T0137924

Lecture Notes in Computer Science 12507

Founding Editors

Gerhard Goos
Karlsruhe Institute of Technology, Karlsruhe, Germany
Juris Hartmanis
Cornell University, Ithaca, NY, USA

Editorial Board Members

Elisa Bertino
Purdue University, West Lafayette, IN, USA
Wen Gao
Peking University, Beijing, China
Bernhard Steffen
TU Dortmund University, Dortmund, Germany
Gerhard Woeginger
RWTH Aachen, Aachen, Germany
Moti Yung
Columbia University, New York, NY, USA

More information about this series at http://www.springer.com/series/7409

Jeff Z. Pan · Valentina Tamma ·
Claudia d'Amato · Krzysztof Janowicz ·
Bo Fu · Axel Polleres · Oshani Seneviratne ·
Lalana Kagal (Eds.)

The Semantic Web – ISWC 2020

19th International Semantic Web Conference
Athens, Greece, November 2–6, 2020
Proceedings, Part II

Springer

Editors
Jeff Z. Pan
University of Edinburgh
Edinburgh, UK

Claudia d'Amato (iD)
University of Bari
Bari, Italy

Bo Fu (iD)
California State University, Long Beach
Long Beach, CA, USA

Oshani Seneviratne (iD)
Rensselaer Polytechnic Institute
Troy, NY, USA

Valentina Tamma (iD)
University of Liverpool
Liverpool, UK

Krzysztof Janowicz
University of California, Santa Barbara
Santa Barbara, CA, USA

Axel Polleres (iD)
Vienna University of Economics
and Business
Vienna, Austria

Lalana Kagal
Massachusetts Institute of Technology
Cambridge, MA, USA

ISSN 0302-9743   ISSN 1611-3349 (electronic)
Lecture Notes in Computer Science
ISBN 978-3-030-62465-1   ISBN 978-3-030-62466-8 (eBook)
https://doi.org/10.1007/978-3-030-62466-8

LNCS Sublibrary: SL3 – Information Systems and Applications, incl. Internet/Web, and HCI

© Springer Nature Switzerland AG 2020, corrected publication 2021, 2022
Chapter "Transparent Integration and Sharing of Life Cycle Sustainability Data with Provenance" is licensed under the terms of the Creative Commons Attribution 4.0 International License (http://creativecommons.org/licenses/by/4.0/). For further details see licence information in the chapter.
This work is subject to copyright. All rights are reserved by the Publisher, whether the whole or part of the material is concerned, specifically the rights of translation, reprinting, reuse of illustrations, recitation, broadcasting, reproduction on microfilms or in any other physical way, and transmission or information storage and retrieval, electronic adaptation, computer software, or by similar or dissimilar methodology now known or hereafter developed.
The use of general descriptive names, registered names, trademarks, service marks, etc. in this publication does not imply, even in the absence of a specific statement, that such names are exempt from the relevant protective laws and regulations and therefore free for general use.
The publisher, the authors and the editors are safe to assume that the advice and information in this book are believed to be true and accurate at the date of publication. Neither the publisher nor the authors or the editors give a warranty, expressed or implied, with respect to the material contained herein or for any errors or omissions that may have been made. The publisher remains neutral with regard to jurisdictional claims in published maps and institutional affiliations.

This Springer imprint is published by the registered company Springer Nature Switzerland AG
The registered company address is: Gewerbestrasse 11, 6330 Cham, Switzerland

Preface

Throughout the years, the International Semantic Web Conference (ISWC) has firmly established itself as the premier international forum to discuss and present the latest advances in fundamental research, innovative technology, and applications of the Semantic Web, Linked Data, Knowledge Graphs, and Knowledge Processing on the Web. Now in its 19th edition, ISWC 2020 aims to bring together researchers and practitioners to present new approaches and findings, share ideas, and discuss experiences. The conference involves researchers with diverse skills and interests, thanks to the increased adoption of semantic technologies. Furthermore, knowledge-driven technologies have become increasingly synergetic in many subfields of artificial intelligence, such as natural language processing and machine learning, and this year's call for papers for the main conference tracks was broadened to include such topics to acknowledge these cooperative research efforts.

This year, the submission process and the conference planning were heavily affected by the COVID-19 pandemic outbreak. Despite the myriad of challenges faced, ISWC has maintained its excellent reputation as a premier scientific conference. As a means of recognizing the difficulties experienced by the community, the submission deadline was postponed by five weeks, and the decision was made to run the conference as a virtual event. We received submissions from 51 different countries with Germany, the USA, China, Italy, and France featuring prominently in the submissions list.

Across the conference, we witnessed a real effort by the community – authors, Senior Program Committee (SPC) members, Program Committee (PC) members, and additional reviewers – all of whom were all incredibly supportive of the changes we had to make to the conference organization, demonstrating remarkable dedication and energy during the whole process. We also saw the pandemic become an opportunity to support the scientific community at large, with multiple papers related to COVID-19 research submitted to the conference.

The Research Track, chaired by Jeff Pan and Valentina Tamma, received 170 submissions and ultimately accepted 38 papers, resulting in an acceptance rate of 22.3%. Continuing with the approach taken last year, we adopted a double-blind review policy, i.e., the authors' identity was not revealed to the reviewers and vice versa. Furthermore, reviewers assigned to a paper were not aware of the identity of their fellow reviewers. We strengthened the composition of the PC, which comprised 34 SPC and 244 regular PC members. An additional 66 sub-reviewers were recruited to support the review process further.

ISWC has traditionally had a very rigorous reviewing process, which was again reflected this year. For every submission, several criteria were assessed by the PC members, including originality, novelty, relevance, and impact of the research contributions; soundness, rigor, and reproducibility; clarity and quality of presentation; and the positioning to the literature. This year, the vast majority of papers were reviewed by four reviewers and an SPC member. All of the reviewers engaged in lively and thorough discussions once the initial reviews had been submitted, and later after the authors' responses were made available. Each paper was then discussed among the Research Track PC chairs and the SPC members to reach a consensus on the final list of accepted papers. As a further measure to recognize the COVID-19 pandemics' challenges, some papers were conditionally accepted, with the SPC members overseeing them and kindly agreeing to shepherd the papers to address the concerns raised by the reviewers. The PC chairs would like to express their gratitude to all SPC members, PC members, and external reviewers for the time, the dedication, and energy they put into the reviewing process, despite these very challenging circumstances.

The In-Use Track continues the tradition to showcase and learn from the growing adoption of Semantic Web technologies in concrete and practical settings, demonstrating the crucial roles that Semantic Web technologies often play in supporting more efficient, effective interoperable solutions in a range of real-world contexts. This year, the track chairs Bo Fu and Axel Polleres received 47 paper submissions, and they accepted 21 papers, leading to an acceptance rate of 44.7%, which reflects a continued increase in the number of submissions as well as acceptances compared to previous years, which indicates a growing maturity and adoption of Semantic Web technologies. The In-Use Track PC consisted of 50 members who engaged in extensive discussions to ensure a high-quality program, where the committee assessed each submission following review criteria including novelty and significance of the application, acceptance and uptake, scalability and technical soundness, as well as the generalizability of the lessons learned regarding the benefits, risks, and opportunities when adopting Semantic Web technologies. Each paper received at least three reviews. The final accepted papers describe successful applications of technologies, including ontologies, Knowledge Graphs, and Linked Data in a diverse range of domains (e.g., digital humanities, pharmaceutics, manufacturing, taxation, and transportation) and highlight the suitability of Semantic Web methods to advance solutions in various challenging areas (e.g., adaptive systems, data integration, collaborative knowledge management, machine learning, and recommendations).

The Resources Track solicited contributions ranging from ontologies and benchmarks to workflows and datasets over software, services, and frameworks. Many of these contributions are research enablers. For instance, ontologies are used to lift data semantically, datasets become core hubs of the Linked Data cloud, and benchmarks enable others to evaluate their research more systematically. In this year's edition, track chairs Claudia d'Amato and Krzysztof Janowicz received 71 submissions, out of which they decided to accept 22. These submissions are well representative of the spirit of the track and the variety of Semantic Web research. They include knowledge graphs related to COVID-19, benchmarks for OWL2 ontologies, web crawlers, and ontologies. The track chairs are incredibly thankful for the timely and high-quality reviews they received and would like to express their gratitude towards the SPC members who provided excellent meta-reviews and engaged in discussions to ensure fair evaluation of all papers.

In light of the reproducibility crisis in natural sciences, we believe that sharing experimental code, data, and setup will benefit scientific progress, foster collaboration, and encourage the exchange of ideas. We want to build a culture where sharing results, code, and scripts are the norm rather than an exception. To highlight the importance in this area, Valentina Ivanova and Pasquale Minervini chaired the second edition of the reproducibility initiative at ISWC. The track's focus was to evaluate submissions from the ISWC Research and Resources Tracks' accepted papers. This year, the ISWC Reproducibility Track extended the evaluation scope, which now includes two assessment lines: Reproducibility Line of Assessment for reproducing systems set ups and computational experiments and Replicability Line of Assessment for evaluating quantitative laboratory experiments with users. For the Reproducibility Line of Assessment, two independent members of the PC interacted with the authors to check the data's availability, source code, documentation, configuration requirements, and reproduce the paper's most important results. For the Replicability Line of Assessment, one member of the PC interacted with the authors to assess if the authors supplied enough materials about their work so an interested researcher could re-run the experiments in question. We received 10 submissions from the Resources Track in the Reproducibility Line of Assessment.

The Industry Track provides industry adopters an opportunity to highlight and share the key learnings and challenges of applying Semantic Web technologies in real-world and scalable implementations. This year, the track chairs Freddy Lecue and Jun Yan received 22 submissions from a wide range of companies of different sizes, and 15 submissions were accepted. The submissions were assessed in terms of quantitative and qualitative value proposition provided, innovative aspects, impact, and lessons learned, as well as business value in the application domain; and the degree to which semantic technologies are critical to their offering. Each paper got one review from an industry Semantic Web expert, which was checked and validated by the Industry Track chairs. The final decision was based on the evidence and impact of industrial applications using/based on Semantic Web technologies.

The Sister Conference Track has been designed as a forum for presentations of significant Semantic Web-related research results that have been recently presented at very well-established conferences other than the ISWC. The goal is to give visibility of these results to the ISWC audience and promote discussions concerning such results. For this first issue, chaired by Jérôme Euzenat and Juanzi Li, we decided to adopt a dual strategy, issuing an open call for papers and actively looking for relevant papers to invite. We invited 22 papers, out of which five applied. Four Additional papers replied to the call for papers. The authors of one other paper asked to submit, but were discouraged. Of these, we retained 8 papers. These were published in the past two year editions of the European Conference on Artificial Intelligence (ECAI), the Association for the Advancement of Artificial Intelligence (AAAI) conference, the International Joint Conferences on Artificial Intelligence (IJCAI), the International Conference on Autonomous Agents and Multi-Agent Systems (AAMAS), and the World Wide Web (WWW) conference. These papers did not undergo a further peer review, nor are they republished in the ISWC proceedings. They complemented and added value to the ISWC 2020 program.

The workshop program, chaired by Sabrina Kirrane and Satya Sahoo, included a mix of established and relatively new topics. Workshops on established topics included ontology matching, ontology design and patterns, scalable knowledge base systems, semantic statistics, querying and benchmarking, evolution and preservation, profiling, visualization, and Semantic Web for health data management. Workshops on relatively new topics included contextualized knowledge graphs, semantics for online misinformation detection, semantic explainability, natural language interfaces, research data management, artificial intelligence technologies for legal documents, the Semantic Web in practice, and Wikidata. Tutorials on a variety of topics such as knowledge graph construction, common sense knowledge graphs, pattern-based knowledge base construction, building large knowledge graphs efficiently, scalable RDF analytics, SPARQL endpoints, Web API, data science pipelines, semantic explainability, shape applications and tools, and building mobile Semantic Web applications complemented the workshop program.

As of ISWC 2020, the Semantic Web Challenges mark their 17th appearance at the conference. Since last year, all proposed challenges need to provide a benchmarking platform, on which participants can have their solution validated using objective measures against fixed datasets. Three exciting challenges were open for submissions: the SeMantic AnsweR Type prediction task (SMART), the Semantic Web Challenge on Tabular Data to Knowledge Graph Matching (SemTab), and the Mining the Web of HTML-embedded Product Data. For SMART, participants focus on predicting the type of answers to English questions, which is essential to the topic of question answering within the natural language processing and information retrieval domain. For SemTab, participants aimed to convert tables into knowledge graphs to better exploit the information contained in them. For the Product Data challenge, participants had to address tasks in the domain of e-commerce data integration, specifically product matching, and product classification. Challenge entries and lessons learned were discussed at ISWC 2020.

The Posters and Demos Track is one of the most vibrant parts of every ISWC. This year, the track was chaired by Kerry Taylor and Rafael Gonçalves, who received a total of 97 submissions: 58 posters and 39 demos. The PC consisting of 97 members and the track chairs, accepted 43 posters and 35 demos. The decisions were primarily based on relevance, originality, and clarity of the submissions.

The conference also included a Doctoral Consortium (DC) Track, chaired by Elena Simperl and Harith Alani. The DC Track was designed to enable PhD students to share their work and initial results with fellow students and senior researchers from the Semantic Web community, gain experience in presenting scientific research, and receive feedback in a constructive and informal environment. This year, the PC accepted 6 papers for oral presentations out of 11 submissions. The DC program focused on allowing the students to work together during multiple activity sessions on joint tasks, such as articulating research questions or forming an evaluation plan. The aim was to increase their interactions and receive hands-on guidance from the ISWC community's senior members. DC Tracks also included a fantastic invited talk, delivered by Prof. Payam Barnaghi.

This year, ISWC offered Student Grant Awards to support the full conference's registration cost. We acknowledge the Semantic Web Science Association (SWSA) and the *Artificial Intelligence Journal* (AIJ) for generously funding this year's student grants. The applications were solicited from students attending a higher education institution, having either an ISWC 2020 paper accepted or just intending to participate in the conference. Preference was given to the students having a first-authored paper in either the main conference, the doctoral consortium, a workshop, the poster/demo session, or the Semantic Web challenge. This year, given the conference's virtual nature and the challenge of increasing student engagement, we planned a unique program for the Student Engagement and Mentoring Session that was open to all the student attendees of the conference. The session included three main parts. First, we hosted career-advising panels, consisting of senior researchers (mentors) with an open Q&A session on research and career advice. Second, a brainstorming group activity was planned to engage students in participatory design to creatively combine and articulate their research ideas for the Semantic Web's future vision. Lastly, a fun-filled social virtual party took place to help students socially engage with their peers.

Our thanks go to Elmar Kiesling and Haridimos Kondylakis, our publicity chairs, and Ioannis Chrysakis and Ioannis Karatzanis, our Web chairs. Together they did an amazing job of ensuring that all conference activities and updates were made available on the website and communicated across mailing lists and on social media. Gianluca Demartini and Evan Patton were the metadata chairs this year, and they made sure that all relevant information about the conference was available in a format that could be used across all applications, continuing a tradition established at ISWC many years ago. We are especially thankful to our proceedings chair, Oshani Seneviratne, who oversaw the publication of this volume alongside a number of CEUR proceedings for other tracks.

Sponsorships are essential to realizing a conference and were even more important this year as additional funds were necessary to put together the virtual conference. Despite numerous hurdles caused by the unusual situation, our highly committed trio of sponsorship chairs, Evgeny Kharlamov, Giorgios Stamou, and Veronika Thost, went above and beyond to find new ways to engage with sponsors and promote the conference to them.

Finally, our special thanks go to the members of the Semantic Web Science Association (SWSA), especially Ian Horrocks, the SWSA President, for their continuing support and guidance and to the organizers of previous ISWC conferences who were a constant source of knowledge, advice, and experience.

September 2020

Jeff Z. Pan
Valentina Tamma
Claudia d'Amato
Krzysztof Janowicz
Bo Fu
Axel Polleres
Oshani Seneviratne
Lalana Kagal

Organization

Organizing Committee

General Chair

Lalana Kagal — Massachusetts Institute of Technology, USA

Local Chairs

Manolis Koubarakis — University of Athens, Greece
Dimitris Plexousakis — ICS-FORTH, University of Crete, Greece
George Vouros — University of Piraeus, Greece

Research Track Chairs

Jeff Z. Pan — The University of Edinburgh, UK
Valentina Tamma — The University of Liverpool, UK

Resources Track Chairs

Claudia d'Amato — University of Bari, Italy
Krzysztof Janowicz — University of California, Santa Barbara, USA

In-Use Track Chairs

Bo Fu — California State University, Long Beach, USA
Axel Polleres — Vienna University of Economics and Business, Austria

Reproducibility Track Chairs

Valentina Ivanova — RISE Research Institutes of Sweden, Sweden
Pasquale Minervini — University College London, UK

Industry Track Chairs

Freddy Lecue — Inria, France, and CortAIx Thales, Canada
Jun Yan — Yidu Cloud Technology Company Ltd., China

Sister Conference Paper Track Chairs

Jérôme Euzenat — Inria, Université Grenoble Alpes, France
Juanzi Li — Tsinghua University, China

Workshop and Tutorial Chairs

Sabrina Kirrane — Vienna University of Economics and Business, Austria
Satya S. Sahoo — Case Western Reserve University, USA

Semantic Web Challenges Track Chairs

Anna Lisa Gentile   IBM Research, USA
Ruben Verborgh   Ghent University – imec, Belgium

Poster and Demo Track Chairs

Rafael Gonçalves   Stanford University, USA
Kerry Taylor   The Australian National University, Australia,
     and University of Surrey, UK

Doctoral Consortium Chairs

Harith Alani   KMI, The Open University, UK
Elena Simperl   University of Southampton, UK

Student Coordination Chairs

Maribel Acosta   Karlsruhe Institute of Technology, Germany
Hemant Purohit   George Mason University, USA

Virtual Conference Chairs

Mauro Dragoni   Fondazione Bruno Kessler, Italy
Juan Sequada   data.world, Austin, USA

Proceedings Chair

Oshani Seneviratne   Rensselaer Polytechnic Institute, USA

Metadata Chairs

Gianluca Demartini   The University of Queensland, Australia
Evan Patton   CSAIL MIT, USA

Publicity Chairs

Elmar Kiesling   Vienna University of Economics and Business, Austria
Haridimos Kondylakis   ICS-FORTH, Greece

Web Site Chairs

Ioannis Chrysakis   ICS-FORTH, Greece, and Ghent University, IDLab,
     imec, Belgium.
Ioannis Karatzanis   ICS-FORTH, Greece

Vision Track Chairs

Natasha Noy   Google Research, USA
Carole Goble   The University of Manchester, UK

Sponsorship Chairs

Evgeny Kharlamov   Bosch Center for Artificial Intelligence, Germany,
and University of Oslo, Norway
Giorgios Stamou   AILS Lab, ECE NTUA, Greece
Veronika Thost   MIT, IBM Watson AI Lab, USA

Research Track Senior Program Committee

Eva Blomqvist	Linköping University, Sweden
Gianluca Demartini	The University of Queensland, Australia
Mauro Dragoni	Fondazione Bruno Kessler, Italy
Achille Fokoue	IBM, USA
Naoki Fukuta	Shizuoka University, Japan
Anna Lisa Gentile	IBM, USA
Birte Glimm	Universität Ulm, Germany
Jose Manuel Gomez-Perez	Expert System, Spain
Olaf Hartig	Linköping University, Sweden
Laura Hollink	Vrije Universiteit Amsterdam, The Netherlands
Andreas Hotho	University of Würzburg, Germany
Wei Hu	Nanjing University, China
Mustafa Jarrar	Birzeit University, Palestine
Ernesto Jimenez-Ruiz	City, University of London, UK
Kang Liu	National Laboratory of Pattern Recognition, Chinese Academy of Sciences, China
Vanessa Lopez	IBM, Ireland
Markus Luczak-Roesch	Victoria University of Wellington, New Zealand
David Martin	Nuance Communications, USA
Thomas Meyer	University of Cape Town, CAIR, South Africa
Boris Motik	University of Oxford, UK
Raghava Mutharaju	IIIT-Delhi, India
Francesco Osborne	The Open University, UK
Matteo Palmonari	University of Milano-Bicocca, Italy
Bijan Parsia	The University of Manchester, UK
Terry Payne	The University of Liverpool, UK
Guilin Qi	Southeast University, China
Simon Razniewski	Max Planck Institute for Informatics, Germany
Marta Sabou	Vienna University of Technology, Austria
Elena Simperl	King's College London, UK
Daria Stepanova	Bosch Center for Artificial Intelligence, Germany
Hideaki Takeda	National Institute of Informatics, Japan
Tania Tudorache	Stanford University, USA
Maria-Esther Vidal	Universidad Simón Bolívar, Venezuela
Kewen Wang	Griffith University, Australia

Research Track Program Committee

Ibrahim Abdelaziz	IBM, USA
Maribel Acosta	Karlsruhe Institute of Technology, Germany
Panos Alexopoulos	Textkernel B.V., The Netherlands
Muhammad Intizar Ali	Insight Centre for Data Analytics, National University of Ireland Galway, Ireland
José-Luis Ambite	University of Southern California, USA
Reihaneh Amini	Wright State University, USA
Grigoris Antoniou	University of Huddersfield, UK
Hiba Arnaout	Max Planck Institute for Informatics, Germany
Luigi Asprino	University of Bologna, STLab, ISTC-CNR, Italy
Nathalie Aussenac-Gilles	IRIT, CNRS, France
Carlos Badenes-Olmedo	Universidad Politécnica de Madrid, Spain
Pierpaolo Basile	University of Bari, Italy
Valerio Basile	University of Turin, Italy
Sumit Bhatia	IBM, India
Christian Bizer	University of Mannheim, Germany
Carlos Bobed	Everis, NTT Data, University of Zaragoza, Spain
Alex Borgida	Rutgers University, USA
Paolo Bouquet	University of Trento, Italy
Zied Bouraoui	CRIL, CNRS, Artois University, France
Alessandro Bozzon	Delft University of Technology, The Netherlands
Anna Breit	Semantic Web Company, Austria
Carlos Buil Aranda	Universidad Técnica Federico Santa María, Chile
Davide Buscaldi	LIPN, Université Paris 13, Sorbonne Paris Cité, France
David Carral	TU Dresden, Germany
Giovanni Casini	ISTI-CNR, Italy
Irene Celino	Cefriel, Italy
Vinay Chaudhri	SRI International, USA
Wenliang Chen	Soochow University, China
Jiaoyan Chen	University of Oxford, UK
Yubo Chen	Institute of Automation, Chinese Academy of Sciences, China
Gong Cheng	Nanjing University, China
Cuong Xuan Chu	Max Planck Institute for Informatics, Germany
Philipp Cimiano	Bielefeld University, Germany
Pieter Colpaert	Ghent University – imec, Belgium
Mariano Consens	University of Toronto, Canada
Olivier Corby	Inria, France
Oscar Corcho	Universidad Politécnica de Madrid, Spain
Luca Costabello	Accenture Labs, Ireland
Isabel Cruz	University of Illinois at Chicago, USA
Philippe Cudre-Mauroux	University of Fribourg, Switzerland
Bernardo Cuenca Grau	University of Oxford, UK
Victor de Boer	Vrije Universiteit Amsterdam, The Netherlands

Daniele Dell'Aglio	University of Zurich, Switzerland
Elena Demidova	L3S Research Center, Germany
Ronald Denaux	Expert System, Spain
Grit Denker	SRI International, USA
Danilo Dessì	University of Cagliari, Italy
Tommaso Di Noia	Politecnico di Bari, Italy
Stefan Dietze	GESIS - Leibniz Institute for the Social Sciences, Germany
Djellel Difallah	New York University, USA
Anastasia Dimou	Ghent University, Belgium
Dejing Dou	University of Oregon, USA
Jianfeng Du	Guangdong University of Foreign Studies, China
Michel Dumontier	Maastricht University, The Netherlands
Shusaku Egami	Electronic Navigation Research Institute, Japan
Fajar J. Ekaputra	Vienna University of Technology, Austria
Alessandro Faraotti	IBM, Italy
Michael Färber	Karlsruhe Institute of Technology, Germany
Catherine Faronzucker	Université Nice Sophia Antipolis, France
Yansong Feng	The University of Edinburgh, UK
Alba Fernandez	Universidad Politécnica de Madrid, Spain
Miriam Fernandez	Knowledge Media Institute, UK
Javier D. Fernández	F. Hoffmann-La Roche AG, Switzerland
Besnik Fetahu	L3S Research Center, Germany
Valeria Fionda	University of Calabria, Italy
Fabian Flöck	GESIS Cologne, Germany
Luis Galárraga	Inria, France
Fabien Gandon	Inria, France
Huan Gao	Microsoft, USA
Raúl García-Castro	Universidad Politécnica de Madrid, Spain
Andrés García-Silva	Expert System
Daniel Garijo	Information Sciences Institute, USA
Manas Gaur	KNO.E.SIS Josh Research Center, Wright State University, USA
Pouya Ghiasnezhad Omran	The Australian National University, Australia
Shrestha Ghosh	Max Planck Institute for Informatics, Germany
Jorge Gracia	University of Zaragoza, Spain
Alasdair Gray	Heriot-Watt University, UK
Dagmar Gromann	University of Vienna, Austria
Paul Groth	University of Amsterdam, The Netherlands
Kalpa Gunaratna	Samsung Research, USA
Peter Haase	metaphacts, Germany
Mohand-Said Hacid	Université Claude Bernard Lyon 1, France
Tom Hanika	Knowledge and Data Engineering, University of Kassel, Germany
Andreas Harth	University of Erlangen-Nuremberg, Fraunhofer IIS-SCS, Germany

Mounira Harzallah	LS2N, France
Oktie Hassanzadeh	IBM, USA
Rinke Hoekstra	University of Amsterdam, The Netherlands
Katja Hose	Aalborg University, Denmark
Lei Hou	Tsinghua University, China
Pan Hu	University of Oxford, UK
Renfen Hu	Beijing Normal University, China
Linmei Hu	Beijing University of Posts and Telecommunications, China
Luis Ibanez-Gonzalez	University of Southampton, UK
Ryutaro Ichise	National Institute of Informatics, Japan
Prateek Jain	Nuance Communications, USA
Tobias Käfer	Karlsruhe Institute of Technology, Germany
Lucie-Aimée Kaffee	University of Southampton, UK
Maulik R. Kamdar	Elsevier Inc., USA
Ken Kaneiwa	The University of Electro-Communications, Japan
Tomi Kauppinen	Department of Computer Science, Aalto University, Finland
Takahiro Kawamura	National Agriculture and Food Research Organization, Japan
Mayank Kejriwal	Information Sciences Institute, USA
Ilkcan Keles	Aalborg University, Denmark
Carsten Keßler	Aalborg University, Denmark
Sabrina Kirrane	Vienna University of Economics and Business, Austria
Pavel Klinov	Stardog Union Inc., USA
Matthias Klusch	DFKI, Germany
Boris Konev	The University of Liverpool, UK
Egor V. Kostylev	University of Oxford, UK
Kouji Kozaki	Osaka Electro-Communication University, Japan
Adila A. Krisnadhi	Universitas Indonesia, Indonesia
Markus Krötzsch	TU Dresden, Germany
Sreenivasa Kumar	Indian Institute of Technology Madras, India
Jose Emilio Labra Gayo	Universidad de Oviedo, Spain
Agnieszka Ławrynowicz	Poznan University of Technology, Poland
Danh Le Phuoc	TU Berlin, Germany
Maurizio Lenzerini	Sapienza University of Rome, Italy
Ru Li	Shanxi University, China
Yuan-Fang Li	Monash University, Australia
Matteo Lissandrini	Aalborg University, Denmark
Ming Liu	Harbin Institute of Technology, China
Carsten Lutz	Universität Bremen, Germany
Ioanna Lytra	Enterprise Information Systems, University of Bonn, Germany
Gengchen Mai	University of California, Santa Barbara, USA
Bassem Makni	IBM, USA
Ioana Manolescu	Inria, Institut Polytechnique de Paris, France

Fiona McNeill	Heriot-Watt University, UK
Albert Meroño-Peñuela	Vrije Universiteit Amsterdam, The Netherlands
Nandana Mihindukulasooriya	Universidad Politécnica de Madrid, Spain
Dunja Mladenic	Jožef Stefan Institute, Slovenia
Ralf Möller	University of Lübeck, Germany
Pascal Molli	University of Nantes, LS2N, France
Gabriela Montoya	Aalborg University, Denmark
Deshendran Moodley	University of Cape Town, South Africa
Isaiah Onando Mulang'	University of Bonn, Germany
Varish Mulwad	GE Global Research, USA
Summaya Mumtaz	University of Oslo, Norway
Hubert Naacke	Sorbonne Université, UPMC, LIP6, France
Shinichi Nagano	Toshiba Corporation, Japan
Yavor Nenov	Oxford Semantic Technologies, UK
Axel-Cyrille Ngonga Ngomo	Paderborn University, Germany
Vinh Nguyen	National Library of Medicine, USA
Andriy Nikolov	AstraZeneca, Germany
Werner Nutt	Free University of Bozen-Bolzano, Italy
Daniela Oliveira	Insight Centre for Data Analytics, Ireland
Fabrizio Orlandi	ADAPT, Trinity College Dublin, Ireland
Magdalena Ortiz	Vienna University of Technology, Austria
Julian Padget	University of Bath, UK
Ankur Padia	UMBC, USA
Peter Patel-Schneider	Xerox PARC, USA
Rafael Peñaloza	University of Milano-Bicocca, Italy
Bernardo Pereira Nunes	The Australian National University, Australia
Catia Pesquita	LaSIGE, Universidade de Lisboa, Portugal
Alina Petrova	University of Oxford, UK
Patrick Philipp	KIT (AIFB), Germany
Giuseppe Pirrò	Sapienza University of Rome, Italy
Alessandro Piscopo	BBC, UK
María Poveda-Villalón	Universidad Politécnica de Madrid, Spain
Valentina Presutti	ISTI-CNR, Italy
Yuzhong Qu	Nanjing University, China
Alexandre Rademaker	IBM, EMAp/FGV, Brazil
David Ratcliffe	Defence, USA
Domenico Redavid	University of Bari, Italy
Diego Reforgiato	Università degli studi di Cagliari, Italy
Achim Rettinger	Trier University, Germany
Martin Rezk	Google, USA
Mariano Rico	Universidad Politécnica de Madrid, Spain
Giuseppe Rizzo	LINKS Foundation, Italy
Mariano Rodríguez Muro	Google
Dumitru Roman	SINTEF, Norway

Oscar Romero	Universitat Politécnica de Catalunya, Spain
Marco Rospocher	Università degli Studi di Verona, Italy
Ana Roxin	University of Burgundy, UMR CNRS 6306, France
Sebastian Rudolph	TU Dresden, Germany
Anisa Rula	University of Milano-Bicocca, Italy
Harald Sack	FIZ Karlsruhe - Leibniz Institute for Information Infrastructure, KIT, Germany
Angelo Antonio Salatino	The Open University, UK
Muhammad Saleem	AKSW, University of Leizpig, Germany
Cristina Sarasua	University of Zurich, Switzerland
Kai-Uwe Sattler	TU Ilmenau, Germany
Uli Sattler	The University of Manchester, UK
Ognjen Savkovic	Free University of Bolzano-Bolzano, Italy
Marco Luca Sbodio	IBM, Ireland
Konstantin Schekotihin	Alpen-Adria Universität Klagenfurt, Austria
Andreas Schmidt	Kasseler Verkehrs- und Versorgungs-GmbH, Germany
Juan F. Sequeda	data.world, USA
Chuan Shi	Beijing University of Posts and Telecommunications, China
Cogan Shimizu	Kansas State University, USA
Kuldeep Singh	Cerence GmbH, Zerotha Research, Germany
Hala Skaf-Molli	University of Nantes, LS2N, France
Sebastian Skritek	Vienna University of Technology, Austria
Kavitha Srinivas	IBM, USA
Biplav Srivastava	AI Institute, University of South Carolina, USA
Steffen Staab	IPVS, Universität Stuttgart, Germany, and WAIS, University of Southampton, UK
Andreas Steigmiller	Universität Ulm, Germany
Nadine Steinmetz	TU Ilmenau, Germany
Armando Stellato	Tor Vergata University of Rome, Italy
Umberto Straccia	ISTI-CNR, Italy
Gerd Stumme	University of Kassel, Germany
Eiichi Sunagawa	Toshiba Corporation, Japan
Pedro Szekely	USC, Information Sciences Institute, USA
Yan Tang	Hohai University, China
David Tena Cucala	University of Oxford, UK
Andreas Thalhammer	F. Hoffmann-La Roche AG, Switzerland
Krishnaprasad Thirunarayan	Wright State University, USA
Steffen Thoma	FZI Research Center for Information Technology, Germany
Veronika Thost	IBM, USA
David Toman	University of Waterloo, Canada
Riccardo Tommasini	University of Tartu, Estonia
Takanori Ugai	Fujitsu Laboratories Ltd., Japan
Jacopo Urbani	Vrije Universiteit Amsterdam, The Netherlands
Ricardo Usbeck	Paderborn University, Germany

Marieke van Erp	KNAW Humanities Cluster, The Netherlands
Miel Vander Sande	Meemoo, Belgium
Jacco Vanossenbruggen	CWI, Vrije Universiteit Amsterdam, The Netherlands
Serena Villata	CNRS, Laboratoire d'Informatique, Signaux et Systèmes de Sophia-Antipolis, France
Domagoj Vrgoc	Pontificia Universidad Católica de Chile, Chile
Hai Wan	Sun Yat-sen University, China
Meng Wang	Southeast University, China
Peng Wang	Southeast University, China
Xin Wang	Tianjin University, China
Zhe Wang	Griffith University, Australia
Haofen Wang	Intelligent Big Data Visualization Lab, Tongji University, China
Weiqing Wang	Monash University, Australia
Gregory Todd Williams	J. Paul Getty Trust, USA
Frank Wolter	The University of Liverpool, UK
Tianxing Wu	Southeast University, China
Honghan Wu	King's College London, UK
Josiane Xavier Parreira	Siemens AG, Austria
Guohui Xiao	KRDB Research Centre, Free University of Bozen-Bolzano, Italy
Yanghua Xiao	Fudan University, China
Bo Xu	Donghua University, China
Ikuya Yamada	Studio Ousia Inc., Japan
Veruska Zamborlini	University of Amsterdam, The Netherlands
Fadi Zaraket	American University of Beirut, Lebanon
Haifa Zargayouna	Université Paris 13, France
Xiaowang Zhang	Tianjin University, China
Yuanzhe Zhang	Institute of Automation, Chinese Academy of Sciences, China
Xin Zhao	Renmin University of China, China
Kenny Zhu	Shanghai Jiao Tong University, China
Antoine Zimmermann	École des Mines de Saint-Étienne, France
Aya Zoghby	Mansoura University, Egypt

Research Track Additional Reviewers

George Baryannis
Sotiris Batsakis
Alberto Benincasa
Akansha Bhardwaj
Russa Biswas
Julia Bosque-Gil
Janez Brank
Tanya Braun

Christoph Braun
Maxime Buron
Ling Cai
Muhao Chen
Francesco Corcoglioniti
Marco Cremaschi
Vincenzo Cutrona
Jiwei Ding

Hang Dong
Dominik Dürrschnabel
Cristina Feier
Maximilian Felde
Herminio García González
Yuxia Geng
Simon Gottschalk
Peiqin Gu
Ryohei Hisano
Fabian Hoppe
Elena Jaramillo
Pavan Kapanipathi
Christian Kindermann
Benno Kruit
Felix Kuhr
Sebastian Lempert
Qiuhao Lu
Maximilian Marx
Qaiser Mehmood
Sylvia Melzer
Stephan Mennicke
Sepideh Mesbah
Payal Mitra
Natalia Mulligan
Anna Nguyen

Kristian Noullet
Erik Novak
Romana Pernischová
Md Rashad Al Hasan Rony
Paolo Rosso
Tarek Saier
Md Kamruzzaman Sarker
Bastian Schäfermeier
Thomas Schneider
Matteo Antonio Senese
Lucia Siciliani
Rita Sousa
Ahmet Soylu
Maximilian Stubbemann
Víctor Suárez-Paniagua
S. Subhashree
Nicolas Tempelmeier
Klaudia Thellmann
Vinh Thinh Ho
Rima Türker
Sahar Vahdati
Amir Veyseh
Gerhard Weikum
Xander Wilcke
Huayu Zhang

Resources Track Senior Program Committee

Irene Celino   Cefriel, Italy
Olivier Curé   Université Paris-Est, LIGM, France
Armin Haller   The Australian National University, Australia
Yingjie Hu   University at Buffalo, USA
Ernesto Jimenez-Ruiz   City, University of London, UK
Maxime Lefrançois   MINES Saint-Etienne, France
Maria Maleshkova   University of Bonn, Germany
Matteo Palmonari   University of Milano-Bicocca, Italy
Vojtěch Svátek   University of Economics Prague, Czech Republic
Ruben Verborgh   Ghent University, Belgium

Resources Track Program Committee

Benjamin Adams   University of Canterbury, New Zealand
Mehwish Alam   FIZ Karlsruhe - Leibniz Institute for Information
                        Infrastructure, AIFB Institute, KIT, Germany
Francesco Antoniazzi   École des Mines de Saint-Etienne, France

Mahdi Bennara	École des Mines de Saint-Étienne, France
Felix Bensmann	GESIS - Leibniz Institute for the Social Sciences, Germany
Fernando Bobillo	University of Zaragoza, Spain
Katarina Boland	GESIS - Leibniz Institute for the Social Sciences, Germany
Elena Cabrio	Université Côte d'Azur, CNRS, Inria, I3S, France
Valentina Anita Carriero	University of Bologna, Italy
Victor Charpenay	Friedrich-Alexander Universität, Germany
Yongrui Chen	Southeast University, China
Andrea Cimmino Arriaga	Universidad Politécnica de Madrid, Spain
Andrei Ciortea	University of St. Gallen, Switzerland
Pieter Colpaert	Ghent University, Belgium
Francesco Corcoglioniti	Free Researcher, Italy
Simon Cox	CSIRO, Australia
Vincenzo Cutrona	University of Milano-Bicocca, Italy
Enrico Daga	The Open University, UK
Jérôme David	Inria, France
Daniele Dell'Aglio	University of Zurich, Switzerland
Stefan Dietze	GESIS - Leibniz Institute for the Social Sciences, Germany
Anastasia Dimou	Ghent University, Belgium
Iker Esnaola-Gonzalez	Fundación Tekniker, Spain
Nicola Fanizzi	University of Bari, Italy
Raúl García-Castro	Universidad Politécnica de Madrid, Spain
Genet Asefa Gesese	FIZ Karlsruhe - Leibniz Institute for Information Infrastructure, Germany
Martin Giese	University of Oslo, Norway
Seila Gonzalez Estrecha	Michigan State University, USA
Rafael S. Gonçalves	Stanford University, USA
Tudor Groza	Garvan Institute of Medical Research, Australia
Christophe Guéret	Accenture Labs, Ireland
Peter Haase	metaphacts, Germany
Armin Haller	The Australian National University, Australia
Karl Hammar	Jönköping AI Lab, Jönköping University, Sweden
Aidan Hogan	DCC, Universidad de Chile, Chile
Antoine Isaac	Europeana, Vrije Universiteit Amsterdam, The Netherlands
Marcin Joachimiak	Lawrence Berkeley National Laboratory, USA
Tomi Kauppinen	Aalto University, Finland
Elmar Kiesling	Vienna University of Economics and Business, Austria
Sabrina Kirrane	Vienna University of Economics and Business, Austria
Tomas Kliegr	University of Economics, Prague, Czech Republic
Adila A. Krisnadhi	Universitas Indonesia, Indonesia
Benno Kruit	University of Amsterdam, The Netherlands

Christoph Lange	Fraunhofer, Technology FIT, RWTH Aachen University, Germany
Paea Le Pendu	University of California, Riverside, USA
Martin Leinberger	Universität Koblenz-Landau, Germany
Weizhuo Li	Southeast University, China
Albert Meroño-Peñuela	Vrije Universiteit Amsterdam, The Netherlands
Pascal Molli	University of Nantes, LS2N, France
Summaya Mumtaz	University of Oslo, Norway
Lionel Médini	LIRIS, University of Lyon, France
Hubert Naacke	Sorbonne Université, UPMC, LIP6, France
Raul Palma	Poznań Supercomputing and Networking Center, Poland
Heiko Paulheim	University of Mannheim, Germany
Catia Pesquita	LaSIGE, Universidade de Lisboa, Portugal
Alina Petrova	University of Oxford, UK
Rafael Peñaloza	University of Milano-Bicocca, Italy
Giuseppe Pirrò	Sapienza University of Rome, Italy
María Poveda-Villalón	Universidad Politécnica de Madrid, Spain
Blake Regalia	University of California, Santa Barbara, USA
Giuseppe Rizzo	LINKS Foundation, Italy
Sergio José Rodríguez Méndez	The Australian National University, Australia
Maria Del Mar Roldan-Garcia	Universidad de Malaga, Spain
Marco Rospocher	Università degli Studi di Verona, Italy
Ana Roxin	University of Burgundy, France
Michael Röder	Paderborn University, Germany
Tzanina Saveta	ICS-FORTH, Greece
Mark Schildhauer	NCEAS, USA
Stefan Schlobach	Vrije Universiteit Amsterdam, The Netherlands
Patricia Serrano Alvarado	LS2N, University of Nantes, France
Cogan Shimizu	Kansas State University, USA
Blerina Spahiu	University of Milano-Bicocca, Italy
Kavitha Srinivas	IBM, USA
Pedro Szekely	USC, Information Sciences Institute, USA
Ruben Taelman	Ghent University, Belgium
Hendrik Ter Horst	CITEC, Bielefeld University, Germany
Matthias Thimm	Universität Koblenz-Landau, Germany
Tabea Tietz	FIZ Karlsruhe, Germany
Jacopo Urbani	Vrije Universiteit Amsterdam, The Netherlands
Ricardo Usbeck	Paderborn University, Germany
Maria-Esther Vidal	Universidad Simón Bolívar, Venezuela
Tobias Weller	Karlsruhe Institute of Technology, Germany
Bo Yan	University of California, Santa Barbara, USA

Ziqi Zhang   The University of Sheffield, UK
Qianru Zhou   University of Glasgow, UK
Rui Zhu   University of California, Santa Barbara, USA

In-Use Track Program Committee

Renzo Angles   Universidad de Talca, Chile
Carlos Buil Aranda   Universidad Técnica Federico Santa María, Chile
Stefan Bischof   Siemens AG, Austria
Irene Celino   Cefriel, Italy
Muhao Chen   University of Pennsylvania, USA
James Codella   IBM, USA
Oscar Corcho   Universidad Politécnica de Madrid, Spain
Philippe Cudre-Mauroux   University of Fribourg, Switzerland
Brian Davis   Dublin City University, Ireland
Christophe Debruyne   Trinity College Dublin, Ireland
Djellel Difallah   New York University, USA
Luis Espinosa-Anke   Cardiff University, UK
Achille Fokoue   IBM, USA
Daniel Garijo   Information Sciences Institute, USA
Jose Manuel Gomez-Perez   Expert System, Spain
Damien Graux   ADAPT Centre, Trinity College Dublin, Ireland
Paul Groth   University of Amsterdam, The Netherlands
Daniel Gruhl   IBM, USA
Peter Haase   metaphacts, Germany
Armin Haller   The Australian National University, Australia
Nicolas Heist   University of Mannheim, Germany
Aidan Hogan   Universidad de Chile, Chile
Tobias Käfer   Karlsruhe Institute of Technology, Germany
Maulik Kamdar   Elsevier Health Markets, USA
Tomi Kauppinen   Aalto University, Finland
Mayank Kejriwal   Information Sciences Institute, USA
Elmar Kiesling   Vienna University of Economics and Business, Austria
Sabrina Kirrane   Vienna University of Economics and Business, Austria
Vanessa Lopez   IBM, USA
Beatrice Markhoff   LI, Université François Rabelais Tours, France
Sebastian Neumaier   Vienna University of Economics and Business, Austria
Andriy Nikolov   AstraZeneca, Germany
Alexander O'Connor   Autodesk, Inc., USA
Declan O'Sullivan   Trinity College Dublin, Ireland
Francesco Osborne   The Open University, UK
Matteo Palmonari   University of Milano-Bicocca, Italy
Josiane Xavier Parreira   Siemens AG, Austria
Catia Pesquita   LaSIGE, Universidade de Lisboa, Portugal
Artem Revenko   Semantic Web Company, Austria
Mariano Rico   Universidad Politécnica de Madrid, Spain

Petar Ristoski	IBM, USA
Dumitru Roman	SINTEF, Norway
Melike Sah	Near East University, Cyprus
Miel Vander Sande	Meemoo, Belgium
Dezhao Song	Thomson Reuters, Canada
Anna Tordai	Elsevier B.V., The Netherlands
Tania Tudorache	Stanford University, USA
Svitlana Vakulenko	Vienna University of Economics and Business, Austria
Xuezhi Wang	Google, USA
Matthäus Zloch	GESIS - Leibniz Institute for the Social Sciences, Germany

In-Use Track Additional Reviewers

Robert David	Semantic Web Company, Austria
Kabul Kurniawan	Vienna University of Economics and Business, Austria
Michael Luggen	eXascale Infolab, Switzerland
Maria Lindqvist	Aalto University, Finland
Nikolay Nikolov	SINTEF, Norway
Mario Scrocca	Cefriel, Italy
Ahmet Soylu	SINTEF, Norway

Sponsors

Platinum Sponsors

http://www.ibm.com

https://inrupt.com

https://www.metaphacts.com

Gold Sponsors

Google

https://google.com

ORACLE

https://www.oracle.com/goto/rdfgraph

Silver Sponsors

Co-inform

https://coinform.eu

https://www.ebayinc.com/careers

https://sirius-labs.no

https://www.accenture.com

Best Resource Paper Sponsor

https://www.springer.com/gp/computer-science/lncs

Student Support and Grants Sponsor

https://www.journals.elsevier.com/artificial-intelligence

Doctoral Consortium Track Sponsor

https://www.heros-project.eu

Contents – Part II

Contents – Part I

Resources Track

CASQAD – A New Dataset for Context-Aware Spatial Question Answering

Jewgeni Rose[1]([⊠])(iD) and Jens Lehmann[2,3](iD)

[1] Volkswagen Innovation Center Europe, Wolfsburg, Germany
jewgeni.rose@volkswagen.de
[2] Computer Science III, University of Bonn, Bonn, Germany
jens.lehmann@cs.uni-bonn.de
[3] Fraunhofer IAIS, Dresden, Germany
jens.lehmann@iais.fraunhofer.de
http://sda.tech

Abstract. The task of factoid question answering (QA) faces new challenges when applied in scenarios with rapidly changing context information, for example on smartphones. Instead of asking who the architect of the "Holocaust Memorial" in Berlin was, the same question could be phrased as "Who was the architect of the many stelae in front of me?" presuming the user is standing in front of it. While traditional QA systems rely on static information from knowledge bases and the analysis of named entities and predicates in the input, question answering for temporal and spatial questions imposes new challenges to the underlying methods. To tackle these challenges, we present the **Context-aware Spatial QA Dataset** ($CASQAD$) with over 5,000 annotated questions containing visual and spatial references that require information about the user's location and moving direction to compose a suitable query. These questions were collected in a large scale user study and annotated semi-automatically, with appropriate measures to ensure the quality.

Keywords: Datasets · Benchmark · Question answering · Knowledge graphs

Resource Type: Dataset
Website and documentation: http://casqad.sda.tech
Permanent URL: https://figshare.com/projects/CASQAD/80897.

1 Introduction

Factoid question answering over static and massive scale knowledge bases ($KBQA$) such as DBpedia [1], Freebase [4] or YAGO [27] are well researched and recent approaches show promising performance [37]. State-of-the-art systems (e.g. [5,11,33,34,37]) perform well for simple factoid questions around a

© Springer Nature Switzerland AG 2020
J. Z. Pan et al. (Eds.): ISWC 2020, LNCS 12507, pp. 3–17, 2020.
https://doi.org/10.1007/978-3-030-62466-8_1

target named entity and revolving predicates. A question like *"Who was the architect of the Holocaust Memorial in Berlin, Germany?"* can be translated into a SPARQL expression to query a KB with the result *Peter Eisenman*[1]. However, in practice question answering is mostly applied in virtual digital assistants on mobile devices, such as Siri, Alexa or Google Assistant. Users address these systems as if they are (physically) present in the situation and their communication changes compared to traditional QA scenarios. Questions contain deictic references such as "there" or "here" that need additional context information (e.g. time and geographic location [17]) to be fully understood. For example, instead of asking who the architect of the "Holocaust Memorial" in Berlin was, the same question could be phrased as *"Who was the architect of the many stelae in front of me?"* presuming the virtual assistant has knowledge about user position and viewing direction. These types of questions require the QA systems to use volatile information sets to generate the answer. Information like location or time change frequently with very different update rates. Instead of using fixed knowledge bases Context-aware QA (*CQA*) systems have to adapt to continuous information flows. That changes not only the structure of the knowledge base itself but also impacts the methodology of how to resolve the correct answer. To answer the aforementioned example question, a QA system would need an additional processing unit that provides external context with location information and matches this information set with spatial and visual signals in the input question, such as *tall* and *in front of me*.

Related works in the field of Spatial Question Answering combined geographic information system modules[2] with a semantic parsing based QA system [18,19]; proposed a system that facilitates crowdsourcing to find users that are likely nearby the according point of interest to answer temporal and location-based questions [22]; or utilizing a QA component to conduct user-friendly spatio-temporal analysis [38]. Latter is achieved by searching the input for temporal or spatial key words, which are mapped to a predefined dictionary. Despite a certain success, a commonality is that no attempt has been made to formalize and systematically combine question answering with external context, e.g. the GPS position where the question was asked. We believe, our dataset will help to close this gap, and tackling one of the main challenges in Question Answering [17].

Contribution. To help bridging the gap between traditional QA systems and Context-aware QA, we offer a new and to the best of our knowledge first Context-aware Spatial QA Dataset (called *CASQAD*) focusing on questions that take spatial context information into account, i.e. visual features, user's location and moving direction. Context has a variety of different meanings and scales, depending on application and research field. We therefore take a look at the concept of context in linguistics first and provide a crisp definition that will be used to annotate the questions. For the task of question collection, we define a case

---

[1] He designed the Memorial to the Murdered Jews of Europe https://www.visitberlin. de/en/memorial-murdered-jews-europe.

[2] A visibility engine computes, which objects are visible from the user's point of view.

study and carry out a user study on Amazon's MTurk crowdsourcing platform. For reproducibility, we provide the source code for the data collection and the resulting dataset[3]. In brief, our question collection and annotation process is as follows.

Raw Data Collection. A crowdworker is presented a Human Intelligence Task (*HIT*[4]) on MTurk, containing the instructions, the project or scenario description and a StreetView panorama embedding. The instructions cover how to control the panorama and how to pose a question with respect to our definition and goals. The scenario describes the purpose and background for the case study, which is as follows: "Imagine driving through a foreign city and ask questions about the surrounding you would usually ask a local guide". The panorama is a StreetView HTML embedding that the user can rotate and zoom in or out but not move freely around the streets, which forces the focus on the presented points of interest. To ensure the quality of the collected questions, we developed comprehensive guidelines, including splitting the batches and monitoring the collection process. We collected over 5,000 questions in sum for 25 panoramas in the German city of Hanover from over 400 different workers.

Question Annotation. For the annotation process we follow a two-step approach. First, we pre-process the raw input automatically to detect named entities, spatial and visual signals[5] and annotate the questions. Second, three human operators evaluate these question-annotation pairs and either approve or correct them.

2 Related Work

In recent years multiple new datasets have been published for the task of QA [2,3,6,9,13–15,20,21,24,25,28,35], including benchmarks provided by the Question Answering over Linked Data (QALD) challenge [29–31]. The datasets and benchmarks differ particularly in size (a few hundreds to hundreds of thousands), complexity (simple facts vs. compositional questions), naturalness (artificially generated from KB triples vs. manually created by human experts), language (mono- vs. multilingual) and the underlying knowledge base (DBpedia, YAGO2, Freebase or Wikidata), in case SPARQL queries are provided.

SimpleQuestions [6] and WebQuestions [3] are the most popular datasets for the evaluation of simple factoid question answering systems, despite the fact that most of the questions can already be answered by standard methods [12,23]. The recent QALD benchmarks contain more complex questions of higher quality with aggregations and additional filter conditions, such as "Name all buildings in London which are higher than 100m" [31]. These questions are hand-written

---

[3] https://casqad.sda.tech/.

[4] A HIT describes the micro tasks a requester posts to the workers on Amazon's platform, also known as a "project".

[5] Using state-of-the-art models from https://spacy.io/.

by the organizers of the challenge and are small in number (up to a few hundred questions).

The SQuAD [25] dataset introduces 100,000 crowdsourced questions for the reading comprehension task. The crowdworkers formulate a question after reading a short text snippet from Wikipedia that contains the answer. The SQuAD 2.0 [24] dataset introduces unanswerable questions to make the systems more robust by penalizing approaches that heavily rely on type matching heuristics. NarrativeQA [21] presents questions which require deep reasoning to understand the narrative of a text rather than matching the question to a short text snippet. The recently published LC-QuAD 2.0 [14] dataset contains 30,000 questions, their paraphrases and corresponding SPARQL queries. The questions were collected by verbalizing SPARQL queries that are generated based on hand-written templates around selected entities and predicates. These verbalizations are then corrected and paraphrased by crowdworkers. For a more detailed description and comparison of standard and recent benchmarks for question answering, we refer to [32,36].

The TempQuestions [20] benchmark contains 1,271 questions with a focus on the temporal dimension in question answering, such as *"Which actress starred in Besson's first science fiction and later married him?"*, which requires changes to the underlying methods regarding question decomposition and reasoning about time points and intervals [20]. The questions were selected from three publicly available benchmarks [2,3,9] by applying hand-crafted rules and patterns that fit the definition of temporal questions, and verified by human operators in the post-processing. However, the processing of questions specifically containing spatial or visual references that require additional context information to be answered was not considered so far.

3   Defining Spatial Questions

There are various different types of questions, which require additional information to be fully understood. Questions can contain a personal aspect, cultural background, or simply visual references to the surrounding location. More formally, in linguistics context is described as a frame that surrounds a (focal) event being examined and provides resources for its appropriate interpretation [8]. This concept is extended by four dimensions, namely setting, behavioral environment, language and extra-situational context. Behavioral environment and language describe how a person speaks and how she presents herself, i.e. the use of gestures, facial expressions, speech emphasis or use of specific words. For instance, this can be used to differentiate between literally or sarcastically meant phrases. The setting describes the social and spatial framework and the extra-situational dimension provides deeper background knowledge about the participants, e.g. the personal relationship and where a conversation is actually held (office vs. home). All dimensions describe important information to process a question properly. To make a first step towards Context-aware QA, in this work we focus on the setting dimension, specifically questions containing spatial and visual references, which require reasoning over multiple data sources. A spatial location is defined by its 2-dimensional geo-coordinate (latitude and longitude). However, users in a real-world scenario rather ask for information about a target object by referring or relate to visually more salient adjacent objects or describe the target visually or both. For this reason, we will define a spatial question by the visual and spatial signals contained in the phrase. The task of spatial question collection is covered in Sect. 4.

3.1 Spatial Signals

We refer to spatial signals as keywords or phrases that modify a question such that it requires a QA system to have additional knowledge about the spatial surrounding of the user. Table 1 shows samples of spatial signals used in the context of spatial question answering applied on mobile assistants. Deictic references are used to point to entities without knowing the name or label, such as *that* building. Positional or vicinity signals reinforce the disambiguation of nearby entities by facilitating the matching between the input question and possible surrounding entities. For example, in the question "What is the column next to the Spanish restaurant?" the signal *next to* is used to point to the column that is next to the more salient object "Spanish restaurant".

Table 1. Spatial signals examples distributed over different categories with according text snippets. Further, all spatial signals can be combined, such as in "What is *that* building *next to* the book store?"

	Spatial Signals	Snippet
Deixis	That, This, There	"over there"
Position	Left, right, in front	"left to me"
Vicinity	Next to, after, at	"right next to the book store"

3.2 Visual Signals

Visual signals are keywords and phrases that specify or filter the questions target entity. Similar to the spatial signals for position and vicinity they facilitate the disambiguation of nearby entities or entities in the same direction from the user's point of view. Visual signals are stronger in terms of filtering visible salient

Table 2. Visual signal examples for different categories with according example snippets.

	Visual Signals	Snippet
Color	Red, green, blue	"yellowish building"
Size	Tall, small, big, long	"tall column"
Shape	Flat, rounded, conical	"rounded corners"
Salience	Flags, brick wall, glass	"flags on the roof"

features and attributes, such as color, shape or unique features. Table 2 shows samples of visual signals for different categories.

3.3 Spatial Questions

Utilizing the described concepts for spatial and visual signals from the Sects. 3.1 and 3.2 we can define a spatial question as follows:

Definition 1. *A spatial question contains at least one spatial signal and requires additional context knowledge to understand the question and disambiguate the target entity. A spatial question can contain multiple visual signals.*

The results of our theoretical considerations disclose challenges to QA systems dealing with spatial questions corresponding to our definition. The QA system requires additional knowledge about user's position, moving or viewing direction and surroundings. The questions contain deictic references (*that*) and location information (*next to*), making it impossible to use traditional approaches based on named entity recognition. The exemplary question taken from our case study that will be presented in Sect. 4, shows the need for new methods for CQA.

Example 1. "What is the white building on the corner with the flags out front?"

Here, we have visual signals *white building* (color) and *with flags out front* (salience) which filter the possible entity candidates for the spatial signal *on the corner*. These filters are important to pinpoint the target entity with a higher probability. Even with distinctive spatial signals such as *on the corner*, we could face four different buildings to choose from – potentially even more, in case there are multiple buildings.

4 CASQAD – Context-Aware Spatial QA Dataset

The main objective of our work is the introduction of a spatial questions corpus that fits the definition in Sect. 3.3, i.e. the questions require the QA system to combine different input sources (at least one for the question and one for the context information) to reason about the question objective and target entity, which is a big step towards Context-aware Question Answering. The most intuitive way to collect natural questions with minimized bias, is to conduct a user study. We use Amazon's MTurk crowdsourcing platform for this task, considering common best practices [10]. MTurk is an efficient option to collect data in a timely fashion [7] and is the de facto standard to collect human generated data for natural language processing. Since we are interested in spatial questions we have to design the collection task accordingly. Therefore, we focus on our motivational scenario that pictures the use of a Context-aware QA system on a mobile device.

4.1   Experimental Setup on MTurk

For the collection task, we first define an appropriate scenario and design to instruct the crowdworkers at MTurk (also called *turkers*). Instead of showing a textual description from Wikipedia containing the answer, we present the task in a more natural way, which also fits our scenario. We show a Google Street View[6] HTML embedding in the survey with the following instructions: "Imagine driving through the German city of Hanover, which is foreign to you. To get to know the city better, you hire a local guide who can answer your questions about surrounding points of interest (POIs). The Street View panorama represents the view from your car". Street View is used on MTurk for various image annotation tasks, for example to support the development of vision-based driver assistance systems [26]. Hara et al. [16] incorporated Street View images in a MTurk survey to identify street-level accessibility problems. In contrast to static images we embed dynamic Street View panoramas in an HTML document, which facilitates an interactive user-system interaction.

Instructions:

- Please ask questions about surrounding POIs visible in the panorama.
- When posing a question, please make sure the current field of view is oriented towards the questions subject.
- General questions such as "Where am I?" will not be awarded.
- You drive through the German city of Hanover.
- The Street View Panorama represents the view from your car.
- You have never been to Hanover and would like to know more about the POIs.
- That's why you hired a local guide to answer your questions about the POI's.

The Route: A crucial part to ensure validity of the experiments is the choice of anchor points for the Street View panoramas. An anchor point is the initial point of view that is presented to the turker. Our goal is to collect spatial questions that people would ask about visible surroundings. For this reason, we picked panoramas containing several POIs from a typical commercial tourist city tour in Hanover[7]. The route consists of 24 different panoramas showing 521 directly visible POIs (buildings, stations, monuments, parks) that have an entry in Open Street Map[8]. Further, some of the panoramas show dynamic objects that were present at the time the pictures were taken, such as pedestrians and vehicles.

---

[6] https://www.google.com/intl/en/streetview/.

[7] https://www.visit-hannover.com/en/Sightseeing-City-Tours/Sightseeing/City-tours.

[8] https://www.openstreetmap.org.

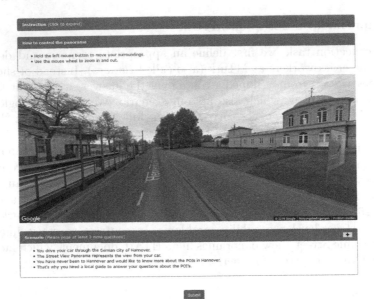

Fig. 1. An example Street View embedding showing the train stop *Hannover Herrenhäuser Gärten* to the left and the *Library Pavilion* with the *Berggarten* to the right.

The HITs: A Human Intelligence Task (HIT) describes the task a crowdworker is supposed to solve in order to earn the reward. The requester has to provide information for the worker including a (unique) title, job description and clear reward conditions. In addition the requester has to specify qualification requirements in the MTurk form to filter desirable from undesirable crowdworkers, such as gender, age, profession, or more specific qualifications like having a driving license or visited places. Here, we specified crowdworkers to be equally distributed over the common age groups and gender. All workers are English speakers and have their residence in the United States, spread proportionally among the population of the individual states[9]. Additionally, we asked the workers if they ever visited the German city of Hanover before, to make sure the scenario of visiting a foreign city holds to minimize the bias. The task for the turkers is to pose at least three different questions to the system, which shows one of the 24 panoramas. When the user submits a question, she has to focus the view on the target object, e.g. the bridge or monument. As a result, we automatically annotate the question with potential context information, by analyzing the position, viewing direction, pitch and zoom level of the panorama[10]. To prevent empty or too short questions, we analyze the input in real time. This is achieved by hosting

---

[9] https://en.wikipedia.org/wiki/List_of_states_and_territories_of_the_United_States_by_population.

[10] All meta information is provided by Google's Street View API https://developers.google.com/maps/documentation/streetview/.

Table 3. The table shows all 24 anchor points from the route including the position (longitude and latitude), the heading as angle degree (0° is north, 90° is east, 180° is south, and 270° is west), and the number of visible points of interest (POIs) and buildings (e.g. office or apartment buildings).

Title	Latitude	Longitude	Heading	Visible POIs	Visible Buildings
Schlosshäuser im Berggarten Hannover	52.3915561	9.7001302	0	8	24
Landesmuseum Hannover und Staatskanzlei Niedersachsen	52.365528	9.7418318	228	20	24
Neues Rathaus	52.3680496	9.7366261	169	10	14
Marktkirche Hannover	52.3722614	9.7353854	188	12	27
Staatstheater Hannover	52.3737913	9.7417629	238	8	31
Landesmuseum Hannover	52.3650037	9.739624	20	23	11
Leibnitz Universität	52.3816144	9.7175591	7	19	12
Musiktheater Bahnhof Leinhausen	52.3963175	9.6770442	8	8	12
Stöckener Friedhof	52.4003496	9.6692401	11	27	28
Marktkirche Hannover	52.372351	9.7352942	194	12	27
Christuskirche (Hannover)	52.3816745	9.7262545	198	9	58
Neues Rathaus	52.3677048	9.7386612	240	15	23
Landesmuseum	52.3655038	9.7397131	93	17	6
Döhrener Turm	52.3467714	9.7605805	5	19	22
Amtsgericht	52.3773252	9.7449014	168	3	5
VW-Tower	52.3798545	9.7419755	285	15	20
Niedersächsisches Finanzministerium	52.3723053	9.7455448	134	3	35
Börse Hannover	52.3723032	9.7417374	127	11	29
Niedersächsisches Wirtschaftsministerium	52.3689436	9.7347076	29	19	16
Ruine der Aegidienkirche	52.3692021	9.738762	64	11	30
Waterloosäule	52.3663441	9.726473	78	8	19
Niedersachsenhalle	52.3770004	9.7693496	198	17	24
Landtag Niedersachsen	52.3707776	9.7336929	218	6	12
Hauptbahnhof	52.3759631	9.7401624	0	24	18

the web page, which is embedded into the MTurk form, on our own servers on Azure. In our experiments every turker is limited up to eight HITs, which is a good trade-off between cost efficiency and diversity of the workers.

The MTurk Form: Figure 1 shows a screenshot from the document presented to the turkers. On top of the figure is the collapsible instructions box with general instructions, for example, what button to click to submit a question, and reward constraints (the turkers are not paid, if we detect spam, fraud attempts or any random input). To lessen the distraction in the view, we don't use control panels in the embedding. There is a small panel with a control description above the Street View embedding, which is the center of the form. Using the mouse, the turker has a 360° view and can change the pitch and zoom level. In contrast to Google's web application, the turkers cannot move freely around in the panorama, i.e. change their position. This is done to force the focus of the turker on the visible objects in the given panorama. After completing a HIT, randomly another HIT containing a different anchor point from the route is offered to the turker. A panorama is never presented twice to the same user.

4.2   Annotation Process

In sum we collected 5,232 valid[11] questions by 472 different turkers. An exploratory analysis shows, that the questions range from questions about salient buildings, like "Is this a government building?" to questions such as "Is the bus station a good spot to pick up girls?". However, most of the questions are about nice places to stay and eat or interesting looking monuments nearby – questions to be expected from a foreigner to ask an assistant or to look up in a city guide. More specific, the turkers asked for information about the cuisine and opening hours of nearby restaurants and theaters, or building dates and architectural styles. The required information to answer these questions is typically available in common knowledge sources such as OpenStreetMap[12], Google Places[13] or Wikidata[14]. More details about the annotation analysis will be presented in Sect. 4.4. We annotated the dataset in a two-step approach that will be presented below.

Automated Processing: The first processing step is normalizing the user input. Sentences containing multiple questions are separated and white space characters normalized first[15]. For example *"What is this building? When can I visit it?"* is separated into two questions. Then, every question is labeled automatically with relevant meta data from Street View, i.e. position, heading, pitch and zoom level, and the according Street View panorama direct HTTP link. Storing and sharing the images is not permitted per terms of use.

Manual Processing: In the second processing step, three local experts within our team annotate the previously processed questions. We prepared a form containing the raw input, the normalized questions, the meta information from Google Street View and the according image. The annotation task was to tag the questions with the objective of the question, such as the age of a building, mark vicinity phrases as explicit spatial references, as well as phrases containing visual signals. We differentiate between vicinity and simple deictic references to express the complexity and difficulty of these questions, such as *"What is across the street from the Borse building?"*. Finally, the annotators have to choose the questions target object, such as a POI, a nearby location (*"Is **this area** safe at night?"*) or something else (e.g. questions such as *"In what direction is the capitol?"*).

4.3   Experiments with Crowd-Based Annotations

In an early experiment with a batch of 200 Hits we attempted to annotate the phrases by the crowdworkers. We created an additional input mask in the MTurk data collection questionnaire, in which the crowdworkers were supposed

---

[11] We removed manually questions such as "Who am I?".

[12] https://www.openstreetmap.org.

[13] https://cloud.google.com/maps-platform/places.

[14] https://www.wikidata.org/.

[15] Even though we instructed the turkers to phrase only one question per input frame, not all followed the instruction.

to annotate their questions themselves. Using the meta information provided by the Google Street View API we approximated the visible objects in a panorama, queried every available information in the aforementioned knowledge sources (Google Places, OSM and Wikidata) and offered a list of possible answers or information for all records. Then we asked the turkers to annotate the questions with the following information:

1. Choose the object of interest from the given list of objects (object displayed including name, type, and a list of all available attributes)
2. Choose the intent of your question (this is basically a record from the list of attributes, such as construction date for buildings or cuisine for restaurants)
3. If there is no appropriate entry, choose "misc" for object or intent

However, our evaluation revealed critical flaws in this process. The number of approximated visible objects was too high for each panorama to disambiguate these correctly, especially for non-locals. Consequently, the same applies for the choice of the right intent. In addition, the missing English terms for German local places made it difficult to understand the meaning or usage of a place or building. The error rate was over 50% (not including the cases when the crowdworkers selected "misc" as the intent or object). We decided not to use the crowdworker annotations, if every annotation had to be checked by experts again anyway.

4.4 Annotation Analysis

The questions length ranges from 3 to 31 words, whereas the average length of the words is 4.4. The average number of words per question is 6.8 and the according median is 6, which is similar to comparable datasets [20]. Table 4 shows the frequency count for the first token of the question at the left columns. The right columns show the frequency count for the question intent. Both lists cover similarly 84% of all questions. Figure 2 shows the comparison of the word distribution with related datasets.

Table 4. Top 10 list with first token and intent frequency count.

First token	Count	Intent	Count
What	2368	Category	1288
is	1017	Construction Date	671
how	502	Name	292
when	414	Usage	281
are	176	Opening Hours	144
do(es)	137	Significance	120
can	113	Offering	116
who	105	Accessibility	91
where	101	Architecture	81
	2952		**3084**

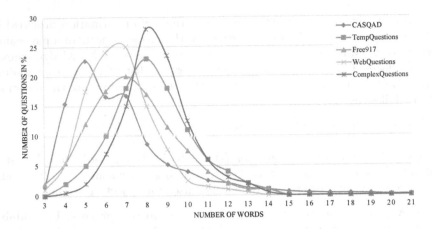

Fig. 2. Distribution of words per question in CASQAD compared to some popular datasets for Question Answering over Knowledge Bases: ComplexQuestions [2], WebQuestions [3], Free917 [9], and TempQuestions [20]

Spatial and Visual Signals: A detailed analysis shows, that the turkers phrase 92% of the questions using simple deictic references to refer to nearby points of interest, otherwise naming the entities (e.g. some businesses have a name on the entrance sign). Questions that contain explicit signals for vicinity or visual information are less frequent. On the other hand, these questions are more complex and challenging, for example *"What is behind the field across from the large building?"*. Table 5 shows the distribution of spatial and visual signals of the annotated questions.

Table 5. Questions distribution by spatial signals. Questions containing named entities usually aren't spatial by our definition.

Signal Type	Example Question	Total
Visual	*What's inside the large stone building?*	490
Vicinity	*Are there any good pubs around here?*	350
Deixis	*What type of architecture is this?*	4839
Size	*How tall is this building?*	260
Color	*What is the building over there with the blue symbol?*	214
Named Entity	*What happens at the Amt-G. Hannover?*	402

5   Conclusion

We published a new dataset containing 5,232 textual questions that are spatial by nature – *CASQAD*. In addition we enriched the questions with according meta context information from Google Street View, such as the GPS position of the point of view, and direct links to the according images. The questions complexity ranges from rather simple questions querying one attribute of a point of interest, to questions about the social and historical background of specific symbols in the images. The versatility of this dataset facilitates the usage for KBQA as well as for text comprehension, or hybrid systems including visual QA. We hope to spur research Context-aware Question Answering systems with this dataset. *CASQAD* is currently being used in multiple internal projects in the Volkswagen Group Innovation[16], in particular in the research field of digital assistants. Future work will include a ready to use end-to-end baseline and an extensive evaluation in a real world scenario with users driving in a car to explore a foreign city, to encourage further research in this direction.

References

1. Auer, S., et al.: DBpedia: a nucleus for a web of open data. In: Aberer, K., et al. (eds.) ASWC/ISWC -2007. LNCS, vol. 4825, pp. 722–735. Springer, Heidelberg (2007). https://doi.org/10.1007/978-3-540-76298-0_52
2. Bao, J., Duan, N., Yan, Z., Zhou, M., Zhao, T.: Constraint-based question answering with knowledge graph. In: Proceedings of COLING 2016, the 26th International Conference on Computational Linguistics: Technical Papers, pp. 2503–2514. The COLING 2016 Organizing Committee, Osaka, Japan (2016)
3. Berant, J., Chou, A., Frostig, R., Liang, P.: semantic parsing on freebase from question-answer pairs. In: Proceedings of EMNLP (October), pp. 1533–1544 (2013)
4. Bollacker, K., Cook, R., Tufts, P.: Freebase: a shared database of structured general human knowledge. In: Proceedings of the National Conference on Artificial Intelligence, vol. 22, no 2, p. 1962 (2007)
5. Bordes, A., Chopra, S., Weston, J.: Question answering with subgraph embeddings. In: Proceedings of the 2014 Conference on Empirical Methods in Natural Language Processing (EMNLP), 25–29 October, pp. 615–620 (2014). https://doi.org/10.3115/v1/D14-1067
6. Bordes, A., Usunier, N., Chopra, S., Weston, J.: Large-scale simple question answering with memory networks (2015). https://doi.org/10.1016/j.geomphys.2016.04.013
7. Buhrmester, M.D., Talaifar, S., Gosling, S.D.: An evaluation of amazon's mechanical turk, its rapid rise, and its effective use. Perspect. Psychol. Sci. **13**(2), 149–154 (2018). https://doi.org/10.1177/1745691617706516
8. Bulcaen, C.: Rethinking context: language as an interactive phenomenon. Lang. Lit. **4**(1), 61–64 (1995). https://doi.org/10.1177/096394709500400105
9. Cai, Q., Yates, A.: Large-scale semantic parsing via schema matching and lexicon extension. In: Proceedings of the 51st Annual Meeting of the Association for Computational Linguistics (Volume 1: Long Papers), pp. 423–433. Association for Computational Linguistics, Sofia, Bulgaria, August 2013
10. Cheung, J.H., Burns, D.K., Sinclair, R.R., Sliter, M.: Amazon mechanical turk in organizational psychology: an evaluation and practical recommendations. J. Bus. Psychol. **32**(4), 347–361 (2016). https://doi.org/10.1007/s10869-016-9458-5

---

[16] https://www.volkswagenag.com/en/group/research---innovations.html.

11. Dhingra, B., Danish, D., Rajagopal, D.: Simple and effective semi-supervised question answering. In: Proceedings of the 2018 Conference of the North American Chapter of the Association for Computational Linguistics: Human Language Technologies, Volume 2 (Short Papers), pp. 582–587. Association for Computational Linguistics, Stroudsburg, PA, USA (2018). https://doi.org/10.18653/v1/N18-2092
12. Diefenbach, D., Lopez, V., Singh, K., Maret, P.: Core techniques of question answering systems over knowledge bases: a survey. Knowl. Inf. Syst. **55**(3), 529–569 (2017). https://doi.org/10.1007/s10115-017-1100-y
13. Diefenbach, D., Tanon, T.P., Singh, K., Maret, P.. Question answering benchmarks for Wikidata. In: CEUR Workshop Proceedings, vol. 1963, pp. 3–6 (2017)
14. Dubey, M., Banerjee, D., Abdelkawi, A., Lehmann, J.: LC-QuAD 2.0: a large dataset for complex question answering over Wikidata and DBpedia. In: Ghidini, C., et al. (eds.) ISWC 2019. LNCS, vol. 11779, pp. 69–78. Springer, Cham (2019). https://doi.org/10.1007/978-3-030-30796-7_5
15. Dunn, M., Sagun, L., Higgins, M., Guney, V.U., Cirik, V., Cho, K.: SearchQA: a new q&a dataset augmented with context from a search engine (2017)
16. Hara, K., Le, V., Froehlich, J.: Combining crowdsourcing and google street view to identify street-level accessibility problems. In: Proceedings of the SIGCHI Conference on Human Factors in Computing Systems - CHI '2013, p. 631 (2013). https://doi.org/10.1145/2470654.2470744
17. Höffner, K., Walter, S., Marx, E., Usbeck, R., Lehmann, J., Ngonga Ngomo, A.C.: Survey on challenges of question answering in the semantic web. Semant. Web **8**(6), 895–920 (2017). https://doi.org/10.3233/SW-160247
18. Janarthanam, S., et al.: Evaluating a city exploration dialogue system combining question-answering and pedestrian navigation. In: 51st Annual Meeting of the Association of Computational Linguistics (October 2015), pp. 1660–1668 (2013). https://doi.org/10.18411/a-2017-023
19. Janarthanam, S., et al.: Integrating location, visibility, and question-answering in a spoken dialogue system for pedestrian city exploration. In: Proceedings of the 13th Annual Meeting of the Special Interest Group on Discourse and Dialogue (July), pp. 134–136 (2012)
20. Jia, Z., Abujabal, A., Roy, R.S., Strötgen, J., Weikum, G.: TempQuestions: a benchmark for temporal question answering. In: WWW (Companion Volume), vol. 2, pp. 1057–1062 (2018)
21. Kočiský, T., et al.: The narrativeQA reading comprehension challenge. Trans. Assoc. Comput. Linguist. **6**, 317–328 (2018). https://doi.org/10.1162/tacl_a_00023
22. Liu, Y., Alexandrova, T., Nakajima, T.: Using stranger [sic] as sensors: temporal and geo-sensitive question answering via social media. In: WWW '13: Proceedings of the 22nd international conference on World Wide Web, pp. 803–813 (2013). https://doi.org/10.1145/2488388.2488458
23. Petrochuk, M., Zettlemoyer, L.: SimpleQuestions nearly solved: a new upperbound and baseline approach. In: Proceedings of the 2018 Conference on Empirical Methods in Natural Language Processing, pp. 554–558. Association for Computational Linguistics, Stroudsburg, PA, USA (2018). https://doi.org/10.18653/v1/D18-1051
24. Rajpurkar, P., Jia, R., Liang, P.: Know what you don't know: unanswerable questions for SQuAD. In: ACL 2018–56th Annual Meeting of the Association for Computational Linguistics, Proceedings of the Conference (Long Papers) 2, pp. 784–789 (2018)

25. Rajpurkar, P., Zhang, J., Lopyrev, K., Liang, P.: SQuAD: 100,000+ questions for machine comprehension of text. In: Proceedings of the 2016 Conference on Empirical Methods in Natural Language Processing (ii), pp. 2383–2392 (2016). https://doi.org/10.18653/v1/D16-1264
26. Salmen, J., Houben, S., Schlipsing, M.: Google street view images support the development of vision-based driver assistance systems. In: Proceedings of the IEEE Intelligent Vehicles Symposium (June 2012), pp. 891–895 (2012). https://doi.org/10.1109/IVS.2012.6232195
27. Suchanek, F.M., Kasneci, G., Weikum, G.: Yago: a core of semantic knowledge. In: Proceedings of the 16th International Conference on World Wide Web (2007). https://doi.org/10.1145/1242572.1242667
28. Trivedi, P., Maheshwari, G., Dubey, M., Lehmann, J.: LC-QuAD: a corpus for complex question answering over knowledge graphs. In: d'Amato, C., et al. (eds.) ISWC 2017. LNCS, vol. 10588, pp. 210–218. Springer, Cham (2017). https://doi.org/10.1007/978-3-319-68204-4_22
29. Unger, C., et al.: Question answering over linked data (QALD-4). In: CEUR Workshop Proceedings, vol. 1180, pp. 1172–1180 (2014)
30. Unger, C., et al.: Question answering over linked data (QALD-5). In: CLEF, vol. 1180, pp. 1172–1180 (2015)
31. Usbeck, R., Ngomo, A.C.N., Haarmann, B., Krithara, A., Röder, M., Napolitano, G.: 7th open challenge on question answering over linked data (QALD-7). Commun. Comput. Inf. Sci. **769**, 59–69 (2017). https://doi.org/10.1007/978-3-319-69146-6_6
32. Usbeck, R., et al.: Benchmarking question answering systems. Semant. Web **1**, 1–5 (2016)
33. Yang, M.C., Duan, N., Zhou, M., Rim, H.C.: Joint relational embeddings for knowledge-based question answering. In: Proceedings of the 2014 Conference on Empirical Methods in Natural Language Processing (EMNLP), pp. 645–650 (2014). https://doi.org/10.3115/v1/D14-1071
34. Yang, Z., Hu, J., Salakhutdinov, R., Cohen, W.: Semi-supervised QA with generative domain-adaptive nets. In: Proceedings of the 55th Annual Meeting of the Association for Computational Linguistics (Volume 1: Long Papers), pp. 1040–1050. Association for Computational Linguistics, Stroudsburg, PA, USA (2017). https://doi.org/10.18653/v1/P17-1096
35. Yang, Z., et al.: HotpotQA: a dataset for diverse, explainable multi-hop question answering. In: Proceedings of the 2018 Conference on Empirical Methods in Natural Language Processing, pp. 2369–2380. Association for Computational Linguistics, Stroudsburg, PA, USA (2019). https://doi.org/10.18653/v1/D18-1259
36. Yatskar, M.: A qualitative comparison of CoQA, SQuAD 2.0 and QuAC. NAACL-HLT, September 2018
37. Yih, W.T., Chang, M.W., He, X., Gao, J.: Semantic parsing via staged query graph generation: question answering with knowledge base. In: ACL, pp. 1321–1331 (2015). https://doi.org/10.3115/v1/P15-1128
38. Yin, Z., Goldberg, D.W., Zhang, C., Prasad, S.: An NLP-based question answering framework for spatio-temporal analysis and visualization. In: ACM International Conference Proceeding Series Part, vol. F1482, pp. 61–65 (2019). https://doi.org/10.1145/3318236.3318240

HDTCat: Let's Make HDT Generation Scale

Dennis Diefenbach[1,2](✉) and José M. Giménez-García[2]

[1] The QA Company SAS, Saint-Étienne, France
dennis.diefenbach@qanswer.eu
[2] Univ Lyon, UJM-Saint-Étienne, CNRS, Laboratoire Hubert Curien, UMR 5516,
42023 Saint-Étienne, France
jose.gimenez.garcia@univ-st-etienne.fr

Abstract. Data generation in RDF has been increasing over the last years as a means to publish heterogeneous and interconnected data. RDF is usually serialized in verbose text formats, which is problematic for publishing and managing huge datasets. HDT is a binary serialization of RDF that makes use of compact data structures, making it possible to publish and query highly compressed RDF data. This allows to reduce both the volume needed to store it and the speed at which it can be transferred or queried. However, it moves the burden of dealing with huge amounts of data from the consumer to the publisher, who needs to serialize the text data into HDT. This process consumes a lot of resources in terms of time, processing power, and especially memory. In addition, adding data to a file in HDT format is currently not possible, whether this additional data is in plain text or already serialized into HDT.

In this paper, we present HDTCat, a tool to merge the contents of two HDT files with low memory footprint. Apart from creating an HDT file with the added data of two or more datasets efficiently, this tool can be used in a divide-and-conquer strategy to generate HDT files from huge datasets with low memory consumption.

Keywords: RDF · Compression · HDT · Scalability · Merge · HDTCat

1 Introduction

RDF (Resource Description Framework)[1] is the format used to publish data in the Semantic Web. It allows to publish and integrate heterogeneous data. There exists a number of standard RDF serializations in plain text (N-triples, RDF/XML, Turtle, ...). While these serializations make RDF easy to process, the resulting files tend to be voluminous. A common solution consists of using a universal compressor (like bzip2) on the data before publication. This solution, however, requires the decompression of the data before using it by the consumer.

---

[1] https://www.w3.org/TR/rdf11-concepts/.

© Springer Nature Switzerland AG 2020
J. Z. Pan et al. (Eds.): ISWC 2020, LNCS 12507, pp. 18–33, 2020.
https://doi.org/10.1007/978-3-030-62466-8_2

HDT (Header-Dictionary-Triples) is a binary serialization format that encodes RDF data in two main components: The Dictionary and the Triples. The Dictionary gives an ID to each term used in the data. These IDs are used in the Triples part to encode the graph structure of the data. Both components are serialized in compressed space using compact data structures that allow the data to be queried without the need to decompress it beforehand. Because of this, HDT has become the center piece of RDF data stores [2,11], public query end-points [12], or systems for query answering in natural language [3,4]. However, the serialization process requires important amounts of memory, hampering its scalability. In addition, the current workflow to serialize RDF into HDT does not cover use cases such as adding data to an existing HDT file or merging two separate HDT files into one. This forces a user to fully decompress the HDT file.

In this paper we present HDTCat, a tool to merge two HDT files. This allows several functionalities: (1) to create an HDT file that combines the data of two HDT files without decompressing them, (2) to add data to an existing HDT file, by compressing this data first into HDT and then merging with the existing file, or (3) compressing huge datasets of RDF into HDT, by the means of splitting the data in several chunks, compressing each one separately and then merging them.

The rest of the paper is organized as follows: Sect. 2 presents background information about RDF and HDT, as well as related work on scalability of HDT serialization. Section 3 describes the algorithms of HDTCat. Section 4 shows how HDT performs against current alternatives. Finally, in Sect. 5 we give some closing remarks and present current and future lines of work for HDTCat.

2 Background

In this section, we provide basic background knowledge about RDF and how it is serialized into HDT. This is necessary to understand the approach to merge two HDT files.

2.1 RDF

RDF is the data model used in the Semantic Web. The data is organized in *triples* in the form (s, p, o), where s (the subject) is the resource being described, p (the predicate) is the property that describes it, and o (the object) is the actual value of the property. An object can be either a resource or a literal value. In a set of triples, resources can appear as subject or object in different triples, forming a directed labeled graph, which is known as *RDF graph*. Formal definitions for RDF triple and RDF graph (adapted from [9]) can be seen in Definition 1 and 2, respectively.

Definition 1 (RDF triple). *Assume an infinite set of terms* $\mathcal{N} = \mathcal{I} \cup \mathcal{B} \cup \mathcal{L}$, *where* $\mathcal{I}$, $\mathcal{B}$, *and* $\mathcal{L}$ *are mutually disjoint, and* $\mathcal{I}$ *are IRI references,* $\mathcal{B}$ *are Blank Nodes, and* $\mathcal{L}$ *are Literals. An RDF triple is a tuple* $(s, p, o) \in (\mathcal{I} \cup \mathcal{B}) \times \mathcal{I} \times (\mathcal{I} \cup \mathcal{B} \cup \mathcal{L})$, *where "s" is the subject, "p" is the predicate and "o" is the object.*

Definition 2 (RDF graph). *An RDF graph G is a set of RDF triples of the form* (s, p, o). *It can be represented as a directed labeled graph whose edges are* $s \xrightarrow{p} o$.

Example 1. The following snippet show an RDF file, that we call RDF_1, in N-Triples format:

```
<so1> <p1> <o1>.
<so1> <p1> <o2>.
<s1>  <p2> <so1>.
```

Moreover we denote as RDF_2 the following RDF file in N-Triples:

```
<so1> <p3> <o2>.
<o2> <p1> <s1>.
```

We will use these two files as running examples and show how they can be compressed and merged using HDTCat.

2.2 HDT

HDT [6] is a binary serialization format for RDF based on compact data structures. Compact data structures are data structures that compress the data as close as possible to its theoretic lower bound, but allow for efficient query operations. HDT encodes an RDF graph as a set of three components: (1) Header, that contains the metadata about the file and the data itself; (2) Dictionary, which assigns an unambiguous ID to each term appearing in the data; and (3) Triples, that replaces the terms by their ID in the dictionary and encodes them in a compressed structure. While HDT allows for different implementations of both Dictionary and Triples components, efficient default implementations are currently published. These implementations are the Four-Section Dictionary and the Bitmap Triples. We provide brief descriptions of those implementations down below.

The *Header* component stores metadata information about the RDF dataset and the HDT serialization itself. This data can be necessary to read the other sections of an HDT file. The *Dictionary* component stores the different IRIs, blank nodes, and literals, and assigns to each one an unambiguous integer ID. The *Triples* component stores the RDF graph, where all the terms are replaced by the ID assigned in the Dictionary component. From now on to represent an HDT file, we write $HDT = (H, D, T)$, where H is the header component, D is the dictionary component, and T is the triples component. In theory, each component allows different encoding. In practice, however, current compression formats are based in sorting lexicographically their elements. We describe thereafter characteristics of current HDT encoding.

In the Four-Section Dictionary an integer ID is assigned to each term (IRI, Blank Node and Literal). The set of terms is divided into four sections: (1) the *Shared* section, that stores the terms that appear at the same time as subjects and objects of triples; the *Subjects* section, which stores the terms that appear exclusively as subjects of triples; the *Objects* section, which contains the terms that appear only as object of triples; and finally the *Predicates* section, storing the terms that appear as predicates of the triples. From now on, we write the Dictionary as a tuple $D = (SO, S, O, P)$, where SO is the shared section, S is the subjects section, O is the objects section, and P is the predicates section. In each section the terms are sorted lexicographically and compressed (*e.g.*, using Plain [1] or Hu-Tucker FrontCoding [10]). The position of each term is then used as its implicit ID in each section. This way to each term an integer is assigned in a space-efficient way. The dictionary needs to provide global IDs for subjects and objects, independently of the section in which they are stored. Terms in P and SO do not change, while IDs for S and O sections are increased by the size of SO (*i.e.*, $ID_S := ID_S + \max(ID_{SO})$ and $ID_O := ID_O + \max(ID_{SO})$).

Example 2. Consider the file RDF_1 from Example 1. We call the corresponding HDT file $HDT[1] = (H_1, D_1, T_1)$ with $D_1 = (SO_1, S_1, O_1, P_1)$. The dictionary sections look as follows (note that the compression is not shown here as it is not important to understand HDTCat):

SO_1			S_1			O_1			P_1	
IRI	ID		IRI	ID		IRI	ID		IRI	ID
<so1>	1		<s1>	2		<o1>	2		<p1>	1
						<o2>	3		<p2>	2

Note that the ids in the S_1 and O_2 section start by 2 since there is one entry in the common section SO_1. Similarly for RDF_2 we get HDT_2 with:

SO_2			S_2		O_2			P_2	
IRI	ID		IRI	ID	IRI	ID		IRI	ID
<o2>	1				<s1>	3		<p1>	1
<so1>	2							<p3>	2

□

In the triples component T, each term in the triples is replaced by the ID from the dictionary and sorted in what is known as *Plain Triples*. The ordering is defined in the following.

Definition 3. If $T_1 = (s_1, p_1, o_1)$ and $T_2 = (s_2, p_2, o_2)$ are two triples then $T_1 \geq T_2$ if and only if:

1. $s_1 \geq s_2$;
2. if $s_1 = s_2$ then $p_1 \geq p_2$;
3. if $s_1 = s_2$ and $p_1 = p_2$ then $o_1 \geq o_2$;

Example 3. The triples from RDF_1 from Example 2 in *Plain Triples* are:

1 1 2
1 1 3
2 2 1

Note that they respect the order defined in Definition 3. The one from RDF_2 are:

1 1 3
2 2 1

Note that the triples were reordered.

$\square$

The triples can be compressed in *Compact Triples*, which uses two coordinated sequences of IDs, Q_P and Q_O, to store the IDs of predicates and objects respectively, in the order they appear in the sorted triples. The first ID in Q_P is assumed to have the subject with $ID_s = 1$. Each following ID is assumed to have the same ID as its predecessor. If the ID 0 appears in the sequence, it means a change to the following ID (*i.e.*, the ID is incremented by one). Respectively, the first ID in Q_O is matched with the property in the first position of Q_P. Each following ID is assumed to have the property as its predecessor, and if the ID 0 appears in the sequence, it means a change to the following ID (that is, the next ID in Q_P). This can be further compressed in *BitMap Triples* by removing the 0 from the ID sequences and adding two bit sequences, B_P and B_O, that mark the position where the change of subject (for Q_P) or predicate (for Q_O) happen. Note that the data-structures described above allow fast retrieval of all triple patterns with fixed subject. Some more indexes are added to resolve fast triple patterns with fixed predicate or object. Moreover, note that due to the global ordering updates are not supported.

2.3   Works on Scalability of HDT

To the best of our knowledge, there are only two publications that deal with scalable HDT generation. The first one is HDT-MR [7], a MapReduce-based tool to serialize huge datasets in RDF into HDT. MDT-MR has proven able to compress more than 5 billion triples into HDT. However, HDT-MR needs a MapReduce cluster to compress the data, while HDTCat can run in a single computer.

A single HDT file containing over 28 billion triples has been published in LOD-a-lot [5]. The aim was to generate a snapshot of all current RDF triples in the LOD cloud. However, both the algorithm and tool used to create this HDT file are not public. HDTCat tries to fill this gap making the algorithm and tool needed to create HDT files of this size open to the public.

3   HDTCat

In this section we describe the HDTCat algorithm. Given two HDT files HDT_1 and HDT_2, HDTCat generates a new HDT file HDT_{cat} that contains the union of the triples in HDT_1 and HDT_2. Its goal is to achieve this in a scalable way, in particular in terms of memory footprint, since this is generally the limited resource on current hardware.

Let's assume two HDT files, $HDT_1 = (H_1, D_1, T_1)$ and $HDT_2 = (H_2, D_2, T_2)$ are given. The current solution to merge these two HDT files is to first decompress them into text. Then, the two text files are concatenated, and the resulting file is serialized again into HDT. Basically, two ordered lists are put one after the other and ordered again without exploiting their initial order. The problem addressed by HDTCat is how to merge the dictionaries D_1, D_2 and the triples T_1, T_2 without decompressing them, so that the resulting HDT file contains the union of the RDF triples. The result of of merging the two HDT files needs to be the same as the serialization of the contatenation of the two uncompressed files, that is rdf2hdt(RDF_1+RDF_2)=hdtcat(rdf2hdt(RDF_1),rdf2hdt(RDF_2)).

The algorithm can be decomposed into three phases:

1. Joining the dictionaries,
2. Joining the triples,
3. Generating the header.

For the two first phases, HDTCat uses merge-sort-based algorithms that take advantage of the initial ordering of the HDT components. The general idea of the algorithms is described in Fig. 1. Briefly, there are two iterators over the two lists. Recursively, the current entries of the two iterators are compared and the lowest entry is added to the final list.

There are two important consequences. Imagine the two components have n respectively m entries. The first consequence is that the time complexity is reduced. If two components are merged by first decompressing and then serializing their union, the time complexity is $O((n + m) \cdot log(n + m))$ because of the need to merge an unsorted set of triples. However, when sorting two already sorted lists, using the algorithm above, the time complexity is $O(n + m)$. The second, and in our eyes the more important, is the memory consumption. The existing approach to serialize RDF into HDT stores every uncompressed triple in memory so that the memory complexity is in the order of $O(n+m)$. Iterating over the sorted lists by letting them compressed, and decompressing only the current entry, reduces the memory complexity to $O(1)$ This explains the main idea behind HDTCat. We are now going to explain more in detail the merging strategy and the data-structures needed.

3.1   Joining the Dictionary

Assume two HDT dictionaries $D_1 = (SO_1, S_1, O_1, P_1)$ and $D_2 = (SO_2, S_2, O_2, P_2)$. The goal is to create a new HDT dictionary $D_{cat} = (SO_{cat}, S_{cat}, O_{cat}, P_{cat})$.

Merging the sections P_1 and P_2 is a simple process. P_1 and P_2 are two arrays of ordered compressed strings. Algorithm 1 assumes that there are two iterators over the two lists. Recursively, the current entries of the two iterators are compared and the lowest entry is added to the final list. To compare the entries they are decompressed, and the new entry is compressed directly and

Data: Two sorted lists a and b
Result: A sorted list c containing all entities in a and b

```
 1  n= length of a; m = length of b
 2  allocate c with length n+m
 3  i = 1; j = 1
 4  while i < n || j < m do
 5    │  if i = n then
 6    │  │  copy rest of b into c
 7    │  │  break
 8    │  end
 9    │  if j = m then
10    │  │  copy rest of a into c
11    │  │  break
12    │  end
13    │  if a[i] < b[j] then
14    │  │  copy a[i] into c
15    │  │  i=i+1
16    │  end
17    │  if b[j] < a[i] then
18    │  │  copy b[j] into c
19    │  │  j=j+1
20    │  end
21    │  if a[i] = b[j] then
22    │  │  copy a[i] into c
23    │  │  i=i+1
24    │  │  j=j+1
25    │  end
26  end
```

Algorithm 1: Algorithm to merge two sorted lists. Note that the algorithm has a time complexity of $O((n+m))$. All computation do not need to be done on RAM but can be performed on disk.

added to P_{cat}. Note that since the strings are uncompressed and compressed directly, the memory footprint remains low.

Example 4. The predicate section of HDT_{cat} is:

P_{cat}

IRI	ID
<p1>	1
<p2>	2
<p3>	3

□

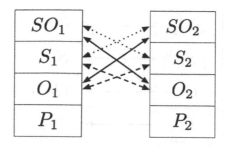

Fig. 1. This figure shows the non-trivial sections that can share an entry. Clearly SO_1 and SO_2, S_1 and S_2, O_1 and O_2, P_1 and P_2 can contain common entries. The other sections that can contain common entries are connected by a double arrow. It is important to take care of these common entries when merging the dictionaries.

Merging the other sections needs to take into account, however, that some terms can move to different sections in the HDT files to be merged. For example, if S_1 contains an IRI that appears also in O_2. Figure 1 shows the sections that can contain common elements (excluding the non-trivial cases). The following cases need to be taken into account:

- If SO_1 and S_2, or S_1 and SO_2 contain common entries, then they must be skipped when joining the S sections.
- If SO_1 and O_2, or O_1 and SO_2 contain common entries, then they must be skipped when joining the O sections.
- If S_1 and O_2, or O_1 and S_2 contain common entries then they must be skipped when joining the S and O sections, and additionally they must be added to the SO_{cat} section.

For this reason, terms can be assigned to different sections in the final HDT dictionary. For the example where S_1 contains as IRI that appears also in O_2, this IRI should be assigned to the section SO_{cat}, since the IRI will appear both in the subject and the object of some triples. Figure 2 shows to which sections the terms can be assigned depending on where they are in the initial dictionaries.

Example 5. The sections of HDT_{cat} different from P_{cat} look like this:

SO_{cat}		S_{cat}		O_{cat}	
IRI	ID	IRI	ID	IRI	ID
<o2>	1			<o1>	4
<so1>	2				
<s1>	3				

Note that the IRI <s1> moved from section S_1 to section SO_{cat}.

□

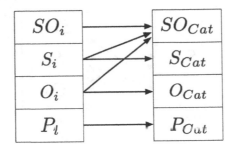

Fig. 2. This figure shows to which sections of the HDT_{cat} dictionary, the entries from the dictionary section of either HDT_1 or HDT_2 can move. The SO section and the P section ids are going to the SO_{cat} and P_{cat} section respectively. If there is an entry that appears both in the S section and the O section, then the corresponding entry will go to the SO_{cat} section. Otherwise the entry goes to the S or O section.

To store the merged sections of HDT_{cat}, since they are written sequentially, data-structures stored on disk can be used, reducing their memory complexity to $O(1)$.

When joining the triples in the next step, it will be necessary to know the correspondence between the IDs in D_1 and D_2, and the IDs in D_{cat}. To keep track of those mappings, we introduce data structures that, for each ID in the section $Sec \in \{SO_1, S_1, O_1, P_1, SO_2, S_2, O_2, P_2\}$,. assign the new ID in the corresponding section $Sec_{cat} \in \{SO_{cat}, S_{cat}, O_{cat}, P_{cat}\}$. For one section Sec the data structure contains two arrays:

1. An array indicating, for each ID of Sec, which is the corresponding section in Sec_{cat}.
2. An array mapping the IDs of Sec to the corresponding ID in the section Sec_{cat}.

We indicate every such mapping as M(Sec). Moreover, we construct also the mappings form SO_{cat}, S_{cat} (note: the IDs of these two sections are consecutive) to SO_1, S_1 and SO_2, S_2 respectively. This consists of two arrays:

1. An array indicating ,for each ID of SO_{cat} or S_{cat}, the corresponding ID in SO_1, S_1 (if it exists).
2. An array indicating for each ID of SO_{cat} or S_{cat}, the corresponding ID in SO_2, S_2 (if it exists).

The arrays are directly written to disk. We indicate the two mappings as M(cat,1) and M(cat,2) .

Example 6. The mappings for HDT_{cat} are as follows:

M(SO_1)		
ID	Sec$_{cat}$	ID$_{cat}$
1	SO_{cat}	2

M(S_1)		
ID	Sec$_{cat}$	ID$_{cat}$
2	SO_{cat}	3

M(O_1)		
ID	Sec$_{cat}$	ID$_{cat}$
2	O_{cat}	4
3	SO_{cat}	1

M(P_1)		
ID	Sec$_{cat}$	ID$_{cat}$
1	P_{cat}	1
2	P_{cat}	2

M(SO_2)		
ID	Sec$_{cat}$	ID$_{cat}$
1	SO_{cat}	1
2	SO_{cat}	2

M(S_2)		
ID	Sec$_{cat}$	ID$_{cat}$

M(O_2)		
ID	Sec$_{cat}$	ID$_{cat}$
3	SO_{cat}	1

M(P_2)		
ID	Sec$_{cat}$	ID$_{cat}$
1	P_{cat}	1
2	P_{cat}	3

M(cat,1)	
ID$_{cat}$	ID$_{old}$
1	-
2	1
3	2

M(cat,2)	
ID$_{cat}$	ID$_{old}$
1	1
2	2
3	-

□

3.2 Joining the Triples

In this section we describe the process to merge the triples T_1, T_2 in HDTCat. This process exploits the fact that the triples are ordered only indirectly. That is, the fact that the HDT files are queriable.

Remember that by Definition 3 the triples need to be ordered first by subjects, then by predicates, and finally by objects. The order of the subjects is given by the subjects section in the merged dictionary S_{cat}. Then, for each ID in S_{cat}, we use the mappings $M(S[cat], S[1])$ and $M(S[cat], S[2])$ (constructed when joining the dictionary sections) to find the IDs ID_1 and ID_2 of the original HDT files HDT_1 and HDT_1 that mapped to ID_{cat} in HDT_{cat}. Since both HDT_1 and HDT_2 are queriable, we can retrieve all triples with subjects ID_1 and ID_2 respectively. By using again the mappings constructed when joining the dictionaries, we can now translate the IDs of these triples used in HDT_1 and HDT_2 to the corresponding IDs in HDT_{cat}. We generate the triples by iterating over the subjects and by writing the triples directly to disk.

Example 7. Let's first join the triples with $ID_{cat} = 1$. According to $M_{Cat,2}$ there are only triples in HDT_2 mapping to it. In fact there is only the triple:
1 2 3
By using the mappings of Example 6 this will become:
1 3 1
For $ID_{cat} = 2$ we search all triples associated to $ID_{cat} = 2$. These triples are:
1 1 2
1 1 3
in HDT_1 and:
2 2 1

in HDT_2. By using the mappings of Example 6 these correspond to the new IDs:
2 1 3
2 1 1
and:
2 3 1
Note that the triples of HDT_1 where initially ordered, while the mapped triples
are not $((2,1,3)>(2,1,1))$. The merged triples for $ID_{cat} = 2$ are then:
2 1 1
2 1 3
2 3 1

□

3.3   Creating the Header

While the dictionary and the triples must be merged from the corresponding
sections of the two HDT files, the header just contain some statistical information
like the number of triples and the number of distinct subjects. This means that
there is nothing to do here except writing the statistics corresponding to D_{cat}
and T_{cat} that have been generated.

4   Experiments

In this section we evaluate the performance of HDTCat. In particular we compare
the scalability of HDTCat when generating HDT files (starting from N-Triples
against (1) the regular HDT serialization, using the command line tool rdf2hdt
that is part of the HDT repository[2], and (2) HDT-MR. We perform three different experiments to compare how HDTCat performs in different situations.

Experiment 1. We use synthetic data generated using LUBM [8]. LUBM is a
benchmark to test the performance of SPARQL queries and contains both a tool
to generate synthetic RDF data and a set of SPARQL queries. The generated
RDF contains information about universities (like departments, students, pro-
fessors and so on). We generated the following LUBM datasets: (1) from 1000
to 8000 universities in steps of 1000, and (2) from 8000 to 40000 universities in
steps of 4000. We used 3 methods to compress these files to HDT:

- **rdf2hdt:** We concatenate the LUBM datasets generated to obtain the
  datasets of increasing size by steps of 1000 universities up to 8000 universities, then we increase the steps by 4000 universities. We then used rdf2hdt
  to generate the corresponding HDT files.
- **HDT-MR:** HDT-MR is used in the same way as rdf2hdt, using the same
  concatenated files, and then converted to HDT.

---

[2] https://github.com/rdfhdt/hdt-java.

- **HDTCat:** We first serialized the generated datasets into HDT, then we used HDTCat to recursively compute the merged HDT files. *I.e.*, we generated lubm.1–2.000.hdt from lubm.1–1.000.hdt and lubm.1001–2.000.hdt; then lubm.1–3.000.hdt from lubm.1–2.000.hdt and lubm.2001–3.000.hdt; and so on.

We run the experiments for rdf2hdt and HDTCat on different hardware configurations:

- **Configuration 1:** A server with 128 Gb of RAM, 8 cores of type Intel(R) Xeon(R) CPU E5–2637 v3 @ 3.50 GHz. RAID-Z3 with 12x HDD 10TB SAS 12Gb/s 7200 RPM. We run hdt2rdf and hdtCat on this configuration. For the results of HDT-MR we report the ones achieved by [7], that where executed on a cluster with a total memory of 128 Gb of RAM. While rdf2hdt and HDTCat are designed to be used on a single server, HDT-MR is designed to be used on a cluster. To make the results comparable we choose a single node and a cluster configuration with the same amount of RAM since this is the limited resource for compressing RDF serializations to HDT.
- **Configuration 2:** A server with 32 Gb of RAM, 16 cores of type Intel(R) Xeon(R) CPU E5–2680 0 @ 2.70 GHz. RAID-Z3 with 12x HDD 10TB SAS 12Gb/s 7200 RPM.
- **Configuration 3:** A desktop computer with 16 Gb of RAM, AMD A8–5600K with 4 cores. 1x HDD 500GB SCSI 6 Gb/s, 7200 RPM.

Note that while the two first configurations have a RAID deployment with 10 drives, the third one is limited to a single HDD. Since HDTCat is I/O intensive, this can affect its performance.

The results obtained by the 3 methods on the 3 hardware configurations are shown in Table 1. It summarizes the comparison between the three methods to generate HDT from other N-Triples of LUBM datasets. **T** indicates the time and **M** the maximal memory consumption of the process. In the case of HDTCat we also report T_{com} the time to compress the N-Triples into HDT and T_{cat} the time to cat the two files together. $\star$ indicates that the experiment failed with an OUT OF MEMORY error. "−" indicates that the experiment was not performed. This has two reasons. Either a smaller experiment failed with an OUT OF MEMORY, or the experiment with HDT-MR was not performed on the corresponding configuration. The experiments in the T_com column are very similar because we compress similar amount of data. We report the average times of these experiments and indicated that with "$\star$".

The results for Configuration 1 show that while hdt2rdf fails to compress lubm-12000, by using HDTCat we are able to compress lubm-40000. This means that one can compress at least as much as the HDT-MR implementation. Note that lubm-40000 does not represent an upper bound for both methods. For lubm-8000, HDT-MR is 121% faster then HDTCat. This is expected since HDT-MR exploits parallelism while HDTCat does not. Moreover while the single node configuration has HDD disks, the cluster configuration used SSD disks. For

Table 1. Comparison between methods to serialize RDF into HDT.

Configuration 1: 128 Gb RAM

LUBM	Triples	hdt2rdf		HDT-MR	HDTCat			
		T (s)	M (Gb)	T (s)	T_com (s)	T_cat (s)	T (s)	M_cat (Gb)
1000	0.13BN	1856	53.4	936	970*	–	–	—
2000	0.27BN	4156	70.1	1706		317	2257	26.9
3000	0.40BN	6343	89.3	2498		468	3695	35.4
4000	0.53BN	8652	105.7	3113		620	5285	33.8
5000	0.67BN	11279	118.9	4065		803	7058	41.7
6000	0.80BN	23595	122.7	4656		932	8960	47.5
7000	0.93BN	78768	123.6	5338		1088	11018	52.9
8000	1.07BN	⋆	⋆	6020		1320	13308	58.7
12000	1.60BN	–	–	9499	4710*	1759	19777	54.7
16000	2.14BN	—	–	13229		2338	26825	73.4
20000	2.67BN	–	–	15720		2951	34486	90.5
24000	3.20BN	–	–	26492		3593	42789	90.6
28000	3.74BN	–	–	36818		4308	51807	84.9
32000	4.27BN	–	–	40633		4849	61366	111.1
36000	4.81BN	–	–	48322		6085	72161	109.4
40000	5.32BN	–	–	55471		7762	84633	100.1

Configuration 2: 32 Gb RAM

LUBM	Triples	HDT		-	HDTCat			
		T (s)	M (Gb)	-	T_com (s)	T_cat (s)	T (s)	M_cat (Gb)
1000	0.13BN	1670	28.3	-	1681*	–	–	–
2000	0.27BN	⋆	⋆	-		454	3816	17.3
3000	0.40BN	—	–	–		660	6366	20.1
4000	0.53BN	–	–	—		869	8916	25.5
5000	0.67BN	–	–	–		1097	11694	29.3
6000	0.80BN	–	–	–		1345	14720	28.5
7000	0.93BN	–	–	–		1584	17985	30.6
8000	1.07BN	–	–	–		1830	21496	30.4
12000	1.60BN	–	–	–	⋆	2748	-	31.0
16000	2.14BN	–	–	–	–	3736	—	31.1
20000	2.67BN	–	–	–	–	5007	–	30.5
24000	3.20BN	–	–	–	–	5514	–	30.8
28000	3.74BN	–	–	–	–	6568	–	30.8
32000	4.27BN	–	–	–	–	7358	–	30.8
36000	4.81BN	–	–	–	–	9126	–	30.6
40000	5.32BN	–	–	–	–	9711	–	30.8

Table 1. (*continued*)

Configuration 3: 16 Gb RAM

LUBM	Triples	HDT			HDTCat			
		T (s)	M (Gb)	–	T_com (s)	T_cat (s)	T (s)	M_cat (Gb)
1000	0.13BN	2206	14.5	–	2239*	–	–	–
2000	0.27BN	★	★	–		517	4995	10.7
3000	0.40BN	–	–	–		848	8082	11.8
4000	0.53BN	–	–	–		1301	11622	11.9
5000	0.67BN	–	–	–		1755	15616	12.7
6000	0.80BN	–	–	–		2073	19928	11.8
7000	0.93BN	–	–	–		2233	24400	12.6
8000	1.07BN	–	–	–		3596	30235	12.2
12000	1.60BN	–	–	–	★	4736	–	14.3
16000	2.14BN	–	–	–	–	6640	–	14.3
20000	2.67BN	–	–	–	–	9058	–	14.4
24000	3.20BN	–	–	–	–	10102	–	14.3
28000	3.74BN	–	–	–	–	13287	–	12.8
32000	4.27BN	–	–	–	–	14001	–	13.9
36000	4.81BN	–	–	–	–	17593	–	14.0
40000	5.32BN	–	–	–	–	19929	–	13.9

lubm-40000 the speed advantage reduces, HDT-MR is 52% faster then HDT-Cat. The results for Configuration 2 show that the speed of hdtCat to compress lubm-40000 in comparison to Configuration 1 is reduced, but only by 25%. The results for Configuration 3 show that it is possible to compress on a 16 Gb machine HDT files containing 5 Billion triples. In particular this means that it is possible to index on a 16Gb machine an RDF file with 5 Billion triples and construct a SPARQL endpoint on top. This is unfeasible for every other SPARQL endpoint implementation we are aware of. Moreover this also shows that for Configuration 1, lubm-4000 is far from being an upper bound so that potentially huge RDF files can be indexed, which was not imaginable before.

Experiment 2. While the above results are using the synthetic data provided by LUBM we also performed an experiment using real datasets. In particular we join the Wikidata dump of the 19-02-2018 (330G in ntriple format) and the 2016 DBpedia dump[3] (169G in ntriple format). This corresponds to 3.5 billion

---

[3] All files retrieved by: wget -r -nc -nH –cut-dirs=1 -np -l1 -A '*ttl.bz2' -A '*.owl'-R '*unredirected*'–tries 2 http://downloads.dbpedia.org/2016-10/core-i18n/en/, i.e. all files published in the english DBpedia. We exclude the following files: nif_page_structure_en.ttl, raw_tables_en.ttl and page_links_en.ttl since they do not contain typical data used in application relying on DBpedia.

triples. We where able to join the corresponding HDT file in 143 min and 36 s using a 32 Gb RAM machine. The maximal memory consumption was 27.05 Gb.

Experiment 3. Note that Wikidata and DBpedia are not sharing many IRIs. So one valid argument is if HDTCat is also performing well when the two joined HDT files contain many common IRIs. To test this we randomly sorted the lubm 2 000 nt file and split it in two files containing the same amount of triples. We then join them using HDTCat. While joining lubm.1–1000.hdt and lubm.1001–2000.hdt took 287 s, joining the randomly sorted files took 431 s. This corresponds to a 66% increase of time which is expected. This shows that HDTCat is still performing well in such a scenario.

Code. The code for HDTCat is currently part of the HDT code repository available under https://github.com/rdfhdt/hdt-java. The code is released under the *Lesser General Public License* as the existing Java code. We also provide a command line tool, called rdf2hdtcat, that allows to compress HDT in a divide and conquer method (pull request #109) to easily serialize big RDF file to HDT.

5   Conclusion and Future Work

In this paper we have presented HDTCat, an algorithm and command line tool to merge two HDT files with improved time and memory efficiency. We have described in detailed how the algorithm works and we have compared our implementation against the other two available alternatives: regular HDT serialization and HDT-MR, a MapReduce-based designed to tackle scalability in HDT serialization. The experiments shows that it is possible to compress 5 billion triples on a 16 Gb machine which was not imaginable before.

Our future work include the creating of a tool that combines rdf2hdt and HDTCat to parallelize RDF serialization into HDT to generate HDT files faster. Moreover we are working on extending HDTCat to be able to merge an arbitrary number of HDT files simultaneously.

In the long term, we plan to work in the use of HDTCat to support updates on HDT-based tools. A strategy is to have a read-only index and to store the updates in a delta structure that is periodically merged (with HDTCat) with the read-only part.

Finally we believe that HDTCat will enable the Semantic Web Community to tackle scenarios which were non feasible before because of scalability.

Acknowledgements. We would like to thank Pedro Migliatti for executing part of the experiments as well as Javier D. Fernández for the helpful discussions with him. We also want to thank Dimitris Nikolopoulos and Wouter Beek from Triply for porting the algorithm to C++. This project has received funding from the European Union's Horizon 2020 research and innovation program under the Marie Sklodowska-Curie grant agreement No 642795.

References

1. Brisaboa, N.R., Cánovas, R., Martínez-Prieto, M.A., Navarro, G.: Compressed string dictionaries. CoRR abs/1101.5506 (2011). http://arxiv.org/abs/1101.5506
2. Curé, O., Blin, G., Revuz, D., Faye, D.C.: WaterFowl: a compact, self-indexed and inference-enabled immutable RDF store. In: Presutti, V., d'Amato, C., Gandon, F., d'Aquin, M., Staab, S., Tordai, A. (eds.) ESWC 2014. LNCS, vol. 8465, pp. 302–316. Springer, Cham (2014). https://doi.org/10.1007/978-3-319-07443-6_21
3. Diefenbach, D., Both, A., Singh, K., Maret, P.: Towards a question answering system over the semantic web. Semant. Web J. 1–19 (2020)
4. Diefenbach, D., Singh, K., Maret, P.: WDAqua-core0: a question answering component for the research community. In: Dragoni, M., Solanki, M., Blomqvist, E. (eds.) SemWebEval 2017. CCIS, vol. 769, pp. 84–89. Springer, Cham (2017). https://doi.org/10.1007/978-3-319-69146-6_8
5. Fernández, J.D., Beek, W., Martínez-Prieto, M.A., Arias, M.: LOD-a-lot. In: d'Amato, C., et al. (eds.) ISWC 2017. LNCS, vol. 10588, pp. 75–83. Springer, Cham (2017). https://doi.org/10.1007/978-3-319-68204-4_7
6. Fernández, J.D., Martínez-Prieto, M.A., Gutiérrez, C., Polleres, A., Arias, M.: Binary RDF representation for publication and exchange (HDT). J. Web Semant. **19**(22–41), 00124 (2013)
7. Giménez-García, J.M., Fernández, J.D., Martínez-Prieto, M.A.: HDT-MR: a scalable solution for RDF compression with HDT and MapReduce. In: Gandon, F., Sabou, M., Sack, H., d'Amato, C., Cudré-Mauroux, P., Zimmermann, A. (eds.) ESWC 2015. LNCS, vol. 9088, pp. 253–268. Springer, Cham (2015). https://doi.org/10.1007/978-3-319-18818-8_16
8. Guo, Y., Pan, Z., Heflin, J.: LUBM: a benchmark for OWL knowledge base systems. J. Web Semant. **3**(2), 158–182 (2005)
9. Gutierrez, C., Hurtado, C.A., Mendelzon, A.O., Pérez, J.: Foundations of semantic web databases. J. Comput. Syst. Sci. **77**(3), 520–541 (2011). https://doi.org/10.1016/j.jcss.2010.04.009
10. Hu, T.C., Tucker, A.C.: Optimal computer search trees and variable-length alphabetical codes. Siam J. Appl. Math. **21**(4), 514–532 (1971)
11. Martínez-Prieto, M., Arias, M., Fernández, J.: Exchange and consumption of huge RDF data. In: Proceeding of ESWC, pp. 437–452 (2012)
12. Verborgh, R., et al.: Querying Datasets on the Web with High Availability. In: Mika, Peter, Tudorache, Tania, Bernstein, Abraham, Welty, Chris, Knoblock, Craig, Vrandečić, Denny, Groth, Paul, Noy, Natasha, Janowicz, Krzysztof, Goble, Carole (eds.) ISWC 2014. LNCS, vol. 8796, pp. 180–196. Springer, Cham (2014). https://doi.org/10.1007/978-3-319-11964-9_12, http://linkeddatafragments.org/publications/iswc2014.pdf

Squirrel – Crawling RDF Knowledge Graphs on the Web

Michael Röder[1,2(✉)] [iD], Geraldo de Souza Jr[1] [iD],
and Axel-Cyrille Ngonga Ngomo[1,2] [iD]

[1] DICE Group, Department of Computer Science, Paderborn University,
Paderborn, Germany
{michael.roeder,gsjunior,axel.ngonga}@upb.de
[2] Institute for Applied Informatics, Leipzig, Germany

Abstract. The increasing number of applications relying on knowledge graphs from the Web leads to a heightened need for crawlers to gather such data. Only a limited number of these frameworks are available, and they often come with severe limitations on the type of data they are able to crawl. Hence, they are not suited to certain scenarios of practical relevance. We address this drawback by presenting SQUIRREL, an open-source distributed crawler for the RDF knowledge graphs on the Web, which supports a wide range of RDF serializations and additional structured and semi-structured data formats. SQUIRREL is being used in the extension of national data portals in Germany and is available at https://github.com/dice-group/squirrel under a permissive open license.

Keywords: Linked data · Crawler · Open data

1 Introduction

The knowledge graphs available on the Web have been growing over recent years both in number and size [4][1]. This development has been accelerated by governments publishing public sector data on the web[2]. With the awareness of the power of 5-star linked open data has come the need for these organizations to (1) make the semantics of their datasets explicit and (2) connect their datasets with other datasets available on the Web. While the first step has been attended to in a plethora of projects on the semantification of data, the second goal has remained a challenge, addressed mostly manually. However, a manual approach to finding and linking datasets is impractical due to the steady growth of datasets provided by both governments and the public sector in both size and

---

[1] See https://lod-cloud.net/ for an example of the growth.
[2] Examples include the European Union at https://ec.europa.eu/digital-single-market/en/open-data and the German Federal Ministry of Transport and Digital Infrastructure with data at https://www.mcloud.de/.

© Springer Nature Switzerland AG 2020
J. Z. Pan et al. (Eds.): ISWC 2020, LNCS 12507, pp. 34–47, 2020.
https://doi.org/10.1007/978-3-030-62466-8_3

number[3]. Different public services have hence invested millions of Euros into research projects aiming to automate the connection of government data with other data sources[4].

An indispensable step towards automating the second goal is the *automated and periodic gathering of information about available open data that can be used for linking to newly published data of the public sector*. A necessary technical solution towards this end is a *scalable crawler for the Web of Data*. While the need for such a solution is already dire, it will become even more pressing to manage the growing amount of data that will be made available each year into the future. At present, the number of open-source crawlers for the web of data that can be used for this task is rather small and all come with several limitations. We close this gap by presenting SQUIRREL—a distributed, open-source crawler for the web of data[5]. SQUIRREL supports a wide range of RDF serializations, decompression algorithms and formats of structured data. The crawler is designed to use Docker[6] containers to provide a simple build and run architecture [13]. SQUIRREL is built using a modular architecture and is based on the concept of dependency injection. This allows for a further extension of the crawler and adaptation to different use cases.

The rest of this paper is structured as follows: we describe related work in Sect. 2 and the proposed crawler in Sect. 3. Section 4 presents an evaluation of the crawler, while Sect. 5 describes several applications of SQUIRREL. We conclude the paper in Sect. 6.

2 Related Work

There are only a small number of open-source Data Web crawlers available that can be used to crawl RDF datasets. An open-source Linked Data crawler to crawl data from the web is LDSpider[7] [10]. It can make use of several threads in parallel to improve the crawling speed, and offers two crawling strategies. The breadth-first strategy follows a classical breadth-first search approach for which the maximum distance to the seed URI(s) can be defined as termination criteria. The load-balancing strategy tries to crawl URIs in parallel without overloading the servers hosting the data. The crawled data can be stored either in files or can be sent to a SPARQL endpoint. It supports a limited amount of RDF serialisations (details can be found in Table 1 in Sect. 3). In addition, it cannot be deployed in a distributed environment. Another limitation of LDSpider is the missing functionality to crawl SPARQL endpoints and open data portals. A detailed comparison of LDSpider and SQUIRREL can be found in Sects. 3 and 4.

---

[3] See, e.g., https://www.mdm-portal.de/, where traffic data from the German Federal Ministry of Transport and Digital Infrastructure is made available.

[4] See, e.g., the German mFund funds at http://mfund.de.

[5] Our code is available at https://github.com/dice-group/squirrel and the documentation at https://w3id.org/dice-research/squirrel/documentation.

[6] https://www.docker.com/.

[7] https://github.com/ldspider/ldspider.

A crawler focusing on structured data is presented in [6]. The authors describe a 5-step pipeline that converts structured data formats like XHTML or RSS into RDF. In [8,9], a distributed crawler is described, which is used to index resources for the Semantic Web Search Engine. To the best of our knowledge, both crawlers are not available as open-source projects.

In [2], the authors present the LOD Laundromat—a framework that downloads, parses, cleans, analyses and republishes RDF datasets. The framework has the advantage of coming with a robust parsing algorithm for various RDF serialisations. However, it solely relies on a given list of seed URLs. In contrast to a crawler, it does not extract new URLs from the fetched data to crawl.

Since web crawling is an established technique, there are several open-source crawlers. An example of a scalable, general web crawler is presented in [7]. However, most of these crawlers cannot process RDF data without further adaptation. A web crawler extended for processing RDF data is the open-source crawler Apache Nutch[8]. Table 1 in Sect. 3 shows the RDF serialisations, compressions and forms of structured data that are supported by the Apache Nutch plugin[9]. However, the plugin stems from 2007, relies on an out-dated crawler version and failed to work during our tests[10].

Overall, the open-source crawlers currently available are either not able to process RDF data, are limited in the types of data formats they can process, or are restricted in their scalability.

3   Approach

3.1   Requirements

Web of Data crawler requirements were gathered from nine organisations within the scope of the projects LIMBO[11] and OPAL[12]. OPAL aims to create an open data portal by integrating the available open data of different national and international data sources[13]. The goal of LIMBO is to collect available mobility data of the ministry of transport, link them to open knowledge bases and publish them within a data portal[14].

To deliver a robust, distributed, scalable and extensible data web crawler, we pursue the following goals with SQUIRREL:

R1: The crawler should be designed to provide a distributed and scalable solution on crawling structured and semi-structured data.

---

[8] http://nutch.apache.org/.

[9] The information has been gathered by an analysis of the plugin's source code.

[10] A brief description of the plugin and its source code can be found at https://issues. apache.org/jira/browse/NUTCH-460.

[11] https://www.limbo-project.org/.

[12] http://projekt-opal.de/projektergebnisse/deliverables/.

[13] See http://projekt-opal.de/en/welcome-project-opal/ and https://www.bmvi.de/ SharedDocs/DE/Artikel/DG/mfund-projekte/ope-data-portal-germany-opal.html.

[14] See https://www.limbo-project.org/ and https://www.bmvi.de/SharedDocs/DE/ Artikel/DG/mfund-projekte/linked-data-services-for-mobility-limbo.html.

R2: The crawler must exhibit "respectful" behaviour when fetching data from servers by following the Robots Exclusion Standard Protocol [11]. This reduces the chance that a server is overloaded by the crawler's request and the chance that the crawler is blocked by a server.

R3: Since not all data is available as structured data, crawlers for the data web should offer a way to gather semi-structured data.

R4: The project should offer easy addition of further functionality (e.g., novel serialisations, other types of data, etc.) through a fully extensible architecture.

R5: The crawler should provide metadata about the crawling process, allowing users to get insights from the crawled data.

In the following, we give an overview of the crawler's components, before describing them in more detail.

3.2   Overview

SQUIRREL comprises two main components: *Frontier* and *Worker* (**R1**). To achieve a fully extensible architecture, both components rely on the dependency injection pattern, i.e., they comprise several modules that implement the single functionalities of the components. These modules can be injected into the components, facilitating the addition of more functionalities (**R4**). To support the addition of the dependency injection, SQUIRREL has been implemented based on the Spring framework[15]. Fig. 1 illustrates the architecture of SQUIRREL.

When executed, the crawler has exactly one Frontier and a number of Workers, which can operate in parallel (**R1**). The Frontier is initialised with a list of seed URIs. It normalises and filters the URIs, which includes a check of whether the URIs have been seen before. Thereafter, the URIs are added to the internal queue. Once the Frontier receives a request from a Worker, it gives a set of URIs to the Worker. For each given URI, the Worker fetches the URI's content, analyses the received data, collects new URIs and forwards the data to its sink. When the Worker is done with the given set of URIs, it sends it back to the Frontier together with the newly identified URIs. The crawler implements the means for a periodic re-evaluation of URIs known to have been crawled in past iterations.

3.3   Frontier

The Frontier has the task of organising the crawling. It keeps track of the URIs to be crawled, and those that have already been crawled. It comprises three main modules:

1. A Normalizer that preprocesses incoming URIs,
2. a Filter that removes already seen URIs
3. a Queue used to keep track of the URIs to be crawled in the future.

---

[15] https://spring.io/.

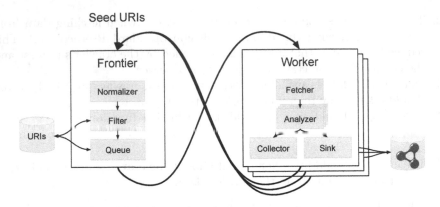

Fig. 1. Squirrel core achitecture

3.3.1 Normalizer

The Normalizer preprocesses incoming URIs by transforming them into a normal form. This reduces the number of URIs that are different but point to the same resources. The URI normalisation comprises the following actions:

- Removal of default ports, e.g., port 80 for HTTP.
- Removal of percentage-encoding for unreserved characters [3].
- Normalization of the URI path, e.g., by removing punctuations [3].
- Removal of the URIs' fragment part.
- Alphanumeric sorting of key-value pairs for query parts that contain several key value pairs.

In addition, the Normalizer tries to find session identifiers or similar parts of the URI that have no influence on the retrieved content. The strings that mark such a part of the URI are configurable.

3.3.2 Filter

The Filter module is mainly responsible for filtering URIs that have already been processed. To achieve this goal, the Frontier makes use of a NoSQL database (i.e., MongoDB in the current implementation), which is used to store all crawled URIs in a persistent way. This ensures that a crawler can be interrupted and restarted later on. Additionally, black or white lists can be used to narrow the search space of the crawler if necessary.

3.3.3 Queue

The Queue is the module that stores the URIs to be crawled. It groups and sorts the URIs, which makes it the main module for implementing crawling strategies. At present, SQUIRREL offers two queue implementations—an IP- and a domain-based first-in-first-out (short: FIFO) queue. Both work in a similar way by grouping URIs based on their IP or their pay-level domain, respectively. The

URI groups are sorted following the FIFO principle. When a Worker requests a new set of URIs, the next available group is retrieved from the queue and sent to the Worker. Internally, this group is marked as blocked, i.e., it remains in the queue and new URIs can be added by the Frontier but it cannot be sent to a different Worker. As soon as the Worker returns the requested URIs, the group is unblocked and the crawled URIs are removed from it. If the group is empty, it is removed from the queue. This implements a load-balancing strategy that aims to crawl the web as fast as possible without overloading single IPs or pay-level domains.

Like the Filter module, the Queue relies on a persistent MongoDB to store the URIs. This enables a restart of the Frontier without a loss of its internal states.

3.4   Worker

The Worker component performs the crawling based on a given set of URIs. Crawling a single URI is done in four steps:

1. URI content is fetched,
2. fetched content is analysed,
3. new URIs are collected, and
4. the content is stored in a sink.

The modules for these steps are described in the following:

3.4.1   Fetcher

The fetcher module takes a given URI and downloads its content. Before accessing the given URI, the crawler follows the Robots Exclusion Standard Protocol [11] and checks the server's `robots.txt` file (**R2**). If the URI's resource can be crawled, one of the available fetchers is used to access it. At present, SQUIRREL uses four different fetchers. Two general fetchers cover the HTTP and the FTP protocol, respectively. Two additional fetchers are used for SPARQL endpoints and CKAN portals, respectively. However, other fetchers can be added by means of the extensible SQUIRREL API if necessary[16].

The Worker tries to retrieve the content of the URI by using the fetchers, in the order in which they were defined, until one of them is successful. The fetcher then stores the data on the disk and adds additional information (like the file's MIME type) to the URI's properties for later usage. Based on the MIME type, the Worker checks whether the file is a compressed or an archive file format. In this case, the file is decompressed and extracted for further processing. In its current release, SQUIRREL supports the formats Gzip, Zip, Tar, 7z and Bzip2[17].

---

[16] Details about implementing a new fetcher can be found at https://dice-group.github.io/squirrel.github.io/tutorials/fetcher.html.

[17] Details regarding the compressions can be found at https://pkware.cachefly.net/Webdocs/APPNOTE/APPNOTE-6.3.5.TXT, https://www.gnu.org/software/gzip/ and http://sourceware.org/bzip2/, respectively.

Table 1. Comparison of RDF serialisations, compressions, methods to extract data from HTML and other methods to access data supported by Apache Nutch (including the RDF plugin), LDSpider and SQUIRREL.

	RDF Serialisations										Comp.					HTML					
	RDF/XML	RDF/JSON	Turtle	N-Triples	N-Quads	Notation 3	JSON-LD	TriG	TriX	HDT	ZIP	Gzip	bzip2	7zip	tar	RDFa	Microdata	Microformat	HTML (scraping)	SPARQL	CKAN
Apache Nutch	✓	–	✓	✓	–	✓	–	–	–	–	✓	✓	–	–	–	✓	✓	✓	–	–	–
LDSpider	✓	–	✓	✓	✓	✓	✓	–	–	–	–	–	–	–	–	✓	✓	✓	–	–	–
SQUIRREL	✓	✓	✓	✓	✓	✓	✓	✓	✓	✓	✓	✓	✓	✓	✓	✓	✓	✓	✓	✓	✓

3.4.2 Analyser

The task of the Analyser module is to process the content of the fetched file and extract triples from it. The Worker has a set of Analysers that are able to handle various types of files. Table 1 lists the supported RDF serialisations, the compression formats and the different ways SQUIRREL can extract data from HTML files. It compares the supported formats with the formats supported by Apache Nutch and LDSpider [10]. Each Analyser offers an isElegible method that is called with a URI and the URI's properties to determine whether it is capable of analysing the fetched data. The first Analyser that returns true receives the file together with a Sink and a Collector, and starts to analyse the data.

The following Analysers are available in the current implementation of SQUIRREL:

1. The RDF Analyser handles RDF files and is mainly based on the Apache Jena project[18]. Thus, it supports the following formats: RDF/XML, N-Triples, N3, N-Quads, Turtle, TRIG, JSON-LD and RDF/JSON.
2. The HDT Analyser is able to process compressed RDF graphs that are available in the HDT file format [5].
3. The RDFa Analyser processes HTML and XHTML Documents extracting RDFa data using the Semargl parser[19].
4. The scraping Analyser uses the Jsoup framework for parsing HTML pages and relies on user-defined rules to extract triples from the parsed page[20].

---

[18] https://jena.apache.org.
[19] https://github.com/semarglproject/semargl.
[20] https://jsoup.org/.

This enables the user to use SQUIRREL to gather not only structured but also semi-structured data from the web (**R3**).

5. The CKAN Analyser is used for the JSON line files generated by the CKAN Fetcher when interacting with the API of a CKAN portal. The analyser transforms the information about datasets in the CKAN portal into RDF triples using the DCAT ontology [1].

6. The Any23-based Analyser processes HTML pages, searching for Microdata or Microformat embedded within the page.

7. In contrast to the other Fetchers, the SPARQL-based Fetcher directly performs an analysis of the retrieved triples.

New analysers can be implemented if the default API does not match the user's needs[21].

3.4.3 Collector

The Collector module collects all URIs from the RDF data. SQUIRREL offers an SQL-based collector that makes use of a database to store all collected URIs. It ensures the scalability of this module for processing large data dumps. For testing purposes, a simple in-memory collector is provided. As soon as the Worker has finished crawling the given set of URIs, it sends all collected URIs to the Frontier and cleans up the collector.

3.4.4 Sink

The Sink has the task to store the crawled data. Currently, a user can choose from three different sinks that are implemented. First, a file-based sink is available. This sink stores given triples in files using the Turtle serialisation for RDF[22]. These files can be further compressed using GZip. The second sink is an extension of the file-based sink and stores triples in the compressed HDT format [5]. It should be noted that both sinks separate the crawled data by creating one file for each URI that is crawled. An additional file is used to store metadata from the crawling process. Both sinks have the disadvantage that each Worker has a local directory in which the data is stored. The third sink uses SPARQL update queries to insert the data in a SPARQL store. This store can be used by several Workers in parallel. For each crawled URI, a graph is created. Additionally, a metadata graph is used to store the metadata generated by the Workers. New sinks can be added by making use of the extensible API[23].

3.4.5 Activity

The Workers of SQUIRREL document the crawling process by writing metadata to a metadata graph (**R5**). This metadata mainly relies on the PROV ontol-

---

[21] Details about implementing a new analyzer can be found at https://dice-group. github.io/squirrel.github.io/tutorials/analyzer.html.

[22] https://www.w3.org/TR/turtle/.

[23] Details about implementing a new sink can be found at https://dice-group.github. io/squirrel.github.io/tutorials/sink.html.

ogy [12] and has been extended where necessary. Figure 2 gives an overview of the generated metadata. The crawling of a single URI is modelled as an activity. Such an activity comes with data like the start and end time, the approximate number of triples received, and a status line indicating whether the crawling was successful. The result graph (or the result file in case of a file-based sink) is an entity generated by the activity. Both the result graph and the activity are connected to the URI that has been crawled.

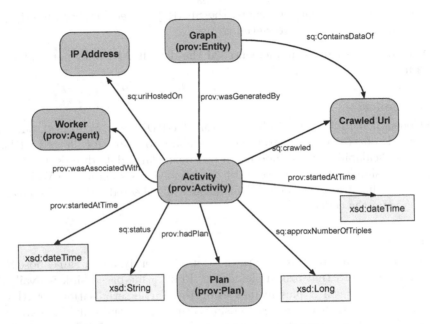

Fig. 2. Squirrel activity, extending the PROV ontology

4   Evaluation

4.1   Benchmark

We carried out two experiments to compare SQUIRREL with the state-of-the-art Data Web crawler, LDSpider [10]. LDSpider was chosen because it is one of the most popular crawlers for the linked web, and is widely used in various projects. All experiments were carried out using the ORCA benchmark for Data Web crawlers [15][24]. ORCA is built upon the HOBBIT benchmarking platform [14] and ensures repeatable and comparable benchmarking of crawlers for the Data Web. To this end, it creates a network of servers from which the crawler should download data. For each server, ORCA generates an RDF dataset with outgoing

---

[24] https://github.com/dice-group/orca.

links to other servers. This allows the crawler to start with one or more seed URIs and crawl the complete network. Note, the benchmark ensures that all triples created for the single servers can be reached by a crawler by traversing the links. ORCA offers five different types of servers:

1. a dump file server,
2. a SPARQL endpoint,
3. a CKAN portal,
4. a server for HTML with embedded RDFa triples
5. a dereferencing server.

The latter can be called via the URI of an RDF resource and answers with the triples that have the resource as subject. The dereferencing server negotiates the RDF serialisation with the crawler based on the crawler's HTTP request. However, the serialisation of each dump file is randomly chosen to be either RDF/XML, Turtle, N-Triples or Notation 3. ORCA measures the completeness of data gathered by the crawler, and its run time.

4.2   Evaluation Setup

We carry out two experiments in which we use ORCA to simulate a network of servers. The first experiment simulates a real-world Data Web and focuses on the effectiveness of the two crawlers, i.e., the amount of correct triples they retrieve. As suggested by the authors of [15], the generated cloud comprises 100 servers with 40% dump file servers, 30% SPARQL servers, 21% dereferencing servers, 5% CKAN servers and 4% servers that offer RDFa within HTML pages. The average degree of each node is set to 20 and the data of each node comprises 1000 triples with an average degree of 9 triples per resource. 30% of the dump file nodes offer their files compressed using zip, gzip or bz2. The results of this experiment are listed in Table 2[25].

The second experiment focuses on the efficiency of the crawler implementations. We follow the suggestion given in [15] for efficiency experiments and solely rely on 200 dereferencing servers. These servers have can negotiate the RDF serialisation with the crawler. Hence, the crawlers are very likely to be able to crawl the complete graph, which eases a comparison of the crawlers with respect to their efficiency. The other parameters are the same as in the first experiment. The results of the second experiment are listed in Table 2[26].

For all experiments, we use a cluster of machines. The crawlers are deployed on 3 machines while the created servers of the ORCA benchmark are hosted on 3 other machines. Each of the machines has 16 cores with hyperthreading

---

[25] The detailed results can be seen at https://w3id.org/hobbit/experiments#158540364 5660,1584545072279,1585230107697,1584962226404,1584962243223,1585574894994, 1585574924888,1585532668155,1585574716469.

[26] Detailed results can be found at https://w3id.org/hobbit/experiments#1586886425 879,1587151926893,1587284972402,1588111671515,1587121394160,1586886364444, 1586424067908,1586374166710,1586374133562.

Table 2. Results for experiment I and II.

Crawler	Experiment I		Experiment II			
	Micro Recall	Run time (in s)	Micro Recall	Run time (in s)	CPU (in s)	RAM (in GB)
LDSpider (T1)	0.31	1798	1.00	2 031	320.0	1.2
LDSpider (T8)	0.30	1792	1.00	2 295	265.0	? 8
LDSpider (T16)	0.31	1858	1.00	1 945	345.4	1.6
LDSpider (T32)	0.31	1847	1.00	2 635	11.6	2.6
LDSpider (T32,LBS)	0.03	66	0.54	765	182.1	7.5
SQUIRREL (W1)	0.98	6 663	1.00	11 821	991.3	3.9
SQUIRREL (W3)	0.98	2 686	1.00	4 100	681.4	8.6
SQUIRREL (W9)	0.98	1 412	1.00	1 591	464.8	18.1
SQUIRREL (W18)	0.97	1 551	1.00	1 091	279.8	22.1

and 256 GB RAM[27]. For both experiments, the usage of `robots.txt` files is disabled. We use several configurations of LDSpider and SQUIRREL. LDSpider (T1), (T8), (T16) and (T32) use a breadth-first crawling strategy and 1, 8, 16 or 32 threads, respectively. Additionally, we configure LDSpider (T32,LSB), which makes use of 32 threads and a load-balancing strategy. Further, we configure SQUIRREL (W1), (W3), (W9) and (W18) to use 1, 3, 9 or 18 Worker instances, respectively.

4.3 Discussion

The results of the first experiment show that LDSpider has a lower recall than SQUIRREL. This difference is due to several factors. LDSpider does not support 1) the crawling of SPARQL endpoints, 2) the crawling of CKAN portals, nor 3) the processing of compressed RDF dump files. In comparison, SQUIRREL comes with a larger set of supported server types, RDF serialisations and compression algorithms. Hence, SQUIRREL was able to crawl nearly all triples. However, not all triples of all CKAN portals and RDFa nodes could be retrieved, leading to a micro recall of up to 0.98.

The second experiment shows that the larger set of features offered by SQUIRREL comes with lower efficiency. LDSpider achieves lower run times using a more economical configuration with respect to consumed CPU time and RAM. With a higher number of workers, SQUIRREL achieves lower run times but consumes much more RAM than LDSpider. At the same time, the experiment reveals that the load-balancing strategy of LDSpider tends to abort the crawling process very early and, hence, achieves only a low recall in both experiments.

---

[27] The details of the hardware setup that underlies the HOBBIT platform can be found at https://hobbit-project.github.io/master#hardware-of-the-cluster.

5 Application

Squirrel is used within several research projects, of which two are of national importance in Germany. The OPAL project creates an integrated portal for open data by integrating datasets from several data sources from all over Europe[28]. At the moment, the project focuses on the portals mCLOUD.de, govdata.de and europeandataportal.eu. In addition, several sources found on OpenDataMonitor.eu are integrated. SQUIRREL is used to regularly gather information about available datasets from these portals. Table 3 lists the number of datasets that are extracted from the portals, the time the crawler needs to gather them, and the way the crawler accesses data. It should be noted that the run times include the delays SQUIRREL inserts between single requests to ensure that the single portals are not stressed. The portals evidently use different ways to offer their data. Two of them are CKAN portals, while mCLOUD.de has to be scraped using SQUIRREL's HTML scraper. Only europeandataportal.eu offers a SPARQL endpoint to access the dataset's metadata. The data integrated by OPAL are to be written back into the mCLOUD.de portal and cater for the needs of private and public organisations requiring mobility data. Users range from large logistic companies needing to plan transport of goods, to single persons mapping their movement with pollen concentration.

Table 3. Crawling statistics of the OPAL project.

	Datasets	Triples	Run time	Type
mCLOUD.de	1 394	19 038	25min	HTML
govdata.de	34 057	138 669	4h	CKAN
europeandataportal.eu	1 008 379	13 404 005	36h	SPARQL
OpenDataMonitor.eu	104 361	464 961	7h	CKAN

Another project that makes use of SQUIRREL to collect data from the web is LIMBO. Its aim is to unify and refine mobility data of the German Federal Ministry of Transport and Digital Infrastructure. The refined data is made available to the general public to create the basis for new, innovative applications. To this end, SQUIRREL is used to collect this and related data from different sources.

6 Conclusion

This paper presented SQUIRREL, a scalable, distributed and extendable crawler for the Data Web, which provides support for several different protocols and data serialisations. Other open-source crawlers currently available are either not able to process RDF data, are limited in the types of data formats they can process, or are restricted in their scalability. SQUIRREL addresses these drawbacks

---

[28] http://projekt-opal.de/.

by providing an extensible architecture adaptable to supporting any format of choice. Moreover, the framework was implemented for simple deployment both locally and at a large scale.

We described the components of the crawler and presented a comparison with LDSpider. This comparison showed the advantages of SQUIRREL with respect to the large amount of supported data and server types. SQUIRREL was able to crawl data from different sources (HTTP, SPARQL and CKAN) and compression formats (zip,gzip,bz2), leading to a higher recall than LDSpider. In addition, we identified SQUIRREL's efficiency as a focus for future development and improvement. SQUIRREL is already used by several projects and we provide tutorials for its usage to empower more people to make use of the data available on the web[29].

Acknowledgments. This work has been supported by the BMVI (Bundesministerium für Verkehr und digitale Infrastruktur) projects LIMBO (GA no. 19F2029C) and OPAL (GA no. 19F2028A).

References

1. Archer, P.: Data catalog vocabulary (dcat) (w3c recommendation), January 2014. https://www.w3.org/TR/vocab-dcat/
2. Beek, W., Rietveld, L., Bazoobandi, H.R., Wielemaker, J., Schlobach, S.: Lod laundromat: a uniform way of publishing other people's dirty data. In: Mika, P., et al. (eds.) The Semantic Web - ISWC 2014, pp. 213–228. Springer International Publishing, Cham (2014)
3. Berners-Lee, T., Fielding, R., Masinter, L.: Uniform Resource Identifier (URI): Generic Syntax. Internet Standard, Internet Engineering Task Force (IETF), January 2005. https://tools.ietf.org/html/rfc3986
4. Fernández, J.D., Beek, W., Martínez-Prieto, M.A., Arias, M.: LOD-a-lot. In: d'Amato, C., et al. (eds.) ISWC 2017. LNCS, vol. 10588, pp. 75–83. Springer, Cham (2017). https://doi.org/10.1007/978-3-319-68204-4_7
5. Fernández, J.D., Martínez-Prieto, M.A., Gutiérrez, C., Polleres, A., Arias, M.: Binary RDF representation for publication and exchange (HDT). Web Semant. Sci. Serv. Agents World Wide Web, **19**, 22–41 (2013). http://www.websemanticsjournal.org/index.php/ps/article/view/328
6. Harth, A., Umbrich, J., Decker, S.: MultiCrawler: a pipelined architecture for crawling and indexing semantic web data. In: Cruz, I., et al. (eds.) ISWC 2006. LNCS, vol. 4273, pp. 258–271. Springer, Heidelberg (2006). https://doi.org/10.1007/11926078_19
7. Heydon, A., Najork, M.: Mercator: a scalable, extensible web crawler. Word Wide Web **2**(4), 219–229 (1999)
8. Hogan, A.: Exploiting RDFS and OWL for Integrating Heterogeneous, Large-Scale, Linked Data Corpora (2011). http://aidanhogan.com/docs/thesis/

---

[29] https://dice-group.github.io/squirrel.github.io/tutorials.html.

9. Hogan, A., Harth, A., Umbrich, J., Kinsella, S., Polleres, A., Decker, S.: Searching and browsing linked data with SWSE: the semantic web search engine. Web Semant. Sci. Serv. Agents World Wide Web, **9**(4), 365–401 (2011). https://doi.org/10.1016/j.websem.2011.06.004. http://www.sciencedirect.com/science/article/pii/S1570826811000473, JWS special issue on Semantic Search

10. Isele, R., Umbrich, J., Bizer, C., Harth, A.: LDspider: an open-source crawling framework for the web of linked data. In: Proceedings of the ISWC 2010 Posters & Demonstrations Track: Collected Abstracts, vol. 658, pp. 29–32. CEUR-WS (2010)

11. Koster, M., Illyes, G., Zeller, H., Harvey, L.: Robots Exclusion Protocol. Internet-draft, Internet Engineering Task Force (IETF), July 2019. https://tools.ietf.org/html/draft-rep-wg-topic-00

12. Lebo, T., Sahoo, S., McGuinness, D.: PROV-O: The PROV Ontology. W3C Recommendation, W3C, April 2013. http://www.w3.org/TR/2013/REC-prov-o-20130430/

13. Merkel, D.: Docker: Lightweight linux containers for consistent development and deployment. Linux J. 2014(239), March 2014. http://dl.acm.org/citation.cfm?id=2600239.2600241

14. Röder, M., Kuchelev, D., Ngonga Ngomo, A.C.: HOBBIT: a platform for benchmarking Big Linked Data. Data Sci. (2019). https://doi.org/10.3233/DS-190021

15. Röder, M., de Souza, G., Kuchelev, D., Desouki, A.A., Ngomo, A.C.N.: Orca: a benchmark for data web crawlers (2019). https://arxiv.org/abs/1912.08026

OBA: An Ontology-Based Framework for Creating REST APIs for Knowledge Graphs

Daniel Garijo[✉][iD] and Maximiliano Osorio[iD]

Information Sciences Institute, University of Southern California, Los Angeles, USA
{dgarijo,mosorio}@isi.edu

Abstract. In recent years, Semantic Web technologies have been increasingly adopted by researchers, industry and public institutions to describe and link data on the Web, create web annotations and consume large knowledge graphs like Wikidata and DBpedia. However, there is still a knowledge gap between ontology engineers, who design, populate and create knowledge graphs; and web developers, who need to understand, access and query these knowledge graphs but are not familiar with ontologies, RDF or SPARQL. In this paper we describe the Ontology-Based APIs framework (OBA), our approach to automatically create REST APIs from ontologies while following RESTful API best practices. Given an ontology (or ontology network) OBA uses standard technologies familiar to web developers (OpenAPI Specification, JSON) and combines them with W3C standards (OWL, JSON-LD frames and SPARQL) to create maintainable APIs with documentation, units tests, automated validation of resources and clients (in Python, Javascript, etc.) for non Semantic Web experts to access the contents of a target knowledge graph. We showcase OBA with three examples that illustrate the capabilities of the framework for different ontologies.

Keywords: Ontology · API · REST · JSON · JSON-LD · Data accessibility · Knowledge graph · OpenAPI · OAS

Resource type: Software
License: Apache 2.0
DOI: https://doi.org/10.5281/zenodo.3686266
Repository: https://github.com/KnowledgeCaptureAndDiscovery/OBA/

1 Introduction

Knowledge graphs have become a popular technology for representing structured information on the Web. The Linked Open Data Cloud[1] contains more than 1200 linked knowledge graphs contributed by researchers and public institutions. Major companies like Google,[2] Microsoft,[3] or Amazon [16] use knowledge graphs

---

[1] https://lod-cloud.net/.
[2] https://developers.google.com/knowledge-graph.
[3] https://www.microsoft.com/en-us/research/project/microsoft-academic-graph/.

© Springer Nature Switzerland AG 2020
J. Z. Pan et al. (Eds.): ISWC 2020, LNCS 12507, pp. 48–64, 2020.
https://doi.org/10.1007/978-3-030-62466-8_4

to represent some of their information. Recently, crowdsourced knowledge graphs such as Wikidata [15] have surpassed Wikipedia in the number of contributions made by users.

In order to create and structure these knowledge graphs, ontology engineers develop vocabularies and ontologies defining the semantics of the classes, object properties and data properties represented in the data. These ontologies are then used in extraction-transform-load pipelines to populate knowledge graphs with data and make the result accessible on the Web to be queried by users (usually as an RDF dump or a SPARQL endpoint). However, consuming and contributing to knowledge graphs exposed in this manner is problematic for two main reasons. First, exploring and using the contents of a knowledge graph is a time consuming task, even for experienced ontology engineers (common problems include lack of usage examples that indicate how to retrieve resources, the ontologies used are not properly documented or without examples, the format in which the results are returned is hard to manipulate, etc.). Second, W3C standards such as SPARQL [12] and RDF [2] are still unknown to a major portion of the web developer community (used to JSON and REST APIs), making it difficult for them to use knowledge graphs even when documentation is available.

In this paper we address these problems by introducing OBA, an Ontology-Based API framework that given an ontology (or ontology network) as input, creates a JSON-based REST API server that is consistent with the classes and properties in the ontology; and can be configured to retrieve, edit, add or delete resources from a knowledge graph. OBA's contributions include:

- A method for automatically creating a **documented REST OpenAPI specification**[4] from an OWL ontology [10], together with the means to customize it as needed (e.g., filtering some of its classes). Using OBA, new changes made to an ontology can be automatically propagated to the corresponding API, making it easy to maintain.
- A framework to create a **server implementation** based on the API specification to handle requests automatically against a target knowledge graph. The implementation will validate posted resources to the API and will deliver the results in a JSON format as defined in the API specification.
- A method for **converting JSON-LD** returned by a SPARQL endpoint [6] into JSON according to the format defined in the API specification.
- A mechanism based on named graphs[5] for users to **contribute** to a knowledge graph through POST requests.
- Automatic generation of **tests** for API validation against a knowledge graph.

OBA uses standards widely used in Web development (JSON) for accepting requests and returning results, while using SPARQL and JSON-LD frames to query knowledge graphs and frame data in JSON-LD. We consider that OBA is a valuable resource for the community, as it helps bridging the gap between

---

[4] http://spec.openapis.org/oas/v3.0.3.
[5] https://www.w3.org/2004/03/trix/.

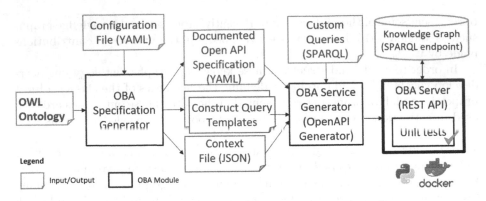

Fig. 1. Overview of the OBA Framework.

ontology engineers who design and populate knowledge graphs and application and service developers who can benefit from them.

The rest of the paper is structured as follows: Sect. 2 describes the architecture and rationale of the OBA framework, while Sect. 3 shows the different features of OBA through three different examples and a performance analysis of the tool. Section 4 discusses adoption, impact and current limitations of the tool, Sect. 5 compares OBA to similar efforts from the Semantic Web community to help developers access knowledge graphs, and Sect. 6 concludes the paper.

2   The Ontology-Based APIs Framework (OBA)

OBA is a framework designed to help ontology engineers create RESTful APIs from ontologies. Given an OWL ontology or ontology network and a knowledge graph (accessible through a SPARQL endpoint), OBA automatically generates a documented standard API specification and creates a REST API server that can validate requests from users, test all API calls and deliver JSON objects following the structure described in the ontology.

Figure 1 shows an overview of the workflow followed by the OBA framework, depicting the target input ontology on the left and the resultant REST API on the right. OBA consists of two main modules: the *OBA Specification Generator*, which creates an API specification template from an input ontology; and the *OBA Service Generator*, which produces a server with a REST API for a target SPARQL endpoint. In this section we describe the different features of OBA for each module, along with the main design decisions and assumptions adopted for configuring the server.

2.1   OBA Specification Generator

One of the drivers for the development of OBA was the need to use standards and file formats commonly used by web developers (who may not necessarily be familiar with Semantic Web technologies). Hence, we decided to use the OpenAPI specification[6] for representing REST APIs and JSON as the main interchange file format.

There are three reasons why we chose the OpenAPI specification (OAS): First, it "defines a standard, programming language-agnostic interface description for REST APIs, which allows both humans and computers to discover and understand the capabilities of a service without requiring access to source code, additional documentation, or inspection of network traffic".[7] Second, OAS is an open source initiative backed up by industry and widely used by the developer community, with more than 17.000 stars in GitHub and over 6.000 forks. Finally, by adopting OAS we gain access to a wide range of tools[8] that can be leveraged and extended (e.g., for generating a server) and are available in multiple programming languages.

2.1.1   Generating an OAS from OWL

OAS describes how to define *operations* (GET, POST, PUT, DELETE) and *paths* (i.e., the different API calls) to be supported by a REST API; together with the information about the *schemas* that define the structure of the objects to be returned by each call. OAS also describes how to provide examples, documentation and customization through parameters for each of the paths declared in an API.

Typically, an OAS would have two paths for each GET operation; and one for POST, PUT and DELETE operations. For instance, let us consider a simple REST API for registering and returning *regions* around the world. An OAS would have the paths '/regions' (for returning all available regions) and '/regions/{id}' (for returning the information about a region in particular) for the GET operation; the '/regions' path for POST;[9] and the '/regions/{id}' path for PUT and DELETE operations.

In OAS, the *schema* to be followed by an object in an operation is described through its properties. For example, we can define a *Region* as a simple object with a *label*, a *type* and a *partOfRegion* property which indicates that a region is part of another region. The associated schema would look as follows in OAS:

---

[6] https://github.com/OAI/OpenAPI-Specification.

[7] http://spec.openapis.org/oas/v3.0.3.

[8] https://github.com/OAI/OpenAPI-Specification/blob/master/
IMPLEMENTATIONS.md.

[9] Alternatively, '/regions/id' may be used to allow developers to post their own ids.

```
Region:
  description: A region refers to an extensive, continuous
        part of a surface or body.
  properties:
   id:
      nullable: false
      type: string
   partOfRegion:
      description: Region where another region is included in
      items:
        $ref: '#/components/schemas/Region'
      nullable: true
      type: array
    label:
      description: Human readable description of the resource
      items:
        type: string
      nullable: true
      type: array
    type:
      description: type of the resource
      items:
        type: string
      nullable: true
      type: array
  type: object
```

Note that the *partOfRegion* property will return objects that follow the *Region* schema (as identified by '#/components/schemas/Region'). The *nullable* parameter indicates that the target property is optional.

The main OAS structure maps naturally to the way classes and properties are specified in ontologies and vocabularies. Therefore, in OBA we support RDFs class expressivity by mapping each ontology class to a different path in the API specification;[10] and adding each object property and data type property in the target ontology to the corresponding schema by looking into its domain and range. Complex class restrictions consisting on multiple unions and intersections are currently not supported. Documentation for each path and property is included in the description field of the OAS by looking at the available ontology definitions (e.g., rdfs:comment annotations on classes and properties). Unions in property domains are handled by copying the property into the respective class schemas (e.g., if the domain of a property is 'Person or Cat', the property will be added in the Person and Cat schemas); and properties declared in superclasses are propagated to their child class schemas. Properties with no domain or range are by default excluded from the API, although this behavior can be configured; and property chains are currently not supported. By default,

---

[10] We follow the best practices for RESTful API design: paths are in non-capital letters and always in plural (e.g., /regions, /persons, etc.).

all properties are *nullable* (optional). The full mapping between OAS and OWL supported by OBA is available online.[11]

Finally, we also defined two filtering features in OBA when generating the OAS to help interacting with the API. First, we allow specifying a subset of classes of interest to include in an API, since ontologies may contain more classes than the ones we may be interested in. Second, by default OBA will define a parameter on each GET path to retrieve entities of a class based on their label.

As a result of executing the OBA specification generator, we create an OAS in YAML format[12] that can be inspected by ontology engineers manually or using an online editor.[13] This specification can be implemented with the OBA server (Sect. 2.2) or by other means (e.g., by implementing the API by hand).

2.1.2 Generating SPARQL Query Templates and JSON-LD Context

The OBA Specification Generator also creates a series of templates with the queries to be supported by each API path. These queries will be used by the server for automatically handling the API calls. For example, the following query is used to return all the information of a resource by its id (`?_resource_iri`):

```
#+ summary: Return resource information by its resource_iri
PREFIX rdfs: <http://www.w3.org/2000/01/rdf-schema#>
CONSTRUCT {
    ?_resource_iri ?predicate ?obj .
    ?obj a ?type .
    ?obj rdfs:label ?label
}
WHERE {
    ?_resource_iri ?predicate ?obj
    OPTIONAL {
        ?obj   a ?type
        OPTIONAL {
            ?obj rdfs:label ?label
        }
    }
}
```

The individual id `?_resource_iri` acts as a placeholder which is replaced with the URI associated with the target path (we reuse GRLC [11] to define parameters in a query[14]). Returned objects will have one level of depth within the graph (i.e., all the outgoing properties of a resource), in order to avoid returning very complex objects. This is useful in large knowledge graphs such as DBpedia [1], where returning all the sub-resources included within an instance may be too costly. However, this default behavior can be customized by editing

---

[11] https://oba.readthedocs.io/en/latest/mapping/.
[12] https://yaml.org/.
[13] https://editor.swagger.io/.
[14] https://github.com/KnowledgeCaptureAndDiscovery/OBA_sparql/.

the proposed resource templates or by adding a custom query (further explained in the next section).

Together with the query templates, OBA will generate a JSON-LD context file from the ontology, which will be used by the server to translate the obtained results back into JSON. We have adapted owl2jsonld [13] for this purpose.

2.2   OBA Service Generator

Once the OAS has been generated, OBA creates a script to set up a functional server with the API. We use OpenAPI generator,[15] one of the multiple server implementations for OAS made available by the community; to generate a server with our API as a Docker image.[16] Currently, we support the Python implementation, but the architecture is flexible enough to change the server implementation in case of need. OBA also includes a mechanism for enabling pagination, which allows limiting the number of resources returned by the server.

OBA handles automatically several aspects that need to be taken into account when setting up a server, including how to validate and insert complex resources in the knowledge graph, how to handle authentication; how to generate unit tests and how to ease the access to the server by making clients for developers. We briefly describe these aspects below.

2.2.1   Converting SPARQL Results into JSON

We designed OBA to generate results in JSON, one of the most popular interchange formats used in web development. Figure 2 shows a sequence diagram with the steps we follow to produce the target JSON in GET and POST requests. For example, for the GET request, we first create a SPARQL CONSTRUCT query to retrieve the result from a target knowledge graph. The query is created automatically using the templates generated by OBA, parametrizing them with information about the requested path. For example, for a GET request to /regions/id, the id will replace ?_resource_iri in the template query as described in Sect. 2.1.2.

As shown in Fig. 2, the construct query returns a JSON-LD file from the SPARQL endpoint. We frame the results to make sure they follow the structure defined by the API and then we transform the resultant JSON-LD to JSON. In order to transform JSON-LD to JSON and viceversa, we keep a mapping file with the API path to ontology class URI correspondence, which is automatically generated from the ontology. The URI structure followed by the instances is stored in a separate configuration file.

2.2.2   Resource Validation and Insertion

OBA uses the specification generated from an input ontology to create a server with the target API. By default, the server is prepared to handle GET, POST,

---

[15] https://github.com/OpenAPITools/openapi-generator.
[16] https://oba.readthedocs.io/en/latest/server/.

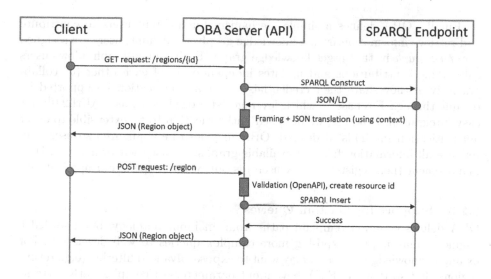

Fig. 2. Sample GET and POST request through the OBA server.

PUT and DELETE requests, which are addressed with CONSTRUCT, INSERT, UPDATE and DELETE SPARQL queries respectively. However, POST, PUT and DELETE requests need to be managed carefully, as they modify the contents of the target knowledge graph.

For POST and PUT, one of the main issues to address is handling *complex objects*, i.e., objects that contain one or multiple references to other objects that do not exist in the knowledge graph yet. Following our above example with regions, what would happen if we received a POST request with a new region where *partOfRegion* points to other regions that do not exist yet in our knowledge graph? For example, let us consider that a developer wants to register a new region `Marina del Rey` that is part of `Los Angeles`, and none of them exist in the knowledge graph. One way would be requiring the developer to issue a new request to register each parent region before registering the child one (e.g., a POST request first to register `Los Angeles` region and then another POST request for `Marina del Rey`); but this makes it cumbersome for developers to use the API. Instead, OBA will recursively validate, insert and generate ids for all subresources that do not have an id already. Hence, in the previous example OBA would register `Los Angeles` first, and then `Marina del Rey`. If a subresource already has an id, it will not be registered as a new resource. When a new resource is created, the server assigns its id with a uuid, and returns it as part of the JSON result.

This recursive behavior is not desirable for DELETE requests, as we could potentially remove a resource referenced by other resources in the knowledge graph. Therefore, OBA currently deletes only the resource identified by its id in the request.

Finally, OBA defines a simple mechanism for different users to contribute and retrieve information from a knowledge graph. By default, users are assigned a *named graph* in the target knowledge graph. Each named graph allows users submitting contributions and updates independently of each other (or collaboratively, if they share their credentials). User authentication is supported by default through Firebase,[17] which leverages standards such as OAUth2.0[18] for easy integration. However, we designed authentication to be extensible to other authentication methods, if desired. OBA may also be configured so users can retrieve all information from all available graphs; or just their own graph. User management (i.e., registering new users) is out of the scope of our application.

2.2.3 Support for Custom Queries

OBA defines common template paths from an input ontology, but knowledge engineers may require exposing more complex queries to web developers. For example, knowledge engineers may want to expose advanced filtering (e.g., return regions that start with "Eu"), have input parameters or complex path patterns (e.g., return only regions that are part of another region). These paths are impossible to predict in advance, as they depend on heterogeneous use cases and requirements. Therefore, in OBA we added a module to allow supporting *custom queries* and let knowledge engineers expand or customize any of the queries OBA supports by default.

To add a custom query, users need to follow two main steps. First, create a CONSTRUCT query in a file; and second, edit the OAS with the path where the query needs to be supported. OBA reuses GRLC's query module [11] to support this feature, as GRLC is an already well established application for creating APIs from SPARQL queries. An example illustrating how to add custom queries to an OAS in OBA can be found online.[19]

2.2.4 Generating Unit Tests

OBA automatically generates unit tests for all the paths specified in the generated OAS (using Flask-Testing,[20] a unit test toolkit for Flask servers[21] in Python). Units tests are useful to check if the data in a knowledge graph is consistent with the classes and properties used in the ontology, and to identify unused classes. By default, OBA supports unit tests for GET requests only, since POST, PUT and DELETE resources may need additional insight of the contents stored in the target knowledge graph. However, this is a good starting point to test the different calls to be supported by the API and detect any inconsistencies. Knowledge engineers may extend the test files with additional tests required

---

[17] https://firebase.google.com/docs/auth/.
[18] https://oauth.net/2/.
[19] https://oba.readthedocs.io/en/latest/adding_custom_queries/.
[20] https://pythonhosted.org/Flask-Testing/.
[21] https://flask.palletsprojects.com/en/1.1.x/.

by their use cases. Unit tests are generated as part of the server, and may be invoked before starting up the API for public consumption.[22]

2.2.5   Generating Clients for API Exploitation

The OpenAPI community has developed tools to generate clients to support an OAS in different languages. We use the OpenAPI generator in OBA to create clients (software packages) to ease API management calls for developers. For example, below is an example of a code snippet using the Python client for one of the APIs we generated with OBA[23] and available through `pip`. The client retrieves the information of a region (with id '*Texas*') and returns the result as a JSON object in a Python dictionary without the need of issuing GET requests or writing SPARQL:

```
import modelcatalog
# modelcatalog is the Python package with our API
api_instance = modelcatalog.ModelApi()
region_id = "Texas"
try:
    # Get a Region by its id. The result is a JSON object
    region = api_instance.regions_id_get(region_id)
    print(region)
except ApiException as e:
    print("Exception when calling ModelApi->regions_id_get: %s\n" % e)
```

3   Usage Examples and Performance Analysis

In this section we demonstrate the full capabilities of OBA in an incremental manner through three different examples of increasing complexity (Sect. 3.1); and a performance analysis to measure the overhead added by OBA (Sect. 3.2). All the resources described in this section are accessible online.[24]

3.1   Illustrating OBA's Features Through Examples

Drafting an API for an Ontology Network: The simplest way in which OBA can be used is by generating a draft OAS from a given ontology (without generating a server). We have tested OBA with ten different ontologies from different domains to generate draft specifications, and we have found this feature very useful in our work. Drafting the API allows knowledge engineers discuss potential errors for misinterpretation, as well as easily detect errors on domains and ranges of properties.

---

[22] https://oba.readthedocs.io/en/latest/test/.

[23] https://model-catalog-python-api-client.readthedocs.io/en/latest/.

[24] https://oba.readthedocs.io/en/latest/examples/.

Generating a GET API for a Large Ontology: DBpedia [1] is a popular knowledge graph with millions of instances over a wide range of categories. The DBpedia ontology[25] contains over 680 classes and 2700 properties; and creating an API manually to support them becomes a time consuming task. We demonstrated OBA by creating two different APIs for DBpedia. The first API contains all the paths associated with the classes in the ontology, showing how OBA can be used by default to generate an API even when the ontology has a considerable size. Since the resultant API is too big to browse manually, we created a Python client[26] and a notebook[27] demonstrating its use. The second API has just a selected group of classes by using a filter, as in some cases not all the classes may need to be supported in the desired API. OBA does a transitive closure on the elements that are needed as part of the API. For example, if the filter only contains "Band", and it has a relationship with "Country" (e.g., *origin*), then by default OBA will also import the schema for "Country" into the API specification to be validated accordingly. In the DBpedia example, selecting just 2 classes (`dbpedia:Genre` and `dbpedia:Band`) led to the inclusion of more than 90 paths in the final specification.

Generating a Full Create, Delete, Update, Delete (CRUD) API: OKG-Soft [5] is an open knowledge graph with scientific software metadata, developed to ease the understanding and execution of complex environmental models (e.g., in hydrology, agriculture or climate sciences). A key requirement of OKG-Soft was for users to be able to contribute with their own metadata in collaborative manner, and hence we used the full capabilities of OBA to support adding, editing and deleting individual resources. OKG-Soft uses two ontologies to structure the knowledge graph, which have evolved over time with new requirements. We used OBA to maintain an API release after each ontology version, generating an OAS, updating it with any required custom queries and generating a server with unit tests, which we executed before deploying the API in production. Having unit tests helped detecting and fixing inconsistencies in the RDF, and improved the overall quality of the knowledge graph. Authenticated users may use the API for POST, PUT and DELETE resources;[28] and we use the contents of the knowledge graph for scientific software exploration, setup and execution in different environments. An application for browsing the contents of the knowledge graph is available online.[29]

The three examples described in this section demonstrate the different features of OBA for different ontologies: the ability to draft API specifications, the capabilities of the tool to be used for large ontologies and to filter classes when

---

[25] https://wiki.dbpedia.org/services-resources/ontology.
[26] https://github.com/sirspock/dbpedia_api_client.
[27] https://github.com/sirspock/dbpedia_example/.
[28] https://model-catalog-python-api-client.readthedocs.io/en/latest/endpoints/.
[29] https://models.mint.isi.edu.

required; and the support for GET, POST, PUT and DELETE operations while following the best practices for RESTful design.

3.2 Performance Analyses

We carried out two main analyses to assess 1) the overhead added by OBA when framing results to JSON and 2) the performance of the API for answering multiple requests per second. Both tests have been performed in two separate machines (one with the API, another one with the SPARQL endpoint) with the same conditions: 8 GB of RAM and 2 CPUs. The analyses retrieve a series of results from a SPARQL endpoint (Fuseki server) by doing SPARQL queries and comparing them against an equivalent request through an OBA-generated API (GET queries, without reverse proxy caching). All requests retrieve individuals of various classes of a knowledge graph (e.g., GET all datasets, get all persons) and not individual resources. The corresponding SPARQL queries use CONSTRUCTs.

Table 1 shows that OBA (without enabling reverse proxy caching) adds an overhead below 150 ms for the majority of the queries with respect to a SPARQL endpoint (below 50ms); and between 150 and 200 ms for 8% of the queries. Overall, this overhead is barely noticeable when using the API in an application.

Table 2 shows the performance of the OBA server when receiving 5, 10 and 60 requests per second. When enabling reverse proxy caching (our recommended option), the API can handle 60 queries/second with a delay of less than 200 ms. Without cache, performance degrades when receiving more than 10 requests per second. This may be improved with an advanced configuration of the Python server used in our implementation.

Table 1. Time delay added by OBA when transforming RDF results into JSON

Time	No. requests		% of requests	
	Endpoint	OBA API	Endpoint	OBA API
[0s, 50ms]	59	0	98.33%	0.00%
[50ms, 100ms]	1	0	1.67%	0.00%
[100ms, 150ms]	0	0	0.00%	0.00%
[150ms, 200ms]	0	55	0.00%	91.67%
[200ms, 250 ms]	0	5	0.00%	8.33%
[250ms, 350 ms]	0	0	0.00%	0.00%
>350ms	0	0	0.00%	0.00%

Table 2. API performance for different number of requests and reverse proxy caching

Time	5 requests/second		10 requests/second		60 requests/second	
	Cache	No cache	Cache	No cache	Cache	No cache
[0s, 100ms]	100%	0.00%	99.83%	0%	99.89%	0%
[100ms, 200ms]	0%	88.67%	0%	0.28%	0.11%	0.24%
[200ms, 300ms]	0%	11.33%	0.17%	1.11%	0%	0%
[300ms, 1s]	0%	0%	0%	3.64%	0%	1.2%
[1s, 5s]	0%	0%	0%	95.00%	0%	6.10%
>5s	0%	0%	0%	0%	0%	92.44%

4   Adoption, Impact and Limitations

We developed OBA to help developers (not familiar with SPARQL) accessing the contents of knowledge graphs structured by ontologies. With OBA, non-expert web developers may use clients in the languages they are more familiar with (e.g., Python, JavaScript, etc.); generated with the OBA Service Generator. Web developers with more knowledge on using APIs may use the API created with the OBA server, while knowledge engineers may choose to query the SPARQL endpoint directly.

OBA builds on the work started by tools like Basil [3] and GRLC [11] - pioneers in exposing SPARQL queries as APIs- to help involve knowledge engineers in the process of data retrieval using the ontologies they designed. In our experience, generating a draft API from an ontology has helped our developer collaborators understand how to consume the information of our knowledge graphs, while helping the ontology engineers in our team detect potential problems in the ontology design.

In fact, similar issues have been raised in the Semantic Web community for some time. For example, the lack of guidance when exploring existing SPARQL endpoints[30] has led to the development of tools such as [9] to help finding patterns in SPARQL endpoints in order to explore them. The Semantic Web community has also acknowledged the difficulties experienced by developers to adopt RDF,[31] which have resulted in ongoing efforts to improve materials and introductory tutorials.[32]

OBA helps addressing these problems by exploiting Semantic Web technologies while exposing the information to developers following the REST standards they are familiar with. OBA allows prototyping APIs from ontologies, helps maintaining APIs (having an API per version of the ontology), helps validating API paths, assists in the creation of unit tests and documents all of the API

---

[30] https://lists.w3.org/Archives/Public/semantic-web/2015Jan/0087.html.
[31] https://lists.w3.org/Archives/Public/semantic-web/2018Nov/0036.html.
[32] https://github.com/dbooth-boston/EasierRDF.

schemas automatically. In addition, the tool is thoroughly documented, with usage tutorials and examples available online.[33]

We end this section by discussing assumptions and limitations in OBA. For instance, OBA assumes that the target endpoint is modeled according to the ontology used to create the API; and changes in the ontology version will lead to a new version of the API (hence keeping track of which version supports which operations). OBA also assumes that two classes in an ontology network don't have the same local name, as each class is assigned a unique path. As per current limitations, OBA simplifies some restrictions in the ontology, such as complex axioms in property domains and ranges (e.g., having unions and intersections at the same time as a property range), to help creating the OAS. In addition, large ontologies may result in extensive APIs, which will work appropriately handling requests, but may be slow to render in a browser (e.g., to see documentation of a path). Finally, by default OBA does not handle reification or blank nodes, although they can be partially supported by creating custom queries.

OBA is proposed as a new resource, and therefore we don't have usage metrics from the community so far.

5   Related Work

The Semantic Web community has developed different approaches for helping developers access and manipulate the contents of knowledge graphs. For instance, the W3C Linked Platform [8] proposes a framework for handling HTTP requests over RDF data; and has multiple implementations such as Apache Marmotta[34] or Trellis.[35] Similarly, the Linked Data Templates specification[36] defines a protocol for read/write Linked Data over HTTP. The difference between these efforts and our approach is that OBA creates an OAS from ontology terms that provides an overview of the contents in a knowledge graph; and also simplifies validating resources with a documented OAS that is popular among developers.

Other efforts like Basil [3], GRLC [11], r4r[37] and RAMOSE[38] create REST APIs from SPARQL queries in order to ease access to knowledge graphs. However, in these efforts there is no validation of posted data according to a schema or ontology; knowledge engineers have to manually define the queries that need to be supported; and additional work is required to transform a result into an easy to use JSON representation. In OBA, posted resources are validated against the OpenAPI schemas, a first version of the API is created automatically from the ontology, and all the results are returned in JSON according to the generated OAS.

---

[33] https://oba.readthedocs.io/en/latest/quickstart/.
[34] http://marmotta.apache.org/.
[35] https://www.trellisldp.org/about.html.
[36] https://atomgraph.github.io/Linked-Data-Templates/.
[37] https://github.com/oeg-upm/r4r.
[38] https://github.com/opencitations/ramose.

Other efforts have attempted to improve the serialization of SPARQL results. For example, SPARQL transformer [7] and SPARQL-JSONLD[39] both present approaches for transforming SPARQL to user-friendly JSON results by using a custom mapping language and JSON-LD frames [6] respectively. In [14] the authors use GraphQL,[40] which is gaining popularity among the developer community, to generate SPARQL queries and serialize the results in JSON. In fact, some triplestores like Stardog have started to natively support interfaces for GraphQL.[41] These approaches facilitate retrieving parseable JSON from a knowledge graph, but developers still need to be familiar with the underlying ontologies used to query the data in those knowledge graphs. In OBA, an OAS with all available calls is generated automatically.

Parallel to the development of OBA, a recent effort has started to map OWL to OAS.[42] However, this approach focuses only on the mapping to OAS, while OBA also provides an implementation for creating an API server.

Finally, [4] proposes to define REST APIs to access the classes and properties of an ontology. This is different from our scope, which uses the ontology as a template to create an API to exploit the contents of a knowledge graph. To the best of our knowledge, our work is the first end-to-end framework for creating REST APIs from OWL ontologies to provide access to the contents of a knowledge graph.

6   Conclusions and Future Work

In this paper we have introduced the Ontology-Based APIs framework (OBA), a new resource for creating APIs from ontologies by using the OpenAPI Specification. OBA has demonstrated to be extremely useful in our work, helping setting up and maintaining API versions, testing and easy prototyping against a target knowledge graph. We believe that OBA helps bridging the knowledge gap between ontology engineers and developers, as it provides the means to create a guide (a documented API) that illustrates how to exploit a knowledge graph using the tools and standards developers are used to.

We are actively expanding OBA to support new features. First, we are working towards supporting additional mappings between OWL and OAS, such as complex domain and range axiomatization. Second, we are working to support accepting and delivering JSON-LD requests (instead of JSON only), which is preferred by some Semantic Web developers. As for future work, we are exploring the possibility of adding support for GraphQL, which has gained popularity lately, as an alternative to using SPARQL to retrieve and return contents. Finally, an interesting approach worth exploring is to combine an ontology with existing tools to mine patterns from knowledge graphs to expose APIs with the most common data patterns.

---

[39] https://github.com/usc-isi-i2/sparql-jsonld.
[40] https://graphql.org/.
[41] https://www.stardog.com/categories/graphql/.
[42] https://github.com/hammar/owl2oas.

Acknowledgements. This work was funded by the Defense Advanced Research Projects Agency with award W911NF-18-1-0027 and the National Science Foundation with award ICER-1440323. The authors would like to thank Yolanda Gil, Paola Espinoza, Carlos Badenes, Oscar Corcho, Karl Hammar and the ISWC anonymous reviewers for their thoughtful comments and feedback.

References

1. Auer, S., Bizer, C., Kobilarov, G., Lehmann, J., Cyganiak, R., Ives, Z.: DBpedia: a nucleus for a web of open data. In: Aberer, K., et al. (eds.) ASWC/ISWC -2007. LNCS, vol. 4825, pp. 722–735. Springer, Heidelberg (2007). https://doi.org/10.1007/978-3-540-76298-0_52
2. Cyganiak, R., Lanthaler, M., Wood, D.: RDF 1.1 concepts and abstract syntax. W3C recommendation, W3C (2014) http://www.w3.org/TR/2014/REC-rdf11-concepts-20140225/
3. Daga, E., Panziera, L., Pedrinaci, C.: A BASILar approach for building Web APIs on top of SPARQL endpoints. In: Proceedings of the Third Workshop on Services and Applications over Linked APIs and Data, co-located with the 12th Extended Semantic Web Conference (ESWC 2015), **1359**, 22–32 (2015)
4. Dirsumilli, R., Mossakowski, T.: RESTful encapsulation of OWL API. In: Proceedings of the 5th International Conference on Data Management Technologies and Applications, DATA 2016, SCITEPRESS - Science and Technology Publications, Lda, Setubal, PRT, pp. 150–157 (2016)
5. Garijo, D., Osorio, M., Khider, D., Ratnakar, V., Gil, Y.: OKG-Soft: an Open Knowledge Graph with Machine Readable Scientific Software Metadata. In: 2019 15th International Conference on eScience (eScience), San Diego, CA, USA, pp. 349–358. IEEE (2019) https://doi.org/10.1109/eScience.2019.00046
6. Kellogg, G., Champin, P.A., Longley, D.: JSON-LD 1.1. Candidate recommendation, W3C (2020) https://www.w3.org/TR/2020/CR-json-ld11-20200417/
7. Lisena, P., Meroño-Peñuela, A., Kuhn, T., Troncy, R.: Easy web API development with SPARQL transformer. In: Ghidini, C., et al. (eds.) ISWC 2019. LNCS, vol. 11779, pp. 454–470. Springer, Cham (2019). https://doi.org/10.1007/978-3-030-30796-7_28
8. Malhotra, A., Arwe, J., Speicher, S.: Linked data platform 1.0. W3C recommendation, W3C (2015) http://www.w3.org/TR/2015/REC-ldp-20150226/
9. Mihindukulasooriya, N., Poveda-Villalón, M., García-Castro, R., Gómez-Pérez, A.: Loupe - an online tool for inspecting datasets in the linked data cloud. In: Proceedings of the ISWC 2015 Posters & Demonstrations Track co-located with the 14th International Semantic Web Conference (ISWC-2015), Bethlehem, PA, USA. CEUR Workshop Proceedings, CEUR-WS.org, **1486** (2015)
10. Patel-Schneider, P., Motik, B., Parsia, B.: OWL 2 web ontology language structural specification and functional-style syntax (2 edn). W3C recommendation, W3C (2012) http://www.w3.org/TR/2012/REC-owl2-syntax-20121211/
11. Meroño-Peñuela, A., Hoekstra, R.: GRLC makes gitHub taste like linked data APIs. In: Sack, H., Rizzo, G., Steinmetz, N., Mladenić, D., Auer, S., Lange, C. (eds.) ESWC 2016. LNCS, vol. 9989, pp. 342–353. Springer, Cham (2016). https://doi.org/10.1007/978-3-319-47602-5_48
12. Seaborne, A., Harris, S.: SPARQL 1.1 query language. W3C recommendation, W3C (2013) http://www.w3.org/TR/2013/REC-sparql11-query-20130321/

13. Soiland-Reyes, S.: Owl2Jsonld 0.2.1 (2014) https://doi.org/10.5281/ZENODO.10565
14. Taelman, R., Vander Sande, M., Verborgh, R.: GraphQL-LD: linked data querying with GraphQL. In: Proceedings of the 17th International Semantic Web Conference: Posters and Demos, pp. 1–4 (2018)
15. Vrandečić, D., Krötzsch, M.: Wikidata: a free collaborative knowledgebase. Commun. ACM **57**(10), 78–85 (2014)
16. Zhu, Q., et al.: Collective multi-type entity alignment between knowledge graphs. In: Proceedings of The Web Conference 2020, WWW '20, Association for Computing Machinery, New York, NY, USA, pp. 2241–2252 (2020)

Schímatos: A SHACL-Based Web-Form Generator for Knowledge Graph Editing

Jesse Wright(✉)(iD), Sergio José Rodríguez Méndez(✉)(iD), Armin Haller(✉)(iD), Kerry Taylor(✉)(iD), and Pouya G. Omran(✉)(iD)

Australian National University, Canberra, ACT 2601, Australia
{jesse.wright,sergio.rodriguezmendez,armin.haller,
kerry.taylor,p.g.omran}@anu.edu.au
https://cecs.anu.edu.au/

Abstract. Knowledge graph creation and maintenance is difficult for naïve users. One barrier is the paucity of user friendly publishing tools that separate schema modeling from instance data creation. The Shapes Constraint Language (SHACL) [12], a W3C standard for validating RDF based knowledge graphs, can help. SHACL enables domain relevant structure, expressed as a set of shapes, to constrain knowledge graphs. This paper introduces *Schímatos*, a form based Web application with which users can create and edit RDF data constrained and validated by shapes. Forms themselves are generated from, and stored as, shapes. In addition, *Schímatos*, can be used to edit shapes, and hence forms. Thus, *Schímatos* enables end-users to create and edit complex graphs abstracted in an easy-to-use custom graphical user interface with validation procedures to mitigate the risk of errors. This paper presents the architecture of *Schímatos*, defines the algorithm that builds Web forms from shape graphs, and details the workflows for SHACL creation and data-entry. *Schímatos* is illustrated by application to Wikidata.

Keywords: Knowledge graph · SHACL · MVC-based Web-Form Generator · RDF editing tool · Linked data platform

1 Introduction

The rapid growth of Knowledge Graphs (KGs) impels the Semantic Web vision [6] of a ubiquitous network of machine readable resources. Popular KGs include the community-driven Wikidata [29], and Google's KG [15] which is largely populated through `schema.org` annotations on websites. An enduring barrier to the development of the machine-readable Web, however, is the lack of tools for authoring semantic annotations [3,14]. While ontology editors such as *Protégé* [16] and Diagrammatic Ontology Patterns [24] are favoured when creating quality assured (TBox) axioms, their primary purpose is to build and validate the model: the schema of an ontology. While some of these editors *can* also create instances, that is, individuals assigned to classes and data values with property

© Springer Nature Switzerland AG 2020
J. Z. Pan et al. (Eds.): ISWC 2020, LNCS 12507, pp. 65–80, 2020.
https://doi.org/10.1007/978-3-030-62466-8_5

relations (ABox), the process is cumbersome and requires detailed knowledge of the RDF(S) and OWL languages. There is limited support to guide users to the correct classes to which an entity can belong and the permissible relationships between entities. Further, as ontology editors do not clearly distinguish schema editing from data editing, data editing users may inadvertently alter the schema.

To address this, some Web publishing tools on top of wikis, microblogs or content management systems have been developed that allow a user to create semantic annotations, that is, instance assertions (e.g. the work discussed in [4,8,13,17,28]). This work has, for example, been incorporated into the semantic MediaWiki software [28] on which Wikidata is based [29]. However, even within this software, and so in Wikidata, there is limited user support for instance assertions, mostly through text auto-completion. Consequently, in order to add instances, users must have a strong understanding of the RDF data model, underlying semantics of the Wikidata ontology, and the typical structure of other instances of the same class.

The Shapes Constraint Language (SHACL) [12] is a recent W3C recommendation developed to express conditions, as shape graphs, for validating RDF based KGs. Thus domain relevant structure can be enforced. For example, Wikidata is an early adopter of shape graphs to define constraints on classes [25]. While Wikidata has chosen to use ShEx [20] to express these constraints[1], *Schímatos* uses the SHACL standard.

At the time of writing, 218 schemas exist under the Wikidata schema entity prefix, but none are enforced within the graph. For instance, the shape available at https://www.wikidata.org/wiki/EntitySchema:E10 [31] defines constraints on entities of type *Human*, but instances of this type (e.g. http://www.wikidata.org/entity/Q88056610 - Sergio José Rodríguez Méndez) are not validated against this shape. Moreover, the authoring tool underlying Wikidata does not use these shapes to guide the user when creating similar entities [25]. However, shapes *could* be used in authoring tools to guide semantic annotations.

This paper proposes *Schímatos*, a SHACL-based Web-Form Generator for Knowledge Graph Editing. Beyond its primary purpose of enabling naïve users to create and edit instance assertions, it also provides means to create and edit shapes. The software is being developed in the Australian Government Records Interoperability Framework (AGRIF) project and has been applied to basic use cases within several Australian Government departments. As these use cases are confidential, the authors cannot report on the knowledge graphs maintained in these cases, so we demonstrate a possible application of *Schímatos* on Wikidata.

The remainder of this paper is structured as follows. First, related systems are presented in Sect. 2. Section 3 describes a motivating use case for this work using an example entity from Wikidata. Then, *Schímatos* and its architecture are presented in Sect. 4, followed by the form display logic in Sect. 5. Section 6 defines formally the execution behaviour of the system with respect to the data entry and the shape creation process. The paper concludes in Sect. 7 with an outlook on the schedule for future work.

---

[1] See https://www.wikidata.org/wiki/Wikidata:WikiProject_Schemas.

2   Related Work

Many mature ontology editors such as Protégé [16], the Neon toolkit [9] or the commercial TopBraid Composer[2], offer ways to create entities based on one or more ontologies. Some of these editors have a Web-based version that can allow ordinary Web users to hand craft ontology instances. However, users must be able to recognise and correctly encode the expected property relationships for each new class instance according to RDF(S)/OWL semantics.

Web-publishing tools on top of wikis, microblogs or content management systems [4,8,13,17,28] allow naïve users to create semantic annotations, that is, instance assertions. However, the user interface of these tools is either fixed to a particular ontology or it has to be manually created based on the desired mapping to a TBox schema.

ActiveRaUL [7,10,11] is closest in functionality to *Schímatos*. It allows the automatic rendering[3] of Web-forms from arbitrary ontologies. The resulting forms can then be used to create instances compliant to the original ontology. The forms themselves become instances of a UI ontology, called RaUL. The major difference to *Schímatos* is that ActiveRaUL predated the introduction of SHACL and, as such, the resulting forms are generated by interpreting ontology assertions as rules. *Schímatos* is using SHACL shapes and therefore does not need to violate the Open World assumption of an underlying ontology, but truthfully renders a form, based on the constraints expressed.

ShEx-form [19] is a recent tool that creates forms from a shape graph. It appears to be inspired by Tim Berners-Lee's draft article, *Linked Data Shapes, Forms and Footprints* [26], which proposes the joint use of Shape Languages, Form Languages (such as the User Interface Ontology [5]), and Footprint Languages which describe where the data from a form is to be stored. To the best of our knowledge, this tool does not enable users to interact directly with external KGs and also does not perform any validation of user input. Furthermore, some features may not suit users unfamiliar with RDF concepts; requiring that they interact directly with the underlying shape serialization in order to generate a form.

3   Motivating Example

Within Wikidata [29] there are many instances of the class *Human*, most of which have missing attributes that are required according to the Wikidata *Human* shape. Of the existing instances of type human, there are only 8,117,293[4] entities which constitute about 0.103% of the current world population [27, Figure A.1.] and 0.007% [18, Table 2] of the total population over time. Whilst there have

---

[2] See https://www.topquadrant.com/products/topbraid-composer/.

[3] The term 'rendering' is used to refer to the generation of a Document Object Model (DOM). This is the same terminology used in the ReactJS framework.

[4] As of 2020-08-19 using the SPARQL query `SELECT (COUNT(?item) AS ?count) WHERE {?item wdt:P31 wd:Q5}`.

```
@prefix ex: <http://example.com/ns#>
@prefix sh: <http://www.w3.org/ns/SHACL#>
@prefix wd: <http://www.wikidata.org/entity/>
@prefix wdt: <http://www.wikidata.org/prop/direct/>
@prefix tp: <http://www.shacl.kg/types/>
@prefix rdfs: <http://www.w3.org/2000/01/rdf-schema#>

ex:humanWikidataShape              sh:path wdt:P734;              sh:property [
  a sh:NodeShape ;                   sh:name "family name" ;        sh:path wdt:P40;
  sh:targetClass wd:Q5 ;             sh:minCount 0 ;                sh:name "children" ;
  rdfs:label "human shape" ;       ] ;                              sh:minCount 0 ;
  sh:property [                    sh:property [                    sh:class wd:Q5 ;
    sh:path wdt:P21 ;                sh:path wdt:P106;            ] ;
    sh:name "gender" ;              sh:name "occupation" ;          sh:property [
    sh:in (                         sh:minCount 0 ,                 sh:path [
      wd:Q6581097 # male          ] ;                                sh:alternativePath (
      wd:Q6581072 # female        sh:property [                        wdt:P1038 [ # relatives
      wd:Q1097630 # intersex        sh:path wdt:P27;                     sh:oneOrMorePath [
      wd:Q1052281 # transgender female (MTF)  sh:name "country of citizenship" ;  sh:alternativePath (
      wd:Q2449503 # transgender male (FTM)    sh:minCount 0 ;            wdt:P22
      wd:Q48270 # non-binary      ] ;                                    wdt:P25
    );                            sh:property [                          wdt:P3373
    sh:maxCount 1 ;                 sh:path wdt:P22 ;                     wdt:P26
  ] ;                               sh:name "father" ;                 )
  sh:property [                     sh:minCount 0 ;                    ]
    sh:path wdt:P19 ;               sh:class wd:Q5 ;                  ]
    sh:name "place of birth" ;      sh:node ex:humanWikidataShape ;  )
    sh:maxCount 1 ;               ] ;                              ] ;
    sh:property [                 sh:property [                    sh:name "relatives" ;
      sh:path wdt:P17 ;             sh:path wdt:P25 ;               sh:minCount 0 ;
      sh:name "country" ;           sh:name "mother" ;              sh:class wd:Q5 ;
      sh:maxCount 1 ;               sh:minCount 0 ;                 sh:node ex:humanWikidataShape ;
    ] ;                             sh:class wd:Q5 ;              ] ;
  ] ;                               sh:node ex:humanWikidataShape ; sh:property [
  sh:property [                   ] ;                                sh:path wdt:P103 ;
    sh:path wdt:P569 ;            sh:property [                      sh:name "native language" ;
    sh:name "date of birth" ;       sh:path wdt:P3373 ;             sh:minCount 0 ;
    sh:maxCount 1 ;                 sh:name "sibling" ;           ] ;
    sh:pattern                      sh:minCount 0 ;                sh:property [
      "^[0-9]{2}\\/[0-9]{2}\\/[0-9]{4}$" ;  sh:class wd:Q5 ;         sh:path wdt:P1412 ;
  ] ;                               sh:node ex:humanWikidataShape ; sh:name "written/spoken language(s)" ;
  sh:property [                   ] ;                                sh:minCount 0 ;
    sh:path wdt:P735 ;           sh:property [                     ] ;
    sh:name "given name" ;         sh:path wdt:P26 ;                sh:property [
    sh:minCount 0 ;                 sh:name "husband|wife" ;         sh:path wdt:P6886 ;
    sh:datatype tp:name            sh:minCount 0 ;                  sh:name "publishing language(s)" ;
  ] ;                               sh:node ex:humanWikidataShape ;  sh:minCount 0 ;
  sh:property [                  ] ;                              ] .
```

Fig. 1. A SHACL shape for the class of human (`wd:Q5`) entities described in the Wikidata ontology.

been attempts to scrape data on existing entities [32], much of the information is either not available online, or not in a machine-readable format. Thus, to effectively complete instances of the class *Human*, widely-accessible tools that enforce logical constraints on the class (i.e. *Human* in this case) are required, whilst also prompting users to enter requisite data for the system. An example is the entity `wd:Q88056610`[5] which appears to have been automatically generated from the ORCID of the researcher [32]. Currently, no other information about this entity is available on Wikidata.

This paper uses the shape graph for *Human* [31] as a running example, and shows how *Schímatos* can be used to enforce constraints on instances of this class. Specifically, the paper demonstrates how the tool can be used to fill out missing information; including gender, birthplace and date of birth; for the entity `wd:Q88056610`[6].

---

[5] This is the identifier for researcher Dr. Sergio José Rodríguez Méndez.

[6] The repo for *Schímatos* at http://schimatos.github.io includes a translation of the *Human* ShEx shape into a SHACL shape which can be obtained by automated means with tools such as RDFShape [21].

4   The *Schímatos* System

Schímatos is an application that automatically generates Web-forms from SHACL shapes. Data from completed forms, including the class and datatype annotations of inputs, can be submitted to a KG over a SPARQL endpoint registered with *Schímatos*. Web-forms may also be edited within the tool and their SHACL definition updated via SPARQL updates. All operations occur in the client so the tool can be packaged as a stand-alone desktop application or served from a website.

Schímatos is built in the ReactJS Framework[7] which compiles to W3C compliant HTML+CSS/JavaScript with cross-browser compatibility. All SPARQL requests are compliant with the LDP standard [23], so that it can read/write data in SPARQL compliant triplestores. The tool also accepts data in the proposed SPARQL 1.1 Subscribe Language [1] so as to receive live updates from SEPA clients [2].

Schímatos is available to download as an HTML+CSS/JavaScript package[8] and it can also be run online[9]. Both of these resources work by default with a local instance of Wikidata and re-use existing ShEx files [30] translated to SHACL files for *Schímatos*. Consider the entity `wd:Q88056610`, which appears to have been automatically generated from the ORCID of the researcher [32]. Currently, no other information about this entity is available on Wikidata. One can apply the SHACL-translated *Human* constraint [31] to this entity in order to create a form that prompts users to fill out the missing information including gender, birthplace and date of birth. Many additional examples are available as translations of the pre-defined ShEx constraints publicly available on the Wikidata platform[10].

4.1   Architecture

At its core, *Schímatos'* design follows the Model-View-Controller (MVC) pattern [22] as depicted in Fig. 2.

Model. The model layer consists of three logical named graphs[11]: (1) SHACL store (shapes), (2) Type store[12], and (3) the KG (data). The SHACL store is a repository of shapes which, when loaded, are translated to form structures by the *View*. The Type store contains constraints for custom class and datatype definitions, such as regular expressions that all literals of a given datatype must match. Examples include: phone numbers, email addresses, and social security numbers.

---

[7] See http://reactjs.org.

[8] http://schimatos.org.

[9] http://schimatos.github.io.

[10] See https://www.wikidata.org/wiki/Wikidata:WikiProject_Schemas.

[11] There is no requirement that these graphs share the same SPARQL endpoint or host.

[12] The current namespace prefix for this graph is http://www.shacl.kg/types/ (`tp:`).

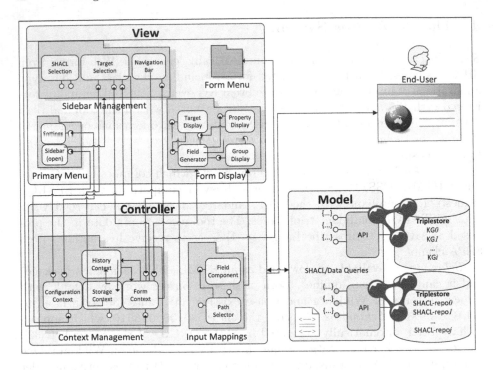

Fig. 2. Model-View-Controller software architectural design pattern for *Schímatos*

These definitions provide validation constraints for any `sh:PropertyShape` with a defined class or datatype. The KG, of course, has the data, which is used to pre-fill the form and is updated whenever a user submits new data.

Schímatos interacts with the model via RESTful API requests. In the current version of the tool, these requests are SPARQL queries and commands made over the SPARQL endpoint of the relevant triplestores.

View. One of the main parts of the view layer in Fig. 2 is the *Form Display* module; which has the components that handle the Web-Form generation from the SHACL structures. This module is explained in detail in Sect. 5. The Sidebar Management module wraps several key components, including the SHACL and Target Selectors, the Navigation Bar, and the SHACL Uploader.

SHACL Selection. Within this component, users can choose one or more constraints they wish to apply to given entities. Shapes can be searched for by properties including `rdfs:label`, `sh:targetClass`, and `sh:targetNode` constructors. Experienced users have the option to write custom SPARQL queries with which they can search for SHACL constraints. Users also have the option to customise many of their selections, including the 'severity' of the constraints they wish to apply and whether they wish to apply several constraints to a single entity. There is also functionality to 'automatically apply' constraints where the target `class/entity/subjectsOf/objectsOf` attributes of SHACL constraints are used to determine whether the constraint applies to the entity a user is targeting.

Target Selection. Within this panel, users can search for entities that they wish to enter into the *Schímatos* form. Similarly to the *SHACL Selection* Component, shapes can be selected using a custom SPARQL query or by querying for properties such as `rdf:type` and `rdfs:label`. This part of the GUI also displays the datatype and object properties of the currently selected entity so that users can verify that they have selected the correct IRI.

Navigation Bar. Within the *Navigation Bar*, users may choose to navigate to, and focus on different components of the form. This is primarily of use when entering data about complex entities for which the form structure may be too overwhelming to view in a single display.

4.2 Controller

Input Mappings. All data entry in *Schímatos* is validated against the shape constraint used to generate the form. The current validation engine for individual inputs is built upon the React Hook Forms package[13]. For validating individual form fields, *Schímatos* uses `sh:PropertyShape` constructors such as `sh:pattern` and `sh:minValue` to validate entry of entities and literals. Each `sh:PropertyShape`, such as that for a human's given name, may have a datatype or class (such as `tp:name`) with an associated set of constraints. If not defined directly in the `sh:PropertyShape`, *Schímatos* can suggest the datatype using the `rdfs:domain` and `rdfs:range` of the predicate (such as `wdt:P735`). These constraints can be defined in a shape and loaded automatically from a designated Type store. For this purpose, SHACL definitions of all pre-defined XSD types are pre-loaded into the *Storage Context*. The controller does not update the *Form Context* with values that fail the validation procedure and alerts users to fix the entry. This means that when a completed form is submitted to the model, no 'invalid' entries are submitted to the KG.

Context Management. *Schímatos* uses three ReactJS contexts[14] as described below, which control and manage the application's state at any given time.

1. The *Configuration Context* defines user settings for interacting with the tool such as the complexity of features available and the prefixes used within the form. In addition, the *Configuration Context* defines the KG, SHACL store and Type store endpoints. If permitted by the user, this is also stored as a browser cookie to maintain settings across sessions.

---

[13] See https://react-hook-form.com/.
[14] See https://reactjs.org/docs/context.html.

2. The *Form Context* captures the information to construct the form, including a representation of the SHACL constraints currently in use. It additionally stores data for all the entities and literals currently entered into the form, the data of children nodes within the supported path length, and the display settings for any *form element*[15]. When multiple shape constraints are applied simultaneously, they are 'merged' within the *Form Context* by applying the most strict set of constraints on each property.

3. The *Storage Context* has a local copy of triples from the KG model which are relevant to the current form. This data can then be used to pre-fill the form, provide suggestions for possible user input, and enable the display of datatype and object properties. Additionally, the *Storage Context* contains datatype constraints which are automatically applied to any property fields that have 'datatype' information in the corresponding SHACL pattern.

5 Form Display Logic

The SHACL standard defines two `sh:Shape` classes which are mapped in *Schímatos* to the DOM as follows: `sh:NodeShapes` are mapped to form elements, and `sh:PropertyShapes` are mapped to form fields that include HTML/JavaScript validators and a label. The following paragraphs provide more detail on this rendering logic.

Rendering Simple Shapes. A `sh:NodeShape` (e.g. `ex:humanWikidataShape`) is a set of `sh:PropertyShapes` (e.g. shapes for `wdt:P1038` - *relatives*, or `wdt:P21` - *gender*) used to generate a form for a chosen focus node (e.g. `wd:Q88056610` - Sergio José Rodríguez Méndez). When rendering a form, *Schímatos* uses the `sh:order` constructor of each `sh:PropertyShape` to determine the position in which it is displayed whilst `sh:PropertyShapes` with the same `sh:group` constructor are grouped in a pane.

Each `sh:PropertyShape` is rendered as a set of one or more HTML inputs that have the same validators, and the values of which follow the same property path to the focus node. A single label is used for each set of inputs. If defined, it is the value of the `sh:name` constructor. Otherwise, if the `sh:path` is a single predicate of length 1 (for instance, `wdt:P1038` rather than `wdt:P22/wdt:P1038*`) then the `rdfs:label` of that predicate is displayed. If `rdfs:label` is undefined, the property path is used as the label. Each set is displayed and validated uniquely depending on the constraints specified in the shape. If the `sh:nodeKind` constraint is `sh:IRI` (e.g. as for `wdt:P1038` - *relatives*, in Fig. 1), then the input must be a valid W3C IRI and the tool presents users with a customised IRI-field where they can use a drop-down input to select the correct namespace prefix before entering the remaining IRI. Users are also presented with suggested inputs, such as known entities of the class, which they can select as the input.

If the `sh:in` constraint is present for a property (e.g. as for `wdt:P21` - *gender*) then the input value must lie within the set of values predefined in the

---

[15] See Sect. 5.

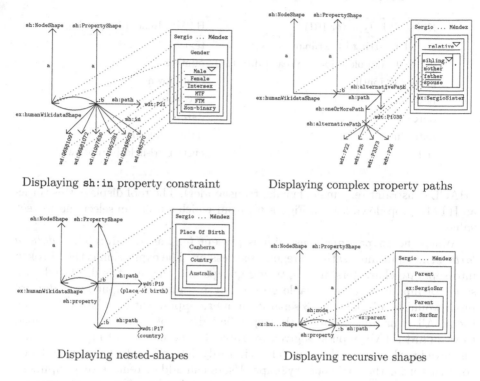

Fig. 3. Rendering of different property constraints in a generated form

Fig. 4. Screenshot of the form generated by *Schímatos* from the *Human* shape

Table 1. XSD/XML and IRI to HTML/ReactJS mapping

XML/XSD type (or IRI)	HTML/ReactJS input type
anyURI | string | hexBinary	text
decimal | double | float | gYear | duration	number
gDay | gMonth	drop-down
boolean	checkbox
date	date
dateTime	datetime-local
IRI	custom-iri-field | dropdown

SHACL constraint (e.g. male, female, transgender). The tool displays this set as an HTML drop-down list (cf. Figure 3a) from which a user can select the correct value.

When the `sh:pattern` constraint is present (e.g. as for `wdt:P569` - *date of birth*), the input must satisfy a regular expression. The expression is then broken into separate characters and capturing groups and the input is displayed as a series of inputs for each variable group. For instance, the input for a date of birth would be displayed as 3 separate numeric inputs with slashes in between them. All remaining inputs with a specified datatype constraint are mapped to standard HTML inputs as given in Table 1. The number of inputs that are displayed to the user in the form field is initially determined by the `sh:minCount` constructor for the `sh:PropertyShape`. Users can add or remove such inputs so long as they remain within the `sh:minCount` and `sh:maxCount` constraint for a property with `sh:severity` of `sh:Violation`, but are by default unbounded for other levels of severity. These inputs are equipped with a set of JavaScript validators for each property constraint; the value is not saved into the form until it passes the validation criteria.

Rendering Property Shapes with Complex Paths. Complex `sh:path` constructors in `sh:PropertyShapes` are displayed as a variable label for the set of inputs. Rather than displaying the `sh:name`, `rdfs:label` or property path IRI as the label, users are presented with a menu from which they can select the path they wish to use to add values to a given property constraint. For instance, the complex property path shape in Fig. 3b shows the `ex:relatives` property of the `ex:humanWikidataShape`. For the label, there is a drop-down where users can select the option `wdt:P1038` (for a generic relative) and another option to create a custom path using the *sibling, mother, father*, and *spouse* properties. For instance, one could enter information about their *grandfather* using the path `wdt:P22/wdt:P22` (*father/father*). Values will be entered into the graph corresponding to the value of the drop-down label at the time of entry. This means, a user could enter her grandfather's name, and then change the path to `wdt:P22/wdt:P25` to enter information about her grandmother.

Rendering Nested Shapes. Nested Shapes are rendered as nested form elements within the DOM. To do this, *Schímatos* first renders the form disregarding any sh:node constructors or nested properties present within a sh:PropertyShape. For each input generated by a given sh:PropertyShape, the nested sh:NodeShape is applied, treating the input as the focus node for the new form element. In our example, *Schímatos* will first generate a form beginning at ex:humanWikidataShape that has a form field for wdt:P19 (*place of birth*). Once the user inputs the IRI denoting place of birth, it becomes the focus node of a nested sh:NodeShape which is a form element containing a single form field for wdt:P17 (*country*). A sample rendering of this form is given in Fig. 3c.

Rendering Recursive Shapes. The SHACL standard specifies that 'validation with recursive shapes is not defined in SHACL and is left to SHACL processor implementations' [12]. In *Schímatos*, this is represented as a nested form structure that responds dynamically to user entry. Recursively defined shapes cannot be implemented in the same manner as nested shapes as doing so would cause a non-terminating loop within the application. Consequently, *Schímatos* loads and stores recursively defined data in the *Storage Context*. In our example, *Schímatos* first loads the ex:humanWikidataShape and renders all form fields within the form element. Since the sh:PropertyShape generating the form field wdt:P22 (*father*) contains a recursive reference to the ex:humanWikidataShape, the rendering of this element terminates and a copy of ex:humanWikidataShape is saved to the *Storage Context*. Once a user submits a value in the form field for wdt:P22 (e.g. wd:SergioSenior), or when the value is pre-filled using data from the KG, *Schímatos* validates the node wd:SergioSenior against the ex:humanWikidataShape. If the validation passes then the form remains unchanged, otherwise, a new ex:humanWikidataShape form element is rendered with wd:SergioSenior as the focus node. The process repeats, as the form element for wd:SergioSenior contains a wdt:P22 form field which references the ex:humanWikidataShape. Users may enter the IRI of wd:Q88056610's *grandfather* (ex:SnrSnr) in this field (cf. Fig. 3d). Users can terminate this process at any point by closing any form element they do not wish to complete.

6   Execution Behavior

The execution behaviour of *Schímatos* can be grouped into two processes, i.e. a data entry process (cf. Fig. 5), and a shape creation process (cf. Fig. 6), that may be executed in an interleaving manner. However, as the intended user of each of these processes is typically distinct, i.e. domain experts that create data and information architects that can change shape constraints, respectively, this paper distinguishes these two processes in the following sections.

6.1   Data Entry

Users may enter data about either a new or an existing entity. In both cases, a user names the entity and selects the shape they wish to use for the process which in turn generates a form for the user to fill in. If there is existing data relating to the named entity, the form will be automatically initialised with information retrieved from the KG. Figure 5 presents a UML sequence diagram for the case where a user wishes to add new data about an existing entity wd:Q88056610 - Sergio José Rodríguez Méndez. Figure 4 depicts a screen capture of this process.

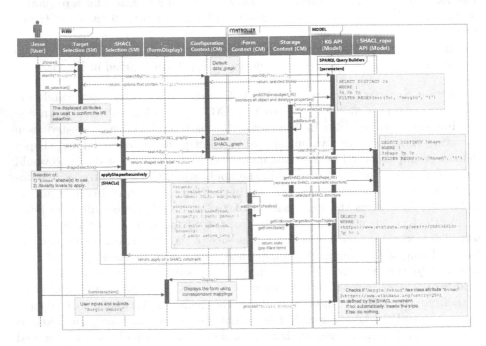

Fig. 5. UML sequence diagram view of *Schímatos* for data creation

To do so, a user would first navigate to the *Target Selection* panel within the sidebar. There is a search bar within the panel where users can search by any datatype (literal) constructor relating to an entity. Since the rdfs:label constructor for the entity is "Sergio José Rodríguez Méndez", the entity will appear in the set of results when a user searches for "Sergio". Once a user selects an entity from the drop-down menu, it is entered into the empty field on the screen. Next users can select which shape (form) they wish to apply to the entity. To do this, they can first navigate to the *SHACL Selection* panel within the sidebar, since the *Human* shape which is to be applied has an rdfs:label "human shape for wikidata instances" (cf. Fig. 1), it will be in the set of results for the search term 'human'. Once a user applies the shape to the entity, a form will be generated via the procedure outlined in Sect. 5. Once the form is rendered, users can choose to add or remove repetitive form components as long as it is allowed by the SHACL constraint underlying that form component. For example, an individual can have between 0 and 2 living biological parents, so the user may choose to add another field under *parent* and enter the names of Sergio's mother *and* Sergio's father. Each user entry is subject to a validation process (based on regular expression patterns, value constraints, etc. as discussed in Sect. 4.2). Users are not able to submit their data for this entry if the validation process fails: for instance, if a user attempts to enter a date of birth, without including the year. When this occurs, a popup will prompt the user to correct the entry. Once the user has 'completed' the form, they can choose to perform a 'final submission' which performs additional validations (including validating certain relationships between nodes, and cardinality of elements). Users will be guided to fix any errors if this validation process fails. If the validation process passes, all changes will be submitted to the KG.

6.2   Shape Creation

Schímatos provides the capacity for information architects to construct SHACL shapes within a form building UI. Figure 6 presents the UML sequence diagram for creating a new *detailed researcher shape*[16]. In this example, there is an existing *Human* shape and a *University professor* shape defined for Wikidata instances. These shapes contain many of the attributes that the user wishes to include in a new shape, say *detailed researcher*. The user can search for the *Human* shape and the *University professor* shape before opting to apply both shapes simultaneously. The user has the option to apply all constraints, or only those with a defined level of violation severity. A severity may be `sh:Info`, `sh:Warning` or `sh:Violation` with an undefined severity defaulting to a `sh:Violation`. These shapes will then be merged within the *Form Context* by the processes outlined in Sect. 4.2. This is displayed to the user as a single form which contains all of the constructors and constraints from both shapes. The user may then manually edit/create form fields using the form building tools. In this use case, the user will add the `wdt:P496` constructor (ORCID). The `sh:maxCount` constraint along with the pattern of the datatype are used to validate that the user entry follows the correct pattern for at most one ORCID. Once the user submits this change, the *Form Context* is updated accordingly. When a user chooses to 'save' a shape back to the SHACL graph, the internal structures storing the shape in the *Form Context* is first serialized to Turtle and then the data is submitted to the SHACL triplestore over the SPARQL endpoint via an INSERT command.

7   Conclusion and Future Work

Schímatos is the first interactive SHACL informed knowledge graph editor. It can be used for knowledge graph completion by domain experts without expertise in RDF(S)/OWL as well as for the development of SHACL shapes by information architects. This paper has shown how *Schímatos* can dynamically transform SHACL shapes to HTML data-entry forms with built-in data validation.

---

[16] This example is described in further detail at https://github.com/schimatos/schimatos.org.

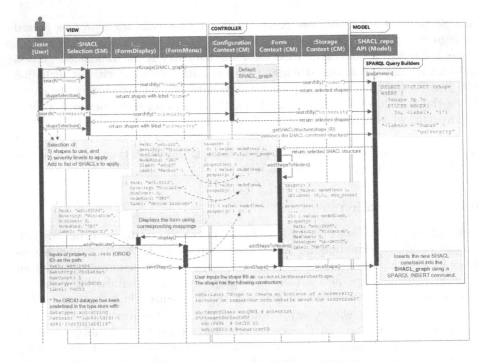

Fig. 6. UML Sequence view of *Schímatos* for SHACL creation

Schímatos is available under the MIT license[17] for download at http://schimatos.org and for Web use at http://schimatos.github.io. The first public release of the software is also available under the DOI https://doi.org/10.5281/zenodo.3988748.

The authors expect that the software will be maintained in the long term by the Australian Government Linked Data Working Group[18] and the Open Source community at large. We would like to invite the Wikidata project to gauge the potential to use the tool in the continuous creation of this public knowledge graph.

In future work, the authors have planned formal user trials in the Australian Government. Currently, *Schímatos* is being improved in the following ways: (1) multi-user support; (2) a Model RESTful service that handles different graph versions (with DELETE operations); (3) support of SHACL generation from TBox axioms; and (4) RDF* support.

---

[17] See https://opensource.org/licenses/MIT.
[18] See http://linked.data.gov.au.

Schímatos should be cited as follows:
Wright, J., Rodríguez Méndez, S.J., Haller, A., Taylor, K., Omran, P.G.: *Schímatos* - A SHACL-based Web-Form Generator (2020), DOI https://doi.org/10.5281/zenodo.3988748.

References

1. Aguzzi, C., Antoniazzi, F., Roffia, L., Viola, F.: SPARQL 1.1 subscribe language. Unofficial Draft 12 October, w3c (2018) http://mml.arces.unibo.it/TR/sparql11-subscribe.html
2. Aguzzi, C., Antoniazzi, F., Roffia, L., Viola, F.: SPARQL event processing architecture (SEPA). Unofficial Draft, w3c (2018) http://mml.arces.unibo.it/TR/sepa.html
3. Barbosa, A., Bittencourt, I.I., Siqueira, S.W., Silva, R.D., Calado, I.: The use of software tools in linked data publication and consumption: a systematic literature review. IJSWIS **13**, 68–88 (2017)
4. Baumeister, J., Reutelshoefer, J., Puppe, F.: KnowWE: a Semantic Wiki for knowledge engineering. Appl. Intell. **35**, 323–344 (2011). https://doi.org/10.1007/s10489-010-0224-5
5. Berners-Lee, T.: User interface ontology. Tech. rep., w3c (2010) https://www.w3.org/ns/ui
6. Berners-Lee, T., Hendler, J., Lassila, O.: The semantic web. Sci. Am. **284**(5), 34–43 (2001)
7. Butt, A.S., Haller, A., Liu, S., Xie, L.: ActiveRaUL: a web form-based user interface to create and maintain RDF data. In: Proceedings of the ISWC 2013 Posters & Demonstrations Track, CEUR-WS.org **1035**, 117–120 (2013)
8. Corlosquet, S., Delbru, R., Clark, T., Polleres, A., Decker, S.: Produce and consume linked data with drupal!. In: Bernstein, A., et al. (eds.) ISWC 2009. LNCS, vol. 5823, pp. 763–778. Springer, Heidelberg (2009). https://doi.org/10.1007/978-3-642-04930-9_48
9. Haase, P., et al.: The neon ontology engineering toolkit. In: Proceedings of the WWW 2008 Developers Track (2008)
10. Haller, A.: Activeraul: a model-view-controller approach for semantic web applications. In: Proceedings of the International Conference on Service-Oriented Computing and Applications (SOCA), pp. 1–8. IEEE (2010)
11. Haller, A., Groza, T., Rosenberg, F.: Interacting with linked data via semantically annotated widgets. In: Pan, J.Z., et al. (eds.) JIST 2011. LNCS, vol. 7185, pp. 300–317. Springer, Heidelberg (2012). https://doi.org/10.1007/978-3-642-29923-0_20
12. Knublauch, H., Kontokostas, D.: Shapes constraint language (SHACL). W3C Recommendation, w3c (2017) https://www.w3.org/TR/shacl
13. Kuhn, T.: Acewiki: Collaborative ontology management in controlled natural language. In: Proceedings of the 3rd Semantic Wiki Workshop (SemWiki 2008), in conjunction with ESWC 2008 (2008)
14. Liao, X., Zhao, Z.: Unsupervised approaches for textual semantic annotation, a survey. ACM Comput. Surv. **52**(4), 1–45 (2019)
15. Noy, N., Gao, Y., Jain, A., Narayanan, A., Patterson, A., Taylor, J.: Industry-scale knowledge graphs: lessons and challenges. ACM Queue **17**(2), 48–75 (2019)

16. Noy, N., Sintek, M., Decker, S., Crubezy, M., Fergerson, R., Musen, M.: Creating semantic web contents with protege-2000. Intelli. Syst., IEEE **16**(2), 60–71 (2001)
17. Passant, A., Breslin, J.G., Decker, S.: Open, distributed and semantic microblogging with SMOB. In: Benatallah, B., Casati, F., Kappel, G., Rossi, G. (eds.) ICWE 2010. LNCS, vol. 6189, pp. 494–497. Springer, Heidelberg (2010). https://doi.org/10.1007/978-3-642-13911-6_36
18. Population Reference Bureau: How many people have ever lived on earth? (2018) https://www.prb.org/howmanypeoplehaveeverlivedonearth/
19. Prud'hommeaux, E.: Play with ShEx-generated forms (2020) https://github.com/ericprud/shex-form
20. Prud'hommeaux, E., Boneva, I., Gayo, J.E.L., Kellogg, G.: Shape expressions language 2.1. Final community group report 8 october 2019, W3C Community Group (2019) http://shex.io/shex-semantics/
21. RDFShape: Parse and convert schema, http://rdfshape.herokuapp.com/schemaConversions
22. Reenskaug, T.: The original MVC reports. Tech. rep., Dept. of Informatics, University of Oslo (2007) http://heim.ifi.uio.no/~trygver/2007/MVC_Originals.pdf
23. Speicher, S., Arwe, J., Malhotra, A.: Linked data platform 1.0. W3C Recommendation, w3c (2015) https://www.w3.org/TR/ldp/
24. Stapleton, G., Howse, J., Taylor, K., Delaney, A., Burton, J., Chapman, P.: Towards diagrammatic ontology patterns. In: Workshop on Ontology and Semantic Web Patterns (WOP 2013), CEUR proceedings, **1188** (2013) http://ceur-ws.org/Vol-1188/paper_4.pdf
25. Thornton, K., et al.: Using shape expressions (ShEx) to share RDF data models and to guide curation with rigorous validation. In: Hitzler, P., et al. (eds.) ESWC 2019. LNCS, vol. 11503, pp. 606–620. Springer, Cham (2019). https://doi.org/10.1007/978-3-030-21348-0_39
26. Tim berners-lee: linked data shapes, forms and footprints. Tech. rep., w3c (2019) https://www.w3.org/DesignIssues/Footprints.html
27. United Nations: World population prospects 2019: Data booklet (2019). https://population.un.org/wpp/Publications/Files/WPP2019_DataBooklet.pdf
28. Völkel, M., Krötzsch, M., Vrandecic, D., Haller, H., Studer, R.: Semantic wikipedia. In: Proceedings of the 15th International Conference on World Wide Web, WWW '06, ACM, New York, NY, USA, pp. 585–594 (2006)
29. Vrandeĉić, D., Krötzsch, M.: Wikidata: a free collaborative knowledgebase. Commun. ACM **57**(10), 78–85 (2014)
30. Wikiproject schemas, https://www.wikidata.org/wiki/Wikidata:WikiProject_Schemas
31. Wikidata: human (E10) (2020) https://www.wikidata.org/wiki/EntitySchema:E10
32. Wikiproject source metadata, https://www.wikidata.org/wiki/Wikidata:WikiProject_Source_MetaData

OWL2Bench: A Benchmark for OWL 2 Reasoners

Gunjan Singh[1], Sumit Bhatia[2(✉)], and Raghava Mutharaju[1]

[1] Knowledgeable Computing and Reasoning Lab, IIIT-Delhi, New Delhi, India
{gunjans,raghava.mutharaju}@iiitd.ac.in
[2] IBM Research AI, New Delhi, India
sumitbhatia@in.ibm.com

Abstract. There are several existing ontology reasoners that span a wide spectrum in terms of their performance and the expressivity that they support. In order to benchmark these reasoners to find and improve the performance bottlenecks, we ideally need several real-world ontologies that span the wide spectrum in terms of their size and expressivity. This is often not the case. One of the potential reasons for the ontology developers to not build ontologies that vary in terms of size and expressivity, is the performance bottleneck of the reasoners. To solve this chicken and egg problem, we need high quality ontology benchmarks that have good coverage of the OWL 2 language constructs, and can test the scalability of the reasoners by generating arbitrarily large ontologies. We propose and describe one such benchmark named OWL2Bench. It is an extension of the well-known University Ontology Benchmark (UOBM). OWL2Bench consists of the following – TBox axioms for each of the four OWL 2 profiles (EL, QL, RL, and DL), a synthetic ABox axiom generator that can generate axioms of arbitrary size, and a set of SPARQL queries that involve reasoning over the OWL 2 language constructs. We evaluate the performance of six ontology reasoners and two SPARQL query engines that support OWL 2 reasoning using our benchmark. We discuss some of the performance bottlenecks, bugs found, and other observations from our evaluation.

Keywords: OWL 2 benchmark · Ontology benchmark · OWL2Bench · Reasoner benchmark

Resource Type: Benchmark
License: Apache License 2.0
Code and Queries: https://github.com/kracr/owl2bench

1 Introduction

OWL 2 [6] has different profiles, namely OWL 2 EL, OWL 2 QL, OWL 2 RL, OWL 2 DL, and OWL 2 Full that vary in terms of their expressivity and reasoning complexity. The first three profiles are tractable fragments of OWL 2

© Springer Nature Switzerland AG 2020
J. Z. Pan et al. (Eds.): ISWC 2020, LNCS 12507, pp. 81–96, 2020.
https://doi.org/10.1007/978-3-030-62466-8_6

having polynomial reasoning time. On the other hand, reasoning over OWL 2 DL ontologies has a complexity of N2EXPTIME and OWL 2 Full is undecidable. Several thousands of ontologies that belong to these OWL 2 profiles are available across repositories such as the NCBO Bioportal[1], AgroPortal[2], and the ORE datasets [12]. In order to compare features and benchmark the performance of different reasoners, one could use a subset of these existing ontologies. However, such an approach is inflexible as most real-world ontologies cover only a limited set of OWL constructs and arbitrarily large, and complex ontologies are seldom available that can be used to test the limits of systems being benchmarked. A synthetic benchmark, on the other hand, offers the flexibility to test various aspects of the system by changing the configuration parameters (such as size and complexity). In particular, it is hard to answer the following questions without a benchmark.

i) Does the reasoner support all the possible constructs and their combinations of a particular OWL 2 profile?
ii) Can the reasoner handle large ontologies? What are its limits?
iii) Can the reasoner handle all types of queries that involve reasoning?
iv) What is the effect of any particular construct, in terms of number or type, on the reasoner performance?

Unless a reasoner can answer these questions, ontology developers will not have the confidence to build large and complex ontologies. Without these ontologies, it will be hard to test the performance and scalability of the reasoner. So an ontology benchmark can fill this gap and help the developers in building better quality reasoners and ontologies.

There are some existing benchmarks such as LUBM [5], UOBM [9], BSBM [3], SP²Bench [15], DBpedia [10], and OntoBench [8]. Some of these are based on RDF (BSBM, SP²Bench, DBpedia) or an older version of OWL (LUBM, UOBM), and those that cover OWL 2 are limited in scope (OntoBench). We propose an OWL 2 benchmark, named OWL2Bench, that focuses on the coverage of OWL 2 language constructs, tests the scalability, and the query performance of OWL 2 reasoners. The main contributions of this work are as follows.

– TBox axioms for each of the four OWL 2 profiles (EL, QL, RL, and DL). They are developed by extending UOBM's university ontology. These axioms are helpful to test the inference capabilities of the reasoners.
– An ABox generator that can generate axioms of varying size over the aforementioned TBox. This is useful for testing the scalability of the reasoners.
– A set of 22 SPARQL queries spanning different OWL 2 profiles that require reasoning in order to answer the queries. These queries also enable benchmarking of SPARQL query engines that support OWL 2 reasoning.

---

[1] https://bioportal.bioontology.org/.
[2] http://agroportal.lirmm.fr/.

Six reasoners, namely, ELK [7], HermiT [4], JFact[3], Konclude [17], Openllet[4], and Pellet [16] were evaluated using OWL2Bench on three reasoning tasks (consistency checking, classification, realisation). SPARQL queries were used to evaluate the performance of Stardog[5] and GraphDB[6]. The results of our evaluation are discussed in Sect. 4.

2 Related Work

There has been limited work on developing benchmarks for OWL 2 reasoners. On the other hand, there are several established benchmarks for RDF query engines and triple stores such as LUBM [5], BSBM [3], SP^{2}Bench [15], WatDiv [1], DBpedia benchmark [10], and FEASIBLE [14]. These benchmarks have been discussed comprehensively in [13]. Since the focus of our work is not on benchmarking the RDF query engines and triple stores, we do not discuss these benchmarks any further.

LUBM [5] provides an ontology for the university domain that covers a subset of OWL Lite constructs, a dataset generator to generate variable size instance data, and a set of 14 SPARQL queries. In the generated dataset, different universities lack necessary interlinks that do not sufficiently test the scalability aspect of the ontology reasoners. To test the scalability of reasoners, we need to increase the size of the generated data. In the case of LUBM, this is achieved by increasing the number of universities. So, if the benchmark lacks necessary interlinks across different universities, the generated instance data would result in multiple isolated graphs rather than a connected large graph. Reasoning over a connected large graph is significantly harder than that on multiple isolated small graphs. Thus, interlinks are necessary to reveal the inference efficiency of the reasoners on scalable datasets. University Ontology Benchmark (UOBM) [9] is an extension of LUBM that has been designed to overcome some of the drawbacks of LUBM. It covers most of the OWL Lite and OWL DL constructs to test the inference capability of the reasoners. Additional classes and properties have been added to create interlinks among the universities. But, UOBM does not support the OWL 2 profiles.

OntoBench [8] covers all the constructs and profiles supported by OWL 2. The primary purpose of OntoBench is to test the coverage of the reasoners rather than their scalability. It provides a web interface[7] for the users to choose the OWL 2 language constructs which are then used to generate an ontology. Thus, OntoBench overcomes the inflexibility of the other static benchmarks. However, it does not support the generation of ABox axioms.

JustBench [2] is a benchmark for ontology reasoners in which the performance is analyzed based on the time taken by the reasoners to generate justifications for

---

[3] http://jfact.sourceforge.net/.
[4] https://github.com/Galigator/openllet.
[5] https://www.stardog.com/.
[6] http://graphdb.ontotext.com/.
[7] http://visualdataweb.de/ontobench/.

Table 1. A comparison of the state-of-the-art benchmarks with OWL2Bench. C indicates coverage in terms of OWL constructs/SPARQL features, S indicates scalability, R indicates the three reasoning tasks (consistency checking, classification, and realisation)

Benchmark	Supported Profile	Evaluated System(s)	Evaluation Type
LUBM	OWL Lite (Partial)	RDFS and OWL Lite Reasoners	C, S
UOBM	OWL Lite and OWL 1 DL	OWL Lite and OWL 1 DL Reasoners	C, S
SP²Bench	RDFS	RDF Stores	C, S
BSBM	RDFS	RDF Stores	C, S
DBpedia	RDFS/OWL (Partial)	RDF Stores	C, S
OntoBench	All OWL 2 Profiles	Ontology Visualization Tool	C
ORE Framework	OWL 2 EL and DL	OWL 2 Reasoners	R
OWL2Bench	All OWL 2 Profiles	OWL 2 Reasoners and SPARQL Query Engines	C, S, R

the entailments. This makes the benchmark independent of the OWL versions and profiles. JustBench does not generate any data (TBox or ABox axioms). So the benchmark, by itself, cannot be used to test the scalability of the reasoners.

Other than the aforementioned benchmarks, there also exists an open-source java based ORE benchmark framework[8] which was a part of OWL Reasoner Evaluation (ORE) Competition [11,12]. The competition was held to evaluate the performance of OWL 2 complaint reasoners over several different OWL 2 EL and OWL 2 DL ontologies. But, the performance evaluation in the context of varying sizes of an ontology was not considered. The ORE competition corpus can be used with the framework for reasoner evaluation. The framework evaluates the reasoners on three reasoning tasks, namely, consistency checking, classification, and realisation. The framework does not cover the evaluation of the SPARQL query engines (with OWL reasoning support) in terms of the coverage of constructs and scalability.

Comparison of OWL2Bench with some of the state-of-the-art benchmarks is provided in Table 1. There is no benchmark that is currently available that can test the OWL 2 reasoners in terms of their coverage, scalability, and query performance. We address this shortcoming by proposing OWL2Bench.

3   OWL2Bench Description

OWL2Bench can be used to benchmark three aspects of the reasoners - support for OWL 2 language constructs, scalability in terms of ABox size, and the query

---

[8] https://github.com/ykazakov/ore-2015-competition-framework.

performance. It consists of three major components, a fixed TBox for each OWL 2 profile, an ABox generator that can generate ABox of varying size with respect to the corresponding TBox, and a fixed set of SPARQL queries that can be run over the generated ontology (combination of TBox and ABox).

3.1 TBox

We built the TBox for each of the four OWL 2 profiles (EL, QL, RL, and DL), by extending UOBM because it has support for OWL Lite and OWL DL profiles[9], along with a mechanism to generate ABox axioms and has a set of SPARQL queries for the generated ABox. UOBM consists of concepts that describe a university (college, department, course, faculty, etc.) and the relationships between them. The TBoxes in OWL2Bench are created by following the steps listed below.

 i) The axioms from UOBM that belong to at least one of the OWL 2 profiles are added to the TBox of the respective profiles and those that did not fit into the corresponding OWL 2 profile due to the syntactic restrictions are restructured to fit into that profile.
 ii) There are several other language constructs from across the four OWL 2 profiles that are not covered by the axioms from UOBM. In such cases, either the UOBM axioms are enriched with additional constructs or new classes and properties are introduced, and appropriate axioms are added to the respective TBoxes.
iii) In order to create a more interconnected graph structures when compared to UOBM, additional axioms that link different universities are created. For example, object properties `hasCollaborationWith`, and `hasAdvisor` connect a `Person` from a `Department` with people across different departments of the same `College`, and across different colleges of the same `University`, or across different universities.

The hierarchy among some of the classes, including the relations between them, is shown in Fig. 1. All the four TBoxes of OWL2Bench consist of classes such as `University`, `College`, `CollegeDiscipline`, `Department`, `Person`, `Program`, and `Course`. They are related to each other through relationships such as `enrollFor`, `teachesCourse`, and `offerCourse`.

Some of the extensions to the axioms, classes, and properties made to UOBM for each OWL 2 profile are listed in Table 2. The complete list of extensions is available on GitHub[10]. Apart from ObjectSomeValuesFrom, the table contains only those OWL 2 constructs that are absent in UOBM. We included ObjectSomeValuesFrom to illustrate the syntactic restrictions imposed on the usage of different constructs in each OWL 2 profile. For example, we renamed the UOBM's `UndergraduateStudent` class to `UGStudent`, added a new class `UGProgram` and a property `enrollFor`, and redefined the class `UGStudent` (UG

---

[9] Profiles of OWL 1.
[10] https://github.com/kracr/owl2bench.

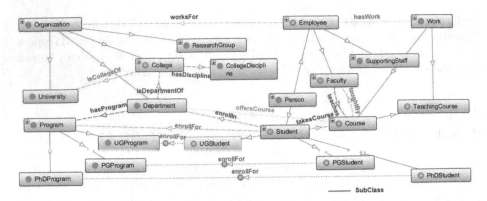

Fig. 1. Partial class hierarchy and relationship among some of the classes in OWL2Bench. The colored labeled (dashed) edges represent different object properties. The unlabeled edges represent the subclass relation. (Color figure online)

student is an undergraduate student who enrolls for any undergraduate program). The class UGStudent is added to all the profiles and the definition changes slightly (given below) depending on the OWL 2 profile.

- In OWL 2 EL, UGStudent ≡ Student ⊓ ∃enrollFor.UGProgram
- In OWL 2 QL, UGStudent ⊑ Student ⊓ ∃enrollFor.UGProgram (existential restrictions are not allowed in the subclass expression)
- In OWL 2 RL, Student ⊓ ∃enrollFor.UGProgram ⊑ UGStudent (existential restrictions are not allowed in the superclass expression)
- In OWL 2 DL, UGStudent ≡ Student ⊓ = 1enrollFor.UGProgram (since qualified exact cardinalities are supported, we could make the axiom more expressive by writing it using exact cardinality)

3.2 ABox

ABox axioms are generated by OWL2Bench based on two user inputs, the number of universities and the OWL 2 profile (EL, QL, RL, DL) of interest. The instance data that is generated complies with the schema defined in the TBox of the selected profile. The size of the instance data depends on the number of universities. The steps to generate the ABox are listed below.

i) Instances (class assertion axioms) for the University class are generated and their number is equal to the number of universities specified by the user.
ii) For each University class instance, instances for College, Department, as well as for all the related classes are generated.
iii) Property assertion axioms are created using these instances. For example, an object property isDepartmentOf links a Department instance to a College instance. Similarly, a data property hasName is used to connect a department name to a Department instance.

Table 2. Examples of TBox axioms from OWL2Bench. Axioms follow the syntactic restriction of the OWL 2 profiles.

OWL 2 EL	
ObjectPropertyChain	worksFor o isPartOf ⊑ isMemberOf
ObjectSomeValuesFrom	UGStudent ≡ Student ⊓ ∃enrollFor.UGProgram
ReflexiveObjectProperty	ReflexiveObjectProperty(knows)
ObjectHasSelf	SelfAwarePerson ≡ Person ⊓ ∃knows.Self

OWL 2 QL	
ObjectSomeValuesFrom	UGStudent ⊑ Student ⊓ ∃enrollFor.UGProgram
IrreflexiveObjectProperty	IrreflexiveObjectProperty(isAdvisedBy)
AsymmetricObjectProperty	AsymmetricObjectProperty(isAffiliatedOrganisationOf)

OWL 2 RL	
ObjectSomeValuesFrom	Student ⊓ ∃enrollFor.UGProgram ⊑ UGStudent
ObjectAllValuesFrom	WomenCollege ⊑ College ⊓ ∀hasStudent.Women
ObjectMaxCardinality	LeisureStudent ⊑ Student ⊓ ≤1 takesCourse.Course

OWL 2 DL	
ObjectAllValuesFrom	WomenCollege ≡ College ⊓ ∀hasStudent.Women
ObjectExactCardinality	UGStudent ≡ Student ⊓ =1enrollFor.UGProgram
ObjectMaxCardinality	LeisureStudent ≡ Student ⊓ ≤1 takesCourse.Course
ObjectMinCardinality	PeopleWithManyHobbies ≡ Person ⊓ ≥3 likes.Interest
DisjointDataProperties	DisjointDataProperties(firstName lastName)
Keys	HasKey(Person hasID)
DisjointObjectProperties	DisjointObjectProperties(likes dislikes)
DisjointUnion	CollegeDiscipline ≡ Engineering ⊔ Management ⊔ Science ⊔ FineArts ⊔ HumanitiesAndSocialScience
DisjointClasses	Engineering ⊓ Management ⊓ Science ⊓ FineArts ⊓ HumanitiesAndSocialScience ⊑ ⊥

iv) The number of instances of each class (other than University) and the number of connections between all the instances are selected automatically and randomly from a range specified in the configuration file. This range (maximum and minimum values of the parameters) can be modified to change the size of the generated ABox as well as to control the density (number of connections between different instances). Moreover, the output ontology format can also be specified in the configuration file. By default, the generated ontology format is RDF/XML.

Since we focus on testing the scalability of the reasoners OWL2Bench, generates (by default) approximately 50,000 ABox axioms for one university and it goes up to 14 million for 200 universities. Table 3 shows the number of TBox axioms and the ABox axioms that are generated by OWL2Bench based on the number of universities.

Table 3. The number of TBox axioms, along with the number of ABox axioms generated by OWL2Bench is given here.

Type		OWL 2 Profile			
		EL	QL	RL	DL
Classes		131	131	134	132
Object Properties		82	85	85	86
Data Properties		12	12	12	13
TBox Axioms		703	738	780	793
ABox Axioms	1 University	50,131	50,125	50,131	50,127
	2 Universities	99,084	99,078	99,084	99,080
	5 Universities	325,412	325,406	325,412	325,408
	10 Universities	711,021	711,015	711,021	711,017
	20 Universities	1,380,428	1,380,422	1,380,428	1,380,424
	50 Universities	3,482,119	3,482,113	3,482,119	3,482,115
	100 Universities	7,260,070	7,260,064	7,260,070	7,260,066
	200 Universities	14,618,828	14,618,822	14,618,828	14,618,824

3.3   Queries

While the generated A-Box and T-Box can be used to benchmark capabilities and performance of various reasoners, OWL2Bench also provides twenty-two SPARQL queries as part of the benchmark. These queries have been designed with the aim of evaluating SPARQL engines that support reasoning (such as Stardog). Each query has a detailed description that captures the intent of the query and the OWL 2 language constructs that the SPARQL engine needs to recognize and process accordingly. Each query makes use of language constructs that belong to at least one OWL 2 profile. Due to the lack of space, we have not included all the queries in the Appendix. They are available on GitHub (see footnote 10).

4   Experiments and Discussions

In this section, we use our benchmark to compare the reasoning and querying performance of six reasoners and two SPARQL query engines. Note that the aim of our experiments is not to present an exhaustive analysis of all the existing reasoners and SPARQL query engines. Instead, we chose a representative subset to demonstrate the utility of OWL2Bench. During our evaluation, we identified possible issues with these systems (some of which have already been communicated with the developers) that need to be fixed and could also pave the way for further research in the development of reasoners and query engines.

All our evaluations ran on a server with an AMD Ryzen Threadripper 2990WX 32-Core Processor and 128GB of RAM. The server ran on a 5.3.0–46-generic #38 18.04.1-Ubuntu SMP operating system. We use Java 1.8 and OWL API 5.1.11 in the experiments. We set the heap space to 24 GB. We run our benchmark on the reasoners ELK 0.4.3 [7], HermiT 1.3.8.1 [4], JFact 5.0.0 (see footnote 3), Konclude 0.6.2 [17], Openllet 2.6.4 (see footnote 4), and Pellet 2.3.6 [16]. ELK supports OWL 2 EL profile and the rest of them are OWL 2 DL reasoners. We use GraphDB 9.0.0 (see footnote 6) and Stardog 7.0.2 (see footnote 5) for running the SPARQL queries. GraphDB supports SPARQL queries and OWL 2 QL, OWL 2 RL profiles (but not OWL 2 EL and OWL 2 DL). Stardog's underlying reasoner is Pellet and hence it supports ontologies of all the profiles including OWL 2 DL. Note that only those reasoners and SPARQL query engines were considered for evaluation that offered full reasoning support with respect to the OWL 2 profiles.

4.1 Evaluating OWL Reasoners

We compare the six ontology reasoners with respect to the time taken for performing three major reasoning tasks, namely, consistency checking, classification, and realisation. We use OWL2Bench to generate ABox axioms for 1, 2, 5, 10, 20, 50, 100, and 200 Universities for the OWL 2 EL, QL, RL, and DL profiles. Table 3 reports the size of each of these datasets. Table 4 reports the time taken by the reasoners on these datasets. The time-out was set to 90 min. We report the average time taken in 5 independent runs for each reasoning task.

We used the OWL API to connect to HermiT, JFact, Pellet, Openllet, and run them. Since the reasoner implementations do quite a bit of work in the reasoner constructor, we included this time as well while calculating the time taken for each reasoning task. For the other two reasoners (ELK and Konclude), parsing and preprocessing times are included by default when run from the command line. Moreover, ELK also includes the ontology loading time in its reported time and ignores some of the axioms (such as Object Property Range, Data Property Assertion, Data Property Domain, and Range, Functional and Equivalent Data Property, Self Restriction and Same Individual) during the preprocessing stage. Since ELK is an OWL 2 EL reasoner, we compare its performance only in the EL profile.

JFact has performed the worst on all the reasoning tasks across all the four OWL 2 profiles. It timed out for every reasoning task except in the OWL 2 QL profile. Even for the QL profile, JFact performed the worst even on the smallest ontology and timed-out when the size of the dataset increased. Konclude outperformed all the other reasoners in terms of time taken for all the three reasoning tasks (except for OWL 2 DL datasets). It was able to complete the reasoning task even when most of the other reasoners had an error or timed out. Although Konclude is faster than most of the reasoners, it requires a lot of memory and thus as the size of the dataset grew larger (100 and 200 universities) it threw a memory error. It is interesting to observe that the ontologies generated by OWL2Bench are reasonably challenging and most of the reasoners were unable to handle the classification and realisation tasks on these ontologies.

For the OWL 2 EL profile, the two concurrent reasoners, ELK and Konclude performed exceptionally well when compared to HermiT, Openllet, and Pellet. Their performance deteriorated with increase in the size of the ontology. On the

Table 4. Time taken in seconds for the three reasoning tasks: Consistency Checking (CC), Classification (CT) and Realisation (RT) by the reasoners on the ontologies from OWL2Bench. j.h.s is Java Heap Space Error, g.c is Garbage Collection Overhead limit error, m.e is Memory Error, and t.o is Timed Out.

Profile	Task	Reasoner	No. of Universities							
			1	2	5	10	20	50	100	200
EL	CC	ELK	1.27	2.47	8.43	12.57	28.32	70.48	106.63	197.68
		HermiT	8.87	35.65	665.12	t.o	t.o	t.o	t.o	t.o
		JFact	t.o	t.o	t.o	t.o	t.o	t.o	t.o	t.o
		Konclude	0.88	1.92	7.67	17.70	35.59	89.68	m.e	m.e
		Openllet*	-	-	-	-	-	-	-	-
		Pellet	2.65	6.93	84.75	608.32	1876.00	t.o	t.o	t.o
	CT	ELK	1.50	2.50	8.51	12.56	30.93	70.77	106.83	177.59
		HermiT	110.13	639.56	t.o	t.o	t.o	t.o	t.o	g.c
		JFact	t.o	t.o	t.o	t.o	t.o	t.o	t.o	t.o
		Konclude	0.88	1.92	7.68	17.70	35.60	89.69	m.e	m.e
		Openllet	9.53	19.41	627.02	1803.05	t.o	t.o	t.o	t.o
		Pellet	11.46	37.72	335.14	t.o	t.o	t.o	t.o	t.o
	RT	ELK	1.60	2.69	8.97	13.42	32.45	74.03	114.48	195.26
		HermiT	t.o	t.o	t.o	t.o	t.o	j.h.s	g.c	g.c
		JFact	t.o	t.o	t.o	t.o	t.o	t.o	t.o	t.o
		Konclude	0.90	1.95	7.77	17.93	36.03	90.70	m.e	m.e
		Openllet	13.88	38.18	281.23	1829.93	t.o	t.o	t.o	t.o
		Pellet	6.08	18.99	234.74	t.o	t.o	t.o	t.o	t.o
QL	CC	HermiT	1.34	3.23	6.56	14.71	29.37	65.81	151.40	392.60
		JFact	15.77	53.98	656.75	3207.62	t.o	t.o	t.o	t.o
		Konclude	0.66	1.41	5.70	13.56	26.72	66.29	m.e	m.e
		Openllet	0.93	1.69	4.81	11.87	25.12	51.23	121.03	335.12
		Pellet	0.96	1.60	5.09	10.93	21.20	51.26	121.866	303.18
	CT	HermiT	1.70	3.37	10.01	23.85	51.61	132.50	300.04	901.10
		JFact	29.88	93.76	935.12	4222.124	t.o	t.o	t.o	t.o
		Konclude	0.66	1.42	5.70	13.57	26.73	66.30	m.e	m.e
		Openllet	1.21	2.20	5.79	12.63	25.84	55.00	127.26	357.696
		Pellet	1.39	2.64	6.43	12.34	24.88	58.87	131.00	340.31
	RT	HermiT	1.70	3.37	10.01	23.85	51.61	132.50	300.04	901.10
		JFact	87.85	364.94	3776.16	t.o	t.o	t.o	t.o	t.o
		Konclude	0.68	1.44	5.79	13.75	27.08	67.20	m.e	m.e
		Openllet	1.95	3.16	6.85	14.43	27.27	65.22	149.11	426.81
		Pellet	1.15	1.91	5.62	12.53	24.63	61.91	141.13	350.85
RL	CC	HermiT	147.17	1782.14	t.o	t.o	t.o	j.h.s	t.o	g.c
		JFact	t.o	t.o	t.o	t.o	t.o	t.o	t.o	t.o
		Konclude	0.75	1.62	6.50	15.08	31.49	74.28	m.e	m.e
		Openllet	9.60	46.11	614.67	2910.94	t.o	t.o	t.o	t.o
		Pellet	3.67	12.51	118.30	467.38	1657.55	t.o	t.o	t.o
	CT	HermiT	2768.19	t.o	t.o	t.o	t.o	t.o	t.o	g.c
		JFact	t.o	t.o	t.o	t.o	t.o	t.o	t.o	t.o
		Konclude	0.76	1.63	6.57	15.08	31.50	74.29	m.e	m.e
		Openllet	13.49	45.20	631.06	2776.39	t.o	t.o	t.o	t.o
		Pellet	8.68	24.49	179.12	1069.02	t.o	t.o	t.o	t.o
	RT	HermiT	209.31	2938.86	t.o	t.o	t.o	j.h.s	t.o	g.c
		JFact	t.o	t.o	t.o	t.o	t.o	t.o	t.o	t.o
		Konclude	0.78	1.66	6.68	15.32	31.91	75.29	m.e	m.e
		Openllet	10.29	49.88	675.60	2996.07	t.o	t.o	t.o	t.o
		Pellet	4.28	15.06	144.64	640.09	3046.57	t.o	t.o	t.o

(continued)

Table 4. (*continued*)

Profile	Task	Reasoner	No. of Universities							
			1	2	5	10	20	50	100	200
DL	CC	HermiT	138.41	2206.51	t.o	t.o	t.o	t.o	t.o	g.c
		JFact	t.o	t.o	t.o	t.o	t.o	t.o	t.o	t.o
		Konclude	78.36	m.e	m.e	m.e	m.e	m.e	m.e	m.e
		Openllet*	-	-	-	-	-	-	-	-
		Pellet	859.61	149.63	1977.04	t.o	t.o	t.o	t.o	t.o
	CT	HermiT	t.o	t.o	t.o	t.o	t.o	t.o	t.o	g.c
		JFact	t.o	t.o	t.o	t.o	t.o	t.o	t.o	t.o
		Konclude	m.e	m.e	m.e	m.e	m.e	m.e	m.e	m.e
		Openllet	824.51	4013.19	t.o	t.o	t.o	t.o	t.o	g.c
		Pellet	t.o	t.o	t.o	t.o	t.o	t.o	t.o	t.o
	RT	HermiT	t.o	t.o	t.o	t.o	t.o	t.o	t.o	g.c
		JFact	t.o	t.o	t.o	t.o	t.o	t.o	t.o	t.o
		Konclude	m.e	m.e	m.e	m.e	m.e	m.e	m.e	m.e
		Openllet	t.o	t.o	t.o	t.o	t.o	t.o	t.o	g.c
		Pellet	t.o	t.o	t.o	t.o	t.o	t.o	t.o	t.o

Table 5. Inconsistent results (in seconds) from Openllet for OWL 2 EL and OWL 2 DL ontologies generated by OWL2Bench for the consistency checking reasoning task.

Profile	No. of Univ.	Iteration				
		1	2	3	4	5
EL	1	7.60	10.48	7.63	7.11	0.87
	2	53.96	5.7	37.98	43.13	2.59
	5	14.08	622.66	579.30	710.91	8.97
	10	45.20	32.42	39.31	52.90	2236.29
	20	t.o	170.55	t.o	t.o	117.09
	50	725.94	733.84	t.o	t.o	668.14
	100	3155.08	t.o	4408.96	t.o	t.o
	200	t.o	t.o	t.o	t.o	t.o
DL	1	1.48	4.09	t.o	t.o	1.57
	2	2.57	3.22	3.21	2.46	2.00
	5	t.o	t.o	t.o	t.o	t.o
	10	13.77	t.o	t.o	t.o	13.76
	20	t.o	t.o	t.o	t.o	t.o
	50	t.o	97.98	1790.08	480.09	t.o
	100	t.o	t.o	t.o	170.39	178.38
	200	428.91	t.o	t.o	342.18	t.o

smaller ontologies (up to 5 universities), Konclude performed better than ELK in all the reasoning tasks. But for the larger ontologies, ELK (a profile specific reasoner) performed better than Konclude (an all profile reasoner) in terms of runtime and memory. We did not include Openllet consistency results in Table 4

Table 6. Load times (in seconds) for Stardog and GraphDB for OWL2Bench data for four profiles, and for 1, 5, and 10 universities. d.s.e denotes Disk Space Error. n.a denotes reasoner not applicable for the corresponding OWL 2 Profile.

	1 University				5 Universities				10 Universities			
	EL	QL	RL	DL	EL	QL	RL	DL	EL	QL	RL	DL
GraphDB	n.a	9.26	2619.53	n.a	n.a	30.66	d.s.e	n.a	n.a	59.30	d.s.e	n.a
Stardog	7.17	6.79	6.86	6.99	8.42	7.97	8.27	7.96	10.08	10.00	10.01	9.60

since we noticed some inconsistencies in the results obtained over different iterations. These are reported in Table 5. We observed two types of inconsistencies. Across the different runs on the same ontology, there is a huge variation in the runtime. For example, for 10 universities in the OWL 2 EL profile, Openllet's runtime varied from 32.42 s to 2236.29 s and for 20 universities, it varied from 117.09 s to a timeout (greater than 90 min). The second inconsistency that we observed is that Openllet takes more time or times out on a smaller ontology but completes the same reasoning task on a larger ontology. A case in point are the column values of "Iteration 2" in the EL profile across the universities (Table 5). The inconsistency in the results was reported to the openllet support[11].

Since Openllet is an extension of Pellet, we ran the same experiments as in Table 5 using Pellet to check for the inconsistencies in the results. We observed only one among the two inconsistencies, which is that consistency checking of a larger ontology sometimes takes more time than for a smaller ontology. Since both these ontologies belong to the same OWL 2 profile and the same TBox was used, the results are surprising.

Most of the reasoners (except JFact) worked well on all the reasoning tasks in the OWL 2 QL profile. Pellet/Openllet perform the best in terms of the runtime, but as expected, the performance of all the reasoners deteriorate with increase in the size of the ontology. The performance of Konclude is comparable to that Pellet and Openllet, except that it was not able to handle ontologies that are very large (100 and 200 universities).

In the OWL 2 RL profile, most reasoners could not handle the larger ontologies. The performance of Pellet and Openllet was comparable in all the other profiles. But in OWL 2 RL, Pellet's performance was significantly better when compared to Openllet. In the case of OWL 2 DL profile, Openllet completed the classification reasoning task for 1 and 2 universities. HermiT and Pellet were able to complete the consistency checking task only (for 2 and 5 universities, respectively). Other than that, most of the reasoners timed out or had memory related errors even for small ontologies (1 and 2 universities).

4.2   Evaluating SPARQL Engines

We evaluated the twenty-two SPARQL queries from OWL2Bench on GraphDB and Stardog. We compare these two systems in terms of their loading time and

---

[11] https://github.com/Galigator/openllet/issues/50.

query response time. We use the ontologies generated by OWL2Bench for 1, 5, and 10 universities (Table 3) for the experiments.

Loading Time Comparison: Table 6 reports the time taken (averaged over 5 independent runs) by the two SPARQL engines. Stardog consistently outperforms GraphDB and is able to load the data an order of magnitude faster than GraphDB. This is because GraphDB performs materialization during load time and thus the size of the GraphDB repository is much more than the Stardog repository. We observed that GraphDB could not load the data from OWL 2 RL ontology for 5 and 10 universities. It ended up using more than 2.5 TB of disk space and gave an insufficient disk space error after about 6 h. This indicates that GraphDB may not be able to handle even medium sized ontologies (around 300k axioms). This issue has been reported to the GraphDB support team.

Table 7. Time taken (in seconds) by Stardog and GraphDB for different SPARQL queries. ABox is generated for 1, 5, and 10 universities for the four OWL 2 profiles. $o.w$ indicates that the system did not produce the expected result due to Open World Assumption. $t.o$ indicates time out (10 min). $n.a$ denotes Not Applicable because datasets could not be loaded for querying. 'x' indicates that the particular query is not applicable to an OWL 2 profile.

Query	EL						QL			RL						DL		
	Stardog			GraphDB			Stardog			GraphDB			Stardog			Stardog		
	1	5	10	1	5	10	1	5	10	1	5	10	1	5	10	1	5	10
Q1	2.95	1.69	3.71	0.50	0.80	1.10	0.91	2.14	3.77	x	x	x	x	x	x	t.o	t.o	t.o
Q2	1.59	4.27	7.60	x	x	x	x	x	x	0.20	n.a	n.a	3.58	14.63	29.01	t.o	t.o	t.o
Q3	0.13	0.22	0.27	x	x	x	x	x	x	0.40	n.a	n.a	0.21	0.65	1.17	t.o	t.o	t.o
Q4	0.34	0.92	1.32	x	x	x	x	x	x	0.70	n.a	n.a	0.21	0.71	0.78	t.o	t.o	t.o
Q5	0.13	0.20	0.11	x	x	x	x	x	x	0.60	n.a	n.a	0.13	0.16	0.18	t.o	t.o	t.o
Q6	1.84	2.08	4.00	x	x	x	x	x	x	x	x	x	x	x	x	t.o	t.o	t.o
Q7	x	x	x	0.20	0.30	0.40	0.62	1.39	1.90	0.20	n.a	n.a	0.20	1.44	1.71	t.o	t.o	t.o
Q8	x	x	x	0.20	0.30	0.40	0.14	0.14	0.14	0.30	n.a	n.a	0.14	0.16	0.17	t.o	t.o	t.o
Q9	x	x	x	o.w	o.w	o.w	o.w	o.w	o.w	o.w	n.a	n.a	o.w	o.w	o.w	t.o	t.o	t.o
Q10	x	x	x	0.30	0.40	0.70	0.22	0.40	0.62	0.60	n.a	n.a	0.26	0.40	0.42	t.o	t.o	t.o
Q11	x	x	x	0.70	0.20	0.20	0.33	0.54	0.91	0.20	n.a	n.a	0.44	0.96	1.36	t.o	t.o	t.o
Q12	x	x	x	x	x	x	x	x	x	0.20	n.a	n.a	205.44	654.00	t.o	t.o	t.o	t.o
Q13	x	x	x	x	x	x	x	x	x	o.w	n.a	n.a	o.w	o.w	o.w	t.o	t.o	t.o
Q14	x	x	x	x	x	x	x	x	x	o.w	n.a	n.a	o.w	o.w	o.w	t.o	t.o	t.o
Q15	x	x	x	x	x	x	x	x	x	0.10	n.a	n.a	0.12	0.16	0.21	t.o	t.o	t.o
Q16	x	x	x	x	x	x	x	x	x	0.10	n.a	n.a	0.14	0.09	0.19	t.o	t.o	t.o
Q17	x	x	x	x	x	x	x	x	x	x	x	x	x	x	x	t.o	t.o	t.o
Q18	x	x	x	x	x	x	x	x	x	x	x	x	x	x	x	t.o	t.o	t.o
Q19	0.09	0.19	0.28	0.20	0.20	0.20	0.11	0.18	0.28	0.20	n.a	n.a	0.09	0.20	0.28	t.o	t.o	t.o
Q20	0.23	0.48	0.84	0.20	0.30	0.90	0.13	0.31	0.47	0.30	n.a	n.a	99.08	t.o	t.o	t.o	t.o	t.o
Q21	0.24	0.37	0.71	0.10	0.20	0.20	0.15	0.35	0.39	0.20	n.a	n.a	0.86	0.65	1.38	t.o	t.o	t.o
Q22	0.18	0.22	0.36	0.10	0.20	0.20	0.19	0.26	0.37	0.20	n.a	n.a	0.17	0.11	0.31	t.o	t.o	t.o

Query Runtime Comparison: We use the twenty-two SPARQL queries from OWL2Bench to compare the query runtime performance of GraphDB and Stardog. Note that, despite the availability of a number of SPARQL engines such as

Virtuoso[12], Blazegraph[13], and RDF-3X[14], we chose these two SPARQL engines because of their support for OWL 2 reasoning, albeit with different expressivity. The time-out for the query execution is 10 min for each query. Table 7 summarizes the query runtimes averaged over 5 independent runs for the two systems. The queries involving constructs from OWL 2 EL and OWL 2 DL have been ignored for GraphDB since it does not support those two profiles. Since GraphDB does inferencing at load time, it is able to answer all the OWL 2 QL and OWL 2 RL related queries in a fraction of a second for even larger datasets. Stardog, on the other hand, performs reasoning at run time and thus timed out for Q12 on the OWL 2 RL 10 universities ontology and Q20 on OWL 2 RL with 5 and 10 universities. For Stardog, we observed that it could not handle any of the queries related to the OWL 2 DL profile (Q1 to Q22) across the three ontologies (1, 5, and 10 Universities). It could, however, handle most of the queries for other profiles. Due to the open world assumption, we do not get the desired results for Q9, Q13, and Q14 since they involve cardinalities, complement, and universal quantifiers.

5   Conclusions and Future Work

We presented an ontology benchmark named OWL2Bench. The focus of our benchmark is on testing the coverage, scalability, and query performance of the reasoners across the four OWL 2 profiles (EL, QL, RL, and DL). To that end, we extended UOBM to create four TBoxes for each of the four OWL 2 profiles. Our benchmark also has an ABox generator and comes with a set of twenty-two SPARQL queries that involve reasoning. We demonstrated the utility of our benchmark by evaluating six reasoners and two SPARQL engines using our benchmark. Inconsistencies in the results were observed in some of the cases and these have been already reported to the appropriate support teams.

We plan to extend this work by making the TBox generation more configurable, i.e., users can select the particular language constructs (across all the four OWL 2 profiles) that they are interested in benchmarking. Another extension that we plan to work on is to provide an option to the users to choose the desired hardness level (easy, medium, and hard) of the ontology with respect to the reasoning time and OWL2Bench will then generate such an ontology. The hardness of an ontology can be measured in terms of the reasoning runtime and the memory needed to complete the reasoning process.

Acknowledgement. The first and the third author would like to acknowledge the partial support received from the Infosys Center for Artificial Intelligence at IIIT-Delhi.

Appendix

Some of the OWL2Bench SPARQL Queries are listed below.

---

[12] https://virtuoso.openlinksw.com/.
[13] https://blazegraph.com/.
[14] https://code.google.com/archive/p/rdf3x/.

Q1. SELECT DISTINCT ?x ?y WHERE { ?x :knows ?y }
Description: Find the instances who know some other instance.
Construct Involved: <u>knows</u> is a Reflexive Object Property.
Profile: EL, QL, DL

Q2. SELECT DISTINCT ?x ?y WHERE { ?x :isMemberOf ?y }
Description: Find Person instances who are member (Student or Employee) of some Organization.
Construct Involved: ObjectPropertyChain
Profile: EL, RL, DL

Q6. SELECT DISTINCT ?x ?y WHERE { ?x rdf:type :SelfAwarePerson }
Description: Find all the instances of class SelfAwarePerson. Self Aware person is a Person who knows themselves.
Construct Involved: ObjectHasSelf
Profile: EL, DL

Q8. SELECT DISTINCT ?x ?y WHERE { ?x :isAffiliatedOrganizationOf ?y }
Description: Find the Affiliations of all the Organizations.
Construct Involved: <u>isAffiliatedOrganizationOf</u> is an Asymmetric Object Property. Domain(Organization), Range(Organization).
Profile: QL, RL, DL

Q11. SELECT DISTINCT ?x ?y WHERE { ?x :isAdvisedBy ?y }
Description: Find all the instances who are advised by some other instance.
Construct Involved: <u>isAdvisedBy</u> is an Irreflexive Object Property. Domain(Person), Range(Person)
Profile: QL, RL, DL

Q17. SELECT DISTINCT ?x WHERE {?x rdf:type :UGStudent}
Description: Find all the instances of class UGStudent. UGStudent is a Student who enrolls in exactly one UGProgram.
Construct Involved: ObjectExactCardinality
Profile: DL

References

1. Aluç, G., Hartig, O., Özsu, M.T., Daudjee, K.: Diversified stress testing of RDF data management systems. In: Mika, P., et al. (eds.) ISWC 2014. LNCS, vol. 8796, pp. 197–212. Springer, Cham (2014). https://doi.org/10.1007/978-3-319-11964-9_13

2. Bail, S., Parsia, B., Sattler, U.: JustBench: a framework for OWL benchmarking. In: Patel-Schneider, P.F., et al. (eds.) ISWC 2010. LNCS, vol. 6496, pp. 32–47. Springer, Heidelberg (2010). https://doi.org/10.1007/978-3-642-17746-0_3

3. Christian, B., Andreas, S.: The berlin SPARQL benchmark. Int. J. Semant. Web Inf. Syst. **5**, 1–24 (2009)

4. Glimm, B., Horrocks, I., Motik, B., Stoilos, G., Wang, Z.: HermiT: an OWL 2 reasoner. J. Autom. Reasoning **53**(3), 245–269 (2014). https://doi.org/10.1007/s10817-014-9305-1

5. Guo, Y., Pan, Z., Heflin, J.: LUBM: a benchmark for OWL knowledge base systems. J. Web Semant. **3**(2–3), 158–182 (2005)
6. Hitzler, P., Krötzsch, M., Parsia, B., Patel-Schneider, P.F., Rudolph, S.: OWL 2 Web Ontology Language Profiles. (Second Edition) (2012) https://www.w3.org/TR/owl2-primer/
7. Kazakov, Y., Krötzsch, M., Simančík, F.: The incredible ELK. J. Autom. Reasoning **53**(1), 1–61 (2014)
8. Link, V., Lohmann, S., Haag, F.: OntoBench: generating custom OWL 2 benchmark ontologies. In: Groth, P., et al. (eds.) ISWC 2016. LNCS, vol. 9982, pp. 122–130. Springer, Cham (2016). https://doi.org/10.1007/978-3-319-46547-0_13
9. Ma, L., Yang, Y., Qiu, Z., Xie, G., Pan, Y., Liu, S.: Towards a complete OWL ontology benchmark. In: Sure, Y., Domingue, J. (eds.) ESWC 2006. LNCS, vol. 4011, pp. 125–139. Springer, Heidelberg (2006). https://doi.org/10.1007/11762256_12
10. Morsey, M., Lehmann, J., Auer, S., Ngomo, A.C.: DBpedia SPARQL benchmark-performance assessment with real queries on real data. Semant. Web ISWC **2011**, 454–469 (2011)
11. Parsia, B., Matentzoglu, N., Gonçalves, R.S., Glimm, B., Steigmiller, A.: The OWL reasoner evaluation (ORE) 2015 resources. In: Groth, P., et al. (eds.) ISWC 2016. LNCS, vol. 9982, pp. 159–167. Springer, Cham (2016). https://doi.org/10.1007/978-3-319-46547-0_17
12. Parsia, B., Matentzoglu, N., Gonçalves, R.S., Glimm, B., Steigmiller, A.: The OWL reasoner evaluation (ORE) 2015 competition report. J. Autom. Reasoning **59**(4), 455–482 (2017)
13. Sakr, S., Wylot, M., Mutharaju, R., Le Phuoc, D., Fundulaki, I.: Linked Data - Storing, Querying, and Reasoning. Springer International Publishing AG, Springer, Cham (2018). https://doi.org/10.1007/978-3-319-73515-3
14. Saleem, M., Mehmood, Q., Ngonga Ngomo, A.-C.: FEASIBLE: a feature-based SPARQL benchmark generation framework. In: Arenas, M., et al. (eds.) ISWC 2015. LNCS, vol. 9366, pp. 52–69. Springer, Cham (2015). https://doi.org/10.1007/978-3-319-25007-6_4
15. Schmidt, M., Hornung, T., Lausen, G., Pinkel, C.: SP2Bench: a SPARQL performance benchmark. In: 2009 IEEE 25th International Conference on Data Engineering, pp. 222–233. IEEE (2010)
16. Sirin, E., Parsia, B., Cuenca Grau, B., Kalyanpur, A., Katz, Y.: Pellet: a practical OWL-DL reasoner. J. Web Semant. **5**(2), 51–53 (2007)
17. Steigmiller, A., Liebig, T., Glimm, B.: Konclude: system description. J. Web Semant. **27**, 78–85 (2014)

RuBQ: A Russian Dataset for Question Answering over Wikidata

Vladislav Korablinov[1] and Pavel Braslavski[2,3,4]($\boxtimes$) (iD)

[1] ITMO University, Saint Petersburg, Russia
vladislav.korablinov@gmail.com
[2] Ural Federal University, Yekaterinburg, Russia
[3] HSE University, Saint Petersburg, Russia
[4] JetBrains Research, Saint Petersburg, Russia
pbras@yandex.ru

Abstract. The paper presents **RuBQ**, the first Russian knowledge base question answering (KBQA) dataset. The high-quality dataset consists of 1,500 Russian questions of varying complexity, their English machine translations, SPARQL queries to Wikidata, reference answers, as well as a Wikidata sample of triples containing entities with Russian labels. The dataset creation started with a large collection of question-answer pairs from online quizzes. The data underwent automatic filtering, crowd-assisted entity linking, automatic generation of SPARQL queries, and their subsequent in-house verification.

The freely available dataset will be of interest for a wide community of researchers and practitioners in the areas of Semantic Web, NLP, and IR, especially for those working on multilingual question answering. The proposed dataset generation pipeline proved to be efficient and can be employed in other data annotation projects.

Keywords: Knowledge base question answering · Semantic parsing · Evaluation · Russian language resources

Resource location: http://doi.org/10.5281/zenodo.3835913
Project page: https://github.com/vladislavneon/RuBQ

1 Introduction

Question answering (QA) addresses the task of returning a precise and concise answer to a natural language question posed by the user. QA received a great deal of attention both in academia and industry. Two main directions within QA are *Open-Domain Question Answering (ODQA)* and *Knowledge Base Question Answering (KBQA)*. ODQA searches for the answer in a large collection of text documents; the process is often divided into two stages: 1) retrieval of potentially relevant paragraphs and 2) spotting an answer span within the paragraph

---

V. Korablinov—Work done as an intern at JetBrains Research.

© Springer Nature Switzerland AG 2020
J. Z. Pan et al. (Eds.): ISWC 2020, LNCS 12507, pp. 97–110, 2020.
https://doi.org/10.1007/978-3-030-62466-8_7

(referred to as *machine reading comprehension, MRC*). In contrast, KBQA uses a *knowledge base* as a source of answers. A knowledge base is a large collection of factual knowledge, commonly structured in subject–predicate–object (SPO) triples, for example (Vladimir_Nabokov, spouse, Véra_Nabokov).

A potential benefit of KBQA is that it uses knowledge in a distilled and structured form that enables reasoning over facts. In addition, knowledge base structure is inherently language-independent – entities and predicates are assigned unique identifiers that are tied to specific languages through labels and descriptions, – which makes KBs more suitable for multilingual QA. The task of KBQA can be formulated as a translation from natural language question into a formal KB query (expressed in SPARQL, SQL, or λ-calculus). In many real-life applications, like in *Jeopardy!* winning IBM Watson [15] and major search engines, hybrid QA systems are employed – they rely on both text document collections and structured knowledge bases.

High-quality annotated data is crucial for measurable progress in question answering. Since the advent of SQuAD [27], a wide variety of datasets for machine reading comprehension have emerged, see a recent survey [39]. We are witnessing a growing interest in multilingual question answering, which leads to the creation of multilingual MRC datasets [1,8,24]. Multilingual KBQA is also an important research problem and a promising application [9,16]. Russian is among top-10 languages by its L1 and L2 speakers[1]; it has a Cyrillic script and a number of grammar features that make it quite different from e.g. English and Chinese – the languages most frequently used in NLP and Semantic Web research.

In this paper we present **RuBQ** (pronounced ['rubik]) – **Ru**ssian Knowledge **B**ase **Q**uestions, a KBQA dataset that consists of 1,500 Russian questions of varying complexity along with their English machine translations, corresponding SPARQL queries, answers, as well as a subset of Wikidata covering entities with Russian labels. To the best of our knowledge, this is the first Russian KBQA and semantic parsing dataset. To construct the dataset, we started with a large collection of trivia Q&A pairs harvested on the Web. We built a dedicated recall-oriented Wikidata entity linking tool and verified the obtained answers' candidate entities via crowdsourcing. Then, we generated paths between possible question entities and answer entities and carefully verified them.

The freely available dataset is of interest for a wide community of Semantic Web, natural language processing (NLP), and information retrieval (IR) researchers and practitioners, who deal with multilingual question answering. The proposed dataset generation pipeline proved to be efficient and can be employed in other data annotation projects.

2   Related Work

Table 1 summarizes the characteristics of KBQA datasets that have been developed to date. These datasets vary in size, underlying knowledge base, presence

---

[1] https://en.wikipedia.org/wiki/List_of_languages_by_total_number_of_speakers.

Table 1. KBQA datasets. Target knowledge base (**KB**): Fb – Freebase, DBp – DBpedia, Wd – Wikidata (MSParS description does not reveal details about the KB associated with the dataset). **CQ** indicates the presence of complex questions in the dataset. Logical form (**LF**) annotations: λ – lambda calculus, S – SPARQL queries, t – SPO triples. Question generation method (**QGM**): M – manual generation from scratch, SE – search engine query suggest API, L – logs, T+PP – automatic generation of question surrogates based on templates followed by crowdsourced paraphrasing, CS – crowdsourced manual generation based on formal representations, QZ – quiz collections, FA – fully automatic generation based on templates.

Dataset	Year	#Q	KB	CQ	LF	QGM	Lang
Free917 [7]	2013	917	Fb	+	λ	M	en
WebQuestions [3]	2013	5,810	Fb	+	–	SE	en
SimpleQuestions [5]	2015	108,442	Fb	–	t	CS	en
ComplexQuestions [2]	2016	2,100	Fb	+	–	L, SE	en
GraphQuestions [30]	2016	5,166	Fb	+	S	T+PP	en
WebQuestionsSP [38]	2016	4,737	Fb	+	S	SE	en
SimpleQuestions2Wikidata [11]	2017	21,957	Wd	–	t	CS	en
30M Factoid QA Corpus [29]	2017	30M	Fb	–	t	FA	en
LC-QuAD [32]	2017	5,000	DBp	+	S	T+PP	en
ComplexWebQuestions [31]	2018	34,689	Fb	+	S	T+PP	en
ComplexSequentialQuestions [28]	2018	1.6M	Wd	+	–	M+CS+FA	en
QALD9 [33]	2018	558	DBp	+	S	L	mult
LC-QuAD 2.0 [13]	2019	30,000	DBp, Wd	+	S	T+PP	en
FreebaseQA [19]	2019	28,348	Fb	+	S	QZ	en
MSParS [12]	2019	81,826	–	+	λ	T+PP	zh
CFQ [21]	2020	239,357	Fb	+	S	FA	en
RuBQ (this work)	2020	1,500	Wd	+	S	QZ	ru

of questions' logical forms and their formalism, question types and sources, as well as the language of the questions.

The questions of the earliest Free917 dataset [7] were generated by two people without consulting a knowledge base, the only requirement was a diversity of questions' topics; each question is provided with its logical form to query Freebase. Berant et al. [3] created WebQuestions dataset that is significantly larger but does not contain questions' logical forms. Questions were collected through Google suggest API: authors fed parts of the initial question to the API and repeated the process with the returned questions until 1M questions were reached. After that, 100K randomly sampled questions were presented to MTurk workers, whose task was to find an answer entity in Freebase. Later studies have shown that only two-thirds of the questions in the dataset are completely correct; many questions are ungrammatical and ill-formed [37,38]. Yih et al. [38] enriched 81.5% of WebQuestions with SPARQL queries and demonstrated that semantic parses substantially improve the quality of KBQA. They also showed that semantic parses can be obtained at an acceptable cost when the task is broken down into smaller steps and facilitated by a handy interface. Annotation was performed by five people familiar with Freebase design, which hints at the fact that the task is still too tough for crowdsourcing. WebQuestions were used in further studies aimed to generate complex questions [2,31]. SimpleQuestions [5] is the largest manually created KBQA dataset to date. Instead of providing logical parses for existing questions, the approach explores the opposite direction: based on formal representation, a natural language question is generated by crowd workers. First, the authors sampled SPO triples from a Freebase subset, favoring non-frequent subject–predicate pairs. Then, the triples were presented to crowd workers, whose task was to generate a question about the subject, with the object being the answer. This approach doesn't guarantee that the answer is unique – Wu et al.[37] estimate that SOTA results on the dataset (about 80% correct answers) reach its upper bound, since the rest of the questions are ambiguous and cannot be answered precisely. The dataset was used for the fully automatic generation of a large collection of natural language questions from Freebase triples with neural machine translation methods [29]. Dieffenbach et al. [11] succeeded in a semi-automatic matching of about one-fifth of the dataset to Wikidata.

The approach behind FreebaseQA dataset [19] is the closest to our study – it builds upon a large collection of trivia questions and answers (borrowed largely from TriviaQA dataset for reading comprehension [20]). Starting with about 130K Q&A pairs, the authors run NER over questions and answers, match extracted entities against Freebase, and generate paths between entities. Then, human annotators verify automatically generated paths, which resulted in about 28 K items marked relevant. Manual probing reveals that many questions' formal representations in the dataset are not quite precise. For example, the question eval-25: *Who captained the Nautilus in 20,000 Leagues Under The Sea?* is matched with the relation *book.book.characters* that doesn't represent its meaning and leads to multiple answers along with a correct one (*Captain Nemo*). Our approach differs from the above in several aspects. We implement a recall-oriented IR-based entity linking since many questions involve general concepts that cannot be recognized by off-the-shelf NER tools. After that, we verify answer entities via crowdsourcing. Finally, we perform careful in-house verification of automatically generated paths between question and answer entities in KB. We can conclude that our pipeline leads to a more accurate representation of questions' semantics.

The questions in the KBQA datasets can be *simple*, i.e. corresponding to a single fact in the knowledge base, or *complex*. Complex questions require a combination of multiple facts to answer them. WebQuestions consists of 85% simple questions; SimpleQuestions and 30M factoid QA Corpus contain only simple questions. Many studies [2,12,13,21,28,31] purposefully target complex questions.

The majority of datasets use Freebase [4] as target knowledge base. Freebase was discontinued and exported to Wikidata [25]; the latest available Freebase dump dates back to early 2016. QALD [33] and both versions of LC-QuAD

[13,32] use DBpedia [22]. LC QuAD 2.0 [13] and ComplexSequentialQuestions [28] use Wikidata [36], which is much larger, up-to-date, and has more multilingual labels and descriptions. The majority of datasets, where natural language questions are paired with logical forms, employ SPARQL as a more practical and immediate option compared to lambda calculus.

Existing KBQA datasets are almost exclusively English, with Chinese MSParS dataset being an exception [12]. QALD-9 [33], the latest edition of QALD shared task,[2] contains questions in 11 languages: English, German, Russian, Hindi, Portuguese, Persian, French, Romanian, Spanish, Dutch, and Italian. The dataset is rather small; at least Russian questions appear to be non-grammatical machine translations.[3]

There are several studies on knowledge base question generation [14,17,21, 29]. These works vary in the amount and form of supervision, as well as the structure and the complexity of the generated questions. However, automatically generated questions are intended primarily for training; the need for high-quality, human-annotated data for testing still persists.

3   Dataset Creation

Following previous studies [19,20], we opted for quiz questions that can be found in abundance online along with the answers. These questions are well-formed and diverse in terms of properties and entities, difficulty, and vocabulary, although we don't control these properties directly during data processing and annotation.

The dataset generation pipeline consists of the following steps: 1) data gathering and cleaning; 2) entity linking in answers and questions; 3) verification of answer entities by crowd workers; 4) generation of paths between answer entities and question candidate entities; 5) in-house verification/editing of generated paths. In parallel, we created a Wikidata sample containing all entities with Russian labels. This snapshot ensures reproducibility – a reference answer may change with time as the knowledge base evolves. In addition, the smaller dataset requires less powerful hardware for experiments with RuBQ. In what follows we elaborate on these steps.

3.1   Raw Data

We mined about 150,000 Q&A pairs from several open Russian quiz collections on the Web.[4] We found out that many items in the collection aren't actual factoid questions, for example, cloze quizzes (*Leonid Zhabotinsky was a champion of Olympic games in ... [Tokyo][5]*), crossword, definition, and multi-choice questions, as well as puzzles (*Q: There are a green one, a blue one, a red one and an*

---

[2] See overview of previous QALD datasets in [34].

[3] We manually verified all the 558 Russian questions in the QALD-9 dataset – only two of them happen to be grammatical.

[4] http://baza-otvetov.ru, http://viquiz.ru, and others.

[5] Hereafter English examples are translations from original Russian questions and answers.

east one in the white one. What is this sentence about? A: The White House). We compiled a list of Russian question words and phrases and automatically removed questions that don't contain any of them. We also removed duplicates and cross-word questions mentioning the number of letters in the expected answer. This resulted in 14,435 Q&A pairs.

3.2   Entity Linking in Answers and Questions

We implemented an IR-based approach for generating Wikidata entity candidates mentioned in answers and questions. First, we collected all Wikidata entities with Russian labels and aliases. We filtered out Wikimedia disambiguation pages, dictionary and encyclopedic entries, Wikimedia categories, Wikinews articles, and Wikimedia list articles. We also removed entities with less than four outgoing relations – we used this simple heuristic to remove less interconnected items that can hardly help solving KBQA tasks. These steps resulted in 4,114,595 unique entities with 5,430,657 different labels and aliases.

After removing punctuation, we indexed the collection with Elasticsearch using built-in tokenization and stemming. Each text string (question or answer) produces three types of queries to the Elasticsearch index: 1) all token trigrams; 2) capitalized bigrams (many named entities follow this pattern, e.g. *Alexander Pushkin, Black Sea*); and 3) free text query containing only nouns, adjectives, and numerals from the original string. N-gram queries (types 1 and 2) are run as phrase queries, whereas recall-oriented free text queries (type 3) are executed as Elasticsearch fuzzy search queries. Results of the latter search are re-ranked using a combination of BM25 scores from Elasticsearch and aggregated page view statistics of corresponding Wikipedia articles.[6] Finally, we combine search results preserving the type order and retain top-10 results for further processing. The proposed approach effectively combines precision- (types 1 and 2) and recall-oriented (type 3) processing.

3.3   Crowdsourcing Annotations

Entity candidates for answers obtained through the entity linking described above were verified on Yandex.Toloka crowdsourcing platform.[7] Crowd workers were presented with a Q&A pair and a ranked list of candidate entities. In addition, they could consult a Wikipedia page corresponding to the Wikidata item, see Fig. 1. The task was to select a single entity from the list or the *None of the above* option. The average number of candidates on the list is 5.43.

Crowd workers were provided with a detailed description of the interface and a variety of examples. To proceed to the main task, crowd workers had to first pass a qualification consisting of 20 tasks covering various cases described in the instruction. We also included 10% of honeypot tasks for live quality monitoring. These results are in turn used for calculating confidence of the annotations obtained so far as a weighted majority vote (see details of the approach in [18]).

---

[6] https://dumps.wikimedia.org/other/pageviews/.
[7] https://toloka.ai/.

Fig. 1. Interface for crowdsourced entity linking in answers: 1 – question and answer; 2 – entity candidates; 3 – Wikpedia page for a selected entity from the list of candidates (in case there is no associated Wikipedia page, the Wikidata item is shown).

Confidence value governs overlap in annotations: if the confidence is below 0.85, the task is assigned to the next crowd worker. We hired Toloka workers from the best 30% cohort according to internal rating. As a result, the average confidence for the annotation is 98.58%; the average overlap is 2.34; average time to complete a task is 19 s.

In total, 9,655 out of 14,435 answers were linked to Wikidata entities. Among the matched entities, the average rank of the correct candidate appeared to be 1.5. The combination of automatic candidate generation and subsequent crowdsourced verification proved to be very efficient. A possible downside of the approach is a lower share of literals (dates and numerical values) in the annotated answers. We could match only a fraction of those answers with Wikidata: Wikidata's standard formatted literals may look completely different even if representing the same value. Out of 1,255 date and numerical answers, 683 were linked to a Wikidata entity such as a particular year. For instance, the answer for *In what year was Immanuel Kant born?* matches *Q6926 (year 1724)*, whereas the corresponding Wikidata value is "1724-04-22"^^xsd:dateTime. Although the linkage is deemed correct, this barely helps generate a correct path between question and answer entities.

3.4 Path Generation and In-House Annotation

We applied entity linking described above to the 9,655 questions with verified answers and obtained 8.56 candidate entities per question on average. Next, we generated candidate subgraphs spanning question and answer entities, restricting the length between them by two hops.[8]

We investigated the option of filtering out erroneous question entities using crowdsourcing analogous to answer entity verification. A pilot experiment on a small sample of questions showed that this task is much harder – we got only 64% correct matches on a test set. Although the average number of generated paths decreased (from 1.9 to 0.9 and from 6.2 to 3.5 for paths of length one and two, respectively), it also led to losing correct paths for 14% of questions. Thus,

---

[8] We examined the sample and found out that there are only 12 questions with distances between question and answer entities in the Wikidata graph longer than two.

we decided to perform an in-house verification of the generated paths. The work was performed by the authors of the paper.

After sending queries to the Wikidata endpoint, we were able to find chains of length one or two for 3,194 questions; the remaining 6,461 questions were left unmatched. We manually inspected 200 random unmatched questions and found out that only 10 of them could possibly be answered with Wikidata, but the required facts are missing in the KB.

Out of 2,809 1-hop candidates corresponding to 1,799 questions, 866 were annotated as correct. For the rest 2,328 questions, we verified 3,501 2 hop candidates, but only 55 of them were deemed correct. 279 questions were marked as answerable with Wikidata. To increase the share of complex questions in the dataset, we manually constructed SPARQL queries for them.

Finally, we added 300 questions marked as non-answerable over Wikidata, although their answers are present in the knowledge base. The majority of them are unanswerable because semantics of the question cannot be expressed using the existing Wikidata predicates, e.g. *How many bells does the tower of Pisa have? (7)*. In some cases, predicates do exist and a semantically correct SPARQL query can be formulated, but the statement is missing in the KG thus the query will return an empty list, e.g. *What circus was founded by Albert Salamonsky in 1880? (Moscow Circus on Tsvetnoy Boulevard)*. These adversarial examples are akin to unanswerable questions in the second edition of SQuAD dataset [26]; they make the task more challenging and realistic.

4   RuBQ Dataset

4.1   Dataset Statistics

Our dataset has 1,500 unique questions in total. It mentions 2,357 unique entities – 1,218 in questions and 1,250 in answers. There are 242 unique relations in the dataset. The average length of the original questions is 7.99 words (median 7); machine-translated English questions are 10.58 words on average (median 10). 131 questions have more than one correct answer. For 1,154 questions the answers are Wikidata entities, and for 46 questions the answers are literals. We consider empty answers to be correct for 300 unanswerable questions and do not provide answer entities for them.

Inspired by a taxonomy of query complexity in LC-QuAD 2.0 [13], we annotated obtained SPARQL queries in a similar way. The query type is defined by the constraints in the SPARQL query, see Table 2. Note that some queries have multiple type tags. For example, SPARQL query for the question *How many moons does Mars have?* is assigned *1-hop* and *count* types and therefore isn't simple in terms of SimpleQuestions dataset.

Taking into account RuBQ's modest size, we propose to use the dataset primarily for testing rule-based systems, cross-lingual transfer learning models, and models trained on automatically generated examples, similarly to recent MRC datasets [1,8,24]. We split the dataset into development (300) and test (1,200) sets in such a way to keep a similar distribution of query types in both subsets.

Table 2. Query types in RuBQ (#D/T – number of questions in development and test subsets, respectively).

Type	#D/T	Description
1-hop	198/760	Query corresponds to a single SPO triple
multi-hop	14/55	Query's constraint is applied to more than one fact
multi-constraint	21/110	Query contains more than one SPARQL constraint
qualifier-answer	1/5	Answer is a value of a qualifier relation, similar to "fact with qualifiers" in LC-QuAD 2.0
qualifier-constraint	4/22	Query poses constraints on qualifier relations; a superclass of "temporal aspect" in LC-QuAD 2.0
reverse	6/29	Answer's variable is a subject in at least one constraint
count	1/4	Query applies COUNT operator to the resulting entities, same as in LC-QuAD 2.0
ranking	3/16	ORDER and LIMIT operators are applied to the entities specified by constraints, same as in LC-QuAD 2.0
0-hop	3/12	Query returns an entity already mentioned in the questions. The corresponding questions usually contain definitions or entity's alternative names
exclusion	4/18	Query contains NOT IN, which excludes entities mentioned in the question from the answer
no-answer	60/240	Question cannot be answered with the knowledge base, although answer entity may be present in the KB

4.2 Dataset Format

For each entry in the dataset, we provide: the original question in Russian, original answer text (may differ textually from the answer entity's label retrieved from Wikidata), SPARQL query representing the meaning of the question, a list of entities in the query, a list of relations in the query, a list of answers (a result of querying the Wikidata subset, see below), and a list of query type tags, see Table 3 for examples. We also provide machine-translated English questions obtained through Yandex.Translate without any post-editing.[9] The reason to include them into the dataset is two-fold: 1) the translations, although not perfectly correct, help understand the questions' meaning for non-Russian speakers and 2) they are ready-to-use for cross-lingual QA experiments (as we did with English QA system QAnswer). RuBQ is distributed under CC BY-SA license and is available in JSON format.

The dataset is accompanied by RuWikidata8M – a Wikidata sample containing all the entities with Russian labels.[10] It consists of about 212M triples with 8.1M unique entities. As mentioned before, the sample guarantees the correctness

---

[9] https://translate.yandex.com/.
[10] https://zenodo.org/record/3751761, project's page on github points here.

Table 3. Examples from the RuBQ dataset. Answer entities' labels are not present in the dataset and are cited here for convenience. Note that the original Q&A pair corresponding to the third example below contains only one answer – *geodesist*.

Question	Who wrote the novel "Uncle Tom's Cabin"?

SPARQL query	`SELECT ?answer`
	`WHERE {`
	`    wd:Q2222 wdt:P50 ?answer`
	`}`

Answers IDs	Q102513 (Harriet Beecher Stowe)
Tags	1-hop
Question	Who played Prince Andrei Bolkonsky in Sergei Bondarchuk's film "War and Peace"?

SPARQL query	`SELECT ?answer`
	`WHERE {`
	`    wd:Q845176 p:P161`
	`    [ ps:P161 ?answer; pq:P453 wd:Q2737140 ]`
	`}`

Answers IDs	Q312483 (Vyacheslav Tikhonov)
Tags	qualifier-constraint
Question	Who uses a theodolite for work?

SPARQL query	`SELECT ?answer`
	`WHERE {`
	`    wd:Q181517 wdt:P366 [ wdt:P3095 ?answer ]`
	`}`

Answers IDs	Q1734662 (cartographer), Q11699606 (geodesist), Q294126 (land surveyor)
Tags	multi-hop

of the queries and answers and makes the experiments with the dataset much simpler. For each entity, we executed a series of CONSTRUCT SPARQL queries to retrieve all the truthy statements and all the full statements with their linked data.[11] We also added all the triples with subclass of (P279) predicate to the sample. This class hierarchy can be helpful for question answering task in the absence of an explicit ontology in Wikidata. The sample contains Russian and English labels and aliases for all its entities.

4.3   Baselines

We provide two RuBQ baselines from third-party systems – DeepPavlov and QAnswer – that illustrate two possible approaches to cross-lingual KBQA.

---

[11] Details about Wikidata statement types can be found here: https://www.mediawiki. org/wiki/Wikibase/Indexing/RDF_Dump_Format#Statement_types.

Table 4. DeepPavlov's and QAnswer's top-1 results on RuBQ's answerable and unanswerable questions in the test set, and the breakdown of correct answers by query type.

	DeepPavlov	QAnswer
Answerable (960)		
correct	129	153
1-hop	123	136
1-hop + reverse	0	3
1-hop + count	0	2
1-hop + exclusion	0	2
Multi-constraint	4	9
Multi-hop	1	0
Qualifier-constraint	1	0
Qualifier-answer	0	1
incorrect/empty	831	807
Unanswerable (240)		
incorrect	65	138
empty/not found	175	102

To the best of our knowledge, the KBQA library[12] from an open NLP framework DeepPavlov [6] is the only freely available KBQA implementation for Russian language. The library uses Wikidata as a knowledge base and implements the standard question processing steps: NER, entity linking, and relation detection. According to the developers of the library, they used machine-translated SimpleQuestions and a dataset for zero-shot relation extraction [23] to train the model. The library returns a single string or *not found* as an answer. We obtained an answer entity ID using reverse ID-label mapping embedded in the model. If no ID is found, we treated the answer as a literal.

QAnswer [10] is a rule-based KBQA system that answers questions in several languages using Wikidata. QAnswer returns a (possibly empty) ranked list of Wikidata item IDs along with a corresponding SPARQL query. We obtain QAnswer's results by sending RuBQ questions machine-translated into English to its API.[13]

QAnswer outperforms DeepPavlov in terms of precision@1 on the answerable subset (16% vs. 13%), but demonstrates a lower accuracy on unanswerable questions (43% vs. 73%). Table 4 presents detailed results. In contrast to DeepPavlov, QAnswer returns a ranked list of entities as a response to the query, and for 23 out of 131 questions with multiple correct answers, it managed to perfectly

---

[12] http://docs.deeppavlov.ai/en/master/features/models/kbqa.html.   The   results
   reported below are as of April 2020; a newer model has been released in June 2020.
[13] https://qanswer-frontend.univ-st-etienne.fr/.

match the set of answers. For eight questions with multiple answers, QAnswer's top-ranked answers were correct, but the lower-ranked ones contained errors. To facilitate different evaluation scenarios, we provide an evaluation script that calculates precision@1, exact match, and precision/recall/F1 measures, as well as the breakdown of results by query types.

5   Conclusion and Future Work

We presented RuBQ – the first Russian dataset for Question Answering over Wikidata. The dataset consists of 1,500 questions, their machine translations into English, and annotated SPARQL queries. 300 RuBQ questions are unanswerable, which poses a new challenge for KBQA systems and makes the task more realistic. The dataset is based on a collection of quiz questions. The data generation pipeline combines automatic processing, crowdsourced and in-house verification, and proved to be very efficient. The dataset is accompanied by a Wikidata sample of 212M triples that contain 8.1M entities with Russian and English labels, and an evaluation script. The provided baselines demonstrate the feasibility of the cross-lingual approach in KBQA, but at the same time indicate there is ample room for improvements. The dataset is of interest for a wide community of researchers in the fields of Semantic Web, Question Answering, and Semantic Parsing.

In the future, we plan to explore other data sources and approaches for RuBQ expansion: search query suggest APIs as for WebQuestions [3], a large question log [35], and Wikidata SPARQL query logs.[14] We will also address complex questions and questions with literals as answers, as well as the creation of a stronger baseline for RuBQ.

Acknowledgments. We thank Mikhail Galkin, Svitlana Vakulenko, Daniil Sorokin, Vladimir Kovalenko, Yaroslav Golubev, and Rishiraj Saha Roy for their valuable comments and fruitful discussion on the paper draft. We also thank Pavel Bakhvalov, who helped collect RuWikidata8M sample and contributed to the first version of the entity linking tool. We are grateful to Yandex.Toloka for their data annotation grant. PB acknowledges support by Ural Mathematical Center under agreement No. 075-02-2020-1537/1 with the Ministry of Science and Higher Education of the Russian Federation.

References

1. Artetxe, M., Ruder, S., Yogatama, D.: On the cross-lingual transferability of monolingual representations. arXiv preprint arXiv:1910.11856 (2019)
2. Bao, J., Duan, N., Yan, Z., Zhou, M., Zhao, T.: Constraint-based question answering with knowledge graph. In: Proceedings of COLING 2016, the 26th International Conference on Computational Linguistics: Technical Papers, pp. 2503–2514 (2016)

---

[14] https://iccl.inf.tu-dresden.de/web/Wikidata_SPARQL_Logs/en.

3. Berant, J., Chou, A., Frostig, R., Liang, P.: Semantic parsing on freebase from question-answer pairs. In: Proceedings of the 2013 conference on empirical methods in natural language processing, pp. 1533–1544 (2013)
4. Bollacker, K., Evans, C., Paritosh, P., Sturge, T., Taylor, J.: Freebase: a collaboratively created graph database for structuring human knowledge. In: Proceedings of the 2008 ACM SIGMOD international conference on Management of data, pp. 1247–1250 (2008)
5. Bordes, A., Usunier, N., Chopra, S., Weston, J.: Large-scale simple question answering with memory networks. arXiv preprint arXiv:1506.02075 (2015)
6. Burtsev, M., et al.: Deeppavlov: Open-source library for dialogue systems. In: Proceedings of ACL 2018, System Demonstrations, pp. 122–127 (2018)
7. Cai, Q., Yates, A.: Large-scale semantic parsing via schema matching and lexicon extension. In: Proceedings of the 51st Annual Meeting of the Association for Computational Linguistics, pp. 423–433 (2013)
8. Clark, J.H., et al.: TyDi QA: a benchmark for information-seeking question answering in typologically diverse languages. arXiv preprint arXiv:2003.05002 (2020)
9. Diefenbach, D., Both, A., Singh, K., Maret, P.: Towards a question answering system over the semantic web. arXiv preprint arXiv:1803.00832 (2018)
10. Diefenbach, D., Giménez-García, J., Both, A., Singh, K., Maret, P.: QAnswer KG: designing a portable question answering system over RDF data. In: Hart, A., et al. (eds.) ESWC 2020. LNCS, vol. 12123, pp. 429–445. Springer, Cham (2020). https://doi.org/10.1007/978-3-030-49461-2_25
11. Diefenbach, D., Tanon, T.P., Singh, K.D., Maret, P.: Question answering benchmarks for wikidata. In: ISWC (Posters & Demonstrations) (2017)
12. Duan, N.: Overview of the NLPCC 2019 shared task: open domain semantic parsing. In: Tang, J., Kan, M.-Y., Zhao, D., Li, S., Zan, H. (eds.) NLPCC 2019. LNCS (LNAI), vol. 11839, pp. 811–817. Springer, Cham (2019). https://doi.org/10.1007/978-3-030-32236-6_74
13. Dubey, M., Banerjee, D., Abdelkawi, A., Lehmann, J.: LC-QuAD 2.0: a large dataset for complex question answering over wikidata and DBpedia. In: Ghidini, C., et al. (eds.) ISWC 2019. LNCS, vol. 11779, pp. 69–78. Springer, Cham (2019). https://doi.org/10.1007/978-3-030-30796-7_5
14. Elsahar, H., Gravier, C., Laforest, F.: Zero-shot question generation from knowledge graphs for unseen predicates and entity types. In: NAACL, pp. 218–228 (2018)
15. Ferrucci, D., et al.: Building watson: an overview of the deepQA project. AI Mag. **31**(3), 59–79 (2010)
16. Hakimov, S., Jebbara, S., Cimiano, P.: AMUSE: multilingual semantic parsing for question answering over linked data. In: d'Amato, C., et al. (eds.) ISWC 2017. LNCS, vol. 10587, pp. 329–346. Springer, Cham (2017). https://doi.org/10.1007/978-3-319-68288-4_20
17. Indurthi, S.R., Raghu, D., Khapra, M.M., Joshi, S.: Generating natural language question-answer pairs from a knowledge graph using a RNN based question generation model. In: Proceedings of the 15th Conference of the European Chapter of the Association for Computational Linguistics, pp. 376–385 (2017)
18. Ipeirotis, P.G., Provost, F., Sheng, V.S., Wang, J.: Repeated labeling using multiple noisy labelers. Data Min. Knowl. Discov. **28**(2), 402–441 (2014)
19. Jiang, K., Wu, D., Jiang, H.: FreebaseQA: a new factoid QA data set matching trivia-style question-answer pairs with Freebase. In: Proceedings of the 2019 Conference of the North American Chapter of the Association for Computational Linguistics: Human Language Technologies, pp. 318–323 (2019)

20. Joshi, M., Choi, E., Weld, D.S., Zettlemoyer, L.: TriviaQA: a large scale distantly supervised challenge dataset for reading comprehension. In: ACL, pp. 1601–1611 (2017)
21. Keysers, D., et al.: Measuring compositional generalization: a comprehensive method on realistic data. In: ICLR (2020)
22. Lehmann, J., et al.: DBpedia-a large-scale, multilingual knowledge base extracted from Wikipedia. Seman. Web 6(2), 167–195 (2015)
23. Levy, O., Seo, M., Choi, E., Zettlemoyer, L.: Zero-shot relation extraction via reading comprehension. In: CoNLL, pp. 333–342 (2017)
24. Lewis, P., Oğuz, B., Rinott, R., Riedel, S., Schwenk, H.: MLQA. evaluating cross lingual extractive question answering. arXiv preprint arXiv:1910.07475 (2019)
25. Pellissier Tanon, T., Vrandečić, D., Schaffert, S., Steiner, T., Pintscher, L.: From freebase to wikidata: the great migration. In: Proceedings of the 25th international conference on world wide web, pp. 1419–1428 (2016)
26. Rajpurkar, P., Jia, R., Liang, P.: Know what you don't know: unanswerable questions for SQuAD. In: ACL, pp. 784–789 (2018)
27. Rajpurkar, P., Zhang, J., Lopyrev, K., Liang, P.: SQuAD: 100,000+ questions for machine comprehension of text. In: EMNLP, pp. 2383–2392 (2016)
28. Saha, A., Pahuja, V., Khapra, M.M., Sankaranarayanan, K., Chandar, S.: Complex sequential question answering: towards learning to converse over linked question answer pairs with a knowledge graph. arXiv preprint (2018)
29. Serban, I.V., et al.: Generating factoid questions with recurrent neural networks: the 30M factoid question-answer corpus. In: ACL, pp. 588–598 (2016)
30. Su, Y., et al.: On generating characteristic-rich question sets for QA evaluation. In: Proceedings of the 2016 Conference on Empirical Methods in Natural Language Processing, pp. 562–572 (2016)
31. Talmor, A., Berant, J.: The Web as a knowledge base for answering complex questions. In: NAACL, pp. 641–651 (2018)
32. Trivedi, P., Maheshwari, G., Dubey, M., Lehmann, J.: LC-QuAD: a corpus for complex question answering over knowledge graphs. In: d'Amato, C., et al. (eds.) ISWC 2017. LNCS, vol. 10588, pp. 210–218. Springer, Cham (2017). https://doi.org/10.1007/978-3-319-68204-4_22
33. Usbeck, R., Gusmita, R.H., Axel-Cyrille Ngonga Ngomo, Saleem, M.: 9th challenge on question answering over linked data (QALD-9). In: SemDeep-4, NLIWoD4, and QALD-9 Joint Proceedings, pp. 58–64 (2018)
34. Usbeck, R., et al.: Benchmarking question answering systems. Semant. Web 10(2), 293–304 (2019)
35. Völske, M., et al.: What users ask a search engine: analyzing one billion Russian question queries. In: Proceedings of the 24th ACM International on Conference on Information and Knowledge Management, pp. 1571–1580 (2015)
36. Vrandečić, D., Krötzsch, M.: Wikidata: a free collaborative knowledgebase. Commun. ACM 57(10), 78–85 (2014)
37. Wu, Z., Kao, B., Wu, T.H., Yin, P., Liu, Q.: PERQ: Predicting, explaining, and rectifying failed questions in KB-QA systems. In: Proceedings of the 13th International Conference on Web Search and Data Mining, pp. 663–671 (2020)
38. Yih, W.T., Richardson, M., Meek, C., Chang, M.W., Suh, J.: The value of semantic parse labeling for knowledge base question answering. In: Proceedings of the 54th Annual Meeting of the Association for Computational Linguistics, pp. 201–206 (2016)
39. Zhang, X., Yang, A., Li, S., Wang, Y.: Machine reading comprehension: a literature review. arXiv preprint arXiv:1907.01686 (2019)

HDGI: A Human Device Gesture Interaction Ontology for the Internet of Things

Madhawa Perera[1,2]($\boxtimes$), Armin Haller[1] (iD), Sergio José Rodríguez Méndez[1] (iD), and Matt Adcock[1,2]

[1] Australian National University, Canberra, ACT 2601, Australia
{madhawa.perera,armin.haller,sergio.rodriguezmendez,
matt.adcock}@anu.edu.au
[2] CSIRO, Canberra, AU 2601, Australia
{madhawa.perera,matt.adcock}@csiro.au
https://cecs.anu.edu.au/
https://csiro.au/

Abstract. Gesture-controlled interfaces are becoming increasingly popular with the growing use of Internet of Things (IoT) systems. In particular, in automobiles, smart homes, computer games and Augmented Reality (AR)/Virtual Reality (VR) applications, gestures have become prevalent due to their accessibility to everyone. Designers, producers, and vendors integrating gesture interfaces into their products have also increased in numbers, giving rise to a greater variation of standards in utilizing them. This variety can confuse a user who is accustomed to a set of conventional controls and has their own preferences. The only option for a user is to adjust to the system even when the provided gestures are not intuitive and contrary to a user's expectations.

This paper addresses the problem of the absence of a systematic analysis and description of gestures and develops an ontology which formally describes gestures used in Human Device Interactions (HDI). The presented ontology is based on Semantic Web standards (RDF, RDFS and OWL2). It is capable of describing a human gesture semantically, along with relevant mappings to affordances and user/device contexts, in an extensible way.

Keywords: Ontology · Gesture · Semantic web · Internet of Things · Gesture interfaces

1 Introduction

Gesture-based systems are becoming widely available hence widely explored as methods for controlling interactive systems. Especially in modern automobiles, smart homes, computer games, and Augmented Reality (AR)/ Virtual Reality (VR) applications, gestures have become prevalent due to their accessibility to

© Springer Nature Switzerland AG 2020
J. Z. Pan et al. (Eds.): ISWC 2020, LNCS 12507, pp. 111–126, 2020.
https://doi.org/10.1007/978-3-030-62466-8_8

everyone. Most of these gesture interactions consist of physical movements of the face, limbs, or body [19] and allow users to express their interaction intentions and send out corresponding interactive information [9] to a device or a system. However, most of the gestural interfaces are built based on a manufacturer's design decision.

Introducing "guessability of a system" in 2005, Wobbrock et al. [18] emphasize that "a user's initial attempts at performing gestures, typing commands, or using buttons or menu items must be met with success despite the user's lack of knowledge of the relevant symbols". Their study enables the collection of end users' preferences for symbolic input, and is considered as the introduction of Gesture Elicitation Studies (GES) [17]. Since then many researches have attempted to define multiple gesture vocabularies. However, a majority of them are limited in their scope and limited to specific uses. As a result, an impressive amount of knowledge has resulted from these GES. At present this knowledge is cluttered. There are multiple studies that show 'best gestures' for the same referent. Here, the referent is the effect of a gesture or the desired effect of an action which the gestural sign refers to [8]. Hence, there are redundant gesture vocabularies. If all the knowledge of GES is properly linked, researchers could find gesture vocabularies that are defined for similar referents. However, a lack of linked data in this area has resulted in researchers conducting new GES whenever they need a particular gesture-referent mapping instead of using existing knowledge. Hence, we see the necessity of a gesture ontology that can describe gestures with its related referents and facilitate automated reasoning.

Further, there currently exist several sensors, such as Microsoft Kinect, allowing out-of-the-box posture or movement recognition which allow developers to define and capture mid-air gestures and to use them in various applications. With the advancements in AR and VR, the use of gestural interfaces has increased as these immersive technologies tend to use more intuitive Human Compute Interaction (HCI) techniques. All these systems have the capability to detect rich gestural inputs. This has resulted in designers, developers, producers and vendors integrating gesture interfaces into their products contributing to a surge in their numbers and causing greater variation in ways of utilizing them. Riener et al. [13] also show that, most of the time, "system designers define gestures based on their own preferences, evaluate them in small-scale user studies, apply modifications, and teach end users how to employ certain gestures". Further, Riener et al. [13] state that this is problematic because people have different expectations of how to interact with an interface to perform a certain task. This could confuse the users who are accustomed to a set of conventional controls.

Most of the time, these systems have either binary or a few choices when it comes to gesture selection. Therefore, users do not have much of a choice even though the manufacturer defined gestures are undesirable or counter-intuitive. For example, if we take Microsoft HoloLens[1], its first version (HoloLens 1) has a 'bloom' gesture to open its 'start' menu. In contrast, in HoloLens 2, a user has to pinch their thumb and index finger together while looking at the start icon

---

[1] https://docs.microsoft.com/en-au/hololens/.

that appears near a user's wrist when they hold out their hand with their palm facing up, to open the start menu. Optionally they can also tap the icon that appears near the wrist using their other hand.

BMW's iDrive infotainment system expects users to point a finger to the BMW iDrive touchscreen to 'accept a call' whereas Mercedes-Benz' User Experience (MBUX) multimedia infotainment system uses the same gesture to select an icon on their touch screen. Further, online search engines currently do not provide sufficient information for gesture related semantics. For example, search query to retrieve 'gestures to answer a call in a car', would not provide relevant gesture vocabularies supported by different vendors. Designers/developers have to find individual studies separately and read/learn necessary data manually. Being able to retrieve semantics and refer to a central location which maps all the available gestures to the affordance of 'answering a call in a car' would be convenient for designers and developers in such situations.

Additionally, understanding semantics of these gestures and inter-mapping them will help to bring interoperability among interfaces increasing User Experience (UX). The problem is how to do this mapping. Our approach is to design an ontology to map existing and increasingly prolific gesture vocabularies and their relationships to systems with the intention of providing the ability to understand and interpret user gestures. Henceforth, users are individually shown the desired effect of an action (called a referent) to their preferred gestures.

Villarreal-Narvaez et al.'s [17] most recent survey paper shows that a majority of gestures are performed using the upper limbs of a human body (i.e. hands). Thereby keeping extensibility in mind, we designed a Human Device Gesture Interaction (HDGI) ontology to describe and map existing and upcoming upper limb related gestures along with relevant device affordances. This allows systems to query the ontology after recognizing the gesture to understand its referents without having to be pre-programmed. This further helps personalisation of gestures for particular sets of users. As such, a user does not have to memorize a particular gesture for each different system, which improves system reliability.

This paper describes the HDGI ontology and its sample usage and state of the art in this area. First, in Sect. 2 we discuss existing approaches to address the problem of ubiquitousness in human device gesture interactions. In Sect. 3, we describe the syntax, semantics, design and formalisation of HDGI v0.1, and the rationale behind such a design. In Sect. 4, we illustrate tools for mapping HDGI v0.1 to leap motion[2] and the Oculus Quest[3] devices. This serves as an evaluation of the expressive power of our ontology and provides developers and designers with a tool on how to integrate the HDGI ontology in their development. We conclude and discuss future work in Sect. 5.

2   Related Work

A large number of studies can be found dealing with the problem of hand gesture recognition and its incorporation into the design and development of gestural

---

[2] See https://www.ultraleap.com/.
[3] See https://www.oculus.com/quest/.

interfaces [16]. In most of these cases, gestures are predefined with their meaning and actions. Yet, the studies do seem to explore the capability of identifying the relationship beyond predefined mappings of a gesture. Thus, we see very few studies that have attempted to define and formalise the relationship between each gesture. A review conducted by Villarreal Narvaez et al. [17] in 2020 shows that GES has not yet reached its peak which indicates that there will be many more gesture-related vocabularies in the future which, consequently increases the need to have interoperability between them.

One approach that has been adopted by researchers is to define taxonomies, enabling designers and manufacturers to use standard definitions when defining gesture vocabularies. Following this path, Scoditti et al. [15] proposed a gestural interaction taxonomy in order to guide designers and researchers, who "need an overall systematic structure that helps them to reason, compare, elicit (and create) the appropriate techniques for the problem at hand." Their intention is to introduce system-wide consistent languages with specific attention for gestures. However, those authors do not map existing gesture vocabularies with semantical relationships. Following this, Choi et al. [3] developed a 3D hand gesture taxonomy and notation method. The results of this study can be used as a guideline to organize hand gestures for enhancing the usability of gesture-based interfaces. This again follows a similar approach to Scoditti et al. [15]. However, this research is restricted to 6 commands (43 gestures) of a TV and blind(s) that were used in the experiment. Therefore, further experiments with an increased number of commands is necessary to see the capability and adaptability of the proposed taxonomy and notation method. Also, this notation is using a numeric terminology which is not easily readable, unless designers strictly follow a reference guide that is provided. In addition, they mention that the size or speed of hand gestures have not been considered in their approach.

Moving beyond taxonomies there is also existing research using ontologies. Osumer et al. [12] have modelled a gesture ontology based on a Microsoft Kinect-based skeleton which aims to describe mid-air gestures of human body. Their ontology mainly focuses on capturing the holistic posture of the human body, hence misses details like the finger pose or movements and a detailed representation of the hand. In addition, the ontology is not openly shared, hence it prevents use and extensibility. In addition, their main contribution is to have a "sensor-independent ontology of body-based contextual gestures, with intrinsic and extrinsic properties" where mapping different gestures with their semantic relationships to affordances is not considered.

Khairunizam et al. [6] have conducted a similar study with the intention of addressing the challenge of how to increase the knowledge level of the computational systems to recognize gestural information with regard to arm movements. In their research, they have tried to describe knowledge of the arm gestures and attempted to recognize it with a higher accuracy. This can be identified as an interesting study where the authors have used Qualisys motion capture (MOCAP) to capture the movement of the user's right arm when they perform an arm gesture. However, their focus was mainly on recognizing geometrical gestures and the gesture set was limited to 5 geometrical shapes. Again, their ontological framework does not consider the mapping of other gestures that carry similar referents.

Overall, the attempts above have a different scope compared to our ontology. Our focus is not on modelling the infinite set of concepts, features, attributes and relationships attached to arm-based gestures. We do not consider gestures that do not carry a referent to a particular affordance of a device. Nonetheless, our ontology is extensible to allow the addition of emerging gestures with a referent to an affordance or to be extended to other body parts, i.e. extending the gestures beyond the upper limbs of the human body. As a best practise we have used existing ontologies whenever they fit and provided mappings to concepts and properties in these ontologies.

3   Human Device Gesture Interaction (HDGI) Ontology

The HDGI ontology models the pose and/or movement of human upper limbs that are used to interact with devices. This ontology; 1) Describes gestures related to device interactions and which are performed using a human's upper limb region; 2) Maps affordances and human gestures to facilitate devices/automated systems to understand different gestures that humans perform to interact with the same affordances; 3) acts as a dictionary for manufacturers, designers and developers to search/identify the commonly used gestures for certain affordances, and/or to understand the shape and dynamics of a certain gesture. The ontology is developed with a strong focus on flexibility and extensibility where device manufacturers, designers and users can introduce new gestures and map its relations to necessary affordances. Most importantly, this does not enforce designers and manufacturers to follow a standard but it maps the ubiquitousness in gesture vocabularies by linking them appropriately. The aim of this study is to define a semantic model of gestures combined with its associated knowledge. As such, GES becomes more permissive, which opens up the opportunity to introduce a shareable and reusable gesture representation that can be mapped according to the relationships introduced in HDGI.

We defined a new namespace https://w3id.org/hdgi with the prefix hdgi (registered entry at http://prefix.cc) for all the classes used in the ontology so as to be independent of external ontologies. However, we have provided relevant mappings to external ontologies where appropriate. We are using w3id.org as the permanent URL service. Furthermore, the relevant code, data and ontology are made available for the community via GitHub[4] allowing anyone interested to join as a contributor.

3.1   Design Rationale

We have arranged the classes and properties of the HDGI ontology[5] to represent human upper limb region gestures with their associated affordances

---

[4] https://github.com/madhawap/human-device-gesture-interaction-ontology.
[5] https://w3id.org/hdgi.

and context. The ontology is designed around a core that consists of seven main classes: hdgi:Gesture, hdgi:BodyPart, hdgi:Pose, hdgi:Movement, hdgi:Affordance, hdgi:Device, and hdgi:Human, establishing the basic relationships between those along with an hdgi:Observer and hdgi:Context classes. Figure 1 depicts this core ontology design pattern[6]. This pattern will be registered in the ontology design pattern initiative[7]. Please note that the ontology introduces all classes and relationships depicted in Fig. 1 in its own namespace, but for illustration purposes we use their equivalent classes/properties from external ontologies, when appropriate. All the classes and properties are expressed in OWL2 and we use Turtle syntax throughout our modelling.

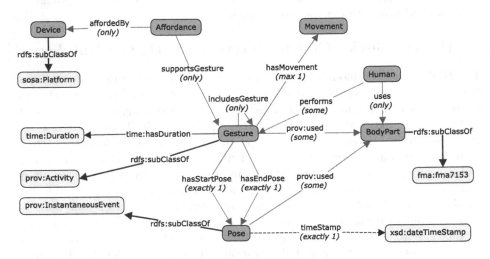

Fig. 1. The core structure of HDGI, showing relationships to external ontologies

We use global domain and range restrictions on properties sparingly, but as much as possible, use guarded local restrictions instead, i.e. universal and existential class restrictions for a specific property such that only for instances of that property with that class as the subject, the range of the property is asserted. This helps in the alignment of the ontology with other external ontologies, in particular, if they also use guarded local restrictions. We provide alignments to these ontologies as separate ontology files in Github at: human-device-interaction-ontology/ontologyAlignments[8].

---

[6] The prefixes denoted in the figure are: sosa: <http://www.w3.org/ns/sosa/>, time: <http://www.w3.org/2006/time#>, prov: <http://www.w3.org/ns/prov#>, fma: <http://purl.org/sig/ont/fma/>.

[7] See http://ontologydesignpatterns.org.

[8] https://github.com/madhawap/human-device-gesture-interaction-ontology/tree/master/v0.1/ontologyAlignments.

3.2   Core Classes and Properties

Gesture A hdgi:Gesture is defined in such a way that it distinguishes two atomic types of gestures, namely static and dynamic gestures. A dynamic gesture consists of exactly one start hdgi:Pose at a given time, exactly one end hdgi:Pose at a given time, an atomic hdgi:Movement, and involves a single hdgi:BodyPart at a time. However, since a gesture can have multiple poses and movements of multiple body parts, we provide a means to define a sequence of gestures. Since, the ontology is designed in a way that it can capture and describe individual body parts separately, a gesture that involves multiple movements and poses of body parts can be described using the object property hdgi:includesGesture that aggregates hdgi:Gesture and through their mapping to Allen time [1] puts them in sequence or concurrent. That is, a gesture can contain one or more gestures.

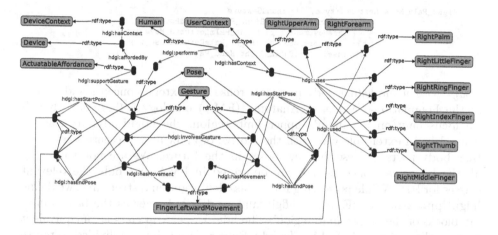

Fig. 2. HDGI dynamic gesture instance example

To give a concrete example of the modelling of a dynamic gesture, we use a 'swipe gesture' performed with the right hand (named 'right hand swipe left') illustrated below in Listing 1.1 and Fig. 2. As per the description above, 'right hand swipe left' consists of eight atomic gestures. Only some of these atomic gestures are shown in Fig. 2 and listed in Listing 1.1 and each of those include a single body part, a start pose and an end pose, with a movement.

For extensibility, we added several possible gesture subclasses such as hdgi:HandGesture, hdgi:ForearmGesture, hdgi:FacialGesture, hdgi:LegGesture, hdgi:UpperArmGesture etc. However, at this moment only hand, forearm, and upper arm gestures are modelled in detail in HDGI.

Listing 1.1. Gesture Right Hand Swipe Left

```
@prefix owl: <http://www.w3.org/2002/07/owl#> .
@prefix hdgi: <https://w3id.org/hdgi#> .

hdgi:Right_Hand_Swipe_Left rdf:type, hdgi:Gesture ;
            hdgi:includesGesture hdgi:Right_Forearm_Move_Left ,
                                 hdgi:Right_IndexFinger_Move_Left ,
                                 hdgi:Right_LittleFinger_Move_Left ,
                                 hdgi:Right_MiddleFinger_Move_Left ,
                                 hdgi:Right_Palm_Move_Left ,
                                 hdgi:Right_RingFinger_Move_Left ,
                                 hdgi:Right_Thumb_Move_Left ,
                                 hdgi:Right_UpperArm_Move_Left .

hdgi:Right_Forearm_Move_Left rdf:type, hdgi:ForearmGesture ;
                hdgi:hasEndPose hdgi:Right_ForearmPose_2 ;
                hdgi:hasMovement hdgi:Right_ForearmLeftward_Move ;
                hdgi:hasStartPose hdgi:Right_ForearmPose_1 ;
                hdgi:used hdgi:Sofia_RightForearm .

hdgi:Right_Palm_Move_Left rdf:type, hdgi:HandGesture ;
                hdgi:hasEndPose hdgi:Right_PalmOutwardPose_2 ;
                hdgi:hasMovement hdgi:Right_PalmLeftward_Move ;
                hdgi:hasStartPose hdgi:Right_PalmOutwardPose_1 ;
                hdgi:used hdgi:Sofia_Right_Hand_Palm .
```

BodyPart For modelling body parts, we reuse and extend concepts and classes in the Foundational Model of Anatomy (FMA) ontology [14]. Again, though we focus only on a human's upper limb region, the hdgi:BodyPart class is defined in an extensible way with the motive of allowing representation of further body parts to describe new poses in the future. We are not modelling all the biological concepts that are described in FMA, but only the relevant classes for HDI. While preserving FMA class definitions and structures, we define hdgi:UpperArm, hdgi:Forearm, hdgi:Palm, and hdgi:Finger as the basic building blocks of the 'upper limb region'. The hdgi:UpperArm class is an equivalent class to the Arm class in FMA. The hdgi:Finger class is further divided to represent each individual finger as hdgi:Thumb, hdgi:IndexFinger, hdgi:MiddleFinger, hdgi:RingFinger, and hdgi:LittleFinger and are mapped to the respective subclasses of a "Region of hand" in FMA. These fingers are further divided into left and right entities. Figure 3 depicts each of these sections of the 'upper limb region'. Thus, we define a gesture as a combination of one or more poses and/or movements involved by one or more of these eight sections of upper limb region.

Pose. Each body part can be involved in a pose. In other words, a Pose must hdgi:involves one hdgi:BodyPart at a point in time. For each body part there is a corresponding, potential pose. Stepping down a layer of abstraction, the hdgi:Pose class describes the exact placement of a pose in a 3D space, by modelling the 'position' and 'rotation' of a pose. The hdgi:hasPosition and hdgi:hasRotation relationships are used for this mapping; e.g. hdgi:ThumbCurled -> hdgi:hasPosition -> xPosition. In order to avoid the problem of having different origin points based on the gesture recognition device configurations, the HDGI ontology always considers relative positions. That is, upper arm positions are always relative to the shoulder joint (Refer to Fig. 3 - point A). The position

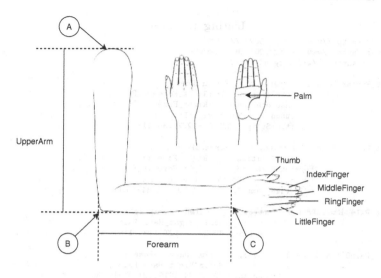

Fig. 3. Body parts of Upper limb region

of a hdgi:ForearmPose is always relative to the elbow joint (cf. Figure 3 – point B). Palm and finger positions are always relative to the wrist (cf. Figure 3 – point C). Further, the hdgi:Position class must describe the local coordinate system that its hdgi:xPosition, hdgi:yPosition, and zPosition values are based on. Thus, every hdgi:Position must have a hdgi:hasLocalCoordinateSystem object property with a hdgi:LocalCoordinateSystem as its range. This is to avoid problems, such as different SDKs/systems using slightly different coordinate systems. For example, Unity3D[9] is using a left-hand rule coordinate system where the Z-axis always points outwards from the users. In contrast, the leap-motion SDK uses a right-hand rule where the Z-axis is pointed inwards. In order to allow either type of modelling and to avoid unnecessary conversions steps, we separately model the hdgi:LocalCoordinateSystem class and hdgi:Position class relationship. The rotation of a pose can be represented in two different ways. Some systems use yaw (angle with y-axis), pitch (angle with x-axis), and roll (angle with z-axis) angles to describe the rotation of a 3D rigid body, whereas some systems use quaternions. By allowing support for both of these representations (yet one at a time), we keep our model flexible and able to model data received from different manufacturers/devices.

Further, a hdgi:Pose represents a static gesture. Thus, similar to the hdgi:Gesture class, a hdgi:Pose can contain one or more poses within itself. A hdgi:Pose always has a time stamp and involves a single body part at a time (thus, individual body parts can be modelled separately). Again, for extensibility, we added several possible poses as subclasses such as hdgi:LegPose, hdgi:FootPose, etc. However, at the moment we only model hdgi:UpperArm, hdgi:Forearm, hdgi:Palm, and each individual hdgi:Finger poses.

---

[9] See https://unity.com/.

Listing 1.2. Pose

```
@prefix owl: <http://www.w3.org/2002/07/owl#> .
@prefix xsd: <http://www.w3.org/2001/XMLSchema#> .
@prefix hdgi: <https://w3id.org/hdgi#> .

hdgi:Right_ForearmPose_1 rdf:type, hdgi:ForearmPose ;
                hdgi:hasPosition hdgi:Right_Forearm_Position_1 ;
                hdgi:hasRotation hdgi:Right_Forearm_Rotation_1 ;
                hdgi:used hdgi:Sofia_RightForearm ;
                hdgi:timeStamp "2020-04-12T21:30:11-10:00"^^xsd:dateTimeStamp .

hdgi:Right_ForearmPose_2 rdf:type, hdgi:ForearmPose ;
                hdgi:hasPosition hdgi:Right_Forearm_Position_2 ;
                hdgi:hasRotation hdgi:Right_Forearm_Rotation_2 ;
                hdgi:used hdgi:Sofia_RightForearm ;
                hdgi:timeStamp "2020-04-12T21:31:31-10:00"^^xsd:dateTimeStamp .

hdgi:Right_PalmOutwardPose_1 rdf:type, hdgi:PalmOutwardPose ;
                hdgi:used hdgi:Sofia_Right_Hand_Palm ;
                hdgi:timeStamp "2020-04-12T21:30:11-10:00"^^xsd:dateTimeStamp .

hdgi:Right_PalmOutwardPose_2 rdf:type, hdgi:PalmOutwardPose ;
                hdgi:used hdgi:Sofia_Right_Hand_Palm ;
                hdgi:timeStamp "2020-04-12T21:31:31-10:00"^^xsd:dateTimeStamp .
```

Listing 1.2 provides an example of a pose modelling related to the gesture "Right Hand Swipe Left". The example models the start pose and the end pose of the right forearm and the right palm. As per the description above, each hdgi:Pose hdgi:used only hdgi:BodyPart and has exactly one hdgi:timestamp with a maximum of one hdgi:Position and a hdgi:Rotation (hdgi:Rotation could be modeled either using Euler angles (hdgi:xRotation (roll), hdgi:yRotation (pitch), hdgi:zRotation (yaw)) or hdgi:Quaternion based on received data). Listing 1.3 further explains the hdgi:Position and hasLocalCoordinateSystem mappings. Each hdgi:Position has maximum of one hdgi:xPosition, hdgi:yPosition and hdgi:zPosition and exactly one hdgi:LocalCoordinateSystem. Notice in hdgi:LocalCoordinateSystem, each axis direction is pre-known (enum), hence for hdgi:xAxisDirection it is either "leftward" or "rightward", for hdgi:yAxisDirection it is either "upward" or "downward", and for hdgi:zAxisDirection it is "outward" or "inward".

Listing 1.3. Position

```
@prefix owl: <http://www.w3.org/2002/07/owl#> .
@prefix xsd: <http://www.w3.org/2001/XMLSchema#> .
@prefix hdgi: <https://w3id.org/hdgi#> .

hdgi:Right_Forearm_Position_1 rdf:type, hdgi:Position ;
                hdgi:hasLocalCoordinateSystem hdgi:
                    ↪ NewForearm_LocalCoordinateSystem ;
                hdgi:hasUnitOfMeasure hdgi:centimeters ;
                hdgi:xPosition "0.0"^^xsd:float ;
                hdgi:yPosition "30.211"^^xsd:float ;
                hdgi:zPosition "19.021"^^xsd:float .

hdgi:Right_Forearm_Position_2 rdf:type, hdgi:Position ;
                hdgi:hasLocalCoordinateSystem hdgi:
                    ↪ NewForearm_LocalCoordinateSystem ;
                hdgi:hasUnitOfMeasure hdgi:centimeters ;
                hdgi:xPosition "16.517"^^xsd:float ;
                hdgi:yPosition "28.456"^^xsd:float ;
                hdgi:zPosition "19.121"^^xsd:float .
```

```
hdgi:NewForearm_LocalCoordinateSystem rdf:type, hdgi:LocalCoordinateSystem ;
                                hdgi:x-axisDirection "leftward" ;
                                hdgi:y-axisDirection "upward" ;
                                hdgi:z-axisDirection "outward" .
```

Movement. The hdgi:Movement class only relates to dynamic gestures and has no relationship to a hdgi:Pose. A hdgi:Movement consists of a predefined set of movements that we identified as sufficient to describe the movements of hdgi:UpperArm, hdgi:Forearm, hdgi:Palm, and hdgi:Finger. This is extensible for designers and developers to include their own new movements. As this is not tightly-coupled with other classes such as hdgi:Gesture, hdgi:Pose, and hdgi:BodyPart, the flexibility is there for customisations. Each hdgi:Movement is atomic (that is related to only one position change or one rotation change) and must have exactly a single hdgi:Duration. This can be derived from hdgi:timestamp difference between start hdgi:Pose and end hdgi:Pose.

Affordances. According to Norman [11] "the term affordance refers to the perceived and actual properties of the thing that determines just how the thing could possibly be used". Later on, this view has become standard in Human Computer Interaction and Design [2]. Further, Maier et al. [7] define affordances to be "potential uses" of a device. This implies that the "human is able to do something using the device" [2]. Hence affordances of a device can be stated as the set of "all potential human behaviours that the device might allow" [2]. Therefore, Brown et al. [2] conclude that "affordances are context dependent action or manipulation possibilities from the point of view of a particular actor". This highlighted the necessity for us to model both an hdgi:Affordance and a hdgi:Context (both hdgi:UserContext and hdgi:DeviceContext) class when modelling Human Device Gesture Interactions. As a user's choice of gestures is heavily based on their context [10], to understand the correct intent it is important that HDGI can map both the context and affordance. This helps systems to understand user specific gesture semantics and behave accordingly.

In gesture interactions, necessary affordances are communicated by the user to a device via a gesture that is supported by the device. If there is an openly accessible gesture affordance mapping with automated reasoning, we could integrate multiple gesture recognition systems to cater for user needs, and thereby increase user experience (UX). For example, assume that Device A has an affordance X and Device B has affordance Y. If a user performs a gesture which can only be detected by Device B but the user's intent is to interact with affordance X, by using the mappings in hdgi-ontology and the use of automated reasoning, Device B would be able to understand the user intent and communicate that to Device A accordingly. This further implies, that it is the affordance that should be mapped to a gesture rather than the device. This is modelled as hdgi:Affordance -> hdgi:supportsGesture -> hdgi:Gesture, where an affordance can have none to many supported gestures. A hdgi:Device can be a host to multiple affordances and the same affordance can be hosted by multiple devices. Hence, hdgi:Affordance -> hdgi:affordedBy -> hdgi:Device has cardinality of many to many. Here, hdgi:Device is a sub class of sosa:Platform. SOSA [5] (Sensor, Observation, Sample, and Actuator) is a lightweight but self-contained core ontology which itself is the core of the new Semantic Sensor Network (SSN) [4] ontology. The SSN ontology describes sensors and their observations, involved procedures, studied features of interest, samples, and observed properties, as well as actuators. We further reuse sosa:Sensor and sosa:Actuator and hdgi:ActuatableAffordance and hdgi:ObservableAffordance are sub-classes of sosa:ActuatableProperty and sosa:ObservableProperty.

In addition, the HDGI ontology models the relationship between hdgi:Device and a hdgi:DeviceManufacturer as there can be the same gesture mapped to different affordances by different vendors or the same affordance can be mapped to different gestures (refer to the BMW and Mercedes-Benz example in Sect. 1). We model this in HDGI through the hdgi:Device hdgi:manufacturedBy a hdgi:DeviceManufacturer relationship where a device must have just one manufacturer.

Listing 1.4 provides an example modelling of hdgi:Affordance, hdgi:Device and hdgi:Context relationships corresponding to the description above above.

Listing 1.4. Affordance and Device Mapping

```
@prefix owl: <http://www.w3.org/2002/07/owl#> .
@prefix rdf: <http://www.w3.org/1999/02/22-rdf-syntax-ns#> .
@prefix hdgi: <https://w3id.org/hdgi#> .

hdgi:go_to_next_channel rdf:type, hdgi:ActuatableAffordance ;
                        hdgi:affordedBy hdgi:Television ;
                        hdgi:supportsGesture hdgi:Right_Palm_Move_Left .

hdgi:Television rdf:type, hdgi:Device ;
                hdgi:hasContext hdgi:Hotel_Room_Type_1 ;
                hdgi:manufacturedBy hdgi:Samsung .

hdgi:Hotel_Room_Type_1 rdf:type, hdgi:DeviceContext .

hdgi:Samsung rdf:type, hdgi:DeviceManufacturer .

hdgi:Hotel_Room_Type_1_Visitor rdf:type, hdgi:UserContext .

hdgi:Sofia rdf:type, hdgi:Human ;
           hdgi:hasContext hdgi:Hotel_Room_Type_1_Visitor ;
           hdgi:performs hdgi:Right_Palm_Move_Left ;
           hdgi:uses hdgi:Sofia_Right_Hand_Index_Finger ,
                     hdgi:Sofia_Right_Hand_Little_Finger ,
                     hdgi:Sofia_Right_Hand_Middle_Finger ,
                     hdgi:Sofia_Right_Hand_Palm ,
                     hdgi:Sofia_Right_Hand_Ring_Finger ,
                     hdgi:Sofia_Right_Hand_Thumb .
```

Most importantly, this is one of the major contributions in this ontology and when correctly modelled, this will help systems to automatically identify the semantics of a user's gesture and perform the necessary affordance mapping through an interconnected knowledge base instead of predefined one to one mappings. This allows gesture recognition systems to run gesture recognition, detection, mappings and communication separately in independent layers.

4   Device Mappings to HDGI

In addition to ontology building and annotating, it is equally important to consider its integration and documentation as a part of ontology engineering. Figure 4 illustrates a proof-of-concept implementation of the HDGI ontology. Here we wrapped a set of predefined SPARQL endpoints with RESTful APIs, in order to make the integration with third party Software Development Kits and Services easier and faster.

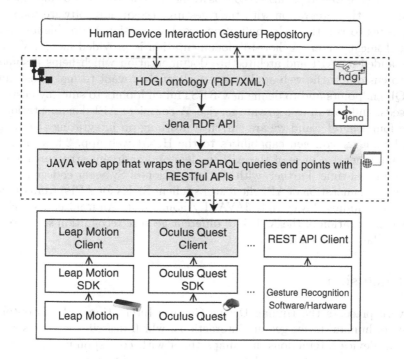

Fig. 4. HDGI ontology data integration workflow

HDGI-Mapping Service, is a fully API-driven RESTful web service, where designers, device manufacturers and developers can refer to one place - the HDGI-gesture repository - to find currently available and contemporary gestures and their relevant mappings to device affordances. In addition, APIs further allow them to define their own gesture vocabularies and map them and upload them to the gesture repository. This means that their gesture vocabularies will be easily accessible to the research community. We anticipate that this will help to reduce the redundant gesture vocabularies and increase the reuse of existing ones, eventually helping to reduce the ubiquitousness currently prominent in gestural interfaces.

In our study, we looked at the gesture vocabularies in the current literature and tried to map them into the ontology as a starting point. This allows using the HDGI-service endpoints to query about available gesture vocabularies. As we have made this an OpenSource project under Apache 2.0 license, anyone can contribute to the open GitHub code repository for further improvements. In addition, they can deploy this service in their own private cloud, if necessary. Either way, adhering to the HDGI ontology mappings will allow universal integration of gesture data instead of having a universal gesture standard that is not yet available and may never emerge.

Further information on HDGI mappings can be explored here[10]. Sample mapping service (the web application) code is available to anyone to download locally, and continue the integration with their gesture recognition software tools. The prerequisites to run the web application are Java version 1.9 or higher and an Apache Tomcat server. A 'how-to' documentation is provided here[11]. We have further added an API and architecture documentation which helps if someone needs to customize the web application itself, if they want to make customised SPARQL endpoints and to define new RESTful endpoints to suit any customizable needs, and to run as a private service. A complete API documentation can also be found here[12], and we are currently working on integrating the swagger UI, and Swagger codegen capabilities to the HDGI web app. Thus, users can get a comprehensive view of the API, understand endpoint structures and try it online in real-time. Further, with the integration of Swagger codegen, we will allow instant generation of API client stubs (client SDKs for APIs) from different languages including JavaScript, JAVA, Python, Swift, Android etc. which will make the integration of the APIs into different gesture recognition software/services even faster and easier.

5   Conclusion

This work presents the Human Device Gesture Interaction (HDGI) ontology: a model of human device gesture interactions; which describes gestures related to human device interactions and maps them with corresponding affordances. This is an initial step towards building a comprehensive human device gesture interaction knowledge base with the ultimate purpose of bringing better user experience. The HDGI ontology can assist gesture recognition systems, designers, manufacturers, and even developers to formally express gestures and to carry automated reasoning tasks based on relationships between gesture and devices affordances. While developing the ontology, we extracted elements observed from existing gesture vocabularies defined in previous studies. We also present a Web service interface (HDGI Mapping service) that can be integrated with existing gesture recognition systems.

The intention and scope of the HDGI ontology can be summarized as follows: 1) describe gestures related to HDI performed using the human upper-limb

---

[10] https://github.com/madhawap/human-device-gesture-interaction-ontology/blob/master/README.md.

[11] https://w3id.org/hdgi/mappings-docs.

[12] https://w3id.org/hdgi/mappings-api.

region; 2) map the relationship between affordances and a particular gesture based on the user context (therefore, devices could understand different gestures that human performs to interact with same affordances); and 3) act as a dictionary and a repository for manufacturers, developers, and designers to identify the commonly used gestures for certain affordances, to specify formally what a certain gesture means, and to introduce new gestures if necessary.

As future work, there are several possible extensions that can be made to the ontology by incorporating more gesture types such as facial gestures, head gestures etc. Further, we are planning to release and deploy the HDGI RESTful service in the Cloud and release API clients to leading hand-gesture supported systems such as Microsoft HoloLens 2, Microsoft Kinect, and Soli. Since gesture interactions in Mixed Reality (MR) are becoming increasingly popular, we plan to conduct several gesture elicitation studies using Microsoft HoloLens 2 especially to map gesture interactions in MR to the HDGI ontology.

References

1. Allen, J.F.: Maintaining knowledge about temporal intervals. Commun. ACM **26**(11), 832–843 (1983)
2. Brown, D.C., Blessing, L.: The relationship between function and affordance. In: ASME 2005 International Design Engineering Technical Conferences and Computers and Information in Engineering Conference, American Society of Mechanical Engineers Digital Collection, pp. 155–160 (2005)
3. Choi, E., Kim, H., Chung, M.K.: A taxonomy and notation method for three-dimensional hand gestures. Int. J. Ind. Ergon. **44**(1), 171–188 (2014)
4. Haller, A., et al.: The modular SSN ontology: a joint W3C and OGC standard specifying the semantics of sensors, observations, sampling, and actuation. Semanti. Web **10**(1), 9–32 (2019). https://doi.org/10.3233/SW-180320
5. Janowicz, K., Haller, A., Cox, S.J., Le Phuoc, D., Lefrançois, M.: Sosa: a lightweight ontology for sensors, observations, samples, and actuators. J. Web Semant. **56**, 1–10 (2019)
6. Khairunizam, W., Ikram, K., Bakar, S.A., Razlan, Z.M., Zunaidi, I.: Ontological framework of arm gesture information for the human upper body. In: Hassan, M.H.A. (ed.) Intelligent Manufacturing & Mechatronics. LNME, pp. 507–515. Springer, Singapore (2018). https://doi.org/10.1007/978-981-10-8788-2_45
7. Maier, J.R., Fadel, G.M.: Affordance-based methods for design. In: ASME 2003 International Design Engineering Technical Conferences and Computers and Information in Engineering Conference, American Society of Mechanical Engineers Digital Collection, pp. 785–794 (2003)
8. McNeill, D.: Hand and Mind: What Gestures Reveal About Thought. University of Chicago press (1992)
9. Mitra, S., Acharya, T.: Gesture recognition: a survey. IEEE Trans. Syst. Man Cybern. Part C Appl. Rev. **37**(3), 311–324 (2007)
10. Morris, M.R., et al.: Reducing legacy bias in gesture elicitation studies. Interact. **21**(3), 40–45 (2014)
11. Norman, D.A.: The Psychology of Everyday Things. Basic books (1988)
12. Ousmer, M., Vanderdonckt, J., Buraga, S.: An ontology for reasoning on body-based gestures. In: Proceedings of the ACM SIGCHI Symposium on Engineering Interactive Computing Systems, pp. 1–6 (2019)

13. Riener, A.: Gestural interaction in vehicular applications. Comput. **45**(4), 42–47 (2012)
14. Rosse, C., Mejino, J.: The Foundational Model of Anatomy Ontology. In: Burger, A., Davidson, D., Baldock, R. (eds.) Anatomy Ontologies for Bioinformatics: Principles and Practice, vol. 6, pp. 59–117. Springer, London (2007). https://doi.org/10.1007/978-1-84628-885-2_4
15. Scoditti, A., Blanch, R., Coutaz, J.: A novel taxonomy for gestural interaction techniques based on accelerometers. In: Proceedings of the 16th international conference on Intelligent user interfaces, pp. 63–72 (2011)
16. Stephanidis, C., et al.: Seven HCI grand challenges. Int. J. Hum. Comput. Interact. **35**(14), 1229–1269 (2019)
17. Villarreal Narvaez, S., Vanderdonckt, J., Vatavu, R.D., Wobbrock, J.O.: A systematic review of gesture elicitation studies: what can we learn from 216 studies? In: ACM conference on Designing Interactive Systems (DIS'20) (2020)
18. Wobbrock, J.O., Aung, H.H., Rothrock, B., Myers, B.A.: Maximizing the guessability of symbolic input. In: CHI'05 extended abstracts on Human Factors in Computing Systems, pp. 1869–1872 (2005)
19. Yang, L., Huang, J., Feng, T., Hong-An, W., Guo-Zhong, D.: Gesture interaction in virtual reality. Virtual Reality Intell. Hardware. **1**(1), 84–112 (2019)

AI-KG: An Automatically Generated Knowledge Graph of Artificial Intelligence

Danilo Dessì[1,2]([✉]), Francesco Osborne[3], Diego Reforgiato Recupero[4], Davide Buscaldi[5], Enrico Motta[3], and Harald Sack[1,2]

[1] FIZ Karlsruhe - Leibniz Institute for Information Infrastructure, Eggenstein-Leopoldshafen, Germany
[2] Karlsruhe Institute of Technology, Institute AIFB, Karlsruhe, Germany
{danilo.dessi,harald.sack}@fiz-karlsruhe.de
[3] Knowledge Media Institute, The Open University, Milton Keynes, UK
{francesco.osborne,enrico.motta}@open.ac.uk
[4] Department of Mathematics and Computer Science, University of Cagliari, Cagliari, Italy
diego.reforgiato@unica.it
[5] LIPN, CNRS (UMR 7030), Université Sorbonne Paris Nord, Villetaneuse, France
davide.buscaldi@lipn.univ-paris13.fr

Abstract. Scientific knowledge has been traditionally disseminated and preserved through research articles published in journals, conference proceedings, and online archives. However, this article-centric paradigm has been often criticized for not allowing to automatically process, categorize, and reason on this knowledge. An alternative vision is to generate a semantically rich and interlinked description of the content of research publications. In this paper, we present the Artificial Intelligence Knowledge Graph (AI-KG), a large-scale automatically generated knowledge graph that describes 820K research entities. AI-KG includes about 14M RDF triples and 1.2M reified statements extracted from 333K research publications in the field of AI, and describes 5 types of entities (tasks, methods, metrics, materials, others) linked by 27 relations. AI-KG has been designed to support a variety of intelligent services for analyzing and making sense of research dynamics, supporting researchers in their daily job, and helping to inform decision-making in funding bodies and research policymakers. AI-KG has been generated by applying an automatic pipeline that extracts entities and relationships using three tools: DyGIE++, Stanford CoreNLP, and the CSO Classifier. It then integrates and filters the resulting triples using a combination of deep learning and semantic technologies in order to produce a high-quality knowledge graph. This pipeline was evaluated on a manually crafted gold standard, yielding competitive results. AI-KG is available under CC BY 4.0 and can be downloaded as a dump or queried via a SPARQL endpoint.

Keywords: Artificial Intelligence · Scholarly data · Knowledge graph · Information Extraction · Natural Language Processing

© Springer Nature Switzerland AG 2020
J. Z. Pan et al. (Eds.): ISWC 2020, LNCS 12507, pp. 127–143, 2020.
https://doi.org/10.1007/978-3-030-62466-8_9

1   Introduction

Scientific knowledge has been traditionally disseminated and preserved through research articles published in journals, conference proceedings, and online archives. These documents, typically available as PDF, lack an explicit machine-readable representation of the research work. Therefore, this article-centric paradigm has been criticized for not allowing to automatically process, categorize, and reason on this knowledge [13]. In recent years, these limitations have been further exposed by the increasing number of publications [6], the growing role of interdisciplinary research, and the reproducibility crisis [20].

An alternative vision, that is gaining traction in the last few years, is to generate a semantically rich and interlinked description of the content of research publications [7,13,24,29]. Integrating this data would ultimately allow us to produce large scale knowledge graphs describing the state of the art in a field and all the relevant entities, e.g., tasks, methods, metrics, materials, experiments, and so on. This knowledge base could enable a large variety of intelligent services for analyzing and making sense of research dynamics, supporting researchers in their daily job, and informing decision-making in funding bodies and governments.

The research community has been working for several years on different solutions to enable a machine-readable representations of research, e.g., by creating bibliographic repositories in the Linked Data Cloud [19], generating knowledge bases of biological data [5], encouraging the Semantic Publishing paradigm [27], formalising research workflows [31], implementing systems for managing nano-publications [14] and micropublications [26], automatically annotating research publications [24], developing a variety of ontologies to describe scholarly data, e.g., SWRC[1], BIBO[2], BiDO[3], SPAR [21], CSO[4] [25], and generating large-scale knowledge graphs, e.g., OpenCitation[5], Open Academic Graph[6], Open Research Knowledge Graph[7] [13], Academia/Industry DynAmics (AIDA) Knowledge Graph[8] [3]. Most knowledge graphs in the scholarly domain typically contain metadata describing entities, such as authors, venues, organizations, research topics, and citations. Very few of them [12–14,26] actually include explicit representation of the knowledge presented in the research papers. A recent example is the Open Research Knowledge Graph [13] that also offers a web interface for annotating and navigating research papers. Typically, these knowledge graphs are populated either by human experts [13,14] or by automatic pipelines based on Natural Language Processing (NLP) and Information Extraction (IE) [16,23]. The first solution usually produces an high-quality

---

[1] http://ontoware.org/swrc.
[2] http://bibliontology.com.
[3] http://purl.org/spar/bido.
[4] http://cso.kmi.open.ac.uk.
[5] https://opencitations.net/.
[6] https://www.openacademic.ai/oag/.
[7] https://www.orkg.org/orkg/.
[8] http://w3id.org/aida/.

outcome, but suffers from limited scalability. Conversely, the latter is able to process very large corpora of publications, but may yield a noisier outcome.

The recent advancements in deep learning architectures have fostered the emergence of several excellent tools that extract information from research publications with a fair accuracy [4,11,16,18]. However, integrating the output of these tools in a coherent and comprehensive knowledge graph is still an open challenge.

In this paper, we present the Artificial Intelligence Knowledge Graph (AI-KG), a large-scale automatically generated knowledge graph that describes 820K research entities in the field of AI. AI-KG includes about 14M RDF triples and 1.2M reified statements extracted from 333K research publications in the field of AI and describes 5 types of entities (research topics, tasks, methods, metrics, materials) linked by 27 relations. Each statement is also associated to the set of publications it was extracted from and the tools that allowed its detection.

AI-KG was generated by applying an automatic pipeline [9] on a corpus of publications extracted from the Microsoft Academic Graph (MAG). This approach extracts entities and relationships using three state of the art tools: DyGIE++ [30], the CSO Classifier [23], and Stanford CoreNLP [2,17]. It then integrates similar entities and relationships and filters contradicting or noisy triples. AI-KG is available online[9] and can be queried via a Virtuoso triplestore or downloaded as a dump. We plan to release a new version of AI-KG every six months, in order to include new entities and relationships from recent publications.

The remainder of this paper is organized as follows. Section 2 discusses the related work, pointing out the existing gaps. Section 3 describes AI-KG, the pipeline used for its generation, and our plan for releasing new versions. Section 4 reports the evaluation. Finally, Sect. 5 concludes the paper, discusses the limitations, and defines future directions of research where we are headed.

2   Related Work

Due to its importance in the automatic and semi-automatic building and maintenance of Knowledge Bases, the area of Information Extraction (IE) comprises a large body of work, which includes a variety of methods for harvesting entities and relationships from text. In many of the proposed solutions, IE relies on Part-Of-Speech (PoS) tagging and various type of patterns, morphological or syntactical [22,28], often complementing themselves to compensate for reduced coverage. The most recent approaches exploit various resources to develop ensemble methodologies [18]. If we consider IE as the combination of two main tasks, extracting entities and identifying relations from text, the latter has proven without doubt the most challenging. The most successful models for relation extraction are either based on knowledge or supervised and, therefore, depend on large annotated datasets, which are rare and costly to produce. Among the

---

[9] http://w3id.org/aikg.

knowledge-based ones, it is worth to cite FRED[10], a machine reader developed by [11] on top of Boxer [8]. However, these tools are built for open-domain extraction and do not usually performs well on research publications that typically use scientific jargon and domain-dependent terms.

For a number of years, researchers have targeted scientific publications as a challenge domain, from which to extract structured information. The extraction of relations from scientific papers has recently raised interest among the NLP research community, thanks also to challenges such as SemEval 2017, scienceIE[11] and SemEval 2018 Task 7 *Semantic Relation Extraction and Classification in Scientific Papers* [10], where participants tackled the problem of detecting and classifying domain-specific semantic relations. Since then, extraction methodologies for the purpose of building knowledge graphs from scientific papers started to spread in the literature [15]. For example, authors in [1] employed syntactical patterns to detect entities, and defined two types of relations that may exist between two entities (i.e., *hyponymy* and *attributes*) by defining rules on noun phrases. Another attempt to build scientific knowledge graphs from scholarly data was performed by [16], as an evolution of the authors' work at SemEval 2018 Task 7. First, authors proposed a Deep Learning approach to extract entities and relations from the scientific literature; then, they used the retrieved triples for building a knowledge graph on a dataset of 110,000 papers. However, they only used a set of six predefined relations, which might be too generic for the purpose of yielding insights from the research landscape. Conversely, we also detected frequent verbs used on research articles and mapped them to 27 semantic relations, making our results more precise and fine-grained.

3   AI-KG

3.1   AI-KG Overview

The Artificial Intelligence Knowledge Graph (AI-KG) includes about 14M RDF triples and describes a set of 1.2M statements and 820K entities extracted from a collection of 333,609 publications in Artificial Intelligence (AI) in the period 1989–2018. In order to interlink AI-KG with other well-known knowledge bases, we also generated 19,704 *owl:sameAs* relationships with Wikidata and 6,481 with CSO. The current version of AI-KG was generated and will be regularly updated through an automatic pipeline that integrates and enriches data from Microsoft Academic Graph, the Computer Science Ontology (CSO), and Wikidata.

The AI-KG ontology is available online[12] and builds on SKOS[13], PROV-O[14], and OWL[15]. Each statement in AI-KG is associated with a triple describing the

---

[10] http://wit.istc.cnr.it/stlab-tools/fred/.
[11] https://scienceie.github.io/.
[12] http://w3id.org/aikg/aikg/ontology.
[13] https://www.w3.org/2004/02/skos/.
[14] https://www.w3.org/TR/prov-o/.
[15] https://www.w3.org/OWL/.

relationship between two entities and a number of relevant metadata. Specifically, a statement is described by the following relationships:

- *rdf:subject*, *rdf:predicate*, and *rdf:object*, which provide the statement in standard triple form;
- *aikg-ont:hasSupport*, which reports the number of publications the statement was derived from;
- *PROV-O:wasDerivedFrom*, which provides provenance information and lists the IDs of the publications from which the statement was extracted;
- *PROV-O:wasGeneratedBy*, which provides provenance and versioning information, listing (i) the tools used to detect the relationship, and (ii) the version of the pipeline that was used;
  aikg-ont:isInverse, which signals if the statement was created by inferring the inverse of a relationship extracted from the text.
- *aikg-ont:isInferredByTransitivity*, which signals if the statement was inferred by other statements (i.e., via transitive closure).

An example of an AI-KG statement is shown in the following:

```
aikg:statement_110533 a aikg-ont:Statement, provo:Entity ;
aikg-ont:hasSupport 4 ;
aikg-ont:isInferredByTransitivity false ;
aikg-ont:isInverse false ;
rdf:subject aikg:learning_algorithm ;
rdf:predicate aikg-ont:usesMethod ;
rdf:object aikg:gradient_descent ;
provo:wasDerivedFrom aikg:1517004310,
    aikg:1973720487,
    aikg:1996503769,
    aikg:2085159862 ;
provo:wasGeneratedBy aikg:DyGIE++,
    aikg:OpenIE,
    aikg:pipeline_V1.2 .
```

The example illustrates the statement `<learning_algorithm, usesMethod, gradient_descent>` and all its relevant information. It declares that this statement was extracted from four publications (using *aikg-ont:hasSupport*) and gives the IDs for these publications (using *provo:wasDerivedFrom*).

It also uses *provo: wasGeneratedBy* to declare the specific tools that were used to identify the statement, and which version of our pipeline was used to process it.

The AI-KG ontology describes five types of research entities (*Method, Task, Material, Metric, OtherEntity*). We focused on those types since they are already supported by several information extraction tools [16] and benchmarks [10].

The relations between the instances of these types were instead crafted analysing the main predicates and triples returned by several tools. We selected the most frequent predicates extracted by NLP tools and generated a set of candidate relations by combining them with the five supported entities.

For example, the predicate *uses* was used to produce *usesMethod, usesTask, usesMaterial, usesMetric, usesOtherEntity*. The *is a* predicate was instead mapped to the *skos:broader* relation, e.g., `<neural_network, skos:broader, machine_learning_technique>`. This draft was revised in subsequent iterations by four domain experts, who eventually selected 27 relations derived from 9 basic verbs (*uses, includes, is, evaluates, provides, supports, improves, requires,* and *predicts*) and defined their characteristics, such as domain, range, and transitiveness. Defining the correct domain for each relationship also enabled us to filter many invalid statements returned by the original tools as discussed in Sect. 3.3. AI-KG is licensed under a Creative Commons Attribution 4.0 International License (CC BY 4.0). It can be downloaded as a dump at http://w3id.org/aikg/ and queried via a Virtuoso triplestore at http://w3id.org/aikg/sparql/.

In the following subsection, we will discuss the automatic generation of AI-KG triples (Sect. 3.2), how it was assembled (Sect. 3.3), and describe it in more details (Sect. 3.4).

3.2   Research Entities and Relations Extraction

This section illustrates the pipeline for extracting entities and relationships from research papers and generating AI-KG. Figure 1 shows the architecture of the pipeline. This approach first detects sub-strings that refer to research entities, links them by using both pre-defined and verb-based relations, and generates three disjoint sets of triples. Then, it applies NLP techniques to remove too generic entities (e.g., "approach", "algorithm") and cleans unusual characters (e.g., hyphens used in text to start a new row). Finally, it merges together triples that have the same subject and object and uses a manually crafted dictionary to generate their relationships.

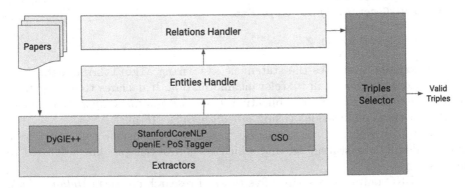

Fig. 1. Schema of our pipeline to extract and handle entities and relations.

Description of Employed Tools and Methods. The following tools were used to extract research entities and their relations:

- *DyGIE++* [30] designed by Wadden et al. was used to perform the first parsing of the input scientific data. It is a framework which exploits BERT embeddings into a neural network model to analyze scientific text. The DyGIE++ framework extracts six types of research entities *Task, Method, Metric, Material, Other-Scientific-Term, and Generic* and seven types of relations (i.e., *Compare, Part-of, Conjunction, Evaluate-for, Feature-of, Used-for, Hyponym-Of*). For the purpose of this work, we discarded all the triples with relation *Conjunction* and *Generic*, since they did not carry sufficient semantic information. DyGIE++ exploits a feed-forward neural network that is applied on span representations of the input texts to compute two scores v_1 and v_2, which measure the probability of span representations to be research entities or relations within the predefined types.
- The *CSO Classifier*[16] [23], is a tool built on top of the Computer Science Ontology, an automatically generated ontology of research areas in the field of Computer Science [25]. It identifies topics by means of two different components, the syntactic module and the semantic module. The *syntactic module* adopts syntactical rules in order to detect topics in the text. In particular, on unigrams, bigrams, and trigrams computed on text, it applies the Levenshtein similarity with the labels of the topics in CSO. If the similarity meets a given threshold the n-gram is recognized as research topic. The *semantic module* exploits the knowledge contained in a Word2Vec model trained on a corpus of scientific papers and a regular expression on PoS tags of the input text to map n-grams to research topics.
- The *Open Information Extraction (OpenIE)* [2] is an annotator provided by the Stanford Core NLP suite. It extracts general entities and relations from a plain text. It detects groups of words (clauses) where there are at least a subject and a verb by exploring a parsing tree of its input. First, clauses that hold this syntactic structure are built. Then, it adopts a multinomial logistic regression classifier to recursively explore the dependency tree of sentences from governor to dependant nodes. The natural logic of clauses is captured by exploiting semantic dictating contexts and, finally, long clauses are segmented into triples. In order to detect only triples that are related to research entities, we removed all OpenIE triples where the string of detected entities did not overlap with the string of the research entities previously found by DyGIE++ and CSO classifier.
- *PoS Tagger of Stanford Core NLP*[17] which annotates PoS tags of an input text. The PoS tags were used to detect all verbs that might represent a relation between two research entities. More precisely, for each sentence s_i we held all the verbs $V = \{v_0, \ldots, v_k\}$ between each pair of research entities (e_m, e_n) to create triples in the form $<e_m, v, e_n>$ where $v \in V$.

From each abstract a_i of the input AI papers, the pipeline extracted entities E_i and relations R_i. More specifically, these sets were firstly extracted by using

---

[16] https://github.com/angelosalatino/cso-classifier.
[17] https://nlp.stanford.edu/software/tagger.shtml.

the DyGIE++ tool[18]. Then, E_i was expanded by using all research topics that were found by the CSO classifier. Subsequently, OpenIE was applied to parse the text, and all triples in the form <*subject, verb, object*> with both subject and object that overlap research entities in E_i were added to R_i. The set R_i was finally expanded by using all triples built by exploiting the PoS tagger. The reader notices that between two entities different relations might be detected by the origin tools, therefore, two entities within AI-KG can be at most linked by 3 different relations.

Handling of Research Entities. Research entities extracted from plain text can contain very generic nouns, noisy elements, and wrong representations due to mistakes in the extraction process. In addition, different text representations might refer to the same research entity. To prevent some of these issues, our approach performed the following steps. First, it cleaned entities from punctuation signs (e.g., hyphens and apostrophes) and stop-words. Then, it exploited a manually built blacklist of entities to filter out ambiguous entities, such as "learning". Then, it applied simple rules to split strings that contained more than one research entity. For example, a research entity like *machine learning and data mining* was split in *machine learning* and *data mining*. Subsequently, acronyms were detected and solved within the same abstract by exploiting the fact that they usually appear in brackets next to the extended form of the related entities e.g., *Support Vector Machine (SVM)*.

In order to discard generic entities (e.g., *approach, method, time, paper*), we exploited the Information Content (IC) score computed on our entities by means of the NLTK[19] library, and a white-list of entities that had to be preserved. Specifically, our white-list was composed by all CSO topics and all keywords coming from our input research papers. Our pipeline discarded all entities that were not in the white-list and that had a IC equal or lower than an empirically and manually defined threshold of 15.

Finally, we merged singular and plural forms of the same entities to avoid that many resulting triples expressed the same information. We transformed plural entities in their singular form using the Wordnet lemmatizer and merged entities that refer to the same research topic (e.g., *ontology alignment* and *ontology matching*) according to the *relevantEquivalent* relation in CSO.

Handling of Relations. In this section, we describe how the pipeline identified specific relations between entities.

Best Relations Selector. Our relations can be divided in three subsets i) R_{D++}: the set of triples derived by the DyGIE++ framework where relations are pre-defined ii) R_{OIE}: the set of triples detected by OpenIE where each relation is a verb, and iii) R_{PoS}: the set of triples that were built on top of the PoS tagger results where each relation is a verb.

---

[18] We thank NVIDIA Corp. for the donation of 1 Titan Xp GPU used in this research.
[19] https://www.nltk.org/howto/wordnet.html.

In order to integrate these triples and identify one relation for each pair of entities we performed the following operations.

- The set of triples R_{D++} containing predefined relations was modified as follows. Let $LR = [r_1, \ldots, r_n]$ the list of relations between a pair of entities e_p, e_q such that $(e_p, r_i, e_q) \in R_{D++}$. Then, the most frequent relation r_{freq} was selected as the most frequent relation in LR and used to build the triple $<e_p, r_{freq}, e_q>$. The set of triples so built generated the set T_{D++}.
- The set of triples R_{OIE} relations was transformed as follows. For each pair of research entities (e_p, e_q) all their relations $LR = [r_1, \ldots, r_n]$ were collected. For each relation r_i, its corresponding word embedding w_i was associated and the list LR_w was built. The word embeddings were built by applying the Word2Vec algorithm over the titles and abstracts of 4,5M English papers in the field of Computer Science from MAG after replacing spaces with underscores in all n-grams matching the CSO topic labels and for frequent bigrams and trigrams. Then, all word embeddings w_i were averaged yielding w_{avg}, and by using the cosine similarity the relation r_j with word embedding w_j nearest to w_{avg} was chosen as best predicate label. Triples like $<e_p, r_j, e_q>$ were used to create the set T_{OIE}. The same process was also applied on the triples R_{PoS} yielding the set T_{PoS}.
- Finally, each triple within sets T_{D++}, T_{OIE}, and T_{PoS} was associated to the list of research papers from which they were extracted in order to preserve the provenance of each statement. Additionally, we refer to the number of research papers as the *support*, which is a confidence value about the consensus of the research community over that specific triple.

Relations Mapping. Many triples presented relations that were semantically similar, but syntactically different, such as *exploit, use,* and *adopt.* Therefore, we reduced the relation space by building a map M for merging similar relations. All verb relations in sets T_{OIE} and T_{PoS} were taken into account. We mapped all verb relations with the corresponding word embeddings and created a hierarchical clustering by exploiting the algorithm provided by the SciKit-learn library. The values $1-$ cosine similarity were used as distance between elements. Then the silhouette-width measure was used to quantify the quality of the clusters for various cuts. Through an empirical analysis the dendrogram was cut when the average silhouette-width was 0.65. In order to remove noisy elements we manually revised the clusters. Finally, using the clusters we created the map M where elements of the same cluster were mapped to the cluster centroid. In addition, M was also manually integrated to map the relations of the set T_{D++} to the same verb space. The map M was used to transform all relations of triples in sets T_{D++}, T_{OIE}, and T_{PoS}.

Triple Selection. In order to preserve only relevant information about the AI field, we adopted a selection process that labels our triples as *valid* and *not-valid.*

Valid Triples. In order to define the set of valid triples we considered which method was used for the extraction and the number of papers associated to each triple. In more details, we used the following criteria to consider triples as *valid*:

- All triples that were extracted by DyGIE++ and OpenIE (i.e., triples of the sets T_{D++} and T_{OIE}) were considered valid since the quality of results of those tools has been already proved by their related scientific publications.
- All triples of the set T_{PoS} that were associated to at least 10 papers with the goal to hold triples with a fair consensus. We refer to this set as T'_{PoS}.

The set T_{valid} was hence composed by the union of T_{D++}, T_{OIE}, and T'_{PoS}. All the other triples were temporarily added to the set $T_{\neg valid}$.

Consistent Triples. Several triples in the set $T_{\neg valid}$ might still contain relevant information even if they are not well-supported. For their detection, we exploited the set T_{valid} as good examples to move triples from the set $T_{\neg valid}$ to T_{valid}. More precisely, we designed a classifier $\gamma : (e_p, e_q) \rightarrow l$ where (e_p, e_q) is a pair of research entities and l is a predicted relation. The idea was that a triple consistent with T_{valid} would have its relation correctly guessed by γ. In more details the following steps were performed:

- A Multi-Perceptron Classifier (MLP) to guess the relation between a couple of entities was trained on the T_{valid} set. The input was made by the concatenation of entity word embeddings e_p, e_q, i.e., w_{e_p}, w_{e_q}. The adopted word embeddings model was the same used to cluster verbs.
- We applied γ on entities for each triple $<e_p, r, e_q> \in T_{\neg valid}$, yielding a relation r'. The relations r and r' were compared. If $r = r'$ then the triple $<e_p, r, e_q>$ was considered consistent and included to T_{valid}. Otherwise we computed the cosine similarity cos_sim similarity between r and r' word embeddings, and the *Wu-Palmer* wup_sim similarity between r and r' Wordnet synsets. If the average between cos_sim and wup_sim was higher than the threshold $th = 0.5$ then the triple $<e_p, r, e_q>$ was considered consistent with T_{valid} and added to this set.

The set T_{valid} after these steps contained 1,493,757 triples.

3.3   AI-KG Generation

In this section, we discuss the generation of AI-KG from the triples of the set T_{valid} and describe how it is driven by the AI-KG ontology introduced in Sect. 3.1. We also report how we materialized several additional statements entailed by the AI-KG schema using inverse and transitive relations. Finally, we describe how we mapped AI-KG to Wikidata and CSO.

Ontology-Driven Knowledge Graph Generation. As discussed in Sect. 3.1, the most frequent predicates of the set T_{valid} were given to four domain experts associated with several examples of triples. After several iteration, the domain experts produced a final set of 27 relations and defined their range, domain, and transitivity. We mapped the relations in T_{valid} to those 27 relations whenever was possible and discarded the inconsistent triples. The latter included both the triples whose predicate was different from the nine predicates adopted for the AI-KG ontology or their synonymous and the ones that did not respect the domain of the relations. For instance, the domain of the relation "includesTask" does not include the class "Material", since materials cannot include tasks. Therefore, all triples stating that a material includes a task, such as `<jaffe_face_database, includesTask, face_detection>`, were filtered out.

This step generated 1,075,655 statements from the 1,493,757 triples in T_{valid}. These statements were then reified using the RDF reification vocabulary[20].

Statement Materialization. In order to support users querying AI-KG via SPARQL and allowing them to quickly retrieve all information about a specific entity, we decided to also materialize some of the statements that could be inferred using transitive and inverse relations. Since we wanted for the resulting statements to have a minimum consensus, we computed the transitive closure of all the statements extracted by at least two research articles. This resulted in additional 84,510 inferred statements.

We also materialized the inverse of each statement, e.g., given the statement `<sentiment_analysis, usesMaterial, twitter>` we materialized the statement `<twitter, materialUsedBy, sentiment_analysis>`. The final version of the KG, including all the inverse statements, counts 27,142,873 RDF triples and 2,235,820 reified statements.

Integration with Other Knowledge Graphs. We mapped the entities in AI-KG to Wikidata, a well-known knowledge base containing more than 85M of data items, and to CSO. In particular, each entity in AI-KG was searched in Wikidata and, when there was just one corresponding valid Wikidata entry, we generated a `owl:sameAs` relation between the two entities. The analysis of the correct mapping and the problem of the correct identification of multiple Wikidata entries for a given entity are considered future works as beyond the scope of this paper. Overall, we found 19,704 of such entities. Similarly, we mapped 6,481 research entities to the research topics in CSO.

3.4   AI-KG Statistics

In this section we briefly discuss some analytics about AI-KG.

Table 1 shows the number of statements derived from each of the basic tools.

---

[20] https://www.w3.org/TR/rdf-mt/#Reif.

Table 1. Contribution of extracting resources in term of number of statements.

Source	Triples number
DyGIE++ ($T_{D++}set$)	1,002,488
OpenIE ($T_{OIE}set$)	53,883
PoS Tagger ($T'_{PoS}set$)	55,900

DyGIE++ provided the highest number of triples (T_{D++}), while the OpenIE tool, and the PoS tagger methodology provided a comparable number of triples (T'_{PoS} + *Cons. Triples*). However, the set T_{D++} contains a large majority of statements that were extracted from a single article.

To highlight this trend, in Fig. 2 we report the distribution of the statements generated by T_{D++}, T_{OIE} and T'_{PoS} + *Cons. Triples* according to their number of associated publications (support). While T_{D++} produces the most sizable part of those statements, most of them have a very low support. For higher support levels, the set T'_{PoS} + *Cons. Triples* contains more statements than T_{D++} and T_{OIE}. This suggests that the inclusion of T'_{PoS} enables to generate more statements in accordance within the AI community consensus.

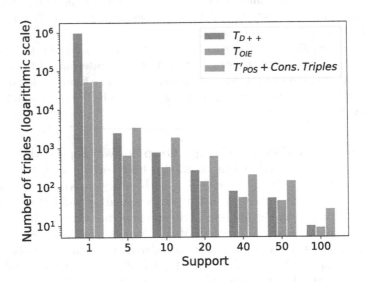

Fig. 2. Distribution of the statements support for each source.

The total number of entities in our KG is 820,732 distributed across the various types as shown by Table 2. The most frequent entities are methods, but we also have a large number of tasks and materials.

The distribution of relations within the AI-KG is shown in Fig. 3. The most frequent relation by a large margin is *usesMethod* that is used to describe the fact that an entity (Task, Method, or OtherEntity) typically uses a method for a certain purpose. This relation has many practical uses. For example it enables to retrieve all the methods used for a certain task (e.g., computer vision). This can in turn support literature reviews and automatic hypotheses generation tools. Other interesting and frequent relations include *usesMaterial*, that could be used to track the usage of specific resources (e.g., DBpedia), *includesMethod*, which enables to assess which are the components of a method, and *evaluatesMethod*, that can be used to determine which metrics are used to evaluate a certain approach. A comprehensive analysis of all the information that can be derived from AI-KG is out of the scope of this paper and will be tackled in future work.

Table 2. Distribution of entities over types

Type	Number of entities
Method	327,079
OtherEntity	298,777
Task	145,901
Material	37,510
Metric	11,465

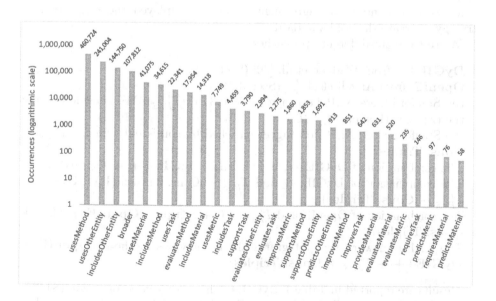

Fig. 3. Degree distribution of relations adopted within our statements.

3.5 Generation of New Versions

The pipeline described in this section will be employed on novel research outcomes in order to keep the AI-KG updated with the latest developments in the AI community. Specifically, we plan to run it every 6 months on a recent corpus of articles and release a new version. We are currently working on ingesting a larger set research papers in the AI domain and further improving our characterization of research entities. As next step, we plan to release a more granular categorization of the materials by identifying entities such as knowledge bases, textual datasets, image datasets, and others.

4 Evaluation

For annotation purposes, we focused only on statements where the underlying subjects and objects covered at least one of the 24 sub-topics of Semantic Web and at least another topic in the CSO ontology. This set includes 818 statements: 401 from T_{D++}, 102 from T_{OIE}, 170 from T'_{PoS} (110 of them returned by the classifier for identifying *Cons. Triples*), and 212 noisy triples that were discarded by the pipeline as described in Sect. 3.2. We included the latter to be able to properly calculate the recall. The total number of triples is slightly less than the sum of the sets because some of them have been derived by more than one tool.

We asked five researchers in the field of Semantic Web to annotate each triple either as *true* or *false*. Their averaged agreement was 0.747 ± 0.036, which indicates a high inter-rater agreement. Then we employed the majority rule strategy to create the gold standard.

We tested eight different approaches:

- **DyGIE++** from Wadden et al. [30] (Sect. 3.2).
- **OpenIE**, from Angeli et al. [2] (Sect. 3.2).
- the Stanford Core NLP PoS tagger (Sect. 3.2). (T'_{PoS}). We considered only the triples with support ≥ 10.
- the Stanford Core NLP PoS tagger enriched by consistent triples. $(T'_{PoS} +$ **Cons. Triples**).
- The combination of DyGIE++ and OpenIE (**DyGIE++ + OpenIE**).
- The combination of DyGIE++ and T'_{PoS} + Cons. Triples (**DyGIE++ +** T'_{PoS} **+ Cons. Triples**).
- The combination of OpenIE and T'_{PoS} + Cons. Triples (**OpenIE +** T'_{PoS} **+ Cons. Triples**).
- The final framework that integrates all the previous methods (**OpenIE + DyGIE++ +** T'_{PoS} **+ Cons. Triples**).

Results are reported in Table 3. DyGIE++ has a very good precision (84.3%) but a relatively low recall, 54.4%. OpenIE and T'_{PoS} yield a good precision but a very low recall. T'_{PoS} + Cons. Triples obtains the highest precision (84.7%) of all the tested methods, highlighting the advantages of using a classifier for selecting consistent triples. Combining the basic methods together raises the recall without losing much precision. DyGIE++ + OpenIE yields a F-measure of 72.8% with a recall of 65.1% and DyGIE++ + T'_{PoS} + Cons. Triples a F-measure of 77.1% with a recall of 71.6%. The final method used to generate AI-KG yields the best recall (80.2%) and F-measure (81.2%) and yields also a fairly good precision (78.7%).

Table 3. Precision, Recall, and F-measure of each method adopted to extract triples.

Triples identified by	Precision	Recall	F-measure
DyGIE++	0.8429	0.5443	0.6615
OpenIE	0.7843	0.1288	0.2213
T'_{PoS}	0.8000	0.0773	0.1410
T'_{PoS} + Cons. Triples	**0.8471**	0.2319	0.3641
DyGIE++ + OpenIE	0.8279	0.6506	0.7286
DyGIE++ + T'_{PoS} + Cons. Triples	0.8349	0.7166	0.7712
OpenIE + T'_{PoS} + Cons. Triples	0.8145	0.3253	0.4649
DyGIE++ + OpenIE + T'_{PoS} + Cons. Triples	0.7871	**0.8019**	**0.8117**

5   Conclusions

In this paper we presented AI-KG, a large-scale automatically generated knowledge graph that includes about 1,2M statements describing 820K research entities in the field of Artificial Intelligence. This novel resource was designed for supporting a variety of systems for analyzing research dynamics, assisting researchers, and informing founding bodies. AI-KG is freely available online and we hope that the scientific community will further build on it. In future, we plan to explore more advanced techniques, e.g., graph embeddings for inferring additional triples and cleaning up wrong statements. Moreover, we intend to perform a comprehensive analysis of AI-KG and assess its ability to support a variety of AI tasks, such as recommending publications, generating new graph embeddings, and detecting scientific trends. We would also like to allow the scientific community to give feedback and suggest edits on AI-KG as we did for CSO[21]. We then plan to apply our pipeline on a even larger set of articles in Computer Science, in order to generate an extensive representation of this field. Finally, we will investigate the application of our approach to other domains, including Life Sciences and Humanities.

References

1. Al-Zaidy, R.A., Giles, C.L.: Extracting semantic relations for scholarly knowledge base construction. In: IEEE 12th ICSC, pp. 56–63 (2018)

---

[21] https://cso.kmi.open.ac.uk/participate.

2. Angeli, G., Premkumar, M.J.J., Manning, C.D.: Leveraging linguistic structure for open domain information extraction. In: Proceedings of the 53rd Annual Meeting of the ACL and the 7th IJCNLP, vol. 1, pp. 344–354 (2015)
3. Angioni, S., Salatino, A.A., Osborne, F., Recupero, D.R., Motta, E.: Integrating knowledge graphs for analysing academia and industry dynamics. In: Bellatreche, L., et al. (eds.) TPDL/ADBIS -2020. CCIS, vol. 1260, pp. 219–225. Springer, Cham (2020). https://doi.org/10.1007/978-3-030-55814-7_18
4. Auer, S., Kovtun, V., Prinz, M., et al.: Towards a knowledge graph for science. In: 8th International Conference on Web Intelligence, Mining and Semantics (2018)
5. Belleau, F., Nolin, M.A., Tourigny, N., Rigault, P., Morissette, J.: Bio2RDF: towards a mashup to build bioinformatics knowledge systems. J. Biomed. Inform. **41**(5), 706–716 (2008)
6. Bornmann, L., Mutz, R.: Growth rates of modern science: a bibliometric analysis based on the number of publications and cited references. J. Assoc. Inf. Sci. Technol. **66**(11), 2215–2222 (2015)
7. Buscaldi, D., Dessì, D., Motta, E., Osborne, F., Reforgiato Recupero, D.: Mining scholarly publications for scientific knowledge graph construction. In: Hitzler, P., et al. (eds.) ESWC 2019. LNCS, vol. 11762, pp. 8–12. Springer, Cham (2019). https://doi.org/10.1007/978-3-030-32327-1_2
8. Curran, J.R., Clark, S., Bos, J.: Linguistically motivated large-scale NLP with C&C and boxer. In: Proceedings of the 45th Annual Meeting of the ACL on Interactive Poster and Demonstration Sessions, pp. 33–36 (2007)
9. Dessì, D., Osborne, F., Reforgiato Recupero, D., Buscaldi, D., Motta, E.: Generating knowledge graphs by employing natural language processing and machine learning techniques within the scholarly domain. Future Gener. Comput. Syst. (2020)
10. Gábor, K., Buscaldi, D., Schumann, A.K., et al.: SemEval-2018 task 7: semantic relation extraction and classification in scientific papers. In: Proceedings of The 12th International Workshop on Semantic Evaluation, pp. 679–688 (2018)
11. Gangemi, A., Presutti, V., Reforgiato Recupero, D., et al.: Semantic web machine reading with FRED. Semant. Web **8**(6), 873–893 (2017)
12. Groth, P., Gibson, A., Velterop, J.: The anatomy of a nanopublication. Inf. Serv. Use **30**(1–2), 51–56 (2010)
13. Jaradeh, M.Y., Oelen, A., Farfar, K.E., et al.: Open research knowledge graph: next generation infrastructure for semantic scholarly knowledge. In: Proceedings of the 10th International Conference on Knowledge Capture, pp. 243–246 (2019)
14. Kuhn, T., Chichester, C., Krauthammer, M., Queralt-Rosinach, N., Verborgh, R., et al.: Decentralized provenance-aware publishing with nanopublications. PeerJ Comput. Sci. **2**, e78 (2016)
15. Labropoulou, P., Galanis, D., Lempesis, A., et al.: OpenMinTeD: a platform facilitating text mining of scholarly content. In: 11th International Conference on Language Resources and Evaluation (LREC 2018), Paris, France (2018)
16. Luan, Y., He, L., Ostendorf, M., Hajishirzi, H.: Multi-task identification of entities, relations, and coreference for scientific knowledge graph construction. In: Proceedings of the EMNLP 2018 Conference, pp. 3219–3232 (2018)
17. Manning, C.D., Surdeanu, M., Bauer, J., Finkel, J.R., et al.: The stanford corenlp natural language processing toolkit. In: Proceedings of 52nd Annual Meeting of the Association for Computational Linguistics: System Demonstrations, pp. 55–60 (2014)

18. Martinez-Rodriguez, J.L., Lopez-Arevalo, I., Rios-Alvarado, A.B.: OpenIE-based approach for knowledge graph construction from text. Expert Syst. Appl. **113**, 339–355 (2018)
19. Nuzzolese, A.G., Gentile, A.L., Presutti, V., Gangemi, A.: Semantic web conference ontology - a refactoring solution. In: Sack, H., Rizzo, G., Steinmetz, N., Mladenić, D., Auer, S., Lange, C. (eds.) ESWC 2016. LNCS, vol. 9989, pp. 84–87. Springer, Cham (2016). https://doi.org/10.1007/978-3-319-47602-5_18
20. Peng, R.: The reproducibility crisis in science: a statistical counterattack. Significance **12**(3), 30–32 (2015)
21. Peroni, S., Shotton, D.: The SPAR ontologies. In: Vrandečić, D., et al. (eds.) ISWC 2018. LNCS, vol. 11137, pp. 119–136. Springer, Cham (2018). https://doi.org/10.1007/978-3-030-00668-6_8
22. Roller, S., Kiela, D., Nickel, M.: Hearst patterns revisited: automatic hypernym detection from large text corpora. In: Proceedings of the 56th Annual Meeting of the ACL, pp. 358–363 (2018)
23. Salatino, A., Osborne, F., Thanapalasingam, T., Motta, E.: The CSO classifier: ontology-driven detection of research topics in scholarly articles (2019)
24. Salatino, A.A., Osborne, F., Birukou, A., Motta, E.: Improving editorial workflow and metadata quality at springer nature. In: Ghidini, C., et al. (eds.) ISWC 2019. LNCS, vol. 11779, pp. 507–525. Springer, Cham (2019). https://doi.org/10.1007/978-3-030-30796-7_31
25. Salatino, A.A., Thanapalasingam, T., Mannocci, A., Osborne, F., Motta, E.: The computer science ontology: a large-scale taxonomy of research areas. In: Vrandečić, D., et al. (eds.) ISWC 2018. LNCS, vol. 11137, pp. 187–205. Springer, Cham (2018). https://doi.org/10.1007/978-3-030-00668-6_12
26. Schneider, J., Ciccarese, P., Clark, T., Boyce, R.D.: Using the micropublications ontology and the open annotation data model to represent evidence within a drug-drug interaction knowledge base (2014)
27. Shotton, D.: Semantic publishing: the coming revolution in scientific journal publishing. Learn. Publish. **22**(2), 85–94 (2009)
28. Snow, R., Jurafsky, D., Ng, A.: Learning syntactic patterns for automatic hypernym discovery. In: Advances in Neural Information Processing Systems, vol. 17, pp. 1297–1304 (2005)
29. Tennant, J.P., Crane, H., Crick, T., Davila, J., et al.: Ten hot topics around scholarly publishing. Publications **7**(2), 34 (2019)
30. Wadden, D., Wennberg, U., Luan, Y., Hajishirzi, H.: Entity, relation, and event extraction with contextualized span representations. In: Proceedings of the 2019 Joint Conference EMNLP-IJCNLP, pp. 5788–5793 (2019)
31. Wolstencroft, K., Haines, R., et al.: The taverna workflow suite: designing and executing workflows of web services on the desktop, web or in the cloud. Nucleic Acids Res. **41**(W1), W557–W561 (2013)

NanoMine: A Knowledge Graph
for Nanocomposite Materials Science

Jamie P. McCusker[1]([✉]) [ID], Neha Keshan[1] [ID], Sabbir Rashid[1] [ID],
Michael Deagen[3] [ID], Cate Brinson[2] [ID], and Deborah L. McGuinness[1] [ID]

[1] Department of Computer Science, Rensselaer Polytechnic Institute,
Troy, NY, USA
{mccusj2,keshan,rashis}@rpi.edu, dlm@cs.rpi.edu
[2] Department of Mechanical Engineering and Materials Science, Duke University,
Durham, NC, USA
cate.brinson@duke.edu
[3] College of Engineering and Mathematical Sciences, University of Vermont,
Burlington, VT, USA
Michael.Deagen@uvm.edu

Abstract. Knowledge graphs can be used to help scientists integrate
and explore their data in novel ways. NanoMine, built with the Whyis
knowledge graph framework, integrates diverse data from over 1,700
polymer nanocomposite experiments. Polymer nanocomposites (polymer materials with nanometer-scale particles embedded in them) exhibit
complex changes in their properties depending upon their composition
or processing methods. Building an overall theory of how nanoparticles interact with the polymer they are embedded in therefore typically
has to rely on an integrated view across hundreds of datasets. Because
the NanoMine knowledge graph is able to integrate across many experiments, materials scientists can explore custom visualizations and, with
minimal semantic training, produce custom visualizations of their own.
NanoMine provides access to experimental results and their provenance
in a linked data format that conforms to well-used semantic web ontologies and vocabularies (PROV-O, Schema.org, and the Semanticscience
Integrated Ontology). We curated data described by an XML schema
into an extensible knowledge graph format that enables users to more
easily browse, filter, and visualize nanocomposite materials data. We
evaluated NanoMine on the ability for material scientists to produce
visualizations that help them explore and understand nanomaterials and
assess the diversity of the integrated data. Additionally, NanoMine has
been used by the materials science community to produce an integrated
view of a journal special issue focusing on data sharing, demonstrating
the advantages of sharing data in an interoperable manner.

Keywords: Knowledge graph · Semantic science · Knowledge
representation

Supported by the National Science Foundation.

© Springer Nature Switzerland AG 2020
J. Z. Pan et al. (Eds.): ISWC 2020, LNCS 12507, pp. 144–159, 2020.
https://doi.org/10.1007/978-3-030-62466-8_10

1   Introduction

Polymer nanocomposites are materials made with particles that have at least one dimension that is less than 100 nm [15] that have been embedded in a polymer base, or matrix. Polymer nanocomposites exhibit complex, nonlinear responses to changes in material composition and processing methods. Additionally, these materials are used in many different applications, which means that a wide array of properties have been investigated, including mechanical, thermal, electrical, and others. This complexity makes it difficult to predict the properties of polymer nanocomposites from the performance of their consituents. The diversity of materials used, the ways in which they can be prepared, and the different kinds of properties that could be measured resulted in a complex XML-based data standard that was difficult to understand and query. The NanoMine project attempts to support researchers in their quest to explore new options for nanomaterials by providing an integrated resource where they can explore existing data about nanomaterials and their interrelationships. The biggest challenge facing NanoMine was how to create an extensible data standard that meets the needs of the present and future, while avoiding over-complexity of that standard so that the data can be queried and explored without expanding the software as new data types are added.

The foundation of materials science is the analysis of the processing, structure, and properties (PSP) interrelationships of materials. Materials scientists follow this paradigm to invent new kinds of materials that have a desired performance in real world applications.

Processing describes the sequence of steps needed to create and prepare a material. These might include mixing, melting, cooling, heating, or other methods, and each step can have many parameters.

Structure describes the composition of a material in terms of its parts, and how those constituent parts are arranged within the material. In the case of polymer nanocomposites, this usually consists of a polymer *matrix*, or base material, that has nanoparticles of different types added to it. These nanoparticles are collectively referred to as *filler*.

Properties are measured values describing how the material responds to mechanical, electrical, thermal, or other observable events. A property can indicate how strong a material is, how well it insulates or conducts electricity, how it stores or transmits heat, etc.

Originally, the NanoMine resource was conceived as a repository of XML files that describe polymer nanocomposite experiments. However, the representation needed to adequately describe experiments was complex and difficult to navigate [18]. Finding a specific piece of data required detailed knowledge of the schema, and typically included a sometimes cumbersome process of navigating built-in hierarchies of properties and different representations of measurements. We needed a way to simplify access and querying of data by providing a consistent representation for material processing methods, material composition or structure, and material properties. As a result, we built the following semantically-enabled resources for polymer nanocomposite materials science:

NanoMine Knowledge Graph: The NanoMine knowledge graph provides access to experimental data, including full PSP data, citation information, and provenance.

NanoMine Ontology: The NanoMine Ontology extends the Semanticscience Integrated Ontology (SIO) [3] and provides classes for 141 material properties, 25 processing methods, and 171 material types, compatible with and extending the NanoMine XML Schema.

NanoMine Whyis Instance: A website to search and visualize curated experiments through faceted search, full text search, a SPARQL endpoint, and user-contributed visualizations.

2   Availability

The NanoMine knowledge graph is published at http://nanomine.org, using the Whyis knowledge graph framework. The NanoMine ontology is published as part of the knowledge graph and is available at http://nanomine.org/ns. All entities in the graph are published as 5 star linked data aligned with SIO, Dublin Core Terms, and the W3C Provenance ontology, PROV-O. A read-only SPARQL API is available at https://materialsmine.org/wi/sparql, providing access to all material data and its provenance, as well as how it was transformed into RDF.

This knowledge graph is currently used by the polymer nanocomposite research community to explore their data, and will be featured in a special data issue of ACS Macro Letters [1].

Table 1. Namespace prefixes used in this paper.

Prefix	URI
nanomine	http://nanomine.org/ns/
sio	http://semanticscience.org/resource/
prov	http://www.w3.org/ns/prov#
np	http://www.nanopub.org/nschema#
setl	http://purl.org/twc/vocab/setl/
bibo	http://purl.org/ontology/bibo/

3   Modeling Materials Science

The NanoMine XML schema describes the parts of PSP for nanomaterials that materials scientists considered essential for understanding those nanomaterials [18]. As we have stated, the resulting XML files were complex and difficult to query. The NanoMine knowledge graph is designed to enable consistent representation across data types for properties, material composition, and material processing that are appropriate for nanocomposite materials research, while providing simple methods for querying and visualizing the data. To support this, we re-use representations for entities, processes, and attributes from the Semanticscience Integrated Ontology (SIO) [3]. SIO has been used in many domains, including epidemiology [10], computational biology [9,13], and sea ice data [2]. SIO provides a representation for the most commonly needed relationships and top-level classes for integrated science applications. This reuse supports interoperability with other tools and applications that also use SIO. The basic SIO classes and relationships are outlined in Fig. 1, adapted from figures in [3]. We build on these fundamental properties and classes to create the models needed for the NanoMine ontology and knowledge graph. The namespace prefixes used in this paper are listed in Table 1.

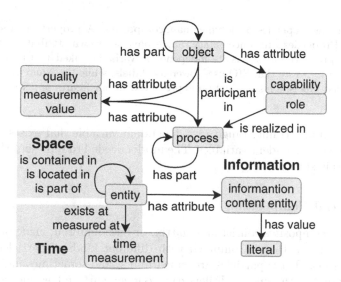

Fig. 1. The basic relationships of SIO include properties for representing attributes, values, parts, and temporal measurements. All classes and properties are part of SIO. Structures in "Space", "Time", and "Information" are the parts of SIO used to model those particular aspects of knowledge. Adapted from [3].

3.1   Properties

The physical properties of materials are at the heart of materials science. Numerous means of measuring and expressing physical properties exist. The NanoMine project represents the common electrical, mechanical, thermal, crystalline, volumetric, and rheological (flow) properties of nanomaterials as instances, following the modeling approach of SIO. Each property is represented as a single value,

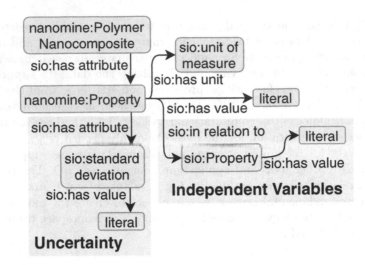

Fig. 2. Modeling properties of polymer nanocomposites. A property is a subclass of *sio:quantity*. Properties with uncertainty ranges have their own attribute of a standard deviation. A property curve has its independent variable linked to the property by *sio:'in relation to'*. Note that SIO uses lowercase labels, while the NanoMine Ontology uses capitalized labels. This distinction is preserved in our figures and visualizations.

a value with an uncertainty range, or a dependent variable that exists in relation to one or more independent variables. The model needed to represent these kinds of properties is shown in Fig. 2.

3.2 Material Composition

Polymer nanocomposites consist of a matrix, or base material, made of polymers like epoxy or latex, that is combined with different kinds of particles, such as graphite or silica. These particles are called fillers, and sometimes multiple filler types are added to the matrix. Fillers can be given surface treatments of different types, which can also change how they interact with the matrix. All of these components have their own properties, like density, volume or mass fraction, and dimensionality to individual particles. Some of these constituent properties are complex, and have a mean and standard deviation (like particle dimension), so we use the same property templates available for the aggregate, nanocomposite representation, as shown in Fig. 2. We represent polymer nanocomposites in terms of their constituents - fillers, matrices, and surface treatments, while linking them to the main nanocomposite using *sio:'has component part'* and *sio:'is surrounded by'*, as shown in Fig. 3. The actual consituent materials are given types, like "silica" or "graphene," based on the kind of material it is. That certain materials are fillers versus matrices is represented using subclasses of *sio:role*.

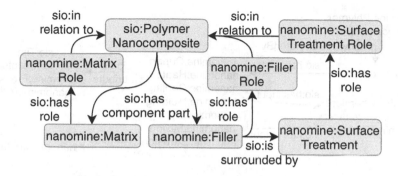

Fig. 3. Modeling composition of polymer nanocomposites. Constituent materials are given a type corresponding to the physical material type (silica, graphene, BPA, etc.), while the roles of filler or matrix is represented as *sio:role* instances.

3.3 Material Processing

The processing methods associated with a polymer nanocomposite are as important to its properties as the materials used to create it. Depending on how the material was prepared, its properties can be wildly different. Analogous to baking a cake, where a precise sequence of recipe steps leads to a tasty result, preparation of polymer nanocomposites requires a sequence of processing steps under controlled conditions in order to achieve the desired material properties. Similarly for polymer nanocomposites, changes in conditions can result in different interactions between the filler and matrix.

The ontology supports the encoding of processing methods using subclasses of *sio:procedure* to represent different kinds of mixing, heating, cooling, curing, molding, drying, and extrusion. Some of these processes, like extrusion, require complex parameters and descriptions of equipment. These are expressed as objects that are used in a process in addition to the process input, and attributes on the process and equipment. This takes advantage of process modeling in both SIO (from a science modeling perspective) and PROV-O (from a provenance perspective) [7]. Figure 4 illustrates this modeling approach for processing polymer nanocomposites.

In the curated XML files, processing steps are able to be expressed as a plain list, but experimental methods often have multiple flows and partial ordering. The use of the SIO hasInput/hasOutput properties allows us to expand the expression of processing workflows beyond the current representations.

3.4 Provenance

Provenance is crucial to an effective openly available knowledge graph with community contributions. The Whyis knowledge graph does not allow additions of knowledge without that addition being wrapped in at least minimal provenance, such as who contributed the knowledge and at what time. In NanoMine, all samples are curated from specific journal articles and unpublished data. Therefore,

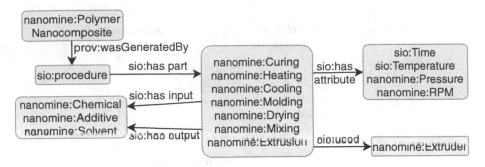

Fig. 4. Modeling processing of polymer nanocomposites. Each sample is marked as being generated by a top level procedure that is composed of numerous steps. Each step has inputs and outputs, and the use of one step's output as another step's input provides partial ordering to the overall workflow.

it is important to track where the sample information comes from. We rely on nanopublications [5] to provide structure for that provenance. This allows us to link a provenance named graph to the general assertion named graph. Provenance includes the processing (how the sample was created) and characterization (how the measurements were made) methods, as well as DOI-compatible metadata for the papers that the samples are curated from. Further details on the representation are shown in Fig. 5.

4   Curating the NanoMine Knowledge Graph

NanoMine has a data curation pipeline (Fig. 6) that takes data from papers and datasets and converts it into a knowledge graph through a collaboration of materials scientists, software engineers, and knowledge graph engineers. Our pipeline is optimized to support contributions by many different collaborators with differing levels of technical skill. Curators on the project generally use a web tool to manually extract information from published papers, and use the WebPlotDigitizer[1] to extract data from figures. Curators and collaborators can fill in an Excel file template, modeled after the XML schema, with these data. The Excel files are uploaded to NanoMine[2] and converted into XML. Once curation is complete, the system uploads the XML files to the NanoMine Whyis instance. This instance is configured to recognize NanoMine XML files and run a Semantic ETL script using SETLr [8] to convert the XML files into the RDF representation described in Sect. 3. The output of this script is added to the knowledge graph. Additionally, we created the NanoMine Ontology using the approach developed for the CHEAR Ontology [10], where we extend SIO, add concepts on-demand, and use a spreadsheet and SETLr to compile the ontology into OWL. Most of the concepts in the NanoMine Ontology were contributed

---

[1] https://automeris.io/WebPlotDigitizer/.
[2] https://materialsmine.org/nm#/XMLCONV.

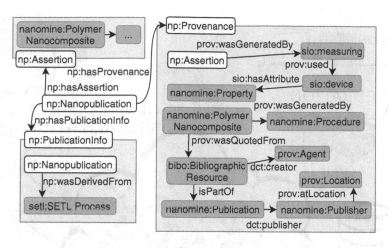

Fig. 5. Modeling the provenance of polymer nanocomposites. Named graphs are indicated as grey boxes. The provenance graph contains information about the methods used to produce the characteristics in the provenance, the processing graph from Fig. 4, and bibliographic information. The Publication Info graph contains statements about how the nanopublication itself was created, usually as a result of a Semantic ETL (SETL) process.

and defined by domain experts, and then integrated into SIO by our ontology experts.

We determined that a knowledge graph was needed instead of an XML file repository because it was difficult to search and visualize the data without custom software. The same properties were stored in many different ways, depending on if they had dependent variables or not, and whether or not there were uncertainty values. It was also difficult to tell if the XML files were curated in a consistent way - as it turns out many chemical names and units of meaures were entered in different ways. Converting the data to a knowledge graph also provides us with an opportunity to enhance the data progressively. For instance, we will be able to search for samples that can be used in different kinds of simulations and computational analyses to produce computed properties. Additionally, we will be providing conversions to preferred units of measure from the ones that were reported in the original papers.

5   Resource Evaluation

We evaluate the ontology by the ability to search, explore, and visualize polymer nanocomposites and their properties from many different experiments. Additionally, we evaluate the knowledge graph itself through the ability for materials scientists to explore nanomaterials using a faceted browser, and the ability of semantics-trained materials scientists to create custom visualizations to share with others.

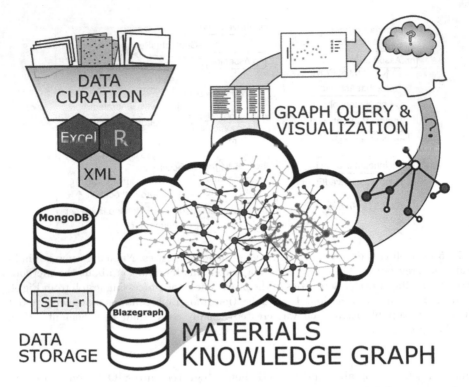

Fig. 6. Overview of the NanoMine curation process. Data is curated from papers by the NanoMine team and other materials science contributors into XML files and Excel spreadsheets (sometimes using R for to improve data quality) and stored in a MongoDB database. The files are copied over to Whyis to be converted into nanopublications that express sample properties. This knowledge graph is then used to visualize properties on demand by users, and further properties are inferred from additional data, including images of the materials.

5.1 Ontology Evaluation

We were able to formally map the NanoMine schema into the NanoMine Ontology using the approaches expressed above. The NanoMine Ontology is fully integrated into SIO and PROV-O. All classes are subsumed by classes in SIO or PROV-O, and no additional properties were needed beyond the ones in those ontologies. This makes it easier for tools that understand SIO or PROV-O to be re-used on NanoMine data, and for NanoMine-specific tools to be used with other scientific or provenance-related data. The NanoMine Ontology is in $ALEOF$, and contains 318 classes and no new Object or Datatype Properties. It is also self-consistent (including the imported ontologies of SIO and PROV-O) as confirmed by both Hermit and Pellet.

The modeling used with the ontology has supported a number of key data exploration features that previously would have been difficult or impossible. For instance, it is now possible to provide summary statistics across groupings of properties, as well as individual property types, as shown in Fig. 8. These views are important for curation quality control, as we can assess if there are any unexpected property names being used in the XML files. We are also able to analyze the use of different materials as fillers and matrices, as shown in Fig. 9. It is important to note that, while the XML Schema and its translation into RDF has evolved over the course of several years, the underlying representation in RDF has not had to change to support those improvements, suggesting that the representation may be durable against future changes. The use of the PROV-O-based partial ordering representation expands on the kinds of experimental procedures we can represent from simple lists of steps to complex workflows. Use of this representation also allows for quick analysis of what kinds of processing steps co-occur, as shown in Fig. 7. Table 2 provides a summary of the total subclasses and instances of the major classes in the knowledge graph.

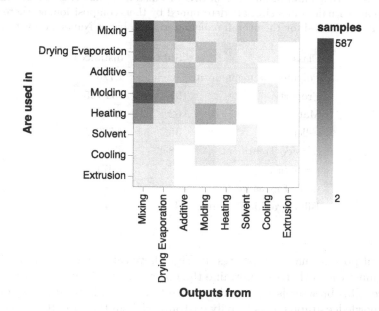

Fig. 7. A co-occurrence matrix of processing steps. Outputs of processing steps from the X axis are fed as inputs into processing steps on the Y axis. This lets users see how common certain methods are and what they tend to be used with, available at http://nanomine.org/viz/b2b74728f1751f2a.

5.2 Visualizing the Knowledge Graph

We provide two primary means of visualizing the knowledge graph - a faceted browser and a custom visualizer. Our general purpose faceted browser for browsing and visualizing samples in the graph allows users to find, select, and visualize

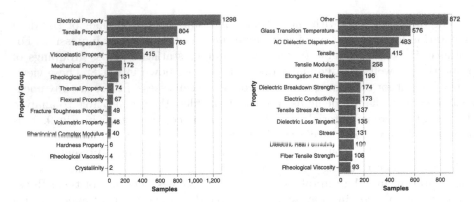

Fig. 8. Material property types by frequency, both by immediate superclass (left) and overall, available at http://nanomine.org/viz/77a5fa51556d064b.

Table 2. Summary of instances and subclasses of the major classes in the NanoMine Ontology. Note that units of measure in SIO are instances, not classes, and that the kinds of polymer nanocomposites are determined by their composition, so there are no current subclasses used for the overall nanomaterial (Polymer Nanocomposite).

Class	Subtypes	Instances
Polymer Nanocomposite	–	1,725
Property	97	733,155
Matrix	62	1,787
Filler	35	1,560
Surface Treatment	88	478
procedure	8	7,471
unit of measure	–	81
Bibliographic Resource	4	183

data about polymer nanocomposites. In Fig. 10, we can see a user plotting properties against each other to determine the effects of certain values on their performance. This browser is a general purpose tool available to any developers of Whyis knowledge graphs. It is heavily customized from the SPARQL Faceter [6], and can autogenerate facet configurations by introspecting the knowledge graph.

Advanced users can also create custom visualizations using the Vega and Vega-Lite tools [14]. In Fig. 12, a materials scientist with training in SPARQL has created a linked visualization using the Vega-Lite framework. In Whyis, the user can publish and share this visualization for others to see their analysis. Figure 11 shows the chart from Fig. 7 being edited. Other plots have been produced using this approach, and are available in the NanoMine visualization gallery.[3]

---

[3] Available from the main NanoMine page at http://nanomine.org.

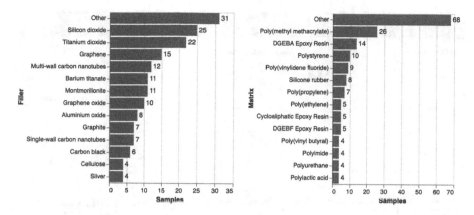

Fig. 9. Particle filler and polymer matrix types in the NanoMine knowledge graph, available at http://nanomine.org/viz/e037b3b61ab26244.

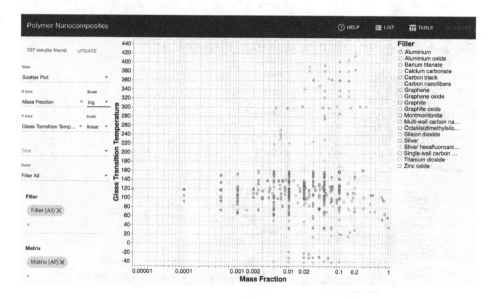

Fig. 10. Users can search for nanomaterial samples of interest and visualize their properties using simple controls. Here, a common view of the data, comparing the mass fraction (amount of nanoparticles by mass) of more than 700 samples to its glass transition temperature (the temperature below which a material becomes glassy and brittle). Users can select from any properties, constituent properties, material types, processing methods, and publication data to filter and visualize.

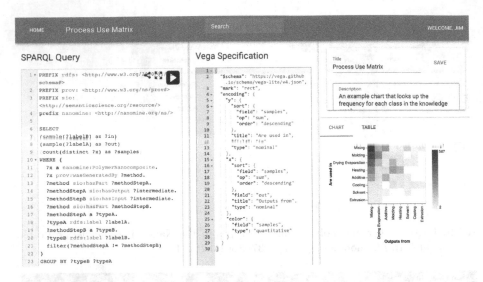

Fig. 11. Editing the Fig. 7 visualization in Whyis. Users can iterate over the SPARQL query and Vega specification to produce their preferred plot.

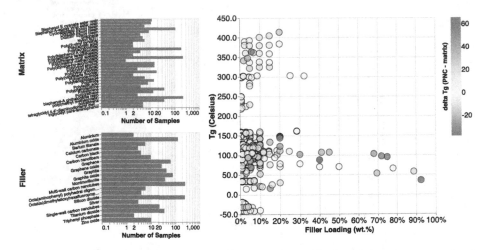

Fig. 12. A user-created custom visualization, created by a materials scientist. In this linked plot example, the heat map dynamically updates as the user selects matrix and filler materials from the bar charts below. Collaborators can use the resulting plot to identify overall trends in how the glass transition temperature, Tg, may change as filler particles are added to the matrix material. Available at http://nanomine.org/viz/10720c80b5b41ab8.

6   Related Work

There are several polymer data resources, but none focus on polymer nanocomposites, nor have publicly accessible APIs. These include the CRC POLYMER-

SnetBASE by the Taylor and Francis Group [4], the Polymer Property Predictor and Database (PPPDB) by the University of Chicago,[4] and the PolyInfo database from NIMS of Japan [12]. All of these data resources distribute curated polymer data from publications and polymer handbooks, with detailed annotations of chemical properties and characteristics. However, there are limitations for those data resources; for PolyInfo and PPPDB, the lack of application program interface (API) access prevents the application of a user-defined search and exploration of data, the CRC POLYMERSnetBASE requires paid access, and PPPDB covers only a few properties (notably the chi parameter and glass transition temperature) and is focused on polymer blends. In all those data resources, very few records are related to composites or nanocomposites, and those data resources rarely contain the complete information needed to describe the nanocomposite processing parameters, microstructure, and properties. Other materials science knowledge graphs include the Metallic Materials Knowledge Graph (MMKG) [17], which subsets DBpedia from seed entities. It does not attempt to manage the properties of those materials, only aggregate terms of interest. Another materials science knowledge graph is propnet [11], which constructs a graph of related materials based on their properties, but does not create a graph that contains representations of those properties, nor does it seem to handle complex composite materials. The Bosch Materials Science Knowledge Base is an internal project to produce a knowledge base of materials produced by the manufacturing company [16], but is not publicly available. Generally, these knowledge graphs provide conceptual models of material types, but do not manage or visualize experimental data from actual or simulated materials.

7 Future Work

There are a number of image analysis tools that the NanoMine team has developed for computing additional properties of nanomaterials. We plan to integrate these as autonomous agents that can perform these analyses automatically when a sample is added with the appropriate image type. Because Whyis allows deep metadata about any kind of file, including the context of how the file was produced, it is simple to identify the relevant image files for analysis. We also plan to introduce quality metrics for sample data that can determine if curation errors have been introduced. This can come from providing additional metadata about specific property types, and then checking for consistency with that metadata. We are also working to provide more consistent naming and identification of the chemicals used in creating the nanomaterials. Many chemicals have multiple names, and no current chemical databases have compete coverage of the materials used. We plan to improve the user experience of the Whyis interface through a usability study. Finally, we are continually seeking to curate more materials into the resource, and are currently working on expanding it to support similar capabilities for metamaterials. This new resource, which will encompass NanoMine, will be called MaterialsMine.

---

[4] https://pppdb.uchicago.edu.

8   Conclusion

We introduced the NanoMine *ontology, knowledge graph, and Whyis instance* as complementary semantically-enabled resources that help materials scientists explore the known effects of polymer nanocomposite design on their many properties. This resource is published as Linked Data and through a SPARQL endpoint, using the Whyis knowledge graph framework. Scientists can filter and visualize polymer nanocomposite samples to find relationships between aspects of those samples to gain further insights into nanocomposite design. Additionally, materials scientists, with some semantic training, can create and have created advanced visualizations using the Vega and Vega-Lite libraries. The use of nanopublications in the knowledge graph allows for the management of the knowledge graph as curation improves. The resources are in use today by a number of material science groups and the resources provide a semantic foundation for a consolidated data resource of data related to a special issue of a material science journal. While we have focused here on nanomaterials, we believe this model can be used by a wide range of efforts that need to model and explore content that rely on the process-structure-property linkages, which is most of material science. We plan to continue to evolve and improve the NanoMine knowledge graph and ontology as our team and collaborators identify and curate more polymer nanocomposite experiments into the graph, and to expand the framework of NanoMine to provide similar support for metamaterials research.

References

1. Brinson, L.C., et al.: Polymer nanocomposite data: curation, frameworks, access, and potential for discovery and design. ACS Macro Lett. **9**, 1086–1094 (2020)
2. Duerr, R.E., et al.: Formalizing the semantics of sea ice. Earth Sci. Inf. **8**(1), 51–62 (2014). https://doi.org/10.1007/s12145-014-0177-z
3. Dumontier, M., et al.: The semantic science integrated ontology (SIO) for biomedical research and knowledge discovery. J. Biomed. Semant. **5**(1), 14 (2014)
4. Ellis, B., Smith, R.: Polymers: A Property Database. CRC Press (2008)
5. Groth, P., Gibson, A., Velterop, J.: The anatomy of a nanopublication. Inf. Serv. Use **30**(1–2), 51–56 (2010)
6. Koho, M., Heino, E., Hyvönen, E., et al.: SPARQL faceter-client-side faceted search based on SPARQL. In: LIME/SemDev@ ESWC (2016)
7. Lebo, T., Sahoo, S., McGuinness, D.: PROV-O: The PROV Ontology (2013). http://www.w3.org/TR/prov-o/
8. McCusker, J.P., Chastain, K., Rashid, S., Norris, S., McGuinness, D.L.: SETLr: the semantic extract, transform, and load-r. PeerJ Preprints 6, e26476v1 (2018)
9. McCusker, J.P., Dumontier, M., Yan, R., He, S., Dordick, J.S., McGuinness, D.L.: Finding melanoma drugs through a probabilistic knowledge graph. PeerJ Comput. Sci. **3**, e106 (2017)
10. McCusker, J.P., et al.: Broad, interdisciplinary science in tela: an exposure and child health ontology. In: Proceedings of the 2017 ACM on Web Science Conference, pp. 349–357 (2017)
11. Mrdjenovich, D., et al.: Propnet: a knowledge graph for materials science. Matter **2**(2), 464–480 (2020). https://doi.org/10.1016/j.matt.2019.11.013

12. Otsuka, S., Kuwajima, I., Hosoya, J., Xu, Y., Yamazaki, M.: Polyinfo: polymer database for polymeric materials design. In: 2011 International Conference on Emerging Intelligent Data and Web Technologies, pp. 22–29. IEEE (2011)
13. Piñero, J., et al.: Disgenet: a comprehensive platform integrating information on human disease-associated genes and variants. Nucleic Acids Res. gkw943 (2016)
14. Satyanarayan, A., Moritz, D., Wongsuphasawat, K., Heer, J.: Vega-lite: a grammar of interactive graphics. IEEE Trans. Vis. Comput. Graph. **23**(1), 341–350 (2016)
15. Schadler, L., Brinson, L.C., Sawyer, W.: Polymer nanocomposites: a small part of the story. JOM **59**(3), 53–60 (2007)
16. Strötgen, J., et al.: Towards the bosch materials science knowledge base. In: ISWC Satellites (2019)
17. Zhang, X., Liu, X., Li, X., Pan, D.: MMKG: an approach to generate metallic materials knowledge graph based on DBpedia and Wikipedia. Comput. Phys. Commun. **211**, 98–112 (2017). https://doi.org/10.1016/j.cpc.2016.07.005. High Performance Computing for Advanced Modeling and Simulation of Materials
18. Zhao, H., et al.: Nanomine schema: an extensible data representation for polymer nanocomposites. APL Mater. **6**(11), 111108 (2018)

G2GML: Graph to Graph Mapping Language for Bridging RDF and Property Graphs

Hirokazu Chiba[1(✉)] [iD], Ryota Yamanaka[2], and Shota Matsumoto[3]

[1] Database Center for Life Science, Chiba 277-0871, Japan
chiba@dbcls.rois.ac.jp
[2] Oracle Corporation, Bangkok 10500, Thailand
ryota.yamanaka@oracle.com
[3] Lifematics Inc., Tokyo 101-0041, Japan
shota.matsumoto@lifematics.co.jp

Abstract. How can we maximize the value of accumulated RDF data? Whereas the RDF data can be queried using the SPARQL language, even the SPARQL-based operation has a limitation in implementing traversal or analytical algorithms. Recently, a variety of database implementations dedicated to analyses on the property graph (PG) model have emerged. Importing RDF datasets into these graph analysis engines provides access to the accumulated datasets through various application interfaces. However, the RDF model and the PG model are not interoperable. Here, we developed a framework based on the Graph to Graph Mapping Language (G2GML) for mapping RDF graphs to PGs to make the most of accumulated RDF data. Using this framework, accumulated graph data described in the RDF model can be converted to the PG model, which can then be loaded to graph database engines for further analysis. For supporting different graph database implementations, we redefined the PG model and proposed its exchangeable serialization formats. We demonstrate several use cases, where publicly available RDF data are extracted and converted to PGs. This study bridges RDF and PGs and contributes to interoperable management of knowledge graphs, thereby expanding the use cases of accumulated RDF data.

Keywords: RDF · Property graph · Graph database

1 Introduction

Increasing amounts of scientific and social data are being published in the form of the Resource Description Framework (RDF) [1], which presently constitutes a large open data cloud. DBpedia [2] and Wikidata [3] are well-known examples of such RDF datasets. SPARQL [4] is a protocol and query language that serves as a standardized interface for RDF data. This standardized data model and interface enables the construction of integrated graph data. However, the lack of

© Springer Nature Switzerland AG 2020
J. Z. Pan et al. (Eds.): ISWC 2020, LNCS 12507, pp. 160–175, 2020.
https://doi.org/10.1007/978-3-030-62466-8_11

an interface for graph-based analysis and performant traversal limits use cases of the graph data.

Recently, the property graph (PG) model [5,6] has been increasingly attracting attention in the context of graph analysis [7]. Various graph database engines, including Neo4j [8], Oracle Database [9], and Amazon Neptune [10] adopt this model. These graph database engines support algorithms for traversing or analyzing graphs. However, few datasets are published in the PG model and the lack of an ecosystem for exchanging data in the PG model limits the application of these powerful engines.

In light of this situation, developing a method to transform RDF into PG would be highly valuable. One of the practical issues faced by this challenge is the lack of a standardized PG model. Another issue is that the transformation between RDF and PG is not straightforward due to the differences in their models. In RDF graphs, all information is expressed by triples (node-edge-node), whereas in PGs, arbitrary information can be contained in each node and edge as the key-value form. Although this issue was previously addressed on the basis of predefined transformations [11], users still cannot control the mapping for their specific use cases.

In this study, we redefine the PG model incorporating the differences in existing models and propose serialization formats based on the data model. We further propose a graph to graph mapping framework based on the Graph to Graph Mapping Language (G2GML). Employing this mapping framework, accumulated graph data described in RDF can be converted into PGs, which can then be loaded into several graph database engines for further analysis. We demonstrate several use cases, where publicly available RDF data is extracted and converted to PGs. Thus, this study provides the foundation for the interoperability between knowledge graphs.

The main contributions of this study are as follows: 1) language design of G2GML and 2) its prototype implementation. Furthermore, we propose 3) the common PG model and its serialization, which are essential to ensure that this mapping framework is independent from the implementations of databases.

2   Graph to Graph Mapping

2.1   Overview

We provide an overview of the graph to graph mapping (G2G mapping) framework (Fig. 1).

In this framework, users describe mappings from RDF to PG in G2GML. According to this G2GML description, the input RDF dataset is converted into a PG dataset. The new dataset can also be optionally saved in specific formats for loading into major graph database implementations.

G2GML is a declarative language comprising pairs of RDF patterns and PG patterns. The core concept of a G2GML description can be represented by a map from RDF subgraphs, which match specified SPARQL patterns, to PG components.

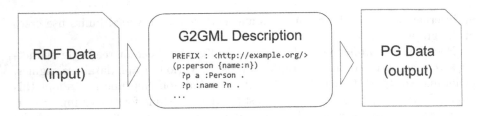

Fig. 1. Overview of mapping from RDF to PG

2.2   G2GML

G2GML is a language to describe maps from RDF graphs to PGs. It represents a domain-specific declarative language, where programmers only have to describe patterns of RDF and PG. The RDF pattern is described in terms of the SPARQL syntax, and the PG pattern is described in a syntax derived from openCypher [12].

Preliminaries. *An RDF triple* (s, p, o) *is an element of* $(I \cup B) \times I \times (I \cup B \cup L)$, *where* I, L, *and* B *are a set of IRIs, a set of literals, and a set of blank nodes, respectively, which are considered pairwise disjoint. For an RDF triple* (s, p, o), *s is the subject, p the predicate, and o the object. An RDF graph is defined as a finite set of RDF triples.*

A G2GML description comprises one or more *node maps* and *edge maps*. A *node map* is a function that maps resources from an RDF graph to a set of *PG nodes*. Similarly, an *edge map* is a function that maps resources from an RDF graph to a set of *PG edges*. Here, we define the syntax of G2GML as follows.

Definition 1 (EBNF notation of G2GML).

```
g2gml    ::= SPARQL_PREFIX* mapping+
mapping ::= ( node_pat | edge_pat ) NEWLINE
            INDENTED_RDF_PATTERN NEWLINE
```

In Definition 1, `SPARQL_PREFIX` and `INDENTED_RDF_PATTERN` are almost the same as *PrefixDecl* and *GroupGraphPattern* in SPARQL syntax (see EBNF notation of SPARQL grammar [4]). The only difference is that `INDENTED_RDF_PATTERN` must have one or more white spaces at the beginning of each line. The symbols `node_pat` and `edge_pat` are described according to Definition 2 and 3. `NEWLINE` is used as an abstract symbol corresponding to the appropriate character to start a new line. Minimal examples of G2GML are shown in Sect. 2.5.

An RDF graph pattern in G2GML specifies the resources that should be mapped to PG components. Variables in RDF graph patterns are embedded into the corresponding location in the preceding PG patterns, which yield resultant PGs.

A node map is described as a pair of a PG node pattern and an RDF graph pattern. Here, we define the syntax of the PG node pattern as follows.

Definition 2 (EBNF notation of PG node pattern).

```
node_pat   ::= "(" NODE   ":" LABEL  ("{" prop_list "}")? ")"
prop_list ::= property "," prop_list | property
property   ::= PROP_NAME ":" PROP_VAL
```

In Definition 2, NODE, LABEL, PROP_NAME, and PROP_VAL are all identifiers. A PG node pattern specifies a node, a label, and properties to be generated by the map. The NODE symbol specifies the ID of the generated node, which is replaced by the resource retrieved from the RDF graph. The LABEL symbol specifies the label of the resultant node. A label also serves as the name of the node map, which is referred from edge maps. Therefore, the identifier of each label must be unique in a G2GML description. Each node pattern contains zero or more properties in a prop_list. A property is a pair of PROP_NAME and PROP_VAL that describes the properties of nodes. The identifiers NODE and PROP_VAL must be names of variables in the succeeding RDF graph pattern, while LABEL and PROP_NAME are regarded as text literals.

In contrast, a PG edge pattern specifies the pattern of the edge itself, its label, and properties, as well as the source and destination nodes. Here, we define the syntax of the PG edge pattern as follows.

Definition 3 (EBNF notation of PG edge pattern).

```
edge_pat ::= "(" SRC   ":" SRC_LAB ") -"
             "[" EDGE?  ":" EDGE_LAB  ("{" prop_list "}")? "]"
             ("->" | "-") "(" DST ":" DST_LAB ")"
```

In Definition 3, SRC, SRC_LAB, EDGE, EDGE_LAB, DST, and DST_LAB are all identifiers. Identifiers of SRC and DST specify the variables in the succeeding RDF graph pattern that should be mapped to the endpoint nodes of resultant edges. A SRC_LAB and a DST_LAB serve not only as labels of resultant nodes, but also as implicit constraints of the edge map. The resultant source and destination nodes must match patterns in the corresponding node maps (described in Sect. 2.6 with a concrete example). For this reason, SRC_LAB and DST_LAB must be defined in other node maps. EDGE, EDGE_LAB, and prop_list can be described in the same manner as their counterparts in node maps.

The resultant edges can be either directed or undirected. If '->' is used as a delimiter between an edge and a destination part, the edge will be directed.

In node and edge maps, a variable bound to PROP_VAL may have multiple candidates for a single node or edge. If multiple candidates exist, they are concatenated into an array in the same manner as GROUP_CONCAT in SPARQL. If EDGE is omitted in an edge map, such candidates are concatenated within groups of tuples of SRC and DST.

2.3   Property Graph Model

We define the PG model independent of specific graph database implementations. For the purpose of interoperability, we incorporate differences in PG models, taking into consideration multiple labels or property values for nodes and edges, as well as mixed graphs with both directed and undirected edges. The PG model that is redefined here requires the following characteristics:

- A PG contains nodes and edges.
- Each node and edge can have zero or more labels.
- Each node and edge can have properties (key-value pairs).
- Each property can have multiple values.
- Each edge can be directed or undirected.

Formally, we define the PG model as follows.

Definition 4 (Property Graph Model).
A property graph *is a tuple* $PG = \langle N, E_d, E_u, e, l_n, l_e, p_n, p_e \rangle$, *where:*

1. N *is a set of nodes.*
2. E_d *is a set of directed edges.*
3. E_u *is a set of undirected edges.*
4. $e : E \rightarrow \langle N \times N \rangle$ *is a function that yields the endpoints of each directed or undirected edge where* $E := E_d \cup E_u$. *If the edge is directed, the first node is the source, and the second node is the destination.*
5. $l_n : N \rightarrow 2^S$ *is a function mapping each node to its multiple labels where* S *is a set of all strings.*
6. $l_e : E \rightarrow 2^S$ *is a function mapping each edge to its multiple labels.*
7. $p_n : N \rightarrow 2^P$ *is a function used to assign nodes to their multiple properties.* P *is a set of properties. Each property assumes the form* $p = \langle k, v \rangle$, *where* $k \in S$ *and* $v \in 2^V$. *Here* V *is a set of values of arbitrary data types.*
8. $p_e : E \rightarrow 2^P$ *is a function used to assign edges to their multiple properties.*

2.4   Serialization of Property Graphs

According to our definition of the PG model, we propose serialization in flat text and JSON. The flat text format (PG format) performs better in terms of human readability and line-oriented processing, while the JSON format (JSON-PG format) is best used for server–client communication.

The PG format has the following characteristics. An example is given in Fig. 2, which is visualized in Fig. 3.

- Each line describes a node or an edge.
- All elements in a line are separated by spaces or tabs.
- The first column of a node line contains the node ID.
- The first three columns of an edge line contain the source node ID, direction, and destination node ID.
- Each line can contain an arbitrary number of labels.
- Each line can contain an arbitrary number of properties (key-value pairs).

More formally, we describe the PG format in the EBNF notation as follows.

```
# NODES
101  :person  name:Alice  age:15  country:"United States"
102  :person  :student  name:Bob  country:Japan  country:Germany

# EDGES
101 -- 102  :same_school  :same_class  since:2002
102 -> 101  :likes  since:2005
```

Fig. 2. Example of PG format

Definition 5 (EBNF notation of the PG format).

```
pg          ::= (node | edge)+
node        ::= NODE_ID labels properties NEWLINE
edge        ::= NODE_ID direction NODE_ID labels properties NEWLINE
labels      ::= label*
properties  ::= property*
label       ::= ":" STRING
property    ::= STRING ":" VALUE
direction   ::= "--" | "->"
```

According to this definition, each property value of PGs is a set of items of any datatype. Meanwhile, our G2G mapping prototype implementation currently supports the three main datatypes (integer, double, and string) as property values, and those types are inferred from the format of serialized values.

Furthermore, we implemented command-line tools to convert formats between the flat PG and JSON-PG, where the latter follows the JSON syntax in addition to the above definition. We further transform them into formats for well-known graph databases such as Neo4j, Oracle Database, and Amazon Neptune. The practical use cases of our tools demonstrate that the proposed data model and formats have the capability to describe PG data used in existing graph databases (see https://github.com/g2glab/pg).

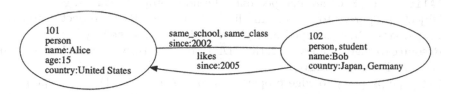

Fig. 3. Example of PG format (Visualization)

2.5 Minimal Example

Figure 5 shows the minimal example of G2G mapping from RDF data (Fig. 4) to PG data (Fig. 6), representing the following five types of typical mapping.

```
@prefix : <http://example.org/> .
:person1 a :Person ;
         :name 'Alice' .
:person2 a :Person ;
         :name 'Bob' .
:person1 :supervised_by :person2 .
[] a :Email ;
     :sender      :person1 ;
     :receiver    :person2 ;
     :year        2017 ;
     :attachment  '01.pdf' .
```

Fig. 4. Example of input RDF data

```
PREFIX : <http://example.org/>
(p:person {name:n})
     ?p a :Person .
     ?p :name ?n .
(p1:person)-[:supervised_by]->(p2:person)
     ?p1 :supervised_by ?p2 .
(p1:person)-[:emailed {year:y, attachment:a}]->(p2:person)
     ?f a :Email ;
        :sender    ?p1 ;
        :receiver  ?p2 ;
        :year      ?y .
     OPTIONAL { ?f :attachment ?a }
```

Fig. 5. Example of G2G mapping definition

```
"http://example.org/person1" :person name:Alice
"http://example.org/person2" :person name:Bob
"http://example.org/person1" -> "http://example.org/person2" :supervised_by
"http://example.org/person1" -> "http://example.org/person2" :emailed year:2017 attachment:"01.pdf"
```

Fig. 6. Example of output PG data

- Resource to node: In lines 2–4, the RDF resources with type :`Person` are mapped into the PG nodes using their IRIs as node IDs.
- Datatype property to node property: In lines 2–4, the RDF datatype property :`name` is mapped onto the PG node property key `name`. The literal objects 'Alice' and 'Bob' are mapped onto the node property values.
- Object property to edge: In lines 5–6, the RDF object property :`supervised_by` is mapped onto the PG edge `supervised_by`.
- Resource to edge: In lines 7–12, the RDF resource with type :`Email` is mapped onto the PG edge `emailed`.
- Datatype property to edge property: In lines 7–12, the RDF datatype property :`year` and :`attachment` are mapped onto the PG edge property `year` and `attachment`. The literal objects 2017 and '01.pdf' are mapped onto the edge property values.

2.6 Mapping Details

The G2G mapping based on G2GML is designed to be intuitive in general as shown above; however, there are several discussions regarding the details of mapping.

```
@prefix : <http://example.org/> .
:person1 a :Person ;
            :name 'Alice' .
:person2 a :Person ;
            :name 'Bob' .
[] a :Email ;
    :sender   :person1 ;
    :receiver :person2 ;
    :year     2017 .
[] a :Email ;
    :sender   :person1 ;
    :receiver :person2 ;
    :year     2018 .
```

Fig. 7. Example of input RDF data (multi-edges)

```
"http://example.org/person1" :person name:Alice
"http://example.org/person2" :person name:Bob
"http://example.org/person1" -> "http://example.org/person2" :emailed year:2017
"http://example.org/person1" -> "http://example.org/person2" :emailed year:2018
```

Fig. 8. Example of output PG data (multi-edges)

Referencing Node Labels. The node labels in edge maps must be defined in node maps (described in Sect. 2.2) to reference the conditions. This means that the conditions for defining `person` are imported into the definition of `supervised_by` and `emailed` relationships. Therefore, the `supervised_by` relationship will be generated only between the nodes satisfying the `person` condition. In this manner, users can retrieve the PG datasets that are intuitively consistent (it is also possible to avoid specifying node labels). In contrast, edge conditions are not inherited to nodes, such that the nodes are retrieved without all relationships, `supervised_by` and `emailed`. Although the nodes with no relationship also satisfy the conditions, the prototype implementation does not retrieve such orphan nodes by default from a practical perspective.

Multi-edges. When a pair of RDF resources has multiple relationships, those will be converted to multiple PG edges instead of a single PG edge with multiple properties. The G2GML in Fig. 5 generates two PG edges for the RDF data in Fig. 7 to maintain the information that Alice emailed Bob twice (Fig. 8).

List of Property Values. Each PG property can assume multiple values. In an RDF that includes the same data property multiple times, the values are assumed to be the members of a list. The G2GML in Fig. 5 generates one PG edge with two properties for the RDF data in Fig. 9, keeping the information that Alice emailed Bob once, and the email contained two attachments (Fig. 10).

3   Use Cases

We present several use cases to show how to apply the G2G mapping framework, where we utilize publicly available RDF datasets through SPARQL endpoints.

```
@prefix : <http://example.org/> .
:person1 a :Person ;
         :name 'Alice' .
:person2 a :Person ;
         :name 'Bob' .
[] a :Email ;
    :sender    :person1 ;
    :receiver  :person2 ;
    :year      2017 ;
    :attachment '01.pdf' ;
    :attachment '02.pdf' .
```

Fig. 9. Example of input RDF data (list of property values)

```
"http://example.org/person1" :person name:Alice
"http://example.org/person2" :person name:Bob
"http://example.org/person1" -> "http://example.org/person2" :emailed year:2017 attachment:"01.pdf" attachment:"02.pdf"
```

Fig. 10. Example of output PG data (list of property values)

3.1 Using Wikidata

Wikidata is a useful source of open data from various domains. We present an example of mapping using a disease dataset from Wikidata, where each disease is associated with zero or more genes and drugs. Figure 11 illustrates the schematic relationships between those entities. This subset of Wikidata can be converted as a PG using the G2G mapping framework by writing a mapping file in G2GML (see Fig. 12). Here, we focus on human genes and specify the necessary conditions in G2GML. Each instance of Q12136 (disease) can have a property P2176 (drug used for treatment), which is thereby linked to items of Q12140 (medication). Further, each disease can have a property P2293 (genetic association), which is thereby linked to items of Q7186 (gene). The resultant PG includes 4696 diseases, 4496 human genes, and 1287 drugs. The total numbers of nodes and edges are summarized in Table 1. As a result of G2G mapping, the numbers of nodes and edges are reduced to almost half and one third, respectively.

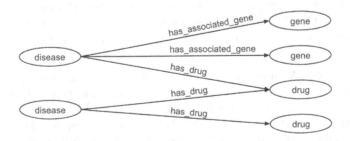

Fig. 11. Schematic relations of Wikidata entities

```
PREFIX wd: <http://www.wikidata.org/entity/>
PREFIX wdt: <http://www.wikidata.org/prop/direct/>

(g:human_gene {symbol: s})
    ?g wdt:P31 wd:Q7187 ;        # "instance of" "gene"
       wdt:P703 wd:Q15978631 ;   # "found in taxon" "Homo sapiens"
       wdt:P353 ?s .             # "HGNC gene symbol"

(d:disease {name: n})
    ?d wdt:P31 wd:Q12136 ;       # "instance of" "disease"
       rdfs:label ?l .
    FILTER(lang(?l) = "en")
    BIND(str(?l) AS ?n)

(m:drug {name: n})
    ?m wdt:P31 wd:Q12140 ;       # "instance of" "medication"
       rdfs:label ?l .
    FILTER(lang(?l) = "en")
    BIND(str(?l) AS ?n)

(d:disease)-[:has_associated_gene]->(g:human_gene)
    ?d wdt:P2293 ?g .            # "genetic association"

(d:disease)-[:has_drug]->(m:drug)
    ?d wdt:P2176 ?m .            # "drug used for treatment"
```

Fig. 12. G2GML for Wikidata mapping

3.2 Using DBpedia

Figure 13 schematically illustrates an example of the G2G mapping to convert
RDF data retrieved from DBpedia into PG data. Focusing on a relationship
where two musicians (?m1 and ?m2) belong to the same group, this information
can be represented in the PG shown on the right side of the figure. The relationships are originally presented as independent resources ?g. In the mapping, we
map ?g, ?n, and ?s onto a PG edge labeled same_group with attributes. Such
compaction by mapping is useful in numerous use cases, for example, when users
are interested only in relationships between musicians.

The mapping rules for this example are specified in the G2GML description
shown in Fig. 14. The G2GML description contains node mapping for musician
entities and edge mapping for same_group relationships. In each specified PG
pattern, {m, n, h} and {m1, m2, n, s} are used as variables to reconstruct
resources and literals extracted from RDF graphs.

Figure 15 shows queries to retrieve the pairs of musicians that are in the same
group in SPARQL and graph query languages [12,13]. The queries in Cypher
and PGQL are more succinct owing to the simple structure of the PG obtained
by G2G mapping.

4 Availability

The prototype implementation of G2G mapping is written in JavaScript and can
be executed using Node.js in the command line. It has an endpoint mode and a
local file mode. The local file mode uses Apache Jena ARQ to execute SPARQL

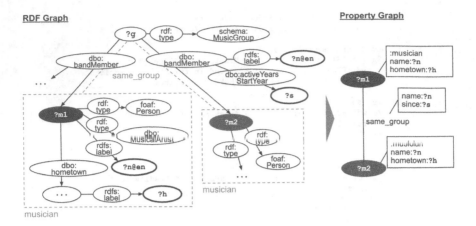

Fig. 13. Schematic example of DBpedia mapping

```
PREFIX rdf: <http://www.w3.org/1999/02/22-rdf-syntax-ns#>
PREFIX rdfs: <http://www.w3.org/2000/01/rdf-schema#>
PREFIX schema: <http://schema.org/>
PREFIX dbo: <http://dbpedia.org/ontology/>
PREFIX foaf: <http://xmlns.com/foaf/0.1/>

# Node mapping
(m:musician {name:n, hometown:h})                        # PG Pattern
    ?m rdf:type foaf:Person , dbo:MusicalArtist ;         # RDF Pattern
       rdfs:label ?n . FILTER(lang(?n) = "en") .
    OPTIONAL { ?m dbo:hometown/rdfs:label ?h . FILTER(lang(?h) = "en") }

# Edge mapping
(m1:musician)-[:same_group {name:n, since:s}]-(m2:musician)   # PG Pattern
    ?g rdf:type schema:MusicGroup ;                          # RDF Pattern
       dbo:bandMember ?m1 , ?m2 . FILTER(?m1 != ?m2)
    OPTIONAL { ?g rdfs:label ?n . FILTER(lang(?n) = "en")}
    OPTIONAL { ?g dbo:activeYearsStartYear ?s }
```

Fig. 14. G2GML for DBpedia mapping

queries internally, whereas the endpoint mode accesses SPARQL endpoints via the Internet. An example of the usage in the endpoint mode is as follows:

$ g2g musician.g2g http://dbpedia.org/sparql

where the first argument is a G2GML description file, and the second argument is the target SPARQL endpoint, which provides the source RDF dataset.

Furthermore, a Docker image (https://hub.docker.com/r/g2glab/g2g) and a demonstration site of the G2G mapping framework (https://purl.org/g2gml) are available (Fig. 16).

In future adoptions of this framework, we expect two scenarios: The use of the standalone G2G mapping tools, such as our implementation, for generating PG datasets from RDF resources. Further, the adoption of the framework within the database management systems, which support both RDF and PG datasets.

```
# SPARQL
PREFIX rdf: <http://www.w3.org/1999/02/22-rdf-syntax-ns#>
PREFIX rdfs: <http://www.w3.org/2000/01/rdf-schema#>
PREFIX schema: <http://schema.org/>
PREFIX dbo: <http://dbpedia.org/ontology/>
PREFIX foaf: <http://xmlns.com/foaf/0.1/>

SELECT DISTINCT ?nam1 ?nam2
WHERE {
    ?mus1 rdf:type foaf:Person , dbo:MusicalArtist .
    ?mus2 rdf:type foaf:Person , dbo:MusicalArtist .
    ?mus1 rdfs:label ?nam1 . FILTER(lang(?nam1) = "ja") .
    ?mus1 rdfs:label ?nam2 . FILTER(lang(?nam2) = "ja") .
    ?grp a schema:MusicGroup ;
        dbo:bandMember ?mus1 , ?mus2 .
    FILTER(?mus1 != ?mus2)
}


# Cypher
MATCH (m1)-[:same_group]-(m2) RETURN DISTINCT m1.name, m2.name

# PGQL
SELECT DISTINCT m1.name, m2.name MATCH (m1)-[:same_group]-(m2)
```

Fig. 15. SPARQL, Cypher, and PGQL

Table 1. Number of nodes and edges in use cases

	RDF nodes	RDF edges	PG nodes	PG edges
Wikidata disease	20,692	36,826	10,477	11,770
DBpedia musician	23,846	32,808	7,069	10,755

5 Related Work

5.1 Property Graph Model and Serialization

Recently, graph data has increasingly attracted attention, leading to a plethora of database implementations. Thus far, different data models and query languages have been available for these graph database implementations. Consequently, there have been community-based activities for standardizing the graph query language for interoperable use of PG databases [14]. Similarly, the standardization of the PG model for various database implementations enhances the interoperable use of graph data. There is indeed a demand for graph standardization in the community, which was recently discussed in a W3C workshop [15]. Another proposal for the PG model and serialization [16] that was similar to ours was presented in the workshop. Notably, that the two independent studies converged to a similar solution. They seem to be interchangeable; however, this remains to be tested. Future studies should address collaboration and standardization of the data model. In particular, our serialization has implementations for some of the major database engines and has the potential to further cover various database engines. The serialization formats that are independent of specific database implementations will increase the interoperability of graph databases and make it easier for users to import accumulated graph data.

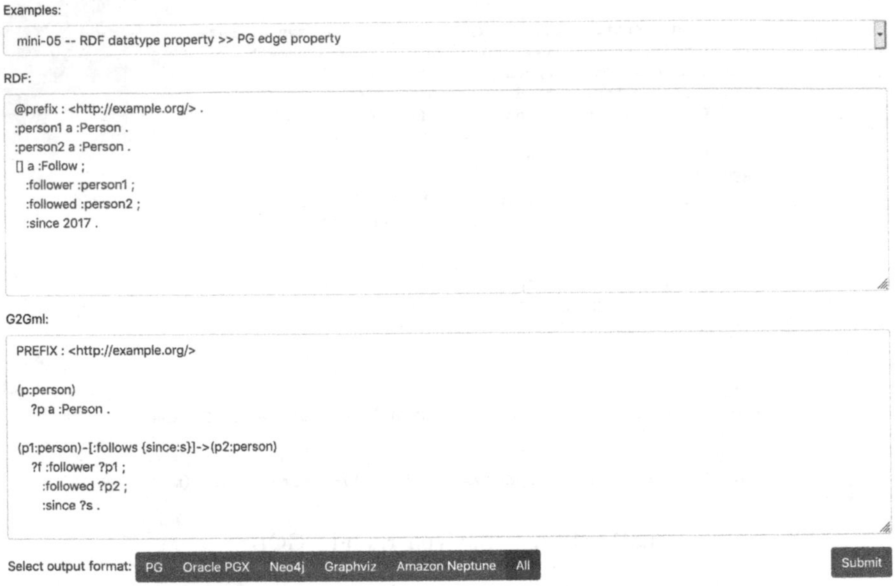

Examples:

mini-05 -- RDF datatype property >> PG edge property

RDF:

```
@prefix : <http://example.org/> .
:person1 a :Person .
:person2 a :Person .
[] a :Follow ;
  :follower :person1 ;
  :followed :person2 ;
  :since 2017 .
```

G2Gml:

```
PREFIX : <http://example.org/>

(p:person)
  ?p a :Person .

(p1:person)-[:follows {since:s}]->(p2:person)
  ?f :follower ?p1 ;
    :followed ?p2 ;
    :since ?s .
```

Select output format: PG   Oracle PGX   Neo4j   Graphviz   Amazon Neptune   All   Submit

Fig. 16. G2G mapping demonstration site

5.2   Graph to Graph Mapping

A preceding study on converting existing data into graph data included an effort
to convert relational databases into graph databases [17]. However, given that
RDF has prevailed as a standardized data model in scientific communities, considering mapping based on the RDF model is crucial. The interoperability of
RDF and PG [11,18–20] has been discussed, and efforts were made to develop
methods to convert RDF into PG [21,22]. However, considering the flexibility
regarding the type of information that can be expressed by edges in property
graphs, a novel method for controlling the mapping is necessary. We discuss the
comparison of controlled mapping and direct mapping later in this section.

To the best of our knowledge, this study presents the first attempt to develop
a framework for controlled mapping between RDF and PG. Notably, the designed
G2GML is a declarative mapping language. As a merit of the declarative description, we can concentrate on the core logic of mappings. In the sense that the
mapping process generates new graph data on the basis of existing graph data,
it has a close relation to the semantic inference. A similar concept is found in
the SPARQL CONSTRUCT queries. While the SPARQL CONSTRUCT clause
defines mapping on the same data model, G2GML defines mapping between different data models. Thus, G2GML is considered as a specific extension of the
SPARQL CONSTRUCT clause for generation or inference of PG data.

A previous study compared the performance of RDF triple stores, such as
Blazegraph and Virtuoso, and Neo4j [23]. They concluded that RDF has an
advantage in terms of performance and interoperability. However, they tested

only Neo4j as an implementation of the PG model. It is necessary to update such benchmarks using additional implementations, which requires improved interoperability and standardization of the PG model. Our study is expected to contribute to the improved interoperability of PG models, although a continuous discussion is necessary for the standardization of query languages for the PG model to achieve an interoperable ecosystem of both RDF and PG, along with a fair benchmarking.

Direct Mapping. Other mapping frameworks, such as Neosemantics (a Neo4j plugins), propose a method to convert RDF datasets without mapping definitions (for convenience, we call such methods direct mapping). However, the following capabilities are essential in practical usage.

1. Filtering data - RDF is designed for the web of data. When the source RDF dataset is retrieved from the public web space, it is inefficient (or even unrealistic) to convert the whole connected dataset. In G2GML, users can specify the resources they need, such that the SPARQL endpoints can return the filtered dataset. (This design is described in Sect. 2.2).
2. Mapping details - There is no common ruleset to uniquely map RDF terms to PG elements (labels, key-value pairs of properties) and to name new PG elements. Further, the conversion rules of multi-edges and lists of property values (discussed in Sect. 2.6) are not always obvious to users. Therefore, defining a method for mapping details is necessary to create precise data in actual use cases.
3. Schema definition - We assume that mapping is often used for developing specific applications on top of PG datasets. In this development, the schema of the dataset should be known, while the original RDF data source could contain more information that is not covered by the schema. G2GML helps developers understand the data schema in its intuitive definition (separately for nodes and edges, and their referencing - discussed in Sect. 2.6), while direct mapping has the potential to generate PGs without defining a schema.

We observe a similar discussion in the conversion from the relational model to RDF, where are two W3C standards, i.e., Direct Mapping [24] and R2RML [25].

6   Conclusion

We designed the G2GML for mapping RDF graphs to PGs and developed its prototype implementation. To ensure a clear definition of this mapping, we defined the PG model independent of specific graph database implementations and proposed its exchangeable serialization formats.

The advantage of using RDF is that different applications can retrieve necessary information from the same integrated datasets owing to the interoperable nature of semantic modeling.

For such increasing RDF resources, graph databases are potentially the ideal data storage as the property graph model can naturally preserve the relationship information semantically defined in RDF. G2GML therefore plays an important role in this data transformation process.

Various graph database implementations are actively developed and the standardization of query languages is currently ongoing. We expect the G2GML or its enhanced mapping framework to be supported by database management systems and other software. We believe that our efforts of generalization and prototype implementation will promote further discussion.

Acknowledgements. Part of the work was conducted in BioHackathon meetings (http://www.biohackathon.org). We thank Yuka Tsujii for helping create the figures. We thank Ramona Röß for careful review of the manuscript and useful comments.

References

1. RDF 1.1 Concepts and Abstract Syntax, W3C Recommendation, 25 February 2014. http://www.w3.org/TR/rdf11-concepts/
2. Lehmann, J., et al.: DBpedia-a large-scale, multilingual knowledge base extracted from Wikipedia. Semant. Web **6**(2), 167–195 (2015)
3. Vrandevcć, D., Krötzsch, M.: Wikidata: a free collaborative knowledgebase. Commun. ACM **57**(10), 78–85 (2014)
4. SPARQL 1.1 Query Language, W3C Recommendation, 21 March 2013. http://www.w3.org/TR/sparql11-query/
5. Angles, R., Gutierrez, C.: An Introduction to Graph Data Management. arXiv preprint arXiv:1801.00036 (2017)
6. Angles, R., Arenas, M., Barceló, P., Hogan, A., Reutter, J., Vrgoc, D.: Foundations of modern query languages for graph databases. ACM Comput. Surv. (CSUR) **50**(5), 68 (2017)
7. Abad-Navarro, F., Bernabé-Diaz, J.A., García-Castro, A., Fernandez-Breis, J.T.: Semantic publication of agricultural scientific literature using property graphs. Appl. Sci. **10**(3), 861 (2020)
8. The Neo4j Graph Platform. https://neo4j.com/
9. Oracle Database Property Graph. https://www.oracle.com/goto/propertygraph
10. Amazon Neptune. https://aws.amazon.com/neptune/
11. Hartig, O.: Reconciliation of RDF* and property graphs. arXiv preprint arXiv:1409.3288 (2014)
12. openCypher. https://www.opencypher.org/
13. van Rest, O., Hong, S., Kim, J., Meng, X., Chafi, H.: PGQL: a property graph query language. In: Proceedings of the Fourth International Workshop on Graph Data Management Experiences and Systems, p. 7. ACM (2016)
14. Angles, R., et al.: G-CORE: a core for future graph query languages. In: Proceedings of the 2018 International Conference on Management of Data, pp. 1421–1432 (2018)
15. W3C Workshop on Web Standardization for Graph Data. https://www.w3.org/Data/events/data-ws-2019/
16. Tomaszuk, D., Angles, R., Szeremeta, L., Litman, K., Cisterna, D.: Serialization for property graphs. In: Kozielski, S., Mrozek, D., Kasprowski, P., Małysiak-Mrozek, B., Kostrzewa, D. (eds.) BDAS 2019. CCIS, vol. 1018, pp. 57–69. Springer, Cham (2019). https://doi.org/10.1007/978-3-030-19093-4_5

17. De Virgilio, R., Maccioni, A., Torlone, R.: Converting relational to graph databases. In: First International Workshop on Graph Data Management Experiences and Systems, p. 1. ACM (2013)
18. Angles, R., Thakkar, H., Tomaszuk, D.: RDF and property graphs interoperability: status and issues. In: Proceedings of the 13th Alberto Mendelzon International Workshop on Foundations of Data Management (2019)
19. Das, S., Srinivasan, J., Perry, M., Chong, E. I., Banerjee, J.: A tale of two graphs: property graphs as RDF in Oracle. In: EDBT, pp. 762–773 (2014)
20. Thakkar, H., Punjani, D., Keswani, Y., Lehmann, J., Auer, S.: A stitch in time saves nine-SPARQL querying of property graphs using gremlin traversals. arXiv preprint arXiv:1801.02911 (2018)
21. Tomaszuk, D.: RDF data in property graph model. In: Garoufallou, E., Subirats Coll, I., Stellato, A., Greenberg, J. (eds.) MTSR 2016. CCIS, vol. 672, pp. 104–115. Springer, Cham (2016). https://doi.org/10.1007/978-3-319-49157-8_9
22. De Virgilio, R.: Smart RDF data storage in graph databases. In: 2017 17th IEEE/ACM International Symposium on Cluster, Cloud and Grid Computing, pp. 872–881. IEEE (2017)
23. Alocci, D., Mariethoz, J., Horlacher, O., Bolleman, J.T., Campbell, M.P., Lisacek, F.: Property graph vs RDF triple store: a comparison on glycan substructure search. PLoS ONE 10(12), e0144578 (2015)
24. A Direct Mapping of Relational Data to RDF, W3C Recommendation, 27 September 2012. https://www.w3.org/TR/r2rml/
25. R2RML: RDB to RDF Mapping Language, W3C Recommendation, 27 September 2012. https://www.w3.org/TR/r2rml/

The International Data Spaces Information Model – An Ontology for Sovereign Exchange of Digital Content

Sebastian Bader[1,7]($\boxtimes$)(ID), Jaroslav Pullmann[2](ID), Christian Mader[3], Sebastian Tramp[1](ID), Christoph Quix[5,9](ID), Andreas W. Müller[6], Haydar Akyürek[5](ID), Matthias Böckmann[1](ID), Benedikt T. Imbusch[7], Johannes Lipp[5,8](ID), Sandra Geisler[5](ID), and Christoph Lange[5,8](ID)

[1] Fraunhofer IAIS, Sankt Augustin, Germany
{sebastian.bader,matthias.boeckmann}@iais.fraunhofer.de
[2] Stardog Union, Arlington, USA
jaro.pullmann@stardog.com
[3] MANZ Solutions GmbH, Vienna, Austria
christian.mader@manz.at
[4] eccenca GmbH, Leipzig, Germany
sebastian.tramp@eccenca.com
[5] Fraunhofer FIT, Sankt Augustin, Germany
{christoph.quix,haydar.akyuerek,johannes.lipp,
sandra.geisler,christoph.lange-bever}@fit.fraunhofer.de
[6] Schaeffler Technologies, Herzogenaurach, Germany
andreas_w.mueller@schaeffler.com
[7] University of Bonn, Bonn, Germany
benedikt.imbusch@uni-bonn.de
[8] RWTH Aachen University, Aachen, Germany
[9] Hochschule Niederrhein, University of Applied Sciences, Krefeld, Germany

Abstract. The International Data Spaces initiative (IDS) is building an ecosystem to facilitate data exchange in a secure, trusted, and semantically interoperable way. It aims at providing a basis for smart services and cross-company business processes, while at the same time guaranteeing data owners' sovereignty over their content. The IDS Information Model is an RDFS/OWL ontology defining the fundamental concepts for describing actors in a data space, their interactions, the resources exchanged by them, and data usage restrictions. After introducing the conceptual model and design of the ontology, we explain its implementation on top of standard ontologies as well as the process for its continuous evolution and quality assurance involving a community driven by industry and research organisations. We demonstrate tools that support generation, validation, and usage of instances of the ontology with the focus on data control and protection in a federated ecosystem.

Keywords: Data model · Digital ecosystems · Data sovereignty · Federated architecture · Ontology

© Springer Nature Switzerland AG 2020
J. Z. Pan et al. (Eds.): ISWC 2020, LNCS 12507, pp. 176–192, 2020.
https://doi.org/10.1007/978-3-030-62466-8_12

1 Introduction: IDS Key Principles

Seamless collaboration and information exchange are the foundations of digital business models. Huge internet-based platforms have emerged, connecting people around the world and exchanging information in unprecedented speed. While end-users got used to such convenient communication and data exchange in their private interactions, they expect similar characteristics in their professional environment. However, data exchange in business-to-business relations faces a significant amount of still unresolved challenges. One example is the typical dilemma of digital strategies – sharing valuable data involves the risk of losing the company's competitive advantage, whereas not participating prevents innovative business models and undermines upcoming revenue opportunities.

There is currently no standardised, widely accepted means for a trustful exchange of business data that ensures traceability, data owner's privacy and sovereignty. Privacy concerns and protection of proprietary information are critical factors of future data infrastructures [7]. Such an infrastructure is a key prerequisite for a secure, standardised and fine-grained sharing of sensitive business data, unlocking the potential for novel value creation chains and the inception of intermediation platforms [9].

The International Data Spaces initiative[1] (IDS; formerly "Industrial Data Space") targets the requirements mentioned above by promoting a standard for virtual data spaces for reliable data exchange among business partners. To achieve the goal of sovereign data exchange, aspects of data management, semantic data integration, and security have to be addressed. The IDS proposes a message-based approach to bridge syntactic differences. Still, a successful exchange of data objects requires sufficient understanding of its content and meaning. A shared information model is therefore needed.

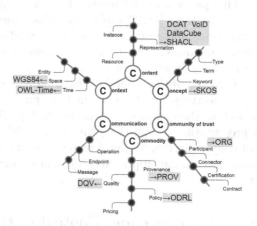

Fig. 1. Partitions of the ontology by concern (pointing to standards reused).

The *IDS Information Model* (IDS IM) is an RDFS/OWL ontology, which defines the general concepts depicted in Fig. 1 along with roles required to describe actors, components, roles and interactions in a data space. This ontology serves two purposes, (1) as a catalogue of machine-readable terms and data schema for IDS components and (2) as a shared language for all stakeholders. Each involved player needs to understand and be able to interpret this set of terms, thus enabling semantic interoperability in federated environments. The

---

[1] https://internationaldataspaces.org.

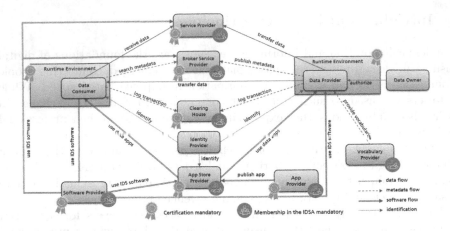

Fig. 2. IDS Reference Architecture with its main roles and interactions.

IDS IM therefore presents the backbone and common denominator for the data-sovereign ecosystem as envisioned by the IDS.

This paper presents version 4.0.0 of the IDS IM. Recent advances over earlier publications [18, 22] especially include the elaborated coverage of enforceable permissions and restrictions as the foundation of data usage policies, a significantly enhanced supply of interaction messages, as well as improved tool support for instance creation and validation.

Section 2 outlines the IDS environment and explains the fundamental concepts. Sections 3 and 4 explain the implementation using standard ontologies as well as the continuous evolution and quality assurance methods, followed by a presentation of tools for generation, validation, and usage of instances of the ontology in Sect. 5. Section 6 reviews current adoption and Sect. 7 reviews related work and similar approaches. Section 8 concludes the paper and outlines next steps.

2   Governance and Context of the IDS Information Model

The IDS has been designed in a systematic process with broad involvement of industrial stakeholders [17]. Its specification and reference implementations are maintained and supported by the International Data Spaces Association (IDSA), a non-profit organisation to disseminate and evolve the IDS views and principles. The IDSA, with more than 100 member organisations meanwhile, serves as the institutional body for promoting the IDS in research projects and industrial applications. In particular, via its sub-working group (SWG) 4 "Information Model", the IDSA ensures the sustainability of the ontology and provides the resources for future extensions (cf. Sect. 3.2 for details).

The IDS Reference Architecture Model (RAM) defines the roles assumed and the responsibilities of organisations interacting in a data space [18]. Figure 2

Table 1. Key facts about the IDS Information Model and related resources.

General	Licence	Apache License 2.0
	Size	278 classes, 149 object properties, 115 data properties, 684 individuals
	Total size	3912 triples
Reuse	Reused ontologies	CC, DCAT, DCMI Terms, FOAF, ODRL, OWL-Time, VoID, etc.
Documentation	Ontology documentation	https://w3id.org/idsa/core/
	Element description	Using rdfs:label, rdfs:comment
Availability	Namespace	ids: https://w3id.org/idsa/core/
		idsc: https://w3id.org/idsa/code/
	Serialisations	Turtle, RDF/XML, JSON-LD, N-Triples
	GitHub	https://github.com/International-Data-Spaces-Association/InformationModel/
	VoCol Instance	http://vocol.iais.fraunhofer.de/ids/

shows, for a broad initial overview, the core *interactions* and *roles* in the IDS. Data Providers exchange messages with Data Consumers via standardised software interfaces, and use multiple services to support this. They can, for example, publish metadata about resources to a directory ("broker") and thus allow others to find these. At the heart of every IDS interaction is the adherence to the usage rules – accomplished by the connection of machine-readable usage policies with each interaction and the application of certified, trustworthy execution environments. The so-called IDS Connectors interpret and enforce the applied policies, thus creating a federated network for a trustworthy data exchange.

The IDS IM specifies the domain-agnostic common language of the IDS. The IM is the essential agreement shared by the participants and components of the IDS, facilitating compatibility and interoperability. It serves the stakeholders' requirement "that metadata should not be limited to syntactical information about data, but also include data ownership information, general usage conditions, prices for data use, and information about where and how the data can be accessed" [17] by supporting the description, publication and identification of (digital) resources. It is, like other elementary IDS software components, available as open source to foster adoption (cf. Table 1). The ontology, the normative implementation of the declarative UML representation in the IDS RAM, was originally created in 2017 and first released in 2018.

2.1 Motivating Example

We use the example of the provider of financial intelligence data, the 'Business Intel Inc.', which collects, verifies, and processes stock market data for investment companies. One of their top seller is a cleared dataset of all Wall Street rates, which high frequency traders use to train their AI models. In order to further automate their selling process, 'Business Intel Inc.' provides their dataset in an

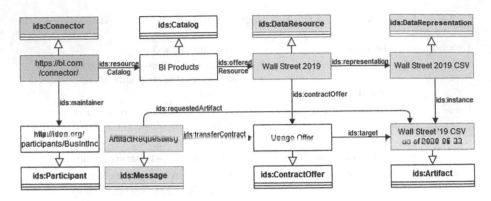

Fig. 3. IDS core classes and their instances in the running example.

IDS ecosystem, through their IDS Connector at https://bi.com/connector/ to ensure to (1) transforming existing data into economic value, while (2) restricting access and subsequent usage, and thus (3) ensuring their sovereignty over their data. These translate to four requirements:

(R1) Describe the data resource to make it discoverable (to potential, still unknown costumers)

(R2) Create business value through data exchange

(R3) Describe intended and prevent unintended usages

(R4) Control the usage over the complete digital life cycle

The description and announcement of the data resource (R1) is shown in Listing 1.[2] The unambiguous metadata is understood by every other participant in the IDS. In addition, as Attard et al. explain, added value from digital data can only be created through value co-creation [2]. Therefore, data resources must be made available at the right time to the right consumer. The requirement (R2) is fulfilled by the IDS infrastructure and the involved components, which are able to interpret a data resource based on its self-description and understand the described relations (cf. Fig. 3).

Listing 1. Stock market data modelled as an IDS DataResource.

```
_:StockData a ids:DataResource ;
  ids:title "Wall Street Stock Prices 2019"@en ;
  ids:description "This dataset contains the complete stock market prices
    of all 2019 Wall Street listed companies by milliseconds."@en ;
  ids:keyword "stock price", "Wall Street", "2019" ;
  ids:publisher <http://idsa.org/participants/BusIntInc>;
  ids:temporalCoverage [ a ids:Interval ;
    ids:begin [ a ids:Instant ;
      ids:dateTime "2019-01-01T00:00:00.000-04:00"^^xsd:dateTimeStamp ];
    ids:end [ a ids:Instant ;
```

---

[2] We abbreviate URIs following http://prefix.cc/.

```
    ids:dateTime "2019-12-31T23:59:59.999-04:00"^^xsd:dateTimeStamp]];
ids:language idsc:EN ;
ids:representation [ ids:instance _:StockDataCSV ; ids:mediaType
  <https://www.iana.org/assignments/media-types/text/csv>];
ids:resourceEndpoint [ a ids:ConnectorEndpoint ; ids:accessURL
  "https://bi.com/connector/reports/2019_wall_street.csv"^^xsd:anyURI];
ids:contractOffer _:StockDataOffer .
```

3 Methodology

3.1 Design Principles

The IDS overall has been designed as an alliance-driven multi-sided platform [17]. The basic process is aligned with the *eXtreme Design* method [21] with a strong focus on agile and collaborative workflows. The role of a customer is filled through a dedicated *ontology owner*, an experienced ontology expert who acts as the link to the developer community. In addition, the IDS IM is driven by the initial requirements originally collected and described in the RAM [18], and later on represented through publicly accessible *issues*. Furthermore, the *eXtreme Design* proposal to use separate ontology modules has let to the partitions shown in Fig. 1. As demanded by [10], the ontology development process needs to be test-driven, which is implemented by an automated syntax validation together with a semi-automated code generation pipeline (cf. Sect. 5.1). This code is integrated into several runtime components, in particular IDS Connectors, serving as test environment for each and every update.

Deep integration with state of the art software development platforms (Git, continuous integration, build agents, sprint-based development) enables an agile, iterative release management. Combining these characteristics with Semantic Web best practices led to the core design principles of the IDS IM:

Reuse: The body of existing work is evaluated and reused by refining terms of standard vocabularies, many of them being W3C Recommendations.
Linked Data: The IM is published under a stable namespace, in common RDF serialisations together with a human-readable documentation and interlinked with external resources.
FAIR: The ontology as a whole follows the FAIR principles (findable, accessible, interoperable, reusable [23]).
Separation of concerns: Each module of the ontology addresses a dedicated concern that applies to a digital resource (cf. Fig. 1).

3.2 Maintenance and Update Process

As stated, the IM's development within the IDSA SWG4 follows an agile methodology involving different stakeholder groups. Interested IDSA members support the core modelling team by supplying domain knowledge, providing use cases

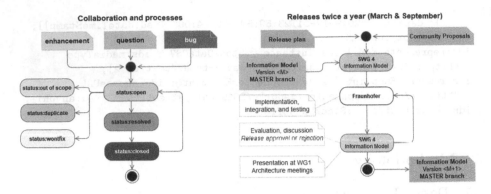

Fig. 4. The IDS IM update and release process.

and validating the model against their requirements. It has followers and contributors from 13 different IDSA member organisations and represents one of the most active IDSA communities.

The IM is provisioned in two parallel ways. The stable releases, reflecting major model updates, are provided once or twice a year (see also the *eXtreme Design's* integration loop). In the meantime, new features and bug fixes, which have been filed in the GitHub issue tracker, are addressed in monthly sprints as part of the module development loop. Those yield, besides incremental versions, nightly builds and snapshots. The overall process is depicted in Fig. 4. The community as a whole initiates change requests by creating tickets and proposing updates. The IDSA SWG4 then acts as the final authority, reviewing and merging the proposals.

3.3 Ontology Build Process and Quality Control

The continuous evolution of the IDS IM is supported by Continuous Integration and Continuous Deployment (CI/CD) mechanisms with automated quality assessment. As the IM is developed in its Turtle source code representation in a Git repository, CI/CD is realised similar to software projects. We currently run three test stages:

- A syntax check validates the syntactical correctness of all component files (currently 220 Turtle files).
- A reasoner[3] checks for logical inconsistencies, such as disjoint classes that subsume each other, or cycles in the class hierarchy.
- A set of RDFUnit[4] [13] test cases is used to find code smells and report common errors.

For new versions and releases, human-readable documentation is generated by semi-automatic invocations of Widoco, as explained below in Sect. 4. The

---

[3] Currently Pellet: https://github.com/stardog-union/pellet
[4] https://github.com/AKSW/RDFUnit.

Widoco process includes further quality checks provided by the OOPS! ontology pitfall scanner web service[5] [20], such as identifying broken links to external vocabularies ("namespace hijacking"). The quality is furthermore ensured by a code review process, where each change request must be evaluated and approved by at least one reviewer not involved in the creation of the change request. Major updates are additionally discussed in the IDSA SWG4 and require unanimous consent.

Instances of the IM can also be validated against its schema by using SHACL shapes. Every class has its corresponding shape, stating the required properties, their cardinality requirements and value types[6]. The shapes are used to (1) validate incoming data objects but also (2) to describe the restrictions on class attributes. Thereby, the SHACL representations are used both as the enabler for instance validation and as a further extension to the schema description, for instance for cardinality restrictions.

4 Implementation and Reuse of Standards

The declarative representation of the IDS IM is provisioned as 5-Star Linked Data and conforms to the FAIR principles. It is accessible under an open license, in a stable namespace and maintained in a public GitHub repository (cf. Table 1). Dereferencing the namespace URLs redirects the client to either a human-readable website supplying the ontology documentation page, generated in W3C style by Widoco[7] [8], or directly to a serialisation in one of the supported formats (RDF/XML, Turtle, JSON-LD, N-Triples). Further documentation is available at a public, read-only instance of the VoCol vocabulary collaboration environment. This instance includes views of the evolution of the ontology and a public SPARQL endpoint[8]. All classes, properties and instances are enhanced with descriptions and, wherever appropriate, links to further information sources.

4.1 Relations to External Ontologies

Terms from external ontologies are individually imported by extending the respective class (using *rdfs:subClassOf*) or property (*rdfs:subPropertyOf*), in order to adapt its axiomatisation (e.g., *rdfs:range*) or insufficient specification from the IDS context (cf. Table 1 and Fig. 1). The adoption of external concepts into the IDS namespace is necessary, as those concepts are further refined according to the IDS characteristics and facets. For instance, the *DigitalContent* class captures the type and semantics of a binary content in an abstract, format-independent way, extending *dcat:Dataset*. Among others, it records the

---

[5] http://oops.linkeddata.es.
[6] https://github.com/International-Data-Spaces-Association/InformationModel/tree/master/testing.
[7] https://zenodo.org/record/3519946.
[8] https://vocol.iais.fraunhofer.de/ids/.

Table 2. Examples for Information about Data Structure and Semantics.

Information	Standard	User Story
Use of vocabularies	VoID (*void:vocabulary, void:classPartition, void: propertyPartition*, etc.)	"This resource mainly contains information about average and minimum temperatures," or "This resource mainly contains instances of the W3C SOSA/SSN sensor data ontology"
Data structure	Data Cube (*qb:structure, qb:component, qb:dimen-sion, qb:attribute*, etc.)	"This resource consists of a three-dimensional matrix with temperature measurements in degrees centigrade in the dimensions 1. time, 2. geo-coordinates, and 3. sensor used"
Graph structure	SHACL (*sh:shapesGraph*)	"This resource contains measurements of average and minimum temperature in a specific place at a specific time, measured by sensor X"

context in terms of spatial, temporal and real-world entity coverage, the (SKOS) concepts related to the content (theme), and the provenance of the content by leveraging the PROV-O[9] vocabulary. IDS components, however, require specific attributes and relations, which are not stated – and not intended to be – in the original vocabularies.

4.2 Expressing Data Structure and Domain-Specific Semantics

The IDS IM is independent of concrete application domains and thus does not provide terminology for the *content* of data resources. However, as the IDS encourages interoperability and extensible ecosystems, it encourages the use of RDF and domain ontologies for Representations (cf. [22] for a sample scenario using a taxonomy of steel grades). In this context, it is desirable to include information about the domain-specific semantics and, similarly, the structure of content into the metadata of a Resource or some of its Representations – for example, to be able to retrieve more relevant data resources. To this end, the IM reuses VoID, the Data Cube Vocabulary, and SHACL[10], as explained in Table 2 and detailed by examples in the GitHub repository.

5   Tool Support

While the IDS IM serves as the shared language throughout a data space, its adoption is usually challenging for component developers not familiar with the Semantic Web. A set of tools therefore supports the implementers and aims at preventing pitfalls as much as possible.

---

[9] https://www.w3.org/TR/prov-o/.
[10] https://www.w3.org/TR/ {void,vocab-data-cube,shacl}/.

```
DataResource metadata = new DataResourceBuilder()
  ._title_(Util.asList(new TypedLiteral("Wall Street ... 2019","en")))
  ._description_(Util.asList(new TypedLiteral("This dataset...", "en")))
  ._keyword_(Util.asList(new PlainLiteral("stock price"), [...]))
  ._publisher_(URI.create("http://idsa.org/participants/BusIntInc"))
  ._temporalCoverage_(Util.asList(new IntervalBuilder().[...].build()))
  ._language_(Util.asList(Language.EN))
  ._representation_(Util.asList(new RepresentationBuilder()[...]))
  ._resourceEndpoint_(Util.asList(new ResourceEndpointBuilder()[...]))
  ._contractOffer_(<data_restrictions>).build();
```

Listing 2. Java representation of the running example.

5.1 Java API to Generate Instances

Instantiating the IDS IM concepts is crucial when running IDS components in practice, e.g., when sending messages, creating metadata descriptions or specifying usage restrictions (cf. Listing 2). Developers of Connectors within the early IDS projects found it inconvenient and error-prone to create these instances directly on the level of RDF data structures. Therefore, a software stack has been developed to transform the declarative representation of the IDS IM into a Java class library. The code generation process takes the ontology's Turtle source files as input and automatically validates, compiles and pushes the Java library files. To the best of our knowledge, the IDS IM is the only ontology with such a representation directly in executable code.

The Java API is publicly deployed via two channels: it is pushed to a Maven repository[11], and a nightly release is made available as a ZIP file on GitHub[12] as a part of the CI/CD pipeline (cf. Sect. 3.3). Besides the Java API, this ZIP file contains the Turtle sources, an UML-like visualisation of the ontology as well as a parser and serializer for IM instances[13]. This package helps to onboard developers faster and to give them as much support as possible. As the adoption in the developer community is crucial for the success of the IDS in general and the IDS IM in particular, we have also created a thin web application which guides the user through the modelling process.

5.2 GUI for Instance Management

This so-called IDS Semantic Instance Manager (cf. Fig. 5) supports non-RDF expert developers and system architects in expressing their required entities by

---

[11] https://maven.iais.fraunhofer.de/artifactory/eis-ids-public/de/fraunhofer/iais/eis/ ids/infomodel/java/.
[12] https://github.com/International-Data-Spaces-Association/InformationModel/ releases.
[13] Demo project available at https://jira.iais.fraunhofer.de/stash/projects/ICTSL/ repos/ids-infomodel-demo/browse.

a template-based GUI driven by the Java API introduced in Sect. 5.1. Instances can be exported to the common RDF serialisations. In the case of Messages, they can be sent directly to the target, thus turning the Semantic Instance Manager into a GUI for interactive control of a data space. Most IDS concepts have required properties, for example, a timestamp when the message was issued or the URI of the issuer. The GUI supports and guides the users to formulate valid IM instances, thus drastically lowering the entry barrier for constructing, e.g., messages, component and data descriptions or usage policies.

Changes in the evolving IM, such as the introduction of new message types, do not require adapting the GUI, as it is dynamically built from the Java library using reflection.

Fig. 5. IDS Semantic Instance Manager GUI.

6 Adoption

This section gives a brief overview of common use cases in which the IDS IM enables semantic interoperability in data spaces. The adoption processes in general are organised in five main verticalisation initiatives, which map the generic IDS specifications with the domain-specific requirements. These initiatives involve industrial manufacturing community, which is strongly related with the Plattform Industrie 4.0 and the Industrial Internet Consortium, the medical, energy, and material data space, as well as IDS in Smart Cities. In addition, at least seven commercially driven implementation processes are known to the authors. For instance, the public tender on re-implementing the German national mobility data platform explicitly enforces the IDS specifications[14] to ensure a self-sovereign landscape of equally empowered participants.

We regard the IDS IM as a reference ontology for trustworthy, data-driven architectures. It is a cornerstone of any IDS-related implementation and thus used in all related publicly funded projects, and impacts several industry platforms. The IDSA highlights 14 real-world use cases, the majority of them being realised with an investment from companies and contributing to their business success; furthermore, 10 EU research projects alone involve the IDSA (plus several of its members).[15] On a more technical level, at least 11 different Connector

---

[14] https://www.evergabe-online.de/tenderdetails.html?0&id=322425.
[15] https://www.internationaldataspaces.org/success-stories/.

implementations with explicit support for a defined IDS IM version are currently known to the authors. Further adoption of the IDS IM among component developers is fostered at the quarterly IDSA Plugfest[16].

6.1 Community of Trust and Usage Control

Technically implemented trust and data sovereignty are at the heart of the IDS. Unambiguous description of usage restrictions and definition of the required attributes are therefore one of the most important use cases of the IDS IM. As the vocabulary presents the shared understanding of all involved parties – combining the different domains to one consistent ecosystem – it connects their security, certification, governance and interoperability models with each other.

Key challenges in the context of data usage control are the formal description of permissions and obligations. In our example, the Business Intel Inc. is able to present its intended restrictions in terms of a machine-readable policy (see R3). The Open Digital Rights Language (ODRL [11]) provides the terms and concepts for these statements. The IDS IM further details these constructs and defines their implications, focusing on their publication, negotiation, acknowledgement and enforcement. These additional steps enhance the solely *descriptive* ODRL vocabulary to legally *binding and enforceable* statements. Thus, the IDS IM not only allows to state permissions, e.g., that a data asset can be read (*ids:Permission* to *idsc:READ*) by certain users, but it can also express them in decidable terms for usage control engines such as MyData[17]. Such tools independently evaluate the agreed usage policies and, for instance, grant or deny access to individual resources. Modelling usage policies, contracts and the mappings between declarative and technically enforceable policies is a crucial prerequisite for the implementation of the IDS value proposition, to maintain the complete *sovereignty* of data owners with regard to their content.

Listing 3 shows a policy for the example data resource. The IDS IM is – to the best of our knowledge – the only vocabulary to cover the actual enforcement of usage restrictions. Established standard languages, for instance XACML, only focus on access control or, as for instance ODRL, only allow the description and exchange of policies. The IDS IM closes this gap with detailed instructions on how to interpret each attribute, how to resolve statements and how to relate given policies to a system environment [3]. This is one aspect of solving R4. Listing 3 further shows how R2 (Business Value) is expressed. The *postDuty* clause describes and enforces a compensation for using the dataset, thereby combining business and data security statements in one representation.

Furthermore, the clear semantic of the allowed Action (READ) tells every interested buyer that the usage in its own IT landscape is covered (R3.1: describe intended usages), while any further distribution or reselling will and must be prohibited by the Usage Control Framework (R3.2: prevent unintended usage). The contract gives the Business Intel Inc. the tool to enforce its business model

---

[16] https://www.internationaldataspaces.org/get-involved/#plugfest.
[17] https://www.mydata-control.de/.

Listing 3. Exemplary policy (*ids:ContractOffer* class; cf. Fig. 3) that grants read access to members of a certain organisation.

```
_:StockDataOffer a ids:ContractOffer ;
    ids:permission [ ids:target _:StockData ;
        ids:action idsc:READ ;
        ids:constraint [ ids:leftOperand idsc:USER;
            ids:operator idsc:MEMBER_OF;
            ids:rightOperandReference <http://whiterock-invest.com/> ];
        ids:postDuty [
            ids:action [ ids:includedIn idsc:COMPENSATE ;
                ids:actionRefinement [ ids:leftOperand idsc:PAY_AMOUNT ;
                    ids:operator idsc:EQ ;
                    ids:rightOperand "5000000"^^xsd:double ] ] ] [...] ] .
```

in the technical landscape of the customer (R4: control over the complete life cycle), which of course must also be supported by the execution environment of the customer. This however is ensured by the signed certification claims and can be checked on the fly.

6.2   Trust Through Certified Attribute Declarations

In order to evaluate these claims, participants and components are subject to a *certification process* – an additional means to establish trust within and across data spaces. Organisational structures, methodologies, and standards underlying that process are detailed in [12]. A normative, tamper-proof reference of certification, security and identity attributes is maintained by IDS infrastructure components, operated as part of an Identity Provider (cf. Fig. 2).

Being part of the Public Key Infrastructure (PKI), this component augments the core attributes of an identity proof by a set of dynamic IDS-specific attributes. These attributes are defined in the IM for purposes of a single-truth maintenance, bridging the gap between the certification process during the *design time* of components, the security onboarding at *deployment time*, and the automated validations during interactions at *runtime*, not only at a data provider but at any intermediary IDS system throughout the complete data life cycle (R4). As Listing 4 shows, the basic JSON Web Token structure from RFC 7519 has been extended with additional attributes for usage control systems. Most relevant is here the *securityProfile* property, which contains crucial information on the trustworthiness of the target system. To the best of our knowledge, no other data model supports such a holistic approach and combines the various requirements.

Listing 4. Serialised Dynamic Attribute Token (DAT) in JSON-LD.

```
{ "@context" : "https://w3id.org/idsa/contexts/context.jsonld",
  "@id" : "http://w3id.org/idsa/DatPayload/A51317560",
  "@type" : "ids:DatPayload",
```

```
"referringConnector" : { "@id": "http://bi.com/connector" },
"iss": "65:43:5D:E8...:keyid:CB:8C:...AE:2F:31:46",
"sub": "65:43:5D:E8...:keyid:CB:8C:...DD:CB:FD:0B",
"iat": 1589982066, "nbf": 1590154866, "exp": 1590759666,
"aud": { "@id": "idsc:IDS_CONNECTOR_ATTRIBUTES_ALL" },
"scope": "ids_connector_attributes",
"securityProfile": { "@id": "idsc:BASE_SECURITY_PROFILE" }}
```

7 Related Work

Several consortia have been formed to standardise (industrial) data exchange. The most prominent ones so far include the German Plattform Industrie 4.0 (PI4.0) and the US-American Industrial Internet Consortium (IIC). The PI4.0 focuses on physical assets and provides an extensive data model, called the Asset Administration Shell [4]. Nevertheless, this model does not sufficiently reflect the requirements of sovereign data interactions. The IIC focuses on the aspects of interoperable systems and architectures but also specifies a brief vocabulary [5], intended to enable discussions between experts but not to serve as a formal information model for a machine to machine interactions.

A huge amount of semantic description languages for interfaces and federated systems has been proposed. The SOAP technology stack and its service description language WSDL has been extended with the WSMO and WSMO-Light ontologies [6]. OWL-S is a similar OWL-based ontology for semantic descriptions of services. Furthermore, description languages for REST APIs have recently gained popularity, most prominently OpenAPI[18]. The IDS IM's definition of an *Interface*, e.g., of a Resource, is technology-agnostic, comparable to Web Service Interfaces in WSDL 2.0[19] and the concept of Service Profiles in OWL-S ontology[20]. Still, the focus is on the functionality of the endpoints itself, disregarding the challenges proposed through data protection and trust requirements.

The Data Catalog Vocabulary (DCAT) [14] is a related W3C Recommendation making use of well-established vocabularies to describe the distribution of (static) data sets. The limited expressivity of DCAT 1 was a major motivation for the IDS IM to extend it by versioning or temporal and spatial context. DCAT also neither includes relations to originating organisations nor allows for the description of data-related service APIs. *dcat:DataService* is just one example of how many of these limitations have recently been addressed with DCAT 2[21].

In addition to plain description languages, several ecosystems have been designed to seamlessly exchange data. bIoTope[22] aims at enabling interoperability between *vertical IoT silos* via a standardised open API. Data integration is supposed to be based on vocabularies to describe the different data sources.

---

[18] https://swagger.io/docs/specification/about/.
[19] https://www.w3.org/TR/wsdl20/#Interface.
[20] https://www.w3.org/Submission/OWL-S/#4.
[21] https://www.w3.org/TR/vocab-dcat-2/.
[22] http://www.biotope-project.eu.

FIWARE[23] provides data through a RESTful API with RDF semantics called NGSI-LD[24]. Besides the claim to reduce JSON payload costs and a full REST adoption, it offers a more powerful query language, especially for geospatial queries. FIWARE has been used to implement the IDS architecture [1]. Being RESTful, NGSI-LD serves a different purpose than the message-based IDS IM; however, they have in common the "Context" concern of data, e.g., in a spatio-temporal sense. Nevertheless, these ecosystems do not sufficiently express the conditions and restrictions imposed through digital information exchange.

The terminology of authorisations, obligations, and conditions introduced by the influential $UCON_{ABC}$ [19] usage control model has been adopted by many later models. Together with RFC 2904 and the introduction of the different *policy points*, these two works form the theoretical foundation of usage control. However, neither proposes a vocabulary to specify distinct permissions or prohibitions. This task is, to some, degree covered by XACML [15,16]. Still, XACML only focuses on *access* control, not on the more holistic usage control.

The Data Privacy Vocabulary (DPV[25]) provides terms to annotate and categorise instances of legally compliant personal data handling according to the GDPR, including the notions of data categories, data controllers, purposes of processing data, etc. We are considering it as a candidate for extending the IDS IM by terminology for describing privacy aspects of data or software resources.

8   Conclusion and Future Work

We introduced an Information Model for data space ecosystems with a focus on supporting data sovereignty. We described how to support model development, documentation, and usage by different representations for various groups of stakeholders. We further demonstrated the usage of design principles that helped us to advance state-of-the-art models underlying our work.

The IDS IM is available openly on GitHub and comprises the patterns and features necessary to describe and implement digital sovereignty in a federated ecosystem. It shows how semantic technologies can be enhanced with security and trust to pave the way for enforceable, self-determined, i.e., sovereign data management across organisations. The comprehensive view on the challenge of addressing data owners' legitimate concerns while enabling productive data usage by other parties is a requirement for upcoming data-driven business cases.

Following the described contribution methodology, the IM is continuously evolved with industry stakeholders via the IDSA. We thus ensure that it is in line with emerging requirements of data ecosystems concerned with maintaining data sovereignty down to implementation specifications. Thus, the IM also promotes Semantic Web standards in disciplines where there is little awareness so far.

---

[23] https://www.fiware.org.
[24] https://fiware-datamodels.readthedocs.io/en/latest/ngsi-ld_howto/.
[25] https://www.w3.org/ns/dpv.

Next steps include developing tools for automated extraction of IDS IM metadata from the content of data resources, and to fully support retrieving data resources with a defined structure or domain-specific semantics.

Acknowledgements. This research was funded by the German Federal Ministries of Education and Research (grant number 01IS17031) and Transport and Digital Infrastructure (19F2083A), the EU H2020 projects BOOST4.0 (780732), DEMETER (857202) and TRUSTS (871481) and the Fraunhofer Cluster of Excellence "Cognitive Internet Technologies". We thank Eva Corsi, Anna Kasprzik, Jörg Langkau and Michael Theß, who have also contributed to the IDS IM via the IDSA SWG4.

References

1. Alonso, Á., Pozo, A., Cantera, J., de la Vega, F., Hierro, J.: Industrial data space architecture implementation using FIWARE. Sensors **18**(7), 2226 (2018)
2. Attard, J., Orlandi, F., Auer, S.: Data value networks: enabling a new data ecosystem. In: International Conference on Web Intelligence (WI). IEEE (2016)
3. Bader, S.R., Maleshkova, M.: Towards Enforceable Usage Policies for Industry 4.0. LASCAR Workshop at ESWC (2019). http://ceur-ws.org/Vol-2489/
4. Boss, B., et al.: Details of the Asset Administration Shell Part 1. Technical report, ZVEI (2019)
5. Bournival, E., et al.: The Industrial Internet of Things Vocabulary. IIC (2019). https://hub.iiconsortium.org/vocabulary. Accessed 28 Nov 2019
6. Domingue, J., Roman, D., Stollberg, M.: Web Service Modeling Ontology (WSMO)-An Ontology for Semantic Web Services (2005)
7. Finn, R., Wadhwa, K., Grumbach, S., Fensel, A.: Byte Final Report and Guidelines. Technical report D7.3, BYTE Project (2017). http://new.byte-project.eu/wp-content/uploads/2014/02/D7.3-Final-report-FINAL.pdf
8. Garijo, D.: WIDOCO: a wizard for documenting ontologies. In: d'Amato, C., et al. (eds.) ISWC 2017. LNCS, vol. 10588, pp. 94–102. Springer, Cham (2017). https://doi.org/10.1007/978-3-319-68204-4_9
9. Grumbach, S.: Intermediation platforms, an economic revolution. ERCIM News **2014**(99) (2014)
10. Hitzler, P., Gangemi, A., Janowicz, K.: Ontology Engineering with Ontology Design Patterns: Foundations and Applications, vol. 25. IOS Press (2016)
11. Ianella, R., Villata, S.: ODRL Information Model 2.2. Technical report, W3C ODRL Community Group (2018). https://www.w3.org/TR/odrl-model/
12. IDSA: Whitepaper certification. Technical report, IDSA (2018). https://www.internationaldataspaces.org/publications/whitepaper-certification/
13. Kontokostas, D., et al.: Test-driven evaluation of linked data quality. In: WWW (2014)
14. Maali, F., Erickson, J., Archer, P.: Data Catalog Vocabulary (DCAT). Technical report, W3C (2014). https://www.w3.org/TR/vocab-dcat/
15. Mazzoleni, P., Crispo, B., Sivasubramanian, S., Bertino, E.: XACML policy integration algorithms. Trans. Inf. Syst. Secur. **11**(1) (2008)
16. Moses, T., et al.: eXtensible Access Control Markup Language (XACML). OASIS Standard, February 2005
17. Otto, B., Jarke, M.: Designing a multi-sided data platform: findings from the International Data Spaces case. Electron. Markets **29**(4), 561–580 (2019). https://doi.org/10.1007/s12525-019-00362-x

18. Otto, B., et al.: Reference Architecture Model. IDSA (2019). https://www.internationaldataspaces.org/ressource-hub/publications-ids/, version 3.0
19. Park, J., Sandhu, R.: The UCON ABC usage control model. ACM Trans. Inf. Syst. Secur. (TISSEC) **7**(1), 128–174 (2004)
20. Poveda-Villalón, M., Gómez-Pérez, A., Suárez-Figueroa, M.C.:Oops!(ontology pitfall scanner!): an on-line Tool for Ontology Evaluation. IJSWIS **10**(2) (2014)
21. Presutti, V., Daga, E., Gangemi, A., Blomqvist, E.: eXtreme design with content ontology design patterns. In: Proceedings of the Workshop on Ontology Patterns (2009)
22. Pullmann, J., Petersen, N., Mader, C., Lohmann, S., Kemeny, Z.: Ontology-based information modelling in the industrial data space. In: ETFA. IEEE (2017)
23. Wilkinson, M.D., et al.: The FAIR guiding principles for scientific data management and stewardship. Sci. Data **3** (2016)

LDflex: A Read/Write Linked Data Abstraction for Front-End Web Developers

Ruben Verborgh$^{(\boxtimes)}$ⓘ and Ruben Taelmanⓘ

IDLab, Department of Electronics and Information Systems,
Ghent University – imec, Ghent, Belgium
{ruben.verborgh,ruben.taelman}@ugent.be

Abstract. Many Web developers nowadays are trained to build applications with a user-facing browser front-end that obtains predictable data structures from a single, well-known back-end. Linked Data invalidates such assumptions, since data can combine several ontologies and span multiple servers with different APIs. Front-end developers, who specialize in creating end-user experiences rather than back-ends, thus need an abstraction layer to the Web of Data that integrates with existing frameworks. We have developed LDflex, a domain-specific language that exposes common Linked Data access patterns as reusable JavaScript expressions. In this article, we describe the design and embedding of the language, and discuss its daily usage within two companies. LDflex eliminates a dedicated data layer for common and straightforward data access patterns, without striving to be a replacement for more complex cases. The use cases indicate that designing a Linked Data developer experience—analogous to a user experience— is crucial for adoption by the target group, who in turn create Linked Data apps for end users. Crucially, simple abstractions require research to hide the underlying complexity.

1 Introduction

Other than in the beginning days of the Semantic Web, user-facing Web applications nowadays are often built by a dedicated group of specialists called *front-end developers*. This specialization resulted from an increasing maturity of the field of Web development, causing a divergence of skill sets among back-end and front-end developers, as well as different technologies and tool stacks. The current Semantic Web technology stack, in contrast, focuses mostly on back-end or full-stack developers, requiring an intimate knowledge about how data is structured and accessed. A dormant assumption is that *others* will build abstractions for front-end developers [29], whereas designing an adequate developer experience requires a deep understanding of Semantic Web technologies.

If we want front-end developers to build applications that read and write Linked Data, we need to speak their language and equip them with abstractions that fit their workflow and tooling [22]. Crucially, we do *not* see this as a matter

© Springer Nature Switzerland AG 2020
J. Z. Pan et al. (Eds.): ISWC 2020, LNCS 12507, pp. 193–211, 2020.
https://doi.org/10.1007/978-3-030-62466-8_13

of "dumbing down" SPARQL or RDF; rather, we believe it revolves around appropriate primitives for the abstraction level at which front-end applications are developed, similar to how SQL and tables are not front-end primitives either. The keyword is *proportionality* rather than convenience: the effort to access a certain piece of data should be justifiable in terms of its utility to the application.

The difficulty lies in finding an abstraction that hides irrelevant RDF complexities, while still exposing the unbounded flexibility that Linked Data has to offer. Abstractions with rigid objects do not suffice, as their encapsulation tends to conceal precisely those advantages of RDF. Tools should instead enable developers to leverage the power and harness the challenges of the open Web. This empowerment is especially important in decentralized environments such as the Solid ecosystem [26], where data is spread across many sources that freely choose their data models. Building for such a multitude of sources is significantly more complex than interfacing with a single, controlled back-end [22].

This article discusses the design, implementation, and embedding of *LDflex*, a domain-specific language that exposes the Web of Linked Data through JavaScript expressions that appear familiar to developers. We discuss its requirements and formal semantics, and show how it integrates with existing front-end development frameworks. Rather than striving for full coverage of all query needs, LDflex focuses on simple but common cases that are not well covered by existing Semantic Web technologies (which remain appropriate for complex scenarios). We examine the usage of LDflex within two companies, and study its usage patterns within production code in order to assess its application in practice.

2   Related Work

Querying Data on the Web. SPARQL queries [9] carry *universal semantics*: each query maintains a well-defined meaning across data sources by using URIs rather than local identifiers, making queries independent of their processing. In theory, this enables reuse across different data sources; in practice, ontological differences need bridging [25,29]. Although very few Web developers have experience with RDF or SPARQL, query-based development has been gaining popularity because of the GraphQL language [11]. While integrating well with existing development practices, GraphQL queries lack universal semantics, so applications remain restricted to specific sources. Several Semantic Web initiatives focused on providing *simpler* experiences. For example, EasierRDF [6] is a broad investigation into targeting the "average developer", whereas the concrete problem of simplifying query writing and result handling is tackled by SPARQL Transformer [13]. However, we argue that the actual need is not primarily simplification of complex cases, since many front-end developers have a sufficient background to learn RDF and SPARQL. The problem is rather a mismatch of abstraction level, because even conceptually simple data access patterns currently require a technology stack that is considered foreign.

Programming Abstractions for RDF. Programming experiences over RDF data generally fall in one of two categories: either a library offers generic interfaces that represent the RDF model (such as `Triple` and `Literal`), or a framework provides an abstraction inside of the application domain (such as `Person` or `BlogPost`). The latter can be realized through object-oriented wrappers, for instance via *Object–Triple Mapping* (OTM) [12], analogous to *Object-Relational Mapping* (ORM) for relational databases. However, whereas a table in a traditional database represents a *closed* object with a *rigid* structure, representations of RDF resources on the Web are *open* and can take *arbitrary* shapes. As such, there exists an *impedance mismatch* between the object-oriented and resource-oriented worlds [5]. Furthermore, local objects do not provide a good abstraction for distributed resources [30], which developers necessarily encounter when dealing with Linked Data on the Web.

Domain-Specific Languages for Querying. A Domain-Specific Language (DSL) is a programming language that, in contrast to general-purpose languages, focuses on a specific purpose or application domain, thereby trading generality for expressiveness [15]. A DSL is *external* if it has a custom syntax, whereas an *internal* DSL is embedded within the syntax of a host language [8]. For example, the scripting language *Ripple* is an external DSL for Linked Data queries [19]. Inside of another language, external DSLs are typically treated as text strings; for instance, a SPARQL query inside of the Java language would typically not be validated at compile time. Internal DSLs instead blend with the host language and reuse its infrastructure [8]. A prominent example of an internal query DSL is *ActiveRecord* within the Ruby language, which exposes application-level methods (such as `User.find_by_email`) through the *Proxy* pattern [16]. The *Gremlin* DSL [18] instead uses generic graph concepts for traversal in different database implementations.

JavaScript and its Frameworks. Since JavaScript can be used for both front-end and back-end Web application development, it caters to a large diversity of developers in terms of skills and tool stacks. A number of different frameworks exist for front-end development. The classic jQuery library [4] enables browser-agnostic JavaScript code for reading and modifying HTML's Document Object Model (DOM) via developer-friendly abstractions. Recently, frameworks such as *React* [17] have been gaining popularity for building browser-based applications. JavaScript is single-threaded; hence, costly I/O operations such as HTTP requests would block program execution if they were executed on the main thread. JavaScript realizes parallelism through *asynchronous* operations, in which the main thread delegates a task to a separate process and immediately resumes execution. When the process has finished the task, it notifies the JavaScript thread though a *callback function*. To simplify asynchronous code, the `Promise` class was recently introduced into the language, with the keywords `async–await` as syntactical sugar [14].

JavaScript and RDF. Because of its ubiquitous embedding in browsers and several servers, the JavaScript programming language lends itself to reusing the same RDF code in server-side and browser-based Web apps. For compatibility, the majority of RDF libraries for JavaScript conform to API specifications [2] created by the W3C Community Group RDF/JS. The modular *Comunica* query engine [23] is one of them, providing SPARQL query processing over a federation of heterogeneous sources. JavaScript also gave birth to the *JavaScript Object Notation* (JSON), a widely used data format even in non-JavaScript environments. The JSON-LD format [20] allows adding universal semantics to JSON documents by mapping them to RDF. JSON-LD allows JSON terms to be interpreted as URIs using a given JSON-LD *context* that describes term-to-URI mappings. Since JSON-LD contexts can exist independently of JSON-LD documents, they can be reused for other purposes. For example, GraphQL-LD [24] leverages them to add universal semantics to GraphQL. The object-oriented abstractions *SimpleRDF*[1] and *RDF Object*[2] provide access to RDF data by applying JSON-LD contexts to regular objects. Instead of per-property access, *Soukai Solid*[3] considers entire RDF data shapes. All of the aforementioned abstractions require preconfiguring the context or shape and preloading an RDF graph in memory before data can be accessed in an object-oriented manner, limiting them to finite graphs.

3   Requirements Analysis

This section lists the main requirements of LDflex for achieving the goal of a read/write Linked Data abstraction for front-end developers.

R1: Separates Data and Presentation. Because of specialization and separation of concerns, front-end developers who work on the *presentation layer* typically should not come in contact with the data storage layer or its underlying database. Instead, they usually retrieve data through a *data access layer*, which is a higher-level abstraction over the storage layer that hides complexities of data storage that are irrelevant to front-end developers. For example, front-end developers could use a framework that exposes the *active record* architectural pattern instead of manually writing SQL queries for accessing relational databases. While SPARQL queries can abstract data access over a large variety of RDF interfaces, repeated SPARQL patterns in the presentation layer can become cumbersome to write. Specifically, we need a solution to capture repeated access to simple data patterns that occur frequently in typical front-end applications.

R2: Integrates into Existing Tooling. The high tempo at which front-end Web development happens is only possible through specific workflows and tools used by front-end developers. Any solution needs to fit into these workflows, and provide compatibility with these tools. For instance, popular front-end frameworks such as React are based on composable, stateful components. In order for an

---

[1] https://github.com/simplerdf/simplerdf.
[2] https://github.com/rubensworks/rdf-object.js.
[3] https://github.com/NoelDeMartin/soukai-solid.

abstraction layer to be useful, it must be able to integrate directly with such frameworks, without requiring further manual work.

R3: Incorporates the Open World. Since relational databases contain a finite number of data elements with a fixed schema, data access layers for relational databases can be *static* and based on a fixed set of properties and methods. In contrast, there is always more RDF data to be found for a given resource, and RDF data shapes can exist in various ontologies. Therefore, a data access layer for RDF must be *dynamic* so that it can handle arbitrarily shaped RDF data and various ontologies at runtime.

R4: Supports Multiple Remote Sources. In addition to interacting with local RDF data, it is crucial that an abstraction can also seamlessly access Linked Data from remote sources, preserving the semantics of data. Furthermore, RDF data for one resource can be spread over *multiple* sources across the Web, especially in decentralized scenarios, so a solution must consider this distribution and its consequences for application development.

R5: Uses Web Standards. It is required to interoperate with different modes of data access and data interfaces. This means that solutions have to be compatible with existing Web standards, such as RDF, SPARQL, and HTTP protocols and conventions regarding caching and authentication. Furthermore, since the solution will need to be deployed inside of browsers, it needs to be written in (or be able to be compiled to) languages and environments that are supported by modern browsers, such as JavaScript or WebAssembly, and corresponding browser APIs.

R6: Is Configurable and Extensible. Since different applications have different data and behavioral demands, it must be possible to configure and extend the interpretation of the abstraction. On the one hand, developers must be able to configure the mapping from the data access layer to the RDF storage layer. On the other hand, developers must be able to customize existing features and to add new functionality, while controlling the correspondence with the storage layer.

4   Syntax and Semantics

Based on the above requirements, we have designed the LDflex DSL for JavaScript. We discuss related Web languages, and explain its syntactical design and formal interpretation.

4.1   Relation to Existing Languages

The LDflex language draws inspiration from existing path-based languages for the Web. The jQuery library [4] introduced a DSL for the traversal of HTML's Document Object Model, following the *Fluent Interface* pattern [7] with method chaining. For example, the expression `$('ol').children().find('a')` `.text()` obtains the anchor text of the first hyperlink in an ordered list. This DSL is *internal*, as it is embedded within its host language JavaScript. Like Gremlin, it is implemented on the *meta*-level, with built-in methods such as `children` and `attr` referring to abstract HTML constructs (child nodes and attributes) rather than concrete cases (such as `<li>`). We call the evaluation of jQuery paths *safe* because supported methods are always defined—even if intermediary results are missing. For example, for elements without child nodes, calling `element.children().children()` will not produce a runtime error but rather yield an empty set.

The JSON-LD format is a subset of JSON, which itself is a subset of JavaScript. When a JSON-LD document is parsed into main memory during the execution of a JavaScript program, the resulting object can be traversed by chaining property accessors into a path. For instance, given a parsed JSON-LD document stored in a `person` variable, the expression `person.supervisor.department.label` could— depending on the object's JSON-LD context [20] and frame [21]—indicate the department label of a person's supervisor. Like jQuery paths, JSON-LD paths are valid JavaScript expressions and thus form an *embedded* DSL. In contrast to jQuery, JSON-LD paths use *data*-level constructs that refer to a concrete case (such as `supervisor` or `department`), rather than common metamodel concepts shared by all cases (such as `subject` or `predicate`). Evaluation is *unsafe*: syntactically valid paths might lead to runtime errors if an intermediate field is missing. For example, missing or incomplete data could lead the JSON-LD path segments `supervisor` or `department` to be `undefined` and therefore cause the evaluation to error.

4.2   Syntactical Design

Generic Syntactical Structure. To achieve the requirements derived in Sect. 3, we combine the data-level approach of JSON-LD with the safe evaluation from jQuery, leveraging the Fluent Interface pattern [7]. LDflex adopts the syntax of JSON-LD paths consisting of consecutive property accesses, whose set of names is defined by a JSON-LD context [20] that also lists the prefixes. It follows the jQuery behavior that ensures each syntactically valid path within a given context results in an errorless evaluation. It provides extension points in the form of custom properties and methods to which arguments can be passed.

The grammar in Listing 1 expresses the syntax of an LDflex path in Backus– Naur form with start symbol ⟨path⟩. The terminal *root* is a JavaScript object provided by an LDflex implementation. *short-name* corresponds to JSON-LD *term*, *prefix* to JSON-LD *prefix*, *local-name* to JSON-LD *suffix* [20], and *arguments* is any valid JavaScript method arguments expression. Listing 2 displays examples of valid grammar productions.

In practice, several variations on this core grammar exist, leveraging the syntactical possibilities of JavaScript. For instance, an LDflex can be assigned to a variable, after which further segments can be added to that variable. In the remainder of this section, we focus on the core fragment of LDflex consisting of path expressions.

$$\langle\text{path}\rangle \models root \langle\text{segments}\rangle$$
$$\langle\text{segments}\rangle \models \epsilon \mid \langle\text{segment}\rangle \mid \langle\text{segment}\rangle \langle\text{segments}\rangle$$
$$\langle\text{segment}\rangle \models \langle\text{property-access}\rangle \mid \langle\text{method-call}\rangle$$
$$\langle\text{property-access}\rangle \models [\text{"} \; full\text{-}URI \; \text{"}] \mid [\text{"} \langle\text{shorthand}\rangle \text{"}] \mid . \; \langle\text{shorthand}\rangle$$
$$\langle\text{shorthand}\rangle \models prefix \_ local\text{-}name \mid short\text{-}name$$
$$\langle\text{method-call}\rangle \models short\text{-}name \; (\; arguments \;)$$

Listing 1. LDflex expressions follow a path-based syntax, detailed here in its Backus–Naur form.

```
 1  const blog = data["https://alice.example/blog/"];
 2  const comments = blog.blogPost.comment;
 3  const blogAuthor = blog.foaf_maker.givenName;
 4  displayItems(blog, comments, blogAuthor);
 5
 6  async function displayItems(topic, items, creator) {
 7    console.log(`Items of ${await topic.name} at URL ${await topic}`);
 8    console.log(`created by  ${await creator}:`);
 9    for await (const item of items)
10      console.log(`- ${item}: ${await item.name}`);
11  }
```

Listing 2. In this code, LDflex paths are used to collect all comments on posts from a given blog. (This interpretation assumes that the Schema.org JSON-LD context and foaf prefix are set.)

Usage as Paths. Multiple LDflex path syntax variations are displayed in lines 1 to 3 of Listing 2. Line 1 assumes the availability of a root object called data, on which we access a property whose name is a full URL. Since LDflex has safe evaluation, it guarantees that blog is not undefined for *any* arbitrary URL (the mechanism for which is explained in Sect. 5). Line 2 contains a continuation of the path from the previous line, using the blogPost and comment shorthands from the Schema.org JSON-LD context (assumed preset), which represent http://schema.org/blogPost and http://schema.org/comment, respectively. The prefix syntax is shown on Line 3, where foaf_maker represents http://xmlns.com/foaf/0.1/maker (assuming the foaf prefix has been preset). LDflex is JSON-LD-compatible, and thus also supports *compact* IRIs [20] in the familiar foaf:maker syntax. However, since property names containing colons need to be surrounded with brackets and quotes in JavaScript (["foaf:maker"]), we offer an alternative syntax that replaces the colon with either an underscore or dollar sign to bypass such escaping needs. Finally, those three paths are passed as arguments to a function call on line 4. Note in particular how these lines are syntactically indistinguishable from regular JavaScript code, even though they interact with Linked Data on the Web instead of local objects.

Resolution to a Value. On lines 1 to 4, LDflex expressions are treated purely as paths, which are created, extended, and passed around. An unresolved LDflex path *points* to values rather than representing those values itself. Obtaining the actual values involves one or more network requests, but since JavaScript is single-threaded, we cannot afford the mere creation of paths to consume such time. Instead, LDflex leverages a syntactical feature of JavaScript to explicitly trigger asynchronous resolution when needed: by placing the `await` keyword in front of an LDflex path, it resolves to the first value pointed to by the expression. Lines 7, 8 and 10 show this mechanism in action, where for instance `creator` (corresponding to `blog.foaf_maker.givenName`) is resolved to its value. Note how `topic` resolves to a URL because it points to a named node within the RDF model. We can retrieve a human-readable label for it by resolving the `list.name` path (line 7), which will resolve to its `http://schema.org/name` property. Importantly, in addition to representing values, resolved LDflex paths obtained using `await` still behave as regular LDflex paths to which additional segments can be added. This is exemplified on line 10, where the resolved path `item` is extended to `item.name` and in turn resolved via `await`.

Resolution to a Series. In several cases, resolving to a single value is preferred. For instance, even if multiple given names are specified for a person, displaying just one might be sufficient in a given context. In other contexts, all of its values might be needed. LDflex offers developers the choice between resolving to a singular value using `await`, or iterating over multiple values using the `for...await` syntactical construct. On line 9, the `items` path (corresponding to `blog.blogPost.comment`) is resolved asynchronously into a series of values. Every resulting `item` is a resolved LDflex path, that can be used as a raw value (`item` on line 10), or a regular LDflex path that is subsequently resolved (`await item.name` on line 10). In this example, the iteration variable `item` points to the URL of a comment, whereas `item.name` resolves to a human-readable label.

4.3 Formal Semantics

Similar to a JSON-LD document, the specific meaning of an LDflex expression is determined by the context in which it occurs. The resulting interpretation carries universal semantics. For LDflex expressions, the interpretation depends on the active LDflex *configuration* set by a domain expert, which includes settings such as:

- the definition and interpretation of the root path;
- the JSON-LD context used for resolving names into URIs;
- the definition of method calls and special properties.

In general, an LDflex expression consists of consecutive property accesses, representing a path from a known subject to an unknown object. For example, the path `data["https://alice.example/blog/"].blogPost.comment.name` formed by lines 1, 2 and 10 of Listing 2 corresponds to the SPARQL query

```
1  SELECT ?name WHERE {
2    <https://alice.example/blog/> <http://schema.org/blogPost> ?post.
3    ?post <http://schema.org/comment> ?comment.
4    ?comment <http://schema.org/name> ?name.
5  }
```

Listing 3. The expression `data["https://alice.example/blog/"].blogPost.comment.name` of Listing 2 is interpreted as a SPARQL query expressing "titles of comments on posts of a given blog".

displayed in Listing 3. This interpretation assumes a configuration that sets Schema.org as the JSON-LD context, and which interprets properties on the root node `data` as the URL of a subject resource. The configuration also expresses how data sources are selected. For instance, it might consider the document HTTPs://alice.example/blog/! as an RDF graph or as a seed for link traversal [10], or might look up a query interface at `https://alice.example/`. The query processing itself is handled entirely by an existing SPARQL query engine, and partial results can be cached and reused across LDflex paths for performance reasons.

We will now introduce a formal semantics for the set of LDflex expressions LDf. It is determined by an interpretation function $I_C\colon LDf \to Q \times S$, where Q is the set of SPARQL queries and S the set of source expressions over which SPARQL queries can be processed. Such an interpretation function can be instantiated by an LDflex configuration $C = \langle ctx, root, other \rangle$, where $ctx\colon \$ \to \mathcal{U}$ represents a JSON-LD context that maps JavaScript strings to RDF predicate URIs, and $root\colon \$ \to \mathcal{U} \times S$ a function that maps strings to a start subject URI and a source expression. The $other$ set is reserved for other interpretation aspects, such as built-in properties or method names (not covered here).

We consider an LDflex expression $e \in LDf$ as a list consisting of a root property $r \in \$$ and $n > 0$ property accessors $k_i \in \$$, such that $e = (r, k_1, \ldots, k_n) \in \$^n$. The result of its interpretation $I_C(e) = \langle q, s \rangle$ is defined as follows. The root property r is resolved to a URI $u_r \in \mathcal{U}$ using $root(r) = \langle u_r, s_r \rangle$. Every property string k_i is resolved to a URI u_i using ctx, such that $\forall i \in [1, n]\colon u_i = ctx(k_i)$. Then, we generate a set of n triple patterns $TP_e \subset (\mathcal{U} \cup \mathcal{V}) \times \mathcal{U} \times (\mathcal{U} \cup \mathcal{V})$, with $tp_i = \langle s_i, p_i, o_i \rangle \in TP_e$ conforming to the constraints:

- the subject of the first pattern is the root property's URI: $s_1 = u_r$
- the predicates correspond to mapped JSON-LD properties: $\forall i \in [1, n]\colon p_i = u_i$
- the objects are unique variables: $\forall i, j \in [1, n]^2\colon o_i \in \mathcal{V} \wedge i \neq j \Rightarrow o_i \neq o_j$
- the objects and subjects form a chain of variables: $\forall i \in [2, n]\colon s_i = o_{i-1}$

These triple patterns form the SPARQL query q returned by I_C, which is a SELECT query that projects bindings of the basic graph pattern TP_e to the variable o_n. The second element s_r from the result of $root$ determines the returned data source $s = s_r$ for query evaluation. The example in Listing 3 is obtained from its LDflex expression with $v \mapsto \langle v, v \rangle$ as $root$.

Importantly, this semantics ensures that the creation of any sequence of path segments always succeeds—even if no actual RDF triples exist for some intermediate predicate (which would cause an error with JSON-LD). This is because the LDflex expression represents a query, not a value. When the expression is prefixed with the `await` keyword, it resolves to an arbitrary RDF term t_i resulting from the evaluation of its underlying SPARQL query over the specified data sources: $\langle t_i \rangle \in [\![q]\!]_s$ (or `undefined` if there is none). With the `for...await` construct, all RDF terms $(\langle t_1 \rangle, \ldots, \langle t_m \rangle) = [\![q]\!]_s$ are returned one by one in an iterative way.

Some LDflex configurations can have additional functionality, which is not covered in the general semantics above. For instance, a JSON-LD context can have *reverse properties*, for which the subject and objects of the corresponding triple pattern switch places.

4.4   Writing and Appending Data

The formal semantics above cover the case where an LDflex path is used for *reading* data. However, the same query generation mechanism can be invoked to execute SPARQL **UPDATE** queries. Since updates require filling out data in a query, they are modeled as *methods* such that arguments can be passed. The following methods are chainable on each LDflex path:

- `.add(...)` appends the specified objects to the triples matching the path.
- `.set(...)` removes any existing objects, and appends the specified objects.
- `.replace(old, ...new` replaces an existing object with one or more new objects.
- `.delete(...)` removes the specified objects, or all objects (if none specified).

5   Implementation and Embedding

In this section, we discuss the implementation of LDflex within JavaScript to achieve the intended syntax and semantics as described in previous section. We first explain the main architecture, followed by an overview of the developed LDflex libraries.

5.1   Loosely-Coupled Architecture

Proxy. Since the LDflex grammar allows for an *infinite* number of root paths and property URIs, we cannot implement them as a finite set of regular object methods. Instead, we make use of the more flexible Proxy pattern [16]. In JavaScript, `Proxy` allows customizing the behavior of the language by intercepting basic built-in constructs such as property lookup and function invocation. Intercepting every property access at one point is sufficient to define the behavior of the infinite number of possible properties.

Handlers and Resolvers. To achieve flexibility in terms of the functionality and logic that happens during LDflex expression evaluation, we make use of a loosely-coupled architecture of standalone *handlers* and *resolvers*. During the creation of the proxy-based LDflex path expression object, different handlers and resolvers can be configured, which allow the functionality of fields and methods on this path expression to be defined. Handlers are attached to a specific field or method name, which are used for implementing specifically-named functionality such as .subject, .add(), and .sort(). Resolvers are more generic, and are invoked on every field or method invocation if no handler was applicable, after which they can optionally override the functionality. For example, a specific resolver will translate field names into URIs using a configured JSON-LD context.

Implementing Multiple Interfaces. Using the handlers and resolvers, the LDflex path expression behaves as an object with chainable properties. To allow path expressions to resolve to a *single value* using the await keyword, LDflex paths implement the JavaScript Promise interface through a handler. To additionally allow resolution to *multiple values*, LDflex paths also implement implement the AsyncIterable contract through another handler. Thanks to the Proxy functionality, an expression can thus simultaneously behave as an LDflex path, a Promise, and an AsyncIterable. The returned values implement the RDF/JS Term interface and thus behave as URIs, literals, or blank nodes. Furthermore, again by using Proxy, every value also behaves as a full LDflex path such that continuations are possible. This complex behavior is exemplified in Listing 2, where the LDflex path items on line 9 is treated as an iterable with for...await , resulting in multiple item values. On line 10, those are first treated an RDF/JS Term (item), then as an LDflex path (item.name), and finally as a Promise (with await item.name).

Query Execution. To obtain result values, path expressions are first converted into SPARQL queries, after which they are executed by a SPARQL engine. By default, SPARQL queries for data retrieval are generated, as described in Sect. 4.3. When update handlers are used, SPARQL UPDATE queries are generated. This SPARQL query engine can be configured within the constructor of the path expression, which allows a loose coupling with SPARQL engines. Since LDflex passes SPARQL queries to existing query engines, it is not tied to any specific processing strategy. For instance, the engine could execute queries over SPARQL endpoints, in-memory RDF graphs, federations of multiple sources, or different query paradigms such as link-traversal-based query processing [10]. The performance of LDflex in terms of time and bandwidth is thus entirely determined by the query engine.

5.2   LDflex Libraries

Core Libraries. The JavaScript implementation of LDflex is available under the MIT license on GitHub at https://github.com/LDflex/LDflex, via the DOI 10.5281/zenodo.3820072, and the persistent URL https://doi.org/10.5281/

zenodo.3820071, and has an associated canonical citation [28]. The LDflex core is independent of a specific query engine, so we offer plugins to reuse the existing query engines *Comunica* (https://github.com/LDflex/LDflex-Comunica) and *rdflib.js* (https://github.com/LDflex/LDflex-rdflib) which enable full client-side query processing. Following best practices, LDflex and all of its related modules are available as packages compatible with Node and browsers on *npm* and are described following the FAIR principles as machine-readable Linked Data in RDF at https://linkedsoftwaredependencies.org/bundles/npm/ldflex. To make usage easy for newcomers, various documentation pages, examples, and tutorials created by ourselves and others are linked from the GitHub page. A live testing environment is at https://solid.github.io/ldflex-playground/. The sustainability plan includes a minimum of 3 years of maintenance by our team, funded by running projects related to Web querying.

Solid Libraries. An important application domain for LDflex is the Solid decentralized ecosystem [26]. In Solid, rather than storing their data collectively in a small number of centralized hubs, every person has their own *personal data vault*. Concretely, personal data such as profile details, pictures, and comments are stored separately for every person. Solid uses Linked Data in RDF, such that people can refer to each other's data, and to enable universal semantics across all data vaults without resorting to rigid data structures.

We created LDflex for Solid (available at https://github.com/solid/query-ldflex/) as an LDflex configuration that reuses a Solid-specific JSON-LD context containing shorthands for many predicates relevant to Solid. It is configured with custom handlers such as *like* and *dislike* actions. As certain data within Solid data pods requires authentication, this configuration includes a Comunica-based query engine that can perform authenticated HTTP requests against Solid data pods. It allows users to authenticate themselves to the query engine, after which the query engine will use their authentication token for any subsequent queries. Because of authentication, LDflex can be context-sensitive: within an expression such as `user.firstName`, `user` refers to the currently logged-in user.

The LDflex for Solid library is the basis for the Solid React Components (available at https://github.com/solid/react-components/), which are reusable software components for building React front-ends on top of Linked Data sources. These components can be used in React's DSL based on JavaScript and HTML to easily retrieve single and multiple values, as can be seen in Listing 4. Since these LDflex *micro-expressions* are regular strings, there is no specific coupling to the React framework. As such, LDflex can be reused analogously in other front-end frameworks.

```
1   <h2>Ruben's name</h2>
2   <Value src='["https://ruben.verborgh.org/profile/#me"].firstName'/>
3   <h2>Ruben's friends</h2>
4   <List src='["https://ruben.verborgh.org/profile/#me"].friends.firstName'/>
```

Listing 4. This example shows how React components, in this case `Value` and `List`, can use LDflex micro-expressions (highlighted) to retrieve Linked Data from the Web.

6   Usage and Validation

This section summarizes interviews we conducted[4] on the usage of LDflex within two companies. We evaluate the usage of LDflex within Janeiro Digital and Startin'blox by validating the requirements set out in Sect. 3.

6.1   Janeiro Digital

Janeiro Digital[5] is a business consultancy company in Boston, MA, USA that counts around 100 employees. They have worked in close collaboration with Inrupt[6], which was founded as a commercial driver behind the Solid initiative. They have developed the Solid React Software Development Kit (SDK), a toolkit for developing high-quality Solid apps without requiring significant knowledge on decentralization or Linked Data.

The employees within Janeiro Digital have a mixed technology background; several of them are dedicated front-end developers. Janeiro Digital makes use of LDflex as the primary data retrieval and manipulation library within the Solid React SDK. LDflex was chosen as it was a less verbose alternative to existing RDF libraries such as rdflib.js. Since most developers had never used RDF or SPARQL before, rdflib.js was very difficult to work with due to the direct contact with RDF triples. Furthermore, front-end developers would have to write SPARQL queries, while they were used to abstraction layers for such purposes. Since LDflex offers an abstraction layer over RDF triples and SPARQL queries, and makes data look like JavaScript objects, it proved to be easier to learn and work with.

The Solid React SDK provides several React components and code generators, which heavily make use of LDflex to meet simple data retrieval and manipulation needs. Because of LDflex, Janeiro Digital has been able to eliminate their previous dependency on the RDF library rdflib.js for building interactive applications over distributed Linked Data. Below, we briefly discuss three representative usages of LDflex within the SDK.

Collecting Files in a Folder. Listing 5 shows how LDflex `for...await` loops are being used to iterate over all resources within a container in a Linked Data Platform interface.

Saving Profile Photos. Listing 6 shows the code that allows users to change their profile picture using the `.set()` method.

Manipulating Access Control for Files. Listing 7 shows how Solid's Web Access Control authorizations for resource access can be manipulated using LDflex. In this case, a new `acl:Authorization` is created for a certain document.

---

[4] The unabridged interview text is at https://ruben.verborgh.org/iswc2020/ldflex/ interviews/.

[5] https://www.janeirodigital.com/.

[6] https://inrupt.com/.

```
1  const folder = data['http://example.org/myfolder'];
2  const paths = [];
3  for await (const path of folder['ldp:contains']) {
4    paths.push(path.value);
5  }
```

Listing 5. Solid React SDK logic for collecting resources within a container.

```
1  await user.vcard_hasPhoto.set(namedNode(uri));
```

Listing 6. Solid React SDK logic for adding or changing a profile image.

The LDflex usage within the Solid React SDK shows a successful implementation of our requirements. Since Janeiro Digital deliberately chose LDflex due to its abstraction layer over RDF and SPARQL, it *separates data and presentation (R1)*. As LDflex can be used within React applications, even in combination with other RDF libraries such as rdflib.js, it achieves the requirement that it *integrates into existing tooling (R2)*. Next, LDflex *incorporates the open world (R3)* because it allows the SDK to make use of any ontology they need. Since LDflex *uses Web standards (R5)*, the SDK can run in client-side Web applications. Furthermore, the SDK can directly interact with any Solid data pod, and even combine multiple of them, which verifies the requirement that it *supports multiple remote sources (R4)*. The Solid configuration of LDflex discussed in Sect. 5 is used within the SDK, which shows that LDflex *is configurable and extensible (R6)*.

```
1  const { acl, foaf } = ACL_PREFIXES;
2  const subject = `${this.aclUri}#${modes.join('')}`;
3  await data[subject].type.add(namedNode(`${acl}Authorization`));
4  const path = namedNode(this.documentUri);
5  await data[subject]['acl:accessTo'].add(path);
6  await data[subject]['acl:default'].add(path);
```

Listing 7. Solid React SDK logic for authorizing access to a certain document.

6.2   Startin' Blox

Startin'blox[7] (SiB) is a company in Paris, France with a team of 25 free-lancers. They develop the developer-friendly SiB framework with Web components that can fetch data from Solid data vaults. Usage of SiB happens within the Happy Dev network[8] (a decentralized cooperative for self-employed developers), the European Trade Union Confederation, the International Cooperative Alliance, Smart Coop, and Signons.fr.

---

[7] https://startinblox.com/.

[8] https://happy-dev.fr/.

```
1  <sib-display
2     data-src="data/list/users.jsonld"
3     fields="username, first_name, last_name, email, profile.city"
4  ></sib-display>
```

Listing 8. An SiB component for displaying the given fields of a list of users.

```
1  data["data/list/users.jsonld"].username
2  data["data/list/users.jsonld"].first_name
3  data["data/list/users.jsonld"].last_name
4  data["data/list/users.jsonld"].email
5  data["data/list/users.jsonld"].profile.city
```

Listing 9. All LDflex expressions that are produced in the SiB component from Listing 8.

The Startin'blox team has a background in Web development, and assembled to support the creation of Solid applications. LDflex was chosen as an internal library for accessing Solid data pods, as opposed to directly writing SPARQL queries for data access, since they consider SPARQL too complex to learn for new developers, and they do not have a need for the full expressiveness that SPARQL has to offer. Most developers had no direct experience with RDF directly, but they knew JSON, which lowered the entry-barrier.

The SiB framework offers Web components in which developers can define source URIs and the fields that need to be retrieved from them, as shown in Listing 8. An LDflex expression will then be produced for each field, as shown in Listing 9. This example is representative for LDflex usage within SiB, where the majority of expressions select just a single property, and some expressions containing a chain of two properties. The LDflex engine can optimize internally such that, for instance, the document is only fetched once.

The usage of LDflex within SiB shows that LDflex meets all of our introduced requirements. As SiB component users only need to define a data source and a set of fields, the data storage layer is fully abstracted for them, which means it *separates data and presentation (R1)*. Furthermore, the integration of LDflex within the SiB components exemplifies how it *integrates into existing tooling (R2)*. Next, any kind of field can be defined within SiB components without this field having to preconfigured, which shows that LDflex *incorporates the open world (R3)*. SiB is a client-side framework, and it works over Linked Data-based Solid data pods over HTTP, which shows how LDflex *uses Web standards (R5)*. Some SiB users—such as the Happy Dev network—access data that is spread over multiple remote documents, and LDflex *supports multiple remote sources (R4)*. Finally, SiB is able to configure its own JSON-LD context. Some of their specific needs, such as the ability to handle pagination, and support for language-based data retrieval, can be implemented and configured as custom hooks into LDflex, which validates that LDflex *is configurable and extensible (R6)* for their purposes.

7   Conclusion

Most Web developers do not care about Semantic Web technologies—and understandably so: the typical problems they tackle are of a less complex nature than what RDF and SPARQL were designed for. When a front-end is built to match a single, well-known back-end, nothing beats the simplicity of JSON and perhaps GraphQL, despite their lack of universal semantics. This, however, changes when accessing *multiple* back-ends—perhaps simultaneously—without imposing central agreement on all data models. Reusing the RDF technology stack might make more sense than reinventing the wheel, which unfortunately has already started happening within, for instance, the GraphQL community [1].

The Semantic Web definitely has a *user experience* problem, and if the rest of the Web can serve as a reliable predictor, neither researchers nor engineers will be the ones solving it. Front-end Web developers possess a unique skill set for translating raw data quickly into attractive applications. They can build engaging end-user interfaces to the Semantic Web, if we can provide them with the right *developer experience* by packaging RDF technology into a relevant abstraction layer. This requires an understanding of what the actual gaps are, and those look different than what is often assumed. During the design of LDflex, we have interacted with several front-end developers. All of them had a sufficiently technical profile to master RDF and SPARQL—and some of them even did. So there is no inherent need to simplify RDF or SPARQL. The point is rather that, in several common cases, those technologies are simply not the right tools for the job at hand.

LDflex is designed to support the adoption of Semantic Web technologies by front-end developers, who can in turn improve the experience and hence adoption for end-users. Rather than aiming to simplify everything, we want to ensure that straightforward tasks require proportionally sized code. For example, greeting the user by their first name is perfectly possible by fetching an RDF document, executing a SPARQL query, and interpreting the results. That code could even be abstracted into a function. However, the fact this code needs to be written in the first place, makes building user-friendly applications more involved. LDflex reduces such tasks to a single expression, removing a burden for building more engaging apps. The LDflex abstraction layer essentially acts as a runtime-generated data layer, such that a lot of glue code can be omitted. In fact, we witnessed at Janeiro Digital how several helper functions were eliminated by LDflex.

Importantly, LDflex purposely does not strive to provide an all-encompassing tool. The path queries that LDflex focuses on do not cover—by far—the entire spectrum of relevant application queries. While the evaluation shows that path queries are applicable to many common scenarios, more expressive languages such as GraphQL-LD or SPARQL remain appropriate for the remaining cases. LDflex rather aims to fulfill the *Rule of Least Power* [3], so developers can choose the expressivity that fits their problem space. Because of its high degree of extensibility, it can be adapted to different use cases via new, existing, or partly reused configurations.

Thereby, in addition to verifying whether our design requirements were met, the evaluation also brings insights into the technological needs of applications. Crucially, many SPARQL benchmarks focus on complex queries with challenging basic graph patterns, whereas some front-end patterns might actually generate rather simple queries—but a tremendously high volume of them. Furthermore, these queries are processed on the public Web, which is sensitive to latency. Typical scientific experiments are not tuned to such contexts and constraints, so the currently delivered performance might lag behind. This makes it clear that delivering simple abstractions is not necessarily a simple task. On the contrary, exposing complex data through a simple interface involves automating the underlying complexity currently residing in handwritten code [29]. Doing so efficiently requires further research into handling the variety and distribution of data on the Web.

Since Solid presents prominent use cases for LDflex, future work will also need to examine how expressions can be distributed across different sources. For example, an expression such as `user.friends.email` could retrieve the list of friends from the user's data vault, whereas the e-mail addresses themselves could originate from the data vault of each friend (to ensure the recency of the data). Technically, nothing stops us from already doing this today: we could process the corresponding SPARQL query with a link-traversal-based query algorithm [10], which would yield those results. However, the actual problem is rather related to *trust*: when obtaining data for display to a user, which parts should come from which sources? A possible solution is *constrained traversal* [27], in which users can explain what sources they trust for what kinds of data.

One of the enlightening experiences of the past couple of months was that, during browser application development, we found ourselves also using LDflex—despite being well-versed in RDF and SPARQL. This is what opened our eyes to write this article: the reason we sometimes preferred LDflex is because it expressed a given application need in a straightforward way. We surely could have tackled every single need with SPARQL, but were more productive if we did not. This led to perhaps the most crucial insight: enabling developers means enabling ourselves.

Acknowledgements. The authors wish to thank Tim Berners-Lee for his suggestion to build a "jQuery for RDF". We thank James Martin and Justin Bingham from Janeiro Digital and Sylvain Le Bon and Matthieu Fesselier from Startin'blox for their participation in the LDflex interviews.

This research received funding from the Flemish Government under the "Onderzoeksprogramma Artificiële Intelligentie (AI) Vlaanderen" program.

References

1. Baxley, III, J.: Apollo Federation - a revolutionary architecture for building a distributed graph (2019). https://blog.apollographql.com/apollo-federation-f260cf525d21

2. Bergwinkl, T., Luggen, M., elf Pavlik, Regalia, B., Savastano, P., Verborgh, R.: RDF/JS: data model specification. Draft community group report, W3C (2019). https://rdf.js.org/data-model-spec/
3. Berners-Lee, T., Mendelsohn, N.: The rule of least power. TAG finding, W3C Technical Architecture Group (2016). https://www.w3.org/2001/tag/doc/leastPower.html
4. Bibeault, B., Kats, Y.: jQuery in action. Manning (2008)
5. Champin, P.A.: RDF-REST: a unifying framework for Web APIs and Linked Data. In: Proceedings of the First Workshop on Services and Applications over Linked APIs and Data (2013)
6. EasierRDF. https://github.com/w3c/EasierRDF
7. Fowler, M.: FluentInterface (2005). https://www.martinfowler.com/bliki/FluentInterface.html
8. Günther, S.: Development of internal domain-specific languages: design principles and design patterns. In: Proceedings of the 18th Conference on Pattern Languages of Programs, pp. 1:1–1:25. ACM (2011)
9. Harris, S., Seaborne, A., Prud'hommeaux, E.: SPARQL 1.1 query language. Recommendation, W3C (2013). https://www.w3.org/TR/2013/REC-sparql11-query-20130321/
10. Hartig, O.: An overview on execution strategies for Linked Data queries. Datenbank-Spektrum **13**(2), 89–99 (2013). https://doi.org/10.1007/s13222-013-0122-1
11. Hartig, O., Pérez, J.: Semantics and complexity of GraphQL. In: Proceedings of the 27th World Wide Web Conference, pp. 1155–1164 (2018)
12. Ledvinka, M., Křemen, P.: A comparison of object-triple mapping libraries. Seman. Web J. (2019)
13. Lisena, P., Meroño-Peñuela, A., Kuhn, T., Troncy, R.: Easy Web API development with SPARQL transformer. In: Ghidini, C., et al. (eds.) ISWC 2019. LNCS, vol. 11779, pp. 454–470. Springer, Cham (2019). https://doi.org/10.1007/978-3-030-30796-7_28
14. Loring, M.C., Marron, M., Leijen, D.: Semantics of asynchronous JavaScript. In: Proceedings of the 13th ACM SIGPLAN International Symposium on on Dynamic Languages (2017)
15. Mernik, M., Heering, J., Sloane, A.M.: When and how to develop domain-specific languages. ACM Comput. Surv. **37**(4), 316–344 (2005)
16. Peck, M.M., Bouraqadi, N., Fabresse, L., Denker, M., Teruel, C.: Ghost: a uniform and general-purpose proxy implementation. Sci. Comput. Program. **98**, 339–359 (2015)
17. React: Facebook's functional turn on writing JavaScript. Commun. ACM **59**(12), 56–62 (2016)
18. Rodriguez, M.A.: The Gremlin graph traversal machine and language. In: Proceedings of the 15th Symposium on Database Programming Languages, pp. 1–10. ACM (2015)
19. Shinavier, J.: Ripple: functional programs as Linked Data. In: Proceedings of the Workshop on Scripting for the Semantic Web (2007). http://ceur-ws.org/Vol-248/
20. Sporny, M., Longley, D., Kellogg, G., Lanthaler, M., Lindström, N.: JSON-LD 1.0. Recommendation, W3C (2014). http://www.w3.org/TR/json-ld/
21. Sporny, M., Longley, D., Kellogg, G., Lanthaler, M., Lindström, N.: JSON-LD 1.1 framing. Working draft, W3C (2019). https://www.w3.org/TR/json-ld11-framing/

22. Staab, S., Scheglmann, S., Leinberger, M., Gottron, T.: Programming the Semantic Web. In: Proceedings of the European Semantic Web Conference, pp. 1–5 (2014)
23. Taelman, R., Van Herwegen, J., Vander Sande, M., Verborgh, R.: Comunica: a modular SPARQL query engine for the Web. In: Vrandečić, D., et al. (eds.) ISWC 2018. LNCS, vol. 11137, pp. 239–255. Springer, Cham (2018). https://doi.org/10. 1007/978-3-030-00668-6_15
24. Taelman, R., Vander Sande, M., Verborgh, R.: GraphQL-LD: Linked Data querying with GraphQL. In: Proceedings of the 17th International Semantic Web Conference: Posters and Demos (2018). https://comunica.github.io/Article-ISWC2018-Demo-GraphQlLD/
25. Verborgh, R.: Piecing the puzzle - self-publishing queryable research data on the Web. In: Proceedings of the 10th Workshop on Linked Data on the Web, vol. 1809 (2017)
26. Verborgh, R.: Re-decentralizing the Web, for good this time. In: Seneviratne, O., Hendler, J. (eds.) Linking the World's Information: A Collection of Essays on the Work of Sir Tim Berners-Lee. ACM (2020). https://ruben.verborgh.org/articles/redecentralizing-the-web/
27. Verborgh, R., Taelman, R.: Guided link-traversal-based query processing (2020), https://arxiv.org/abs/2005.02239
28. Verborgh, R., Taelman, R., Van Herwegen, J.: LDflex - A JavaScript DSL for querying Linked Data on the Web. Zenodo (2020). https://doi.org/10.5281/zenodo. 3820071
29. Verborgh, R., Vander Sande, M.: The Semantic Web identity crisis: in search of the trivialities that never were. Semant. Web J. 11(1), 19–27 (2020)
30. Waldo, J., Wyant, G., Wollrath, A., Kendall, S.: A note on distributed computing. Technical report, TR-94-29, Sun Microsystems Laboratories, Inc. (1994)

An Ontology for the Materials Design Domain

Huanyu Li[1,3], Rickard Armiento[2,3], and Patrick Lambrix[1,3]

[1] Department of Computer and Information Science, Linköping University,
581 83 Linköping, Sweden
[2] Department of Physics, Chemistry and Biology, Linköping University,
581 83 Linköping, Sweden
[3] The Swedish e-Science Research Centre, Linköping University,
581 83 Linköping, Sweden
{huanyu.li,rickard.armiento,patrick.lambrix}@liu.se

Abstract. In the materials design domain, much of the data from materials calculations are stored in different heterogeneous databases. Materials databases usually have different data models. Therefore, the users have to face the challenges to find the data from adequate sources and integrate data from multiple sources. Ontologies and ontology-based techniques can address such problems as the formal representation of domain knowledge can make data more available and interoperable among different systems. In this paper, we introduce the Materials Design Ontology (MDO), which defines concepts and relations to cover knowledge in the field of materials design. MDO is designed using domain knowledge in materials science (especially in solid-state physics), and is guided by the data from several databases in the materials design field. We show the application of the MDO to materials data retrieved from well-known materials databases.

Keywords: Ontology · Materials science · Materials design · OPTIMADE · Database

Resource Type: Ontology
IRI: https://w3id.org/mdo/full/1.0/

1 Introduction

More and more researchers in the field of materials science have realized that data-driven techniques have the potential to accelerate the discovery and design of new materials. Therefore, a large number of research groups and communities have developed data-driven workflows, including data repositories (for an overview see [14]) and task-specific analytical tools. Materials design is a technological process with many applications. The goal is often to achieve a set of desired materials properties for an application under certain limitations in e.g.,

© Springer Nature Switzerland AG 2020
J. Z. Pan et al. (Eds.): ISWC 2020, LNCS 12507, pp. 212–227, 2020.
https://doi.org/10.1007/978-3-030-62466-8_14

avoiding or eliminating toxic or critical raw materials. The development of condensed matter theory and materials modeling, has made it possible to achieve quantum mechanics-based simulations that can generate reliable materials data by using computer programs [17]. For instance, in [1] a flow of databases-driven high-throughput materials design in which the database is used to find materials with desirable properties, is shown. A global effort, the Materials Genome Initiative[1], has been proposed to govern databases that contain both experimentally-known and computationally-predicted material properties. The basic idea of this effort is that searching materials databases with desired combinations of properties could help to address some of the challenges of materials design. As these databases are heterogeneous in nature, there are a number of challenges to using them in the materials design workflow. For instance, retrieving data from more than one database means that users have to understand and use different application programming interfaces (APIs) or even different data models to reach an agreement. Nowadays, materials design interoperability is achieved mainly via file-based exchange involving specific formats and, at best, some partial metadata, which is not always adequately documented as it is not guided by an ontology. The second author is closely involved with another ongoing effort, the Open Databases Integration for Materials Design (OPTIMADE[2]) project which aims at making materials databases interoperational by developing a common API. Also this effort would benefit from semantically enabling the system using an ontology, both for search as well as for integrating information from the underlying databases.

These issues relate to the FAIR principles (Findable, Accessible, Interoperable, and Reusable), with the purpose of enabling machines to automatically find and use the data, and individuals to easily reuse the data [23]. Also in the materials science domain, recently, an awareness regarding the importance of such principles for data storage and management is developing and research in this area is starting [6].

To address these challenges and make data FAIR, ontologies and ontology-based techniques have been proposed to play a significant role. For the materials design field there is, therefore, a need for an ontology to represent solid-state physics concepts such as materials' properties, microscopic structure as well as calculations, which are the basis for materials design. Thus, in this paper, we present the Materials Design Ontology (MDO). The development of MDO was guided by the schemas of OPTIMADE as they are based on a consensus reached by several of the materials database providers in the field. Further, we show the use of MDO for data obtained via the OPTIMADE API and via database-specific APIs in the materials design field.

The paper is organized as follows. We introduce some well-known databases and existing ontologies in the materials science domain in Sect. 2. In Sect. 3 we present the development of MDO and introduce the concepts, relations and the axiomatization of the ontology. In Sect. 4 we introduce the envisioned usage of

---

[1] https://www.mgi.gov/.
[2] https://www.optimade.org/.

MDO as well as a current implementation. In Sect. 5 we discuss such things as the impact, availability and extendability of MDO as well as future work. Finally, the paper concludes in Sect. 6 with a small summary.

Availability: MDO is developed and maintained on a GitHub repository[3], and is available from a permanent w3id URL[4].

2   Related Work

In this section we discuss briefly well-known databases as well as ontologies in the materials science field. Further, we briefly introduce OPTIMADE.

2.1   Data and Databases in the Materials Design Domain

In the search for designing new materials, the calculation of electronic structures is an important tool. Calculations take data representing the structure and property of materials as input and generate new such data. A common crystallographic data representation that is widely used by researchers and software vendors for materials design, is CIF[5]. It was developed by the International Union of Crystallography Working Party on Crystallographic Information and was first online in 2006. One of the widely used databases is the Inorganic Crystal Structure Database (ICSD)[6]. ICSD provides data that is used as an important starting point in many calculations in the materials design domain.

As the size of computed data grows, and more and more machine learning and data mining techniques are being used in materials design, frameworks are appearing that not only provide data but also tools. Materials Project, AFLOW and OQMD are well-known examples of such frameworks that are publicly available. **Materials Project** [13] is a central program of the Materials Genome Initiative, focusing on predicting the properties of all known inorganic materials through computations. It provides open web-based data access to computed information on materials, as well as tools to design new materials. To make the data publicly available, the Materials Project provides open Materials API and an open-source python-based programming package (pymatgen). **AFLOW** [4] (Automatic Flow for Materials Discovery) is an automatic framework for high-throughput materials discovery, especially for crystal structure properties of alloys, intermetallics, and inorganic compounds. AFLOW provides a REST API and a python-based programming package (aflow). **OQMD** [19] (The Open Quantum Materials Database) is also a high-throughput database consisting of over 600,000 crystal structures calculated based on density functional theory[7]. OQMD is designed based on a relational data model. OQMD supports a REST API and a python-based programming package (qmpy).

---

[3] https://github.com/huanyu-li/Materials-Design-Ontology.
[4] https://w3id.org/mdo.
[5] Crystallographic Information Framework, https://www.iucr.org/resources/cif.
[6] https://icsd.products.fiz-karlsruhe.de/.
[7] http://oqmd.org.

2.2   Ontologies and Standards

Within the materials science domain, the use of semantic technologies is in its infancy with the development of ontologies and standards. The ontologies have been developed, focusing on representing general materials domain knowledge and specific sub-domains respectively.

Two ontologies representing general materials domain knowledge and to which our ontology connects are ChEBI and EMMO. **ChEBI** [5] (Chemical Entities of Biological Interest) is a freely available data set of molecular entities focusing on chemical compounds. The representation of such molecular entities as atom, molecule ion, etc. is the basis in both chemistry and physics. The ChEBI ontology is widely used and integrated into other domain ontologies. **EMMO** (European Materials & Modelling Ontology) is an upper ontology that is currently being developed and aims at developing a standard representational ontology framework based on current knowledge of materials modeling and characterization. The EMMO development started from the very bottom level, using the actual picture of the physical world coming from applied sciences, and in particular from physics and material sciences. Although EMMO already covers some sub-domains in materials science, many sub-domains are still lacking, including the domain MDO targets.

Further, a number of ontologies from the materials science domain focus on specific sub-domains (e.g., metals, ceramics, thermal properties, nanotechnology), and have been developed with a specific use in mind (e.g., search, data integration) [14]. For instance, the Materials Ontology [2] was developed for data exchange among thermal property databases, and MatOnto ontology [3] for oxygen ion conducting materials in the fuel cell domain. NanoParticle Ontology [21] represents properties of nanoparticles with the purpose of designing new nanoparticles, while the eNanoMapper ontology [11] focuses on assessing risks related to the use of nanomaterials from the engineering point of view. Extensions to these ontologies in the nanoparticle domain are presented in [18]. An ontology that represents formal knowledge for simulation, modeling, and optimization in computational molecular engineering is presented in [12]. Further, an ontology design pattern to model material transformation in the field of sustainable construction, is proposed in [22]. All the materials science domain ontologies above target different sub-domains from MDO.

There are also efforts on building standards for data export from databases and data integration among tools. To some extent the standards formalize the description of materials knowledge and thereby create ontological knowledge. A recent approach is Novel Materials Discovery (NOMAD[8]) [7] of which the metadata structure is defined to be independent of specific material science theory or methods that could be used as an exchange format [9].

---

[8] https://www.nomad-coe.eu/externals/european-centres-of-excellence.

2.3 Open Databases Integration for Materials Design

OPTIMADE is a consortium gathering many database providers. It aims at enabling interoperability between materials databases through a common REST API. During the development OPTIMADE takes widely used materials databases such as those introduced in Sect. 2.1 into account. OPTIMADE has a schema that defines the specification of the OPTIMADE REST API and provides essentially a list of terms for which there is a consensus from different database providers. The OPTIMADE API is taken into account in the development of MDO as shown in Sect. 3.

3 The Materials Design Ontology (MDO)

3.1 The Development of MDO

The development of MDO followed the NeOn ontology engineering methodology [20]. It consists of a number of scenarios mapped from a set of common ontology development activities. In particular, we focused on applying scenario 1 (*From Specification to Implementation*), scenario 2 (*Reusing and re-engineering non-ontological resources*), scenario 3 (*Reusing ontological resources*) and scenario 8 (*Restructuring ontological resources*). We used OWL2 DL as the representation language for MDO. During the whole process, two knowledge engineers, and one domain expert from the materials design domain were involved. In the remainder of this section, we introduce the key aspects of the development of MDO.

Requirements Analysis. During this step, we clarified the requirements by proposing Use Cases (UC), Competency Questions (CQ) and additional restrictions.

The use cases, which were identified through literature study and discussion between the domain expert and the knowledge engineers based on experience with the development of OPTIMADE and the use of materials science databases, are listed below.

- UC1: MDO will be used for representing knowledge in basic materials science such as solid-state physics and condensed matter theory.
- UC2: MDO will be used for representing materials calculation and standardizing the publication of the materials calculation data.
- UC3: MDO will be used as a standard to improve the interoperability among heterogeneous databases in the materials design domain.
- UC4: MDO will be mapped to OPTIMADE's schema to improve OPTIMADE's search functionality.

The competency questions are based on discussions with domain experts and contain questions that the databases currently can answer as well as questions that experts would want to ask the databases. For instance, CQ1, CQ2, CQ6, CQ7, CQ8 and CQ9 cannot be asked explicitly through the database APIs, although the original downloadable data contains the answers.

- CQ1: What are the calculated properties and their values produced by a materials calculation?
- CQ2: What are the input and output structures of a materials calculation?
- CQ3: What is the space group type of a structure?
- CQ4: What is the lattice type of a structure?
- CQ5: What is the chemical formula of a structure?
- CQ6: For a series of materials calculations, what are the compositions of materials with a specific range of a calculated property (e.g., band gap)?
- CQ7: For a specific material and a given range of a calculated property (e.g., band gap), what is the lattice type of the structure?
- CQ8: For a specific material and an expected lattice type of output structure, what are the values of calculated properties of the calculations?
- CQ9: What is the computational method used in a materials calculation?
- CQ10: What is the value for a specific parameter (e.g., cutoff energy) of the method used for the calculation?
- CQ11: Which software produced the result of a calculation?
- CQ12: Who are the authors of the calculation?
- CQ13: Which software or code does the calculation run with?
- CQ14: When was the calculation data published to the database?

Further, we proposed a list of additional restrictions that help in defining concepts. Some examples are shown below. The full list of additional restrictions can be found at the GitHub repository[9].

- A materials property can relate to a structure.
- A materials calculation has exactly one corresponding computational method.
- A structure corresponds to one specific space group.
- A materials calculation is performed by some software programs or codes.

Reusing and Re-engineering Non-ontological Resources. To obtain the knowledge for building the ontology, we followed two steps: (1) the collection and analysis of non-ontological resources that are relevant to the materials design domain, and (2) discussions with the domain expert regarding the concepts and relationships to be modeled in the ontology. The collection of non-ontological resources comes from: (1) the dictionaries of CIF and International Tables for Crystallography; (2) the APIs from different databases (e.g., Materials Project, AFLOW, OQMD) and OPTIMADE.

Modular Development Aiming at Building Design Patterns. We identified a pattern related to provenance information in the repository of Ontology Design Patterns (ODPs) that could be reused or re-engineered for MDO. This has led to the reuse of entities in PROV-O [15]. Further, we built MDO in modules considering the possibility for each module to be an ontology design pattern, e.g., the calculation module.

---

[9] https://github.com/huanyu-li/Materials-Design-Ontology/blob/master/requiremen ts.md.

Connection and Integration of Existing Ontologies. MDO is connected to EMMO by reusing the concept 'Material', and to ChEBI by reusing the concept 'atom'. Further, we reuse the concepts 'Agent' and 'SoftwareAgent' from PROV-O. In terms of representation of units we reuse the 'Quantity', 'QuantityValue', 'QuantityKind' and 'Unit' concepts from QUDT (Quantities, Units, Dimensions and Data Types Ontologies) [10]. We use the metadata terms from the Dublin Core Metadata Initiative (DCMI)[10] to represent the metadata of MDO.

3.2   Description of MDO

MDO consists of one basic module, *Core*, and two domain-specific modules, *Structure* and *Calculation*, importing the *Core* module. In addition, the *Provenance* module, which also imports *Core*, models provenance information. In total, the OWL2 DL representation of the ontology contains 37 classes, 32 object properties, and 32 data properties. Figure 9 shows an overview of the ontology. The ontology specification is also publicly accessible at w3id.org[11]. The competency questions can be answered using the concepts and relations in the different modules (CQ1 and CQ2 by *Core*, CQ3 to CQ8 by *Structure*, CQ9 and CQ10 by *Calculation*, and CQ11 to CQ14 by *Provenance*).

The **Core** module as shown in Fig. 1, consists of the top-level concepts and relations of MDO, which are also reused in other modules. Figure 2 shows the description logic axioms for the *Core* module. The module represents general information of materials calculations. The concepts *Calculation* and *Structure* represent materials calculations and materials' structures, respectively, while *Property* represents materials properties. *Property* is specialized into the disjoint concepts *CalculatedProperty* and *PhysicalProperty* (Core1, Core2, Core3). *Property*, which can be viewed as a quantifiable aspect of one material or materials system, is defined as a sub concept of *Quantity* from QUDT (Core4). *Properties* are also related to *structures* (Core5). When a calculation is applied on materials structures, each *calculation* takes some *structures* and *properties* as input, and may output *structures* and *calculated properties* (Core6, Core7). Further, we use EMMO's concept *Material* and state that each *structure* is related to some *material* (Core8).

The **Structure** module as shown in Fig. 3, represents the structural information of materials. Figure 4 shows the description logic axioms for the *Structure* module. Each *structure* has exact one *composition* which represents what chemical elements compose the structure and the ratio of elements in the *structure* (Struc1). The *composition* has different representations of chemical formulas. The *occupancy* of a structure relates the *sites* with the *species*, i.e. the specific chemical elements, that occupy the *site* (Struc2 - Struc5). Each *site* has at most one representation of coordinates in Cartesian format and at most one in fractional format (Struc6, Struc7). The spatial information regarding structures is essential to reflect physical characteristics such as melting point and strength of

---

[10] http://purl.org/dc/terms/.
[11] https://w3id.org/mdo/full/1.0/.

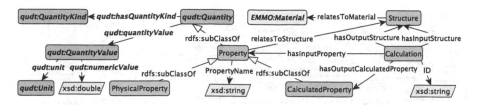

Fig. 1. Concepts and relations in the Core module.

(Core1) CalculatedProperty ⊑ Property
(Core2) PhysicalProperty ⊑ Property
(Core3) CalculatedProperty ⊓ PhysicalProperty ⊑ ⊥
(Core4) Property ⊑ Quantity
(Core5) Property ⊑ ∀ relatesToStructure.Structure
(Core6) Calculation ⊑ ∃ hasInputStructure.Structure ⊓ ∀ hasInputStructure.Structure
*   ⊓ ∀ hasOutputStructure.Structure*
(Core7) Calculation ⊑ ∃ hasInputProperty.Property ⊓ ∀ hasInputProperty.Property
*   ⊓ ∀ hasOutputCalculatedProperty.CalculatedProperty*
(Core8) Structure ⊑ ∃ relatesToMaterial.Material ⊓ ∀ relatesToMaterial.Material

Fig. 2. Description logic axioms for the Core module.

materials. To represent this spatial information, we state that each *structure* is represented by some *bases* and a (periodic) *structure* can also be represented by one or more *lattices* (Struc8). Each *basis* and each *lattice* can be identified by one *axis-vectors* set or one *length triple* together with one *angle triple* (Struc9, Struc10). An *axis-vectors* set has three connections to *coordinate vector* representing the coordinates of three translation vectors respectively, which are used to represent a (minimal) repeating unit (Struc11). These three translation vectors are often called a, b, and c. Point groups and space groups are used to represent information of the symmetry of a structure. The *space group* represents a symmetry group of patterns in three dimensions of a *structure* and the *point group* represents a group of linear mappings which correspond to the group of motions in space to determine the symmetry of a *structure*. Each *structure* has one corresponding *space group* (Struc12). Based on the definition from International Tables for Crystallography, each *space group* also has some corresponding *point groups* (Struc13).

The **Calculation** module as shown in Fig. 5, represents the classification of different computational methods. Figure 6 shows the description logic axioms for the *Calculation* module. Each *calculation* is achieved by a specific *computational method* (Cal1). Each *computational method* has some *parameters* (Cal2). In the current version of this module, we represent two different methods, the *density functional theory method* and the *HartreeFock method* (Cal3, Cal4). In particular, the density functional theory method is frequently used in materials design to investigate the electronic structure. Such method has at least one

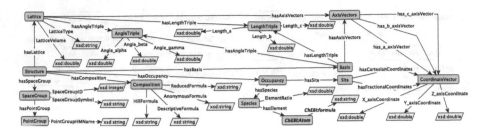

Fig. 3. Concepts and relations in the Structure module.

(Struc1) *Structure* ⊑ = 1 *hasComposition.Composition*
         ⊓ ∀ *hasComposition.Composition*
(Struc2) *Structure* ⊑ ∃ *hasOccupancy.Occupancy* ⊓ ∀ *hasOccupancy.Occupancy*
(Struc3) *Occupancy* ⊑ ∃ *hasSpecies.Species* ⊓ ∀ *hasSpecies.Species*
(Struc4) *Occupancy* ⊑ ∃ *hasSite.Site* ⊓ ∀ *hasSite.Site*
(Struc5) *Species* ⊑ = 1 *hasElement.Atom*
(Struc6) *Site* ⊑ ≤ 1 *hasCartesianCoordinates.CoordinateVector*
         ⊓ ∀ *hasCartesianCoordinates.CoordinateVector*
(Struc7) *Site* ⊑ ≤ 1 *hasFractionalCoordinates.CoordinateVector*
         ⊓ ∀ *hasFractionalCoordinates.CoordinateVector*
(Struc8) *Structure* ⊑ ∃ *hasBasis.Basis* ⊓ ∀ *hasBasis.Basis* ⊓ ∀ *hasLattice.Lattice*
(Struc9) *Basis* ⊑ = 1 *hasAxisVectors.AxisVectors* ⊔
         (= 1 *hasLengthTriple.LengthTriple* ⊓ = 1 *hasAngleTriple.AngleTriple*)
(Struc10) *Lattice* ⊑ = 1 *hasAxisVectors.AxisVectors* ⊔
         (= 1 *hasLengthTriple.LengthTriple* ⊓ = 1 *hasAngleTriple.AngleTriple*)
(Struc11) *AxisVectors* ⊑ = 1 *has_a_axisVector.CoordinateVector*
         ⊓ = 1 *has_b_axisVector.CoordinateVector*
         ⊓ = 1 *has_c_axisVector.CoordinateVector*
(Struc12) *Structure* ⊑ = 1 *hasSpaceGroup.SpaceGroup* ⊓ ∀ *hasSpaceGroup.SpaceGroup*
(Struc13) *SpaceGroup* ⊑ ∃ *hasPointGroup.PointGroup* ⊓ ∀ *hasPointGroup.PointGroup*

Fig. 4. Description logic axioms for the Structure module.

corresponding *exchange correlation energy functional* (Cal5) which is used to
calculate the exchange-correlation energy of a system. There are different kinds
of functionals to calculate exchange–correlation energy (Cal6–Cal11).

The **Provenance** module as shown in Fig. 7, represents the provenance information of materials data and calculation. Figure 8 shows the description logic
axioms for the *Provenance* module. We reuse part of PROV-O and define a new
concept *ReferenceAgent* as a sub-concept of PROV-O's agent (Prov1). We state
that each *structure* and *property* can be published by *reference agents* which
could be databases or publications (Prov2, Prov3). Each *calculation* is produced
by a specific *software* (Prov4).

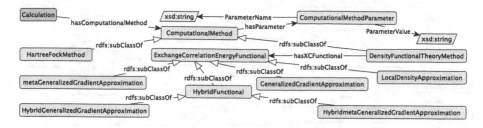

Fig. 5. Concepts and relations in the Calculation module.

(Cal1) Calculation ⊑ = 1 hasComputationalMethod.ComputationalMethod
(Cal2) ComputationalMethod ⊑ ∃ hasParameter.ComputationalMethodParameter
        ⊓ ∀ hasParameter.ComputationalMethodParameter
(Cal3) DensityFunctionalTheoryMethod ⊑ ComputationalMethod
(Cal4) HartreeFockMethod ⊑ ComputationalMethod
(Cal5) DensityFunctionalTheoryMethod ⊑
        ∃ hasXCFunctional.ExchangeCorrelationEnergyFunctional
        ⊓ ∀ hasXCFunctional.ExchangeCorrelationEnergyFunctional
(Cal6) GeneralizedGradientApproximation ⊑ ExchangeCorrelationEnergyFunctional
(Cal7) LocalDensityApproximation ⊑ ExchangeCorrelationEnergyFunctional
(Cal8) metaGeneralizedGradientApproximation ⊑
        ExchangeCorrelationEnergyFunctional
(Cal9) HybridFunctional ⊑ ExchangeCorrelationEnergyFunctional
(Cal10) HybridGeneralizedGradientApproximation ⊑ HybridFunctional
(Cal11) HybridmetaGeneralizedGradientApproximation ⊑ HybridFunctional

Fig. 6. Description logic axioms for the Calculation module.

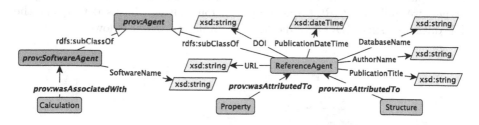

Fig. 7. Concepts and relations in the Provenance module.

(Prov1) ReferenceAgent ⊑ Agent
(Prov2) Structure ⊑ ∀ wasAttributedTo.ReferenceAgent
(Prov3) Property ⊑ ∀ wasAttributedTo.ReferenceAgent
(Prov4) Calculation ⊑ ∃ wasAssociatedwith.SoftwareAgent

Fig. 8. Description logic axioms for the Provenance module.

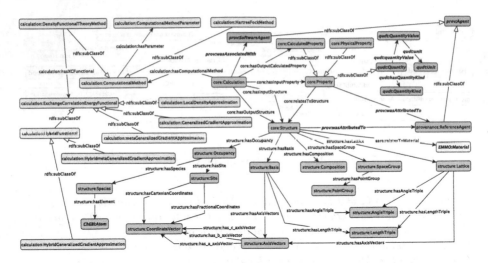

Fig. 9. An overview of MDO.

4   MDO Usage

In Fig. 10, we show the vision for the use of MDO for semantic search over OPTI-MADE and materials science databases. By generating mappings between MDO and the schemas of materials databases, we can create MDO-enabled query interfaces. The querying can occur, for instance, via MDO-based query expansion, MDO-based mediation or through MDO-enabled data warehouses.

As a proof of concept (full lines in the figure), we created mappings between MDO and the schemas of OPTIMADE and part of Materials Project. Using the mappings we created an RDF data set with data from Materials project. Further, we built a SPARQL query application that can be used to query the RDF data set using MDO terminology. Examples are given below.

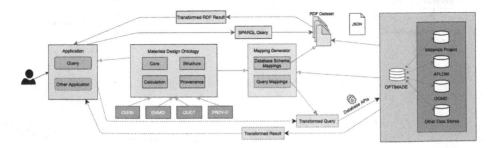

Fig. 10. The vision of the use of MDO. The full-lined components in the figure are currently implemented in a prototype.

Instantiating a Materials Calculation Using MDO. In Fig. 11 we exemplify the use of MDO to represent a specific materials calculation and related data in an instantiation. The example is from one of the 85 stable materials published in Materials Project in [8]. The calculation is about one kind of elpasolites, with the composition $Rb_2Li_1Ti_1Cl_6$. To not overcrowd the figure, we only show the instances corresponding to the calculation's output structure, and for multiple calculated properties, species and sites, we only show one instance respectively. Connected to the instances of the Core module's concepts, are instances representing the structural information of the output structure, the provenance information of the output structure and calculated property, and the information about the computational method used for the calculation.

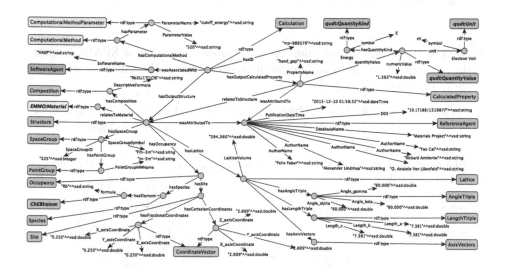

Fig. 11. An instantiated materials calculation.

Mapping the Data from a Materials Database to RDF Using MDO. As presented in Sect. 2.1, data from many materials databases are provided through the providers' APIs. A commonly used format is JSON. Our current implementation mapped all JSON data related to the 85 stable materials from [8] to RDF. We constructed the mappings by using SPARQL-Generate [16]. Listing 1.1 shows a simple example on how to write the mappings on 'band gap' which is a *CalculatedProperty*. The result is shown in Listing 1.2. The final RDF dataset contains 42,956 triples. The SPARQL-generate script and the RDF dataset are available from the GitHub repository[12]. This RDF dataset is used for executing SPARQL queries such as the one presented below.

---

[12] https://github.com/huanyu-li/Materials-Design-Ontology/tree/master/mapping_generator.

Listing 1.1. A simple example of mapping

```
BASE <https://w3id.org/mdo/data/1.0/>
PREFIX fun: <http://w3id.org/sparql-generate/fn/>
PREFIX core: <https://w3id.org/mdo/core/>
PREFIX qudt: <http://qudt.org/schema/qudt/
PREFIX qudt_unit: <http://qudt.org/vocab/unit/>

GENERATE {
    ?band_gap_node a core:CalculatedProperty;
    qudt:quantityValue ?band_gap_quantity_value;
    core:hasPropertyName "band_gap"
    GENERATE {
        ?band_gap_quantity a qudt:QuantityValue;
        qudt:unit qudt_unit:EV;
        qudt:numericValue "band_gap"
    }.
}
SOURCE <http://example.com/mp-989579_Rb2LiT1C16.json>
    AS ?source
WHERE {
    BIND(fun:JSONPath(?source,"$.band_gap") AS ?band_gap)
    BIND(BNODE() AS ?band_gap_node)
    BIND(BNODE() AS ?band_gap_quantity_value)
}
```

Listing 1.2. RDF data

```
@prefix core: <https://w3id.org/mdo/core/> .
@prefix qudt: <http://qudt.org/schema/qudt/ .
@prefix qudt_unit:
<http://qudt.org/vocab/unit/> .

<https://w3id.org/mdo/data/1.0/mp-989579_band_gap>
    a core:CalculatedProperty ;
    core:hasPropertyName "band_gap" ;
    qudt:quantityValue [ a qudt:QuantityValue ;
    qudt:numericValue 1.5623e0 ;
    qudt:unit qudt_unit:EV
    ];
```

A SPARQL Query Example. As an example, we show a SPARQL query related to CQ6 in Listing 1.3. The result contains 7 records, which are shown in Table 1. The query is:

– "What are the materials of which the value of band gap is higher than 5eV?" (The result should contain the formula, and the value of band gap.)

Listing 1.3. A SPARQL query example on Materials Project's dataset

```
PREFIX rdf: <http://www.w3.org/1999/02/22-rdf-syntax-ns#>
PREFIX core: <https://w3id.org/mdo/core/>
PREFIX structure: <https://w3id.org/mdo/structure/>
PREFIX qudt: <http://qudt.org/schema/qudt/>

SELECT ?formula ?value WHERE {
    ?calculation rdf:type core:Calculation;
            core:hasOutputCalculatedProperty ?property;
            core:hasOutputStructure ?output_structure.
    ?property qudt:quantityValue ?quantity_value;
            core:hasPropertyName ?name.
    ?quantity_value rdf:type qudt:QuantityValue;
            qudt:numericValue ?value.
    ?output_structure structure:hasComposition ?composition.
    ?composition structure:hasDescriptiveFormula ?formula.
    FILTER (?value>5 && ?name="band_gap")
}
```

Table 1. The result of the query

Formula	Value
$Cs_2Rb_1In_1F_6$	5.3759
$Cs_2Rb_1Ga_1F_6$	5.9392
$Cs_2K_1In_1F_6$	5.4629
$Rb_2Na_1In_1F_6$	5.2687
$Cs_2Rb_1Ga_1F_6$	5.5428
$Rb_2Na_1Ga_1F_6$	5.9026
$Cs_2K_1Ga_1F_6$	6.0426

We show more SPARQL query examples and the corresponding result in the GitHub repository[13].

5   Discussion and Future Work

To our knowledge, MDO is the first OWL ontology representing solid-state physics concepts, which are the basis for materials design.

The ontology fills a need for semantically enabling access to and integration of materials databases, and for realizing FAIR data in the materials design field. This will have a large impact on the effectiveness and efficiency of finding relevant

---

[13] https://github.com/huanyu-li/Materials-Design-Ontology/tree/master/sparql_query.

materials data and calculations, thereby augmenting the speed and the quality of the materials design process. Through our connection with OPTIMADE and because of the fact that we have created mappings between MDO and some major materials databases, the potential for impact is large.

The development of MDO followed well-known practices from the ontology engineering point of view (NeOn methodology and modular design). Further, we reused concepts from PROV-O, ChEBI, QUDT and EMMO. A permanent URL is reserved from w3id.org for MDO. MDO is maintained on a GitHub repository from where the ontology in OWL2 DL, visualizations of the ontology and modules, UCs, CQs and restrictions are available. It is licensed via an MIT license[14].

Due to our modular approach MDO can be extended with other modules, for instance, regarding different types of calculations and their specific properties. We identified, for instance, the need for an *X Ray Diffraction* module to model the experimental data of the diffraction used to explore the structural information of materials, and an *Elastic Tensor* module to model data in a calculation that represents a structure's elasticity. We may also refine the current ontology. For instance, it may be interesting to model workflows containing *multiple calculations*.

6 Conclusion

In this paper, we presented MDO, an ontology which defines concepts and relations to cover the knowledge in the field of materials design and which reuses concepts from other ontologies. We discussed the ontology development process showing use cases and competency questions. Further, we showed the use of MDO for semantically enabling materials database search. As a proof of concept, we mapped MDO to OPTIMADE and part of Materials Project and showed querying functionality using SPARQL on a dataset from Materials Project.

Acknowledgements. This work has been financially supported by the Swedish e-Science Research Centre (SeRC), the Swedish National Graduate School in Computer Science (CUGS), and the Swedish Research Council (Vetenskapsrådet, dnr 2018-04147).

References

1. Armiento, R.: Database-driven high-throughput calculations and machine learning models for materials design. In: Schütt, K.T., Chmiela, S., von Lilienfeld, O.A., Tkatchenko, A., Tsuda, K., Müller, K.-R. (eds.) Machine Learning Meets Quantum Physics. LNP, vol. 968, pp. 377–395. Springer, Cham (2020). https://doi.org/10.1007/978-3-030-40245-7_17
2. Ashino, T.: Materials ontology: an infrastructure for exchanging materials information and knowledge. Data Sci. J. **9**, 54–61 (2010). https://doi.org/10.2481/dsj.008-041

---

[14] https://github.com/huanyu-li/Materials-Design-Ontology/blob/master/LICENSE.

3. Cheung, K., Drennan, J., Hunter, J.: Towards an ontology for data-driven discovery of new materials. In: AAAI Spring Symposium: Semantic Scientific Knowledge Integration, pp. 9–14 (2008)
4. Curtarolo, S., et al.: AFLOW: an automatic framework for high-throughput materials discovery. Comput. Mater. Sci. **58**, 218–226 (2012). https://doi.org/10.1016/j.commatsci.2012.02.005
5. Degtyarenko, K., et al.: ChEBI: a database and ontology for chemical entities of biological interest. Nucleic Acids Res. **36**(suppl_1), D344–D350 (2008). https://doi.org/10.1093/nar/gkm791
6. Draxl, C., Scheffler, M.: NOMAD: the FAIR concept for big data-driven materials science. MRS Bull. **43**(9), 676–682 (2018). https://doi.org/10.1557/mrs.2018.208
7. Draxl, C., Scheffler, M.: The NOMAD laboratory: from data sharing to artificial intelligence. J. Phys.: Mater. **2**(3), 036001 (2019). https://doi.org/10.1088/2515-7639/ab13bb
8. Faber, F.A., Lindmaa, A., Von Lilienfeld, O.A., Armiento, R.: Machine learning energies of 2 million elpasolite (a b c 2 d 6) crystals. Phys. Rev. Lett. **117**(13), 135502 (2016). https://doi.org/10.1103/PhysRevLett.117.135502
9. Ghiringhelli, L.M., et al.: Towards a common format for computational materials science data. PSI-K Scientific Highlights (2016)
10. Haas, R., Keller, P.J., Hodges, J., Spivak, J.: Quantities, units, dimensions and data types ontologies (QUDT). http://qudt.org. Accessed 03 Aug 2020
11. Hastings, J., et al.: eNanoMapper: harnessing ontologies to enable data integration for nanomaterial risk assessment. J. Biomed. Semant. **6**(1), 10 (2015). https://doi.org/10.1186/s13326-015-0005-5
12. Horsch, M.T., et al.: Semantic interoperability and characterization of data provenance in computational molecular engineering. J. Chem. Eng. Data **65**(3), 1313–1329 (2020). https://doi.org/10.1021/acs.jced.9b00739
13. Jain, A., et al.: The materials project: a materials genome approach to accelerating materials innovation. APL Mater. **1**(1), 011002 (2013). https://doi.org/10.1063/1.4812323
14. Lambrix, P., Armiento, R., Delin, A., Li, H.: Big semantic data processing in the materials design domain. In: Sakr, S., Zomaya, A.Y. (eds.) Encyclopedia of Big Data Technologies, pp. 1–8. Springer, Heidelberg (2019). https://doi.org/10.1007/978-3-319-63962-8_293-1
15. Lebo, T., et al.: PROV-O: the PROV ontology. In: W3C Recommendation, W3C (2013). https://www.w3.org/TR/prov-o/. Accessed Apr 2020
16. Lefrançois, M., Zimmermann, A., Bakerally, N.: A SPARQL extension for generating RDF from heterogeneous formats. In: Blomqvist, E., Maynard, D., Gangemi, A., Hoekstra, R., Hitzler, P., Hartig, O. (eds.) ESWC 2017. LNCS, vol. 10249, pp. 35–50. Springer, Cham (2017). https://doi.org/10.1007/978-3-319-58068-5_3
17. Lejaeghere, K., et al.: Reproducibility in density functional theory calculations of solids. Science **351**(6280), aad3000 (2016). https://doi.org/10.1126/science.aad3000
18. Li, H., Armiento, R., Lambrix, P.: A method for extending ontologies with application to the materials science domain. Data Sci. J. **18**(1), 1–21 (2019). https://doi.org/10.5334/dsj-2019-050
19. Saal, J.E., Kirklin, S., Aykol, M., Meredig, B., Wolverton, C.: Materials design and discovery with high-throughput density functional theory: the open quantum materials database (OQMD). JOM **65**(11), 1501–1509 (2013). https://doi.org/10.1007/s11837-013-0755-4

20. Suárez-Figueroa, M.C., Gómez-Pérez, A., Fernández-López, M.: The NeOn methodology for ontology engineering. In: Suárez-Figueroa, M.C., Gómez-Pérez, A., Motta, E., Gangemi, A. (eds.) Ontology Engineering in a Networked World, pp. 9–34. Springer, Heidelberg (2012). https://doi.org/10.1007/978-3-642-24794-1_2

21. Thomas, D.G., Pappu, R.V., Baker, N.A.: Nanoparticle ontology for cancer nanotechnology research. J. Biomed. Inform. 44(1), 59–74 (2011). https://doi.org/10.1016/j.jbi.2010.03.001

22. Vardeman II, C.F., et al.: An ontology design pattern and its use case for modeling material transformation. Semant. Web 8(5), 719–731 (2017). https://doi.org/10.3233/SW-160231

23. Wilkinson, M.D., et al.: The FAIR guiding principles for scientific data management and stewardship. Sci. Data 3(160018), 1–9 (2016). https://doi.org/10.1038/sdata.2016.18

Explanation Ontology: A Model of Explanations for User-Centered AI

Shruthi Chari[1], Oshani Seneviratne[1], Daniel M. Gruen[1],
Morgan A. Foreman[2], Amar K. Das[2], and Deborah L. McGuinness[1(✉)]

[1] Rensselaer Polytechnic Institute, Troy, NY, USA
{charis,senevo,gruend2}@rpi.edu, dlm@cs.rpi.edu
[2] IBM Research, Cambrdige, MA, USA
morgan.foreman@us.ibm.com, amardasmdphd@gmail.com

Abstract. Explainability has been a goal for Artificial Intelligence (AI) systems since their conception, with the need for explainability growing as more complex AI models are increasingly used in critical, high-stakes settings such as healthcare. Explanations have often added to an AI system in a non-principled, post-hoc manner. With greater adoption of these systems and emphasis on user-centric explainability, there is a need for a structured representation that treats explainability as a primary consideration, mapping end user needs to specific explanation types and the system's AI capabilities. We design an explanation ontology to model both the role of explanations, accounting for the system and user attributes in the process, and the range of different literature-derived explanation types. We indicate how the ontology can support user requirements for explanations in the domain of healthcare. We evaluate our ontology with a set of competency questions geared towards a system designer who might use our ontology to decide which explanation types to include, given a combination of users' needs and a system's capabilities, both in system design settings and in real-time operations. Through the use of this ontology, system designers will be able to make informed choices on which explanations AI systems can and should provide.

Keywords: Explainable AI · Explanation ontology · Modeling of explanations and explanation types · Supporting explainable ai in clinical decision making and decision support

Resource: https://tetherless-world.github.io/explanation-ontology

1 Introduction

Explainability has been a key focus area of Artificial Intelligence (AI) research, from expert systems, cognitive assistants, the Semantic Web, and more recently, in the machine learning (ML) domain. In our recent work [5], we show that advances in explainability have been coupled with advancements in the sub-fields of AI. For example, explanations in second-generation expert systems typically

© Springer Nature Switzerland AG 2020
J. Z. Pan et al. (Eds.): ISWC 2020, LNCS 12507, pp. 228–243, 2020.
https://doi.org/10.1007/978-3-030-62466-8_15

address *What, Why, and How* questions [6,23]. With ML methods, explainability has focused on interpreting the functioning of black-box models, such as identifying the input features that are associated the most with different outputs [13,15]. However, while explanations of the "simplified approximations of complex decision-making functions" [17] are important, they do not account for "specific context and background knowledge" [18] that users might possess, and hence, are often better suited for experts or debugging purposes. Several researchers have written about this shortcoming [13,17], and the fallacy of associating explainability to be solely about model transparency and interpretability [9]. Given this shift in focus of explainable AI, due to the adoption of AI in critical and user-facing fields such as healthcare and finance, researchers are drawing from adjacent "explanation science" fields to make explainable AI more usable [16]. The term "explanation sciences" was introduced by Mittlestadt et al. to collectively refer to the fields of "law, cognitive science, philosophy, and the social sciences." [17]

Recent review papers [3,17] point out that explainability is diverse, serving and addressing different purposes, with user-specific questions and goals. Doshi et al. [7] propose a set of questions beyond the *What, Why, How* questions that need to be addressed by explanations: "What were the main factors in a decision?", "Would changing a certain factor have changed the decision?" and "Why did two similar-looking cases get different decisions or vice versa?" Other researchers, like Wang et al. [26], in their conceptual framework linking human reasoning methods to explanations generated by systems, support various explainability features, such as different "intelligibility queries." Lim and Dey observed these intelligibility queries during their user study where they were studying for mechanisms to improve the system's intelligibility (comprehensibility) by looking to gain user's trust and seeking to avoid situations that could lead to a "mismatch between user expectation and system behavior" [12]. Support for such targeted provisions for explanations would enable the creation of "explainable knowledge-enabled systems", a class of systems we defined in prior work [5] as, "AI systems that include a representation of the domain knowledge in the field of application, have mechanisms to incorporate the *users' context*, are *interpretable*, and host *explanation facilities* that generate *user-comprehensible, context-aware, and provenance-enabled* explanations of the mechanistic functioning of the AI system and the knowledge used."

Currently, a single class of AI models, with their specific focus on particular problems that tap into specific knowledge sources, cannot wholly address these broad and diverse questions. The ability to address a range of user questions points to the need for providing explanations as a service via a framework that interacts with multiple AI models with varied strengths. To achieve this flexibility in addressing a wide range of user questions, we see a gap in semantic support for the generation of explanations that would allow for explanations to be a core component of AI systems to meet users' requirements. We believe an ontology, a machine-readable implementation, can help system designers, and eventually a service, to identify and include methods to generate a variety of explanations that suit users' requirements and questions. While there have been a few efforts to establish connections between explanation types and the mechanisms that are capable of generating them [1, 26], these efforts are either not made available in machine-readable formats or not represented semantically. The lack of a semantic representation that would offer support for modeling explanations makes it difficult for a system designer to utilize their gathered user requirements as a means to build in explanation facilities into their systems.

We first present related work on taxonomies of explanation (Sect. 2). We then introduce the design of an explanation ontology (Sect. 3) that treats explanations as a primary consideration of AI system design and that can assist system designers to capture and structure the various components necessary to enable the generation of user-centric explanations computationally. These components and the attributes of explanations are gathered from our literature review of AI [4, 5], associated "explanation sciences" domains including social sciences and philosophy. We use the results of a previously conducted user-centered design study, which used a prototype decision-support system for diabetes management, to demonstrate the usage of our ontology for the design and implementation of an AI system (Sect. 4). Finally, we evaluate our ontology's competency in assisting system designers by our ontology's ability to support answering a set of questions aimed at helping designers build "explainable, knowledge-enabled," AI systems (Sect. 5).

2 Related Work

While there have been several taxonomies proposed within explainable AI, we limit our review to taxonomies that are closest to our focus, in that they catalog AI models and the different types of explanations they achieve [1, 2], or ones that capture user-centric aspects of explainability [26]. Recently, Arya et al. [2] developed a taxonomy for explainability, organizing ML methods and techniques that generate different levels of explanations, including *post-hoc* (contain explanations about the results or model functioning), *local* (about a single prediction), and *general* (describes behavior of entire model) explanations, across various modalities (interactive/visual, etc.). Their taxonomy has been used as a base for the AI Explainability 360 (AIX 360) toolkit [20] to recommend applicable, explainable AI methods to users. Their taxonomy only considers attributes of explanations from an implementation and data perspective and does not account for end-user-specific requirements. Further, their taxonomy, implemented as a decision tree, lacks a semantic mapping of the terms involved, which makes it hard for system designers to extend this taxonomy flexibly or to understand the interaction between the various entities involved in the generation of explanations. In our ontology, we provide a semantic representation that would help system designers support and include different explanation types in their system, while accounting for both system and user attributes.

Similarly, Arrieta et al. [1] have produced a taxonomy, mapping ML models (primarily deep learning models) to the explanations they produced and the features within these models that are responsible for generating these explanations. Their taxonomy covers different types of explanations that are produced by ML models, including *simplification, explanation by examples, local explanations, text explanations, visual explanations, feature relevance, and explanations by transparent models.* However, in their structural taxonomy, due to the lack of a semantic representation, they often refer to explanation types and the modalities in which they are presented interchangeably. In addition, the explanations they cover are tightly coupled with the capabilities of ML models and they do not explore other aspects that could be included in explanations, such as different forms of knowledge, to make them amenable to end users. Through our ontology, we address this gap by incorporating diverse aspects of explanaibility that are relevant to supporting the generation of user-centric explanation types (e.g., counterfactual, contrastive, scientific, trace based explanations, etc.) that address different user goals beyond model interpretability.

Wang et al. have developed a "Conceptual Framework for Reasoned Explanations" to describe "how human reasoning processes inform" explainable AI techniques [26]. Besides supporting connections between how humans reason and how systems generate an explanation, this conceptual framework also draws parallels between various aspects of explainability, such as between explanation goals and explanation types, human reasoning mechanisms and AI methods, explanation types and intelligibility queries. While they cover some explanation types and point to the need for "integrating multiple explanations," we support a broader set of literature-derived explanation types via our ontology. Also, it remains unclear as to whether their framework is in a machine-readable format that can be used to support system development. Within our ontology, we model some of these explainability aspects that are captured in their framework, including explanation types, explanation goals, and different modalities.

Tiddi et al. [24] created an ontology design pattern (ODP) to represent explanations, and showcased the ability of the ODP to represent explanations across "explanation sciences," such as linguistics, neuroscience, computer science, and sociology. In this ODP, they associate explanations with attributes, including the situation, agents, theory, and events. Additionally, they provide support for the association of explanations with *explanandum* (the fact or event to be explained) and *explanan* (that which does the explaining) to associate explanations with some premise to the *explanandum*. Their contribution is a general-purpose ODP, however, it cannot be applied as is in practice, due to the difficulty in condensing explanations to their suggested form of *<explanans (A), posterior explanandum (P), theory (T), and situational context (C)>*, without the background understanding of how these entities were generated in their field of application. In our ontology, we reuse classes and properties from this ODP, where applicable, and expand on their mappings to support the modeling of explanations generated via computational processes and that address the users' questions, situations, and contexts.

3   Explanation Ontology

As we have discussed in Sect. 1 and 2, explainability serves different purposes. Given this diversity, there is a need for a semantic representation of explanations that models associations between entities and attributes directly and indirectly related to explanations from the system as well as user standpoints. In designing our "Explanation Ontology," (EO) we have used both bottom-up and top-down modeling approaches. We undertook a bottom-up literature review to primarily identify different explanation types and their definitions in the literature. We utilize our literature review as a base for our modeling and use a top-down approach to refine the modeling by analyzing the usage of different explanation types by clinicians during a requirements gathering session we conducted. In Sect. 3.1, we describe our modeling, then, in Sect. 3.2, we showcase our representation of literature-derived explanation types using this modeling.

3.1 Ontology Composition

We design our ontology around the central explanation class (ep:Explanation) and include entities and attributes that we see occurring often in the literature as explanation components. In Fig. 1, we present a conceptual overview of our ontology and depict associations necessary to understand its coverage. In Table 1, we list ontology prefixes that we use to refer to classes and properties.

Table 1. List of ontology prefixes used in the paper.

Ontology prefix	Ontology	URI
sio	SemanticScience Integrated Ontology	http://semanticscience.org/resource/
prov	Provenance Ontology	http://www.w3.org/ns/prov-o
eo	Explanation Ontology	https://purl.org/heals/eo
ep	Explanation Patterns Ontology	http://linkedu.eu/dedalo/explanationPattern.owl

In our ontology, we build on the class and property hierarchies provided by Tiddi et al.'s explanation ODP [24] and the general-purpose SemanticScience Integrated Ontology (SIO) [8]. When referencing classes and properties from the SemanticScience Integrated Ontology (SIO) that use numeric identifiers, we follow the convention used in their paper [8] by referring to classes and properties via their labels. E.g., sio:'in relation to'.

We introduce classes and properties as necessary to construct a model of explanations that supports property associations between explanations (ep:Explanation), the AI Task (eo:AITask) that generated the recommendation (eo:SystemRecommendation) upon which the explanation is based (ep:isBasedOn) and the end-user (eo:User) who consumes them. In our modeling, we note that explanations are dependent on (ep:isBasedOn) both system recommendations as well as implicit/explicit knowledge (eo:knowledge) available to systems or possessed by users. In addition, we also model that the knowledge available to the system can be in relation to (sio:'in relation to') various entities, such as the domain knowledge, situational knowledge of the end-user (eo:User), and knowledge about a record of the central object (eo:ObjectRecord) fed into the system (e.g., as we will see in Sect. 4, a patient is a central object in a clinical decision support system). We also model that explanations address a ques-

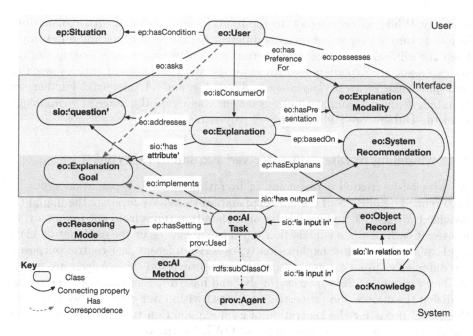

Fig. 1. A conceptual overview of our explanation ontology, capturing attributes of explanations to allow them to be assembled by an AI Task, used in a system interacting with a user. We depict user-attributes of explanations in the upper portion (green highlight), system-attributes in the lower portion (blue highlight), and attributes that would be visible in a user interface are depicted in the middle portion in purple. (Color figure online)

tion (sio:'question') posed by end-users, and these questions are implemented (eo:implements) by AI Tasks that generate recommendations.

AI Tasks (eo:AITask) can be thought of as analogous with different reasoning types (i.e., inductive, deductive, abductive or a hybrid reasoning strategy) and that are implemented by different AI methods (eo:AIMethod) (e.g., similarity algorithms, expert systems) to arrive at recommendations. This decomposition of an AI Task to methods is inspired by Tu et al.'s research in the problem-solving domain [25] of creating domain-independent AI systems that can be instantiated for specific domains. Besides capturing the interplay between an eo:AITask and its implementation, an eo:AIMethod, we also model that an AI Task is implemented in a particular reasoning mode (eo:ReasoningMode) of the system which dictates the overall execution strategy. We believe our approach to supporting the different granularities of work separation within an AI system can be valuable to building AI systems with hybrid reasoning mechanisms capable of generating different explanation types. In addition to capturing the situational context of the user, we also support modeling their existing knowledge and preferences for different forms for presentations of explanation (eo:ExplanationModality).

In the rest of this paper, we refer to ontology classes within single quotes and italicize them (e.g., *'knowledge'*) to give readers an idea of the coverage of our ontology. While, we only depict the top-level classes associated with explanations in Fig. 1, through representations of different explanation types and an example from the clinical requirements gathering session, we show how system designers can associate explanations with more specific subclasses of entities, such as with particular forms of *'knowledge'*, *'AI Task's*, and *'AI methods'*. Further, we maintain definitions and attributions for our classes via the usage of terminology from the DublinCore [19] and Prov-O [10] ontologies.

3.2   Modeling of Literature-Derived Explanation Types

We previously created a taxonomy of literature-derived, explanation types [4] with refined definitions of the nine explanation types. We leverage the mappings provided within EO, and knowledge of explanation types from our taxonomy, to represent each of these explanation types as subclasses of the explanation class (ep:Explanation). These explanation types serve different user-centric purposes, are differently suited for users' *'situations,'* context and *'knowledge,'* are generated by various *'AI Task'* and *'methods,'* and have different informational needs. Utilizing the classes and properties supported within our ontology, we represent the varied needs for the generation of each explanation type, or the sufficiency conditions for each explanation type, as OWL restrictions.

In Table 2, we present an overview of the different explanation types along with their descriptions and sufficiency conditions. In Listing 1.1, we present an RDF representation of a *'contextual explanation,'* depicting the encoding of sufficiency conditions on this class.

Listing 1.1. OWL expression of the representation of a *'contextual explanation'* (whose sufficiency conditions can be referred to from Table 2) in Manchester syntax. In this snippet, we show the syntax necessary to understand the composition of the *'contextual explanation'* class in reference to the classes and properties introduced in Fig. 1.

```
 1 Class: eo:ContextualExplanation
 2   EquivalentTo:
 3     (isBasedOn some eo:'System Recommendation')
 4     and (
 5     (ep:isBasedOn some
 6       (eo:'Contextual Knowledge'
 7         and (sio:'in relation to' some ep:Situation))) or
 8     (ep:isBasedOn some ('Contextual Knowledge'
 9         and (sio:'in relation to' some eo:'Object Record'))))
10   SubClassOf:
11     ep:Explanation
12
13 Class: ep:Explanation
14   SubClassOf:
15       sio:'computational entity',
16       ep:isBasedOn some eo:Knowledge,
17       ep:isBasedOn some eo:SystemRecommendation,
18       ep:isConceptualizedBy some eo:AITask
```

Table 2. An overview of explanation types against simplified descriptions of their literature-synthesized definitions, and natural language descriptions of sufficiency conditions. Within the explanation type description we also include a general prototypical question that can be addressed by each explanation type. Further, within the sufficiency conditions, we highlight ontology classes using single quotes and italics.

Explanation type	Description	Sufficiency conditions
Case based	Provides solutions that are based on actual prior cases that can be presented to the user to provide compelling support for the system's conclusions, and may involve analogical reasoning, relying on similarities between features of the case and of the current situation. **"To what other situations has this recommendation been applied?"**	Is there at least one other prior case (*'object record'*) similar to this situation that had an *'explanation'*? Is there a similarity between this case, and that other case?
Contextual	Refers to information about items other than the explicit inputs and output, such as information about the user, situation, and broader environment that affected the computation. **"What broader information about the current situation prompted the suggestion of this recommendation?"**	Are there any other extra inputs that are not contained in the *'situation'* description itself? And by including those, can better insights be included in the *'explanation'*?
Contrastive	Answers the question "Why this output instead of that output," making a contrast between the given output and the facts that led to it (inputs and other considerations), and an alternate output of interest and the foil (facts that would have led to it). **"Why choose option A over option B that I typically choose?"**	Is there a *'system recommendation'* that was made (let's call it A)? What facts led to it? Is there another *'system recommendation'* that could have happened or did occur, (let's call it B)? What was the *'foil'* that led to B? Can A and B be compared?
Counterfactual	Addresses the question of what solutions would have been obtained with a different set of inputs than those used. **"What if input A was over 1000?"**	Is there a different set of inputs that can be considered? If so what is the alternate *'system recommendation'*?
Everyday	Uses accounts of the real world that appeal to the user, given their general understanding and knowledge. **"Why does option A make sense"**	Can accounts of the real world be simplified to appeal to the user based on their general understanding and *'knowledge'*?
Scientific	References the results of rigorous scientific methods, observations, and measurements. **"What studies have backed this recommendation?"**	Are there results of rigorous *'scientific methods'* to explain the situation? Is there *'evidence'* from the literature to explain this *'situation'*?
Simulation Based	Uses an imagined or implemented imitation of a system or process and the results that emerge from similar inputs. **"What would happen if this recommendation is followed?"**	Is there an *'implemented'* imitation of the *'situation'* at hand? Does that other scenario have inputs similar to the current *'situation'*?
Statistical	Presents an account of the outcome based on data about the occurrence of events under specified (e.g., experimental) conditions. Statistical explanations refer to numerical evidence on the likelihood of factors or processes influencing the result. **"What percentage of people with this condition have recovered?"**	Is there *'numerical evidence'*/likelihood account of the *'system recommendation'* based on data about the occurrence of the outcome described in the recommendation?
Trace Based	Provides the underlying sequence of steps used by the system to arrive at a specific result, containing the line of reasoning per case and addressing the question of why and how the application did something. **"What steps were taken by the system to generate this recommendation?"**	Is there a record of the underlying sequence of steps (*'system trace'*) used by the *'system'* to arrive at a specific *'recommendation'*?

4   Clinical Use Case

We demonstrate the use of EO in the design and operations of an AI system to support treatment decisions in the care of patients with diabetes. We previously conducted a two-part user-centered design study that focused on determining which explanations types are needed within such a system. In the first part of the study, we held an expert panel session with three diabetes specialists to understand their decision support needs when applying guideline-based recommendations in diabetes care. We then used the requirements gathered from this session to design a prototype AI system. In the second part, we performed cognitive walk-throughs of the prototype to understand what reasoning strategies clinicians used and which explanations were needed when presented with a complex patient.

In modeling the reasoning strategies that need to be incorporated into this system design, we found that the Select-Test (ST) model by Stefanoli and Ramoni [22] mirrored the clinician's approach. Applying their ST model, we can organize the clinical reasoning strategy within the system design based on types of *'reasoning mode,'* such as differential diagnosis or treatment planning. Each of these modes can be associated with AI tasks, such as *ranking* that creates a preferential order of options like diagnoses or treatments, *deduction* that predicts consequences from hypotheses, *abstraction* that identifies relevant clinical findings from observations, and *induction* that selects the best solution by matching observations to the options or requests new information where necessary. Each of these AI tasks can generate system recommendations that requires explanations from the clinicians.

We discovered that, of the types of explanations listed in Table 1, everyday and contextual explanations were required more than half the time. We noted that clinicians were using a special form of everyday explanations, specifically their experiential knowledge or *'clinical pearls'* [14] to explain the patient's case. We observed concrete examples of the explanation components being used in the explanations provided by clinicians, such as *'contextual knowledge'* of a patient's condition being used for diagnosis and drug *'recommendations'*. Other examples of explanation types needed within the system design include *trace-based explanations* in a treatment planning mode, to provide an algorithmic breakdown of the guideline steps that led to a drug recommendation; *'scientific explanations'* in a plan critiquing mode, to provide references to studies that support the drug, as well as *'counterfactual explanations'*, to allow clinicians to add/edit information to view a change in the recommendation; and *'contrastive explanations'* in a differential diagnosis mode, to provide an intuition about which drug is the most recommended for the patient. The results of the user studies demonstrated the need for a diverse set of explanation types and that modeling explanation requires various components to support AI system design.

An example of our ontology being used to represent the generation process for a *'contrastive explanation'*, while accounting for the *'reasoning mode,'* *'AI Task'* involved, can be viewed in Listing 1.2. In the RDF representation of a *'contrastive explanation'* used by a clinician, we depict how our ontology would be useful to guide a system to provide an explanation in real-time to the question, "Why Drug B over Drug A?" In our discussion of the Listing 1.2 hereafter, we include the entity IRIs from the listing in parantheses. Upon identifying what explanation type would best suit this question, our ontology would guide the system to access different forms of *'knowledge'* and invoke the corresponding *'AI tasks'* that are suited to generate *'contrastive explanations'*. In this example, a deductive AI task (:AITaskExample) is summoned and generates a system recommendation (:SystemRecExampleA) that Drug A is insufficient based on contextual knowledge of the patient record (:ContextualKnowledgePatient). In addition, the deductive task is also fed with guideline evidence that Drug B is a preferred drug, which results in the generation of a recommendation (:SystemRecExampleB) in favor of Drug B. Finally, our ontology would help guide a system to populate the components of a *'contrastive explanation'* from *'facts'* that supported the hypothesis, "Why Drug B?" and its *'foil'*, "Why not Drug A?," or the facts that ruled out Drug A. We note that the annotation of granular content, such as patient and drug data within these explanations, would require the usage of domain-specific ontologies, whose concepts would need to be inferred into classes supported within our ontology. We defer the granular content annotation effort to future work.

Listing 1.2. Turtle representation of the process a system would undergo to generate a *'contrastive explanation'*, such as the one presented during our cognitive walk-through to address, "Why drug B over drug A?"

```
1  :ContrastiveQuestion
2      a sio:'question';
3      rdfs:label ''Why Drug B over Drug A?'' .
4
5  :ContrastiveExpInstance
6      a eo:ContrastiveExplanation;
7      ep:isBasedOn :SystemRecExampleA, :SystemRecExampleB;
8      rdfs:label ''Guidelines recommend Drug B for this patient'';
9      :addresses :ContrastiveQuestion .
10
11 :SystemRecExampleA
12     a eo:SystemRecommendation;
13     prov:used :ContextualKnowledgePatient;
14     rdfs:label ''Drug A is not sufficient for the patient'' .
15
16 :SystemRecExampleB
17     a eo:SystemRecommendation;
18     prov:used :GuidelineEvidence;
19     rdfs:label ''Drug B is recommended by the guidelines'' .
20
21 :AITaskExample
22     a eo:DeductiveTask;
23     sio:'has output' :SystemRecExampleA, :SystemRecExampleB;
24     ep:hasSetting [a eo:ReasoningMode; rdfs:label ''Treatment Planning''];
25     prov:used :ContextualKnowledgePatient, :GuidelineEvidence;
26     rdfs:label ''Deductive task'' .
27
28 :ContextualKnowledgePatient
29     a eo:ContextualKnowledge, eo:Foil;
30     sio:'in relation to' [a sio:'patient'];
```

```
31    sio:'is input in' :AITaskExample;
32    rdfs:label ''patient has hyperglycemia'' .
33
34 :GuidelineEvidence
35    a eo:ScientificKnowledge, eo:Fact;
36    sio:'is input in' :AITaskExample;
37    rdfs:label ''Drug B is the preferred drug'' .
```

5   Evaluation

We evaluate our ontology via a set of competency questions posed from the perspective of a system designer who may need to design a system that includes appropriate explanation support and may hope to use our ontology. These competency questions are designed to aid system designers in their planning of resources to include for generating explanations that are suitable to the expertise level of the end-user, the scenario for which the system is being developed, etc. The resources that a system designer would need to consider could include the *'AI method'* and *'tasks'* capable of generating the set of explanations that best address the user's *'question'* [11], the reasoning *'modes'* that need to be supported within the system, and the *'knowledge'* sources. These competency questions would be ones that, for example, we, as system designers looking to implement a clinical decision-support system, such as the prototype described in Sect. 4, would ask ourselves upon analyzing the requirements of clinicians gathered from a user study. Contrarily, if the specifications were to live in documentation, versus an explanation ontology, it would be cumbersome for a system designer to perform a lookup to regenerate the explanation requirements for every new use case. Through EO we support system designers to fill in what can be thought of as "slots" (i.e., instantiate classes) for explanation generation capabilities.

In Table 3, we present a list of competency questions along with answers. The first three questions can be addressed before or during system development, when the system designer has gathered user requirements, and the last two questions need to be answered in real-time, based on the system and a user's current set of attributes. During system development and after the completion of a requirements gathering session, if a system designer learns that a certain set of explanation types would be suitable to include, through answers to the first three questions, they can be made aware of the AI methods capable of generating explanations of the type (Q1), the components to build into their system to ensure that the explanations can be generated (Q3), and some of the questions that have been addressed by the particular explanation type (Q2). When a system is running in real-time, an answer to (Q4) would help a system designer decide what pre-canned explanation type already supported by the system would be the best current set of system, and user attributes and an answer to (Q5) would help a system designer decide on whether their system, with its current capabilities, can generate an explanation type that is most suitable to the form of the question being asked. While we address these competency

Table 3. A catalog of competency questions and candidate answers produced by our ontology. These questions can be generalized to address queries about other explanation types supported within our ontology.

Competency question	Answer
Q1. Which AI model (s) is capable of generating this explanation type (e.g. trace-based)?	Knowledge-based systems, Machine learning model: decision trees
Q2. What example questions have been identified for counterfactual explanations?	What other factors about the patient does the system know of? What if the major problem was a fasting plasma glucose?
Q3. What are the components of a scientific explanation?	Generated by an AI Task, Based on recommendation, and based on evidence from study or basis from scientific method
Q4. Given the system was performing abductive reasoning and has ranked specific recommendations by comparing different medications, what explanations can be provided for that recommendation?	Contrastive explanation
Q5. Which explanation type best suits the user question, "Which explanation type can expose numerical evidence about patients on this drug?," and how will the system generate the answer?	Explanation type: statistical System: run 'Inductive' AI task with 'Clustering' method to generate numerical evidence

questions, and specifically ask questions Q4 and Q5 in the setting of our clinical requirements gathering session, we expect that these questions can be easily adapted for other settings to be addressed with the aid of EO. In addition to presenting natural-language answers to sample competency questions, we depict a SPARQL query used to address the third question in Listing 1.3.

Listing 1.3. A SPARQL query that retrieves the sufficiency conditions encoded on an explanation type to answer a competency question of the kind, "What are the components of a scientific explanation?"

```
1 prefix rdfs:<http://www.w3.org/2000/01/rdf-schema#>
2 prefix owl:<http://www.w3.org/2002/07/owl#>
3
4 select ?class ?restriction
5 where {
6     ?class (rdfs:subClassOf|owl:equivalentClass) ?restriction .
7     ?class rdfs:label ''Scientific Explanation'' .
8 }
```

Table 4. Results of the SPARQL query to retrieve the sufficiency conditions defined on the *'scientific explanations'* class. These results depict the flexibility that we allow so that *'scientific explanations'* can either be directly based on *'scientific knowledge'* or on system recommendations that use *'scientific knowledge'*.

Subject	Restriction
Scientific explanation	(ep:isBasedOn some (eo:'Scientific Knowledge' and ((prov:wasGeneratedBy some 'Study') or (prov:wasAssociatedWith some eo:'Scientific Method'))) and (isBasedOn some eo:'System Recommendation')) or (ep:isBasedOn some (eo:'System Recommendation' and (prov:used some (eo:'Scientific Knowledge' and ((prov:wasGeneratedBy some 'Study') or (prov:wasAssociatedWith some eo:'Scientific Method'))))))

6   Resource Contributions

We contribute the following publicly available artifacts: **Explanation Ontology** with the **logical formalizations of the different explanation types** and **SPARQL queries** to evaluate the competency questions, along with the applicable documentation available on our resource website. These resources, listed in Table 5, are useful for anyone interested in building explanation facilities into their systems. The ontology has been made available as an open-source artifact under the Apache 2.0 license [21] and we maintain an open source Github repository for all our artifacts. We also maintain a persistent URL for our ontology hosted on the PURL service (Table 5).

Table 5. Links to resources we have released and refer to in the paper.

Resource	Link to resource
Resource website	http://tetherless-world.github.io/explanation-ontology
EO PURL URL	https://purl.org/heals/eo
Github repository	https://github.com/tetherless-world/explanation-ontology

7   Discussion and Future Work

To address the gap in a semantic representation that can be used to support the generation of different explanation types, we designed an OWL ontology, an explanation ontology, that can be used by system designers to incorporate different explanation types into their AI-enabled systems. We leverage and maintain compatibility with an existing explanation patterns ontology [24] and we expand on it to include representational primitives needed for modeling explanation types for system designers. We also leveraged the widely-used general-purpose

SIO ontology, and introduce the classes and properties necessary for a system to generate a variety of user-centric explanations. During our modeling, we make certain decisions that are inspired by our literature-review and knowledge of the usage of explanation types in clinical practice, to include classes (e.g., *'system recommendation,'* *'knowledge,'* *'user'*) that we deem as central to generating explanations. We include other classes that would indirectly be needed to generate explanations and reason about them, hence, capturing the process involved in explanation generation. However, through our ontology we do not generate natural language explanations, and rather provide support to fill in "slots" that will be included in them.

Our explanation ontology is comprehensive and flexible as it was designed from requirements gathered from a relatively extensive literature review along with a requirements gathering session in a clinical domain. In this paper, we have described how the ontology can be used to represent literature-derived explanation types and then how those explanation types address the questions posed by clinicians. The ontology is also designed to be extensible, as with all ontologies, representational needs may arise as applications arise and evolve. Our competency questions provide guidance to system designers as they make their plans for providing explanations within their decision support systems.

In the future, we plan to build a middleware framework (such as the service we alluded to in Sect. 1) that would interact with the system designer, take a user's *'question'* as input, and apply learning techniques on a combination of the user's *'question'* and the inputs available to the AI system, which could include the user's *'situation'* and context, and the system's *'reasoning mode'* to decide on the most suitable *'explanation'* type. Upon identifying the appropriate *'explanation'* type, the framework would leverage the sufficiency conditions encoded in our ontology to gather different forms of *'knowledge'* to generate the suitable explanation type and summon the AI *'tasks'* and *'methods'* that are best suited to generating the explanation. Such a framework would then be capable of working in tandem with hybrid AI reasoners to generate hybrid explanations [4,17] that serve the users' requirements.

We have represented the explainability components that we deem necessary to generate explanations with a user focus. However, there are other aspects of user input that may be harder to capture. Wang et al. [26] have shown that there is a parallel between one of these user aspects, a user's reasoning strategies, and the reasoning types that an *'AI Task'* uses to support the generation of explanations to address these situations. We are investigating how to include classes, such as a user's reasoning strategies, that are harder to capture/infer from a system perspective and would be hard to operationalize in a system's model. The user-centric focus of AI has been emphasized recently by Liao et al. in their question bank of various user questions around explainability that suggests "user needs should be understood, prioritized and addressed" [11]. As we start to build more user attributes into our ontology, we believe that our model will evolve to support more human-in-the-loop AI systems. We are using EO as a foundation for generating different explanation types in designing a clinical decision support system, and we will publish updates to our ontology as we make edits to support new terms.

8   Conclusion

We have built an ontology aimed at modeling explanation primitives that can support user-centered AI system design. Our requirements came from a breadth of literature and requirements gathered by prospective and actual users of clinical decision support systems. We encode sufficiency conditions encapsulating components necessary to compose hybrid explanation types that address different goals and expose different forms of knowledge in our ontology. Through a carefully crafted set of competency questions, we have exposed and evaluated the coverage of our ontology in helping system designers make decisions about explanations to include in their systems. We believe our ontology can be a valuable resource for system designers as they plan for what kinds of explanations their systems will need to support. We are continuing to work towards supporting the generation of different explanation types and designing a service that would use our explanation ontology as a base to generate explanation types that address users' questions.

Acknowledgements. This work is done as part of the HEALS project, and is partially supported by IBM Research AI through the AI Horizons Network. We thank our colleagues from IBM Research, including Ching-Hua Chen, and from RPI, Sabbir M. Rashid, Henrique Santos, Jamie P. McCusker and Rebecca Cowan, who provided insight and expertise that greatly assisted the research.

References

1. Arrieta, A.B., et al.: Explainable artificial intelligence (XAI): concepts, taxonomies, opportunities and challenges toward responsible AI. Inf. Fus. **58**, 82–115 (2020)
2. Arya, V., et al.: One explanation does not fit all: a toolkit and taxonomy of AI explainability techniques. arXiv preprint arXiv:1909.03012 (2019)
3. Biran, O., Cotton, C.: Explanation and justification in machine learning: a survey. In: IJCAI 2017 Workshop on Explainable AI (XAI), vol. 8, p. 1 (2017)
4. Chari, S., Gruen, D.M., Seneviratne, O., McGuinness, D.L.: Directions for explainable knowledge-enabled systems. In: Tiddi, I., Lecue, F., Hitzler, P. (eds.) Knowledge Graphs for Explainable AI - Foundations, Applications and Challenges. Studies on the Semantic Web. IOS Press (2020, to appear)
5. Chari, S., Gruen, D.M., Seneviratne, O., McGuinness, D.L.: Foundations of Explainable Knowledge-Enabled Systems. In: Tiddi, I., Lecue, F., Hitzler, P. (eds.) Knowledge Graphs for Explainable AI - Foundations, Applications and Challenges. Studies on the Semantic Web. IOS Press (2020, to appear)
6. Dhaliwal, J.S., Benbasat, I.: The use and effects of knowledge-based system explanations: theoretical foundations and a framework for empirical evaluation. Inf. Syst. Res. **7**(3), 342–362 (1996)
7. Doshi-Velez, F., Kim, B.: Towards a rigorous science of interpretable machine learning. arXiv preprint arXiv:1702.08608 (2017)

8. Dumontier, M., et al.: The Semanticscience integrated ontology (SIO) for biomedical research and knowledge discovery. J. Biomed. Semant. **5**(1), 14 (2014)
9. Holzinger, A., Biemann, C., Pattichis, C.S., Kell, D.B.: What do we need to build explainable AI systems for the medical domain? arXiv preprint arXiv:1712.09923 (2017)
10. Lebo, T., et al.: Prov-o: the prov ontology. W3C recommendation (2013)
11. Liao, Q.V., Gruen, D., Miller, S.: Questioning the AI: informing design practices for explainable AI user experiences. arXiv preprint arXiv:2001.02478 (2020)
12. Lim, B.Y., Dey, A.K., Avrahami, D.: Why and why not explanations improve the intelligibility of context-aware intelligent systems. In: Proceedings of the SIGCHI Conference on Human Factors in Computing Systems, pp. 2119–2128. ACM (2009)
13. Lipton, Z.C.: The mythos of model interpretability. arXiv preprint arXiv:1606.03490 (2016)
14. Lorin, M.I., Palazzi, D.L., Turner, T.L., Ward, M.A.: What is a clinical pearl and what is its role in medical education? Med. Teach. **30**(9–10), 870–874 (2008)
15. Lou, Y., Caruana, R., Gehrke, J.: Intelligible models for classification and regression. In: Proceedings of the 18th ACM SIGKDD International Conference on Knowledge Discovery and Data Mining, pp. 150–158. ACM (2012)
16. Miller, T.: Explanation in artificial intelligence: insights from the social sciences. Artif. Intell. **267**, 1–38 (2019)
17. Mittelstadt, B., Russell, C., Wachter, S.: Explaining explanations in AI. In: Proceedings of the Conference on Fairness, Accountability, and Transparency, pp. 279–288. ACM (2019)
18. Páez, A.: The pragmatic turn in explainable artificial intelligence (XAI). Mind. Mach. **29**(3), 441–459 (2019). https://doi.org/10.1007/s11023-019-09502-w
19. Dublin Core Metadata Initiative: Dublin Core Metadata Terms. Resource Page. https://www.dublincore.org/specifications/dublin-core/dcmi-terms/2006-12-18/. Accessed 18 Aug 2020
20. IBM Research Trusted AI: AI Explainability 360 Open Source Toolkit. Resource Page. https://aix360.mybluemix.net/. Accessed 18 Aug 2020
21. The Apache Software Foundation: Apache 2.0 License. License Description Page. https://www.apache.org/licenses/LICENSE-2.0. Accessed 18 Aug 2020
22. Stefanelli, M., Ramoni, M.: Epistemological constraints on medical knowledge-based systems. In: Evans, D.A., Patel, V.L. (eds.) Advanced Models of Cognition for Medical Training and Practice. NATO ASI Series (Series F: Computer and Systems Sciences), vol. 97, pp. 3–20. Springer, Heidelberg (1992). https://doi.org/10.1007/978-3-662-02833-9_1
23. Swartout, W.R., Moore, J.D.: Explanation in second generation expert systems. In: David, J.M., Krivine, J.P., Simmons, R. (eds.) Second Generation Expert Systems, pp. 543–585. Springer, Heidelberg (1993). https://doi.org/10.1007/978-3-642-77927-5_24
24. Tiddi, I., d'Aquin, M., Motta, E.: An ontology design pattern to define explanations. In: Proceedings of the 8th International Conference on Knowledge Capture, pp. 1–8 (2015)
25. Tu, S.W., Eriksson, H., Gennari, J.H., Shahar, Y., Musen, M.A.: Ontology-based configuration of problem-solving methods and generation of knowledge-acquisition tools: application of PROTEGE-II to protocol-based decision support. Artif. Intell. Med. **7**(3), 257–289 (1995)
26. Wang, D., Yang, Q., Abdul, A., Lim, B.Y.: Designing theory-driven user-centric explainable AI. In: Proceedings of the 2019 CHI Conference on Human Factors in Computing Systems, pp. 1–15 (2019)

An SKOS-Based Vocabulary on the Swift Programming Language

Christian Grévisse$^{(\boxtimes)}$ⓘ and Steffen Rothkugelⓘ

Department of Computer Science, University of Luxembourg,
2, avenue de l'Université, 4305 Esch-sur-Alzette, Luxembourg
{christian.grevisse,steffen.rothkugel}@uni.lu

Abstract. Domain ontologies about one or several programming languages have been created in various occasions, mostly in the context of Technology Enhanced Learning (TEL). Their benefits range from modeling learning outcomes, over organization and annotation of learning material, to providing scaffolding support in programming labs by integrating relevant learning resources. The Swift programming language, introduced in 2014, is currently gaining momentum in different fields of application. Both its powerful syntax as well as the provided type safety make it a good language for first-year computer science students. However, it has not yet been the subject of a domain ontology. In this paper, we present an SKOS-based vocabulary on the Swift programming language, aiming at enabling the benefits of previous research for this particular language. After reviewing existing ontologies on other programming languages, we present the modeling process of the Swift vocabulary, its integration into the LOD Cloud and list all of its resources available to the research community. Finally, we showcase how it is being used in different TEL tools.

Keywords: Swift · Vocabulary · SKOS · E-Learning

Resource Type: Ontology/Vocabulary
License: CC BY-SA 4.0
Permanent URL: http://purl.org/lu/uni/alma/swift

1 Introduction

Programming languages have been the subject of domain ontologies in several occasions. Such ontologies have typically covered both syntactic and semantic elements of one or several programming languages. The modeled domain knowledge was then used mostly in the context of Technology Enhanced Learning (TEL). The benefits have been manifold: modeling learning outcomes [16], creation of learning paths [8,13,16], semi-automatic annotation of learning resources [7,11,13], organization of learning objects (LOs) [14,19], scaffolding support in

© Springer Nature Switzerland AG 2020
J. Z. Pan et al. (Eds.): ISWC 2020, LNCS 12507, pp. 244–258, 2020.
https://doi.org/10.1007/978-3-030-62466-8_16

programming labs [7,12,21], reusable knowledge beyond the boundaries of systems [11,14,16] as well as visual and linguistic [14,20] aids. As some approaches realize a mapping between the abstract syntax tree (AST) returned by a parser and the elements of their respective ontologies [4,11,13], it became possible to perform static code analysis through SPARQL queries [4] or even retrieve information on a piece of source code through natural language questions [9].

One of the more recent programming languages is Swift, which was only introduced in 2014. First a proprietary language, it became open source and can nowadays be used within Apple's ecosystem, on Linux and even on Windows (using *Windows Subsystem for Linux*). Often considered a niche language for iOS development, it can be used for server-side development (with frameworks like *Vapor*[1]) and has more recently been considered for deep learning and differentiable computing in *Swift for TensorFlow*[2]. While its first versions often broke backward compatibility, the language has, in its 6 years of existence, reached a certain maturity, with Swift 5 guaranteeing ABI (application binary interface) stability. As of May 2020, Swift is among the top 9 programming languages according to the *PYPL (PopularitY of Programming Language)* Index[3] which measures how often language tutorials were searched for on Google, and among the top 11 on the *TIOBE* index[4]. Furthermore, the *Developer Survey 2019* carried out by the popular question-and-answer site *Stack Overflow* shows Swift among the top 6 most "loved" languages[5]. From an educational point of view, Swift is, similar to Python, less verbose than Java, but, as a statically-typed language, provides type safety, which can be beneficial to both new and experienced programmers.

In this paper, we present an SKOS-based vocabulary on the Swift programming language. Building on the growing popularity of the language, this vocabulary aims at enabling the previously mentioned benefits of modeling domain ontologies on programming languages for Swift. Currently, its main goal is to provide a controlled set of concepts used in the annotation of related learning material, as well as to enable scaffolding support in programming labs by integrating relevant resources.

The remainder of this paper is organized as follows. In Sect. 2, we present an overview of previous programming language ontologies. The modeling process and implementation of the Swift vocabulary is described in Sect. 3. Section 4 shows how the vocabulary is currently used. A general discussion is given in Sect. 5. We conclude in Sect. 6, along some ideas for future work.

---

[1] https://vapor.codes.
[2] https://www.tensorflow.org/swift/.
[3] https://pypl.github.io/PYPL.html.
[4] https://tiobe.com/tiobe-index/.
[5] https://insights.stackoverflow.com/survey/2019#most-loved-dreaded-and-wanted.

2   Related Work

There have been several contributions in research covering one or more programming languages in domain ontologies, of differing complexity and construction methodologies.

Most of these attempts focused on the Java programming language. In [16], a Java ontology is created to model learning outcomes of a distance education course, effectively determining learning paths. The concepts comprised in this ontology were collected from the *Java Tutorials*[6]. The *QBLS* intelligent tutoring system (ITS) presented in [7] uses a similar Java ontology to annotate and provide learning resources as a support in programming labs. While the initial set of concepts was retrieved again from the Java Tutorials, the authors needed to add further concepts while annotating learning material, which shows a gap between the initial domain ontology design and the intention of the annotating teacher. The *Protus* ITS also uses an ontology on Java to provide semantic recommendations of resources [21]. Ivanova built a bilingual Java ontology to help Bulgarian students understand English terminology [14]. This ontology goes beyond pure language features, but also includes application types, such as applets or servlets. The author states that the visualization of such an ontology provides students a quick overview of the concepts and terminology to be learned. The *JavaParser* [13] is a fine-grained concept indexing tool for Java problems. Mapping AST elements to entities in a custom Java ontology enables sequencing of code examples. The *CodeOntology* [4], while mainly tailored to Java, provides a formal representation of object-oriented programming (OOP) languages, and serializes Java source code to RDF triples. This ultimately makes it possible to perform static code analysis through SPARQL queries. Here, the Java language specification was used to retrieve the comprised concepts.

There has also been work on ontologies about the C programming language, used in University courses [18,20]. The ontology presented in [20] was constructed based on glossary terms from the reference book by Kernighan and Ritchie (often abbreviated as *K&R*) [15]. Similar to [14], the authors state that visualizing their ontology can give students a navigation interface across learning objects.

There is also work covering more than one concrete programming language. Pierrakeas et al. constructed two ontologies on Java and C in order to organize LOs [19]. They again used a glossary-based approach, using the Java Tutorials and the K&R book. The *ALMA Ontology* in [11] combines both languages under one SKOS-based umbrella ontology, separating language-specific concepts in different modules, while aligning common elements across the modules. A similar approach focusing on OOP languages is done in *OOC-O* [1], which realizes a consensual conceptualization between common elements from Smalltalk, Eiffel, C++, Java and Python to foster *polyglot programming*.

Finally, there have also been ontologies on other languages not yet mentioned. *CSCRO* is an ontology for semantic analysis of C# source code [9]. Based on the *Source Code Representation Ontology (SCRO)* [2], an ontology solution

---

[6] https://docs.oracle.com/javase/tutorial/.

for detecting programming patterns in Java code, concepts from MSDN references on C# language features were added. This ontology was used to retrieve information from source code through natural language questions. *PasOnto*, an ontology on the Pascal language, was created to describe prerequisites necessary to understand exercise solutions [8]. Again, a glossary approach was used, extracting terms from existing C ontologies and Pascal programming courses.

While mostly TEL applications were the target of the previously mentioned ontologies, a semantic representation of code can enable code analysis and queries [4,9]. The *Function ontology* [6] was created to declare and describe functions, problems and algorithms, independently of a concrete programming language. As opposed to a specification, which defines how to use a function (e.g., on a web service), using this ontology an unambiguous description of what a function does can be realized, while omitting any language-specific implementation details. Similarly, in model-driven development, platform ontologies can be used to define platform dependencies of model transformations and reason about their applicability to specific platforms [22].

3   Swift Vocabulary

The use of ontologies generally fosters the interoperability, discovery, reusability and integration of data and resources within and beyond the boundaries of heterogeneous systems in an unambiguous and machine-readable way. To enable the advantages related to these objectives as realized in the related work for other languages, our main requirement for the Swift vocabulary was to cover the major features of the language. The level of granularity should be reasonable, considering its main application domain, i.e., the annotation and retrieval of learning material. In fact, the resulting vocabulary shall not overwhelm a human annotator. A too fine-grained knowledge representation, as returned by a parser, would violate *Occam's razor*[7], whereas a too coarse-grained one would not allow to annotate learning material in the necessary detail.

For annotation purposes, we can stay in the spectrum of *lightweight* ontologies [3,18] that (i) comprise a controlled vocabulary of entities, (ii) organize these entities in a hierarchical structure, and (iii) enable relations among them. Similar to [7,11,18], we decided to use *SKOS (Simple Knowledge Organization System)*[8], a W3C recommendation for KOS using RDF. SKOS organizes *concepts* within *concept schemes*, provides *labels* to concepts and realizes both hierarchical (e.g., skos:broader) and associative (skos:related) links.

The vocabulary described in this paper is based on version *5.2* of the Swift programming language, released in March 2020. Nevertheless, upcoming releases of the language will hardly change a lot of the existing language features, rather than add new ones, making this initial version of the Swift vocabulary a solid starting point for future versions of the ontology.

---

[7] *"Pluralitas non est ponenda sine necessitate."*
[8] https://www.w3.org/2004/02/skos/.

3.1 Ontology Design

For the Swift Vocabulary, we decided to go for a glossary-based approach that was used most of the time in related work. Our previous experience described in [11], where we proceeded with a bottom-up approach by modeling the syntactic elements returned by a parser, resulted in too many purely parser-related (and parser-dependent) elements, requiring a significant filtering and renaming effort, as well as the addition of missing high-level concepts. While a glossary-based approach might still suffer from a gap between the resulting ontology and the concepts needed for annotations as mentioned in [7], this gap would be even bigger when relying on a parser-based entity extraction method.

As a main reference, we used *The Swift Programming Language*[9] (henceforth called the "Swift book"), the authoritative reference comprising a language guide and its formal specification. A vocabulary created based on the content of this reference book can reasonably be considered as a vocabulary about the language itself, comprising concepts that fit the purpose for our TEL use case. However, this vocabulary shall not be seen as a full-fledged, formal ontology covering the semantics of the language itself, which was not the goal of this work.

The Swift book is available for free, either in the browser or as an ePub. A Python script using the popular *Beautiful Soup*[10] library crawled and parsed the online version. Due to the lack of an index, our working assumption here is that relevant concepts are mentioned in the headings, i.e., the HTML tags h1 to h6. The heading strings were then passed through *spaCy*[11], a natural language processing library. After stopword removal and lemmatization, the encountered noun phrases (if any) were considered as candidate concepts. In case no noun phrase was contained (e.g., headings simply stating the name of a keyword like break or continue), the verbatim text was taken as a candidate. All headings comprise a permalink anchor (a tag) to the respective section.

The resulting set contained 458 candidate concepts, mapped to their respective permalinks (some concepts were mentioned at several occasions). This high number is due to the fact that the book includes many in-depth discussions and examples, such that not all headings represent a new, dedicated concept. To filter such headings and avoid a too fine-grained representation, we manually selected a subset of 139 concepts, however added 31 additional ones representing important keywords and access level modifiers that were not subject of a dedicated heading, yet mentioned in the text. For the remaining 170 concepts, we cleansed the concept names (removing non-alphabetical characters, applying camel case), gave them a preferred label (skos:prefLabel) and selected the best-matching resource from the reference book. A first representation in RDF comprising only the concepts and their resource was created using the *RDFLib*[12] package. The skos:definition property was chosen to map each concept to the URL of the corresponding excerpt in the Swift book.

---

[9] https://docs.swift.org/swift-book/.
[10] https://www.crummy.com/software/BeautifulSoup/.
[11] https://spacy.io.
[12] https://rdflib.readthedocs.io/en/stable/.

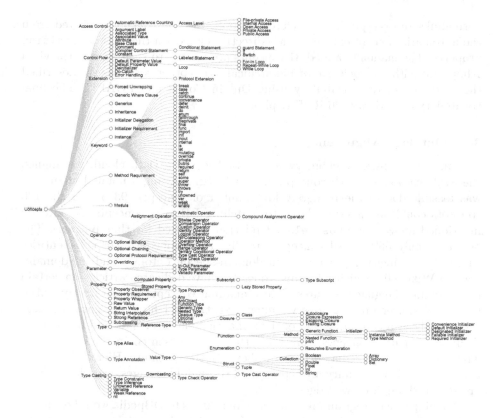

Fig. 1. Dendrogram of the Swift vocabulary

Using *Protégé*[13], we then created both hierarchical and associative links among the concepts. As the number of relevant concepts about a programming language is typically below 200, this manual process is feasible. An automatic approach relying on cross-references in the text to establish associative links would have been possible, but would have been incomplete (not enough cross-references) and yielded false positives (e.g., the section on *Property Observers* mentions *Overriding*, which a priori is not a related concept). This again would have required some manual curation.

Hierarchical links were modeled using 120 `skos:broader` relations. A dendrogram representation of the resulting hierarchy, generated using the *Ontospy*[14] library, is shown in Fig. 1.

Associative links were modeled using 218 `skos:related` relations, of which 55% were internal relations, the remainder being used for ontology alignment purposes (see Sect. 3.2). All concepts are covered by at least one associative or hierarchical link, effectively avoiding any orphan concepts. With respect to the

---

[13] https://protege.stanford.edu.
[14] https://lambdamusic.github.io/Ontospy/.

permalinks to excerpts in the reference book, the 139 concepts selected from the automatic concept extraction are directly related via the skos:definition property. The manually added 31 concepts are related to such an excerpt via a 1-hop link (either associative or hierarchical) to another concept. We described the metadata of the vocabulary using Dublin Core[15] and FOAF[16]. The final vocabulary comprises 1226 RDF triples.

3.2 Ontology Alignment

To connect the Swift vocabulary to the Linked Open Data cloud, we aligned the Swift vocabulary with concepts from DBpedia. Again, a manual alignment was feasible due to the relatively low number of concepts. 93 out of the 170 concepts could be directly related to matching DBpedia concepts. As already mentioned above, the skos:related relation was used for that purpose. The fact that only 55% of all concepts are directly mapped to DBpedia entities shows the specificity gap between our domain-specific ontology and the domain-general Wikipedia, which oftentimes does not have dedicated articles on certain programming language features. The remaining concepts are indirectly linked to DBpedia, over up to 2 hops via skos:related or 1 hop via skos:broader. Figure 2 shows a chord diagram of the semantic relations inside and beyond the Swift vocabulary. Concepts from the latter are represented in the orange arc, whereas mapped DBpedia entities are shown in the green arc. Hierarchical links inside the Swift vocabulary are represented through blue edges, associative links are shown through brown edges.

Aligning other programming language ontologies to DBpedia would establish an indirect relation to the Swift vocabulary, which ultimately could enable the retrieval of related learning material across different programming languages. This can be particularly useful, if a student has already seen a concept present in a language A in another language B, and wants to review or compare the resources on the concept from language B to better understand the concept in language A, fostering higher level thinking skills of students. In addition, collaboration among researchers could be improved this way, to avoid recreating an ontology on the same programming language, as shown in Sect. 2.

3.3 Available Resources

Table 1 lists all available resources related to the Swift vocabulary. It has been serialized in RDF/XML and Turtle format. The permanent URL (PURL) leads to the documentation, which has been generated using WIDOCO [10] and the aforementioned OntoSpy library. Through content negotiation, the RDF/Turtle representations can also be retrieved through the PURL, e.g.,

```
curl -sH "Accept: application/rdf+xml" \
     -L http://purl.org/lu/uni/alma/swift
```

---

[15] https://www.dublincore.org/specifications/dublin-core/dcmi-terms/.
[16] http://xmlns.com/foaf/spec/.

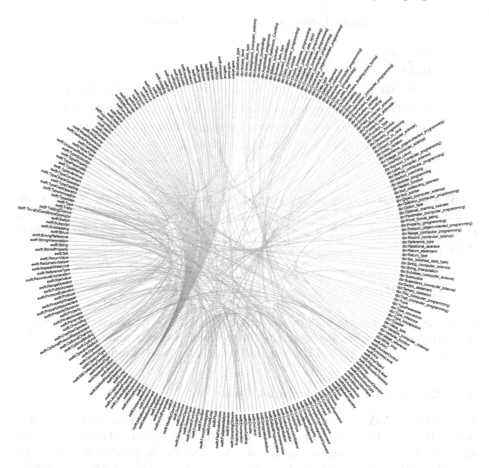

Fig. 2. Chord diagram showing semantic relations inside and beyond the Swift Vocabulary (Color figure online)

For the SPARQL endpoint, we use the *Virtuoso Open-Source Edition*[17]. It allows to retrieve, e.g., the resource from the Swift reference book on the "For-In Loop":

```
SELECT DISTINCT ?resource
WHERE {
    ?concept skos:prefLabel "For-In Loop" .
    ?concept skos:definition ?resource
}
```

Further learning resources could be indexed using an INSERT query, the necessary access rights provided.

---

[17] http://vos.openlinksw.com.

Table 1. Swift Vocabulary-related resources

Resource	URL
PURL	http://purl.org/lu/uni/alma/swift
Turtle file	https://alma.uni.lu/ontology/swift/5.2/swift.ttl
RDF file	https://alma.uni.lu/ontology/swift/5.2/swift.rdf
GitHub repository	https://github.com/cgrevisse/swift-vocabulary
SPARQL endpoint	https://alma.uni.lu/sparql

The vocabulary is released under the Attribution-ShareAlike 4.0 International (CC BY-SA 4.0)[18] license. As a canonical citation, please use: "Grévisse, C. and Rothkugel, S. (2020). *Swift Vocabulary.* http://purl.org/lu/uni/alma/swift".

4   Applications

With the main focus on building a vocabulary suitable for annotating, retrieving and integrating learning material on the Swift programming language, this section will now present how this vocabulary is being used in different TEL tools.

4.1   Extension for Visual Studio Code

Visual Studio Code is a cross-platform source-code editor. Released in 2015, it is among the top 4 IDEs according to the *Top IDE index*[19], as of May 2020, and the most popular development environment according to the previously mentioned Stack Overflow development survey. We created the *ALMA 4 Code* extension[20], which enables the user to select a piece of code and be provided with related learning material. The extension currently supports Swift in its version 5.2. The learning resources are retrieved from the ALMA repository [12], which is hosted on the same server as our SPARQL endpoint. The extension focuses on providing resources to the select syntactical element(s), which can be useful to a student in a programming course to understand, e.g., code examples or exercise solutions.

Under the hood, the extension uses `swift-semantic`[21], a command-line utility we wrote that, for a given `.swift` file and selection range, returns the semantic concept from the Swift vocabulary of the top-most AST node. The underlying parser is given by the *SwiftSyntax*[22] library.

---

[18] https://creativecommons.org/licenses/by-sa/4.0/.
[19] https://pypl.github.io/IDE.html.
[20] https://marketplace.visualstudio.com/items?itemName=cgrevisse.alma4code.
[21] https://github.com/cgrevisse/swift-semantic.
[22] https://github.com/apple/swift-syntax.

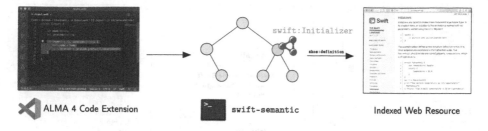

Fig. 3. Example workflow of the "ALMA 4 Code" extension for Visual Studio Code

An example workflow is shown in Fig. 3. The user has selected the initializer of a structure. Through a simple keyboard shortcut, the ALMA 4 Code extension is activated, which passes the source code as well as the selection parameters (start/end line/column) to the `swift-semantic` utility. The latter then builds the AST using the SwiftSyntax library and retrieves the top-most node inside of the selection. Through a mapping between AST node types and Swift vocabulary concepts, the corresponding concept is returned to the extension. Finally, the ALMA repository is searched for learning material indexed for this concept. In this example, the part on initializers from the Swift book is returned.

4.2 Annotation of Learning Material

The previously mentioned ALMA repository is not only used by the ALMA 4 Code extension. The ecosystem described in [12] also comprises Yactul, a gamified student response platform. The Yactul app can be used to replay quizzes from the class, keep track of the learning progress for an individual user and direct her to learning material related to a quiz activity. As shown in Fig. 4, if a player wants to know more about the concepts behind a question (e.g., when her answer was wrong), she can consult related resources, due to the fact that Yactul activities are tagged with semantic concepts. In this example, a question targeting the `mutating` keyword (required by methods that change the state of a Swift structure) is tagged with the concept `swift:mutating` from this vocabulary, such that the related excerpt from the Swift book can be retrieved and shown directly in the app.

Finally, we also use the concepts from our vocabulary as hashtags on our lab sheets (Fig. 5). Hashtags are a popular type of folksonomy used in many social networking sites. The LaTeX-generated lab sheets include these hashtags both for the learner (human agent) to understand the topic of an exercise, as well as for software agents in the form of machine-readable metadata of the PDF file. Clicking on a hashtag leads students to resources indexed in the ALMA repository, similar to the VS Code extension or the Yactul app. Our previous research [12] has shown that all 3 approaches of learning material integration, i.e., in an IDE, in the Yactul app and as hashtags on lab sheets are considered useful by first-year computer science students.

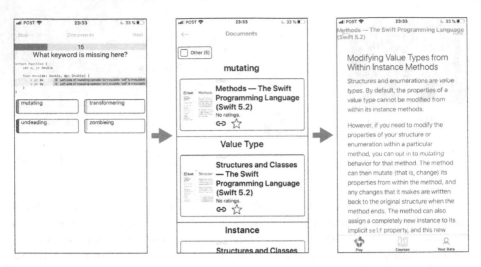

Fig. 4. Retrieval of learning material in the Yactul app

Exercise 31 – Hospital   **#Closure**

The goal of this exercise is to write a 🔶 **Swift** console application that models a patient management system for hospitals. On Moodle, you can find some boilerplate code, which defines a set of symptoms and hospital units, as well as permits the generation of random patients. You can also simply print a patient instance, a meaningful string representation has been provided.

Using closures, you will print patient lists according to certain criteria. Common operations like `filter`, `map`, `reduce`, `sorted` and `forEach` will be heavily used throughout this exercise. It is not permitted to use any explicit loops in this exercise.

Fig. 5. Use of a vocabulary concept in form of a hashtag on a lab sheet

During the summer semester 2020 (February - July), in which instruction was mostly delivered in a remote way due to the COVID-19 pandemic, the Swift vocabulary was first used in the context of a Swift course taught at the University of Luxembourg. At the time of writing, we have observed that learning resources from the Swift book were consulted 84 times through clicking on hashtags on the lab sheets. The most frequently visited resource was the excerpt on "Computed Properties"[23] (concept URI: `swift:ComputedProperty`), a feature unknown to the students from their prior knowledge on Java.

5   Discussion

Impact. The Swift vocabulary presented in this paper is, to the best of our knowledge, the first semantic knowledge representation about the Swift programming language. Similar to related work, which was mainly focusing on Java, this vocabulary was modeled using a glossary-based approach, relying on the Swift reference book. The benefits that related work has brought, mainly in the context of e-learning systems, can now be implemented for this relatively new, yet

---

[23] https://docs.swift.org/swift-book/LanguageGuide/Properties.html#ID259.

stable language, using this vocabulary. As previously mentioned, it has already been adopted in a Swift course for annotating and retrieving learning material, which is actively used by students.

Extensibility. Although the language has become rather stable by now, new language features are constantly integrated, and developers can submit proposals on a dedicated GitHub repository of Apple[24]. The architecture of SKOS, used in this vocabulary, follows the principle of making *minimal ontological commitment* [5], hence makes it easy to extend the vocabulary by integrating new language features.

Availability. As listed in Table 1, all resources are available through a permanent URL and on GitHub. The VS Code extension described in Sect. 4.1 is available on the Visual Studio Marketplace. The Swift Vocabulary has also been registered in the LOD Cloud[25].

Reproducibility. The code used in the ontology creation process described in Sect. 3 is available on the GitHub repository mentioned in Table 1. Apart from the reference serializations, it contains a Python script to generate the initial set of candidate concepts by extraction from the Swift book, along a roadmap for the manual curation.

Reusability. The only semantic commitment is given by the few formal requirements imposed by SKOS. No further domains or ranges have been defined as purely SKOS properties were used. Concepts can thus be easily reused in other ontologies, either directly or indirectly using the alignment with DBpedia.

Sustainability. As the Swift vocabulary is being actively used in the context of a Swift programming course, it is in the developing group's very own interest to keep it up-to-date with respect to further language features that would be introduced to students, either in a pure Swift programming course or an iOS development-related one. The vocabulary could also be pitched to receive its own sub-category in the "Related Projects" forum[26] to gain the recognition and support of the Swift open source community.

Maintenance. Should new language features be introduced or missing, the authors are always welcoming pull requests and issues on GitHub. We are also open for collaboration with other researchers in aligning their programming language ontologies with ours, either directly or indirectly through DBpedia.

---

[24] https://github.com/apple/swift-evolution.
[25] https://lod-cloud.net/dataset/Swift.
[26] https://forums.swift.org/c/related-projects/25.

Quality and Validation. The RDF serialization of the Swift vocabulary validated successfully according to the W3C RDF Validation Service[27]. We also used *qSKOS* [17] to check the quality of the SKOS vocabulary, with no *errors* found. The encountered warnings can be justified as follows: As previously mentioned, the 31 concepts manually added do not include a documentation property, but are linked to an excerpt of the Swift book through a 1-hop relation. The concepts `swift:Attribute` and `swift:Comment` are indeed not connected to other concepts in the Swift vocabulary, which is due to their true nature, but they are linked to matching DBpedia entities. Finally, reciprocal relations are not included, as `skos:related` is an `owl:SymmetricProperty`, whereas `skos:broader` and `skos:hasTopConcept` declare inverse properties (`owl:inverseOf`).

6   Conclusion and Future Work

In this paper, we presented an SKOS-based vocabulary on the Swift programming language. Previous work on other languages has shown the advantages of such domain ontologies, mainly in the context of TEL applications. Using a glossary-based modeling approach, we created a controlled set of concepts covering the majority of both syntactical and semantical elements for this increasingly popular language. Our vocabulary is aligned with DBpedia and registered in the LOD Cloud, making it 5-star open data. Furthermore, we have showcased applications of the vocabulary, both in an extension for the popular Visual Studio Code editor, as well as in form of annotations on learning material, such as quiz activities or hashtags on lab sheets. The latter, which is being actively used in our Swift course enables scaffolding support by integrating relevant resources from the Swift book. The Swift vocabulary can be beneficial to fellow programming teachers and Semantic Web researchers alike.

For future work, we plan on integrating a separate concept scheme on SwiftUI, a new framework for declarative user interface design for Apple's platforms. Furthermore, prerequisite relations could be introduced in the vocabulary, to determine learning paths. These relations could be deduced from the structure of the Swift book, and could ultimately be used to sequence other learning resources (e.g., videos) likewise. Finally, Swift code on GitHub could be indexed for concepts from the Swift vocabulary, to provide code examples on certain topics to students.

Acknowledgments. We would like to thank Rubén Manrique from the Universidad de los Andes (Bogotá, Colombia) for his valuable input on Virtuoso.

---

[27] https://www.w3.org/RDF/Validator/.

References

1. de Aguiar, C.Z., de Almeida Falbo, R., Souza, V.E.S.: OOC-O: a reference ontology on object-oriented code. In: Laender, A.H.F., Pernici, B., Lim, E.-P., de Oliveira, J.P.M. (eds.) ER 2019. LNCS, vol. 11788, pp. 13–27. Springer, Cham (2019). https://doi.org/10.1007/978-3-030-33223-5_3
2. Alnusair, A., Zhao, T.: Using ontology reasoning for reverse engineering design patterns. In: Ghosh, S. (ed.) MODELS 2009. LNCS, vol. 6002, pp. 344–358. Springer, Heidelberg (2010). https://doi.org/10.1007/978-3-642-12261-3_32
3. Andrews, P., Zaihrayeu, I., Pane, J.: A classification of semantic annotation systems. Semant. Web **3**(3), 223–248 (2012). https://doi.org/10.3233/SW-2011-0056
4. Atzeni, M., Atzori, M.: CodeOntology: RDF-ization of source code. In: d'Amato, C., et al. (eds.) ISWC 2017. LNCS, vol. 10588, pp. 20–28. Springer, Cham (2017). https://doi.org/10.1007/978-3-319-68204-4_2
5. Baker, T., Bechhofer, S., Isaac, A., Miles, A., Schreiber, G., Summers, E.: Key choices in the design of Simple Knowledge Organization System (SKOS). J. Web Semant. **20**, 35–49 (2013). https://doi.org/10.1016/j.websem.2013.05.001
6. De Meester, B., Dimou, A., Verborgh, R., Mannens, E.: An ontology to semantically declare and describe functions. In: Sack, H., Rizzo, G., Steinmetz, N., Mladenić, D., Auer, S., Lange, C. (eds.) ESWC 2016. LNCS, vol. 9989, pp. 46–49. Springer, Cham (2016). https://doi.org/10.1007/978-3-319-47602-5_10
7. Dehors, S., Faron-Zucker, C.: QBLS: a semantic web based learning system. In: Proceedings of EdMedia: World Conference on Educational Media and Technology 2006, pp. 2795–2802. Association for the Advancement of Computing in Education (AACE) (2006)
8. Diatta, B., Basse, A., Ouya, S.: PasOnto: ontology for learning Pascal programming language. In: 2019 IEEE Global Engineering Education Conference (EDUCON), pp. 749–754 (2019). https://doi.org/10.1109/EDUCON.2019.8725092
9. Epure, C., Iftene, A.: Semantic analysis of source code in object oriented programming. A case study for C#. Romanian J. Human-Comput. Interact. **9**(2), 103–118 (2016)
10. Garijo, D.: WIDOCO: a wizard for documenting ontologies. In: d'Amato, C., et al. (eds.) ISWC 2017. LNCS, vol. 10588, pp. 94–102. Springer, Cham (2017). https://doi.org/10.1007/978-3-319-68204-4_9
11. Grévisse, C., Botev, J., Rothkugel, S.: An extensible and lightweight modular ontology for programming education. In: Solano, A., Ordoñez, H. (eds.) CCC 2017. CCIS, vol. 735, pp. 358–371. Springer, Cham (2017). https://doi.org/10.1007/978-3-319-66562-7_26
12. Grévisse, C., Rothkugel, S., Reuter, R.A.P.: Scaffolding support through integration of learning material. Smart Learn. Environ. **6**(1), 1–24 (2019). https://doi.org/10.1186/s40561-019-0107-0
13. Hosseini, R., Brusilovsky, P.: JavaParser: a fine-grain concept indexing tool for Java problems. In: First Workshop on AI-supported Education for Computer Science (AIEDCS) at the 16th Annual Conference on Artificial Intelligence in Education, pp. 60–63 (2013)
14. Ivanova, T.: Bilingual ontologies for teaching programming in Java. In: Proceedings of the International Conference on Information Technologies, pp. 182–194 (2014). https://doi.org/10.13140/2.1.2126.4967
15. Kernighan, B.W., Ritchie, D.M.: The C Programming Language. Prentice Hall, Englewood Cliffs (1988)

16. Kouneli, A., Solomou, G., Pierrakeas, C., Kameas, A.: Modeling the knowledge domain of the Java programming language as an ontology. In: Popescu, E., Li, Q., Klamma, R., Leung, H., Specht, M. (eds.) ICWL 2012. LNCS, vol. 7558, pp. 152–159. Springer, Heidelberg (2012). https://doi.org/10.1007/978-3-642-33642-3_16

17. Mader, C., Haslhofer, B., Isaac, A.: Finding quality issues in SKOS vocabularies. In: Zaphiris, P., Buchanan, G., Rasmussen, E., Loizides, F. (eds.) TPDL 2012. LNCS, vol. 7489, pp. 222–233. Springer, Heidelberg (2012). https://doi.org/10.1007/978-3-642-33290-6_23

18. Miranda, S., Orciuoli, F., Sampson, D.G.: A SKOS-based framework for subject ontologies to improve learning experiences. Comput. Hum. Behav. **61**, 609–621 (2016). https://doi.org/10.1016/j.chb.2016.03.066

19. Pierrakeas, C., Solomou, G., Kameas, A.: An ontology-based approach in learning programming languages. In: 2012 16th Panhellenic Conference on Informatics, pp. 393–398 (2012). https://doi.org/10.1109/PCi.2012.78

20. Sosnovsky, S., Gavrilova, T.: Development of educational ontology for C-programming. Int. J. Inf. Theor. Appl. **13**(4), 303–308 (2006)

21. Vesin, B., Ivanović, M., Klašnja-Milićević, A., Budimac, Z.: Protus 2.0: ontology-based semantic recommendation in programming tutoring system. Expert Syst. Appl. **39**(15), 12229–12246 (2012). https://doi.org/10.1016/j.eswa.2012.04.052

22. Wagelaar, D., Van Der Straeten, R.: Platform ontologies for the model-driven architecture. Eur. J. Inf. Syst. **16**(4), 362–373 (2007). https://doi.org/10.1057/palgrave.ejis.3000686

The Virtual Knowledge Graph System Ontop

Guohui Xiao[1,2]($\boxtimes$) (iD), Davide Lanti[1] (iD), Roman Kontchakov[3] (iD),
Sarah Komla-Ebri[2] (iD), Elem Güzel-Kalaycı[4] (iD), Linfang Ding[1] (iD),
Julien Corman[1] (iD), Benjamin Cogrel[2] (iD), Diego Calvanese[1,2,5] (iD),
and Elena Botoeva[6] (iD)

[1] Free University of Bozen-Bolzano, Bolzano, Italy
{xiao,lanti,ding,corman,calvanese}@inf.unibz.it
[2] Ontopic s.r.l., Bolzano, Italy
{guohui.xiao,sarah.komla-ebri,benjamin.cogrel,
diego.calvanese}@ontopic.biz
[3] Birkbeck, University of London, London, UK
roman@dcs.bbk.ac.uk
[4] Virtual Vehicle Research GmbH, Graz, Austria
elem.guezelkalayci@v2c2.at
[5] Umeå University, Umeå, Sweden
diego.calvanese@umu.se
[6] Imperial College London, London, UK
e.botoeva@imperial.ac.uk

Abstract. *Ontop* is a popular open-source virtual knowledge graph system that can expose heterogeneous data sources as a unified knowledge graph. *Ontop* has been widely used in a variety of research and industrial projects. In this paper, we describe the challenges, design choices, new features of the latest release of *Ontop* v4, summarizing the development efforts of the last 4 years.

1 Introduction

The Virtual Knowledge Graph (VKG) approach, also known in the literature as Ontology-Based Data Access (OBDA) [16,23], has become a popular paradigm for accessing and integrating data sources [24]. In such approach, the data sources, which are normally relational databases, are *virtualized* through a mapping and an ontology, and presented as a unified *knowledge graph*, which can be queried by end-users using a vocabulary they are familiar with. At query time, a VKG system translates user queries over the ontology to SQL queries over the database system. This approach frees end-users from the low-level details of data organization, so that they can concentrate on their high-level tasks. As it is gaining more and more importance, this paradigm has been implemented in several systems [3,4,18,21], and adopted in a large range of use cases. Here, we present the latest major release, *Ontop* v4, of a popular VKG system.

The development of *Ontop* has been a great adventure spanning the past decade. Developing such a system is highly non-trivial. It requires both a

© Springer Nature Switzerland AG 2020
J. Z. Pan et al. (Eds.): ISWC 2020, LNCS 12507, pp. 259–277, 2020.
https://doi.org/10.1007/978-3-030-62466-8_17

theoretical investigation of the semantics, and strong engineering efforts to implement all the required features. *Ontop* started in 2009, only one year after the first version of the SPARQL specification had been standardized, and OWL 2 QL [14] and R2RML [9] appeared 3 years later in 2012. At that time, the VKG research focused on union of conjunctive queries (UCQs) as a query language. With this target, the v1 series of *Ontop* was using Datalog as the core for data representation [20], since it was fitting well a setting based on UCQs. The development of *Ontop* was boosted during the EU FP7 project Optique (2013–2016). During the project, the compliance with all the relevant W3C recommendations became a priority, and significant progress has been made. The last release of *Ontop* v1 was v1.18 in 2016, which is the result 4.6K git commits. A full description of *Ontop* v1 is given in [3], which has served as the canonical citation for *Ontop* so far.

A natural requirement that emerged during the Optique project were aggregates introduced in SPARQL 1.1 [12]. The *Ontop* development team spent a major effort, internally called *Ontop* v2, in implementing this query language feature. However, it became more and more clear that the Datalog-based data representation was not well suited for this implementation. Some prototypes of *Ontop* v2 were used in the Optique project for internal purposes, but never reached the level of a public release. We explain this background and the corresponding challenges in Sect. 2.

To address the challenges posed by aggregation and others that had emerged in the meantime, we started to investigate an alternative core data structure. The outcome has been what we call *intermediate query* (IQ), an algebra-based data structure that unifies both SPARQL and relational algebra. Using IQ, we have rewritten a large fragment of the code base of *Ontop*. After two beta releases in 2017 and 2018, we have released the stable version of *Ontop* v3 in 2019, which contains 4.5K commits with respect to *Ontop* v1. After *Ontop* v3, the development focus was to improve compliance and add several major features. In particular, aggregates are supported since *Ontop* v4-beta-1, released late 2019. We have finalized *Ontop* v4 and released it in July 2020, with 2.3K git commits. We discuss the design of *Ontop* v4 and highlight some benefits of IQ that VKG practitioners should be aware of in Sect. 3.

Ontop v4 has greatly improved its compliance with relevant W3C recommendations and provides good performance in query answering. It supports almost all the features of SPARQL 1.1, R2RML, OWL 2 QL and SPARQL entailment regime, and the SPARQL 1.1 HTTP Protocol. Two recent independent evaluations [7,15] of VKG systems have confirmed the robust performance of *Ontop*. When considering all the perspectives, like usability, completeness, and soundness, *Ontop* clearly distinguishes itself among the open source systems. We describe evaluations of *Ontop* in Sect. 4.

Ontop v4 is the result of an active developer community. The number of git commits now sums up to 11.4K. It has been downloaded more 30K times from Sourceforge. In addition to the research groups, *Ontop* is also backed by a commercial company, Ontopic s.r.l., born in April 2019. *Ontop* has been adopted in many academic and industrial projects [24]. We discuss the community effort and the adoption of *Ontop* in Sect. 5.

2 Background and Challenges

A Virtual Knowledge Graph (VKG) system provides access to data (stored, for example, in a relational database) through an *ontology*. The purpose of the ontology is to define a vocabulary (classes and properties), which is convenient and familiar to the user, and to extend the data with background knowledge (e.g., subclass and subproperty axioms, property domain and range axioms, or disjointness between classes). The terms of the ontology vocabulary are connected to data sources by means of a *mapping*, which can be thought of as a collection of database queries that are used to construct class and property assertions of the ontology (RDF dataset). Therefore, a VKG system has the following components: (*a*) queries that describe user information needs, (*b*) an ontology with classes and properties, (*c*) a mapping, and (*d*) a collection of data sources. W3C published recommendations for languages for components *(a)–(c)*: SPARQL, OWL 2 QL, and R2RML, respectively; and SQL is the language for relational DBMSs.

A distinguishing feature of VKG systems is that they retrieve data from data sources only when it is required for a particular user query, rather than extracting all the data and materializing it internally; in other words, the Knowledge Graph (KG) remains *virtual* rather than materialized. An advantage of this approach is that VKG systems expose the actual up-to-date information. This is achieved by delegating query processing to data sources (notably, relational DBMSs): user queries are translated into queries to data sources (whilst taking account of the ontology background knowledge). And, as it has been evident from the early days [4,19,21,22], performance of VKG systems critically depends on the sophisticated query optimization techniques they implement.

Ontop v1. In early VKG systems, the focus was on answering conjunctive queries (CQs), that is, conjunctions of unary and binary atoms (for class and property assertions respectively). As for the ontology language, OWL 2 QL was identified [2,5] as an (almost) maximal fragment of OWL that can be handled by VKG systems (without materializing all assertions that can be derived from the ontology). In this setting, a *query rewriting* algorithm compiles a CQ and an OWL 2 QL ontology into a union of CQs, which, when evaluated over the data sources, has the same answers as the CQ mediated by the OWL 2 QL ontology. Such algorithms lend themselves naturally to an implementation based on non-recursive Datalog: a CQ can be viewed as a clause, and the query rewriting algorithm transforms each CQ (a clause) into a union of CQs (a set of clauses). Next, in the result of rewriting, query atoms can be replaced by their 'definitions' from the mapping. This step, called *unfolding*, can also be naturally represented in the Datalog framework: it corresponds to partial evaluation of non-recursive Datalog programs, provided that the database queries are SELECT-PROJECT-JOIN (SPJ) [16]. So, Datalog was the core data structure in *Ontop* v1 [20], which translated CQs mediated by OWL 2 QL ontologies into SQL queries. The success of *Ontop* v1 heavily relied on the semantic query optimization (SQO) techniques [6] for simplifying non-recursive Datalog programs. One of the most important lessons learnt in that implementation is that rewriting and unfolding, even though they are separate steps from a theoretical point of view, should be considered together in practice: a mapping can be combined with the subclass and subproperty relations of the ontology, and the resulting *saturated mapping* (or *T-mapping*) can be constructed and optimized before any query is processed, thus taking advantage of performing the expensive SQO only once [19].

Ontop Evolution: From Datalog to Algebra. As *Ontop* moved towards supporting the W3C recommendations for SPARQL and R2RML, new challenges emerged.

- In SPARQL triple patterns, variables can occur in positions of class and property names, which means that there are effectively only two underlying 'predicates': `triple` for triples in the RDF dataset default graph, and `quad` for named graphs.
- More importantly, SPARQL is based on a rich algebra, which goes beyond expressivity of CQs. Non-monotonic features like OPTIONAL and MINUS, and cardinality-sensitive query modifiers (DISTINCT) and aggregation (GROUP BY with functions such as SUM, AVG, COUNT) are difficult to model even in extensions of Datalog.
- Even without SPARQL aggregation, cardinalities have to be treated carefully: the SQL queries in a mapping produce *bags* (multisets) of tuples, but their induced RDF graphs contain no duplicates and thus are *sets* of triples; however, when a SPARQL query is evaluated, it results in a bag of solutions mappings.

These challenges turned out to be difficult to tackle in the Datalog setting. For example, one has to use three clauses and negation to model OPTIONAL, see e.g., [1,17]. Moreover, using multiple clauses for nested OPTIONALs can result in an exponentially large SQL query, if the related clauses are treated independently. On the other hand, such a group of clauses could and ideally *should* be re-assembled into a single LEFT JOIN when translating into SQL, so that the DBMS can take advantage of the structure [25]. Curiously, the challenge also offers a solution because most SPARQL constructs have natural counterparts in SQL: for instance, OPTIONAL corresponds to LEFT JOIN, GROUP BY to GROUP BY, and so on. Also, both SPARQL and SQL have bag semantics and use 3-valued logic for boolean expressions.

As a consequence of the above observations, when redesigning *Ontop*, we replaced Datalog by a relational-algebra-type representation, discussed in Sect. 3.

SPARQL vs SQL. Despite the apparent similarities between SPARQL and SQL, the two languages have significant differences relevant to any VKG system implementation.

Typing Systems. SQL is *statically typed* in the sense that all values in a given relation column (both in the database and in the result of a query) have the same datatype. In contrast, SPARQL is *dynamically typed*: a variable can have values of different datatypes in different solution mappings. Also, SQL queries with values of unexpected datatypes in certain contexts (e.g., a string as an argument for '+') are simply deemed incorrect. In contrast, SPARQL treats such *type errors* as legitimate and handles them similarly to NULLs in SQL. For example, the basic graph pattern ?s ?p ?o FILTER (?o < 4) retrieves all triples with a numerical object whose value is below 4 (but *ignores* all triples with strings or IRIs, for example). Also, the output datatype of a SPARQL function depends on the types or language tags of its arguments (e.g., if both arguments of '+' are xsd:integer, then so is the output, and if both arguments are xsd:decimal, then so is the output). In particular, to determine the output datatype of an aggregate function in SPARQL, one has to look at the datatypes of values in the group, which can vary from one group to another.

Order. SPARQL defines a fixed order on IRIs, blank nodes, unbound values, and literals. For multi-typed expressions, this general order needs to be combined with the orders defined for datatypes. In SQL, the situation is significantly simpler due to its static typing: apart from choosing the required order modifier for the datatype, one only needs to specify whether NULLs come first or last.

Implicit Joining Conditions. SPARQL uses the notion of solution mapping compatibility to define the semantics of the JOIN and OPTIONAL operators: two solution mappings are *compatible* if both map each shared variable to the same RDF term (sameTerm), that is, the two terms have the same type (including the language tag for literals) and the same lexical value. The sameTerm predicate is also used for AGGREGATEJOIN. In contrast, equalities in SQL are satisfied when their arguments are *equivalent*, but not necessarily of the same datatype (e.g., 1 in columns of type INTEGER and DECIMAL), and may even have different lexical values (e.g., timestamps with different timezones). SPARQL has a similar equality, denoted by '=', which can occur in FILTER and BIND.

SQL Dialects. Unlike SPARQL with its standard syntax and semantics, SQL is more varied as DBMS vendors do not strictly follow the ANSI/ISO standard. Instead, many use specific datatypes and functions and follow different conventions, for example, for column and table identifiers and query modifiers; even a common function CONCAT can behave differently: NULL-rejecting in MySQL, but not in PostgreSQL and Oracle. Support for the particular SQL dialect is thus essential for transforming SPARQL into SQL.

3   Ontop v4: New Design

We now explain how we address the challenges in *Ontop* v4. In Sect. 3.1, we describe a variant of relational algebra for representing queries and mappings. In Sect. 3.2, we concentrate on translating SPARQL functions into SQL. We discuss query optimization in Sect. 3.3, and post-processing and dealing with SQL dialects in Sect. 3.4.

3.1 Intermediate Query

Ontop v4 uses a variant of relational algebra tailored to encode SPARQL queries along the lines of the language described in [25]. The language, called *Intermediate Query*, or IQ, is a uniform representation both for user SPARQL queries and for SQL queries from the mapping. When the query transformation (rewriting and unfolding) is complete, the IQ expression is converted into SQL, and executed by the underlying relational DBMS.

In SPARQL, an *RDF dataset* consists of a default RDF graph (a set of triples of the form *s-p-o*) and a collection of named graphs (sets of quadruples *s-p-o-g*, where *g* is a graph name). In accordance with this, a ternary relation `triple` and a quaternary relation `quad` model RDF datasets in IQ. We use atomic expressions of the form

$$\texttt{triple}(s, p, o) \qquad \text{and} \qquad \texttt{quad}(s, p, o, g),$$

where s, p, o, and g are either constants or variables—in relational algebra such expressions would need to be built using combinations of SELECT (σ, to deal with constants and matching variables in different positions) and PROJECT (π, for variable names), see, e.g., [8]: for example, triple pattern `:ex :p ?x` would normally be encoded as $\pi_{x/o}\sigma_{s=":ex",p=":p"}\texttt{triple}$, where s, p, o are the attributes of `triple`. We chose a more concise representation, which is convenient for encoding SPARQL triple patterns.

We illustrate the other input of the SPARQL to SQL transformation (via IQ) using the following mapping in a simplified syntax:

$$T_1(x, y) \rightsquigarrow \texttt{:b}\{x\} \quad \texttt{:p} \quad y, \ T_2(x, y) \rightsquigarrow \texttt{:b}\{x\} \quad \texttt{:p} \quad y,$$
$$T_3(x, y) \rightsquigarrow \texttt{:b}\{x\} \quad \texttt{:p} \quad y, \ T_4(x, y) \rightsquigarrow \texttt{:b}\{x\} \quad \texttt{:q} \quad y,$$

where the triples on the right-hand side of $\rightsquigarrow$ represent `subjectMaps` together with `predicateObjectMaps` for properties `:p` and `:q`. In database tables T_1, T_2, T_3, and T_4, the first attribute is the primary key of type TEXT, and the second attribute is non-nullable and of type INTEGER, DECIMAL, TEXT, and INTEGER, respectively. When we translate the mapping into IQ, the SQL queries are turned into atomic expressions $T_1(x, y)$, $T_2(x, y)$, $T_3(x, y)$, and $T_4(x, y)$, respectively, where the use of variables again indicates the π operation of relational algebra. The translation of the right-hand side is more elaborate.

Remark 1. IRIs, blank nodes, and literals can be constructed in R2RML using *templates*, where placeholders are replaced by values from the data source. Whether a template function is *injective* (yields different values for different values of parameters) depends on the shape of the template. For IRI templates, one would normally use *safe separators* [9] to ensure injectivity of the function. For literals, however, if a template contains more than one placeholder, then the template function may be non-injective. On the other hand, if we construct literal values of type `xsd:date` from three separate database INTEGER attributes (for day, month, and year), then the template function is injective because the separator of the three components, -, is 'safe' for numerical values.

Non-constant RDF terms are built in IQ using the binary function rdf with a TEXT lexical value and the term type as its arguments. In the example above, all triple subjects are IRIs of the same template, and the lexical value is constructed using template function :b{}(x), which produces, e.g., the IRI :b1 when $x = 1$. The triple objects are literals: the INTEGER attribute in T_1 and T_4 is mapped to xsd:integer, the DECIMAL in T_2 is mapped to xsd:decimal, and the TEXT in T_3 to xsd:string. Database values need to be cast into TEXT before use as lexical values, which is done by unary functions i2t and d2t for INTEGER and DECIMAL, respectively. The resulting IQ representation of the mapping assertions is then as follows:

$$T_1(x,y) \rightsquigarrow \texttt{triple}(\texttt{rdf}(\texttt{:b\{\}}(x), \texttt{IRI}), \texttt{:p}, \texttt{rdf}(\texttt{i2t}(y), \texttt{xsd:integer})),$$
$$T_2(x,y) \rightsquigarrow \texttt{triple}(\texttt{rdf}(\texttt{:b\{\}}(x), \texttt{IRI}), \texttt{:p}, \texttt{rdf}(\texttt{d2t}(y), \texttt{xsd:decimal})),$$
$$T_3(x,y) \rightsquigarrow \texttt{triple}(\texttt{rdf}(\texttt{:b\{\}}(x), \texttt{IRI}), \texttt{:p}, \texttt{rdf}(y, \texttt{xsd:string})),$$
$$T_4(x,y) \rightsquigarrow \texttt{triple}(\texttt{rdf}(\texttt{:b\{\}}(x), \texttt{IRI}), \texttt{:q}, \texttt{rdf}(\texttt{i2t}(y), \texttt{xsd:integer})).$$

To illustrate how we deal with multi-typed functions in SPARQL, we now consider the following query (in the context of the RDF dataset discussed above):

```
SELECT ?s WHERE { ?x :p ?n .   ?x :q ?m .
                 BIND ((?n + ?m) AS ?s) FILTER (bound(?s))}
```

It involves an arithmetic sum over two variables, one of which, ?n, is multi-typed: it can be an xsd:integer, xsd:decimal, or xsd:string. The translation of the SPARQL query into IQ requires the use of most of the algebra operations, which are defined next.

A *term* is a variable, a constant (including NULL), or a functional term constructed from variables and constants using SPARQL function symbols such as numeric-add, SQL function symbols such as +, and our auxiliary function symbols such as IF, etc. (IF is ternary and such that IF$(\texttt{true}, x, y) = x$ and IF$(\texttt{false}, x, y) = y$). We treat predicates such as = and sameTerm as function symbols of *boolean* type; boolean connectives $\neg$, $\wedge$, and $\vee$ are also boolean function symbols. Boolean terms are interpreted using the *3-valued logic*, where NULL is used for the 'unknown value.' An *aggregate term* is an expression of the form $agg(\tau)$, where agg is a SPARQL or SQL aggregate function symbol (e.g., SPARQL_Sum or SUM) and τ a term. A *substitution* is an expression of the form $x_1/\eta_1, \ldots, x_n/\eta_n$, where each x_i is a variable and each η_i either a term (for PROJ) or an aggregate term (for AGG). Then, IQs are defined by the following grammar:

$$\phi := P(\mathbf{t}) \mid \text{PROJ}_\tau^{\mathbf{x}} \phi \mid \text{AGG}_\tau^{\mathbf{x}} \phi \mid \text{DISTINCT } \phi \mid \text{ORDERBY}_{\mathbf{x}} \phi \mid \text{SLICE}_{i,j} \phi \mid$$
$$\text{FILTER}_\beta \phi \mid \text{JOIN}_\beta(\phi_1, \ldots, \phi_k) \mid \text{LEFTJOIN}_\beta(\phi_1, \phi_2) \mid \text{UNION}(\phi_1, \ldots, \phi_k),$$

where P is a relation name (triple, quad, or a database table name), $\mathbf{t}$ a tuple of terms, $\mathbf{x}$ a tuple of variables, τ a substitution, $i, j \in \mathbb{N} \cup \{0, +\infty\}$ are values for the offset and limit, respectively, and β is a boolean term. When presenting our examples, we often omit brackets and use indentation instead. The algebraic operators above operate on bags of tuples, which can be thought of as *total* functions from sets of variables to values, in contrast to *partial* functions in SPARQL (such definitions are natural from the SPARQL-to-SQL translation point of view; see [25] for a discussion). Also, JOIN and LEFTJOIN are similar to NATURAL (LEFT) JOIN in SQL, in the sense that the tuples are joined (compatible) if their shared variables have the same values. All the algebraic operators are interpreted using the bag semantics, in particular, UNION preserves duplicates (similarly to UNION ALL in SQL).

In our running example, the SPARQL query is translated into the following IQ:

$$\text{PROJ}^{?s}_{?s/\text{numeric-add}(?n,?m)}$$

$$\text{JOIN}_{\neg isNull(\text{numeric-add}(?n,?m))}(\text{triple}(?x,\ \text{:p},\ ?n),\ \text{triple}(?x,\ \text{:q},\ ?m)),$$

where the **bound** filter is replaced by $\neg isNull()$ in the JOIN operation, and the **BIND** clause is reflected in the top-level PROJ. When this IQ is unfolded using the mapping given above, occurrences of **triple** are replaced by unions of appropriate mapping assertions (with matching predicates, for example). Note that, in general, since the RDF dataset is a *set* of triples and quadruples, one needs to insert DISTINCT above the union of mapping assertion SQL queries; in this case, however, the DISTINCT can be omitted because the first attribute is a primary key in the tables and the values of $?n$ are disjoint in the three branches (in terms of **sameTerm**). So, we obtain the following:

$$\text{PROJ}^{?s}_{?s/\text{numeric-add}(?n,?m)}\ \text{JOIN}_{\neg isNull(\text{numeric-add}(?n,?m))}$$

$$\text{UNION}$$

$$\text{PROJ}^{?x,?n}_{?x/\text{rdf}(:b\{\}(y_1),\text{IRI}),\ ?n/\text{rdf}(i2t(z_1),\text{xsd:integer})}\ T_1(y_1,z_1)$$

$$\text{PROJ}^{?x,?n}_{?x/\text{rdf}(:b\{\}(y_2),\text{IRI}),\ ?n/\text{rdf}(d2t(z_2),\text{xsd:decimal})}\ T_2(y_2,z_2)$$

$$\text{PROJ}^{?x,?n}_{?x/\text{rdf}(:b\{\}(y_3),\text{IRI}),\ ?n/\text{rdf}(z_3,\text{xsd:string})}\ T_3(y_3,z_3)$$

$$\text{PROJ}^{?x,?m}_{?x/\text{rdf}(:b\{\}(y_4),\text{IRI}),\ ?m/\text{rdf}(i2t(z_4),\text{xsd:integer})}\ T_4(y_4,z_4).$$

This query, however, cannot be directly translated into SQL because, for example, it has occurrences of SPARQL functions (**numeric-add**).

3.2 Translating (Multi-typed) SPARQL Functions into SQL Functions

Recall the two main difficulties in translating SPARQL functions into SQL. First, when a SPARQL function is not applicable to its argument (e.g., **numeric-add** to **xsd:string**), then the result is the *type error*, which, in our example, means that the variable remains unbound in the solution mapping; in SQL, such a query would be deemed invalid (and one would have no results). Second, the type of the result may depend on the types of the arguments: numeric-add yields an xsd:integer on xsd:integers and an xsd:decimal on xsd:decimals. Using the example above, we illustrate how an IQ with a multi-typed SPARQL operation can be transformed into an IQ with standard SQL operations.

First, the substitutions of PROJ operators are lifted as high in the expression tree as possible. In the process, functional terms may need to be decomposed so that some of their arguments can be lifted, even though other arguments are blocked. For example, the substitution entries for $?n$ differ in the three branches of the UNION, and each needs to be decomposed: e.g., $?n/\text{rdf}(\text{i2t}(z_1), \text{xsd:integer})$ is decomposed into $?n/\text{rdf}(v, t)$, $v/\text{i2t}(z_1)$, and $t/\text{xsd:integer}$. Variables v and t are re-used in the other branches, and, after the decomposition, all children of the UNION share the same entry $?n/\text{rdf}(v, t)$ in their PROJ constructs, and so, this entry can be lifted up to the top. Note, however, that the entries for v and t remain blocked by the UNION. Observe that one child of the UNION can be pruned when propagating the JOIN conditions down: the condition is unsatisfiable as applying numeric-add to xsd:string results in the SPARQL type error, which is equivalent to false when used as a filter. Thus, we obtain

$$\text{PROJ}^{?s}_{?s/\text{numeric-add}(\text{rdf}(v,t),\ \text{rdf}(\text{i2t}(z_4),\ \text{xsd:integer}))}$$
$$\text{JOIN}_{\neg isNull(\text{numeric-add}(\text{rdf}(v,t),\ \text{rdf}(\text{i2t}(z_4),\ \text{xsd:integer})))}$$
$$\text{UNION}\big(\text{PROJ}^{y,v,t}_{v/\text{i2t}(z_1),t/\text{xsd:integer}} T_1(y, z_1), \text{PROJ}^{y,v,t}_{v/\text{d2t}(z_2),t/\text{xsd:decimal}} T_2(y, z_2)\big)$$
$$T_4(y, z_4).$$

However, the type of the first argument of numeric-add is still unknown at this point, which prevents transforming it into a SQL function. So, first, the substitution entries for t are replaced by $t/f(1)$ and $t/f(2)$, respectively, where f is a freshly generated dictionary function that maps 1 and 2 to xsd:integer and xsd:decimal, respectively. Then, f can be lifted to the JOIN and PROJ by introducing a fresh variable p:

$$\text{PROJ}^{?s}_{?s/\text{numeric-add}(\text{rdf}(v,\boldsymbol{f(p)}),\ \text{rdf}(\text{i2t}(z_4),\ \text{xsd:integer}))}$$
$$\text{JOIN}_{\neg isNull(\text{numeric-add}(\text{rdf}(v,\boldsymbol{f(p)}),\ \text{rdf}(\text{i2t}(z_4),\ \text{xsd:integer})))}$$
$$\text{UNION}\big(\text{PROJ}^{y,v,\boldsymbol{p}}_{v/\text{i2t}(z_1),\ \boldsymbol{p/1}} T_1(y, z_1), \quad \text{PROJ}^{y,v,\boldsymbol{p}}_{v/\text{d2t}(z_2),\ \boldsymbol{p/2}} T_2(y, z_2)\big)$$
$$T_4(y, z_4),$$

where the changes are emphasized in boldface.

Now, the type of the first argument of numeric-add is either xsd:integer or xsd:decimal, and so, it can be transformed into a complex functional term with SQL +: the sum is on INTEGERs if p is 1, and on DOUBLEs otherwise. Observe that these sums are cast back to TEXT to produce RDF term lexical values. Now, the JOIN condition is equivalent to true because the numeric-add does not produce NULL without invalid input types and nullable variables. A similar argument applies to PROJ, and we get

$$\text{Proj}^{?s}_{?s/\text{rdf}(\text{IF}(p=1,\ \text{i2t}(\text{t2i}(v)+z_4),\ \text{d2t}(\text{t2d}(v)+\text{i2d}(z_4))),\ \text{IF}(p=1,\ \text{xsd:integer},\ \text{xsd:decimal}))}$$

$$\text{Join}$$

$$\text{Union}\big(\text{Proj}^{y,v,p}_{v/\text{i2t}(z_1),\ p/1}\ T_1(y,z_1),\quad \text{Proj}^{y,v,p}_{v/\text{d2t}(z_2),\ p/2}\ T_2(y,z_2)\big)$$

$$T_4(y,z_4),$$

which can now be translated into SQL. We would like to emphasize that only the SPARQL variables can be multi-typed in IQs, while the variables for database attributes will always have a unique type, which is determined by the datatype of the attribute.

As a second example, we consider the following aggregation query:

    SELECT ?x (SUM(?n) AS ?s) WHERE { ?x :p ?n . } GROUP BY ?x

This query uses the same mapping as above, where the values of data property :p can belong to xsd:integer, xsd:decimal, and xsd:string from INTEGER, DECIMAL, and TEXT database attributes. The three possible ranges for :p require careful handling because of GROUP BY and SUM: in each group of tuples with the same x, we need to compute (separate) sums of all INTEGERs and DECIMALs, as well as indicators of whether there are any TEXTs and DECIMALs: the former is needed because any string in a group results in a *type error* and undefined sum; the latter determines the type of the sum if all values in the group are numerical. The following IQ is the final result of the transformations:

$$\text{Proj}^{?x,?s}_{?x/\text{rdf}(:b\{\}(y),\text{IRI}),}$$
$$\scriptstyle ?s/\text{rdf}(\text{IF}(ct>0,\ \text{NULL},\ \text{IF}(cd=0,\ \text{i2t}(coalesce(si,0)),\ \text{d2t}(sd+\text{i2d}(coalesce(si,0)))))),$$
$$\scriptstyle \text{IF}(ct>0,\ \text{NULL},\ \text{IF}(cd=0,\ \text{xsd:integer},\ \text{xsd:decimal})))$$

$$\text{Agg}^{y}_{cd/\text{COUNT}(d),\ ct/\text{COUNT}(t),\ si/\text{SUM}(i),\ sd/\text{SUM}(d)}$$

$$\text{Union}\big(\text{Proj}^{y,i,d,t}_{d/\text{NULL},\ t/\text{NULL}}\ T_1(y,i),\quad \text{Proj}^{y,i,d,t}_{i/\text{NULL},\ t/\text{NULL}}\ T_2(y,d),\quad \text{Proj}^{y,i,d,t}_{i/\text{NULL},\ d/\text{NULL}}\ T_3(y,t)\big).$$

Note that the branches of Union have the same projected variables, padded by NULL.

3.3 Optimization Techniques

Being able to transform SPARQL queries into SQL ones is a must-have requirement, but making sure that they can be efficiently processed by the underlying DBMS is essential for the VKG approach. This topic has been extensively studied during the past decade, and an array of optimization techniques, such as redundant join elimination using primary and foreign keys [6,18,19,22] and pushing down Joins to the data-level [13], are now well-known and implemented by many systems. In addition to these, *Ontop* v4 exploits several recent techniques, including the ones proposed in [25] for optimizing left joins due to OPTIONALs and MINUSes in the SPARQL queries.

Self-join Elimination for Denormalized Data. We have implemented a novel self-join elimination technique to cover a common case where data is partially denormalized. We illustrate it on the following example with a single database table `loan` with primary key `id` and all non-nullable attributes. For instance, `loan` can contain the following tuples:

id	amount	organisation	branch
10284124	5000	Global Bank	Denver
20242432	7000	Trade Bank	Chicago
30443843	100000	Global Bank	Miami
40587874	40000	Global Bank	Denver

The mapping for data property `:hasAmount` and object properties `:grantedBy` and `:branchOf` constructs, for each tuple in `loan`, three triples to specify the loan amount, the bank branch that granted it, and the head organisation for the bank branch:

$$\text{loan}(x, a, \_, \_) \rightsquigarrow \text{triple}(\text{rdf}(:l\{\}(x), \text{IRI}), :\text{hasAmount}, \text{rdf}(\text{i2t}(a), \text{xsd}:\text{integer})),$$
$$\text{loan}(x, \_, o, b) \rightsquigarrow \text{triple}(\text{rdf}(:l\{\}(x), \text{IRI}), :\text{grantedBy}, \text{rdf}(:b\{\}/\{\}(o, b), \text{IRI})),$$
$$\text{loan}(\_, \_, o, b) \rightsquigarrow \text{triple}(\text{rdf}(:b\{\}/\{\}(o, b), \text{IRI}), :\text{branchOf}, \text{rdf}(:o\{\}(o), \text{IRI}))$$

(we use underscores instead of variables for attributes that are not projected). Observe that the last assertion is not 'normalized': the same triple can be extracted from many different tuples (in fact, it yields a copy of the triple for each loan granted by the branch). To guarantee that the RDF graph is a set, these duplicates have to be eliminated.

We now consider the following SPARQL query extracting the number and amount of loans granted by each organisation:

```
SELECT ?o (COUNT(?l) AS ?c) (SUM(?a) AS ?s) WHERE {
    ?l :hasAmount ?a. ?l :grantedBy ?b. ?b :branchOf ?o } GROUP BY ?o
```

After unfolding, we obtain the following IQ:

$$\text{AGG}^{?o}_{?c/\text{SPARQL\_Count}(l), \, ?s/\text{SPARQL\_Sum}(a)} \ \text{JOIN}$$
$$\quad \text{PROJ}^{?l,?a}_{?l/\text{rdf}(:l\{\}(x_1),\text{IRI}), \, ?a/\text{rdf}(\text{i2t}(a_1),\text{xsd}:\text{integer})} \ \text{loan}(x_1, a_1, \_, \_)$$
$$\quad \text{PROJ}^{?l,?b}_{?l/\text{rdf}(:l\{\}(x_2),\text{IRI}), \, ?b/\text{rdf}(:b\{\}/\{\}(o_2,b_2),\text{IRI})} \ \text{loan}(x_2, \_, o_2, b_2)$$
$$\quad \text{DISTINCT} \ \text{PROJ}^{?b,?o}_{?b/\text{rdf}(:b\{\}/\{\}(o_3,b_3),\text{IRI}), \, ?o/\text{rdf}(:o\{\}(o_3),\text{IRI})} \ \text{loan}(\_, \_, o_3, b_3).$$

Note that the DISTINCT in the third child of the JOIN is required to eliminate duplicates (none is needed for the other two since `id` is the primary key of table `loan`).

The first step is lifting the PROJ. For the substitution entries below the DISTINCT, some checks need to be done before (partially) lifting their functional terms. The `rdf` function used by `?o` and `?b` is injective by design and can always be lifted. Their first arguments are IRI template functional terms. Both IRI templates, :o{} and :b{}/{}, are injective (see Remark 1): the former is unary, the latter has a safe separator / between its arguments. Consequently, both can be lifted. Note that these checks only concern functional terms, as constants can always be lifted above DISTINCTs. The substitution entry for $?l$ is lifted above the AGG because it is its group-by variable. Other entries are used for substituting the arguments of the aggregation functions. Here, none of the variables is multi-typed. After simplifying the functional terms, we obtain the IQ

$$\text{PROJ}_{?o/\text{rdf}(:o\{\}(o_2),\text{IRI}),\ ?c/\text{rdf}(\text{i2t}(n),\text{xsd:integer}),\ ?s/\text{rdf}(\text{i2t}(m),\text{xsd:integer})}^{?o,?c,?s}$$

$$\text{AGG}_{n/\text{COUNT}(x_1),\ m/\text{SUM}(a_1)}^{o_2}$$

$$\text{JOIN}\big(\text{loan}(x_1,a_1,\_,\_),\ \text{loan}(x_1,\_,o_2,b_2),\ \text{DISTINCT loan}(\_,\_,o_2,b_2)\big).$$

Next, the well-known self-join elimination is applied to the first two children of the JOIN (which is over the primary key). Then, the DISTINCT commutes with the JOIN since the other child of JOIN is also a set (due to the primary key), obtaining the sub-IQ

$$\text{DISTINCT}\ \ \text{JOIN}\big(\text{loan}(x_1,a_1,o_2,b_2),\text{loan}(\_,\_,o_2,b_2)\big),$$

on which our new self-join elimination technique can be used, as the two necessary conditions are satisfied. First, the JOIN does not need to preserve cardinality due to the DISTINCT above it. Second, all the variables projected by the second child (o_2 and b_2) of the JOIN are also projected by the first child. So, we can eliminate the second child, but have to insert a filter requiring the shared variables o_2 and b_2 to be non-NULL:

$$\text{DISTINCT}\ \ \text{FILTER}_{\neg isNull(o_2)\wedge\neg isNull(b_2)}\ \ \text{loan}(x_1,a_1,o_2,b_2).$$

The result can be further optimized by observing that the attributes for o_2 and b_2 are non-nullable and that the DISTINCT has no effect because the remaining data atom produces no duplicates. So, we arrive at

$$\text{PROJ}_{?o/\text{rdf}(:o\{\}(o_2),\text{IRI}),\ ?c/\text{rdf}(\text{i2t}(n),\text{xsd:integer}),\ ?s/\text{rdf}(\text{i2t}(m),\text{xsd:integer})}^{?o,?c,?s}$$

$$\text{AGG}_{n/\text{COUNT}(x_1),\ m/\text{SUM}(a_1)}^{o_2}\ \text{loan}(x_1,a_1,o_2,\_),$$

where b_2 is replaced by $\_$ because it is not used elsewhere.

3.4 From IQ to SQL

In the VKG approach almost all query processing is delegated to the DBMS. *Ontop* v4 performs only the top-most projection, which typically transforms database values into RDF terms, as illustrated by the last query above. The subquery under this projection must not contain any RDF value nor any SPARQL function. As highlighted above, our IQ guarantees that such a subquery is not multi-typed.

In contrast to SPARQL, the ANSI/ISO SQL standards are very loosely followed by DBMS vendors. There is very little hope for generating reasonably rich SQL that would be interoperable across multiple vendors. Given the diversity of the SQL ecosystem, in *Ontop* v4 we model each supported SQL dialect in a fine-grained manner: in particular, we model *(i)* their datatypes, *(ii)* their conventions in terms of attribute and table identifiers and query modifiers, *(iii)* the semantics of their functions, *(iv)* their restrictions on clauses such as WHERE and ORDER BY, and *(v)* the structure of their data catalog. *Ontop* v4 directly uses the concrete datatypes and functions of the targeted dialect in IQ by means of Java factories whose dialect-specific implementations are provided through a dependency injection mechanism. Last but not least, *Ontop* v4 allows IQ to contain arbitrary, including user-defined, SQL functions from the queries of the mapping.

4   Evaluation

Compliance of *Ontop* v4 with relevant W3C recommendations is discussed in Sect. 4.1, and performance and comparison with other systems in Sect. 4.2.

4.1   Compliance with W3C Recommendations

Since the relevant W3C standards have very rich sets of features, and they also interplay with each other, it is difficult to enumerate all the cases. The different behaviors of DBMSs make the situation even more complex and add another dimension to consider. Nevertheless, we describe our testing infrastructure and do our best to summarize the behavior of *Ontop* with all the different standards.

Testing Infrastructure. To ensure the correct behavior of the system, we developed a rich testing infrastructure. The code base includes a large number of unit test cases. To test against different database systems, we developed a Docker-based infrastructure for creating DB-specific instances for the tests[1]. It uses docker-compose to generate a cluster of DBs including MySQL, PostgreSQL, Oracle, MS SQL Server, and DB2.

SPARQL 1.1 [12]. In Table 1, we present a summary of *Ontop* v4 compliance with SPARQL 1.1, where rows correspond to sections of the WC3 recommendation. Most of the features are supported, but some are unsupported or only partially supported.

- Property paths are not supported: the `ZeroOrMorePath` (*) and `OneOrMorePath` (+) operators require linear recursion, which is not part of IQ yet. An initial investigation of using SQL Common Table Expressions (CTEs) for linear recursion was done in the context of SWRL [26], but a proper implementation would require dedicated optimization techniques.

---

[1] https://github.com/ontop/ontop-dockertests

Table 1. SPARQL Compliance: unsupported features are ~~crossed out~~.

Section in SPARQL 1.1 [12]	Features	Coverage
5–7. Graph Patterns, etc.	BGP, FILTER, OPTIONAL, UNION	4/4
8. Negation	MINUS, ~~FILTER [NOT] EXISTS~~	1/2
9. Property Paths	~~PredicatePath, InversePath, ZeroOrMorePath,~~ ...	0
10. Assignment	BIND, VALUES	2/2
11. Aggregates	COUNT, SUM, MIN, MAX, AVG, GROUP_CONCAT, SAMPLE	6/6
12. Subqueries	Subqueries	1/1
13. RDF Dataset	GRAPH, ~~FROM [NAMED]~~	1/2
14. Basic Federated Query	~~SERVICE~~	0
15. Solution Seqs. & Mods	ORDER BY, SELECT, DISTINCT, REDUCED, OFFSET, LIMIT	6/6
16. Query Forms	SELECT, CONSTRUCT, ASK, DESCRIBE	4/4
17.4.1. Functional Forms	BOUND, ~~IF~~, COALESCE, ~~EXISTS~~, ~~NOT EXISTS~~, ||, &&, =, sameTerm, ~~IN~~, ~~NOT IN~~	6/11
17.4.2. Fns. on RDF Terms	isIRI, isBlank, isLiteral, isNumeric, str, lang, datatype, ~~IRI~~, ~~BNODE~~, ~~STRDT~~, ~~STRLANG~~, UUID, STRUUID	9/13
17.4.3. Fns. on Strings	STRLEN, SUBSTR, UCASE, LCASE, STRSTARTS, STRENDS, CONTAINS, STRBEFORE, STRAFTER, ENCODE_FOR_URI, CONCAT, langMatches, REGEX, REPLACE	14/14
17.4.4. Fns. on Numerics	abs, round, ceil, floor, RAND	5/5
17.4.5. Fns. on Dates&Times	now, year, month, day, hours, minutes, seconds, ~~timezone~~, tz	8/9
17.4.6. Hash Functions	MD5, SHA1, SHA256, SHA384, SHA512	5/5
17.5. XPath Constructor Fns	~~casting~~	0
17.6. Extensible Value Testing	~~user-defined functions~~	0

- [NOT] EXISTS is difficult to handle due to its non-compositional semantics, which is not defined in a bottom-up fashion. Including it in IQ requires further investigation.
- Most of the missing SPARQL functions (Section 17.4) are not so challenging to implement but require a considerable engineering effort to carefully define their translations into SQL. We will continue the process of implementing them gradually and track the progress in a dedicated issue[2].
- The 5 hash functions and functions REPLACE and REGEX for regular expressions have limited support because they heavily depend on the DBMS: not all DBMSs provide all hash functions, and many DBMSs have their own regex dialects. Currently, the SPARQL regular expressions of REPLACE and REGEX are simply sent to the DBMS.
- In the implementation of functions STRDT, STRLANG, and langMatches, the second argument has to a be a constant: allowing variables will have a negative impact on the performance in our framework.

R2RML [9]. *Ontop* is fully compliant with R2RML. In particular, the support of rr:GraphMap for RDF datasets and blank nodes has been introduced in *Ontop* v4. The optimization hint rr:inverseExpression is ignored in the

---

[2] https://github.com/ontop/ontop/issues/346

current version, but this is compliant with the W3C recommendation. In the combination of R2RML with OWL, however, ontology axioms (a TBox in the Description Logic parlance) could also be constructed in a mapping: e.g., $T_1(x, y) \rightsquigarrow :\{x\}$ rdfs:subClassOf :$\{y\}$. Such mappings are not supported in online query answering, but one can materialize the triples offline and then include them in the ontology manually.

OWL 2 QL [14] **and SPARQL 1.1 Entailment Regimes** [11]. These two W3C recommendations define how to use ontological reasoning in SPARQL. *Ontop* supports them with the exception of querying the TBox, as in SELECT * WHERE { ?c rdfs:subClassOf :Person. ?x a ?c }. Although we have investigated this theoretically and implemented a prototype [13], a more serious implementation is needed for IQ, with special attention to achieving good performance. This is on our agenda.

SPARQL 1.1 Protocol [10] **and SPARQL Endpoint.** We reimplemented the new SPARQL endpoint from scratch and designed a new command-line interface for it. It is stateless and suitable for containers. In particular, we have created a Docker image for the *Ontop* SPARQL endpoint[3], which has greatly simplified deployment. The endpoint is also packed with several new features, like customization of the front page with predefined SPARQL queries, streaming query answers, and result caching.

4.2   Performance and Comparison with Other VKG Systems

Performance evaluation of *Ontop* has been conducted since *Ontop* v1 by ourselves and others in a number of scientific papers. Here we only summarize two recent independent evaluations of *Ontop* v3. Recall that the main focus of *Ontop* v4 compared to v3 has been the extension with new features. Hence, we expect similar results for *Ontop* v4.

Chaloupka and Necasky [7] evaluated four VKG systems, namely, Morph, *Ontop*, SparqlMap, and their own EVI, using the Berlin SPARQL Benchmark (BSBM). D2RQ and Ultrawrap were not evaluated: D2RQ has not been updated for years, and Ultrawrap is not available for research evaluation. Only *Ontop* and EVI were able to load the authors' version of the R2RML mapping for BSBM. EVI supports only SQL Server, while *Ontop* supports multiple DBMSs. In the evaluation, EVI outperformed *Ontop* on small datasets, but both demonstrated similar performance on larger datasets, which can be explained by the fact that *Ontop* performs more sophisticated (and expensive) optimizations during the query transformation step.

Namici and De Giacomo [15] evaluated *Ontop* and Mastro on the NPD and ACI benchmarks, both of which have complex ontologies. Some SPARQL queries had to be adapted for Mastro because it essentially supports only unions of CQs. In general *Ontop* was faster on NPD, while Mastro was faster on ACI.

---

[3] https://hub.docker.com/r/ontop/ontop-endpoint

Both independent evaluations confirm that although *Ontop* is not always the fastest, its performance is very robust. In the future, we will carry out more evaluations, in particular for the new features of *Ontop* v4.

It is important to stress that when choosing a VKG system, among many different criteria, performance is only one dimension to consider. Indeed, in [17], also the aspects of usability, completeness, and soundness have been evaluated. When considering all of these, *Ontop* is a clear winner. In our recent survey [24], we have also listed the main features of popular VKG systems, including D2RQ, Mastro, Morph, *Ontop*, Oracle Spatial and Graph, and Stardog. Overall, it is fair to claim that *Ontop* is the most mature *open source* VKG system currently available.

5 Community Building and Adoption

Ontop is distributed under the Apache 2 license through several channels. Ready-to-use binary releases, including a command line tool and a Protégé bundle with an *Ontop* plugin, are published on Sourceforge since 2015. There have been **30K+ downloads** in the past 5 years according to Sourceforge[4]. The *Ontop* plugin for Protégé is available also in the plugin repository of Protégé, through which users receive auto-updates. A Docker image of the SPARQL endpoint is available at Docker Hub since the *Ontop* v3 release, and it has been 1.1K times. The documentation, including tutorials, is available at the official website[5].

Ontop is the product of a hard-working developer community active for over a decade. Nowadays, the development of *Ontop* is backed by different research projects (at the local, national, and EU level) at the Free University of Bozen-Bolzano and by Ontopic s.r.l. It also receives regular important contributions from Birkbeck, University of London. As of 13 August 2020, the GitHub repository[6] consists of 11,511 git commits from 25 code contributors, among which 10 have contributed more than 100 commits each. An e-mail list[7] created in August 2013 for discussion currently includes 270 members and 429 topics. In Github, 312 issues have been created and 270 closed.

To make *Ontop* sustainable, it needed to be backed up by a commercial company, because a development project running at a public university cannot provide commercial support to its users, and because not all developments are suitable for a university research group. So, Ontopic s.r.l.[8] was born in April 2019, as the first spin-off of the Free University of Bozen-Bolzano[9]. It provides commercial support for the *Ontop* system and consulting services that rely on it, with the aim to push the VKG technology to industry. Ontopic has now become the major source code contributor of *Ontop*.

---

[4] https://sourceforge.net/projects/ontop4obda/files/stats/timeline
[5] https://ontop-vkg.org/
[6] https://github.com/ontop/ontop/
[7] https://groups.google.com/forum/#!aboutgroup/ontop4obda
[8] http://ontopic.biz/
[9] https://www.unibz.it/it/news/132449 (in Italian).

Ontop has been adopted in many academic and industrial use cases. Due to its liberal Apache 2 license, it is essentially impossible to obtain a complete picture of all use cases and adoptions. Indeed, apart from the projects in which the research and development team is involved directly, we normally learn about a use case only when the users have some questions or issues with *Ontop*, or when their results have been published in a scientific paper. Nevertheless, a few significant use cases have been summarized in a recent survey paper [24]. Below, we highlight two commercial deployments of *Ontop*, in which Ontopic has been involved.

UNiCS[10] is an open data platform for research and innovation developed by SIRIS Academic in Spain. Using *Ontop*, the UNiCS platform integrates a large variety of data sources for decision and policy makers, including data produced by government bodies, data on the higher education & research sector, as well as companies' proprietary data. For instance, the Toscana Open Research (TOR) portal[11] is one such deployment of UNiCS. It is designed to communicate and enhance the Tuscan regional system of research, innovation, and higher education and to promote increasingly transparent and inclusive governance. Recently, Ontopic has also been offering dedicated training courses for TOR users, so that they can autonomously formulate SPARQL queries to perform analytics, and even create VKGs to integrate additional data sources.

Open Data Hub-Virtual Knowledge Graph is a joint project between NOI Techpark and Ontopic for publishing South Tyrolean tourism data as a Knowledge Graph. Before the project started, the data was accessible through a JSON-based Web API, backed by a PostgreSQL database. We created a VKG over the database and a SPARQL endpoint[12] that is much more flexible and powerful than the old Web API. Also, we created a Web Component[13], which can be embedded into any web page like a standard HTML tag, to visualize SPARQL query results in different ways.

6   Conclusion

Ontop is a popular open-source virtual knowledge graph system. It is the result of an active research and development community and has been adopted in many academic and industrial projects. In this paper, we have presented the challenges, design choices, and new features of the latest release v4 of *Ontop*.

Acknowledgements. This research has been partially supported by the Wallenberg AI, Autonomous Systems and Software Program (WASP) funded by the Knut and Alice Wallenberg Foundation, by the Italian Basic Research (PRIN) project HOPE, by the EU H2020 project INODE, grant agreement 863410, by the CHIST-ERA project PACMEL, by the Free Uni. of Bozen-Bolzano

---

[10] http://unics.cloud/

[11] http://www.toscanaopenresearch.it/

[12] https://sparql.opendatahub.bz.it/

[13] https://github.com/noi-techpark/webcomp-kg

through the projects QUADRO, KGID, and GeoVKG, and by the project IDEE (FESR1133) through the European Regional Development Fund (ERDF) Investment for Growth and Jobs Programme 2014–2020.

References

1. Arenas, M., Gottlob, G., Pieris, A.: Expressive languages for querying the semantic web. ACM Trans. Database Syst. 43(3), 13:1–13:45 (2018)
2. Artale, A., Calvanese, D., Kontchakov, R., Zakharyaschev, M.: The DL-Lite family and relations. J. Artif. Intell. Res. 36, 1–69 (2009)
3. Calvanese, D., et al.: Ontop: answering SPARQL queries over relational databases. SWJ 8(3), 471–487 (2017)
4. Calvanese, D., et al.: The MASTRO system for ontology-based data access. SWJ 2(1), 43–53 (2011)
5. Calvanese, D., De Giacomo, G., Lembo, D., Lenzerini, M., Rosati, R.: Tractable reasoning and efficient query answering in description logics: the DL-Lite family. JAR 39, 385–429 (2007)
6. Chakravarthy, U.S., Grant, J., Minker, J.: Logic-based approach to semantic query optimization. ACM TODS 15(2), 162–207 (1990)
7. Chaloupka, M., Necasky, M.: Using Berlin SPARQL benchmark to evaluate relational database virtual SPARQL endpoints. Submitted to SWJ (2020)
8. Chebotko, A., Lu, S., Fotouhi, F.: Semantics preserving SPARQL-to-SQL translation. DKE 68(10), 973–1000 (2009)
9. Das, S., Sundara, S., Cyganiak, R.: R2RML: RDB to RDF mapping language. W3C recommendation, W3C (2012)
10. Feigenbaum, L., Williams, G.T., Clark, K.G., Torres, E.: SPARQL 1.1 protocol. W3C recommendation, W3C (2013)
11. Glimm, B., Ogbuji, C.: SPARQL 1.1 entailment regimes. W3C recommendation (2013)
12. Harris, S., Seaborne, A., Prud'hommeaux, E.: SPARQL 1.1 query language. W3C recommendation, W3C (2013)
13. Kontchakov, R., Rezk, M., Rodriguez-Muro, M., Xiao, G., Zakharyaschev, M.: Answering SPARQL queries over databases under OWL 2 QL entailment regime. In: Proceedings of ISWC (2014)
14. Motik, B., et al.: OWL 2 Web Ontology Language: Profiles. W3C Recommendation, W3C (2012)
15. Namici, M., De Giacomo, G.: Comparing query answering in OBDA tools over W3C-compliant specifications. In: Proceedings of DL, vol. 2211. CEUR-WS.org (2018)
16. Poggi, A., Lembo, D., Calvanese, D., De Giacomo, G., Lenzerini, M., Rosati, R.: Linking data to ontologies. J. Data Sem. 10, 133–173 (2008)
17. Polleres, A.: From SPARQL to rules (and back). In: WWW, pp. 787–796. ACM (2007)
18. Priyatna, F., Corcho, Ó., Sequeda, J.F.: Formalisation and experiences of R2RML-based SPARQL to SQL query translation using morph. In: WWW, pp. 479–490 (2014)
19. Rodríguez-Muro, M., Kontchakov, R., Zakharyaschev, M.: Ontology-based data access: ontop of databases. In: Alani, H., et al. (eds.) ISWC 2013. LNCS, vol. 8218, pp. 558–573. Springer, Heidelberg (2013). https://doi.org/10.1007/978-3-642-41335-3_35

20. Rodriguez-Muro, M., Rezk, M.: Efficient SPARQL-to-SQL with R2RML mappings. J. Web Sem. **33**, 141–169 (2015)
21. Sequeda, J.F., Arenas, M., Miranker, D.P.: Ontology-based data access using views. In: Krötzsch, M., Straccia, U. (eds.) RR 2012. LNCS, vol. 7497, pp. 262–265. Springer, Heidelberg (2012). https://doi.org/10.1007/978-3-642-33203-6_29
22. Unbehauen, J., Stadler, C., Auer, S.: Optimizing SPARQL-to-SQL rewriting. In: Proceedings of IIWAS, pp. 324–330. ACM (2013)
23. Xiao, G., et al.: Ontology-based data access: a survey. In: Proceedings of IJCAI (2018)
24. Xiao, G., Ding, L., Cogrel, B., Calvanese, D.: Virtual knowledge graphs: an overview of systems and use cases. Data Intell. **1**, 201–223 (2019)
25. Xiao, G., Kontchakov, R., Cogrel, B., Calvanese, D., Botoeva, E.: Efficient handling of SPARQL OPTIONAL for OBDA. In: Proceedings of ISWC, pp. 354–373 (2018)
26. Xiao, G., Rezk, M., Rodríguez-Muro, M., Calvanese, D.: Rules and ontology based data access. In: Kontchakov, R., Mugnier, M.-L. (eds.) RR 2014. LNCS, vol. 8741, pp. 157–172. Springer, Cham (2014). https://doi.org/10.1007/978-3-319-11113-1_11

KGTK: A Toolkit for Large Knowledge Graph Manipulation and Analysis

Filip Ilievski[1]([✉]) [iD], Daniel Garijo[1] [iD], Hans Chalupsky[1] [iD],
Naren Teja Divvala[1], Yixiang Yao[1] [iD], Craig Rogers[1] [iD], Rongpeng Li[1] [iD],
Jun Liu[1], Amandeep Singh[1] [iD], Daniel Schwabe[2] [iD], and Pedro Szekely[1]

[1] Information Sciences Institute, University of Southern California, Los Angeles, USA
{ilievski,dgarijo,hans,divvala,yixiangy,rogers,
rli,junliu,amandeep,pszekely}@isi.edu
[2] Department of Informatics, Pontificia Universidade Católica Rio de Janeiro,
Rio de Janeiro, Brazil
dschwabe@inf.puc-rio.br

Abstract. Knowledge graphs (KGs) have become the preferred technology for representing, sharing and adding knowledge to modern AI applications. While KGs have become a mainstream technology, the RDF/SPARQL-centric toolset for operating with them at scale is heterogeneous, difficult to integrate and only covers a subset of the operations that are commonly needed in data science applications. In this paper we present KGTK, a data science-centric toolkit designed to represent, create, transform, enhance and analyze KGs. KGTK represents graphs in tables and leverages popular libraries developed for data science applications, enabling a wide audience of developers to easily construct knowledge graph pipelines for their applications. We illustrate the framework with real-world scenarios where we have used KGTK to integrate and manipulate large KGs, such as Wikidata, DBpedia and ConceptNet.

Keywords: Knowledge graph · Knowledge graph embedding · Knowledge graph filtering · Knowledge graph manipulation

Resource type: Software
License: MIT
DOI: https://doi.org/10.5281/zenodo.3828068
Repository: https://github.com/usc-isi-i2/kgtk/

1 Introduction

Knowledge graphs (KGs) have become the preferred technology for representing, sharing and using knowledge in applications. A typical use case is building a new knowledge graph for a domain or application by extracting subsets of several existing knowledge graphs, combining these subsets in application-specific ways, augmenting them with information from structured or unstructured sources, and

---

The original version of this chapter was revised: An error in the name of the author Rongpeng Li has been corrected. The correction to this chapter is available at https://doi.org/10.1007/978-3-030-62466-8_44

© Springer Nature Switzerland AG 2020, corrected publication 2021
J. Z. Pan et al. (Eds.): ISWC 2020, LNCS 12507, pp. 278–293, 2020.
https://doi.org/10.1007/978-3-030-62466-8_18

computing analytics or inferred representations to support downstream applications. For example, during the COVID-19 pandemic, several efforts focused on building KGs about scholarly articles related to the pandemic starting from the CORD-19 dataset provided by the Allen Institute for AI [25].[1] Enhancing these data with KGs such as DBpedia [1] and Wikidata [24] to incorporate gene, chemical, disease and taxonomic information, and computing network analytics on the resulting graphs, requires the ability to operate these KGs at scale.

Many tools exist to query, transform and analyze KGs. Notable examples include graph databases, such as RDF triple stores and Neo4J;[2] tools for operating on RDF such as graphy[3] and RDFlib[4], entity linking tools such as WAT [18] or BLINK [26], entity resolution tools such as MinHash-LSH [14] or MFIBlocks [12], libraries to compute graph embeddings such as PyTorch-BigGraph [13] and libraries for graph analytics, such as graph-tool[5] and NetworkX.[6]

There are three main challenges when using these tools together. First, tools may be challenging to set up with large KGs (e.g., the Wikidata RDF dump takes over a week to load into a triple store) and often need custom configurations that require significant expertise. Second, interoperating between tools requires developing data transformation scripts, as some of them may not support the same input/output representation. Third, composing two or more tools together (e.g., to filter, search, and analyze a KG) includes writing the intermediate results to disk, which is time and memory consuming for large KGs.

In this paper, we introduce the Knowledge Graph Toolkit (KGTK), a framework for manipulating, validating, and analyzing large-scale KGs. Our work is inspired by Scikit-learn [17] and SpaCy,[7] two popular toolkits for machine learning and natural language processing that have had a vast impact by making these technologies accessible to data scientists and software developers. KGTK aims to build a comprehensive library of tools and methods to enable easy composition of KG operations (validation, filtering, merging, centrality, text embeddings, etc.) to build knowledge-based AI applications. The contributions of KGTK are:

- The **KGTK file format**, which allows representing KGs as hypergraphs. This format unifies the Wikidata data model [24] based on items, claims, qualifiers, and references; property graphs that support arbitrary attributes on nodes and edges; RDF-Schema-based graphs such as DBpedia [1]; and general purpose RDF graphs with various forms of reification. The KGTK format uses tab-separated values (TSV) to represent edge lists, making it easy to process with many off-the-shelf tools.

---

[1] https://github.com/fhircat/CORD-19-on-FHIR/wiki/CORD-19-Semantic-Annotation-Projects.
[2] https://neo4j.com.
[3] https://graphy.link/.
[4] https://rdflib.readthedocs.io/en/stable/.
[5] https://graph-tool.skewed.de/.
[6] https://networkx.github.io/.
[7] https://spacy.io/.

- A comprehensive **validator and data cleaning** module to verify compliance with the KGTK format, and normalize literals like strings and numbers.
- **Import modules** to transform different formats into KGTK, including N-Triples [21], Wikidata qualified terms, and ConceptNet [22].
- **Graph manipulation modules** for bulk operations on graphs to validate, clean, filter, join, sort, and merge KGs. Several of these are implemented as wrappers of common, streaming Unix tools like awk[8], sort, and join.
- **Graph querying and analytics modules** to compute centrality measures, connected components, and text-based graph **embeddings** using state-of-the-art language models: RoBERTa [15], BERT [5], and DistilBERT [19]. Common queries, such as computing the set of nodes reachable from other nodes, are also supported.
- **Export modules** to transform KGTK format into diverse standard and commonly used formats, such as RDF (N-Triples), property graphs in Neo4J format, and GML to invoke tools such as graph-tool or Gephi.[9]
- A **framework for composing multiple KG operations**, based on Unix pipes. The framework uses the KGTK file format on the standard input and output to combine tools written in different programming languages.

KGTK provides an implementation that integrates all these methods relying on widely used tools and standards, thus allowing their composition in pipelines to operate with large KGs like Wikidata on an average laptop.

The rest of the paper is structured as follows. Section 2 describes a motivating scenario and lists the requirements for a graph manipulation toolkit. Section 3 describes KGTK by providing an overview of its file format, supported operations, and examples on how to compose them together. Next, Sect. 4 showcases how we have used KGTK on three different real-world use cases, together with the current limitations of our approach. We then review relevant related work in Sect. 5, and we conclude the paper in Sect. 6.

2   Motivating Scenario

The 2020 coronavirus pandemic led to a series of community efforts to publish and share common knowledge about COVID-19 using KGs. Many of these efforts use the COVID-19 Open Research Dataset (CORD-19) [25], compiled by the Allen Institute for AI. CORD-19 is a free resource containing over 44,000 scholarly articles, including over 29,000 with full text, about COVID-19 and the coronavirus family of viruses. Having an integrated KG would allow easy access to information published in scientific papers, as well as to general medical knowledge on genes, proteins, drugs, and diseases mentioned in these papers, and their interactions.

In our work, we integrated the CORD-19 corpus with gene, chemical, disease, and taxonomic knowledge from Wikidata and CTD databases,[10] as well as entity

---

[8] https://linux.die.net/man/1/awk.

[9] https://gephi.org/.

[10] http://ctdbase.org/.

extractions from Professor Heng Ji's BLENDER lab at UIUC.[11] We extracted all the items and statements for the 30,000 articles in the CORD-19 corpus [25] that were present in Wikidata at the time of extraction, added all Wikidata articles, authors, and entities mentioned in the BLENDER corpus, homogenized the data to fix inconsistencies (e.g., empty values), created nodes and statements for entities that were absent in Wikidata, incorporated metrics such as PageRank for each KG node, and exported the output in both RDF and Neo4J.

This use case exhibited several of the challenges that KGTK is designed to address. For example, extracting a subgraph from Wikidata articles is not feasible using SPARQL queries as it would have required over 100,000 SPARQL queries; using RDF tools on the Wikidata RDF dump (107 GB compressed) is difficult because its RDF model uses small graphs to represent each Wikidata statement; using the Wikidata JSON dump is possible, but requires writing custom code as the schema is specific to Wikidata (hence not reusable for other KGs). In addition, while graph-tool allowed us to compute graph centrality metrics, its input format is incompatible with RDF, requiring a transformation.

Other efforts employed a similar set of processing steps [25].[12] These range from mapping the CORD-19 data to RDF,[13] to adding annotations to the articles in the dataset pointing to entities extracted from the text, obtained from various sources [8].[14] A common thread among these efforts involves leveraging existing KGs such as Wikidata and Microsoft Academic Graph to, for example, build a citation network of the papers, authors, affiliations, etc.[15] Other efforts focused on extraction of relevant entities (genes, proteins, cells, chemicals, diseases), relations (causes, upregulates, treats, binds), and linking them to KGs such as Wikidata and DBpedia. Graph analytics operations followed, such as computing centrality measures in order to support identification of key articles, people or substances,[15] or generation of various embeddings to recommend relevant literature associated with an entity.[16] The resulting graphs were deployed as SPARQL endpoints, or exported as RDF dumps, CSV, or JSON files.

These examples illustrate the need for composing sequences of integrated KG operations that extract, modify, augment and analyze knowledge from existing KGs, combining it with non-KG datasets to produce new KGs. Existing KG tools do not allow users to seamlessly run such sequences of graph manipulation tasks in a pipeline. We propose that an effective toolkit that supports the construction of modular KG pipelines has to meet the following criteria:

---

[11] https://blender.cs.illinois.edu/.

[12] A list of such projects can be found in https://github.com/fhircat/CORD-19-on-FHIR/wiki/CORD-19-Semantic-Annotation-Projects.

[13] https://github.com/nasa-jpl-cord-19/covid19-knowledge-graph,   https://github.com/GillesVandewiele/COVID-KG/.

[14] http://pubannotation.org/collections/CORD-19.

[15] https://scisight.apps.allenai.org/clusters.

[16] https://github.com/vespa-engine/cord-19/blob/master/README.md.

1. **A simple representation format** that all modules in the toolkit operate on (the equivalent of *datasets* in Scikit-learn and *document model* in SpaCy), to enable tool integration without additional data transformations.
2. **Ability to incorporate mature existing tools**, wrapping them to support a common API and input/output format. The scientific community has worked for many years on efficient techniques for manipulation of graph and structured data. The toolkit should be able to accommodate them without the need for a new implementation.
3. **A comprehensive set of features** that include import and export modules for a wide variety of KG formats, modules to select, transform, combine, link, and merge KGs, modules to improve the quality of KGs and infer new knowledge, and modules to compute embeddings and graph statistics. Such a rich palette of functionalities would largely support use cases such as the ones presented in this section.
4. **A pipeline mechanism** to allow composing modules in arbitrary ways to process large public KGs such as Wikidata, DBpedia, or ConceptNet.

3   KGTK: The Knowledge Graph Toolkit

KGTK helps manipulating, curating, and analyzing large real-world KGs, in which each statement may have multiple qualifiers such as the statement source, its creation date or its measurement units. Figure 1 shows an overview of the different capabilities of KGTK. Given an input file with triples (either as tab-separated values, Wikidata JSON, or N-Triples), we convert it to an internal representation (the *KGTK file format*, described in Sect. 3.1) that we use as main input/output format for the rest of the features included in the toolkit. Once data is in KGTK format, we can perform operations for curating (data validation and cleaning), transforming (sort, filter, or join) and analyzing the contents of a KG (computing embeddings, statistics, node centrality). KGTK also provides export operations to common formats, such as N-Triples, Neo4J, and JSON. The KGTK operations are described in Sect. 3.2, whereas their composition into pipelines is illustrated in Sect. 3.3.

3.1   KGTK File Format

KGTK uses a tab-separated column-based text format to describe any attributed, labeled or unlabeled hypergraph. We chose this format instead of an RDF serialization for three reasons. First, tabular formats are easy to generate and parse by standard tools. Second, this format is self-describing and easy to read by humans. Finally, it provides a simple mechanism to define hypergraphs and edge qualifiers, which may be more complicated to describe using Turtle or JSON.

KGTK defines KGs as a set of nodes and a set of edges between those nodes. All concepts of meaning are represented via an edge, including edges themselves, allowing KGTK to represent generalized hypergraphs (while supporting the representation of RDF graphs). The snippet below shows a simple example of a KG

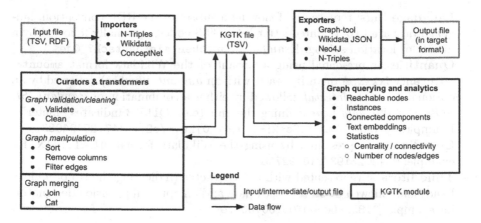

Fig. 1. Overview of the usage workflow and features included in KGTK.

in KGTK format with three people (*Moe, Larry* and *Curly*), the creator of the statements (*Hans*) and the original source of the statements (*Wikipedia*):

node1	label	node2	creator	source	id
"Moe"	rdf:type	Person	"Hans"	Wikipedia	E1
"Larry"	rdf:type	Person	"Hans"	Wikipedia	E2
"Curly"	rdf:type	Person		Wikipedia	E3
"Curly"	hasFriend	"Moe"		Wikipedia	E4

The first line of a KGTK file declares the headers for the document. The reserved words *node1*, *label* and *node2* are used to describe the subject, property and object being described, while *creator* and *source* are optional qualifiers for each statement that provide additional provenance information about the creator of a statement and the original source. Note that the example is not using namespace URIs for any nodes and properties, as they are not needed for local KG manipulation. Still, namespace prefixes (e.g., `rdf`) may be used for mapping back to RDF after the KG manipulations with KGTK. Nodes and edges have unique IDs (when IDs are not present, KGTK generates them automatically).

The snippet below illustrates the representation of qualifiers for individual edges, and shows how the additional columns in the previous example may be represented as edges about edges:

node1	label	node2	id
"Moe"	rdf:type	Person	E1
E1	source	Wikipedia	E5
E1	creator	"Hans"	E6
"Larry"	rdf:type	Person	E2

KGTK is designed to support commonly-used typed literals:

- **Language tags**: represented following a subset of the RDF convention, language tags are two- or three-letter ISO 639-3 codes, optionally followed by a dialect or location subtag. Example: `'Sprechen sie deutsch?'@de`.
- **Quantities**: represented using a variant of the Wikidata format `amount~toleranceUxxxx`. A quantity starts with an *amount* (number), followed by an optional *tolerance interval*, followed by either a combination of *standard (SI) units* or a *Wikidata node* defining the unit (e.g., Q11573 indicates "meter"). Examples include `10m`, `-1.2e+2[-1.0,+1.0]kg.m/s2` or `+17.2Q494083`
- **Coordinates**: represented by using the Wikidata format `@LAT/LON`, for example: `@043.26193/010.92708`
- **Time literals**: represented with a `^` character (indicating the tip of a clock hand) and followed by an ISO 8601 date and an optional precision designator, for example: `^1839-00-00T00:00:00Z/9`

The full KGTK file format specification is available online.[17]

3.2   KGTK Operations

KGTK currently supports 13 operations (depicted in Fig. 1),[18] grouped into four modules: importing modules, graph manipulation modules, graph analytics modules, and exporting modules. We describe each of these modules below.

3.2.1   Importing and Exporting from KGTK

1. The `import` operation transforms an external graph format into KGTK TSV format. KGTK supports importing a number of data formats, including N-Triples, ConceptNet, and Wikidata (together with qualifiers).
2. The `export` operation transforms a KGTK-formatted graph to a wide palette of formats: TSV (by default), N-Triples, Neo4J Property Graphs, graph-tool and the Wikidata JSON format.

3.2.2   Graph Curation and Transformation

3. The `validate` operation ensures that the data meets the KGTK file format specification, detecting errors such as nodes with empty values, values of unexpected length (either too long or too short), potential errors in strings (quotation errors, incorrect use of language tags, etc.), incorrect values in dates, etc. Users may customize the parsing of the file header, each line, and the data values, as well as choose the action taken when a validation rule fails.
4. The `clean` operation fixes a substantial number of errors detected by `validate`, by correcting some common mistakes in data encoding (such as not escaping 'pipe' characters), replacing invalid dates, normalizing values for quantities, languages and coordinates using the KGTK convention for literals. Finally,

---

[17] https://kgtk.readthedocs.io/en/latest/specification/.
[18] https://kgtk.readthedocs.io/en/latest.

it removes rows that still do not meet the KGTK specification (e.g., rows with empty values for required columns or rows with an invalid number of columns).

5. `sort` efficiently reorders any KGTK file according to one or multiple columns. `sort` is useful to organize edge files so that, for example, all edges for `node1` are contiguous, enabling efficient processing in streaming operations.

6. The `remove-columns` operation removes a subset of the columns in a KGTK file (`node1` (source), `node2` (object), and `label` (property) cannot be removed). This is useful in cases where columns have lengthy values and are not relevant to the use case pursued by a user, e.g., removing edge and graph identifiers when users aim to compute node centrality or calculate embeddings.

7. The `filter` operation selects edges from a KGTK file, by specifying constraints ("patterns") on the values for node1, label, and node2. The `pattern` language, inspired by graphy.js, has the following form: "`subject-pattern; predicate-pattern; object-pattern`". For each of the three columns, the filtering pattern can consist of a list of symbols separated using commas. Empty patterns indicate that no filter should be performed for a column. For instance, to select all edges that have property P154 or P279, we can use the pattern "; P154, P279;". Alternatively, a common query of retrieving edges for all humans from Wikidata corresponds to the filter "; P31; Q5".

8. The `join` operation will join two KGTK files. Inner join, left outer join, right outer join, and full outer join are all supported. When a join takes place, the columns from two files are merged into the set of columns for the output file. By default, KGTK will join based on the `node1` column, although it can be configured to join by edge `id`. KGTK also allows the `label` and `node2` columns to be added to the join. Alternatively, the user may supply a list of join columns for each file giving them full control over the semantics of the result.

9. The `cat` operation concatenates any number of files into a single, KGTK-compliant graph file.

3.2.3 Graph Querying and Analytics

10. `reachable-nodes`: given a set of nodes N and a set of properties P, this operation computes the set of reachable nodes R that contains the nodes that can be reached from a node $n \in N$ via paths containing any of the properties in P. This operation can be seen as a (joint) closure computation over one or multiple properties for a predefined set of nodes. A common application of this operation is to compute a closure over the subClassOf property, which benefits downstream tasks such as entity linking or table understanding.

11. The `connected-components` operation finds all connected components (communities) in a graph (e.g., return all the communities connected via an `owl:sameAs` edge in a KGTK file).

12. The `text-embeddings` operation computes embeddings for all nodes in a graph by computing a sentence embedding over a lexicalization of the neighborhood of each node. The lexicalized sentence is created based on a template whose simplified version is:

{label-properties}, {description-properties} is a {isa-properties}, has {has-properties}, and {properties:values}.

The properties for label-properties, description-properties, isa-properties, has-properties, and property-values pairs are specified as input arguments to the operation. An example sentence is "Saint David, patron saint of Wales is a human, Catholic priest, Catholic bishop, and has date of death, religion, canonization status, and has place of birth Pembrokeshire". The sentence for each node is encoded into an embedding using one of 16 currently supported variants of three state-of-the-art language models: BERT, DistilBERT, and RoBERTa. Computing similarity between such entity embeddings is a standard component of modern decision making systems such as entity linking, question answering, or table understanding.

13. The graph-statistics operation computes various graph statistics and centrality metrics. The operation generates a graph summary, containing its number of nodes, edges, and most common relations. In addition, it can compute graph degrees, HITS centrality, and PageRank values. Aggregated statistics (minimum, maximum, average, and top nodes) for these connectivity/centrality metrics are included in the summary, whereas the individual values for each node are represented as edges in the resulting graph. The graph is assumed to be directed, unless indicated differently.

3.3   Composing Operations into Pipelines

KGTK has a pipelining architecture based on Unix pipes[19] that allows chaining most operations introduced in the previous section by using the standard input/output and the KGTK file format. Pipelining increases efficiency by avoiding the need to write files to disk and supporting parallelism allowing downstream commands to process data before upstream commands complete. We illustrate the chaining operations in KGTK with three examples from our own work. Note that we have implemented a shortcut pipe operator "/", which allows users to avoid repeating kgtk in each of their operations. For readability, command arguments are slightly simplified in the paper. Jupyter Notebooks that implement these and other examples can be found online.[20]

Example 1: Alice wants to import the English subset of ConceptNet [22] in KGTK format to extract a filtered subset where two concepts are connected with a more precise semantic relation such as /r/Causes or /r/UsedFor (as opposed to weaker relations such as /r/RelatedTo). For all nodes in this subset, she wants to compute text embeddings and store them in a file called emb.txt.

To extract the desired subset, the sequence of KGTK commands is as follows:

```
kgtk import-conceptnet --english_only conceptnet.csv / \
   filter -p "; /r/Causes,/r/UsedFor,/r/Synonym,/r/DefinedAs,/r/IsA ;" / \
   sort -c 1,2,3 > sorted.tsv
```

---

[19] https://linux.die.net/man/7/pipe.
[20] https://github.com/usc-isi-i2/kgtk/tree/master/examples.

To compute embeddings for this subset, she would use `text-embeddings`:

```
kgtk text-embeddings --label-properties "/r/Synonym" \
  --isa-properties "/r/IsA" --description-properties "/r/DefinedAs" \
  --property-value "/r/Causes" "/r/UsedFor" \
  --model bert-large-nli-cls-token -i sorted.tsv \
  > emb.txt
```

Example 2: Bob wants to extract a subset of Wikidata that contains only edges of the 'member of' (P463) property, and strip a set of columns that are not relevant for his use case (`$ignore_col`), such as id and rank. While doing so, Bob would also like to clean any erroneous edges. On the clean subset, he would compute graph statistics, including PageRank values and node degrees. Here is how to perform this functionality in KGTK (after Wikidata is already converted to a KGTK file called `wikidata.tsv` by `import-wikidata`):

```
kgtk filter -p ' ; P463 ; ' / clean_data /
    remove-columns -c "$ignore_cols" wikidata.tsv > graph.tsv
kgtk graph-statistics --directed --degrees --pagerank graph.tsv
```

Example 3: Carol would like to concatenate two subsets of Wikidata: one containing occupations for several notable people: Sting, Roger Federer, and Nelson Mandela; and the other containing all 'subclass of' (P279) relations in Wikidata. The concatenated file needs to be sorted by subject, after which she would compute the set of reachable nodes for these people via the properties 'occupation' (P106) or 'subclass of' (P279). To achieve this in KGTK, Carol first needs to extract the two subsets with the `filter` operation:

```
kgtk filter -p 'Q8023,Q483203,Q1426;P106;' wikidata.tsv > occupation.tsv
kgtk filter -p ' ; P279 ; ' wikidata.tsv > subclass.tsv
```

Then, she can merge the two files into one, sort it, and compute reachability:

```
kgtk cat occupation.tsv subclass.tsv / \
    sort -c node1 > sorted.tsv
kgtk reachable-nodes --props P106,P279 --root "Q8023,Q483203,Q1426" \
    sorted.tsv > reachable.tsv
```

4   Discussion

Validating, merging, transforming and analyzing KGs at scale is an open challenge for knowledge engineers, and even more so for data scientists. Complex SPARQL queries often time out on online endpoints, while working with RDF dumps locally takes time and expertise. In addition, popular graph analysis tools do not operate with RDF, making analysis complex for data scientists.

The KGTK format intentionally does not distinguish attributes or qualifiers of nodes and edges from full-fledged edges. Tools operating on KGTK graphs can instead interpret edges differently when desired. In the KGTK file format,

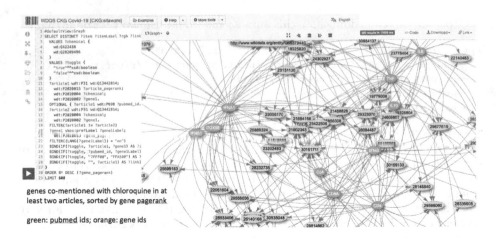

Fig. 2. SPARQL query and visualization of the CORD-19 use case, illustrating the use of the Wikidata infrastructure using our KG that includes a subset of Wikidata augmented with new properties such as "mentions gene" and "pagerank".

everything can be a node, and every node can have any type of edge to any other node. To do so in RDF requires adopting more complex mechanisms such as reification, typically leading to efficiency issues. This generality allows KGTK files to be mapped to most existing DBMSs, and to be used in powerful data transformation and analysis tools such as Pandas.[21]

We believe KGTK will have a significant impact within and beyond the Semantic Web community by helping users to easily perform usual data science operations on large KGs. To give an idea, we downloaded Wikidata (truthy statements distribution, 23.2 GB[22]) and performed a test of filtering out all Qnodes (entities) which have the P31 property (instance of) in Wikidata. This filter took over 20 h in Apache Jena and RDFlib. In graphy, the time went down 4 h 15 min. Performing the same operation in KGTK took less 1 h 30 min.

We have been using KGTK in our own work to integrate and analyze KGs:

- **CORD-19:** As described in Sect. 2, we used KGTK to combine extracted information from the papers in the CORD-19 dataset (such as entities of interest) with metadata about them, and general medical and biology knowledge, all found in Wikidata, CTD and the BLENDER datasets. A notebook illustrating the operations used in this use case is available online.[23] Figure 2 shows the CORD-19 KGTK KG loaded in Wikidata SPARQL query interface. The KGTK tools exported the CORD-19 KG to RDF triples in a format compatible with Wikidata.

---

[21] https://pandas.pydata.org.

[22] https://dumps.wikimedia.org/wikidatawiki/entities/latest-truthy.nt.bz2.

[23] https://github.com/usc-isi-i2/CKG-COVID-19/blob/dev/build-covid-kg.ipynb.

- **Commonsense Knowledge Graph (CSKG):** Commonsense knowledge is dispersed across a number of (structured) knowledge sources, like ConceptNet and ATOMIC [20]. After consolidating several such sources into a single CSKG [11], we used KGTK to compute graph statistics (e.g., number of edges or most frequent relations), HITS, PageRank, and node degrees, in order to measure the impact of the consolidation on the graph connectivity and centrality. We also created RoBERTa-based embeddings of the CSKG nodes, which we are currently using for downstream question answering. A notebook illustrating the operations in this use case is available online.[24]
- **Integrating and exporting Ethiopian quantity data:** We are using KGTK to create a custom extension of Wikidata with data about Ethiopia,[25] by integrating quantity indicators like crime, GDP, population, etc.

The heterogeneity of these cases shows how KGTK can be adopted for multipurpose data-science operations over KGs, independently of the domain. The challenges described in them are common in data integration and data science. Given the rate at which KGs are gaining popularity, we expect KGTK to fill a key gap faced by many practitioners wanting to use KGs in their applications.

The primary limitation of KGTK lies in its functionality coverage. The main focus so far has been on supporting basic operations for manipulating KGs, and therefore KGTK does not yet incorporate powerful browsing and visualization tools, or advanced tools for KG identification tasks such as link prediction, entity resolution, and ontology mapping. Since KGTK is proposed as a new resource, we have no usage metrics at the time of writing this paper.

5 Related Work

Many of the functionalities in KGTK for manipulating and transforming KGs (i.e., join operations, filtering entities, general statistics, and node reachability) can be translated into queries in SPARQL. However, the cost of these queries over large endpoints is often too high, and they will time out or take too long to produce a response. In fact, many SPARQL endpoints have been known to have limited availability and slow response times for many queries [4], leaving no choice but to download their data locally for any major KG manipulation. Additionally, it is unclear how to extend SPARQL to support functionalities such as computing embeddings or node centrality. A scalable alternative to SPARQL is Linked Data Fragments (LDF) [23]. The list of native operations in LDF boils down to triple pattern matching, resembling our proposed `filter` operation. However, operations like merging and joining are not trivial in LDF, while more complex analytics and querying, like embedding computation, are not supported.

Other works have proposed offline querying. LOD Lab [3] and LOD-a-lot [6] combine LDF with an efficient RDF compression format, called Header Dictionary Triples (HDT) [7,16], in order to store a LOD dump of 30–40B statements.

---

[24] https://github.com/usc-isi-i2/kgtk/blob/master/examples/CSKG.ipynb.
[25] https://datamart-upload.readthedocs.io/en/latest/REST-API-tutorial/.

Although the LOD Lab project also employed mature tooling, such as Elastic Search and bash operations, to provide querying over the data, the set of available operations is restricted by employing LDF as a server, as native LDF only supports pattern matching queries. The HDT compression format has also been employed by other efforts, such as sameAs.cc [2], which performs closure and clustering operations over half a billion identity (same-as) statements. However, HDT cannot be easily used by existing tools (e.g., graph-tool or pandas), and it does not describe mechanisms for supporting qualifiers (except for using reification on statements, which complicates the data model).

The recent developments towards supporting triple annotations with RDF* [9] provide support for qualifiers; yet, this format is still in its infancy and we expect it to inherit the challenges of RDF, as described before.

Several RDF libraries exist for different programming languages, such as RDFLib in Python, graphy in JavaScript, and Jena in Java. The scope of these libraries is different from KGTK, as they focus on providing the building blocks for creating RDF triples, rather than a set of operators to manipulate and analyze large KGs (validate, merge, sort, statistics, etc.).

Outside of the Semantic Web community, prominent efforts perform graph operations in graph databases like Neo4J or libraries like graph-tool, which partially overlap with the operations included in KGTK. We acknowledge the usefulness of these tools for tasks like pattern matching and graph traversal, and therefore we provide an export to their formats to enable users to take advantage of those capabilities. However, these tools also have limitations. First, Neo4J "only allows one value per attribute property" and it "does not currently support queries involving joins or lookups on any information associated with edges, including edge ids, edge types, or edge attributes" [10]. The KGTK representation does not have these limitations, and the tasks above can be performed using KGTK commands or via export to SPARQL and graph-tool. Second, while Neo4J performs very well on traversal queries, it is not optimized to run on bulk, relational queries, like "who are the latest reported sports players?" Similarly, [10] shows that Neo4J performs worse and times out more frequently than Postgres and Virtuoso on atomic queries and basic graph patterns, even after removing the labels to improve efficiency. KGTK supports bulk and simple table queries, complex queries are handled by exporting to RDF and Postgres.

Graph-tool provides rich functionality and can natively support property graphs. However, it needs to be integrated with other tools for operations like computation of embeddings or relational data operations, requiring additional expertise.

Finally, the KGX toolkit[26] has a similar objective as KGTK, but it is scoped to process KGs aligned with the Biolink Model, a datamodel describing biological entities using property graphs. Its set of operations can be regarded as a subset of the operations supported by KGTK. To the best of our knowledge, there is no existing toolkit with a comprehensive set of operations for validating, manipulating, merging, and analyzing knowledge graphs comparable to KGTK.

---

[26] https://github.com/NCATS-Tangerine/kgx.

Fig. 3. SQID visualization of local KGTK data (using the CORD-19 example).

6   Conclusions and Future Work

Performing common graph operations on large KGs is challenging for data scientists and knowledge engineers. Recognizing this gap, in this paper we presented the Knowledge Graph ToolKit (KGTK): a data science-centric toolkit to represent, create, transform, enhance, and analyze KGs. KGTK represents graphs in tabular format, and leverages popular libraries developed for data science applications, enabling a wide audience of researchers and developers to easily construct KG pipelines for their applications. KGTK currently supports thirteen common operations, including import/export, filter, join, merge, computation of centrality, and generation of text embeddings. We are using KGTK in our own work for three real-world scenarios which benefit from integration and manipulation of large KGs, such as Wikidata and ConceptNet.

KGTK is actively under development, and we are expanding it with new operations. Our CORD-19 use case indicated the need for a tool to create new edges, which will also be beneficial in other domains with emerging information and many long-tail/emerging new entities. Our commonsense KG use case, which combines a number of initially disconnected graphs, requires new operations that will perform de-duplication of edges in flexible ways. Additional import options are needed to support knowledge sources in custom formats, while new export formats will allow us to leverage a wider span of libraries, e.g., the GraphViz format enables using existing visualization tooling. We are also looking at converting other existing KGs to the KGTK format, both to enhance existing KGTK KGs, and to identify the need for additional functionality. In the longer term, we plan to extend the toolkit to support more complex KG operations, such as entity resolution, link prediction, and entity linking.

We are also working on enhancing further the user experience with KGTK. We are adapting the SQID[27] KG browser (as shown in Fig. 3), which is part of the Wikidata tool ecosystem. To this end, we are using the KGTK export operations to convert any KGTK KG to Wikidata format (JSON and RDF as required by SQID), and are modifying SQID to remove its dependencies on Wikidata. The current prototype can browse arbitrary KGTK files. Remaining work includes computing the KG statistics that SQID requires, and automating deployment of the Wikidata infrastructure for use with KGTK KGs.

Acknowledgements. This material is based on research sponsored by Air Force Research Laboratory under agreement number FA8750-20-2-10002. The U.S. Government is authorized to reproduce and distribute reprints for Governmental purposes notwithstanding any copyright notation thereon. The views and conclusions contained herein are those of the authors and should not be interpreted as necessarily representing the official policies or endorsements, either expressed or implied, of Air Force Research Laboratory or the U.S. Government.

References

1. Auer, S., Bizer, C., Kobilarov, G., Lehmann, J., Cyganiak, R., Ives, Z.: DBpedia: a nucleus for a web of open data. In: Aberer, K., et al. (eds.) ASWC/ISWC -2007. LNCS, vol. 4825, pp. 722–735. Springer, Heidelberg (2007). https://doi.org/10. 1007/978-3-540-76298-0_52

2. Beek, W., Raad, J., Wielemaker, J., van Harmelen, F.: sameAs.cc: the closure of 500M `owl:sameAs` statements. In: Gangemi, A., et al. (eds.) ESWC 2018. LNCS, vol. 10843, pp. 65–80. Springer, Cham (2018). https://doi.org/10.1007/978-3-319-93417-4_5

3. Beek, W., Rietveld, L., Ilievski, F., Schlobach, S.: LOD lab: scalable linked data processing. In: Pan, J.Z., et al. (eds.) Reasoning Web 2016. LNCS, vol. 9885, pp. 124–155. Springer, Cham (2017). https://doi.org/10.1007/978-3-319-49493-7_4

4. Buil-Aranda, C., Hogan, A., Umbrich, J., Vandenbussche, P.-Y.: SPARQL web-querying infrastructure: ready for action? In: Alani, H., et al. (eds.) ISWC 2013. LNCS, vol. 8219, pp. 277–293. Springer, Heidelberg (2013). https://doi.org/10. 1007/978-3-642-41338-4_18

5. Devlin, J., Chang, M.W., Lee, K., Toutanova, K.: BERT: pre-training of deep bidirectional transformers for language understanding. arXiv preprint arXiv:1810.04805 (2018)

6. Fernández, J.D., Beek, W., Martínez-Prieto, M.A., Arias, M.: LOD-a-lot. In: d'Amato, C., et al. (eds.) ISWC 2017. LNCS, vol. 10588, pp. 75–83. Springer, Cham (2017). https://doi.org/10.1007/978-3-319-68204-4_7

7. Fernández, J.D., Martínez-Prieto, M.A., Polleres, A., Reindorf, J.: HDTQ: managing RDF datasets in compressed space. In: Gangemi, A., et al. (eds.) ESWC 2018. LNCS, vol. 10843, pp. 191–208. Springer, Cham (2018). https://doi.org/10.1007/ 978-3-319-93417-4_13

8. Gazzotti, R., Michel, F., Gandon, F.: CORD-19 named entities knowledge graph (CORD19-NEKG) (2020). https://github.com/Wimmics/cord19-nekg, University Côte d'Azur, Inria, CNRS

---

[27] https://tools.wmflabs.org/sqid/.

9. Hartig, O.: RDF* and SPARQL*: an alternative approach to annotate statements in RDF. In: International Semantic Web Conference (Posters, Demos & Industry Tracks) (2017)

10. Hernández, D., Hogan, A., Riveros, C., Rojas, C., Zerega, E.: Querying Wikidata: comparing SPARQL, relational and graph databases. In: Groth, P., et al. (eds.) ISWC 2016. LNCS, vol. 9982, pp. 88–103. Springer, Cham (2016). https://doi.org/10.1007/978-3-319-46547-0_10

11. Ilievski, F., Szekely, P., Cheng, J., Zhang, F., Qasemi, E.: Consolidating common-sense knowledge. arXiv preprint arXiv:2006.06114 (2020)

12. Kenig, B., Gal, A.: MFIBlocks: an effective blocking algorithm for entity resolution. Inf. Syst. **38**(6), 908–926 (2013)

13. Lerer, A., et al.: PyTorch-BigGraph: a large-scale graph embedding system. arXiv preprint arXiv:1903.12287 (2019)

14. Leskovec, J., Rajaraman, A., Ullman, J.D.: Mining of Massive Data Sets. Cambridge University Press, Cambridge (2020)

15. Liu, Y., et al.: RoBERTa: a robustly optimized BERT pretraining approach. arXiv preprint arXiv:1907.11692 (2019)

16. Martínez-Prieto, M.A., Arias Gallego, M., Fernández, J.D.: Exchange and consumption of huge RDF data. In: Simperl, E., Cimiano, P., Polleres, A., Corcho, O., Presutti, V. (eds.) ESWC 2012. LNCS, vol. 7295, pp. 437–452. Springer, Heidelberg (2012). https://doi.org/10.1007/978-3-642-30284-8_36

17. Pedregosa, F., et al.: Scikit-learn: machine learning in Python. J. Mach. Learn. Res. **12**, 2825–2830 (2011)

18. Piccinno, F., Ferragina, P.: From TagME to WAT: a new entity annotator. In: Proceedings of the First International Workshop on Entity Recognition & Disambiguation, pp. 55–62 (2014)

19. Sanh, V., Debut, L., Chaumond, J., Wolf, T.: DistilBERT, a distilled version of BERT: smaller, faster, cheaper and lighter. arXiv preprint arXiv:1910.01108 (2019)

20. Sap, M., et al.: ATOMIC: an atlas of machine commonsense for if-then reasoning. In: Proceedings of the AAAI Conference on Artificial Intelligence, vol. 33, pp. 3027–3035 (2019)

21. Seaborne, A., Carothers, G.: RDF 1.1 N-triples. W3C recommendation, W3C, February 2014. http://www.w3.org/TR/2014/REC-n-triples-20140225/

22. Speer, R., Chin, J., Havasi, C.: ConceptNet 5.5: an open multilingual graph of general knowledge (2016)

23. Verborgh, R., Vander Sande, M., Colpaert, P., Coppens, S., Mannens, E., Van de Walle, R.: Web-scale querying through linked data fragments. In: LDOW. Citeseer (2014)

24. Vrandečić, D., Krötzsch, M.: Wikidata: a free collaborative knowledgebase. Commun. ACM **57**(10), 78–85 (2014)

25. Wang, L.L., et al.: CORD-19: The COVID-19 open research dataset. ArXiv abs/2004.10706 (2020)

26. Wu, L., Petroni, F., Josifoski, M., Riedel, S., Zettlemoyer, L.: Zero-shot entity linking with dense entity retrieval. arXiv preprint arXiv:1911.03814 (2019)

Covid-on-the-Web: Knowledge Graph and Services to Advance COVID-19 Research

Franck Michel[(✉)], Fabien Gandon, Valentin Ah-Kane, Anna Bobasheva,
Elena Cabrio, Olivier Corby, Raphaël Gazzotti, Alain Giboin, Santiago Marro,
Tobias Mayer, Mathieu Simon, Serena Villata, and Marco Winckler

University Côte d'Azur, Inria, CNRS, I3S (UMR 7271), Biot, France
franck.michel@cnrs.fr, {fabien.gandon,anna.bobasheva,elena.cabrio,
olivier.corby,raphael.gazzotti,alain.giboin,
santiago.marro,tobias.mayer,mathieu.simon,serena.villata}@inria.fr,
valentin.ah-kane@etu.univ-cotedazur.fr, winckler@i3s.unice.fr

Abstract. Scientists are harnessing their multi-disciplinary expertise
and resources to fight the COVID-19 pandemic. Aligned with this mind-
set, the Covid-on-the-Web project aims to allow biomedical researchers
to access, query and make sense of COVID-19 related literature. To do so,
it adapts, combines and extends tools to process, analyze and enrich the
"COVID-19 Open Research Dataset" (CORD-19) that gathers 50,000+
full-text scientific articles related to the coronaviruses. We report on the
RDF dataset and software resources produced in this project by leveraging skills in knowledge representation, text, data and argument mining,
as well as data visualization and exploration. The dataset comprises two
main knowledge graphs describing (1) named entities mentioned in the
CORD-19 corpus and linked to DBpedia, Wikidata and other BioPortal
vocabularies, and (2) arguments extracted using ACTA, a tool automating the extraction and visualization of argumentative graphs, meant to
help clinicians analyze clinical trials and make decisions. On top of this
dataset, we provide several visualization and exploration tools based on
the Corese Semantic Web platform, MGExplorer visualization library,
as well as the Jupyter Notebook technology. All along this initiative, we
have been engaged in discussions with healthcare and medical research
institutes to align our approach with the actual needs of the biomedical
community, and we have paid particular attention to comply with the
open and reproducible science goals, and the FAIR principles.

Keywords: COVID-19 · Arguments · Visualization · Named entities ·
Linked data

1 Bringing COVID-19 Data to the LOD: Deep and Fast

In March 2020, as the Coronavirus infection disease (COVID-19) forced us to
confine ourselves at home, the Wimmics research team[1] decided to join the effort

---

[1] https://team.inria.fr/wimmics/.

© Springer Nature Switzerland AG 2020
J. Z. Pan et al. (Eds.): ISWC 2020, LNCS 12507, pp. 294–310, 2020.
https://doi.org/10.1007/978-3-030-62466-8_19

of many scientists around the world who harness their expertise and resources to fight the pandemic and mitigate its disastrous effects. We started a new project called *Covid-on-the-Web* aiming to make it easier for biomedical researchers to access, query and make sense of the COVID-19 related literature. To this end, we started to adapt, re-purpose, combine and apply tools to publish, as thoroughly and quickly as possible, a maximum of rich and actionable linked data about the coronaviruses.

In just a few weeks, we deployed several tools to analyze the *COVID-19 Open Research Dataset* (CORD-19) [20] that gathers 50,000+ full-text scientific articles related to the coronavirus family. On the one hand, we adapted the ACTA platform[2] designed to ease the work of clinicians in the analysis of clinical trials by automatically extracting arguments and producing graph visualizations to support decision making [13]. On the other hand, our expertise in the management of data extracted from knowledge graphs , both generic or specialized, and their integration in the HealthPredict project [9,10], allowed us to enrich the CORD-19 corpus with different sources. We used DBpedia Spotlight [6], Entity-fishing[3] and NCBO BioPortal Annotator [12] to extract Named Entities (NE) from the CORD-19 articles, and disambiguate them against LOD resources from DBpedia, Wikidata and BioPortal ontologies. Using the Morph-xR2RML[4] platform, we turned the result into the *Covid-on-the-Web RDF dataset*, and we deployed a public SPARQL endpoint to serve it. Meanwhile, we integrated the Corese[5] [5] and MGExplorer [4] platforms to support the manipulation of knowledge graphs and their visualization and exploration on the Web.

By integrating these diverse tools, the Covid-on-the-Web project (sketched in Fig. 1) has designed and set up an integration pipeline facilitating the extraction and visualization of information from the CORD-19 corpus through the production and publication of a continuously enriched linked data knowledge graph. We believe that our approach, integrating argumentation structures and named entities, is particularly relevant in today's context. Indeed, as new COVID-19 related research is published every day, results are being actively debated, and moreover, numerous controversies arise (about the origin of the disease, its diagnosis, its treatment...) [2]. What researchers need is tools to help them get convinced that some hypotheses, treatments or explanations are indeed relevant, effective, etc. Exploiting argumentative structures while reasoning on named entities can help address these user's needs and so reduce the number of controversies.

The rest of this paper is organized as follows. In Sect. 2, we explain the extraction pipeline set up to process the CORD-19 corpus and generate the RDF dataset. Then, Sect. 3 details the characteristics of the dataset and services made available to exploit it. Sects. 4 and 5 illustrate the current exploitation and visualization tools, and discuss future applications and potential impact of the dataset. Section 6 draw a review of and comparison with related works.

---

[2] http://ns.inria.fr/acta/.
[3] https://github.com/kermitt2/entity-fishing.
[4] https://github.com/frmichel/morph-xr2rml/.
[5] https://project.inria.fr/corese/.

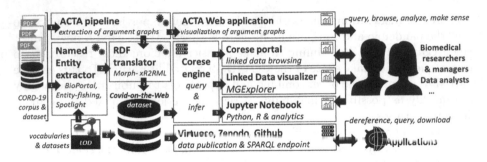

Fig. 1. Covid-on-the-Web overview: pipeline, resources, services and applications

2   From CORD-19 to the Covid-on-the-Web Dataset

The COVID-19 Open Research Dataset [20] (CORD-19) is a corpus gathering scholarly articles (ranging from published scientific publications to pre-prints) related to the SARS-Cov-2 and previous works on the coronavirus family. CORD-19's authors processed each of the 50,000+ full text articles, converted them to JSON documents, and cleaned up citations and bibliography links.

This section describes (Fig. 1) how we harnessed this dataset in order to *(1)* draw meaningful links between the articles of the CORD-19 corpus and the Web of Data by means of NEs, and *(2)* extract a graph of argumentative components discovered in the articles, while respecting the Semantic Web standards. The result of this work is referred to as the *Covid-on-the-Web dataset.*

2.1   Building the CORD-19 Named Entities Knowledge Graph

The *CORD-19 Named Entities Knowledge Graph* (CORD19-NEKG), part of the Covid-on-the-Web dataset, describes NEs identified and disambiguated in the articles of the CORD-19 corpus using three tools:

- DBpedia Spotlight [6] can annotate text in eight different languages with DBpedia entities. Disambiguation is carried out by entity linking using a generative model with maximum likelihood.
- Entity-fishing[6] identifies and disambiguates NEs against Wikipedia and Wikidata at document level. It relies on FastText word embeddings to generate candidates and ranks them with gradient tree boosting and features derived from relations and context.
- NCBO BioPortal Annotator [12] annotates biomedical texts against vocabularies loaded in BioPortal. Patterns are identified using the Mgrep method. Annotator+ [19] extends its capabilities with the integration of a lemmatizer and the Context/NegEx algorithms.

---

[6] https://github.com/kermitt2/entity-fishing.

```
@prefix covidpr: <http://ns.inria.fr/covid19/property/>.
@prefix dct:     <http://purl.org/dc/terms/>.
@prefix oa:      <http://www.w3.org/ns/oa#>.
@prefix schema:  <http://schema.org/>.

[] a                     oa:Annotation;
   schema:about          <http://ns.inria.fr/covid19/f74923b3ce82c...>;
   dct:subject           "Engineering", "Biology";
   covidpr:confidence    "1"^^xsd:decimal;
   oa:hasBody            <http://wikidata.org/entity/Q176996>;
   oa:hasTarget [
      oa:hasSource       <http://ns.inria.fr/covid19/f74923b3ce82c...#abstract>;
      oa:hasSelector     [ a oa:TextPositionSelector, oa:TextQuoteSelector;
         oa:exact "PCR"; oa:start "235"; oa:end "238" ]];
```

Listing 1.1. Representation of the "polymerase chain reaction" (PCR) named entity as an annotation of an article's abstract from offset 235 to 238.

To ensure reusability, CORD19-NEKG leverages well-known, relevant terminological resources to represent articles and NEs in RDF. Below, we outline the main concepts of this RDF modeling. More details and examples are available on the project's Github repository.[7]

Article metadata (e.g., title, authors, DOI) and content are described using DCMI[8], Bibliographic Ontology (FaBiO)[9], Bibliographic Ontology[10], FOAF[11] and Schema.org[12]. NEs are modelled as annotations represented using the Web Annotation Vocabulary[13]. An example of annotation is given in Listing 1.1. The annotation body is the URI of the resource (e.g., from Wikidata) linked to the NE. The piece of text recognized as the NE itself is the annotation target. It points to the article part wherein the NE was recognized (title, abstract or body), and locates it with start and end offsets. Provenance information is also provided for each annotation (not shown in Listing 1.1) using PROV-O[14], that denotes the source being processed, the tool used to extract the NE, the confidence of extracting and linking the NE, and the annotation author.

2.2   Mining CORD-19 to Build an Argumentative Knowledge Graph

The **A**rgumentative **C**linical **T**rial **A**nalysis (ACTA) [13] is a tool designed to analyse clinical trials for argumentative components and PICO[15] elements. Originally developed as an interactive visualization tool to ease the work of clinicians

---

[7] https://github.com/Wimmics/covidontheweb/dataset.

[8] https://www.dublincore.org/specifications/dublin-core/dcmi-terms/.

[9] https://sparontologies.github.io/fabio/current/fabio.html.

[10] http://bibliontology.com/specification.html.

[11] http://xmlns.com/foaf/spec/.

[12] https://schema.org/.

[13] https://www.w3.org/TR/annotation-vocab/.

[14] https://www.w3.org/TR/prov-o/.

[15] PICO is a framework to answer health-care questions in evidence-based practice that comprises patients/population (P), intervention (I), control/comparison (C) and outcome (O).

in analyzing clinical trials, we re-purposed it to annotate the CORD-19 corpus. It goes far beyond basic keyword-based search by retrieving the main claim(s) stated in the trial, as well as the evidence linked to this claim, and the PICO elements. In the context of clinical trials, a *claim* is a concluding statement made by the author about the outcome of the study. It generally describes the relation of a new treatment (intervention arm) with respect to existing treatments (control arm). Accordingly, an observation or measurement is an *evidence* which supports or attacks another argument component. Observations comprise side effects and the measured outcome. Two types of relations can hold between argumentative components. The *attack* relation holds when one component is contradicting the proposition of the target component, or stating that the observed effects are not statistically significant. The *support* relation holds for all statements or observations justifying the proposition of the target component.

Each abstract of the CORD-19 corpus was analyzed by ACTA and translated into RDF to yield the *CORD-19 Argumentative Knowledge Graph*. The pipeline consists of four steps: *(i)* the detection of argumentative components, i.e. claims and evidence, *(ii)* the prediction of relations holding between these components, *(iii)* the extraction of PICO elements, and *(iv)* the production of the RDF representation of the arguments and PICO elements.

Component Detection. This is a sequence tagging task where, for each word, the model predicts if the word is part of a component or not. We combine the BERT architecture[16] [7] with an LSTM and a Conditional Random Field to do token level classification. The weights in BERT are initialized with specialised weights from SciBERT [1] and provides an improved representation of the language used in scientific documents such as in CORD-19. The pre-trained model is fine-tuned on a dataset annotated with argumentative components resulting in .90 f_1-score [14]. As a final step, the components are extracted from the label sequences.

Relation Classification. Determining which relations hold between the components is treated as a three-class sequence classification problem, where the sequence consists of a pair of components, and the task is to learn the relation between them, i.e. *support, attack* or *no relation*. The SciBERT transformer is used to create the numerical representation of the input text, and combined with a linear layer to classify the relation. The model is fine-tuned on a dataset for argumentative relations in the medical domain resulting in .68 f_1-score [14].

PICO Element Detection. We employ the same architecture as for the component detection. The model is trained on the EBM-NLP corpus [17] to jointly predict the participant, intervention[17] and outcome candidates for a given input. Here, the f_1-score on the test set is .734 [13]. Each argumentative component is annotated with the PICO elements it contains. To facilitate structured queries,

---

[16] BERT is a self-attentive transformer models that uses language model (LM) pre-training to learn a task-independent understanding from vast amounts of text in an unsupervised fashion.

[17] The intervention and comparison label are treated as one joint class.

```
@prefix prov:   <http://www.w3.org/ns/prov#>.
@prefix schema: <http://schema.org/>.
@prefix aif:    <http://www.arg.dundee.ac.uk/aif#>.
@prefix amo:    <http://purl.org/spar/amo/>.
@prefix sioca:  <http://rdfs.org/sioc/argument#>.

<http://ns.inria.fr/covid19/arg/4f8d24c531d2c33496...>
    a              amo:Argument;
    schema:about   <http://ns.inria.fr/covid19/4f8d24c531d2c33496...>;
    amo:hasEvidence <http://ns.inria.fr/covid19/arg/4f8d24c531d2c33496.../0>;
    amo:hasClaim   <http://ns.inria.fr/covid19/arg/4f8d24c531d2c33496.../6>.

<http://ns.inria.fr/covid19/arg/4f8d24c531d2c33496.../0>
    a amo:Evidence, sioca:Justification, aif:I-node;
    prov:wasQuotedFrom <http://ns.inria.fr/covid19/4f8d24c531d2c33496...>;
    aif:formDescription "17 patients discharged in recovered condition...";
    sioca:supports <http://ns.inria.fr/covid19/arg/4f8d24c531d2c33496.../6>;
    amo:proves     <http://ns.inria.fr/covid19/arg/4f8d24c531d2c33496.../6>.
```

Listing 1.2. Example representation of argumentative components and their relation.

PICO elements are linked to Unified Medical Language System (UMLS) concepts with ScispaCy [16].

Argumentative Knowledge Graph. The *CORD-19 Argumentative Knowledge Graph* (CORD19-AKG) draws on the Argument Model Ontology (AMO)[18], the SIOC Argumentation Module (SIOCA)[19] and the Argument Interchange Format[20]. Each argument identified by ACTA is modelled as an `amo:Argument` to which argumentative components (claims and evidence) are connected. The claims and evidences are themselves connected by support or attack relations (`sioca:supports`/`amo:proves` and `sioca:challenges` properties respectively). Listing 1.2 sketches an example. Furthermore, the PICO elements are described as annotations of the argumentative components wherein they were identified, in a way very similar to the NEs (as exemplified in Listing 1.1). Annotation bodies are the UMLS concept identifiers (CUI) and semantic type identifiers (TUI).

2.3   Automated Dataset Generation Pipeline

From a technical perspective, the CORD-19 corpus essentially consists of one JSON document per scientific article. Consequently, yielding the Covid-on-the-Web RDF dataset involves two main steps: process each document of the corpus to extract the NEs and arguments, and translate the output of both treatments into a unified, consistent RDF dataset. The whole pipeline is sketched in Fig. 1.

Named Entities Extraction. The extraction of NEs with DBpedia Spotlight, Entity-fishing and BioPortal Annotator produced approximately 150,000 JSON documents ranging from 100 KB to 50 MB each. These documents were loaded into a MongoDB database, and pre-processed to filter out unneeded or

---

[18] http://purl.org/spar/amo/.
[19] http://rdfs.org/sioc/argument#.
[20] http://www.arg.dundee.ac.uk/aif#.

invalid data (e.g., invalid characters) as well as to remove NEs that are less than three characters long. Then, each document was translated into the RDF model described in Sect. 2.1 using Morph-xR2RML,[21] an implementation of the xR2RML mapping language [15] for MongoDB databases. The three NE extractors were deployed on a Precision Tower 5810 equipped with a 3.7 GHz CPU and 64 GB RAM. We used Spotlight with a pre-trained model[22] and Annotator's online API[23] with the Annotator+ features to benefit from the whole set of ontologies in BioPortal. To keep the files generated by Annotator+ of a manageable size, we disabled the options negation, experiencer, temporality, display_links and display_context. We enabled the longest_only option, as well as the lemmatization option to improve detection capabilities. Processing the CORD-19 corpus with the NEs extractors took approximately three days. MongoDB and Morph-xR2RML were deployed on a separate machine equipped with 8 CPU cores and 48GB RAM. The full processing, i.e., spanning upload in MongoDB of the documents produced by the NE extractors, pre-processing and RDF files generation, took approximately three days.

Argumentative Graph Extraction. Only the abstracts longer than ten sub-word tokens[24] were processed by ACTA to ensure meaningful results. In total, almost 30,000 documents matched this criteria. ACTA was deployed on 2.8 GHz dual-Xeon node with 96 GB RAM, and processing the articles took 14 h. Like in the NEs extraction, the output JSON documents were loaded into MongoDB and translated to the RDF model described in Sect. 2.2 using Morph-xR2RML. The translation to RDF was carried out on the same machine as above, and took approximately 10 min.

3   Publishing and Querying Covid-on-the-Web Dataset

The Covid-on-the-Web dataset consists of two main RDF graphs, namely the *CORD-19 Named Entities Knowledge Graph* and the *CORD-19 Argumentative Knowledge Graph*. A third, transversal graph describes the metadata and content of the CORD-19 articles. Table 1 synthesizes the amount of data at stake in terms of JSON documents and RDF triples produced. Table 2 reports some statistics against the different vocabularies used.

Dataset Description. In line with common data publication best practices [8], we paid particular attention to the thorough description of the Covid-on-the-Web dataset itself. This notably comprises (1) licensing, authorship and provenance information described with DCAT[25], and (2) vocabularies, interlinking

---

[21] https://github.com/frmichel/morph-xr2rml/.

[22] https://sourceforge.net/projects/dbpedia-spotlight/files/2016-10/en/.

[23] http://data.bioontology.org/documentation.

[24] Inputs were tokenized with the BERT tokenizer, where one sub-word token has a length of one to three characters.

[25] https://www.w3.org/TR/vocab-dcat/.

Table 1. Statistics of the Covid-on-the-Web dataset.

Type of data	JSON data	Resources produced	RDF triples
Articles metadata and textual content	7.4 GB	n.a.	1.27 M
CORD-19 Named Entities Knowledge Graph			
NEs found by DBpedia Spotlight (titles, abstracts)	35 GB	1.79 M	28.6 M
NEs found by Entity-fishing (titles, abstracts, bodies)	23 GB	30.8 M	588 M
NEs found by BioPortal Annotator (titles, abstracts)	17 GB	21.8 M	52.8 M
CORD-19 Argumentative Knowledge Graph			
Claims/evidence components (abstracts)	138 MB	53 K	545 K
PICO elements		229 K	2.56 M
Total for Covid-on-the-Web (including articles metadata and content)			
	82 GB	54 M named entities 53 K claims/evidence 229 K PICO elements	674 M

and access information described with VOID[26]. The interested reader may look up the dataset URI[27] to visualize this information.

Dataset Accessibility. The dataset is made available by means of a DOI-identified RDF dump downloadable from Zenodo, and a public SPARQL endpoint. All URIs can be dereferenced with content negotiation. A Github repository provides a comprehensive documentation (including licensing, modeling, named graphs and third-party vocabularies loaded in the SPARQL endpoint). This information is summarized in Table 3.

Reproducibility. In compliance with the open science principles, all the scripts, configuration and mapping files involved in the pipeline are provided in the project's Github repository under the terms of the Apache License 2.0, so that anyone may rerun the whole processing pipeline (from articles mining to loading RDF files into Virtuoso OS).

Dataset Licensing. Being derived from the CORD-19 dataset, different licences apply to the different subsets of the Covid-on-the-Web dataset. The subset corresponding to the CORD-19 dataset translated into RDF (including articles metadata and textual content) is published under the terms of the CORD-19 license.[28] In particular, this license respects the sources that are copyrighted.

---

[26] https://www.w3.org/TR/void/.

[27] Covid-on-the-Web dataset URI: http://ns.inria.fr/covid19/covidontheweb-1-1.

[28] CORD-19 license https://www.kaggle.com/allen-institute-for-ai/CORD-19-research -challenge/.

Table 2. Selected statistics on properties/classes/resources.

Property URI	nb of instances	Comments
http://purl.org/dc/terms/subject	62,922,079	Dublin Core subject property
http://www.w3.org/ns/oa#hasBody	54,760,308	Annotation Ontology body property
http://schema.org/about	34,971,108	Schema.org about property
http://www.w3.org/ns/prov#wasGeneratedBy	34,971,108	PROV-O "generated by" property
http://purl.org/dc/terms/creator	34,741,696	Dublin Core terms creator property
http://purl.org/spar/cito/isCitedBy	207,212	CITO citation links
http://purl.org/vocab/frbr/core#partOf	114,021	FRBR part of relations
http://xmlns.com/foaf/0.1/surname	65,925	FOAF surnames

Class URI	nb of instances	Comments
http://www.w3.org/ns/oa#Annotation	34,950,985	Annotations from the Annotation Ontology
http://www.w3.org/ns/prov#Entity	34,721,578	Entities of PROV-O
http://purl.org/spar/amo/Claim	28,140	Claims from AMO
http://rdfs.org/sioc/argument#Justification	25,731	Justifications from SIOC Argument

Resource URI	nb of uses	Comments
http://www.wikidata.org/entity/Q103177	209,183	severe acute respiratory syndrome (Wikidata)
http://purl.obolibrary.org/obo/NCBITaxon_10239	11,488	Virus in NCBI organismal classification
http://dbpedia.org/resource/Severe_acute_respiratory_syndrome	5,280	severe acute respiratory syndrome (DBpedia)
http://www.ebi.ac.uk/efo/EFO_0005741	8,753	Infectious disease in Experimental Factor Ontology

The subset produced by mining the articles, either the NEs (CORD19-NEKG) or argumentative components (CORD19-AKG) is published under the terms of the Open Data Commons Attribution License 1.0 (ODC-By).[29]

Sustainability Plan. In today's context, where new research is published weekly about the COVID-19 topic, the value of Covid-on-the-Web, as well as other related datasets, lies in the ability to keep up with the latest advances and ingest new data as it is being published. Towards this goal, we've taken care of producing a documented, repeatable pipeline, and we have already performed such an update thus validating the procedure. In the middle-term, we intend to improve the update frequency while considering *(1)* the improvements delivered by CORD-19 updates, and *(2)* the changing needs to be addressed based on the expression of new application scenarios (see Sect. 5). Furthermore,

---

[29] ODC-By license: http://opendatacommons.org/licenses/by/1.0/.

Table 3. Dataset availability.

Dataset DOI	10.5281/zenodo.3833753
Downloadable RDF dump	https://doi.org/10.5281/zenodo.3833753
Public SPARQL endpoint	https://covidontheweb.inria.fr/sparql
Documentation	https://github.com/Wimmics/CovidOnTheWeb
URIs namespace	http://ns.inria.fr/covid19/
Dataset URI	http://ns.inria.fr/covid19/covidontheweb-1-1
Citation	Wimmics Research Team. (2020). Covid-on-the-Web dataset (Version 1.1). Zenodo. http://doi.org/10.5281/zenodo.3833753

we have deployed a server to host the SPARQL endpoint that benefits from a high-availability infrastructure and 24/7 support.

4 Visualization and Current Usage of the Dataset

Beyond the production of the Covid-on-the-Web dataset, our project has set out to explore ways of visualizing and interacting with the data. We have developed a tool named *Covid Linked Data Visualizer*[30] comprising a query web interface hosted by a node.js server, a transformation engine based on the Corese Semantic Web factory [5], and the MGExplorer graphic library [4]. The web interface enables users to load predefined SPARQL queries or edit their own queries, and execute them against our public SPARQL endpoint. The queries are parameterized by HTML forms by means of which the user can specify search criterion, e.g., the publication date. The transformation engine converts the JSON-based SPARQL results into the JSON format expected by the graphic library. Then, exploration of the result graph is supported by MGExplorer that encompasses a set of specialized visualization techniques, each of them allowing to focus on a particular type of relationship. Figure 2 illustrates some of these techniques: node-edge diagram (left) shows an overview of all the nodes and their relationships; ClusterVis (top right) is a cluster-based visualization allowing the comparison of node attributes while keeping the representation of the relationships among them; IRIS (bottom right) is an egocentric view for displaying all attributes and relations of a particular node. The proposed use of information visualization techniques is original in that it provides users with interaction modes that can help them explore, classify and analyse the importance of publications. This is a key point for making the tools usable and accessible, and get adoption.

During a meeting with some health and medical research organisations (i.e., Inserm and INCa), an expert provided us with an example query that researchers would be interested in solving against a dataset like the one we generated: "find the articles that mention both a type of cancer and a virus of the corona family". Taking that query as a first competency question, we used Covid Linked Data

---

[30] Covid Linked Data Visualizer can be tested at: http://covid19.i3s.unice.fr:8080.

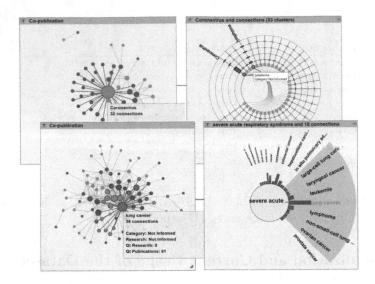

Fig. 2. Covid Linked Data Visualizer: visualization of the subset of articles that mention both a type of cancer (blue dots) and a virus of the corona family (orange dots). (Color figure online)

Visualizer whose results are visualized with the MGExplorer library (Fig. 2). We also created several Python and R Jupyter notebooks[31] to demonstrate the transformation of the result into structures such as Dataframes[32] for further analysis (Fig. 3).

Let us finally mention that, beyond our own uses, the Covid-on-the-Web dataset is now served by the LOD Cloud cache hosted by OpenLink Software.[33]

5   Potential Impact and Reusability

To the best of our knowledge, the Covid-on-the-Web dataset is the first one integrating NEs, arguments and PICO elements into a single, coherent whole. We are confident that it will serve as a foundation for Semantic Web applications as well as for benchmarking algorithms and will be used in challenges. The resources and services that we offer on the COVID-19 literature are of interest for health organisations and institutions to extract and intelligently analyse information on a disease which is still relatively unknown and for which research is constantly evolving. To a certain extent, it is possible to cross-reference information to have a better understanding of this matter and, in particular, to initiate research into unexplored paths. We also hope that the openness of the data and code will allow contributors to advance the current state of knowledge on this disease

---

[31] https://github.com/Wimmics/covidontheweb/tree/master/notebooks.

[32] Dataframes are tabular data structures widely used in Python and R for the data analysis.

[33] https://twitter.com/kidehen/status/1250530568955138048.

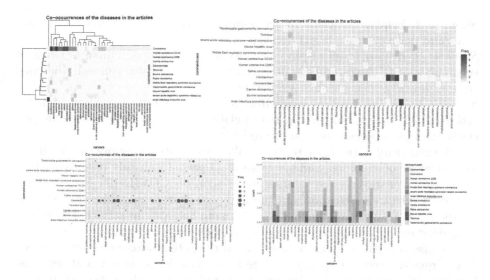

Fig. 3. Visualizations of Jupyter Notebook query results: four different representations of the number of articles that co-mention cancer types and viruses of corona family.

which is impacting the worldwide society. In addition to being interoperable with central knowledge graphs used within the Semantic Web community, the visualizations we offer through MGExplorer and Notebooks show the potential of these technologies in other fields, e.g., the biomedical and medical ones.

Interest of Communities in Using the Dataset and Services. Several biomedical institutions have shown interest in using our resources, eithet direct project partners (French Institute of Medical Research - Inserm, French National Cancer Institute - INCa) or indirect (Antibes Hospital, Nice Hospital). For now, these institutions act as potential users of the resources, and as co-designers. Furthermore, given the importance of the issues at stake and the strong support that they can provide in dealing with them, we believe that other similar institutions could be interested in using the resources.

Documentation/Tutorials. For design rationale purposes, we keep records of the methodological documents we use during the design of the resources (e.g., query elicitation documents), the technical documentation of the algorithms and models[34], the best practices we follow (FAIR, Cool URIs, five-star linked data, etc.) and the end users help (e.g., demonstration notebooks).

Application Scenarios, User Models, and Typical Queries. Our resources are based on generic tools that we are adapting to the COVID-19 issue. Precisely, having a user-oriented approach, we are designing them according to three main motivating scenarios identified through a need analysis of the biomedical institutions with whom we collaborate:

---

[34] https://github.com/Wimmics/covidontheweb/blob/master/doc/01-data-modeling.md.

Scenario 1: Helping clinicians to get argumentative graphs to analyze clinical trials and make evidence-based decisions.

Scenario 2: Helping hospital physicians to collect ranges of human organism's substances (e.g., cholesterol) from scientific articles, to determine if the substances' levels of their patients are normal or not.

Scenario 3: Helping missions heads from a Cancer Institute to collect scientific articles about cancer and coronavirus to elaborate research programs to deeper study the link between cancer and coronavirus.

The genericity of the basic tools will allow us to later on apply the resources to a wider set of scenarios, and our biomedical partners already urge us to start thinking of scenarios related to other issues than the COVID-19.

Besides the scenarios above, we are also eliciting representative user models (in the form of personas), the aim of which is to help us – as service designers – to understand our users' needs, experiences, behaviors and goals.

We also elicited meaningful queries from the potential users we interviewed. These queries serve to specify and test our knowledge graph and services. For genericity purposes, we elaborated a typology from the collected queries, using dimensions such as: Prospective vs. Retrospective queries or Descriptive (requests for description) vs. Explanatory (requests for explanation) vs. Argumentative (requests for argumentation) queries. Here are examples of such queries:

- *Prospective descriptive queries:* What types of cancers are likely to occur in COVID-19 victims in the next years? In what types of patients? Etc.

- *Descriptive retrospective queries:* What types of cancers appeared in [SARSCoV1 | MERS-CoV] victims in the [2|3|n] years that followed? What was the rate of occurrence? In what types of patients? Etc. What are the different sequelae related to Coronaviruses? Which patients cured of COVID-19 have pulmonary brosis?

- *Retrospective explanatory queries:* Did [SARS-CoV1 | MERS-CoV] cause cancer? Was [cell transformation | cancer development] caused directly by coronavirus infection? Or was it caused indirectly through [inflammation | metabolic changes] caused by this infection? Which coronavirus-related sequelae are responsible for the greatest potential for cell transformation? - *Argumentative retrospective queries:* What is the evidence that [SARSCoV1 | MERS-CoV] caused cancer? What experiments have shown that the pulmonary brosis seen in patients cured of COVID-19 was caused by COVID-19?

These queries are a brief illustration of an actual (yet non-exhaustive) list of questions raised by users. It is worthy of notice that whilst some questions might be answered by showing the correlation between components (e.g., types of cancer), others require the representation of trends (e.g., cancer likely to occur in the next years), and analysis of specific attributes (e.g., details about metabolic changes caused by COVID-19). Answering these complex queries requires exploration of the CORD-19 corpus, and for that we offer a variety of analysis and visualization tools. These queries and the generic typology shall be reused in further extensions and other projects.

The Covid Linked Data Visualizer (presented in Sect. 4) supports the visual exploration of the Covid-on-the-Web dataset. Users can inspect the attributes of elements in the graph resulting from a query (by positioning the mouse over elements) or launch a chained visualization using any of the interaction techniques available (ex. IRIS, ClusterVis, etc). These visualization techniques are meant to help users understand the relationships available in the results. For example, users can run a query to visualize a co-authorship network; then use IRIS and ClusterVis to understand who is working together and on which topics. They can also run a query looking for papers mentioning the COVID-19 and diverse types of cancer. Finally, the advanced mode makes it possible to add new SPARQL queries implementing other data exploration chains.

6 Related Works

Since the first release of the CORD-19 corpus, multiple initiatives, ranging from quick-and-dirty data releases to the repurposing of existing large projects, have started analyzing and mining it with different tools and for different purposes. Entity linking is usually the first step to further processing or enriching. Hence, not surprisingly, several initiatives have already applied these techniques to the CORD-19 corpus. **CORD-19-on-FHIR**[35] results of the translation of the CORD-19 corpus in RDF following the HL7-FHIR interchange format, and the annotation of articles with concepts related to conditions, medications and procedures. The authors also used Pubtator [21] to further enrich the corpus with concepts such as gene, disease, chemical, species, mutation and cell line. **KG-COVID-19**[36] seeks the lightweight construction of KGs for COVID-19 drug repurposing efforts. The KG is built by processing the CORD-19 corpus and adding NEs extracted from COVIDScholar.org and mapped to terms from biomedical ontologies. **Covid9-PubAnnotation**[37] is a repository of text annotations concerning CORD-19 as well as LitCovid and others. Annotations are aggregated from multiple sources and aligned to the canonical text that is taken from PubMed and PMC. The **Machine Reading for COVID-19 and Alzheimer's**[38] project aims at producing a KG representing causal inference extracted from semantic relationships between entities such as drugs, biomarkers or comorbidities. The relationships were extracted from the Semantic MEDLINE database enriched with CORD-19. **CKG-COVID-19**[39] seeks the discovery of drug repurposing hypothesis through link prediction. It processed the CORD-19 corpus with state of the art machine reading systems to build a KG where entities such as genes, proteins, drugs, diseases, etc. are linked to their Wikidata counterparts.

---

[35] https://github.com/fhircat/CORD-19-on-FHIR.
[36] https://github.com/Knowledge-Graph-Hub/kg-covid-19/.
[37] https://covid19.pubannotation.org/.
[38] https://github.com/kingfish777/COVID19.
[39] https://github.com/usc-isi-i2/CKG-COVID-19.

When comparing Covid-on-the-Web with these other initiatives, several main differences can be pointed out. First, they restrict processing to the title and abstract of the articles, whereas we process the full text of the articles with Entity-fishing, thus providing a high number of NEs linked to Wikidata concepts. Second, these initiatives mostly focus on biomedical ontologies. As a result, the NEs identified typically pertain to genes, proteins, drugs, diseases, phenotypes and publications. In our approach, we have not only considered biomedical ontologies from BioPortal, but we have also extended this scope with two general knowledge bases that are major hubs in the Web of Data: DBpedia and Wikidata. Finally, to the best our knowledge, our approach is the only one to integrate argumentation structures and named entities in a coherent dataset.

Argument(ation) Mining (AM) [3] is the research area aiming at extracting and classifying argumentative structures from text. AM methods have been applied to heterogeneous types of textual documents. However, only few approaches [11,14,22] focused on automatically detecting argumentative structures from textual documents in the medical domain, e.g., clinical trials, guidelines, Electronic Health Records. Recently, transformer-based contextualized word embeddings have been applied to AM tasks [14,18]. To the best of our knowledge, Covid-on-the-Web is the first attempt to apply AM to the COVID-19 literature.

7   Conclusion and Future Works

In this paper, we described the data and software resources provided by the Covid-on-the-Web project. We adapted and combined tools to process, analyze and enrich the CORD-19 corpus, to make it easier for biomedical researchers to access, query and make sense of COVID-19 related literature. We designed and published a linked data knowledge graph describing the named entities mentioned in the CORD-19 articles and the argumentative graphs they include. We also published the pipeline we set up to generate this knowledge graph, in order to (1) continue enriching it and (2) spur and facilitate reuse and adaptation of both the dataset and the pipeline. On top of this knowledge graph, we developed, adapted and deployed several tools providing Linked Data visualizations, exploration methods and notebooks for data scientists. Through active interactions (interviews, observations, user tests) with institutes in healthcare and medical research, we are ensuring that our approach is guided by and aligned with the actual needs of the biomedical community. We have shown that with our dataset, we can perform documentary research and provide visualizations suited to the needs of experts. Great care has been taken to produce datasets and software that meet the open and reproducible science goals and the FAIR principles.

We identified that, since the emergence of the COVID-19, the unusual pace at which new research has been published and knowledge bases have evolved raises critical challenges. For instance, a new release of CORD-19 is published weekly, which challenges the ability to keep up with the latest advances. Also, the extraction and disambiguation of NEs was achieved with pre-trained models produced before the pandemic, typically before the SARS-Cov-2 entity was even created in Wikidata. Similarly, it is likely that existing terminological resources are being/will be released soon with COVID-19 related updates. Therefore, in the middle term, we intend to engage in a sustainability plan aiming to routinely ingest new data and monitor knowledge base evolution so as to reuse updated models. Furthermore, since there is no reference CORD-19 subset that has been manually annotated and could serve as ground truth, it is hardy possible to evaluate the quality of the machine learning models used to extract named entities and argumentative structures. To address this issue, we are currently working on the implementation of data curation techniques, and the automated discovery of frequent patterns and association rules that could be used to detect mistakes in the extraction of named entities, thus allowing to come up with quality enforcing measures.

References

1. Beltagy, I., Lo, K., Cohan, A.: SciBERT: a pretrained language model for scientific text. In: Proceedings of the 2019 Conference on Empirical Methods in Natural Language Processing and the 9th International Joint Conference on Natural Language Processing (EMNLP-IJCNLP), pp. 3615–3620 (2019)
2. Bersanelli, M.: Controversies about COVID-19 and anticancer treatment with immune checkpoint inhibitors. Immunotherapy **12**(5), 269–273 (2020)
3. Cabrio, E., Villata, S.: Five years of argument mining: a data-driven analysis. Proc. IJCAI **2018**, 5427–5433 (2018)
4. Cava, R.A., Freitas, C.M.D.S., Winckler, M.: Clustervis: visualizing nodes attributes in multivariate graphs. In: Seffah, A., Penzenstadler, B., Alves, C., Peng, X. (eds.) Proceedings of the Symposium on Applied Computing, SAC 2017, Marrakech, Morocco, 3–7 April 2017, pp. 174–179. ACM (2017)
5. Corby, O., Dieng-Kuntz, R., Faron-Zucker, C.: Querying the semantic web with Corese search engine. In: Proceedings of the 16th European Conference on Artificial Intelligence (ECAI), Valencia, Spain, vol. 16, p. 705 (2004)
6. Daiber, M. Jakob, C. Hokamp, J., Mendes, P.N.: Improving efficiency and accuracy in multilingual entity extraction. In: Proceedings of the 9th International Conference on Semantic Systems, pp. 121–124 (2013)
7. J. Devlin, M.-W. Chang, K.L., Toutanova, K.: BERT: pre-training of deep bidirectional transformers for language understanding. In: Proceedings of the 2019 Conference of the North American Chapter of the Association for Computational Linguistics: Human Language Technologies, pp. 4171–4186 (2019)
8. Farias Lóscio, B., Burle, C., Calegari, N.: Data on the Web Best Practices. W3C Recommandation (2017)
9. Gazzotti, R., Faron-Zucker, C., Gandon, F., Lacroix-Hugues, V., Darmon, D.: Injecting domain knowledge in electronic medical records to improve hospitalization prediction. In: Hitzler, P., et al. (eds.) ESWC 2019. LNCS, vol. 11503, pp. 116–130. Springer, Cham (2019). https://doi.org/10.1007/978-3-030-21348-0_8
10. Gazzotti, R., Faron-Zucker, C., Gandon, F., Lacroix-Hugues, V., Darmon, D.: Injection of automatically selected DBpedia subjects in electronic medical records to boost hospitalization prediction. In: Hung, C., Cerný, T., Shin, D., Bechini, A. (eds.) The 35th ACM/SIGAPP Symposium on Applied Computing, SAC 2020, online event, 30 March–3 April 2020, pp. 2013–2020. ACM (2020)

11. Green, N.: Argumentation for scientific claims in a biomedical research article. In: Proceedings of ArgNLP 2014 Workshop (2014)
12. Jonquet, C., Shah, N.H., Musen, M.A.: The open biomedical annotator. Summit Transl. Bioinf. **2009**, 56 (2009)
13. Mayer, T., Cabrio, E., Villata, S.: ACTA a tool for argumentative clinical trial analysis. In: Proceedings of the 28th International Joint Conference on Artificial Intelligence (IJCAI), pp. 6551–6553 (2019)
14. Mayer, T., Cabrio, E., Villata, S.: Transformer-based argument mining for healthcare applications. In: Proceedings of the 24th European Conference on Artificial Intelligence (ECAI) (2020)
15. Michel, F., Djimenou, L., Faron-Zucker, C., Montagnat, J.: Translation of relational and non-relational databases into RDF with xR2RML. In: Proceeding of the 11th International Conference on Web Information Systems and Technologies (WebIST), Lisbon, Portugal, pp. 443–454 (2015)
16. Neumann, M., King, D., Beltagy, I., Ammar, W.: ScispaCy: fast and robust models for biomedical natural language processing. In: Proceedings of the 18th BioNLP Workshop and Shared Task, Florence, Italy, pp. 319–327. Association for Computational Linguistics, August 2019
17. Nye, B., Li, J.J., Patel, R., Yang, Y., Marshall, I., Nenkova, A., Wallace, B.: A corpus with multi-level annotations of patients, interventions and outcomes to support language processing for medical literature. In: Proceedings of 56th Annual Meeting of the Association for Computational Linguistics (ACL), pp. 197–207 (2018)
18. Reimers, N., Schiller, B., Beck, T., Daxenberger, J., Stab, C., Gurevych, I.: Classification and clustering of arguments with contextualized word embeddings. Proc. ACL **2019**, 567–578 (2019)
19. Tchechmedjiev, A., Abdaoui, A., Emonet, V., Melzi, S., Jonnagaddala, J., Jonquet, C.: Enhanced functionalities for annotating and indexing clinical text with the NCBO annotator+. Bioinformatics **34**(11), 1962–1965 (2018)
20. Wang, L.L., et al.: Cord-19: the COVID-19 open research dataset. ArXiv, abs/2004.10706 (2020)
21. Wei, C.-H., Kao, H.-Y., Lu, Z.: PubTator: a web-based text mining tool for assisting biocuration. Nucleic Acids Res. **41**(W1), W518–W522 (2013)
22. Zabkar, J., Mozina, M., Videcnik, J., Bratko, I.: Argument based machine learning in a medical domain. Proc. COMMA **2006**, 59–70 (2006)

ODArchive – Creating an Archive for Structured Data from Open Data Portals

Thomas Weber[1]([✉])[iD], Johann Mitöhner[1][iD], Sebastian Neumaier[1][iD],
and Axel Polleres[1,2][iD]

[1] Vienna University of Economics and Business, Vienna, Austria
thomas.weber@wu.ac.at
[2] Complexity Science Hub Vienna, Vienna, Austria

Abstract. We present ODArchive, a large corpus of structured data collected from over 260 Open Data portals worldwide, alongside with curated, integrated metadata. Furthermore we enrich the harvested datasets by heuristic annotations using the type hierarchies in existing Knowledge Graphs. We both (i) present the underlying distributed architecture to scale up regular harvesting and monitoring changes on these portals, and (ii) make the corpus available via different APIs. Moreover, we (iii) analyse the characteristics of tabular data within the corpus. Our APIs can be used to regularly run such analyses or to reproduce experiments from the literature that have worked on static, not publicly available corpora.

Keywords: Open data · Archiving · Profiling · Reference tables

1 Introduction

The Open Data (OD) movement, mainly driven by public administrations in the form of Open Government Data has over the last years created a rich source of structured data published on the Web, in various formats, covering different domains and typically available under liberal licences. Such OD is typically being published in a decentralized fashion, directly by (governmental) publishing organizations, with data portals, often operated on a national level as central entry points. That is, while OD portals provide somewhat standardized meta-data descriptions and (typically rudimentary, i.e. restricted to metadata only) search functionality, the data resources themselves are available for download on separate locations, as files on specific external download URLs or through web-APIs, again accessible trough a separate URL.

In order to provide unified access to this rich data source, we have been harvesting, integrating and monitoring meta-data from over 260 OD portals for several years now in the Portal Watch project [11,14]. Underlining the increasing importance of providing unified access to structured data on the Web, Google

© Springer Nature Switzerland AG 2020
J. Z. Pan et al. (Eds.): ISWC 2020, LNCS 12507, pp. 311–327, 2020.
https://doi.org/10.1007/978-3-030-62466-8_20

recently started a dataset search [3] facility, which likewise indexes and unifies portal metadata adhering to the Schema.org [5] vocabulary in order to make such metadata searchable at Web scale. In fact, in our earlier works on Portal Watch we demonstrated how the harvested metadata from different OD portal software frameworks, for instance CKAN,[1] can be uniformly mapped to standard formats like DCAT [8] and Schema.org, thereby making the metadata from all indexed portals in Portal Watch available on-the-fly to Google's dataset search.

Yet, while OD *metadata* is well investigated in terms of searchability or quality, the underlying referenced *datasets*, i.e. the actual structured data resources themselves, and their characteristics are still not well understood: What kinds of data are published as OD? How do the datasets themselves develop over time? How do the characteristics of datasets vary between portals? How can search facilities and indexes be built that allow searching within the data and not only the metadata? In order to enable answering such questions, our goal in the present paper is to provide a resource in terms of a dynamically updated corpus of datasets from OD portals, with unified access and filtering capabilities, that shall allow both profiling and scientific analyses of these datasets. To this end we have created, on top of the Portal Watch framework, a dataset crawler and archiver which regularly crawls and indexes OD resources, performs basic data cleansing on known formats, and provides unified access to a large corpus of structured data from OD portals through APIs that allow flexible filtering, e.g. through SPARQL queries over the meta-data, for on-the-fly generation of specific sub-corpora for experiments. We deem this project particularly useful as a resource for experiments on real-world structured data: to name an example, while large corpora of tabular data from Web tables have been made available via CommonCrawl [6], the same is not true for tabular data from OD Portals, for which we expect different characteristics. Indeed, most works on structured OD and its semantics focus on metadata, whereas the structure, properties and linkability of the datasets themselves is, apart from isolated investigations and profiling of adhoc created subcorpora (restricted, for instance, to single data portals), still largely unexplored.

We fill this gap by presenting the Open Dataset Archiver (ODArchive), an infrastructure to crawl, index, and serve a large corpus of regularly crawled structured data from (at the moment) 137 active portals.[2] We describe the challenges that needed to be overcome to build such an infrastructure, including for instance automated change frequency detection in datasets, and make the resource available via various APIs. Moreover, we demonstrate and discuss how these APIs can be used to conduct and regularly update/reproduce various experiments from the literature that have worked on static, not publicly available corpora; as an example we present a detailed profiling analysis on the tabular CSV data in the corpus. Specifically, we make the following concrete contributions:

---

[1] https://ckan.org/, accessed 2020-08-17.

[2] Overall, historically we monitor and have monitored over 260 portals, however, several of those have gone offline in the meantime or are so-called "harvesting" portals that merely replicate metadata from other portals, for details cf. [14].

- We present a detailed architecture of a distributed and scalable Dataset Archiver. The archiver is deployed at https://archiver.ai.wu.ac.at, and the software is openly available on Github[3] under the MIT license.
- Using the introduced archiver, we regularly collect and archive – based on an approximation of the change rates – a large corpus of datasets from OD sources, and make the whole corpus, including the archived versions available via different APIs, incl. download access to subsets of the corpus configurable via SPARQL queries.
- We focus on the prevalent format in our corpus – tabular data – by presenting a detailed profiling analysis of the CSV files in this corpus, and discuss their characteristics. We heuristically annotate columns in these CSVs, using the CSVWeb metadata standard [16], with base datatypes and type hierarchies in existing Knowledge Graphs (DBpedia or Wikidata). Further, we present an approach to scale finding reference columns in this corpus, i.e. tables that contain one or more columns whose values likely reference foreign keys from another reference table: as we can show, there are significantly more reference tables than links to existing KGs in OD, suggesting that such reference tables in OD tables themselves could be the basis for a knowledge graph on its own.

The remainder of this paper is structured as follows: In the next Sect. 2 we present the architecture of our crawling and archiving framework; in Sect. 3 we discuss how to access and query the archived datasets; after an overview of overall corpus characteristics of our archive (Sect. 4), we present experiments on dataset profiling and analysis of identifying reference tables specifically on tabular (CSV) data in Sect. 5. We discuss related and complementary works in Sect. 6, and eventually conclude in Sect. 7.

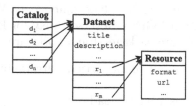

Fig. 1. High-level structure of a data portal.

2 Open Dataset Archive

The datasets that we collect and archive come from the OD Portal Watch project [14]: Portal Watch is a framework for monitoring and quality assessment of (governmental) OD portals, see http://data.wu.ac.at/portalwatch. It monitors

---

[3] https://github.com/websi96/datasetarchiver.

and archives metadata descriptions from (governmental) portals, however, not the actual datasets. The structure of such a data portal (or *catalog*) is similar to digital libraries (cf. Fig. 1): a *dataset* is associated with corresponding *metadata*, i.e. basic descriptive information in structured format, about these resources, for instance, about the authorship, provenance or licensing of the dataset. Each such dataset description typically aggregates a group of data files (referred to as *resources* or distributions) available for download in one or more formats (e.g., CSV, PDF, spreadsheet, etc.) The focus of the present paper is on how to collect, archive, and profile these data files.

2.1   Architecture

Based on our earlier findings on crawling and profiling a static snapshot of CSV data [9] from OD and about capturing and preserving changes on the Web of Data [18], we expect a large amount of dynamically changing source datasets, spread across different domains: as mentioned above the datasets to be crawled themselves, which are cataloged at OD portals, typically reside on different URLs on servers operated by many different data publishers per portal indexed by Portal Watch and accessible by its SPARQL endpoint residing at http://data.wu.ac.at/portalwatch/sparql.

In order to scalably and regularly crawl these datasets from their sources, we therefore designed an infrastructure with three layers to distribute the workload in an extensible manner: (i) a network layer (handled by Kubernetes and Ingress); (ii) a storage layer (using MongoDB), as well as (iii) a Scheduling and Crawling layer handled by specific components written in JavaScript.

That is, the whole system is deployed on an extensible Kubernetes-Cluster with an NGINX Ingress Controller, with currently three server nodes, running our Data storage and Crawling/Scheduling components. We additionally use one external server node with a NGINX Reverse Proxy for load-balancing external traffic. The following software packages/frameworks are used:

1. *Kubernetes:* orchestrates our containerized software on the cluster.
2. *NGINX:*
   - *Ingress Controller:* is a HTTP load balancer for applications, represented by one or more services on different nodes.
   - *Reverse Proxy:* is responsible for load-balancing HTTP requests and database connections from external IPs.
3. *MongoDB:* stores all datasets as chunked binaries along with their associated metadata.
4. *Scheduler:* crawling and scheduling component written in Node.js.

In order to scale the system, it is possible to not only plug in additional server nodes but also whole clusters and spread the workload of the datastore and crawling over their provided nodes. Section 2.1 illustrates the architecture components and their interplay in more detail.

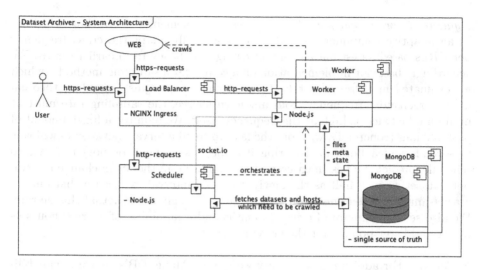

Fig. 2. Archiver architecture

Scheduler. The scheduling component regularly feeds the MongoDB with Resource URLs from the Portal Watch Sparql endpoint. Then it fetches the least-recently downloaded Resource URLs one by each Resource URLs Domain to ensure that our Scheduler does not enforce a denial of service of a domain while distributing the workload to our Crawling Workers via the Load Balancer.

Load Balancer. The Ingress Controller orchestrates the crawling requests by assigning them to worker instances distributed over different nodes/clusters to work in parallel in a Round-Robin fashion. I.e., if there are 3 Workers (w) and 5 requests (r) queued, the requests will be handled by the Workers in following order: $r1 \Rightarrow w1; r2 \Rightarrow w2; r3 \Rightarrow w3; r4 \Rightarrow w1; r5 \Rightarrow w2$

The Crawling Workers then download or crawl the requested resources from the Web and store them in MongoDB.

Database. The MongoDB database instances consist of five collections: *datasets, datasets.chunks, datasets.files, hosts* and *sources*. The *datasets* collection stores essential meta- and crawling-information, e.g., available versions, crawl interval, etc. In the *sources* collection we store information about the source of the datasets, e.g., the data portal (obtained from Portal Watch). The remaining collections organize the storage and chunks of the actual files.

2.2 Workload-Management and Scalability

To ensure our system does not overstrain single hosts nor our own underlying network infrastructure, we make use of the "robots.txt" files and also implemented other strategies to distribute the workload and avoid unnecessary re-crawls.

Dynamic Crawl Frequency. [18] proposes to implement the crawling scheduler as an adaptive component in order to dynamically adapt the crawl frequency per URLs based on estimated content change frequency from earlier crawls. We accordingly base our implementation on a comparison sampling method – which we evaluated in [12] – and take into account the Nyquist sampling theorem [20]: to recreate a frequency from unknown source, the sampling rate must at minimum be twice as high as the frequency itself. We monitor a fixed amount of past versions (concretely, we store the last up to 10 intervals between downloads in seconds and a boolean declaring if a file has changed or not) in order to schedule/predict the best next crawl timestamp. From the mean change interval per dataset we pick half as the newly proposed interval, to ensure that our re-crawl/sampling rate remains on average twice as high as the actual change rate. We also set a maximum of every 6 months and a minimum of every 6 hours as upper and lower bounds for the crawl rate.

Scalability. For additional scalability we rely on MongoDB's sharding capabilities[4] and Kubernetes' container orchestration functionality[5] to horizontally scale across multiple machines: we currently use three nodes totaling 377 GB of memory and 72 CPU cores to distribute all our workload. Each shard contains a subset of the data and each query is routed by a MongoDB instance, providing an interface between client applications and the sharded cluster. A shard key defines which node stores file chunks: we shard by dataset id plus the version number as shard key to keep all chunks of single files on the same node.

The combination of Ingress, Kubernetes and MongoDB connected through micro-services can by extended dynamically, by adding more nodes, when needed.

```
PREFIX arc: <https://archiver.ai.wu.ac.at/ns/csvw#>
PREFIX xsd: <http://www.w3.org/2001/XMLSchema#>
PREFIX csvw: <http://www.w3.org/ns/csvw#>
PREFIX dcat: <http://www.w3.org/ns/dcat#>
PREFIX dc: <http://purl.org/dc/elements/1.1>
INSERT {
  <https://offenedaten.de/dataset/be8c1bf6-50cf-4fab-8ea3-179ca947652a>
      dcat:accessURL <https://www.berlin.de/daten/liste-der-kfz-kennzeichen/kfz-kennz-d.csv> .
  <https://www.berlin.de/daten/liste-der-kfz-kennzeichen/kfz-kennz-d.csv>
      dcat:mediaType "text/csv" ;
      dc:title "kfz-kennz-d.csv" ;
      dc:hasVersion <https://archiver.ai.wu.ac.at/api/v1/get/file/id/5e863ee2b511a4001191dcf8_0> ;
      dc:hasVersion <https://archiver.ai.wu.ac.at/api/v1/get/file/id/5e863ee2b511a4001191dcf8_1> .
  <https://archiver.ai.wu.ac.at/api/v1/get/file/id/5e863ee2b511a4001191dcf8_0>
      dc:identifier "0eec56f69acbda76b375ee982dbd4d7e" ;
      dc:issued "2020-04-06T22:09:56.336Z" ;
      dcat:byteSize 12642 .
  <https://archiver.ai.wu.ac.at/api/v1/get/file/id/5e863ee2b511a4001191dcf8_1>
      dc:identifier "74f78308cb653142663c057744cde84b" ;
      dc:issued "2020-04-12T22:09:56.336Z" ;
      dcat:byteSize 12642 . }
```

Fig. 3. Example `INSERT` statement to add the dataset meta-information.

---

[4] https://docs.mongodb.com/manual/sharding/#shard-keys, accessed 2020-05-22.
[5] https://kubernetes.io/, accessed 2020-05-22.

3   Data Access and Client Interface

SPARQL Endpoint. We make the metadata of the collected and archived datasets queryable over SPARQL by providing the corresponding meta-information in a triple store; the endpoint is available at https://archiver. ai.wu.ac.at/sparql. To describe the datasets we make use of the Data Catalog vocabulary (DCAT) [8] for all crawled datasets (dcat:Dataset) to specify links to the portal (dcat:Catalog) where datasets were published, as well as dcat:accessURLs of *resources* and their respective format (dcat:mediaType). Additionally, for tabular data resources, we provide metadata using the CSV on the Web vocabulary (CSVW) [16]: CSVW provides table-specific properties, such as csvw:tableSchema and csvw:datatypes per column. Figure 3 shows an example of the meta-information stored for an archived dataset.

In this case, as the dataset is a CSV, we also insert CSVWeb metadata as shown in Fig. 4: for these CSVs we heuristically detect the encoding, delimiters, as well as column datatypes of a CSV table, and provide this information using the csvw:dialect property. We further try to detect if the CSV provides a header row, to extract column labels. Details on these heuristic annotations are given in our preliminary work [9]. Additionally, as discussed in more detail in Sect. 5 below, we annotate – where possible – column types as well as basic statistics such as selectivity per table column.

```
INSERT {
    _:csv csvw:url <https://archiver.ai.wu.ac.at/api/v1/get/file/id/5e863ee2b511a4001191dcf8_0> ;
        arc:rows 403 ;
        arc:columns 3 .
    _:csv csvw:dialect [
        csvw:encoding "utf-8" ;
        csvw:delimiter "," ;
        csvw:header true ] .
    _:csv csvw:tableSchema [
        csvw:column <https://archiver.ai.wu.ac.at/api/v1/get/file/id/5e863ee2b511a4001191dcf8_0#1> ;
        csvw:column <https://archiver.ai.wu.ac.at/api/v1/get/file/id/5e863ee2b511a4001191dcf8_0#2> ] .
    <https://archiver.ai.wu.ac.at/api/v1/get/file/id/5e863ee2b511a4001191dcf8_0#1>
        csvw:name "Stadt bzw. Landkreis" ;
        csvw:datatype "string" ;
        rdfs:range <http://dbpedia.org/ontology/Place> .
    <https://archiver.ai.wu.ac.at/api/v1/get/file/id/5e863ee2b511a4001191dcf8_0#2>
        csvw:name "Bundesland" ;
        csvw:datatype "string" ;
        rdfs:range <http://dbpedia.org/ontology/PopulatedPlace> . }
```

Fig. 4. INSERT statement of example CSV meta-information.

API Endpoints. We provide the following API endpoints to interact with the Dataset Archiver. The API is devided into a publicly available API for searching and retrieving our crawled OD resources and a private API used for maintanance, requiring resp. credentials.

Public API.

/stats/basic – Basic statistics on the data stored in the crawler's database.
/get/dataset/{URL} – Returns a JSON object of a dataset description by its referencing URL.
/get/datasets/{domain} – Returns a JSON object of all dataset descriptions provided by the same domain.
/get/dataset/type/{TYPE} – Returns a JSON object of all dataset descriptions which offer resources with the specified filetype e.g. "text/csv" or just "csv".
/get/file/{URL} – Returns a resource (crawled file) by its referencingURL (i.e., for dc:accessURLs the latest downloaded version is retrieved, or, resp. a concrete ds:hasVersion URL can be provided directly).
/get/files/type/{TYPE} – Returns a zip file containing all versions of the specified filetype e.g. "text/csv" or just "csv".
/get/files/sparql?q={QUERY} – Returns a zip file of the resource versions specified by a SPARQL query, that is, all the files corresponding to (version or dataset) URLs that appear in the SPARQL query result cf. detailed explanations below.

Private API.

/post/resource?secret=SECRET – Adds a new resource to the crawler by posting a JSON object containing the URL of the resource, the URL of the portal and the format e.g. 'text/csv' or 'csv'. Only the URL of the resource is mandatory and a secret key credential is needed to post resources.
/post/resources?secret=SECRET – Adds several resources at once in batch, using the same parameters as above.
/crawl?id=ID&domain=DOMAIN&secret=SECRET – Tells the workers which resource has to be crawled. It is used by the master scheduler; a crawl can also be enforced with this endpoint.

Detailed usage examples of the different APIs are documented on our Webpage at https://archiver.ai.wu.ac.at/api-doc.

Data Download via SPARQL. Besides the APIs to directly access files from our crawler and the SPARQL interface to query metadata, we also offer a way of directly downloading data parameterized by SPARQL queries, i.e., for queries that include any URLs from the subject (*datasetURL*) or object (*versionURL*) of the dc:hasVersion property in our triple store, we provide a direct, zipped, download of the data: here *versionURLs* will directly refer to concrete dowloaded file versions, whereas any *datasetURL* will retrieve the resp. latest available version in our corpus.

For instance, the query in Fig. 5 selects all archived resources from a specific *data portal* (data.gv.at),[6] collected after a certain *time stamp*, with a specific

---

[6] To filter datasets by certain data portals we enriched the descriptions by information collected in the Portal Watch (https://data.wu.ac.at/portalwatch/): we use

HTTP media type (in this case CSV files); executing this query at our SPARQL user interface (https://archiver.ai.wu.ac.at/yasgui) gives an additional option to retrieve the specific matching versions directly as a zip file. Alternatively, given this query to the `/get/files/sparql?q={QUERY}` API mentioned above, will retrieve these without the need to use the UI.

```
SELECT ?versionURL WHERE {
    ?datasetURL arc:hasPortal ?Portal ;   # ?datasetURL: the download URL of a specific resource
                                          # ?Portal: a dcat:catalog indexed in Portal Watch
    dc:hasVersion ?versionURL ;           # ?versionURL: a crawled version of the resource
    dcat:mediaType ?mediaType .           # ?mediaType: media type as per HTTP response.
    ?versionURL dc:issued ?dateVersion .  # ?dateVersion: crawl time.

    FILTER (?Portal = <http://data.gv.at> &&
            ?mediaType = "text/csv"       &&
            strdt(?dateVersion, xsd:dateTimeStamp) >= "2020-05-10T00:00Z"^^xsd:dateTimeStamp) }
```

Fig. 5. Example query to get a set of URLs of archived datasets.

4 Overall Corpus Characteristics

Table 1 shows an overview of our overall current ODArchive corpus – as of week 21 in 2020: we regularly crawl a total of ~800k resource URLs of datasets from 137 OD portals; over a time of 8 weeks we collected a total of 4.8 million versions of these datasets. Resource URLs origin from 6k different (pay-level) domains, collected from 137 OD portals, which demonstrates the spread of actual data-providing servers and services indexed by OD portal catalogs. The latest crawled versions of all datasets amount to a total of 1.2 TB uncompressed, and the total of all stored versions sums up to around 5.5 TB. Additionally, Table 1 shows the top-5 most common data formats across the most recent crawl.

Table 1. Total number of URLs of datasets, archived versions, domains/portals, and size of the corpus (left); top-5 most frequent HTTP media types (right).

#Resource URLs	798,091
#Versions	4,833,271
#Domains	6,001
#Portals	137
Latest Versions Corpus Size	1.2 TB
Total Corpus Size	5.5 TB

Media type	Count
text/html	187,196
text/csv	116,922
application/json	102,559
application/zip	93,352
application/xml	76,862

`arc:hasPortal` to add this reference. More sophisticated federated queries could be formulated by including the Portal Watch endpoint [14] which contains additional metadata.

Table 2 shows the main sources of our data corpus: in the left table we provide the ten most frequent domains of the resource URLs in the corpus, whereas the right table shows the top-10 data portals.

Table 2. Top-10 most frequent domains of the resource URLs in the archiver, and the most frequent source portals.

Resource URL Domain	Count	Data Portal	Count
wab.zug.ch	77,436	europeandataportal.eu	282,541
data.opendatasoft.com	63,481	open.canada.ca	118,949
clss.nrcan.gc.ca	59,519	data.opendatasoft.com	63,481
services.cuzk.cz	40,504	offenedaten.de	38,348
abstimmungen.gr.ch	36,604	datamx.io	32,202
www.geoportal.rlp.de	26,275	dados.gov.br	31,961
www150.statcan.gc.ca	20,295	data.gov.ie	20,826
archiv.transparenz.hamburg.de	19,321	hubofdata.ru	19,783
cdn.ruarxive.org	17,242	edx.netl.doe.gov	19,379
www.dati.lombardia.it	15,743	data.gov.gr	18,687

Note that these numbers can easily be computed through our SPARQL endpoint in an always up-to-date manner, and also over time, by restricting to the most recent versions before a certain date, with queries analogous to those shown in Sect. 3. For instance, the following query produces the statistics given in Table 1: https://short.wu.ac.at/odarchiverquery1.

5   CSVs: Column Types from KGs and Reference Tables

In order to demonstrate the potential use of our data collection, we herein discuss reproducible profiling experiments we conducted on a subcorpus of tabular data: the experiments focus on CSV files (116,922, as per Table 1 in the most recent crawl) from our corpus, as the most prominent structured format in OD portals. Also, as mentioned in the introduction, while tabular data on the Web is a popular subject of investigations, the particular characteristics of tabular data from OD portals have thus far not been the main focus of these investigations.

Tables	Header	Columns	Avg Rows	Avg Cols
67974	53279	685276	195.5	10.1

5.1   Labelling Columns with Types from KGs

In the first experiment section, we focus on scalably annotating columns in our CSV table corpus to classes in existing knowledge graphs (KGs), specifically DBpedia [2] and Wikidata [19]. To this end, we distinguish by column datatypes between "textual" and "numeric" columns; we herein specifically focus on scaling named entity recognition (NER) by textual labels in columns to our corpus. As for numeric colums, we note that labeling numeric data in tabular OD corpora with references to KGs has its own challenges and remains a topic of active research, cf. for instance [13] for our own work in this space.

Our basic idea here is to build a NE gazetteer from DBpedia and Wikidata labels, along with references of labels to their associated types (i.e., rdf:type links in DPpedia, or wdt:P31 in Wikidata, resp.). The base assumption here is that, despite potential ambiguities, columns containing labels of predominantly same-typed entities, can be associated with the respective DBpedia/Wikidata type(s). To this end, we extracted label and type information and as well as the transitive class hierarchy (using rdfs:subClassOf, or wdt:P279 links, resp.) from both KGs.[7]

In order to scale, rather than relying on SPARQL, we have constructed our gazetteer by simply compiling the extracted data into huge Python dictionaries (one for the types of a given label, and another for the labels of a given type). This conceptually simple approach is further complicated by two main scalability challenges, which we discuss in the following.

1. Synonyms and Homonyms: A given entity in the NE sources is often associated with a number of ambiguous labels and a given label can be associated with a number of entities, and therefore an even larger number of types; e.g., the entity Abraham Lincoln in the sense of the 16th president of the United States is assigned the types dbpedia.org/ontology/Person, xmlns.com/foaf/0.1/Person, www.w3.org/2002/07/owl#Thing, schema.org/Person, dbpedia.org/ontology/-Agent, and many others within DBpedia.

However, since many more entities share the name of the great president, including bridges, ships, high schools, and universities, the number of types that can be assigned to the *label* 'Abraham Lincoln' is in fact much larger. In addition, the president is also known under a number of other labels, such 'Abe Lincoln', 'Honest Abe', and 'President Lincoln', each of which is also assigned (among others) the types listed above.

2. Multi-linguality: Labels and types are available in various languages; at the moment, we limit ourselves to English and German labels, implementing multi-linguality only in principle and not in any sense exhaustively which would of course still be limited to the language versions available in the NE sources. Restricting to those two languages was also useful to significantly reduce size of the extracted gazetteer from the raw DBPedia and Wikidata HDT dump files containing all language labels. Still, while English labels and types form the

---

[7] The resp. information has been extracted from the most recent DBpedia and Wikidata HDT [4] dumps available at http://www.rdfhdt.org/datasets/.

largest part of the NE sources, we assume many other languages in e.g. nationally operated OD portals, which we do not cover yet. This label-type information was then imported into Python dictionaries for efficient access to all types of labels and vice versa, fitting in memory, provided that the available RAM is sufficient; in this case roughly 30 GB.

CSV Table Pre-Processing. We assume that very large files would not significantly contribute to the results, and therefore only consider files <100 KB in the analysis, resulting in 71,787 CSVs (~60% of all CSV files currently in our corpus). The number of usable tables is further reduced due to import errors (essentially empty, "headers only", CSV files), to 67,974 tables with overall 685,276 columns.

As mentioned above we restrict our comparison to (tables with) columns with textual content only, which further significantly reduces the number of columns to be analysed: overall, the reduced corpus for the experiment considers 61,110 tables and a total of 294,485 textual columns (i.e., an avg. of 4.8 textual columns per table) to be annotated. Looking at the individual values within the remaining data set we find that only around 19% (and only 2% among the unique values) of those values can be associated with at least one DBPedia or Wikidata type:[8]

	Total	With type	Fraction
Values	28,442,981	5,278,327	0.186
Unique	5,299,125	104,985	0.02

Obvious additional measures to be taking into consideration for the type annotations of table columns are the total number of values, the number of distinct values, and the selectivity (the number of distinct values divided by the number of total values): it proved useful to only look at columns with a minimum number of distinct values: For instance, in order to rule out "essentially boolean" attributes,[9] we only considered columns with at least three values. The listing shows the selectivity of columns with at least three values.

Columns	Avg number of values	Avg distinct values	Selectivity
233,416	121.5	46.0	0.28

---

[8] While this needs further investigation, and obviously more sophisticated matching techniques (substrings- or similarity-based), we note that this low percentage seems to hint at the specific textual information in OD tables not necessarily being covered by the more general, encyclopedic knowledge typical in public KGs.

[9] E.g., "Ja" and "Nein" (German for "yes" and "no"), are labels for entities in Wikidata.

Another measure for annotating columns with KG types is the fraction of types covered in the value labels of a given column. For our column-type annotation we consider the following threshold: type coverage for columns with at least one common type with a fraction of 0.8 or greater.

Among the finally remaining **74,467** columns, we collect intersecting types per colums and add those as column annotations (using the CSVWeb vocabulary) to the corpus, cf. the example in Fig. 4. Other column characteristics, such as selectivity, are also added as annotations; via our API one could for instance only consider a specific subcorpus based on these annotations.

The most often identified types are associated with organizations, locations, and various types of media. Note that types from various knowledge graphs are overlapping to varying degrees.

Type	Number
wd/group	10,383
dbpedia.org/ontology/Location	8,601
schema.org/Place	8,601
wd/entity	8,405
wd/intellectual work	7,285
wd/series	7,057
dbpedia.org/ontology/Place	6,799
wd/creative work	6,749
wd/information	5,991
wd/communication medium	5,989

5.2   Finding Reference Tables

Apart from class annotations per column, which serve to link OD datasets with KGs, we also analyzed potential interlinkage between OD tables in our corpus by looking for potential references between tables; whereas e.g. [7] used semantic and machine learning methods to find relations among web tables; here we apply a basic but scalable approach (by again modularly restricting the number of columns to be compared): a reference table contains one or more key columns whose values are referenced i.e. are identical with values in other tables. Our basic approach to identifying reference tables simply compares values in candidate key columns with the complete set of columns from other tables by going through all columns in all tables. We limit this brute force approach as described in the following.

Overall, we compare two approaches for determining a possible reference:

- *Strict* i.e. all values of column A must be present in column B
- *Lax* i.e. at least a fraction of 0.9 of the values in A must be present in B

Both computations are done on the *sets* of column values rather than the original *list* of values.

To further limit the amount of processing we put the following restrictions on reference candidates: (i) the number of distinct values in B must be at least 10, (ii) the selectivity must be 1 in the referenced column B (i.e., we only consider single attribute candidate keys as references).

The "brute force" approach consists in checking every column of every table with each column of every other table satisfying the restrictions (28,524 candidate reference tables), where the Python library ray [10] used for parallelization allowed us to scale this pairwise comparison between candidate tables and every same-typed column of every other table. Applying this reference search by doing a lax or strict check on each column with each reference candidate results in the following number of actually (at least once) referenced tables, where we see that applying a strict check does not decrease the number dramatically:

Reference tables	
lax	15,977
strict	15,052

That is, more than half of the candidate reference tables are actually referenced from other tables, according to these heuristics.

Indeed, some tables are referenced very frequently, for instance reference tables with regionally important area codes, such as US state codes, national ISO country codes, or – as a less obvious example – the following table was cited in 1,811 other tables in the corpus; it has 402 rows showing area codes for German car license plates:

Kennzeichen, Juli 2012	Stadt bzw. Landkreis	Bundesland
A	Augsburg	Bayern
AA	Aalen Ostalbkreis	Baden-Württemberg
AB	Aschaffenburg	Bayern
ABI	Anhalt-Bitterfeld	Sachsen-Anhalt
ABG	Altenburger Land	Thüringen
AC	Aachen	Nordrhein-Westfalen

Overall, while we defer a more detailed analysis to future work, these results hint at a large number of possible additional inter-dataset links, apart from only considering links to existing KGs. We explicitly invite usage of our resource to enable further large scale respective experiments by the community.

---

[10] https://github.com/ray-project/ray, accessed 2020-08-17.

6 Related Work

While OD tables are rarely covered (due to the lack of a readily available corpus as we presented it herein) there are various works on Web tables; we see our resource as a valuable contribution (i) to compare to an alternative evaluation corpus, and (ii) to research the potential of applying existing approaches from these works.

Lehmberg et al. [6] – comparable to our work – presented a large corpus of (typically much smaller) Web tables, consisting of 233 million content tables which they classified as either relational, entity, or matrix tables depending on the orientation and structure of a table, detecting sub-header rows/multi-tables and subject columns in a dataset. In future work, we want to apply this classification to our corpus of tabular resources, in order to highlight and compare the differences of a corpus of Web/HTML and CSV tables. A survey on profiling relational data can be found in [1].

As for related work on entity recognition and semantic interpretation on Web tables [15,17,22] our working hypothesis (partially confirmed by our experiments herein) is that relational data as found on OD repositories are fundamentally different from such Web table corpora, and we will have to leverage additional non-textual/numerical cues in the datasets in order to facilitate linkage to existing KGs. Our archived resources will allow us to test this hypothesis further by applying and reproducing existing works on Web tables as future work. A survey on Web table extraction approaches can be found in [21].

Related to our work on Knowledge Graph types and reference tables, [10] studied the table union problem on a dataset from several OD portals using Locality Sensitive Hashing (LSH) among other approaches, reporting a precision of 0.9005 and a recall of 0.8377 [10, p. 823] on sample queries for unionable tables. Since the subset operation over all columns is essentially a n^2 operation LSH was implemented as an additional alternative approach using minHash LSH Ensemble.[11] Originally designed for queries over large sets of documents the fraction $|Q \cap X|/|Q|$ is the required intersection of query Q with document X which can be specified as threshold when querying the LSH ensemble. This corresponds with the required fraction of elements in column Q present in parent table column X for a referencing relationship as defined in this work.

In our own experiments using this LSH approach we achieved recall and precision values that are somewhat higher than the figures reported in [10] with a speedup of about 5–10 times, but still with a significant number of false positives and negatives, compared to our own brute force set intersection results. A more detailed comparison is on our agenda.

7 Conclusions

ODArchive is set to provide easy access to a large, up-to-date corpus of datasets from OD portals: we archive regularly re-crawled versions of underlying data

---

[11] http://ekzhu.com/datasketch/lshensemble.html, accessed 2020-08-17.

resources for datasets from these portals, based on an adaptive, heuristically estimated crawl rate and have presented a scalable extensible infrastructure to sustainably run such an archive. Apart from overall characteristics of the crawled corpus, in order to demonstrate its use, we presented two experiments in terms of linking tabular OD datasets to existing KGs as well as interlinking them amongst each other by finding reference tables within the corpus. Our initial results clearly suggest that the characteristics of the structured data found on OD portals and readily provided in our corpus are quite different from other available copora, such as Web Tables. In future work we plan to also analyze and attempt to interlink other structured formats in our corpus; additionally, as our framework keeps on running, it shall also enable temporal analyses over the evolution of OD resources. The infrastructure shall allow detailed analyses overall, but also with a narrower scope, restricting to data from particular portals or regions. Last, but not least, we invite the community to use ODArchive and provide feedback (e.g., in terms of additional API feature requests).

References

1. Abedjan, Z., Golab, L., Naumann, F.: Profiling relational data: a survey. VLDB J. **24**(4), 557–581 (2015). https://doi.org/10.1007/s00778-015-0389-y
2. Auer, S., Bizer, C., Kobilarov, G., Lehmann, J., Cyganiak, R., Ives, Z.: DBpedia: a nucleus for a web of open data. In: Aberer, K., et al. (eds.) ASWC/ISWC -2007. LNCS, vol. 4825, pp. 722–735. Springer, Heidelberg (2007). https://doi.org/10.1007/978-3-540-76298-0_52
3. Brickley, D., Burgess, M., Noy, N.F.: Google dataset search: building a search engine for datasets in an open web ecosystem. In: The World Wide Web Conference, WWW 2019, San Francisco, CA, USA, 13–17 May 2019, pp. 1365–1375. ACM (2019). https://doi.org/10.1145/3308558.3313685
4. Fernández, J.D., Martínez-Prieto, M.A., Gutiérrez, C., Polleres, A., Arias, M.: Binary RDF representation for publication and exchange (HDT). In: Web Semantics: Science, Services and Agents on the World Wide Web 2019, pp. 22–41 (2013). http://www.websemanticsjournal.org/index.php/ps/article/view/328
5. Guha, R.V., Brickley, D., Macbeth, S.: Schema.org: evolution of structured data on the web. Commun. ACM **59**(2), 44–51 (2016). https://doi.org/10.1145/2844544
6. Lehmberg, O., Ritze, D., Meusel, R., Bizer, C.: A large public corpus of web tables containing time and context metadata. In: Proceedings of the 25th International Conference Companion on World Wide Web, pp. 75–76 (2016). https://doi.org/10.1145/2872518.2889386
7. Limaye, G., Sarawagi, S., Chakrabarti, S.: Annotating and searching web tables using entities, types and relationships. Proc. VLDB Endow. **3**(1–2), 1338–1347 (2010). https://doi.org/10.14778/1920841.1921005
8. Maali, F., Erickson, J.: Data Catalog Vocabulary (DCAT). W3C Recommendation, January 2014. http://www.w3.org/TR/vocab-dcat/
9. Mitloehner, J., Neumaier, S., Umbrich, J., Polleres, A.: Characteristics of open data CSV files. In: Proceedings - 2016 2nd International Conference on Open and Big Data, OBD 2016 (2016). https://doi.org/10.1109/OBD.2016.18
10. Nargesian, F., Zhu, E., Pu, K.Q., Miller, R.J.: Table union search on open data. Proc. VLDB Endow. **11**(7), 813–825 (2018). https://doi.org/10.14778/3192965.3192973, http://www.vldb.org/pvldb/vol11/p813-nargesian.pdf

11. Neumaier, S.: Semantic enrichment of open data on the Web - or: how to build an open data knowledge graph. Ph.D. thesis, Technische Universität Wien, Vienna, Austria (2019). https://permalink.catalogplus.tuwien.at/AC15550378

12. Neumaier, S., Umbrich, J.: Measures for assessing the data freshness in open data portals. In: 2nd International Conference on Open and Big Data, OBD 2016, Vienna, Austria, 22–24 August 2016, pp. 17–24. IEEE Computer Society (2016). https://doi.org/10.1109/OBD.2016.10

13. Neumaier, S., Umbrich, J., Parreira, J.X., Polleres, A.: Multi-level semantic labelling of numerical values. In: Groth, P., et al. (eds.) ISWC 2016. LNCS, vol. 9981, pp. 428–445. Springer, Cham (2016). https://doi.org/10.1007/978-3-319-46523-4_26

14. Neumaier, S., Umbrich, J., Polleres, A.: Automated quality assessment of metadata across open data portals. J. Data Inf. Qual. **8**(1), 21–229 (2016). https://doi.org/10.1145/2964909

15. Oulabi, Y., Bizer, C.: Extending cross-domain knowledge bases with long tail entities using web table data. In: Advances in Database Technology - 22nd International Conference on Extending Database Technology, EDBT 2019 (2019). https://doi.org/10.5441/002/edbt.2019.34

16. Pollock, R., Tennison, J., Kellogg, G., Herman, I.: Metadata Vocabulary for Tabular Data. W3C Recommendation, December 2015. https://www.w3.org/TR/2015/REC-tabular-metadata-20151217/

17. Sarma, A.D., et al.: Finding related tables. In: Proceedings of the ACM SIGMOD International Conference on Management of Data, SIGMOD 2012, Scottsdale, AZ, USA, 20–24 May 2012, pp. 817–828. ACM (2012). https://doi.org/10.1145/2213836.2213962

18. Umbrich, J., Mrzelj, N., Polleres, A.: Towards capturing and preserving changes on the Web of data. In: CEUR Workshop Proceedings (2015). https://pdfs.semanticscholar.org/971b/178200a0bc14735116ace49a0b164e68a926.pdf

19. Vrandecic, D., Krötzsch, M.: Wikidata: a free collaborative knowledgebase. Commun. ACM **57**(10), 78–85 (2014). https://doi.org/10.1145/2629489

20. Weik, M.H.: Nyquist Theorem, p. 1127. Springer, Boston (2001). https://doi.org/10.1007/1-4020-0613-6_12654

21. Zhang, S., Balog, K.: Web table extraction, retrieval, and augmentation: a survey. ACM Trans. Intell. Syst. Technol. **11**(2), 13:1–13:35 (2020). https://doi.org/10.1145/3372117

22. Zhang, Z.: Effective and efficient semantic table interpretation using tableminer+. Semantic Web **8**(6), 921–957 (2017). https://doi.org/10.3233/SW-160242

Tough Tables: Carefully Evaluating Entity Linking for Tabular Data

Vincenzo Cutrona[1] , Federico Bianchi[2](✉) , Ernesto Jiménez-Ruiz[3,4] ,
and Matteo Palmonari[1]

[1] University of Milano - Dicocca, Milan, Italy
{vincenzo.cutrona,matteo.palmonari}@unimib.it
[2] Bocconi University, Milan, Italy
f.bianchi@unibocconi.it
[3] City, University of London, London, UK
ernesto.jimenez-ruiz@city.ac.uk
[4] University of Oslo, Oslo, Norway

Abstract. Table annotation is a key task to improve querying the
Web and support the Knowledge Graph population from legacy sources
(tables). Last year, the SemTab challenge was introduced to unify different efforts to evaluate table annotation algorithms by providing a common interface and several general-purpose datasets as a ground truth.
The SemTab dataset is useful to have a general understanding of how
these algorithms work, and the organizers of the challenge included some
artificial noise to the data to make the annotation trickier. However, it is
hard to analyze specific aspects in an automatic way. For example, the
ambiguity of names at the entity-level can largely affect the quality of
the annotation. In this paper, we propose a novel dataset to complement
the datasets proposed by SemTab. The dataset consists of a set of high-
quality manually-curated tables with non-obviously linkable cells, i.e.,
where values are ambiguous names, typos, and misspelled entity names
not appearing in the current version of the SemTab dataset. These challenges are particularly relevant for the ingestion of structured legacy
sources into existing knowledge graphs. Evaluations run on this dataset
show that ambiguity is a key problem for entity linking algorithms and
encourage a promising direction for future work in the field.

Keywords: Entity linking · Instance-level matching · Cell entity
annotation · Semantic labeling · Table annotation

1 Introduction

Tables are one of the most used formats to organize data. Every day, both data
practitioners and business people have to handle tables that have been extracted
from databases of sales, pricing, and more. Using these tables to build a new
knowledge graph (KG), populate an existing one [10], or enrich the data in
the table with additional information available in existing KGs [3] requires the

© Springer Nature Switzerland AG 2020
J. Z. Pan et al. (Eds.): ISWC 2020, LNCS 12507, pp. 328–343, 2020.
https://doi.org/10.1007/978-3-030-62466-8_21

source data to be manipulated, interpreted within a graph-based schema (e.g., an ontology), transformed and linked to a reference or to existing KGs. The latter step, in particular, consists in an entity linking task, that is, in connecting cells to reference identifiers (e.g., URIs) that are used to describe, in larger KGs, the entity referred to the cell.

The task of interpreting the table under a graph-based schema and link cells to entities described in a reference KG is referred in the literature to as Table Annotation, Table Interpretation, or Semantic Labeling. It requires the introduction of semantic algorithms, namely semantic table interpreters, that link cells to elements in a KG. Recently, the SemTab 2019 [7] challenge was introduced to unify the community efforts towards the development of performing annotations. The challenge consists of different rounds in which tables of various difficulties have to be annotated. However, it is often difficult to understand what are the shortcoming of each algorithm and how difficult the tables are. For example, are algorithms able to correctly annotate tables that contain homonymic names of people? In Fig. 1, we show a case of ambiguity. In the Semantic Web community, this issue has already been highlighted, and it becomes necessary to have tables that resemble real use-cases [7].

Fig. 1. Ambiguities make table annotation more difficult.

In this paper, we propose a manually curated dataset for table annotation useful for evaluating specific aspects of the annotation task, and, in particular, of the entity linking task in table interpretation. This dataset is complementary to the datasets already introduced in the SemTab challenge. We manually checked the tables with the following question in mind: "Would a human annotator be able to disambiguate this table?". Considering the intrinsic ambiguity of references appearing in tables, we want to ensure that the dataset includes those tables that can be effectively disambiguated, based on the information available in the table by a human annotator. In fact, we report some cases where the human annotators found very hard to match cells of some tables, due to the high ambiguity of their content; when the correct link was hard to be decided based on the table content (the context supporting the disambiguation), we opted for the conservative approach and decided to not annotate the cell (e.g., for the table with the list of bank failures since 2,000, it was not always clear who acquired the bank: whether it was a bank or its holding company).

A good table interpretation algorithm should be able to balance different aspects that need to be considered in the linking process: if the algorithm weights too much the evidence provided by string matching, it will fail to recognize nicknames and different names for things; if it expects clean text, it will fail to identify misspelled entities; if it relies on popularity for disambiguation entities, it will give more weight to popular cities than to the homonymic cities with the same name; if it allows too much fuzziness in the search of the candidate entities, i.e., a pool of entities that are selected as potential links usually based on string matching criteria, it will generate a considerable amount of possible candidates, making the search for the correct one more difficult and prone to errors. We believe that it is helpful to consider these aspects separately in order to evaluate the real power of different entity linking methods in handling the different challenges that data in the tables present. Creating a dataset that features all the aforementioned aspects requires to collect non-artificial tables; in fact, building a dataset via generators (e.g., tables created by querying a SPARQL endpoint) has the advantage of creating a multitude of different tables quickly, but for example it is not possible to create tables with new content (i.e., with facts missing in the reference KG).

Compared to previous benchmarks, this new dataset has the following distinguishing features, which make it a very valuable resource for the fundamental task of table interpretation:

1. **Real Tables** - useful for testing how the table interpretation algorithms deal with the knowledge gap due to novel facts. It can be often the case that some cells in a table refer to entities that are described in the reference KG, for which the algorithm is expected to link the correct entity, and some cells refer to entities not described in the reference KG, for which the algorithm should decide not to link any of the existing entities.
2. **Tables with Ambiguous Mentions** - useful to test the algorithms' capability to handle the ambiguity and link also to non-popular entities (tail entities).
3. **Tables with Misspelled Mentions** - useful for testing the weight of lexical features used by the algorithms. We used the misspelled words as a generator to add controlled noise to other tables.
4. **Tables from Various Sources** - useful to understand which are the most difficult tables to deal with for an algorithm.
5. **Manually Verified Tables** - useful to prevent false positives while evaluating the algorithms; the dataset is of high quality and all the annotations have been manually verified.

2   Background

In this section we define more precisely the terminology we use along the paper. We refer to table annotation as the task of linking cells of a table to elements of a KG. We distinguish from instance-level matching and schema-level matching:

the first aims at linking mentions in the table to entities of the KG (e.g., "10. Alex Del Piero" and "Juventus" can be mapped to dbr:Alessandro_Del_Piero and dbr:Juventus_F.C. in DBpedia), while the latter is dedicated to the columns and their headers, i.e., the table schema, linking the heading cells to elements of the ontology provided by the KG (e.g., the columns "Name" and "Team" could be mapped to types dbo:SoccerPlayer and dbo:Team, and they can be linked with the property "dbo:team"). While the instance-level matching can always rely on the cell content, which represents a mention of the entity to be linked, sometimes the schema-level matching must deal with more challenging tables where the first row is empty (no headers cells), or contains meaningless cells (e.g., codes from the legacy sources). Recently, the SemTab challenge introduced a new terminology, splitting the table annotation task into three sub tasks:

- **CTA** (Column-Type Annotation), that is the schema-level matching focused only on linking columns to ontology types;
- **CEA** (Cell-Entity Annotation), that is the instance-level matching; this is what is also referred to as entity linking, as we do in this paper.
- **CPA** (Columns-Property Annotation), that is the schema-level matching focused only on linking columns of the table through ontology properties.

Even if the challenge evaluates these three tasks separately, they are usually solved together, possibly, iteratively. A fully-automated table annotation approach can start by solving the CEA task, e.g., by finding some entities linked with a certain confidence, then it can use these entities to find the right type for the column, solving CTA, and finally it may go back to refine CEA, e.g., by using the inferred type to filter irrelevant entities and support the disambiguation. Finally, the CPA task is typically solved by combining the information collected during the other two tasks and selecting the properties that fit the pairs of types/entities that have been found. Due to the importance of the CEA task, and to the fact that this task is useful for data enrichment even if a full table to graph transformation is not required, we decided to focus our dataset on the evalation of the entity linking (CEA). By using these annotations then we have also derived type-level annotations of the table headers, thus providing also ground truth for the CTA task (further details in the Appendix A.2). We left the ground truth for the CPA task as future work.

3   Limitations in Related Datasets

In the last decade different benchmark datasets have been proposed in the literature. The most important and used are T2Dv2[1] (also referred to as T2D in this paper), Limaye [8] and W2D [4]. They come with different capabilities. As an example, while T2D provides tables with properties for the CPA task, W2D does not cover this task very well. We can then observe that, even if the above datasets differ in size (T2D has 200 tables, while W2D counts more than 485k),

---

[1] T2Dv2: http://webdatacommons.org/webtables/goldstandardV2.html.

all of them are focused on web tables, i.e., small tables scraped from the web; these datasets are particularly suitable to benchmark table interpretation algorithms whose objective is to support query answering over the large amount of tables published on the web [11]. However, tables considered for the population of KGs are usually quite different from web tables, e.g., they are much larger, and these kind of tables are not representated in the available datasets. We can shortly summarize the difference between these two kinds of tables as follows:

- Legacy tables usually have many rows, while tables in existing benchmark datasets are small (the average number of rows per table is 123 for T2D, 29 for Limaye and only 15 for W2D, according to [4]). Large tables may prevent algorithms from using heuristics that consider the full table (e.g., infer the column type by looking at the whole column).
- Legacy tables, especially CSVs, usually contain de-normalised tables with several columns; this aspect is not well represented in the considered datasets (each table contains on average ~1.77 columns with entities).
- Because of the usual de-normalization, legacy tables contain many columns with entity matches, but tables in existing benchmarks are mostly focused on "entity tables", i.e., tables where each row represents only one entity; in such a table, one column refers to the entity (it is also called *subject* column), and all the other contain attributes of the main entity; this scenario does not fit the case of de-normalised tables. We also report that in some cases (e.g., T2D), if the table contains more than one entity column, they are disregarded and not annotated.
- Entities in web tables are usually mentioned using their canonical name (e.g., Barack Obama is mentioned as "Barack Obama" - it is very unusual to see "B. Obama"); in legacy sources, we find acronyms, abbreviations, misspelled words that considerably increase the ambiguity of the table. For example, the misspelling of drug names is a very important problem in the health domain [5,6].

A recent study reported that many of the approaches tested on such datasets are focused on "obviously linkable" cells [14], showing that a tool like T2K [10] manages to match only 2.85% of a large corpus of Web tables to DBpedia. However, the performance of T2K evaluated on T2D relatively to the CEA task is very high (F1: 0.82, Precision: 0.90, Recall: 0.76). This suggests that many tables in T2D are easy to annotate. The authors themselves specify that the entity linking can be wrong sometimes, especially when the mention is ambiguous, spotlighting that there are a few ambiguous mentions in T2D. Recently, SemTab (Semantic Web Challenge on Tabular Data to Knowledge Graph Matching) was created to conduct a systematic evaluation of table annotation algorithms [7]. The organizers presented new datasets built by automatically generating tables from the results of SPARQL queries. The first rounds of SemTab 2019 also included tables from T2D and W2D datasets. The success of the challenge and the analysis of the results achieved by different competing approaches have revealed some challenges that should be addressed in future work:

- Wikipedia and DBpedia lookup mechanisms are effective for candidate selection tasks, but they might fail when the cell content is misspelled (e.g., Winsconsin vs Wisconsin) or is different from the most common one (e.g., football player nicknames: La Pulga vs Lionel Messi).
- Real world tables are noisy and in general not well formed. Algorithms should be evaluated also with respect to their ability to deal with such tables.
- Missing data can affect the results of the algorithms and thus this aspect should be correctly evaluated.
- Although the overall quality of the SemTab 2019 dataset is higher compared with the other datasets, a manual inspection of the tables in the SemTab brought to the surface some malformed and wrongly annotated tables, like empty rows mapped to an entity and long descriptions (with mentions of different entities) mapped to a single entity.[2]

Finally, existing datasets have contributed to the benchmarking of table interpretation algorithms, however, none of them provide fine grained information about the achievement of a certain score. Some tools have been developed to at least highlight which are the main error patterns, but those patterns must be manually inspected (e.g., [2]). Our dataset has been built to facilitate the understanding of the main limitations of a table interpretation system, since it provides a footprint about the uncovered aspects that led to certain performance score.

4   The 2T Dataset

In this paper we present Tough Tables (2T), a dataset designed to evaluate table annotation approaches on the CEA and CTA tasks. All the annotations are based on DBpedia 2016-10.[3] The structure of the dataset (depicted in Fig. 2) allows the user to know which aspects of the entity linking task are handled better/worse by different approaches. Indeed, the dataset comes with two main categories of tables:

- the *control* group (CTRL), which contains tables that are easy to solve; a table annotation algorithm should at least annotate these tables with relatively high performance.
- the *tough* group (TOUGH), which features only tables that are hard to annotate.

A complete algorithm should solve both the categories, because otherwise i) solving only the CTRL group means that the algorithm is able to only cope with obvious entities and ii) solving only the TOUGH tables highlights that the algorithm is too complex and cannot deal with the simpler cases.

---

[2] See Tables *53822652_0_5767892317858575530* and *12th_Goya_Awards#1* from Round 1 and Round 2, respectively. These errors come from the T2D and W2D datasets used in SemTab 2019.

[3] We checked our annotations against a private replica of the online DBpedia SPARQL endpoint in a local instance, loading the 2016-10 datasets listed at https://wiki.dbpedia.org/public-sparql-endpoint.

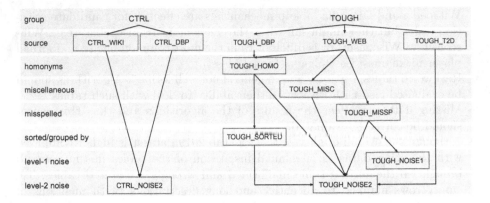

Fig. 2. 2T profile at a glance. Boxed categories are those considered during the evaluation phase. Each category is composed by some of/all the tables from the parent categories.

4.1 Dataset Profile

Figure 2 provides an overview of the different table flavours included within the 2 T dataset. The *control* group (CTRL) contains tables generated by querying the DBpedia SPARQL endpoint[4] (CTRL_DBP), leaving the label as extracted from the KG, and tables collected from Wikipedia (CTRL_WIKI), manually revised and annotated (further details about this procedure are available in Appendix A.1). Like the DBP ones, the WIKI tables are pretty clear and easy to annotate, since most of the mentions refer to entities in DBpedia with just slightly different labels from the ones contained in this KG.

The *tough* group (TOUGH) contains mainly tables scraped from the web. This group contains also a small subset of T2D (TOUGH_T2D), which we re-annotated considering the entities appearing in all the columns, and not only entities in the "subject" column like T2D. In addition, we collected tables from the web that contain nicknames or homonyms (TOUGH_HOMO), and misspelled words (TOUGH_MISSP). We enriched the TOUGH_HOMO category by adding a few tables generated via ad-hoc SPARQL queries. Along with these tables, we included other web tables (TOUGH_MISC), like non-cleaned Wikipedia tables and tables available as Open Data (in a limited quantity, due to motivations stated in Appendix A.1).

We selected some specific tables from the TOUGH_HOMO category and sorted them in a specific order to make the task of detecting the type more difficult (TOUGH_SORTED). If we are creating a table describing athletes we would probably intuitively organize them by category, having the soccer players in the first part, then all the basket players, then all the golf players, etc. Sorting tables might pose a problem for those algorithms that infer the column type by

---

[4] We used the online version at http://dbpedia.org/sparql.

only looking at the first n rows (with n usually small), and then use the inferred type as a filter for the entity linking.[5]

Noisy Tables. We used the already collected tables to generate other noisy tough tables. Starting from tables in TOUGH_MISSP, we generated a new category (TOUGH_NOISE1) by adding a *level 1* noise: for each table, 10 additional tables with noise have been generated, where each of them contains an incremental percentage of misspelled mentions (increasing by 10% at a time). This noise resembles real noise since we use lists of real-world misspelled words and use them to generate noise. From the tables in the CTRL and TOUGH categories (excluding the TOUGH_NOISE1 category), we also created two new categories (CTRL_NOISE2 and TOUGH_NOISE2) via a *level-2* noise, i.e., random noise that changes a bit the labels of randomly selected columns/rows (e.g., it randomly duplicates a symbol). Tables in this new category feature a noise that is random and artificial, thus it does not always resemble a real world scenario.

Novel Facts. One of the main applications of the table interpretation is to extract new facts from tables, especially for KG population/completion tasks. In data integration pipelines, entity linking and new triples generation play an equally important role. Novel facts detection is not considered in the standard CEA evaluation,[6] but we outline that our dataset can be used to test algorithms in finding new facts. 2T tables contain 3,292 entity mentions across 42 tables without a corresponding entity in DBpedia 2016-10. In the CEA asset, a good table annotation algorithm is expected to decide that such cells should not be linked (similar to the *NIL* prediction in Named Entity Linking/Recognition). In more comprehensive assets, like KG construction/population, we expect the algorithm to generate a new triple (with rdf:type) using the discovered column type. Depending on the context, such particular cells might be used in the future to test novel knowledge discovery algorithms.

Overview. Benchmarking a table annotation system using 2 T allows the developers to understand why their algorithms achieved a certain score as follows:

- if the algorithm performs well only on the control tables, then it relies too much on the performance of simple string matching strategies like label lookup (i.e., it looks only for exact matches, or considers only the canonical name of entities);
- if the performance is good also on the control tables with level-2 noise, then the algorithm adopts a kind of fuzziness in its lookup phase (e.g., edit distance), which is still not enough to solve the tough tables;

---

[5] This strategy, that might look naive, is the same implemented in OpenRefine, where the first 10 rows are used to suggest the possible types of the current column.

[6] The SemTab 2019 challenge provided the *target* file with the full list of cells to annotate, disregarding novel facts.

– if the algorithm performs well on some tough tables and bad on others, then the user can better understand the weaknesses of the algorithm by looking at the performance on different categories of tables:
  - if the tables with columns sorted by type are annotated in the wrong way, then the entity linking algorithm is constrained by the type inferred looking at the first n rows, with n too small;
  - if homonyms or nicknames have been wrongly matched, that means that the algorithm employs popularity mechanisms (e.g., page rank), or it is based on a lookup service that returns the most popular entities first (e.g., DBpedia Lookup).[7] Annotating nicknames requires the algorithms to cover aspects of semantics that go a bit beyond simple heuristics;[8]
  - if the tables with level-1 noise are not properly annotated, then the algorithm cannot deal with real-world noise (that can be trickier than the artificial level-2 noise);
  - if the annotations are wrong for the tables containing nicknames, it might be the case the algorithm only focuses on the canonical names of the entities.

Tables 1 and 2 show statistics for 2 T and existing benchmark datasets. Compared to existing datasets, 2T has a higher average number of rows per table, pushing the size of individual tables towards the size of real legacy sources; the number of matches is slightly grater than the number available in SemTab 2019 Rounds 2 and 3 (ST19 - R2 and ST19 - R3 in Table 2), considering that 2 T comes with a number of tables that is up to two orders of magnitude smaller. Since some tables in 2 T are built starting from the same core table, we observe a small number of unique entities. Finally, 2T tables have a lower average number of columns per table, but the highest number of columns with at least a match: this aspect helps in having more columns to annotate in the CTA task, and it is also a starting point for future extensions of 2T, i.e., covering CPA task.

4.2   Evaluation

We ran experiments to evaluate the toughness of our dataset and its capability of spotting the weaknesses of an annotator. We setup an environment that resembles the CEA task of SemTab 2019, i.e., target cells are known and extra annotations are disregarded. We also used the available code to score the algorithms in the same way.[9] We introduce two simple baselines, DBLookup and WikipediaSearch, which run a query against the corresponding online lookup

---

[7] We point out that some homonyms are very easy to solve using DBpedia (e.g., US cities are easy to find, since just appending the state of a city to its canonical name points directly to the right city, e.g., the Cambridge city in Illinois is dbr:Cambridge,_Illinois in DBpedia).

[8] Note that it is possible to solve this problem using a mapping dictionary if available, but this is not a desired solution: this will not make the algorithm *smart*; the same is true for looking up on Google Search.

[9] https://github.com/sem-tab-challenge/aicrowd-evaluator.

Table 1. Detailed statistics for 2T. Values are given as $avg \pm st.dev.$ *(total, min, max)*

Category	Cols	Rows	Matches	Entities	Cols with Matches	Tables
ALL	4.47±1.90 (804, 1, 8)	1080.38±2805.25 (194468, 5, 15477)	3687.94±10142.48 (663830, 6, 61908)	438.77±1241.97 (16464, 6, 7032)	2.99±1.17 (538, 1, 6)	180
CTRL WIKI	5.73±1.28 (86, 4, 7)	66.00±81.39 (990, 10, 263)	241.93±333.21 (3629, 20, 1041)	157.93±224.06 (1940, 15, 771)	3.47±1.41 (52, 2, 6)	15
CTRL DBP	4.40±0.91 (66, 3, 6)	709.60±717.65 (10644, 120, 2408)	2510.67±2573.68 (37660, 360, 7820)	343.40±217.08 (4976, 68, 618)	3.53±0.64 (53, 3, 5)	15
CTRL NOISE2	5.07±1.28 (152, 3, 7)	387.77±599.15 (11633, 10, 2408)	1375.93±2140.04 (41278, 20, 7820)	250.53±236.31 (6745, 15, 770)	3.50±1.07 (105, 2, 6)	30
TOUGH T2D	5.82±1.83 (64, 3, 8)	78.09±77.26 (859, 6, 232)	165.36±150.28 (1819, 6, 464)	94.45±80.05 (991, 6, 248)	2.09+0.94 (23, 1, 4)	11
TOUGH HOMO	3.36±1.12 (37, 2, 5)	1648.82±3272.12 (18137, 13, 8302)	6422.55±13168.36 (70648, 25, 33208)	1469.27±2719.02 (8331, 24, 7032)	3.00±0.77 (33, 2, 4)	11
TOUGH MISC	6.50±1.31 (78, 4, 8)	122.25±162.86 (1467, 11, 561)	366.33±416.95 (4396, 22, 1215)	222.92±261.32 (2374, 16, 770)	3.67±1.44 (44, 2, 6)	12
TOUGH MISSP	3.50±1.29 (14, 2, 5)	4175.50±7549.48 (16702, 52, 15477)	16386.50±30381.91 (65546, 178, 61908)	204.00±350.86 (775, 13, 730)	3.25±0.96 (13, 2, 4)	4
TOUGH SORTED	3.50±2.12 (7, 2, 5)	4215.00±5779.89 (8430, 128, 8302)	16732.00±23300.58 (33464, 256, 33208)	3602.00±4850.75 (7201, 172, 7032)	3.00±1.41 (6, 2, 4)	2
TOUGH NOISE1	2.50±1.13 (100, 1, 4)	2000.30±3701.62 (80012, 5, 14008)	5738.15±11197.12 (229526, 15, 42024)	204.00±307.73 (775, 13, 730)	2.25±0.84 (90, 1, 3)	40
TOUGH NOISE2	5.00±1.97 (200, 2, 8)	1139.88±3183.53 (45595, 6, 15477)	4396.75±12774.85 (175870, 6, 61908)	697.33±1823.97 (11820, 6, 7032)	2.98±1.21 (119, 1, 6)	40

services using only the actual cell content.[10] Both the baselines use the first result returned by the online service as the candidate annotation. Alongside the baselines, we looked for a real algorithm to test among the ones that participated in the SemTab 2019. We contacted the authors of MTab [9], CVS2KG [13], and Tabularisi [12], the tools that performed the best, but their tools were not publicly available. Only the authors of MantisTable [1], winner of the outstanding improvement award at SemTab 2019 (CEA task), provided us with a prototype of their tool.[11] Since the performance obtained by all the tools were similar to each other, we think that MantisTable is a good representative for the evaluation. We run MantisTable and the two baselines on 2T, obtaining the results depicted in Fig. 3. We underline here that our dataset adopts the same standard format defined in SemTab2019, making it compatible with all the systems that participated in the challenge.[12]

---

[10] We used the WikipediaSearch online service available at https://en.wikipedia.org/w/api.php, while we recreated the DBLookup online instance on a dedicated virtual machine.

[11] A fork of the original code repository is available at https://bitbucket.org/vcutrona/mantistable-tool.py.

[12] The standard format introduced in SemTab2019 is directly derived from the T2Dv2 one, thus the number of algorithms that can be tested is potentially greater.

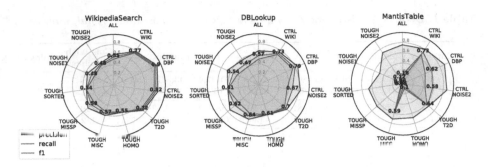

Fig. 3. Precision, Recall and F1 measures for the considered algorithms.

Results. In some cases, MantisTable could not annotate all the table cells (e.g., it failed to annotate tables with thousands of rows); this might be due to some general limits of the MantisTable prototype for which the solution is out of the scope of the paper. Indeed, the algorithm is based on several heuristics, and processing big tables might lead to processing errors. For example, processing 180 tables took more than 24 h. However, this does not compromise our evaluation, proving again that MantisTable can effectively annotate tables as shown in the SemTab 2019 challenge. The reader should take into account that the reported results may not reflect the full performance of MantisTable as they represent the output of a dry-run test without the involvement of the developers. Nevertheless, the results suggest that tough tables are effectively difficult to annotate for state-of-the-art algorithms. For the baselines, their precision and recall are quite similar since the online lookup services almost always return a result, and we set it as the annotation without further processing; focusing on the F1 measure, we observe that WikipediaSearch reaches 0.83 F1 on the CTRL group, which is high compared with state-of-the-art models, considering that the process only relies on the lookup service. The performance of DBLookup is good as well, but it decreases due to the CTRL_NOISE2 subcategory, reaching an average F1 of 0.73 on the CTRL group. Both algorithms are not able to annotate the TOUGH tables, with DBLookup doing a bit better; this might be due to the fact that some of the tough tables have been built with SPARQL queries, giving some advantages to the DBLookup service. In general, the performance is low as expected (0.63 overall F1 for both the baselines), given that these algorithms do not use any kind of sophisticated semantic techniques. Looking at MantisTable results, we see that the tool is focusing on those cells that can be more easily linked to the KG. The semantic techniques employed in the algorithm pushes the precision on the control group, but due to the low recall, the algorithm performs worse (0.32 overall F1) than the previous baselines. Since the precision on CTRL_DBP tables is higher than CTRL_WIKI ones, we can assume that the lookup phase of the algorithm heavily depends on DBpedia (as a lookup service or SPARQL endpoint). The same is confirmed by the low recall on the TOUGH_NOISE1 tables, which are the ones with real-world misspelled men-

Table 2. Comparison with existing benchmark datasets.

Dataset	Cols (avg)	Rows (avg)	Matches	Entities	Cols with matches (avg)	Tables
T2Dv2	1,157 (4.95)	27,996 (123)	26,124	–	–	234
Limaye	–	–	142,737	–	–	6,522
W2D	– (5.58)	7,437,606 (15)	4,453,329	–	–	485,096
ST19 - R1	323 (5.05)	9,153 (143.02)	8,418	6,225	64 (1.00)	64
ST19 - R2	66,734 (5.60)	311,315 (26.12)	463,796	249,216	15,335 (1.29)	11,920
ST19 - R3	9,736 (4.51)	154,914 (71.69)	406,827	174,459	5,762 (2.67)	2,161
ST19 - R4	3,564 (4.36)	52,066 (63.73)	107,352	54,372	1,732 (2.12)	817
2T (ours)	804 (4.47)	194,468 (1080.38)	663,830	16,464	538 (2.99)	180

tions. For the subset of T2D tables that we chose and re-annotated, we spot a F1 score lower than the 0.98 obtained by MantisTable during the Round 1 of the SemTab challenge;[13] this confirms that the T2D tables are focused on obvious entities, disregarding the more difficult ones. The F1 score drastically decreases on the misspelled tables, highlighting that this aspect is still not fully covered in sophisticated state-of-the-art approaches like MantisTable.

4.3 Availability and Long-Term Plan

Resources should be easy accessible to allow data applicability and validation. We follow FAIR (Findable, Accessible, Interoperable and Reusable) guidelines to release our contributions.[14]

We release our dataset in Zenodo (DOI: https://doi.org/10.5281/zenodo. 3840646), in such a way that researchers in the community can benefit from this. Our dataset is released under the Attribution 4.0 International (CC BY 4.0) License.[15] Together with the dataset we release the code that was used to collect the data and create it.[16] This should favor replicability and subsequent extensions.

5 Conclusions and Future Work

In this paper, we presented a novel dataset for benchmarking table annotation approaches on the entity linking and column type annotation tasks, equivalent to the CEA and CTA tasks as defined in current benchmarks [7]. The dataset comes with a mix of real and constructed tables, which resemble many real-world scenarios. We tested our dataset using a state-of-the-art approach, MantisTable, and two baselines. These baselines are represented by online lookup services, usually adopted as a building block of many table annotation approaches. We demonstrate that our tables are tough, and solving them requires the algorithms to

---

[13] https://www.cs.ox.ac.uk/isg/challenges/sem-tab/2019/results.html.
[14] https://www.nature.com/articles/sdata201618.
[15] https://creativecommons.org/licenses/by/4.0/.
[16] https://github.com/vcutrona/tough-tables.

implement sophisticated mechanisms that take into consideration many semantics aspects. In the near future, we intend to extend this dataset to cover also the CPA task; indeed, we observed that the 2T's profile fits well the benchmarking of this task because it contains many columns with matches, making the CPA more challenging (having averagely three possible columns to annotate instead of 2 increases by a degree of freedom the number of possible property annotations). The authors are also analysing the inclusion of this dataset in one of the rounds of SemTab 2020 [17]

Acknowledgments. We would like to thank the authors of MantisTable for sharing the prototype source code. This work was partially supported by the Google Cloud Platform Education Grant. EJR was supported by the SIRIUS Centre for Scalable Data Access (Research Council of Norway). FB is member of the Bocconi Institute for Data Science and Analytics (BIDSA) and the Data and Marketing Insights (DMI) unit.

A   2T Ground Truth Generation Details

A.1   CEA Table Generation and Preprocessing

2T has been built using real tables. Here we clarify that as a "real table" we intend a table, also artificially built, which resembles a real table. Examples are "list of companies with their market segment", or "list of Italian merged political parties", which look like queries that a manager or a journalist could make against a database. The main reasons behind this choice are: *(i)* it is difficult to get access to real databases; *(ii)* open data make available a lot of tables, but mostly always tables are in an aggregated form that makes it difficult to annotate them with entities from a general KG like DBpedia. When the data are fine-grained enough, almost all the entities mentioned are not available in the reference KG. For example, in the list of bank failures got from the U.S. Open Data Portal, only 27 over 561 failed banks are represented in DBpedia.

In this section we describe the processes we adopted to collect real tables, or build tables that resembles real ones.

DBpedia Tables. We used the DBpedia SPARQL endpoint as a table generator (SPARQL results are tables). We run queries to generate tables that include:

- entity columns: columns with DBpedia URIs that represent entities.
- "label columns": columns with possible mentions for the corresponding entities in the entity column. Given an entity column, the corresponding label column has been created by randomly choosing between `rdfs:label`, `foaf:name`, or `dbo:alias` properties.
- literal columns: other columns, with additional information.

---

[17] The SemTab 2020 challenge is still in progress and it is organized to provide tables without known ground truth. For this reason, we will publish the full 2 T dataset, including the ground truth files, at the end of SemTab 2020.

Wikipedia Tables. We browsed Wikipedia looking for pages containing tables of interest (e.g., list of presidents, list of companies, list of singers, etc.). We generated different versions of the collected Wikipedia tables, applying different cleaning steps. The following steps have been applied to Wikipedia tables in the *TOUGH_MISC* category:

- Merged cells have been split in multiple cells with the same value.
- Multi-value cells (slash-separated values, e.g., Pop / Rock, or multi-line values, e.g., Barbados
 United States, or in-line lists, e.g., <ul>, <li>) have been exploded into several lines. If two or more multi-value cells are on the same line, we exploded all the cells (cartesian product of all the values). If a cell contains the same information in more languages (e.g., anthem song titles), we exploded the cell in two or more columns (the creation of new lines would basically represent duplicates).

Wikipedia tables in the *CTRL_WIKI* group underwent the next additional cleaning steps:

- "Note", "Description", and similar long-text columns have been removed.
- Cells with "None", "null", "N/A", "Unaffiliated", and similar values have been emptied.
- Columns with only images (e.g., List of US presidents) have been removed.
- All HTML tags have been deleted from cells (e.g., country flag icons);
- Notes, footnotes, and any other additional within-cell information (e.g., birthYear and deathYear for U.S. presidents) have been removed.

Most of all the tables values are already hyperlinked to their Wikipedia page. We used the hyperlinks as the correct annotations (we trust Wikipedia as a correct source of information), following these criteria:

- If a cell content has several links, we took the most relevant annotation, given the column context (e.g., in table https://en.wikipedia.org/ wiki/List_of_presidents_of_the_United_States#Presidents the "U.S. senator from Tennessee" cell in the "Prior office" column contains two annotations: https://en.wikipedia.org/wiki/U.S._senator and https://en.wikipedia. org/wiki/Tennessee; in this case we took only the https://en.wikipedia.org/ wiki/U.S._senator annotation, as the column is about "Prior offices", not about places).
- Sometimes it happens that if the same value appears several times in the same column (e.g., music genres), only one instance has the hyperlink to the Wikipedia page. In these cases we copied the same hyperlink to all the instances.
- When the hyperlink is missing (e.g., Hard Rock labels in the table https://en.wikipedia.org/wiki/List_of_best-selling_music_artists#250_million_ or_more_records), we manually added the right links by visiting the main entity page (e.g., https://en.wikipedia.org/wiki/Led_Zeppelin) and looking for the missing piece of information (e.g., under the "Genre" section on the Led Zeppelin page we can find Hard Rock linked to https://en.wikipedia.org/

wiki/Hard_rock). In case when the information is missing in the main page (e.g., in the same table, Michael Jackson genres include "Dance", while on his Wiki page the genre is Dance-pop), we manually annotated the value with the most related entity in Wikipedia (in this case, the music genre Dance https://en.wikipedia.org/wiki/Dance_music).

Finally, we converted the Wikipedia links to their DBpedia correspondent links, by replacing https://en.wikipedia.org/wiki/ with http://dbpedia.org/resource in the decoded URL, e.g., https://en.wikipedia.org/wiki/McDonald%27s → dbr:McDonald's, if available, otherwise we manually looked for the right dbpedia link (e.g., https://en.wikipedia.org/wiki/1788-89_United_States_presidential_election →
dbr:United_States_presidential_election,_1788-89). If this attempt also failed, we left the cell blank (no annotations available in DBpedia).

A.2   CTA Ground Truth Construction

Automatic CTA Annotations from CEA. The 2T dataset focus is mainly on the entities because, in our opinion, the CEA task is the core task: with good performance in CEA, it is possible to approximate the CTA task easily. We exploited this observation to automatically construct the CTA annotations starting from the CEA ones, which we trust. For each annotated column, we collected all the annotated entities from the CEA dataset and retrieved the most specific type for all the entities from the DBpedia 2016-10 dump.[18] We then annotate the column with the most specific supertype, i.e., the lowest common ancestor of all the types in the DBpedia 2016-10 ontology.[19]

References

1. Cremaschi, M., De Paoli, F., Rula, A., Spahiu, B.: A fully automated approach to a complete semantic table interpretation. Fut. Gener. Comput. Syst. **112**, 478–500 (2020). https://doi.org/10.1016/j.future.2020.05.019
2. Cremaschi, M., Siano, A., Avogadro, R., Jimenez-Ruiz, E., Maurino, A.: STILTool: a semantic table interpretation evaluation tool. In: ESWC P&D (2020)
3. Cutrona, V., et al.: Semantically-enabled optimization of digital marketing campaigns. In: Ghidini, C., et al. (eds.) ISWC 2019. LNCS, vol. 11779, pp. 345–362. Springer, Cham (2019). https://doi.org/10.1007/978-3-030-30796-7_22
4. Efthymiou, V., Hassanzadeh, O., Rodriguez-Muro, M., Christophides, V.: Matching web tables with knowledge base entities: from entity lookups to entity embeddings. In: d'Amato, C., et al. (eds.) ISWC 2017. LNCS, vol. 10587, pp. 260–277. Springer, Cham (2017). https://doi.org/10.1007/978-3-319-68288-4_16

---

[18] The *instance types* file at http://downloads.dbpedia.org/2016-10/core-i18n/en/instance_types_en.ttl.bz2 contains entities most specific types.
[19] http://downloads.dbpedia.org/2016-10/dbpedia_2016-10.nt.

5. Hussain, F., Qamar, U.: Identification and correction of misspelled drugs names in electronic medical records (EMR). In: ICEIS 2016 - Proceedings of the 18th International Conference on Enterprise Information Systems, vol. 2, pp. 333–338 (2016). https://doi.org/10.5220/0005911503330338

6. Jiang, K., Chen, T., Huang, L., Calix, R.A., Bernard, G.R.: A data-driven method of discovering misspellings of medication names on twitter. In: Building Continents of Knowledge in Oceans of Data: The Future of Co-Created eHealth - Proceedings of MIE 2018, Medical Informatics Europe. Studies in Health Technology and Informatics, vol. 247, pp. 136–140 (2018). https://doi.org/10.3233/978-1-61499-852-5-136

7. Jiménez-Ruiz, E., Hassanzadeh, O., Efthymiou, V., Chen, J., Srinivas, K.: SemTab 2019: resources to benchmark tabular data to knowledge graph matching systems. In: Harth, A., et al. (eds.) ESWC 2020. LNCS, vol. 12123, pp. 514–530. Springer, Cham (2020). https://doi.org/10.1007/978-3-030-49461-2_30

8. Limaye, G., Sarawagi, S., Chakrabarti, S.: Annotating and searching web tables using entities, types and relationships. VLDB 3(1), 1338–1347 (2010). 10.14778/1920841.1921005

9. Nguyen, P., Kertkeidkachorn, N., Ichise, R., Takeda, H.: MTab: matching tabular data to knowledge graph using probability models. In: SemTab@ISWC. CEUR Workshop Proceedings, vol. 2553, pp. 7–14 (2019)

10. Ritze, D., Lehmberg, O., Oulabi, Y., Bizer, C.: Profiling the potential of web tables for augmenting cross-domain knowledge bases. In: Proceedings of the 25th International Conference on World Wide Web, WWW 2016. pp. 251–261. ACM (2016). https://doi.org/10.1145/2872427.2883017

11. Sun, H., Ma, H., He, X., Yih, W.T., Su, Y., Yan, X.: Table cell search for question answering. In: WWW, International World Wide Web Conferences Steering Committee, pp. 771–782 (2016). https://doi.org/10.1145/2872427.2883080

12. Thawani, A., et al.: Entity linking to knowledge graphs to infer column types and properties. In: SemTab@ISWC. CEUR Workshop Proceedings, vol. 2553, pp. 25–32 (2019)

13. Vandewiele, G., Steenwinckel, B., Turck, F.D., Ongenae, F.: CVS2KG: transforming tabular data into semantic knowledge. In: SemTab@ISWC. CEUR Workshop Proceedings, vol. 2553, pp. 33–40 (2019)

14. Zhang, S., Meij, E., Balog, K., Reinanda, R.: Novel entity discovery from web tables. In: Proceedings of The Web Conference 2020, WWW 2020, pp. 1298–1308 (2020). https://doi.org/10.1145/3366423.3380205

Facilitating the Analysis of COVID-19 Literature Through a Knowledge Graph

Bram Steenwinckel(✉) ⓘ, Gilles Vandewiele ⓘ, Ilja Rausch ⓘ,
Pieter Heyvaert ⓘ, Ruben Taelman ⓘ, Pieter Colpaert ⓘ, Pieter Simoens ⓘ,
Anastasia Dimou ⓘ, Filip De Turck ⓘ, and Femke Ongenae ⓘ

IDLab, Ghent University – imec, Technologiepark-Zwijnaarde 126, Ghent, Belgium
{bram.steenwinckel,gilles.vandewiele,ilja.rausch,pieter.heyvaert,
ruben.taelman,pieter.colpaert,pieter.simoens,
anastasia.dimou,filip.deturck,femke.ongenae}@ugent.be

Abstract. At the end of 2019, Chinese authorities alerted the World Health Organization (WHO) of the outbreak of a new strain of the coronavirus, called SARS-CoV-2, which struck humanity by an unprecedented disaster a few months later. In response to this pandemic, a publicly available dataset was released on Kaggle which contained information of over 63,000 papers. In order to facilitate the analysis of this large mass of literature, we have created a knowledge graph based on this dataset. Within this knowledge graph, all information of the original dataset is linked together, which makes it easier to search for relevant information. The knowledge graph is also enriched with additional links to appropriate, already existing external resources. In this paper, we elaborate on the different steps performed to construct such a knowledge graph from structured documents. Moreover, we discuss, on a conceptual level, several possible applications and analyses that can be built on top of this knowledge graph. As such, we aim to provide a resource that allows people to more easily build applications that give more insights into the COVID-19 pandemic.

Keywords: COVID-19 · Knowledge graph creation · Network analysis · Graph embeddings

1 Introduction

In 2019, the World Health Organization (WHO) was alerted that an infectious disease was identified in Wuhan, Central China. Now, in 2020 this disease caused by severe acute respiratory syndrome coronavirus 2 (SARS-CoV-2) has spread globally, resulting in the commonly known COVID-19 pandemic [2].

This virus spread itself easily. Over 20,000,000 people, from all over the world, were infected in a short amount of time [5]. In response to this pandemic, on March 16th, 2020, researchers and leaders from the Allen Institute for AI, Chan Zuckerberg Initiative (CZI), Georgetown University's Center for Security and Emerging Technology (CSET), Microsoft, and the National Library of

© Springer Nature Switzerland AG 2020
J. Z. Pan et al. (Eds.): ISWC 2020, LNCS 12507, pp. 344–357, 2020.
https://doi.org/10.1007/978-3-030-62466-8_22

Medicine (NLM) at the National Institutes of Health released a freely available dataset of scholarly literature about COVID-19, SARS-CoV-2, and the coronavirus group [1]. The goal of releasing such a dataset was to apply recent advances in Natural Language Processing (NLP) and other Artificial Intelligence (AI) techniques to generate new insights in support of the on-going fight against this infectious disease.

The goal of this study was to transform this original dataset into a knowledge graph. Having this data in a graph-based format allows us to reap several benefits. First, by linking concepts to external resources, the dataset can be enriched with knowledge that was initially not available. As an example, linking the studies in the dataset to DBpedia resources of their respective country allows us to explore potential correlations with, for example, geographic and demographic data. Second, the edges in the graph explicitly represent a relation between pairs of entities, which can be taken into account during analysis of the dataset. These edges can result in more precious insights.

This advantage has already been illustrated in several studies. It has been shown that taking into account citation information of a paper and its content can produce useful representations that adopt the idea and benefits of linked data [19]. In this paper, we show the full pipeline to construct such a graph and explain how we've made this resource publicly available, taking into account the Findable, Accessible, Interoperable, and Reusable (FAIR) principles. We also illustrate the advantages of having a knowledge graph of this data by conducting several preliminary analyses and giving potential directions for further applications.

The remainder of this paper is organized as follows. We first discuss some related initiatives that built upon this dataset in Sect. 2. Next, we provide an overview of our architecture used to transform the original dataset into a semantic representation in Sect. 3. In Sect. 4, we then list all resources which were integrated and linked to this knowledge graph, and what type of new information they bring to the knowledge graph. Finally, we discuss some potential applications and provide some preliminary analyses in Sect. 5. We conclude our work in Sect. 6.

2   Related Work

Based on the provided dataset, several other initiatives started to build a knowledge graph, build applications on top of them or used them in order to facilitate the analysis of other researchers of this vast amount of information. The Covid Graph project [17], led by a diverse team based in Germany, is probably the largest initiative. They created a COVID-19 knowledge graph by mining the CORD-19 dataset, linking it to the NCBI Gene Database and other gene ontologies to enable scientific analysis. They currently provide a visual graph explorer and a NEO4J browser as applications. Another notable initiative is the CORD-19-on-FHIR dataset [10]: a Linked Data version of the CORD-19 represented in FHIR RDF. Here, the titles and abstracts were parsed, and more than

180,000 Condition, 32,000 medication and 100,000 procedure instances could be identified and linked. Similar to our initiative, the Dice research group started to build a knowledge graph by creating triples using RDFLIB[1] in Python [23]. The COVID19DS knowledge graph mainly links the papers, authors, refs and cites together in one knowledge graph without looking into the actual content of the papers.

Other knowledge graphs were designed with a particular task in mind. The CORD-SEMANTICTRIPLES initiative derived knowledge from the Semantic MEDLINE database (SemMedDB), reflecting documents also in the COVID 19 corpus[2]. SemMedDB contains concept-relation-concept semantic triples, or predications. After extracting 106 K semantic predications, they imported these into a network and applied network centrality metrics (degree, closeness, betweenness) to identify and substantiate association factors related to COVID-19 for biological plausibility.

The CORD-ReDrugS project enhanced ReDrugS [14] to use the concepts and relations from extracted entities of the COVID dataset to repurpose potential therapies. A knowledge graph to define a cause-and-effect knowledge model of COVID-19 pathophysiology comprising information encoded in Biological Expression Language (BEL) was made for a selected corpus around COVID-19. Mappings, mainly for viral proteins, were made to the NCBI database [8].

Additional efforts incorporated knowledge extracted from the COVID dataset into already existing knowledge graphs. COVID related research findings were added into the Open Research Knowledge Graph [4]. Covid-19 and associated publications were also made available in the Microsoft Academic Knowledge Graph (MAKG) [9].

3   Creating a Knowledge Graph of the CORD-19 Dataset

To create a knowledge graph from an already existing data source, we have to combine both a transparent architecture to generate the linked data and have a correct idea about the originally used data format. In this section, we first describe the original dataset and underlying data format. Next, we give detailed information about the knowledge graph modeling procedure.

3.1   COVID 2019 Open Research Dataset (CORD-19)

At the beginning of March 2020, Kaggle released the COVID 2019 Open Research Dataset (CORD-19) dataset in collaboration with several research groups, such as Microsoft Research. The dataset contains information of over 63,000 papers concerning COVID-19, SARS-CoV-2, or any other related coronaviruses. The papers stem from various sources, such as PubMed Central (PMC) and medRxiv, and are from different research domains. Information about these

---

[1] https://github.com/RDFLib/rdflib.
[2] https://github.com/kingfish777/COVID19.

papers is provided in the form of a CSV file. For more than 51,000 of these papers, a JSON file is provided that contains detailed information about the authors, the content and the other studies that were cited. A schematic overview of such a JSON file is visualized in Fig. 1.

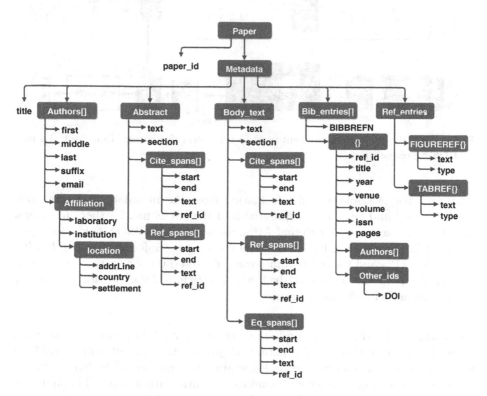

Fig. 1. JSON structure of a given paper. Five main components were nested: abstract, author, body, cites and reference information.

3.2   Knowledge Modeling

In order to facilitate the (meta-)analysis of this significant body of literature, we semantically enriched the data by mapping it to the Resource Description Framework (RDF). Since the data was available in structured formats (JSON and CSV), we can define rules that map chunks of structured information to RDF triples. The RDF Mapping Language (RML) [7] allows intuitively specifying these rules. We now discuss the two main steps of our conversion procedure, which is visualized in Fig. 2.

Extending the JSON. First, each JSON file representing a paper is loaded by a Python script in order to extend it with the information provided by external

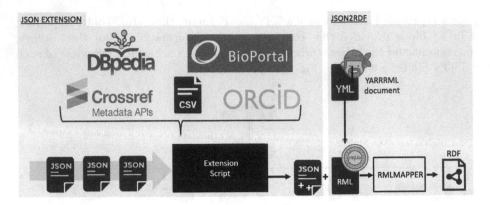

Fig. 2. Overview of each component used to transform the dataset from Sect. 3.1 into an RDF representation.

resources. For example, useful information from the metadata CSV file, such as the journal and DOI, is incorporated into the JSON file. Further on, several modules, interacting with external APIs, use some of the JSON fields displayed in Fig. 1 in their request bodies to acquire even more information. The full list of external APIs and the result values used to extend these papers are described in Sect. 4. When this step is finished, a new extended JSON file for each paper in the original dataset has been made.

Mapping to RDF. In a second step, the extended JSON files are converted to RDF. To make this transformation adaptable, transparent and reusable, a mapping document was created that contains rules on how each element in the JSON can be mapped on a corresponding semantic output value. The mapping document was created with YARRRML, a human-readable text-based representation that can be used to represent RDF Mapping Language (RML) rules [12]. A part of the YARRRML document is being displayed in Listing 1.1. Such a YARRRML document usually consists of two main parts: a part listing all the used prefixes (lines 1 till 6) followed by a part describing the actual semantic mapping (lines 7 till 25). Within this mapping section, we describe 1) the source of the input file (e.g., the file path to the JSON-formatted paper at line 10), 2) the subject mapping (here in lines 11 till 15, we took the DOI from the JSON file to create a unique identifier in and 3) all possible predicate-object relations. In the YARRRML example in Listing 1.1, two such predicate-object relations are defined. One specifying each paper as a `fabio:Work` at line 17, and a second predicate-object relation described by a function which checks whether or not the paper is a `fabio:JournalArticle` (lines 19 till 24). This simple example shows how concepts of fabio (FRBR-aligned Bibliographic Ontology), describing entities that are published or potentially publishable [20], can be mapped using the values of the extended JSON files. The eventually used mapping script outlines a lot more concepts from different domains and ontologies.

As this YARRRML document is only a human-readable text-based representation of RML rules, we have to convert this YARRRML document to an RML document by using the YARRRML Parser[3]. Note that it is possible to write RML rules in this setup directly, but by using YARRRML, we created the ability to let others extend the mapping documents with reduced human effort and without requiring a lot of specific knowledge about semantic web formats.

```
1   prefixes:
2     idlab-fn: "http://example.com/idlab/function/"
3     fabio: "http://purl.org/spar/fabio/"
4     grel: "http://users.ugent.be/~bjdmeest/function/grel.ttl#"
5     ...
6
7   mappings:
8     Realization:
9       sources:
10        - ['tmp/transform_data.json~jsonpath', '$']
11      s:
12        function: grel:array_join
13        parameters:
14          - [grel:p_array_a, "http://dx.doi.org/"]
15          - [grel:p_array_a, "$(doi)"]
16      po:
17        - [a, fabio:Work]
18        - p: a
19          o:
20            - function: idlab-fn:decide
21              parameters:
22                - [idlab-fn:str, $(type)]
23                - [idlab-fn:expectedStr, "journal-article"]
24                - [idlab-fn:result, fabio:JournalArticle]
25      ...
```

Listing 1.1. YAML script to represent the relation between already existing ontological concepts and the JSON values.

The RMLMapper [6] takes both the extended JSON files and the RML document generated using the above YARRRML document as input and produces a set of N-Triples for each paper. Finally, all these N-Triple files were concatenated together to form a single large knowledge graph. A snippet extracted from such an N-Triple file, but represented in a turtle format to improve readability, is provided in Listing 1.2. As defined in the YARRRML subject mapping, all papers are described by a single URI, which is the DOI. All the code used and the input files required in order to perform this conversion are available open-source on Github[4].

3.3 Knowledge Graph Availability

In order to make the data FAIR, we have set up a Linked Data Fragments (LDF) server with a HDT back-end to expose a Triple Pattern Fragments (TPF) interface at the following URL: https://query-covid19.linkeddatafragments.org/. This allows users to query the constructed knowledge graph in a comfortable

---

[3] https://github.com/rmlio/yarrrml-parser.
[4] https://github.com/GillesVandewiele/COVID-KG.

```
@prefix doi: <http://dx.doi.org/> .
@prefix fabio: <http://purl.org/spar/fabio/> .
@prefix COVID19: <http://idlab.github.io/covid19#> .
@prefix orcid: <https://orcid.org/> .
@prefix spar: <http://purl.org/spar/>
@prefix foaf: <http://xmlns.com/foaf/0.1>
@prefix dbr: <http://dbpedia.org/resource/>

doi:10.3390/molecules21121629 a fabio:Work.
doi:10.3390/molecules21121629 a fabio:JournalArticle.
doi:10.3390/molecules21121629 COVID19:hasConcept dbr:Amide.
doi:10.3390/molecules21121629 COVID19:hasConcept dbr:3i.
doi:10.3390/molecules21121629 spar:pro/creator orcid:0000−0002−8523−6340.
orcid:0000−0002−8523−6340 a foaf:Person.
orcid:0000−0002−8523−6340 foaf:surname "Jane".
orcid:0000−0002−8523−6340 foaf:firstName "Smith".
doi:10.3390/molecules191219292 spar:cito/isCitedBy doi:10.3390/molecules21121629.
```

Listing 1.2. Turtle representation of the N-TRiple file for http://dx.doi.org/10.3390/molecules21121629 extracted by the RML mapper.

and scalable fashion. Moreover, the use of TPF guarantees the availability of the resource [25]. We used Comunica [22] to set up the LDF server. Comunica is a query engine platform that offers a plethora of modules for users to design a query engine that fits their needs. The example query which searches for the concept "protein" inside our knowledge graph returns 100 results containing both the DOI identifier and publisher in 12 s.

4   External Resources

As shown in Fig. 2, in addition to the information available in the provided JSON and CSVs, we link the papers to external resources to enrich our dataset with more information. In this section, we give more details on these external resources, as well which fields of the JSON schema visualized in Fig. 1 were used to obtain this additional information.

DBpedia [3]. DBpedia resources were linked to several concepts of each paper. On the one hand, the country, institute and research labs of each of the authors were linked to their respective DBpedia resources. Heuristics were used to check if the DBpedia URI exists by concatenating the domain name with the JSON value or comparing the JSON value with the results of DBpedia lookup [21]. On the other hand, DBPedia Spotlight [15] was used to identify general terms in the title, abstract and body of the paper. A new JSON key `hasConcept` was added for each text block, with all the found values in a list. This list indicated all the DBpedia concepts that were detected within that block.

BioPortal [16]. The title, abstract and body text were processed with the BioPortal annotation tool in order to identify concepts. This annotator returns annotations, especially for biomedical text with classes from biomedical ontologies. We limited the scope of concepts to the COVID-19 surveillance

(COVID19), Coronavirus Infectious Disease (CIDO) and Influenza (FLU) ontologies. Similar to the DBpedia concepts, a `hasConcept` JSON key was added in each text block to list all these newfound concepts.

CrossRef [13]. The citation information was often provided in the form of titles and the authors' abbreviated names. In order to link these papers to their respective DOI, the CrossRef API was used. Moreover, the CrossRef API provides additional metadata, such as the authors' full name and the journal in which it was published. These papers were then linked together using the `isCitedBy` or `cites` predicate between two DOIs.

ORCID [11]. As the author information both obtained by the CrossRef API and provided in the original dataset was sometimes limited or missing, each of the authors was linked to their respective ORCID identifier, when possible. Using this identifier, the ORCID API was used to provide additional information, such as the institution or lab they are working for, which, in turn, could be linked to their DBpedia resources.

5   Applications

Having the original structured data formats in the form of a Knowledge Graph, with additional knowledge linked from external resources, allows for the creation of applications that were a priori not possible. In this section, we discuss some potential applications that could facilitate the analysis of researchers studying the body of COVID-19 literature. All the examples used in this application section are just chosen for information purposes. They merely illustrate the possible applications that can be used by experts within this field to get more insights within the COVID domain.

5.1   Network Analysis

Network analysis is a powerful tool that can reveal interesting patterns hidden in graph datasets. In order to perform such an analysis, we converted our knowledge graph in a regular directed graph by retaining only citation information. This removes the multi-relational aspect of the graph. The conversion is needed as the current network analysis tools can not deal with different labeled edges. The newly constructed graph consists of nodes that represent the papers and edges between these nodes that represent citations from one paper to the other. Below we describe how network analysis on a graph consisting of information on COVID-19 literature could allow, for example, to find communities of related publications or to identify influential scientific contributions. More detailed examples of network analysis, network visualizations and analysis results highlighting central papers can be found online[5].

---

[5] www.kaggle.com/iljara/covid-19-knowledge-graph-a-network-analysis.

Detecting Communities. Clustering reveals information on how tightly some groups of publications are interconnected through citations. Such groups often create communities or even cliques, where every paper cites every other paper within that group. The clustering coefficient can be measured on a local and global scale. In the former case, clustering is considered within each node's neighborhood separately, while the latter is an average over the entire network. The article with the highest local clustering is: *Human Bocavirus infection in hospitalized children during winter*. The total global clustering coefficient for our citation network is 0.024009 ($\pm$0.007078).

Identifying Influential Publications. In order to identify the most influential publications, we study the node centrality. Centrality can be interpreted as a measure of a node's importance. Hence, we conjecture that an influential paper is highly connected, thus, *central* in one way or another. Several metrics can be used to estimate the centrality of a node.

A first, straightforward, metric uses the node degree as a proxy for centrality. The in-degree indicates the number of articles citing a specific article. The out-degree counts the number of articles cited from a specific article. Figure 3 shows the distribution of the total degree (i.e., in-degree plus out-degree) of a part of our knowledge graph and a visualization of the corresponding network. In this example, the highest connected paper stands out visibly. It should be noted that these degrees do not correspond to the number of citations or cited articles provided by, for example, Google Scholar, but rather to the number of citations within the body of COVID-19 literature. The article with the highest number of links is: *Human rhinoviruses: the cold wars resume*. PageRank and Hyperlink-Induced Topic Search (HITS) can be seen as more sophisticated variants of these measures. They also evaluate the importance of a node based on the rank of its neighbors.

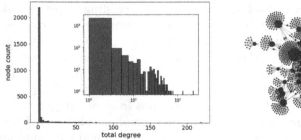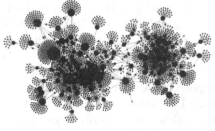

Fig. 3. (a) Distribution of the total degree within the sample network. The inset shows the same data on a double-log scale. (b) Network visualization, the node size is proportionate to the node's total degree.

One other exciting metric is closeness centrality, which can be used to identify influential publications within the network clusters or communities, as defined earlier. Related to the concept of closeness is that of betweenness, which measures how much a certain node acts as a connector or joint. Nodes with high betweenness centrality are often review papers or interdisciplinary works that bridge between several research areas. The article with the highest betweenness centrality in our knowledge graph is: *A novel pancoronavirus RT-PCR assay: frequent detection of human coronavirus NL63 in children hospitalized with respiratory tract infections in Belgium.*

5.2 Embedding Concepts

The knowledge graph created from the CORD-19 dataset excels in representing structured data. However, the underlying symbolic nature of this triple-based format usually makes knowledge graphs hard to manipulate and impractical for machine learning systems. To tackle this issue, knowledge graph embeddings have been proposed, where components of a knowledge graph, including entities and relations, are embedded into continuous vector spaces. The most common technique to build such embeddings is RDF2Vec [18]. In this section, we highlight two applications which benefit from creating these embeddings.

Retrieving Nearest Neighbors. When creating embeddings for the paper nodes within our knowledge graph, RDF2Vec ensures that papers that are related or similar have closely related embedded vectors. If we display these vectors in a two-dimensional space, we see that that vectors of similar or related papers are visually near each other. This allows us to search for papers that are close to or highly associated with a given paper of interest. Merely searching the nearest neighbors of a given paper embedding can already be a useful application. For example, from the generated RDF2Vec embeddings, we have searched for the nearest neighbors of the following paper: *SARS-related Virus Predating SARS Outbreak, Hong Kong* (http://dx.doi.org/10.3201/eid1002.030533) and found that based on the embeddings the following papers are closely related:

```
Development and Evaluation of Novel Real-Time Reverse Transcription-PCR
Assays with Locked Nucleic Acid Probes Targeting Leader Sequences of
Human-Pathogenic Coronaviruses
(http://dx.doi.org/10.1128/jcm.01224-15)

Crystal structure and mechanistic determinants of SARS coronavirus
nonstructural protein 15 define an endoribonuclease family
(http://dx.doi.org/10.1073/pnas.0601708103)

Antigenic and Immunogenic Characterization of Recombinant Baculovirus-
Expressed Severe Acute Respiratory Syndrome Coronavirus Spike Protein:
Implication for Vaccine Design
(http://dx.doi.org/10.1128/jvi.00083-06)
```

They are all highly correlated due to the performed Sars-Cov experiments.

Advanced Clustering. By searching for the nearest neighbors of all given papers, clusters can be identified that indicate groups of papers related to each other, e.g., referring or citing one specific paper or by sharing similar defined concepts. Some experiments with k-Means clustering were performed to show this application potential. The results with a predefined k of 20 are visualized in Fig. 4(a). Papers closely related to each other concerning their created embedding will have the same cluster label. Clustering is, therefore, a lot more informative because the embeddings take into account the whole neighborhood of the node. In the previous section, the network analysis only considered the citation links between nodes to define possible clusters.

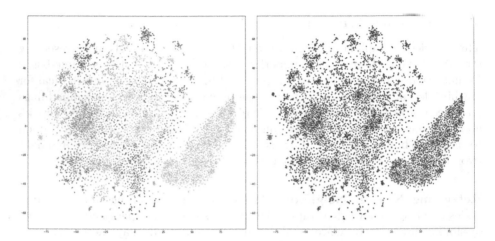

Fig. 4. (a) Clustering outcome by k-means (k = 20) on the generated RDF2Vec embeddings, (b) Example of a selected cluster.

Clustering papers together is beneficial to find groups of related research. However, the embeddings give limited insight into why these papers are grouped together. To increase the understanding of the generated clusters, we experimented with interpretable node classification techniques such as MIND-WALC [24]. Concrete labels based on the defined clusters are assigned to the papers to shift the dataset to a binary classification task. In such a task, we try to discriminate the clustered papers from all other papers in our dataset using neighborhood graph walks. Based on the example visualized in Fig. 4(b), MINDWALC outputs the walks, which are the most discriminating in classifying the selected cluster concerning all other papers. The output is visualized in Listing 1.3 and shows that for the specified cluster, the DBpedia concepts *Aerosol*, *Airborne_disease* and *Hand_washing* can be found following four links starting from any of the papers within this cluster.

```
{
  ('http://dbpedia.org/resource/Aerosol', 4),
  ('http://dbpedia.org/resource/Airborne_disease', 4),
  ('http://dbpedia.org/resource/Airborne_disease', 6),
  ('http://dbpedia.org/resource/Antiseptic', 6),
  ('http://dbpedia.org/resource/Direct_Contact', 4),
  ('http://dbpedia.org/resource/Engineering_controls', 6),
  ('http://dbpedia.org/resource/Hand_sanitizer', 6),
  ('http://dbpedia.org/resource/Hand_sanitizer', 8),
  ('http://dbpedia.org/resource/Hand_washing', 4),
```

```
('http://dbpedia.org/resource/Hand_washing', 6),
('http://dbpedia.org/resource/Hospital-acquired_infection', 4),
('http://dbpedia.org/resource/Huy', 8),
('http://dbpedia.org/resource/Hypochlorite', 6),
('http://dbpedia.org/resource/Methicillin-resistant_Staphylococcus_aureus', 4),
('http://dbpedia.org/resource/Methicillin-resistant_Staphylococcus_aureus', 6),
('http://dbpedia.org/resource/Mycobacterium_tuberculosis', 4),
('http://dbpedia.org/resource/Nebulizer', 6),
('http://dbpedia.org/resource/Nebulizer', 8),
('http://dbpedia.org/resource/Particulates', 2),
('http://dbpedia.org/resource/Particulates', 4),
('http://dbpedia.org/resource/Personal_protective_equipment', 4),
('http://dbpedia.org/resource/Personal_protective_equipment', 6),
('http://dbpedia.org/resource/Respirator_fit_test', 6),
('http://dbpedia.org/resource/Seto_Inland_Sea', 6),
('http://dbpedia.org/resource/Transmission_(medicine)', 6),
('http://dbpedia.org/resource/Tuberculosis_management', 6)
}
```

Listing 1.3. MINDWALC results for the classification of the nodes in the cluster defined in Fig. 4(b).

6 Conclusion

The original CORD-19 dataset delivered a mass of information regarding the COVID-19 pandemic. By transforming the data and available metadata into a knowledge graph, a wide range of useful applications are made possible. The procedure used in this study is generic in such a way that it can be used as a guideline to enrich any structured dataset and transform it into a knowledge graph. New information can be integrated quickly and the whole procedure is transparent as minimal knowledge is required to extend the currently available graph further.

Some potential directions were provided in this paper to show the graph's application potential. Hence, precise research questions must be defined for such applications as this is an essential condition to have insightful results. The created knowledge graph is only as good as the applications built on top of it. Besides the availability of an endpoint, there is still a need for front-ends that allow non-technical people, which many biomedical researchers are, to interact with this resource and to reveal its connected knowledge.

7 Code and Dataset Availability

Both the knowledge graph and the code to enhance and transform the original CORD-19 dataset are made available and is summarized on the resource website http://covid-kg.tools:

- The dataset is available on Kaggle and can be reached by the following: http://doi.org/10.34740/kaggle/dsv/1166450. Tutorial notebooks on how to interact with the knowledge graph using python, how to generate embeddings and how to apply network analysis are also available under the kernels tab.
- The scripts on how the knowledge graph was constructed can be found on Github: https://github.com/GillesVandewiele/COVID-KG
- The TPF interface through which the created knowledge graph can be easily accessed: https://query-covid19.linkeddatafragments.org/

- All RML and YARRRML tools are publicly available: https://rml.io/tools/
- additionally, the embeddings where generated using pyRDF2Vec, which is available open-source: https://github.com/IBCNServices/pyRDF2Vec, as well MINDWALC: https://github.com/IBCNServices/MINDWALC.

References

1. Allen Institute For AIF: Covid-19 open research dataset challenge (cord-19). https://www.kaggle.com/allen-institute-for-ai/CORD-19-research-challenge
2. Andersen, K.G., Rambaut, A., Lipkin, W.I., Holmes, E.C., Garry, R.F.: The proximal origin of sars-cov-2. Nat. Med. **26**(4), 450–452 (2020)
3. Auer, S., Bizer, C., Kobilarov, G., Lehmann, J., Cyganiak, R., Ives, Z.: DBpedia: a nucleus for a web of open data. In: Aberer, K., et al. (eds.) ASWC/ISWC -2007. LNCS, vol. 4825, pp. 722–735. Springer, Heidelberg (2007). https://doi.org/10.1007/978-3-540-76298-0_52
4. Auer, S., Kovtun, V., Prinz, M., Kasprzik, A., Stocker, M., Vidal, M.E.: Towards a knowledge graph for science. In: Proceedings of the 8th International Conference on Web Intelligence, Mining and Semantics, pp. 1–6 (2018)
5. Johns Hopkins Coronavirus Resource Center: Covid-19 dashboard by the center for systems science and engineering (csse) at johns hopkins university (jhu). https://gisanddata.maps.arcgis.com/apps/opsdashboard/index.html#/bda7594740fd40299423467b48e9ecf6
6. Dimou, A., De Meester, B., Heyvaert, P., Verborgh, R., Latré, S., Mannens, E.: RMLMapper: a tool for uniform Linked Data generation from heterogeneous data
7. Dimou, A., Vander Sande, M., Colpaert, P., Verborgh, R., Mannens, E., Van de Walle, R.: RML: a generic language for integrated RDF mappings of heterogeneous data. In: Proceedings of the 7th Workshop on Linked Data on the Web, vol. 1184 (2014)
8. Domingo-Fernandez, D., et al.: Covid-19 knowledge graph: a computable, multimodal, cause-and-effect knowledge model of covid-19 pathophysiology. BioRxiv (2020)
9. Färber, M.: The microsoft academic knowledge graph: a linked data source with 8 billion triples of scholarly data. In: Ghidini, C., et al. (eds.) ISWC 2019. LNCS, vol. 11779, pp. 113–129. Springer, Cham (2019). https://doi.org/10.1007/978-3-030-30796-7_8
10. Guoqian, J., Harold Solbrig, F.T.: Cord-19-on-fhir - semantics for covid-19 discovery. https://github.com/fhircat/CORD-19-on-FHIR
11. Haak, L.L., Fenner, M., Paglione, L., Pentz, E., Ratner, H.: Orcid: a system to uniquely identify researchers. Learn. Publ. **25**(4), 259–264 (2012)
12. Heyvaert, P., De Meester, B., Dimou, A., Verborgh, R.: Declarative rules for linked data generation at your fingertips!. In: Gangemi, A., et al. (eds.) ESWC 2018. LNCS, vol. 11155, pp. 213–217. Springer, Cham (2018). https://doi.org/10.1007/978-3-319-98192-5_40
13. Lammey, R.: Crossref text and data mining services. Science Editing (2015)
14. McCusker, J.P., Dumontier, M., Yan, R., He, S., Dordick, J.S., McGuinness, D.L.: Finding melanoma drugs through a probabilistic knowledge graph. PeerJ Comput. Sci. **3**, e106 (2017)
15. Mendes, P.N., Jakob, M., García-Silva, A., Bizer, C.: Dbpedia spotlight: shedding light on the web of documents. In: Proceedings of the 7th International Conference on Semantic Systems, pp. 1–8 (2011)

16. Noy, N.F., et al.: Bioportal: ontologies and integrated data resources at the click of a mouse. Nucleic Acids Res. **37**, W170–W173 (2009)
17. Preusse, M.: COVID-19 Knowledge Graph (2020). https://covidgraph.org
18. Ristoski, P., Paulheim, H.: RDF2Vec: RDF graph embeddings for data mining. In: Groth, P., et al. (eds.) ISWC 2016. LNCS, vol. 9981, pp. 498–514. Springer, Cham (2016). https://doi.org/10.1007/978-3-319-46523-4_30
19. Schlichtkrull, M., Kipf, T.N., Bloem, P., van den Berg, R., Titov, I., Welling, M.: Modeling relational data with graph convolutional networks. In: Gangemi, A., et al. (eds.) ESWC 2018. LNCS, vol. 10843, pp. 593–607. Springer, Cham (2018). https://doi.org/10.1007/978-3-319-93417-4_38
20. Shotton, D., Peroni, S.: Fabio, the FRBR-aligned bibliographic ontology (2011)
21. Steenwinckel, B., Vandewiele, G., De Turck, F., Ongenae, F.: Csv2kg: Transforming tabular data into semantic knowledge. SemTab, ISWC Challenge (2019)
22. Taelman, R., Van Herwegen, J., Vander Sande, M., Verborgh, R.: Comunica: A Modular SPARQL Query Engine for the Web. In: Vrandečić, D., et al. (eds.) ISWC 2018. LNCS, vol. 11137, pp. 239–255. Springer, Cham (2018). https://doi.org/10.1007/978-3-030-00668-6_15
23. DSG-UPB: Covid19ds: RDF file generation is based on papers related to the covid-19 and coronavirus-related research (2020)
24. Vandewiele, G., Steenwinckel, B., Ongenae, F., De Turck, F.: Inducing a decision tree with discriminative paths to classify entities in a knowledge graph. In: SEPDA2019, the 4th International Workshop on Semantics-Powered Data Mining and Analytics, pp. 1–6 (2019)
25. Verborgh, R., et al.: Triple pattern fragments: a low-cost knowledge graph interface for the web. J. Web Semant. **37**, 184–206 (2016)

In-Use Track

Crime Event Localization
and Deduplication

Federica Rollo$^{(\boxtimes)}$ and Laura Po

'Enzo Ferrari' Engineering Department, Via Vivarelli, 10, 42125 Modena, Italy
{federica.rollo,laura.po}@unimore.it

Abstract. Crime analysis is an approach for identifying patterns and trends in crime events, while information extraction is the task of extracting relevant information from unstructured data. If crime reports are not directly available to the public, a possible solution is to derive crime information published in newspaper articles.

This paper aims at extracting, localizing, deduplicating, and visualizing crime events from online news articles. This work demonstrates how crime-related information can be obtained from newspapers and exploited to create a consistent database of crime events with an automatic process. The approach employs a Named Entity Recognition (NER) algorithm to retrieve locations, organizations and persons and a mapping phase to link entities to Linked Data resources. The date of the event is retrieved through the temporal expressions extraction and normalization. For duplicate detection, an approach analyses and combines crime category, description, location, and crime event date to identify which news articles refer to the same event. The approach has been successfully applied in the Modena province (Italy), focusing on eleven types of crime happen from 2011 till now. The flexibility of the approach allows it to be easily adapted to other cities, regions, or countries and also to other domains.

Keywords: Crime analysis · Crime mapping · News extraction · Document similarity · Duplicate detection

1 Introduction

Focusing resources on high-crime places, high-rate offenders, and repeated victims can help police effectively reducing the crime rate in their communities. Police can take advantage of knowing when, where, and how to focus its resources, as well as how to evaluate the effectiveness of their strategies. Sound crime analysis is paramount to this success. Crime analysis is not merely crime events counting; it is an in-depth examination of the different criminogenic factors (e.g., time, place, socio-demographics) that help understanding why the crime occurs. Data-driven policing and associated crime analysis are still dawning. The use of Geographic Information System (GIS) techniques helps crime

© Springer Nature Switzerland AG 2020
J. Z. Pan et al. (Eds.): ISWC 2020, LNCS 12507, pp. 361–377, 2020.
https://doi.org/10.1007/978-3-030-62466-8_23

analysis and allows localizing crimes to identify the high-risk areas. Unfortunately, in some countries (e.g. in Italy), authorities do not provide free access to updated datasets containing information about crimes happening in the cities. Extracting crime events from news articles published on the web by local newspapers can help overcome the lack of crime up-to-date information.

Several countries provide statistics on crime, but they are often available with some delay. In most of the cases, they are provided as aggregated data, not as single crime events. In the UK, open data about crime and policing are available at street-level with a delay of 2 months[1]. In Italy, the reports of the Italian National Institute of Statistics (ISTAT)[2] provide a clear picture of the types of crime happen in each province during the year. The information provided is aggregated by time and space and become available after (at least) one year from the crime event happening. Based on these official statistics, it is not possible to perform an up-to-date analysis of the local situation in each neighborhood. Newspapers instead provide reliable, localized, and timely information (the time delay between the occurrence of the event and the publication of the news does not exceed 24/48 h). The main drawback is that newspapers do not collect and publish all the facts related to crimes, but only the most relevant, i.e., the ones that arouse the interest of the readers. Therefore, there is a percentage of police reports that will not be turned into news and is lost.

This paper presents a data ingestion approach for extracting crime data from news and enriching them with semantic information. The strategy employs several techniques: crime categorization, named entity extraction, linked data mapping, geolocalization, time expression normalization, and de-duplication. From the best of our knowledge, the integration of multiple techniques, previously used in different contexts, for solving various sub-problems into a common framework in order to perform crime analysis is a novelty.

The method has been applied and tested in the Modena province, a 2,688 Km^2 area populated by more than 700 000 inhabitants and located in the Emilia-Romagna region in Italy. On 13000 reports, collected from 2011 to now (May 2020), the approach was able to geolocalize almost 100% of the crime events and normalize the time expressions on 83% of the news articles. The results produced allows performing crime mapping studies and the identification of crime hot spots in semi real-time. Some preliminary visualizations of these results are shown through a web application.

The paper is organized as follows: Sect. 2 introduces related work; Sect. 3 describes the general approach to extract and analyze news articles; in Sect. 4 the method implemented in the Modena province is described in detail and then evaluated in Sect. 5. Section 6 shows the effectiveness of our approach and demonstrates its scalability. In the end, in Sect. 7, conclusions and possible future work are sketched.

---

[1] On the police open data portal https://data.police.uk/, it was possible to download data about March 2020 on the 21st May 2020.

[2] https://www.istat.it/en/.

2   Related Work

In the last decades, methods for news content extraction have gained relevant interest [4,20], and several online platforms have been developed to visualize the results of the extraction, such as the Europe Media Monitor (EMM) News-Explorer[3] and NewsBrief[4], the Thomson Reuters Open Calais[5], and the Event Registry[6]. These platforms download news articles from the web and exploit different language processing algorithms to detect entities, group news articles into clusters according to their topics [7,12]. In particular, a lot of scientific research is devoted to crime data mining, and new software applications have been created for detecting and analyzing crime data. In [8], an approach is described to extract important entities from police narrative reports written in plain text by using a SOM (self-organizing map) clustering method. Crime analysis methods are applied to find trends [19] or to predict crime events [9], by using neural networks, Bayesian networks, and algorithms as K-nearest-neighbour, boosted decision tree, K-means. An interesting example of crime analysis and mapping in the city of Chicago[7] is explained in [3]. In this case, crime data are extracted from the Chicago Police Department's CLEAR (Citizen Law Enforcement Analysis and Reporting) system, composed of relational databases that allow law enforcement officials to cross-reference available information in investigations and to analyze crime patterns using a geographic information system (GIS). To protect the privacy of crime victims, addresses are shown at the block level only, and specific locations are not identified.

In Italy, two cities provide updated crime datasets on their open data portals. Torino AperTO Open Data Portal includes information about the type of crime, the location and the date of the crime events occurred, but with a delay of two years[8], while Trento provides the annual burglary rate of the previous year[9]. In both cases, data are not timely, and, in Trento, they are also aggregated. In Italy, if it is not possible to get direct access to police reports, the only timely sources of crime data are newspapers that are freely available online for everyone. Our approach aims at extracting crime events from news articles, structuring the information, geolocating them, and linking them to Linked Data resources. The collected data are then published online in a web application to provide a real-time overview and some analysis of the crime situation in the Modena province.

3   The Approach

The approach that we have selected to extract semantic information starting from the news articles published on the web consists of 7 phases. It is a general

---

[3] https://emm.newsexplorer.eu/.
[4] https://emm.newsbrief.eu/.
[5] https://www.crunchbase.com/organization/opencalais.
[6] https://eventregistry.org/.
[7] https://data.cityofchicago.org/Public-Safety/Crimes-Map/dfnk-7re6.
[8] http://aperto.comune.torino.it/.
[9] European Data Portal https://www.europeandataportal.eu/it.

approach that can be applied to any information content that describes events. The phases should be executed in sequence for each news, but different news articles can be processed in parallel.

The first phase is the data extraction in which information of interest that is published on the web is harvested. Then this content is labeled and structured to be stored in a database and to be semantically annotated. Some web content may already expose a structure. For example, HTML pages encoded with the Document Object Model (DOM), have a tree structure wherein each node is an object representing a part of the document.

The second phase is the Named Entity Extraction. This phase is crucial since it allows us to identify persons, organizations, places, and temporal expressions in the text of the news. The correct identification and extraction of entities are of great importance not only for enriching the crime description but also because identified entities act as annotations for the crime event.

The third phase is the Linked Data Mapping. This phase maps the entities into Linked Datasets. In particular, persons, organizations, and locations are linked to URI.

The fourth phase is the categorization of the criminal event. This phase is crucial to map a news w.r.t. a type of crime. Given some pre-categorized news, i.e., annotated training data, machine learning algorithms can be applied on uncategorized news to assign them a type of crime. The entities extracted in the previous two phases can be exploited to enhance the results of this phase.

The fifth phase is the geographical localization; in this phase, the entities that have been identified as locations are processed to be georeferenced. Different methods can be applied; moreover, if a location is not specified in the news, organizations can also be exploited to geolocate the event.

The sixth phase is the normalization of temporal expressions. In this phase, the date of the news published on the web can be revised to identify the exact time of the crime. By analyzing the news text, temporal expressions can be identified that allow normalizing the date (for example, words like "yesterday", "two days ago", "this morning" are identified as temporal expressions).

The last phase is the identification of duplicates. This phase should be applied not only in the case that the input sources are more than one since it is possible to find the same event described in more news articles also within the same newspaper where updates about one crime event are published along time. The duplicate detection phase is carried out by identifying possible duplicates and then making a comparison on the news text to confirm that they are duplicates. At the end of this phase, after identifying the duplicates, it is also possible to merge the information of the criminal events to enrich the information that will be stored on the database.

The use of semantic technologies is a key point in the presented approach for adding knowledge about the crime events, geolocalizing crimes, and performing deduplication. The NER combined with Linked Data mapping allows retrieving entities and associating them to stable URIs. Besides, also the deduplication takes advantage of the semantic information extracted in the previous phases.

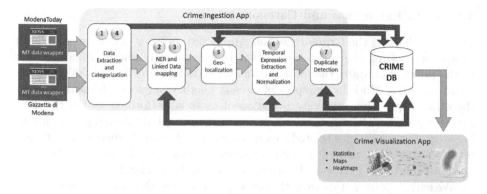

Fig. 1. The method implemented to extract, store, analyze, and visualize the news articles for Modena province (the numbers in the circles represent the phases described in Sect. 3).

4 Modena Crime Ingestion

The approach described in Sect. 3 has been implemented to extract crime-related information in the Modena province. We extended our previous work [13,15] that collected and analyzed the news articles extracted from one newspaper related to thefts. Currently, we ingest news coming from the two most popular local newspapers, "ModenaToday" (MT) and "Gazzetta di Modena" (GM), related to eleven types of crimes: theft, attack, drug dealing, evasion, fraud, abuse, murder, robbery, money laundering, kidnapping, and sexual violence. More than 13000 reports have been collected from 2011 to now (May 2020) using this approach. There exist other 3 minor online newspapers, however integrating them will not change substantially the results since they cover around the 5% of the total news. Since the scope of the news published in these newspapers is to provide information related to single events, a single-crime-event-per-document assumption is made.

Figure 1 displays the phases of our approach applied in the Modena scenario; some phases are performed together because of the particular structure of the data taken into consideration. The Crime Ingestion App aims at extracting, parsing and storing information of the news articles into a PostgreSQL database, called CRIME DB, whose structure is discussed in detail in the following; while, the Crime Visualization App displays the information stored in the DB trough interactive crime maps, heatmaps to discover the high-crime areas, and statistics to identify trends.

The Crime Ingestion App has been implemented through a Java application by using suited libraries and APIs, while, the Crime visualization App is a web application implemented in Python[10].

---

[10] The code of both applications is open source and is available in a github repository https://github.com/federicarollo/Crime-event-localization-and-deduplication.

4.1   Data Extraction and Categorization

The first phase of the Crime Ingestion App is data extraction. This phase is in charge of crawling the web page content to extract semantic information such as the publication date, the location, and a textual description of the event.

In our case, Modena newspapers already classify news articles according to the crime type. Thus, the categorization of news is performed within the data extraction phase. Data extraction takes in input the url of the web page containing a list of news articles related to a specific type of crime[11], then automatically retrieves the URLs related to each news, accesses each URL and extracts information from the HTML tags by using the java web crawler Jsoup.

We extract information from the news exploiting the Document Object Model and taking advantage of the HTML tags of the web page. In the Modena newspapers taken into consideration, the type of crime, the title, the subtitle (description), the date and time of the publication of the news, and the textual information can be harvested directly from the newspapers with specific HTML tags. In some cases, also the location is reported; this information is usually the name of the city or may contain the area and the address. All this information is stored in the "news" table and then refines in the following phases. The structure of the CRIME DB used to store crime-related news is displayed in Fig. 2. The database is a PostgreSQL database that uses the extension PostGIS to store geospatial data. In the "news" table all the information extracted from the news is stored.

4.2   NER and Linked Data Mapping

We implemented the NER phase by using Tint[12] (The Italian NLP Tool) [10], an open-source tool for NLP of Italian texts. Tint is a collection of modules customized for the analysis of text in Italian language and based on the Stanford CoreNLP. In [10] Tint was compared with three other NLP tools for Italian language, (Tanl [5], TextPro [11] and TreeTagger [16]) and it reported the higher values of speed, precision, and F-measure for the NER task (in particular, recognition of entities including persons, organizations, and locations, which is our scope). The NER module of Tint uses the Conditional Random Field (CRF) sequence tagger included in the Stanford CoreNLP, and it is trained on the ICAB dataset, which contains 180 K words taken from the Italian newspaper "L'Adige" [10].

We used Tint to detect persons, organizations, locations, and time expressions. Each entity recognized by the NER algorithm is stored as an instance of the "entity" table and linked to the news in the "news" table (see Figure 2). For the entities retrieved, we perform a mapping w.r.t. linked data, in particular DBpedia and Linked Geo Data. For each person, organization and location retrieved, we search for a corresponding entity in DBpedia. An http request

---

[11] An example is available at https://gazzettadimodena.gelocal.it/ricerca?query=furti where *furti* (theft) is a type of crime.
[12] http://tint.fbk.eu/.

is sent to the link obtained concatenating the base URI of DBpedia[13] to the name of the entity formatted with mixed case with underscores. If the request is successful, the URI is stored in the "entity" table. For each location and organization, we search for the corresponding resource in Linked Geo Data [17]: a SPARQL query selects all the spatial resources that are located in the area of Modena province and looks for the best matches. If a municipality is defined for the news, the query performed retrieves the Linked Geo Data resources of that municipality.

4.3 Geolocalization

The GPS coordinates of the crime locations are retrieved by using the Open-StreetMap API, and in particular, Nominatim[14]. Geolocation is an iterative process; it starts by evaluating whether municipality and addresses are populated in the DB and trying to geolocalize them. If the municipality or address is not present or if their geolocalization failed[15], the process starts exploring the entities that have been extracted with NER and are stored in the entity table. When multiple locations are detected, we consider the one with higher frequency. If the geolocalization of locations failed, organizations are explored. If more locations/organizations with the same frequency are available, we take the first occurrence in the news. This does not affect significantly the result of the geolocalization; indeed, we checked manually some cases, and we discovered that, in the majority of the cases, locations in a news were all close together. In the end, the GPS coordinates and the related address are stored in the "news" table.

4.4 Temporal Expression Extraction and Normalization

To detect when the crime event described in the news happen, an algorithm of temporal expression extraction and normalization is applied to the concatenation of title, sub-title, and text attributes. We use the HeidelTime temporal tagger [18] and, in particular, its implementation included in Tint for the news document type. This tool can extract temporal expressions from documents in natural language and normalize them according to the TIMEX3 annotation standard. Using the date of publication of the news ("publication_datetime") and the result of the temporal expression normalization, the date of the crime event is calculated and stored in the "event_datetime" attribute in the "news" table. When multiple event dates are detected, we select the one with higher frequency.

---

[13] http://dbpedia.org/resource/.

[14] https://nominatim.openstreetmap.org/.

[15] We use the function *OpenStreetMapUtils.getInstance().getCoordinates(location)* where *location* is a string that can be generated by the municipality and the address, or the entity name retrieved from the DB. This function provides the latitude and the longitude of the location. The success of this function depends on how the address is stored in Open Street Map and how the location is reported in the news.

In case of multiple event dates with the same frequency, we consider the date closest to the publication date.

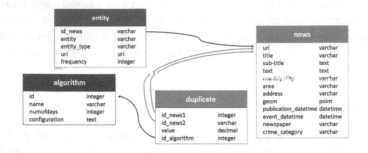

Fig. 2. The structure of the CRIME DB.

4.5 Duplicate Detection

Considering two or more newspapers of the same geographic area, there is a high probability that the news articles describing the same event are published in both newspapers the same day or a few days later. Besides, the same crime-related event could be described more times in the same newspaper to provide updates. Duplicate detection (also known as *deduplication*) is a fundamental task when we are ingesting news articles. All the news articles identified as *duplicates* are collected: in the "duplicate" table, the links to the identifiers of the two news articles, the value of their similarity and the identifier of the algorithm used to find them are stored. In the "algorithm" table, instead, the features and the parameters used by the duplicate detection algorithm are stored in the "configuration" attribute. The duplicate detection is performed in two steps: (1) the reduction of the number of news to be compared, and (2) the comparison of the news articles through similarity algorithms. After the identification of *duplicates*, a merge of the information of the two news is made.

Reduction of the Number of Pairwise Comparisons. This task is also known with the name of blocking. The identification of the event date allows us to compare only the news with the same event date to search for duplicates. If no event date is available, the comparisons are performed by using a date slot which considers the publication date. We assume that news articles related to the same crime event might be published with a few days of delay in different newspapers or the same newspaper. Therefore, we do not need to compare a huge amount of news articles. Besides, the blocking technique we applied compares only news articles classified as the same type of crime and related to crimes that happened in the same municipality, including cases in which municipality is not specified.

Comparison of News Articles Through Similarity Algorithms. The similarity between two news articles takes into account the semantic information extracted in the previous phases and, in particular, the municipality and the event date, in addition to the title, sub-title, and text. In particular, the similarity is computed by using the following multi-variables formula:

$$similarity = \alpha * X + \beta * Y + \gamma * Z + A + B \qquad (1)$$

where X is the similarity between titles of the two news articles, Y is the similarity between sub-titles, Z is the similarity between texts, and α, β, γ are configuration parameters representing the weights to be assigned according to the importance of each similarity; in the end, A is a value added only if the two news articles have the same municipality and B if they have the same event date or the same publication date if the event date was not extracted. The resulted value is normalized and compared to the threshold T. The latter is a configuration parameter that determines if the two compared news articles are related to the same crime event or not. If the similarity computed by the multi-variables formula is equal or greater than T the news articles are labeled as duplicates, otherwise they are considered as related to different events.

For duplicate detection, we apply the shingling technique [6]. A shingle is a portion of the text, also known as "n-gram", where n is the number of consecutive words in the text. In the sentence *"The mugger escaped with the bag"*, if we use the 3-Shingling technique, the shingles are "the mugger escaped", "mugger escaped with", "escaped with the", and "with the bag". We applied the Jaccard index and the cosine similarity methods [1] on a set of 50 news articles manually labeled, and we analyzed the obtained results. We concluded that the cosine similarity computed by using the shingling technique is the best option since it provides better results of recall, precision, and accuracy. This result could be expected since the Jaccard index is not affected by the number of word occurrences. The cosine similarity instead takes into account how many times the shingle appears in specific news and its "importance" thanks to the TF-IDF (Term Frequency - Inverse Document Frequency) function. The "importance" of shingle is made by how many times it appears in the news (i.e., *term frequency*), compared to how many times it appears in all the collection of news articles to be compared (i.e., *document frequency*). The duplicate detection exports data from the DB, applies the similarity algorithm, stores the results, and merges the duplicates. We use the library "java-string-similarity"[16], a Java library implementing different string similarity and distance measures, including shingle based algorithms with cosine similarity.

To determine the best values for the configuration parameters, a validation test has been made on a dataset containing 358 news articles related to robberies published between February 2019 and May 2019 on both newspapers (in particular, 196 news articles are from "ModenaToday" and 162 news articles from "Gazzetta di Modena"). We have read and analyzed the news articles manu-

---

[16] https://github.com/tdebatty/java-string-similarity.

Table 1. Validation test of the *3-days slot* duplicate detection

T	Accuracy	Precision	Recall	F1-score
0.71	0.92	1.00	0.36	0.53
0.70	0.94	0.79	0.68	0.73
0.69	0.92	0.68	0.68	0.68
0.68	0.91	0.60	0.70	0.67
0.67	0.91	0.60	0.82	0.68
0.66	0.87	0.49	0.84	0.62
0.65	0.86	0.47	0.98	0.64

Table 2. Validation test of the *5-days slot* duplicate detection

T	Accuracy	Precision	Recall	F1-score
0.71	0.91	0.89	0.37	0.52
0.70	0.93	0.72	0.70	0.71
0.69	0.92	0.66	0.70	0.68
0.68	0.91	0.61	0.72	0.67
0.67	0.90	0.57	0.84	0.68
0.66	0.86	0.46	0.86	0.60
0.65	0.84	0.43	1.00	0.60

ally and have found 44 duplicates. Then, we have listed the identifiers of these duplicates to be compared with the results of our algorithm.

Tests have been performed between news articles with the same event date, and also considering news articles without event date published at most three or five days before (respectively, *3-days slot* or *5-days slot*). The news articles pairs to be compared selected by the reduction algorithm are 376. At this point, we have to choose the dimension of the shingles. Obviously, the smaller the shingle size, the greater the amount of storage capacity and space complexity required. In [2], Alonso et al. discovered that the optimal shingle dimension without losing precision was between 1 and 3. Based on these tests, we set the length of the shingle to 2 for the title and the sub-title and 3 for the body of the news. Then, we tried to determine the weights α, β, γ, and the values of A and B. After tests with different values, we chose the values $\alpha = 1$, $\beta = 1.25$, $\gamma = 4$, $A = 0.025$, and $B = 0.03$ since they provided the higher values of precision, recall, and accuracy. Therefore, we selected these as the best values. Table 1 shows the evaluation of tests using the slot of 3-days with different values of threshold T. Experiments described in Table 2, instead, are made using the slot of 5-days. In both cases, the best results are obtained with the threshold set to 0.70. The differences in accuracy, precision, recall, and F1-measure are minimal. The time required by the duplicate detection algorithm is a little more than one minute.

5   Evaluation

To evaluate the efficiency of the approach, we performed three tests: one test to check the efficiency and effectiveness of the NER algorithm and the use of Linked Geo Data for geolocalization, another one to evaluate the temporal expression extraction and normalization phase and the last test to access the deduplication algorithm. The data extraction phase implements a high precision method and always extracts the information of interest (title, sub-title, text, location, date and time of publication, type of crime) in the right way since the approach is based on the HTML tags of the source of the news.

5.1 NER Effectiveness

The NER was evaluated on 530 news articles manually labeled, published on both "ModenaToday" and "Gazzetta di Modena" newspapers from January 2020 to March 2020 related to all the types of crime taken into consideration. The manual annotation of the news identifies only one location for each news article.

To prove that the adoption of the NER and the mapping w.r.t. Linked Geo Data enhance the geolocalization, we check the data we obtained without the application of these methodologies. Without NER, we can only make use of the tags on the HTML documents. In a newspaper like "ModenaToday", only the city or the area within the city to which the news refers are provided. In this case, we can establish a correspondence with the municipality for 95% of news articles. This reference is very loose because it does not allow us to locate the crime event to a specific point but only to refer it to an area/city. On the other hand, by using NER, we can extract persons, organizations and locations from each news article. The geolocalization of the location entities allows us to retrieve the GPS coordinates in 479 news (90%). If no location is discovered, the geolocalization is applied to the organizations, finding 34 geolocated entities. Searching in Linked Geo Data the locations/organizations, a location for each news is found. Moreover, with Linked Geo Data, each crime event is enriched and linked to a URI referencing its location. Table 3 shows how the application of NER and the integration with Linked Geo Data increase the possibility of retrieving the GPS coordinates. In the table, the first column explains the method used, the second column is the number of news articles where the location reference is found, the third column shows how many of these locations are geolocated in the area of Modena province, and in the end, the last column is the number of news articles where a location reference is not found. As can be seen, Linked Geo Data allows finding a location reference for all news in the dataset and revises the coordinates geolocated out from the area of Modena province. Without the use of Linked Geo Data, the locations and organizations, identified with the NER, allow finding location reference in 96% of the news articles in the dataset.

Table 3. Evaluation of the coverage in detecting location reference through the application of NER and the integration of Linked Geo Data

Method	Location references found	References localized in Modena province	Location references not found
NER locations	474	454	55
+ NER organizations	+37 (511)	+46 (500)	−37 (18)
+ Linked Geo Data	+18 (529)	+29 (529)	−18 (0)

All the locations retrieved are geolocated in Modena province thanks to the integration with Linked Geo Data. Due to the incompleteness of some addresses found by the NER locations and organizations, the OpenStreetMap API is not able to retrieve the GPS coordinates. In some cases, the crime events are geolocated out from the Modena province. Linked Geo Data revises these errors. Comparing the results with our manual annotations, the geolocalization performs with a 90% precision.

Considering all the news in our dataset discovered by our approach, 12482 news articles among the 13751 news articles have been geolocated exploiting the locations found by NER, 515 of these GPS coordinates were out of the Modena province. On the remaining news, 13153 news articles have been located through organizations (293 out of the Modena province) and 13725 thanks to the integration with Linked Geo Data (1 out of the Modena province). In conclusion, almost 100% of news articles have been geolocated in the province of Modena.

5.2 Temporal Expression Normalization Effectiveness

The use of the HeidelTime tagger allowed extracting temporal expressions in 11426 news articles (83% of the total number of stored news articles). To evaluate the precision of such an approach, a manual evaluation was performed on the same dataset used for the evaluation of the NER effectiveness. One event date has been manually identified for each news. The algorithm has extracted and normalized time reference in 440 news articles (83%). All the time references found are correctly normalized. In the remaining 90 news articles, the identification of time reference was a bit challenging since it was embedded in expressions like "after six months", or "at dawn", or other similar expressions. In other cases, the news is about some updates of an event very far in time; therefore, only the reference to the year is found.

5.3 Duplicate Detection Effectiveness

The duplicate detection algorithm was evaluated on 470 news articles coming from the two different newspapers, and related to different types of crime (165 thefts, 127 robberies, and 178 attacks). The dataset was manually labeled, and 49 duplicates were found. The reduction algorithm selected 261 news articles pairs to be compared (43 about robberies, 100 attacks, and 118 thefts); the time required to find duplicates was two seconds. The deduplication has been applied using three alternatives: 3-days and 5-days time slots on the news date, and the event date. Using the time slots performed similarly with the threshold of 0.70. The results are shown in Table 4. The precision is a bit higher for the 3-days slot, where it reaches the 91%, while the recall is better for the 5-days slot, where it is 88%. The third evaluation was done exploiting the results of the Temporal Expression Normalization phase. On the test set, we found 440 news articles (94%) with a time expression that has been extracted, and normalized, and used to define the event date. The duplicate detection considering the event date in the pairwise comparison allowed discovering 95% of the duplicates. Thus, improving the results of the duplicate detection algorithm.

Table 4. Duplicate detection algorithm on different *day slots*

T	*day slot*	Accuracy	Precision	Recall	F1-measure
0.70	3-days	0.96	0.91	0.65	0.76
0.70	5-days	0.97	0.84	0.88	0.86

6   Impact and Scalability of This Research

In Sect. 4, we took into consideration two local newspapers to ingest crime events. However, newspapers do not publish news for each crime happened in Modena since some crimes are not of public interest. To evaluate the impact of our method, we made a comparison between the number of crimes in the CRIME DB and the number of crimes published in the official report of ISTAT (i.e. the crimes reported to the police). The latest report available[17] contains the crime rates per province from 2014 till 2018. The information is only quantitative; the types of crime are reported per province as a rate out of 100000 inhabitants. No information about where the crime happened (the municipality, the district, or the address), and the period of the year (season, month, or date) is provided. The classification of crimes includes 50 types of crimes and is very exhaustive, including categories that are not taken into consideration in our approach. For providing a comparison between the two datasets, we consider only the crime categories in common. The number of crimes in our approach is considered after the deduplication. Since a location-based comparison was not possible because ISTAT provides a unique report for the entire province, we calculated the number of crimes for the city of Modena in each year by using the total population registered in the same year (from 184700 till 186300 inhabitants). With the total number of crimes in the city of Modena of 7107 from 2014 till 2018, our approach covers around 21% of the crimes reported by ISTAT. A hypothesis on this low coverage can be attributed to the fact that not all the criminal events recorded by ISTAT, and therefore in the police reports, are of high impact. Therefore, not all of them are reported in local news articles.

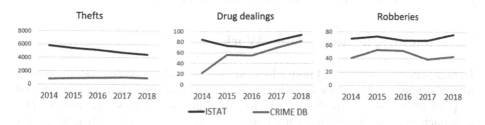

Fig. 3. Distribution of crime reports from 2011 to 2019 in different neighborhoods of Modena.

---

[17] http://dati.istat.it/Index.aspx?QueryId=25097&lang=en.

The most frequent crimes in both datasets are thefts for each year from 2014 to 2018. Figure 3 shows the total number of the top three types of crimes recorded in the report of ISTAT and compared to the number collected with our approach. As can be seen, the lower coverage is reported in thefts (the percentage average is 18%), while the higher one is in drug dealings (70%). It is very interesting to see that the trend over time of drug dealing is very similar in the two distributions if we do not consider the year 2014. Robberies have quite a similar trend in both the datasets (we were able to find 64% of the total robberies).

The approach here described is currently in use to identify the crimes occurring in Modena and its province in real-time. The visualization of crime-related data is available online[18]. This application is a Python web application which uses Tornado as a web framework. It is still an in-progress application since we want to integrate heatmaps to detect hot spots and some statistics. Everybody wants to avoid finding ourselves involved in unpleasant incidents. The citizens and the visitors of a city need to be aware of crime statistics; this may affect where they will stay, go for a walk, live, and work. On the other hand, city councils have the responsibility of identifying crime hot spots to employ appropriate monitoring and controlling mechanisms. Regarding the impact of this approach, even if the approach has been applied in a medium area, it highlights its potentiality. As reported in Sect. 1, in Italy it is not possible to collect real-time crime information from official sources, since official criminal statistics are reported annually with a delay of 6 months. The approach we have implemented can be applied everywhere also in small or medium cities/areas since there will be always one or more newspapers that report the main crimes to happen in that place.

A first scalability test has been executed to ingest all the news articles related to crimes happened in the entire Emilia-Romagna region. We selected other 9 newspapers which publish news related to the 9 provinces of the Emilia-Romagna region. We collected all the available news articles, from 2011 till now, which refer to the eleven crime types. The total number of news articles is 35000 (on average 3900 news articles for each province). The crime ingestion can be run in parallel for different newspapers and for different crime type. Therefore, we executed 99 ingestion processes in parallel to extract, analyze and store data of the region. The total loading time, that depends on the loading time of the province with the higher number of news from 2011, is 3 h for 35000 news[19].

7   Conclusion and Future Work

In this paper, we integrated multiple techniques for solving various sub-problems into a common framework in order to perform crime ingestion that is the first step to allow further analysis and visualization related to crime events. Our approach aims at extracting crime events from news articles, structuring the information, geolocating them, and linking them to Linked Data resources.

---

[18] Crime Visualization App - https://dbgroup.ing.unimore.it/crimemap.

[19] The test has been performed on a Microsoft Windows 10 Pro with 16 GB RAM.

The collected data are deduplicated and published online in a web application to provide a semi real-time overview and some analysis of the crime situation in the Modena province.

The paper described a general procedure that consists of 7 phases and its application in the context of Italian crime news for the Modena province. The approach has made it possible to create a consistent dataset with more than 13000 news articles concerning 11 types of crime, to identify crime events date and location, so allowing a time-space analysis unveiling this critical data to the citizen. A comparison with the official data gave way to discover that this approach allows collecting about 20% of the crime events available in the official reports. The use of the NER algorithm combined with the Linked Data mapping has enhanced the semantic information of each crime and the geolocalization of the events. The time expression normalization has improved the performance of the duplicate detection algorithm. From the best of our knowledge, this is the first case in which a citizen of a medium Italian city can have a look at the real-time crime data and recent statistics on crime trends. The approach is domain-independent. In this paper it is applied on the context of crime-related articles; however, it is possible to apply it to any kind of news. For new sources, an individual wrapper/extractor has to be built to retrieve title, subtitle, and so on. This is the only part that needs to be built (and in some cases only adapted from other wrappers) in order to connect a new source with the application. All the other phases can be re-used as they are provided in the open-source code. The NER algorithm can be applied to new sources provided that the text is in Italian, while the temporal expression extraction and normalization can also be performed on text of other languages. The geolocation process can geolocate addresses all over the world. The duplicate detection can be applied to text in different languages since the similarity measure is not affected by the language. In the end, the Crime Visualization App can show different types of events stored in a database with the structure of the CRIME DB.

In the near future, we will add in the process *Keyphrase/Keyword Extraction* that extracts the main phrases that categorize the text [14]. Moreover, an additional phase will be integrated to detect the key elements of each news. News articles related to the same type of crime have common characteristics. For example, all news articles related to thefts refer to a stolen object (car, money, and other objects); while in the news related to the attacks there is always the mention to who was attacked, and sometimes the reference to who was the attacker and the reason for the attack. This information is specified in a phrase that is characteristic of each type of crime.

References

1. Agarwal, N., Rawat, M., Maheshwari, V.: Comparative analysis of jaccard coefficient and cosine similarity for web document similarity measure. Int. J. Adv. Res. Eng. Technol. **2**(X), 18–21 (2014)

2. Alonso, O., Fetterly, D., Manasse, M.: Duplicate news story detection revisited. In: Banchs, R.E., Silvestri, F., Liu, T.-Y., Zhang, M., Gao, S., Lang, J. (eds.) AIRS 2013. LNCS, vol. 8281, pp. 203–214. Springer, Heidelberg (2013). https://doi.org/10.1007/978-3-642-45068-6_18

3. Alqahtani, A., Garima, A., Alaiad, A.: Crime analysis in Chicago city. In: International Conference on Information and Communication Systems (ICICS), pp. 166–172 (2019). https://doi.org/10.1109/IACS.2019.8809142

4. Arya, C., Dwivedi, S.K.: Content extraction from news web pages using tag tree. Int. J. Auton. Comp. 3(1), 04 51 (2018) https://doi.org/10.1504/IJAC.2018.10013755

5. Attardi, G., Dei Rossi, S., Simi, M.: The tanl pipeline. In: Proceedings of the Workshop on Web Services and Processing Pipelines in HLT, co-located LREC (2010)

6. Broder, A.Z., Glassman, S.C., Manasse, M.S., Zweig, G.: Syntactic clustering of the web. Comput. Netw. ISDN Syst. 29(8–13), 1157–1166 (1997)

7. Chaulagain, B., Shakya, A., Bhatt, B., Newar, D.K.P., Panday, S.P., Pandey, R.K.: Casualty information extraction and analysis from news. In: Proceedings of the International Conference on Information Systems for Crisis Response and Management. ISCRAM Association (2019)

8. Keyvanpour, M.R., Javideh, M., Ebrahimi, M.R.: Detecting and investigating crime by means of data mining: a general crime matching framework. Proc. Comput. Sci. 3, 872–880 (2011)

9. Oatley, G., Zeleznikow, J., Ewart, B.: Matching and predicting crimes. In: International Conference on Innovative Techniques and Applications of Artificial Intelligence, pp. 19–32. Springer (2004). https://doi.org/10.1007/1-84628-103-2_2

10. Palmero Aprosio, A., Moretti, G.: Italy goes to Stanford: a collection of CoreNLP modules for Italian. ArXiv e-prints (2016)

11. Pianta, E., Girardi, C., Zanoli, R., Kessler, F.B.: The textpro tool suite. In: Proceedings of LREC-08 (2008)

12. Piskorski, J., Zavarella, V., Atkinson, M., Verile, M.: Timelines: entity-centric event extraction from online news. In: Proceedings of Text2Story - Third Workshop on Narrative Extraction From Texts. CEUR Workshop Proceedings, vol. 2593, pp. 105–114. CEUR-WS.org (2020)

13. Po, L., Rollo, F.: Building an urban theft map by analyzing newspaper crime reports. In: 13th International Workshop on Semantic and Social Media Adaptation and Personalization, SMAP Zaragoza, Spain, pp. 13–18 (2018). https://doi.org/10.1109/SMAP.2018.8501866

14. Po, L., Rollo, F., Lado, R.T.: Topic detection in multichannel Italian newspapers. In: Semantic Keyword-Based Search on Structured Data Sources - COST Action IC1302 Second International KEYSTONE Conference, IKC, Cluj-Napoca, Romania, pp. 62–75 (2016). https://doi.org/10.1007/978-3-319-53640-8_6

15. Rollo, F.: A key-entity graph for clustering multichannel news: student research abstract. In: Proceedings of the Symposium on Applied Computing, SAC 2017, Marrakech, Morocco, pp. 699–700 (2017). https://doi.org/10.1145/3019612.3019930

16. Schmid, H.: Probabilistic part-of-speech tagging using decision trees. In: International Conference on New Methods in Language Processing, Manchester, UK (1994)

17. Stadler, C., Lehmann, J., Höffner, K., Auer, S.: Linkedgeodata: a core for a web of spatial open data. Semant. Web 3, 333–354 (2012)

18. Strötgen, J., Gertz, M.: Heideltime: high quality rule-based extraction and normalization of temporal expressions. In: Proceedings of the International Workshop on Semantic Evaluation, SemEval@ACL, pp. 321–324 (2010)
19. Wang, T., Rudin, C., Wagner, D., Sevieri, R.: Learning to detect patterns of crime. In: Machine Learning and Knowledge Discovery in Databases, pp. 515–530 (2013)
20. Zhang, K., Zhang, C., Chen, X., Tan, J.: Automatic web news extraction based on DS theory considering content topics. In: Proceedings of International Conference. LNCS, vol. 10860, pp. 194–207. Springer (2018). https://doi.org/10.1007/978-3-319-93698-7_15

Transparent Integration and Sharing of Life Cycle Sustainability Data with Provenance

Emil Riis Hansen[1]([✉])(iD), Matteo Lissandrini[1](iD), Agneta Ghose[2](iD),
Søren Løkke[2](iD), Christian Thomsen[1](iD), and Katja Hose[1](iD)

[1] Department of Computer Science, Aalborg University, Aalborg, Denmark
{emilrh,matteo,chr,khose}@cs.aau.dk
[2] Department of Planning, Aalborg University, Aalborg, Denmark
{agneta,loekke}@plan.aau.dk

Abstract. Life Cycle Sustainability Analysis (LCSA) studies the complex processes describing product life cycles and their impact on the environment, economy, and society. Effective and transparent sustainability assessment requires access to data from a variety of heterogeneous sources across countries, scientific and ecsonomic sectors, and institutions. Moreover, given their important role for governments and policymakers, the results of many different steps of this analysis should be made freely available, alongside the information about how they have been computed in order to ensure accountability. In this paper, we describe how Semantic Web technologies in general and PROV-O in particular, are used to enable transparent sharing and integration of datasets for LCSA. We describe the challenges we encountered in helping a community of domain experts with no prior expertise in Semantic Web technologies to fully overcome the limitations of their current practice in integrating and sharing open data. This resulted in the first nucleus of an open data repository of information about global production. Furthermore, we describe how we enable domain experts to track the provenance of particular pieces of information that are crucial in higher-level analysis.

Keywords: Open data · Provenance · Sustainability analysis

1 Introduction

Sustainability is increasingly becoming a key aspect both for policy making and commercial positioning. Its importance is expected to increase with the global socioeconomic impacts of climate change [11,15]. Life Cycle Sustainability Analysis (LCSA) studies the impacts of products along their life cycle, from the extraction of raw materials to their production, and till their disposal [3]. This enables enterprises and organizations to assess the impact of their current production chain and to find more sustainable means of production, also in line

The original version of this chapter was revised: this chapter was previously published non-open access. The correction to this chapter is available at
https://doi.org/10.1007/978-3-030-62466-8_45

© The Author(s) 2020, corrected publication 2022

J. Z. Pan et al. (Eds.): ISWC 2020, LNCS 12507, pp. 378–394, 2020.
https://doi.org/10.1007/978-3-030-62466-8_24

with the goal of sustainable development [15]. Despite this crucial role, large variations in assumptions and origins of data embedded in the assessments hinder the reliability of the outcome of such analyses. Given the complex nature of the production chain of any product, to perform reliable LCSA, analysts need access to data from a variety of heterogeneous sources across countries, scientific and economic sectors, and institutions. To enable the integration of diverse data sources, previous efforts [6] designed an ontology and corresponding open database to allow multiple organizations and researchers to share LCSA data and to make use of such data to produce analysis and models. These efforts lay the foundations of a platform where domain experts can both freely access data to compute and produce new models, but also re-share their results within the same framework. LCSA involves heterogeneous data sources and actors, hence, it is important to assure transparency, verifiability, and reproducibility of the contents of any data involved in the process. This is achieved by tracking the provenance of the information employed. Information about provenance (also called lineage [16]) allows scientists to track their data through all transformations, analyses, and interpretations [1]. *In this work, we share our experience of opening up datasets from non-open formats.* This will both help any party interested in accessing and sharing LCSA data, as well as provide useful insights to any organization willing to publish their own data to foster open science.

Contributions: *This work presents an account of how Semantic Web (SW) technologies are "in use" in an Open Source Database for Product Life Cycle Sustainability Analysis,* in direct collaboration with domain experts and associations involved in open sustainability assessment (http://bonsai.uno). We first provide an introduction to the domain of Life Cycle Sustainability Analysis and its links with Semantic Web technologies (Sect. 2). We then describe how the BONSAI Open Database for Product Footprinting is tackling the problem of integrating heterogenous LCSA data within a single open knowledge base (Sect. 3). Further, we detail how we represent, keep track of, and allow querying for the provenance of each piece of information in our open data repository. In particular, we describe the data integration workflow and how this is supported by the current open LCSA ontology (Sect. 4). We then detail how the BONSAI Open Database allows modeling all the core data required to develop economic input-output models used in LCSA (Sect. 5). The workflow we implemented allows for integrating datasets from different sources and to republish them as Linked Open Data. Further, we explain how these datasets, once converted, are annotated with provenance information adopting the PROV-O [22] vocabulary, allowing to verify the lineage of the source data (Sect. 6). Finally, we present some important lessons learned while overcoming the challenges of employing SW technologies in this domain (Sect. 7).

2   Background and Domain

Data used to perform LCSA originates from multiple sources such as national statistics, environmental reports, and supply chain reports [20]. In addition, to diversity in data sources, the data models also differ. For example, data on the production of goods or services can be reported in mass or monetary units. Therefore, to perform LCSA, domain experts need to integrate data sets from heterogeneous sources. Usually, LCSA relies on large databases (e.g., Ecoinvent[1]) that contain data at different levels of granularity about many human activities. Practitioners use these databases to compute specific models of the systems and processes they study. *Yet, many of these databases provide little to no access to the techniques used to collect and integrate the data. Moreover, in many cases, these databases are proprietary, expensive to access, and lack inter-operability.* Therefore, given the crucial role of LCSA, the BONSAI organization set out to overcome the current lack of accessibility and transparency with an Open Database for LCSA and developed, as a first step, an appropriate ontology [6]. *Here, we present how SW technologies have been adopted for the first time in LCSA to implement and materialize this open database.*

Availability and accessibility of up-to-date data, as well as legal and technical openness, are important elements that make Open Data the de-facto solution both in open science and in a more transparent government. In this spirit, other efforts have been taken in the direction of creating a database for LCSA analysis. The most notable are Exiobase, YSTAFDB, and Trase Earth. Exiobase [10] is a multi-regional Input-Output database that contains data on 200 product types that are transacted between 164 industries. Moreover, it contains records for 39 resource types, 5 land types, and 66 emission types related to the production and consumption of goods and services in the entire global economy. The Yale Stocks and Flows Database (YSTAFDB) combines material stocks and flows (STAF) data generated since early 2000 and collected by researchers at Yale [12]. Trase Earth[2] is another LCSA initiative that maps supply chain information system for land and forest use in Latin America.

Yet, while legal openness can be provided by applying an appropriate open license, technical openness requires us to ensure that there are no technical barriers to using the data. In particular, the aforementioned databases do not make use of Semantic Web technologies, which limits their ability to seamlessly integrate with other new datasets accessible on the Web. On the other hand, the success of many open-data resources in other domains such as GeoNames and Bio2RDF [13], motivates the decision to adopt Semantic Web technologies and the Linked Open Data format as a more appropriate solution. *Thus, while other efforts provided (legally) open datasets for product footprinting, this work is the first open database for LCSA on the Semantic Web.*

Nonetheless, while a common ontology and data format is the first step towards integrating and publishing free and open data for LCSA [6], *in this paper we focus on the next crucial step: integrating and sharing different Life Cycle Sustainability datasets.* Among others, we describe how we need not only to achieve full interoperability between different datasets, but also how we record and track the *data lifecycle* through provenance to ensure transparency,

---

[1] https://www.ecoinvent.org/.
[2] https://trase.earth/.

verifiability, and traceability of the original datasets and the computed results. To this end, we make use of the W3C PROV-O standard for modeling of provenance [22]. The standard has been widely used in different systems and contexts in the last couple of years. In general, the PROV-O vocabulary is highly flexible and enables the recording of lineage for any collection of data, recording activities (e.g., who gathered what, where, and when), which can be used to evaluate, among others, the reliability of the data. For instance, the W3C PROV-O standard has been used to expose provenance information regarding version control systems (VCS) [4], to enable the publication of VCS provenance on the Web and subsequent integration with other systems that make use of PROV-O. Moreover, it has been used for a Semantic Web-based representation of provenance concerning volunteered geographic information (VGI) [2] and to enhance the quality of an RDF Cube regarding European air quality [5].

3   Life Cycle Sustainability Data

Supply and Use Tables (SUTs) are one of the primary data sources for LCSA. They are comprehensive, non-proprietary data sources, covering the environmental, social, and economic spheres. In practice, the SUT records show what was the *total production* from a specific industrial sector and which other industrial sectors or markets consumed this product in which proportion. For instance, SUT records show that in 2011, ~1237 megatonnes of *steel* were produced in *China* [10]. Furthermore, ~92.7% of the domestic steel production in China was also used in China. Hence, a national SUT database encapsulates production and consumption of products and services for the entire national economy. Global Multi Regional (MR) SUTs are a combination of national SUTs and provide data on the global economy, which includes the transaction of goods and services between countries [10]. Among others, to measure the economic impacts of a change in demand of a specific product or service, Input-Output (IO) models are constructed from SUTs by applying one of the multiple algorithms existing in the literature [7]. IO models represent inter-industry relationships within an economy, showing how output from one industrial sector becomes an input to another industrial sector. As a result, an IO model, obtained from a set of MR SUTs, links flows of productions within and among national markets. If the SUTs include additional data on environmental emissions or social performance (e.g., employment levels), the IO models can be further used to perform environmental or social footprinting (e.g., the impact on carbon emissions by an increase in demand of a product).

A Model for Interoperable LCSA Data. Data available in multi-regional environmentally-extended IO models (EEIO) is aggregated for each industrial sector. Granularity in the analysis can be increased by combining the EEIO with detailed data of the product or service to be analyzed expanding it with different sources [17]. *However, what hinders this process is the lack of access to the relevant datasets and their limited interoperability.* To address this problem, we developed an Ontology for Product Footprinting to ease and promote the

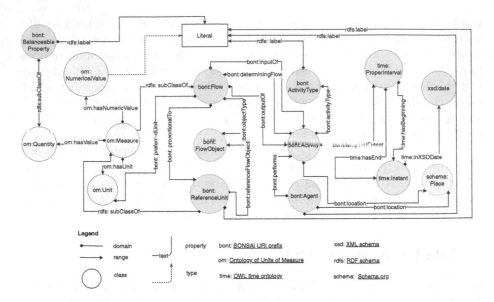

Fig. 1. The BONSAI ontology for LCSA [6]

exchange and integration of diverse LCSA data sources [6]. The proposed ontology (Fig. 1) follows a well-established model around three main concepts: *Activity* (any production activity, e.g., *steel production*), *Flow* (a quantity of product that is either produced or consumed by an activity, e.g., *tonnes of steel produced in China*), and *Flow Object* (the kind of product that is produced or consumed, e.g., *Stainless steel*) [21]. This ontology has been designed to model both economic production and environmental emissions. While the ontology presents a crucial first step for different stakeholders to agree on a common vocabulary and data model, additional tasks are required for the realization of a common open database. In the following, we describe such tasks. In particular, we have established a data integration workflow where multiple data sources are integrated to expand the granularity of the information and allow the construction of more detailed EEIO models. In the following, we adopt the Exiobase dataset and the YSTAFDB as prototypical example resources to demonstrate how we achieve the desired interoperability. *In particular, we describe how we enable integration and sharing of multiple SUTs within the common BONSAI Open Database.*

4   Data Integration Workflow

The integration workflow we established starts when a new dataset is identified for inclusion and terminates with the output of RDF named graphs representing the (annotated) information that was extracted from the identified dataset (see Fig. 2). The graphs are then published as Linked Open Data resources. We note that we tackle explicitly the task of integrating datasets with different formats within a unique repository with a common data model. That is, we enforce (both manually and automatically) syntactic data quality, but we do not tackle the issue of fixing data quality issues in the content of the data we integrate. *This is on purpose since our goal is to collect and store multiple datasets as they are.* Inspecting and solving data quality issues is an orthogonal task that domain experts can carry out only when they have open access to different datasets to compare and cross-reference. This means that, without our open database, ensuring the quality of the data used in LCSA would not be easy (or not feasible at all). In Sect. 7, we provide an example of such a case.

Integration of Multiple Classifications. Different datasets might have distinct classifications for the same concept. To align those datasets, *correspondence tables* systematically encode the semantic correspondence between those concepts within the BONSAI classification. Correspondence tables, hence constitute a reference taxonomy being developed by BONSAI to keep track of conceptual linkage between various datasets. For example, the Exiobase dataset introduces 163 different instances of *Activity Types*, 200 *Flow Objects*, and 43 *Locations*. One of the instances is the *Activity Type* of *cultivation of paddy rice*. In this case, the new concept is added in the BONSAI classification (Fig. 2, top dashed arrow) recording that *cultivation of paddy rice* is an *Activity Type* in the BONSAI classification extracted from the Exiobase dataset.

Moreover, in Exiobase there is a special *Flow Object* labeled *"Other emissions"*. Within the BONSAI classification, this concept is also linked to a set of more specific emissions listed by the United States Environmental Protection Agency (US EPA). This correspondence is hence recorded via the `partOf` relation to make data within the two classifications interoperable. Establishing semantic equivalence requires some domain knowledge, hence correspondence tables are manually created. *Create Correspondence Table* is the first process in the data workflow (Fig. 2). Then we perform the process of *Correspondence Mapping*, which produces the new enhanced dataset containing the updated correspondence information (in Fig. 2 labeled *Correspondence Mapped Dataset*).

Intermediate Data Transformation. In the process of integrating new LCSA datasets, we faced the technical issue of many LCSA datasets being shared in various non-normative formats. As an example, the Exiobase dataset is shared as a set of spreadsheets, without an associated ontology. Similarly, YSTAFDB datasets are provided as plain CSV files. The data structure, even within the same file format (e.g., CSV files), might however also differ from dataset to dataset, due to lack of standardization between LCSA datasets [8]. To allow automatic transformation and integration of new datasets by a common set of data converters, we defined a common intermediate CSV format. The *Formalization Transformation* activity represents the conversion of the specific dataformats to the common one (in Fig. 2 with output *Formalized Dataset*). The formalized datasets will contain a separate list of *Flows*, *Flow Objects*, *Activity Types*, and *Locations*. Finally, this formalization task could also be carried out by any data provider who wants to include their dataset in the BONSAI database.

RDF Data Extraction. The final step in the integration of a new dataset is the actual conversion of the formalized data into an RDF graph coherent with the BONSAI Ontology. Custom scripts are used in this process (called *Data Extraction*) to create named graphs from the formalized data. The result is one or more named graphs with instances of *Flow Objects*, *Activity Types*, *Locations*, and *Flows* (*Named Graphs* in Fig. 2). Our convention is to create a named graph for each class of instances. Thus, if a new dataset presents *Locations*, *Flow Objects*, *Activities*, and *Activity Types* we create four new named graphs, one for each of the four classes. Furthermore, this convention tries to avoid duplicating concepts by storing them only once in their dedicated named graph. Since the same information usually appears in several datasets, the other datasets, when integrated, will just reference the information in the predefined named graph avoiding redundancies. Finally, the newly generated graphs can be published via a SPARQL endpoint. Moreover, while the BONSAI classification is expanded since new named graphs are produced and integrated in the database, the intermediate resources (in the dashed ovals) can be discarded. Finally, since the conversion script is automatic (due to the formalization step), we can ensure its conformity to the proposed ontology and also identify missing information. In our future work, we aim to also adopt shape expressions [14] for syntactic validation of extracted information.

Integration of new Models. After a new dataset is integrated and published, the database is used as a source of information to compute new or updated IO models. Development of IO models from MR SUTs varies depending on the algorithm used for IO Modeling [7,18]. Nonetheless, users of the BONSAI database can apply their own or predefined IO Modeling Algorithms to some or part of the data published in the database by querying only the required data. For instance, given that both Exiobase and YSTAFDB comply with the flow-activity model encoded in the ontology [6], data from both can be processed altogether or a user can select a portion of them for IO Modeling in a specific sector. This step is illustrated in Fig. 2 as the process *IO Modeling* using the named graphs in the database along with an IO Modeling Algorithm. The result of this process is a new named graph representing the *Flows* and the corresponding information in the IO model. This means that the database allows also the insertion of the IO models into the dataset (illustrated with a dashed line between the *IO Models* and the *Named Graphs*).

Metadata Annotation. For all systems that incorporate data from multiple diverse sources, keeping provenance information about individual pieces of data is crucial. For new datasets this corresponds to the information of their origin, especially the organization and the time at which they have been produced. For IO models this also includes the portion of the dataset used to compute them and the metadata about how they have been computed. Therefore, during the integration processes described above, the output datasets are also annotated with provenance information, as described in Sect. 5.

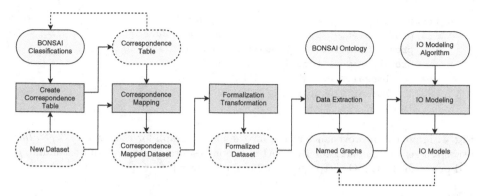

Fig. 2. Integration workflow of a new LCSA dataset. Squares represent processes, ovals represent data, arrows indicate the flow of data, dashed ovals represent data which is not saved after having been used in all their respective processes.

Handling Updates. The pipeline is rerun whenever a new dataset is integrated, or when a new version of an already integrated dataset is available. All steps of the pipeline must be rerun for the integration of new datasets, but changes to existing datasets often do not require the initial manual step of *Create Correspondence Table*, since the schema between versions of a dataset is rarely changed.

5 Support for Provenance

Provenance information is used to determine how an artifact was produced, and from where it origins. This allows, among others, to verify whether correct

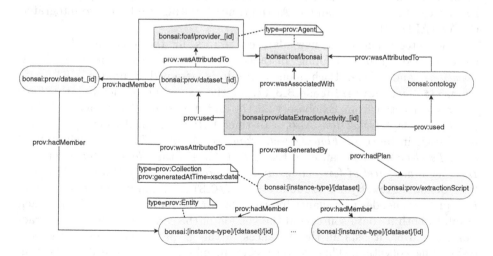

Fig. 3. Implementation of Provenance in the LCSA integration workflow. Pentagons, ellipses, and squares describe PROV-O *Agents*, *Entities*, and *Activities* respectively. Arrows represent PROV-O provenance relationships between model constituents.

```
bonsai:flowobject/exiobase_3_3_17 a prov:Collection ;
    prov:generatedAtTime "2019-11-28"^^xsd:date ;
    prov:wasGeneratedBy bonsai:prov/dataExtractionActivity_0 ;
    prov:hadMember bonsai:flowobject/C_ADDC,
        bonsai:flowobject/C_STEL,
        bonsai:flowobject/C_ALUM,
        ...
```

Fig. 4. Fragment of provenance record for the named graph for Exiobase v3.3.17 *Flow Objects*. Prefixes: **bonsai:** for BONSAI common resources (https://rdf.bonsai.uno/), and **prov:** for PROV-O (http://www.w3.org/ns/prov#)

methods have been utilized to obtain a result and hence whether artifacts can be trusted [22]. Thus, data in the BONSAI database is enhanced with provenance information annotations for all resources integrated through time.

In practice, the BONSAI provenance model is implemented by referencing and subclassing concepts from the W3C PROV-O vocabulary [9]. PROV-O uses the concepts of *Agents*, *Entities*, and *Activities*, to describe objects and their life cycle. In PROV-O *Entities* can be physical, digital, or conceptual objects of which we want to keep track. *Activities* are records of how entities come into existence and how existing *Entities* are changed to become new *Entities*. *Agents* can be a person, a piece of software, an organization, or other entities that may be ascribed responsibility [22]. Hence, PROV-O defines concepts to relate *Agents*, *Entities*, and *Activities* used in the production, delivery, or in other ways influencing an object [22]. Provenance information is automatically produced during the *Data Extraction* process in the data integration workflow described in Sect. 4, following the specific implementation as illustrated in Fig. 3. In the following, we explain how this is materialized using the integration of the Exiobase dataset as an example. At the time of writing, we have also integrated the YSTAFDB [12].

The integration of a new dataset results in the creation of one or more named graphs defining instances of *Flow Objects*, *Activity Types*, *Locations*, and *Flows*. Each named graph is assigned a unique URI (e.g., bonsai:flowobject/exiobase_3_3_17, for the named graph defining the *Flow Objects* extracted from Exiobase v3.3.17) and it is defined both as a distinct *Entity* and as a *Collection*. Also, a distinct URI is assigned to every instance (e.g., each instance of *Flow Object*, *Flow*, or *Activity Type*) in each *dataset* (e.g., *Exiobase*). For instance, in Exiobase (v3.3.17) the *Flow Object* describing *Basic iron and steel of ferro-alloys and first products thereof* (code C_STEL) has URI bonsai:flowobject/exiobase_3_3_17/C_STEL. Finally, the *Data Extraction* process encodes provenance information for the named graphs, linking each graph to both its data source and the version of the script used for data extraction. Moreover, it lists all the instances in the graph as members of the corresponding collection. That is, we record membership to a specific collection for each resource within each graph (lowest level of Fig. 3). In practice, the

PROV-O relation `prov:hadMember` is used to relate instances of data to the same `prov:collection`. This explicit link is materialized to improve accessibility to users and automatic analysis tools. The result of this model is an annotated resource associated with provenance information about the creation time of the named graph, which activity was used in its generation, and the list of its members. A fragment of a concrete example of such a record is shown in Fig. 4 (non-provenance metadata has been omitted from the figure for clarity).

As explained above, named graphs are created during the process *Data Extraction*. Since data extraction is a crucial activity for the creation of the named graphs, we encode information about this step using a PROV-O *Activity*. The *Activity* encodes information about what entities were used in the creation of the named graphs, which was associated with the activity, and references the actual implementation (e.g., the script used for data extraction). Each activity is assigned a unique URI (e.g., `bonsai:prov/dataExtractionActivity_[id]`). Hence, this record links the usage of a set of resources, along with a plan of execution, to a specific data extraction activity. A concrete example of such a record is illustrated in Fig. 5. The record shows how the BONSAI ontology and a dataset were used in the activity referred to as `dataExtractionActivity`, along with the plan `extractionScript`, linking to the version of the script used in the PROV-O *Activity*. As illustrated in Fig. 4, the PROV-O relation `prov:wasGeneratedBy` is used to relate the content of a named graph, to an extraction activity. Hence, we maintain a consistent link between individual instances of extracted data and their respective origin datasets.

```
bonsai:prov/dataExtractionActivity_0 a prov:Activity ;
    prov:hadPlan bonsai:prov/extractionScript ;
    prov:wasAssociatedWith bonsai:foaf/bonsai ;
    prov:used <http://ontology.bonsai.uno/core>,
        bonsai:prov/dataset_0 .
```

Fig. 5. Provenance record of a data extraction activity referring to the BONSAI ontology (`ontology.bonsai.uno/core`), Exiobase v3.3.17 (`bonsai:prov/dataset_0`), and the extraction script identified by `bonsai:prov/extractionScript`.

Finally, we record the specific data extraction activity (e.g., `bonsai:prov/dataExtractionActivity_0`) that extracted data from the specific dataset (e.g., `bonsai:prov/dataset_0`). Hence, each dataset integrated into the BONSAI database is given a unique URI (e.g., `bonsai:prov/dataset_0`, for the Exiobase dataset v3.3.17). Furthermore, the dataset provider (e.g., an organization, a government, or an individual) is also given a unique URI (e.g., `bonsai:foaf/provider_0`, for the Exiobase Consortium maintaining Exiobase). For instance, the provenance record for Exiobase dataset v3.3.17 is illustrated using the turtle format in Fig. 6. The record contains metadata about the dataset (e.g., the version 3.3.17), a link to the organization responsible for it (e.g., Exiobase Consortium), and a date for the latest dataset update before integration into the BONSAI database (e.g., 2019-03-12).

```
bonsai:prov/dataset_0 a prov:Entity ;
        dc:title "Exiobase"
        rdfs:label "LCSA dataset by the EXIOBASE-Consortium, version 3_3_17";
        dc:date "2019-03-12"^^xsd:date;
        dc:license <https://www.exiobase.eu/index.php/terms-of-use> ;
        dc:rights "Copyright©2015 - EXIOBASE Consortium" ;
        owl:versionInfo "3.3.17" ;
        prov:wasAttributedTo bonsai:foaf/provider_0.
```

Fig. 6. Provenance record of the Exiobase dataset v3.3.17. The PROV-O *Entity* records the specific version of Exiobase (i.e., v3.3.17), and the attribution to the EXIOBASE Consortium using the W3 PROV-O predicate prov:wasAttributedTo.

6   BONSAI Database In-Use

Currently, we have published linked open data obtained from the integration of the Exiobase and YSTAFDB datasets. This includes 15.3 M 49 K flows, 164 and 9 flow objects, 49 and 1686 activities, and 200 and 525 locations for Exiobase and YSTAFDB, respectively. In the following, we shortly describe how the BONSAI database can be used in practice for the calculation of environmental emissions. We also describe how the availability of provenance annotations allows us to inspect and assess the reliability of the provided information.

Example Use Case. A typical use of LCSA is the calculation of environmental emissions for industries and products of interest. For example, to estimate the environmental emissions related to production and consumption of steel products in China. In the following, we show which information we have access to by querying the BONSAI database. The queries are executed on the BONSAI SPARQL endpoint[3]. From Exiobase we obtain that ~1237 megatonnes of steel were produced in China in the year 2011. Then we inspect how that steel was used and by whom: China uses approximately 92.7% of its domestic steel production. The database currently records that out of the whole of China's steel production ~617 megatonnes (~50%) were consumed by manufacturers of steel products, ~127 megatonnes (~10% of production) were consumed by manufacturers of electrical machinery, and ~168 megatonnes (~14% of production) were consumed by manufacturers of fabricated metal products. The database also contains information about environmental emissions, such as carbon dioxide, particulate matter (PM 2.5, PM 10), and other such emissions. The database shows that the Chinese steel production contributes to ~1617 megatonnes of carbon dioxide and ~1.77 megatonnes of particulate matter. Similar to the above

---

[3] Available at http://odas.aau.dk.

example, data on many other products of major industrial and agricultural sector can be extracted from Exiobase for the year 2011. Instead, YSTAFDB provides data on a smaller set of 62 elements and various engineering materials, e.g., steel, but on more granular spatial scales and timeframes, ranging from cities to global and from the 1800s to 2013. Therefore, the combined information from the two datasets can be used to make more qualified environmental decisions regarding the environmental performance of steel production, such as comparing the emissions from Chinese steel production to the national environmental emission quotas. Similarly, it can compare the impacts of Chinese steel production to other countries.

Provenance. In the above example, we found that China produced ~1926 tonnes of steel in 2011. One typical question is to verify whether the most current version of the data is available as well as what is the source of such a datapoint. To address this, we query the provenance of our data to find the origin of the information on which we calculated the Chinese *steel* production. The first query finds the named graphs with *Flows* used in the calculation of the Chinese production of *steel*. The query is illustrated in Fig. 7. It finds all *Flows* which are the output of an *Activity*, where the *Activity* has an *Activity Type* labeled as *Manufcature of basic iron and steel*, and the *Activity* was performed in *China* (*Location*). To identify the origin of this information, it finds all named graphs to which such *Flows* belong. In our example, we find that all extracted *Flows* of the product *Iron* used in the calculation of the Chinese production origin from the named graph `bonsai:data/exiobase_3_3_17/hsup`. The SPARQL DESCRIBE query can now be used to describe the resources, which for this example results in the record illustrated in Fig. 8.

```
SELECT DISTINCT ?collection
FROM ...
WHERE {
  ?flows a bont:Flow .
  ?flows bont:outputOf ?activity .
  ?activity bont:activityType / rdfs:label "Manufacture of steel...".
  ?activity bont:location / rdfs:label "CN".
  ?collection prov:hadMember ?flows }
```

Fig. 7. Query fragment (reduced for space constraints) for finding the collections (e.g., named graphs) where flows regarding Chinese steel production origin from.

As illustrated in the figure, the provenance relation `prov:wasGeneratedBy` shows that the graph was generated by a data extraction activity identified by `bonsai:prov/dataExtractionActivity_0`. We further query the provenance of the database to investigate this data extraction activity (also in Fig. 5). The record has a provenance relation `prov:used` to the dataset located at `bonsai:prov/dataset_0`, which means that this dataset was used in the activity to create the named graph located at `bonsai:data/exiobase_3_3_17/hsup`.

```
bonsai:data/exiobase3_3_17/hsup a prov:Collection ;
    prov:generatedAtTime "2019-12-02"^^xsd:date ;
    prov:wasAttributedTo bonsai:foaf/bonsai ;
    prov:wasGeneratedBy bonsai:prov/dataExtractionActivity_0 ;
    prov:hadMember  bonsai:data/exiobase3_3_17/hsup/f_9109,
        bonsai:data/exiobase3_3_17/hsup/f_9096,
    ...
```

Fig. 8. Record for the collection bonsai:data/exiobase_3_3_17/hsup. It encodes provenance metadata for *Flows*, *Activities* and *Locations* extracted from the dataset exiobase_3_3_17. Non-provenance metadata has been omitted from the figure.

Hence, when provenance of the dataset is queried (illustrated in Fig. 6), we find the PROV-O relation prov:wasAttributedTo, which is an attribution from the dataset *Entity* to the *Agent* responsible for its delivery. Hence, we query the database again to find information about the *Agent* with URI bonsai:foaf/provider_0, as referred to in the relation. The resulting record is illustrated in Fig. 9. This allows us to reach directly the source of the dataset and its publisher to verify the currently available information. Moreover, the provenance information about the extraction activity also has a prov:hadPlan relation to an extraction script entity identified with bonsai:prov/extractionScript (see Fig. 5). This entity points to a specific version of a GitHub repository containing the version of code used for the extraction of *Flows*, *Activity Types*, *Flow Objects*, and *Locations*. Hence, provenance for the data extraction script is also maintained.

```
bonsai:foaf/provider_0 a prov:Agent, org:Organization ;
    dc:description "Consortium creating datasets for LCSA" ;
    dc:title "Exiobase Consortium" ;
    foaf:homepage <https://www.exiobase.eu/> .
```

Fig. 9. Provenance record for the PROV-O *Agent* bonsai:foaf/provider_0. The record contains information about the Exiobase Consortium.

Triplestore Performance. To publish the data we collected through an open SPARQL endpoint, we first deployed Jena as our triple store. Yet, during initial tests with different amounts of data, we quickly witnessed that uploading all our triples ($\sim$15 M) proved unfeasible. In particular, as we tried to store more data, the disk space required by Jena was increasing faster than expected. Furthermore, queries to address the competency questions [6] were requiring many minutes to compute, even when processing only a subset of the data. Therefore, we investigated alternative options in terms of triplestore performance for our domain specific case of LCSA. In particular, we compared Jena and Open Virtuoso using the LITMUS benchmark framework [19]. The benchmark was run

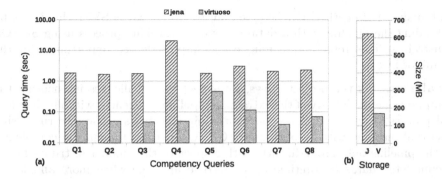

Fig. 10. Comparison of query time and disk footprint. The label under each column, corresponds to queries related to the competency questions [6].

on a virtual server with 8 cores and 64 GB of RAM. Our benchmark adopted the SPARQL queries converted from the competency questions over a subset of our full dataset. In our results (see Fig. 10), Open Virtuoso greatly outperformed Jena in all queries by an order of magnitude in running time. Moreover, Jena storage files on disk had 3× larger space footprint than Virtuoso. Therefore, we concluded that Open Virtuoso was the best choice for our needs and is now used as the DBMS for our SPARQL endpoint.

7 Lessons Learned and Future Work

In this work we have presented how we employed Linked Open Data and Semantic Web technologies for achieving the goal of integrating LCSA datasets. This allowed us to establish the first Linked Open Data database for product footprinting (See footnote 3). The current implementation overcomes several limitations in previous similar efforts. In the following, we present a brief summary of both the advantages and challenges that we encountered in this process.

Advantages of Semantic Web. We demonstrate the benefits of employing Semantic Web technologies to support open and transparent LCS Analysis. To achieve its full potential, LCSA requires the collaboration and sharing of information at many different levels both from governments and organizations. Their interoperability is of crucial importance for the effective computation of IO models. These models are required to investigate global and local impacts due to the change in demand for products and services. In particular, the adoption of a common ontology alongside established standards for data interoperability enables not only researchers and practitioners to have open access to environmental information, but also facilitates other providers to contribute to the database by sharing their own data.

Moreover, we prove the advantages of Semantic Web technologies in the domain of LCSA, by successfully integrating the two datasets: Exiobase and YSTAFDB. These datasets are now fully interoperable and can be queried and analyzed altogether. Further, we plan to exploit novel SW technologies by extending the pipeline with a data consistency checking process using SHACL constraints. This represents a unique and unprecedented opportunity for the future of LCSA.

Challenges. As the project grows, we expect new challenges in ensuring the computational scalability of the extraction pipeline. Currently, without the manual process for correspondence tables, the pipeline takes 8 hours to run, using a virtual machine with 8 cores and 64 GB RAM. Even though the complexity of the pipeline only grows linearly with respect to the number of triples in the accumulated datasets, runtime could become an issue when more and larger datasets are integrated. We plan to cope with this problem using parallelization techniques since the main processes in the pipeline are highly parallelizable.

Choice of Triple Store. As described in the previous section, the choice of data-management system was crucial to allow the necessary scalability of our database. While we first deployed Jena as our triple store, influenced by its popularity along with its open-source license, this choice revealed to be unfeasible. Open Virtuoso instead revealed to be a more solid choice. This was an important practical lesson for us.

User Interface. The use of SW technologies allows open access to the data for both humans and machines, thanks among others to the adoption of the RDF standard and SPARQL query language. Yet, SPARQL is hard to use for non-expert users. To bridge this gap we have deployed a simplified user interface by adapting the yasGUI client (See footnote 3). Also, we enhanced the GUI with query templates for easy access to common LCSA SPARQL queries. This interface presents a list of query templates, among which are present the examples adopted in this work and the competency questions defined with the domain experts. Despite the simplicity of the current GUI, it enabled LCSA experts to query the data in new ways. This led to find a flaw in a fundamental data assumption they were relying on in their handling of the data. Hence, Semantic Web technologies exploited through our GUI, allowed us to open the data enough for this assumption to be tested false by the domain experts and enabled the experts to design corrections in the data processing step. In future, we plan to implement advanced GUIs and as well as a Python library to be used within a data-science notebook to further empower domain experts in their analysis.

Provenance. In this work, we describe the data integration workflow established for the conversion of new datasets into interoperable Linked Open Data. This data will be used to derive complex models, hence we also require to verify the source of the data and the algorithms used in the calculation of the models. Hence, we employed the PROV-O ontology to implement provenance modeling of the entities, activities, and agents involved in the construction and updates of the database. Enabling the tracking of provenance information was one of the most important goals of this work and a key enabler for transparent and reliable LCSA. Nonetheless, while the PROV-O model is easy on the surface, the flexibility of the model presented a non-obvious challenge when deciding how to adapt its vocabulary to our domain. In particular, it is not straightforward to decide the most convenient level of granularity at which to record provenance information. Finally, it was challenging to determine whether a specific provenance model meets all the necessary requirements. To this end, we designed a set of basic provenance competency questions, which we plan to expand in the future. The implementation of a model of provenance was done through multiple iterations and cross-referenced with the competency questions. In its current implementation, we focused on adopting the viewpoint of the domain experts, who are used to handle datasets in terms of files and data-providers.

Acknowledgments. This work is supported by the European Union's Horizon 2020 research and innovation programme under the Marie Skłodowska-Curie grant agreement no. 838216, the Independent Research Fund Denmark under grant agreement no. DFF-8048-00051B, and strategic research funding from Aalborg University Tech Faculty: ODA - Open Data for sustainability Assessment.

References

1. Buneman, P., Khanna, S., Tan, W.-C.: Data provenance: some basic issues. In: International Conference on Foundations of Software Technology and Theoretical Computer Science, pp. 87–93 (2000)
2. Celino, I.: Human computation VGI provenance: semantic web-based representation and publishing. IEEE Trans. Geosc. Remote Sensing **51**(11), 5137–5144 (2013)
3. Curran, M.A.: Environmental life-cycle assessment. Int. J. Life Cycle Assess. **1**(3), 179–179 (1996)
4. De Nies, T., et al.: Git2PROV: exposing version control system content as W3C PROV. In: ISWC, pp. 125–128 (2013)
5. Galárraga, L., Mathiassen, K.A.M., Hose, K.: QBOAirbase: the European air quality database as an RDF cube. In: International Semantic Web Conference (Posters, Demos & Industry Tracks) (2017)
6. Ghose, A., Hose, K., Lissandrini, M., Weidema, B.P.: An open source dataset and ontology for product footprinting. In: ESWC, pp. 75–79 (2019)
7. Jansen, P.K., Raa, T.T.: The choice of model in the construction of input-output coefficients matrices. Int. Econ. Rev. **31**(1), 213–227 (1990)
8. Kuczenski, B., Davis, C.B., Rivela, B., Janowicz, K.: Semantic catalogs for life cycle assessment data. J. Clean. Prod. **137**, 1109–1117 (2016)
9. Lebo, T., et al.: PROV-O: The PROV ontology. W3C recommendation (2013)
10. Merciai, S., Schmidt, J.: Methodology for the construction of global multi-regional hybrid supply and use tables for the EXIOBASE v3 database. J. Ind. Ecol. **22**(3), 516–531 (2018)
11. Messerli, P., et al.: Global Sustainable Development Report 2019: The Future is Now-Science for Achieving Sustainable Development (2019)
12. Myers, R.J., Reck, B.K., Graedel, T.: YSTAFDB, a unified database of material stocks and flows for sustainability science. Sci. Data **6**(1), 1–13 (2019)
13. Nolin, M.-A., et al.: Bio2RDF network of linked data. In: Semantic Web Challenge; ISWC 2008 (2008)
14. Prud'hommeaux, E., Labra Gayo, J.E., Solbrig, H.: Shape expressions: an RDF validation and transformation language. In: Proceedings of the 10th International Conference on Semantic Systems, pp. 32–40 (2014)

15. Sala, S., Ciuffo, B., Nijkamp, P.: A systemic framework for sustainability assessment. Ecol. Econ. **119**, 314–325 (2015)
16. Simmhan, Y.L., Plale, B., Gannon, D.: A survey of data provenance in e-science. ACM Sigmod Rec. **34**(3), 31–36 (2005)
17. Suh, S., Huppes, G.: Methods for life cycle inventory of a product. J. Clean. Prod. **13**(7), 687–697 (2005)
18. Suh, S., Weidema, B., Schmidt, J.H., Heijungs, R.: Generalized make and use framework for allocation in life cycle assessment. J. Ind. Ecol. **14**(2), 335–353 (2010)
19. Thakkar, H.: Towards an open extensible framework for empirical benchmarking of data management solutions: LITMUS. In: Blomqvist, E., Maynard, D., Gangemi, A., Hoekstra, R., Hitzler, P., Hartig, O. (eds.) ESWC 2017. LNCS, vol. 10250, pp. 256–266. Springer, Cham (2017). https://doi.org/10.1007/978-3-319-58451-5_20
20. Visotsky, D., Patel, A., Summers, J.: Using design requirements for environmental assessment of products: a historical based method. Proc. CIRP **61**, 69–74 (2017)
21. Weidema, B.P., Schmidt, J., Fantke, P., Pauliuk, S.: On the boundary between economy and environment in life cycle assessment. Int. J. Life Cycle Assess. **23**(9), 1839–1846 (2018)
22. Yolanda Gil, S.M.: PROV Model Primer. https://www.w3.org/TR/2013/NOTE-prov-primer-20130430/. Accessed 12 May 2019

Open Access This chapter is licensed under the terms of the Creative Commons Attribution 4.0 International License (http://creativecommons.org/licenses/by/4.0/), which permits use, sharing, adaptation, distribution and reproduction in any medium or format, as long as you give appropriate credit to the original author(s) and the source, provide a link to the Creative Commons licence and indicate if changes were made.

The images or other third party material in this chapter are included in the chapter's Creative Commons licence, unless indicated otherwise in a credit line to the material. If material is not included in the chapter's Creative Commons licence and your intended use is not permitted by statutory regulation or exceeds the permitted use, you will need to obtain permission directly from the copyright holder.

A Knowledge Graph for Assessing Agressive Tax Planning Strategies

Niklas Lüdemann[1], Ageda Shiba[1], Nikolaos Thymianis[1], Nicolas Heist[1], Christopher Ludwig[2,3], and Heiko Paulheim[1(✉)]

[1] Data and Web Science Group, University of Mannheim, Mannheim, Germany
{nluedema,agshiba,nthymian}@mail.uni-mannheim.de,
{nico,heiko}@informatik.uni-mannheim.de
[2] Area Accounting and Taxation, University of Mannheim, Mannheim, Germany
[3] Corporate Taxation and Public Finance Department, Leibniz Centre for European Economic Research, Mannheim, Germany
christopher.ludwig@zew.de

Abstract. The taxation of multi-national companies is a complex field, since it is influenced by the legislation of several states. Laws in different states may have unforeseen interaction effects, which can be exploited by allowing multinational companies to minimize taxes, a concept known as *tax planning*. In this paper, we present a knowledge graph of multinational companies and their relationships, comprising almost 1.5 M business entities. We show that commonly known tax planning strategies can be formulated as subgraph queries to that graph, which allows for identifying companies using certain strategies. Moreover, we demonstrate that we can identify anomalies in the graph which hint at potential tax planning strategies, and we show how to enhance those analyses by incorporating information from Wikidata using federated queries.

Keywords: International taxation · Tax haven · Tax planning · Knowledge graph · Graph anomaly · Federated query

1 Introduction

Multinational corporations (MNCs), such as Google, IKEA, and Apple, have been scrutinized in the recent decade for so-called "aggressive" tax planning strategies. Taxes have a considerable effect on the net income of corporations, and it is in principle in the best interest of MNCs to reduce their worldwide tax burden by relocating profits within their group to lower-taxed affiliates.

The increasing internationalization of business activities in combination with the growing importance of the digital economy can create conflicts for the taxation of business profits by local governments [21]. For cross-border businesses' activities, an appropriate allocation of foreign and domestic profits – and the underlying capital – to the involved jurisdictions is necessary, in accordance with the principle of economic allegiance [13]. MNCs represent an economic entity,

© Springer Nature Switzerland AG 2020
J. Z. Pan et al. (Eds.): ISWC 2020, LNCS 12507, pp. 395–410, 2020.
https://doi.org/10.1007/978-3-030-62466-8_25

but they are usually organized as a conglomerate of legally independent separate legal entities or permanent establishments. The direct method to allocate profits and costs follows the separate entity approach and requires corporate divisions to behave as independent market participants, whereas the indirect method follows the unitary entity approach and allocates profits to affiliates by a formulary apportionment. The prevailing method in the international tax system is both for separate legal entities and permanent establishments the direct method which requires the application of the arm's length principle to intragroup transactions [20]. However, for many intermediary goods, services, and license contracts within MNCs, no independent reference market is observable and the implementation of the arm's length principle can be difficult.

Intuitively, MNCs have an incentive to allocate profits and costs in a tax-efficient way to reduce the overall tax burden of the corporation [7]. Tax reduction has a positive effect on the consolidated net income of MNCs which increases shareholder value. Efficient tax systems are – in theory – required to be neutral regarding any investment decision, but the diverse application of international taxation principles leads to a considerable heterogeneity between national tax systems [2]. Taxes represent costs for corporations, thus, MNCs usually consider tax effects intensively and pursue substantive and formal tax planning activities to change and structure economic activities in a tax-efficient way.

The term *tax planning* refers to generally accepted strategies to minimize tax liabilities of MNCs. Up to now, it is not precisely defined which tax planning strategies are considered as "aggressive". The Organization for Economic Co-operation and Development (OECD) defines them as planning activities with "unintended and unexpected tax revenue consequences" [19]. In general, "aggressive" tax planning strategies are said to be in line with legal provisions but these strategies might be able to considerably reduce the tax burden of MNCs in some regions. In the following, the term "aggressive" refers to legal tax planning strategies of MNCs that lead to a substantial reduction of their tax liabilities [11]. Tax planning has to be differentiated from the terminology of *tax avoidance* and *tax evasion*. Tax avoidance strategies exploit loopholes in the tax law to reduce the tax liability. Tax evasion refers to any illegal activities to minimize the tax burden (e.g. misstatements in the tax declaration) [7].

MNCs are usually not one business entity, but a network of parent and child companies and holdings across different countries. Therefore, they can be directly represented in a knowledge graph (KG) [5], i.e., a graph describing entities and their relations [23]. In such a KG, companies can be connected among each other as well as to the countries they belong to, and further information (such as companies' legal forms, countries' populations and GDP etc.) can be added. Such a KG allows for two kinds of analyses: First, companies using certain aggressive tax planning strategies can be identified in the graph, since they correspond to characteristic subgraph patterns. Second, the graph can be analyzed for anomalies, which might hint at tax avoidance strategies, which are not yet known.

The rest of this paper is structured as follows. Section 2 describes the knowledge graph used for our analysis and its sources. Section 3 demonstrates the above mentioned use cases, i.e., the identification of aggressive tax planning strategies and the search for graph anomalies. Section 4 discusses relevant related work, and Sect. 5 closes with a summary and an outlook on future work.

2 Knowledge Graph

For our analysis, we combine data from different sources into a knowledge graph, which can then be queried for analytics purposes.

2.1 Data Sources

The main source of our KG is the Global Legal Entity Identifier Foundation[1]. GLEIF collects data from different legal entity identifier (LEI) issuers and provides a consolidated collection of that data. For each legal entity, different data fields (such as address, legal form, etc.) are collected. GLEIF has two levels of data: level 1 data (*who is who*) contains data about the companies as such, whereas level 2 data (*who owns whom*) provides information about the relationships between companies.

The level 2 data contains both direct as well as ultimate subsidiaries, i.e., child companies of child companies and so on. The latter is, in theory, equivalent to following the transitive closure of the subsidiary relation, however, in some cases, there are subsidiaries missing in between in the data for various reasons (e.g., country specific regulations for disclosing that information).

For further analyses, we include economic and geographic data for the entities at hand. To that end, country-specific data from the World Bank[2] and Wikidata [31] is collected. Those country-wide indicators include population and GDP. Moreover, we included the statutory corporate tax rate for each country from the OECD corporate tax database[3].

Since some data was imported from Wikidata, we also provide interlinks to Wikidata. Countries and companies were trivial to match, since for the former, the GLEIF dataset uses ISO codes also present in Wikidata[4], whereas for the latter, GLEIF identifiers are also used in Wikidata.[5]. Using that approach, we could interlink all countries and a total of 20,734 companies to Wikidata.

For matching cities, first, candidates are retrieved from Wikidata based on postal codes. To that end, a list of all entities with postal codes was retrieved from Wikidata, and attribute values with ranges are preprocessed to get an actual map of postal codes to entities (e.g., Berlin has only one value for the postal code attribute with value 10115-14199[6]). To deal with entities that do not represent a city (e.g., streets or libraries) and with cases where multiple candidates exist

---

[1] https://www.gleif.org/en/.
[2] https://data.worldbank.org/.
[3] https://www.oecd.org/tax/tax-policy/corporate-tax-statistics-database.htm.
[4] https://www.wikidata.org/wiki/Property:P297.
[5] https://www.wikidata.org/wiki/Property:P1278.
[6] https://www.wikidata.org/wiki/Q64.

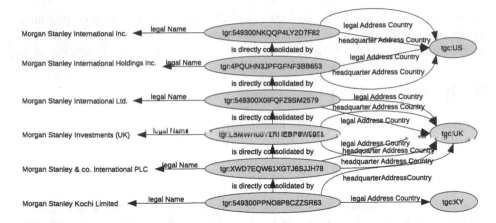

Fig. 1. Example representation of a company and its direct parents in the taxation graph

(e.g., 1000 is the postal code for Brussels, Sofia, Ljubljana, among others), the matching was made based on edit distance, with a maximum threshold of 0.3. Using that approach, we were able to link 43,832 cities to Wikidata.[7]

One basic design decision is collecting the data in one knowledge graph, vs. using SPARQL federated queries for Wikidata and Worldbank data. After some initial experiments with Virtuoso's query federation functionality, we found that federated queries are possible, but significantly slower than local queries. Hence, we follow a mixed approach: data about central entities (such as the population and GDP for countries) are included in our knowledge graph, while still maintaining the possibility to use the full data in Wikidata via federation.

2.2 Resulting Graph

The resulting graph contains about 1.5 M companies and 180k relationships between those companies, as shown in Table 1. An example representation of a company is shown in Fig. 1. Overall, the graph has 22,839,123 triples and is stored in a Virtuoso RDF store [8]. The knowledge graph is available online for browsing, download, and querying via a SPARQL endpoint.[8]

As depicted in Fig. 3, the distribution of direct and ultimate children follows a power law distribution. There are a few companies with very high number of ultimate children, as shown in Table 2, whereas the majority has only one or no ultimate children, as shown in Fig. 11. Companies with children have on average 2.6 direct children and 4.1 ultimate children (i.e., members of the transitive closure of the child relation). The longest chains of subsidiaries that we find

---

[7] The full code for generating the knowledge graph is available online at https://github.com/tax-graph/taxgraph.

[8] http://taxgraph.informatik.uni-mannheim.de/.

Table 1. Contents of the Knowledge Graph

Class	Count
Company	1,491,143
Country	225
City	95,306
Legal Form	1,286
Relation	Count
Direct subsidiary	87,020
Ultimate subsidiary	96,465

Table 2. Top 10 Companies with the most ultimate children

Company	No. of ultimate children
The Goldman Sachs Group, Inc.	2,534
Deutsche Bank Aktiengesellschaft	885
Morgan Stanley	793
Citigroup Inc.	686
Lloyds Banking Group PLC	680
Aegon N.V	629
The Royal Bank of Scotland Group Public Limited Company	496
HSBC Holdings PLC	472
Siemens Aktiengesellschaft	455
Societe Generale	429

spans across six companies, as shown in Fig. 1: Here, the ultimate child has a legal address in the Cayman Islands.

Figure 4 shows the distribution of legal and headquarter addresses. While the distribution among the top legal and headquarter addresses is similar, we can observe that two tax havens, i.e., Cayman Islands (KY) and British Virgin Islands (VG), appear among the top legal addresses, but not among the top headquarter addresses. For 36,400 of all companies in the graph (2.4%), the headquarter and legal address country differ; the majority of legal addresses in this set are the Cayman Islands (9,838), British Virgin Islands (5,878), Ireland (2,496), and Luxembourg (2,389). The most common combination is a headquarter address in the USA and a legal address in the Cayman Islands, as depicted in Fig. 2.

When comparing the corporate tax rates in the legal and headquarter addresses' countries, it can be observed that the corporate tax rate in the legal address country is, on average, 0.24 % points lower than in the headquarter's country. When considering only the 36,400 companies with differing addresses, that difference is even 10.5 % points. As depicted in Fig. 6, companies having their headquarter and legal address in different countries have a higher tendency of using a legal address in a lower-tax country.

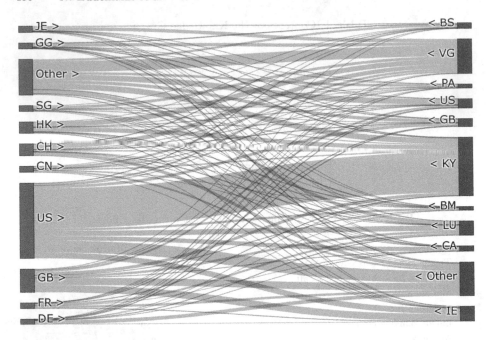

Fig. 2. Most frequent headquarter (left) and legal address country (right) for companies where headquarter and legal address are in different countries.

For subsidiary relations between companies, 35.7% of those are multinational, i.e., the legal address country of the subsidiary and its affiliate differ. Figure 5 depicts the most common relations for such multinational relationships. It can be observed that Ireland, India, and Singapore appear among the top 10 subsidiaries, but not among the top 10 parents.

When looking at the corporate tax rates for multinational companies, it can again be observed that the tax rate in which the subsidiary is located is typically lower than the one of the consolidating company. Across all subsidiary relations, the corporate tax rate in the child company's country is by 0.62 % points lower than in the parent company's country; if restricting this to multinational relations (i.e., where the parent and child company have their legal address in different countries), the difference is 2.46 % points.

3   Usage Examples

The knowledge graph can be used both for finding evidence for well-known tax avoidance strategies, as well as for searching for anomalies in the graph which hint at avoidance strategies not yet known.

3.1   Tax Avoidance Strategies

Well-known strategies for tax avoidance can be observed in the graph and formulated as query patterns and graph queries.

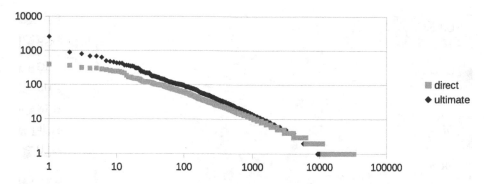

Fig. 3. Distribution of the number of direct and ultimate children per company

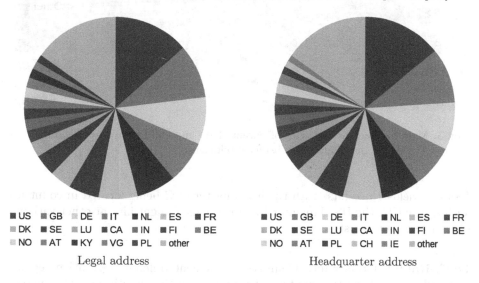

■ US ■ GB ■ DE ■ IT ■ NL ■ ES ■ FR
■ DK ■ SE ■ LU ■ CA ■ IN ■ FI ■ BE
■ NO ■ AT ■ KY ■ VG ■ PL ■ other

Legal address

■ US ■ GB ■ DE ■ IT ■ NL ■ ES ■ FR
■ DK ■ SE ■ LU ■ CA ■ IN ■ FI ■ BE
■ NO ■ AT ■ PL ■ CH ■ IE ■ other

Headquarter address

Fig. 4. Distribution of legal and headquarter addresses

Double Irish with a Dutch Arrangement. The Double Irish with a Dutch Arrangement uses in essence three companies: Two companies are located in Ireland (company A and C) and a conduit entity in the Netherlands (company B). Yet, the Irish fiscal authority considers only company A as taxable in Ireland, the second company is tax resident in a tax haven (company C). This allows to attribute all revenues to a tax haven (company C) [10,14].

Figure 7 depicts the query for a Double Irish with a Dutch Arrangement. Note that since further intermediate companies might be involved, we allow for chains of ownership by using `tgp:isDirectlyConsolidatedBy+`. Since the data in our knowledge graph is not complete, we could not find direct evidence for the Double Irish with a Dutch Arrangement construct. However, removing the last condition of the query (i.e., that company C has to have its headquarter in

Fig. 5. Most frequent headquarter of parent (left) and child company's legal address country (right) for multinational subsidiary relations.

Ireland) yields 19 results (with the headquarter of C being located in countries such as the UK, the US, Japan, or Finland), which might hint at other variants of that tax planning strategy.

Duck-Rabbit Construct. Countries implement different legislative regulations which can have the unintended consequence that hybrid entities emerge. The OECD considers hybrid entities as firms with a dual residency and no country recognizes the entity as taxable [16,22]. These constructs are called *duck-rabbit construct* in the following, named after the optical illusion in which some people see a duck, and some see a rabbit[9]. The structure can be as follows: a company C in the Netherlands having the legal form of a BV (a private limited partnership) is the child of a company B in a tax haven, which in turn is the ultimate child of some international company A, usually located in the US. In that case, the Dutch laws consider B a company under US tax legislation, while the US laws consider B a company under Dutch tax legislation, which ultimately leads to the company being taxed in none of the two countries.

The corresponding graph pattern and query are shown in Fig. 8. Running this query against the graph returns three constructs using the Bermudas and one using the Cayman Islands as an offshore tax haven. Among the former, there

---

[9] https://en.wikipedia.org/wiki/Rabbit-duck_illusion.

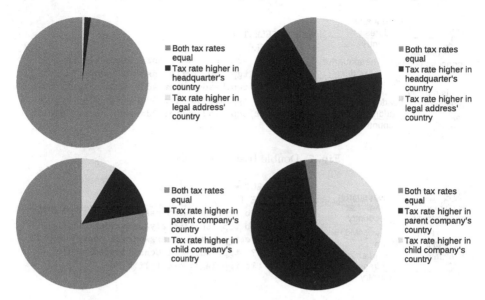

Fig. 6. Corporate tax differences between headquarter and legal address (upper part) and parent and child companies' addresses (lower part). The left hand side diagrams depict all companies, the right hand side diagrams are filtered to those where the two countries are different.

is also the game company *Activision*, which has become one of the well-known examples for this kind of tax avoidance strategy.[10]

3.2   Graph Anomalies

Since we included additional data about countries in our graph, we can use this as background information for further interesting observations [27]. One of those observations is the density of companies per state.

Table 3 depicts the top 10 countries by companies per capita and companies per GDP. It can be observed that many known tax havens appear in the top positions, with some values being clearly out of range (e.g., Liechtenstein lists one company per three inhabitants).

In the table of companies per capita, Denmark appears to be a bit of an outlier at first glance. Digging a bit deeper, we found that private holding companies – so called *Anpartselskab* – in Denmark are not taxed under certain conditions, and the creation of such companies is even advertised as a means for tax planning.[11] While this finding was new to the domain experts in the team, and we have not been able to fully explain the Denmark anomaly, we can, as of today, only find that "something is rotten in the state of Denmark" [29].

---

[10] https://thecorrespondent.com/6942.

[11] See, e.g., https://www.offshorecompany.com/company/denmark-holding/.

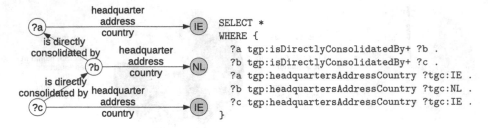

Fig. 7. Double Irish Arrangement

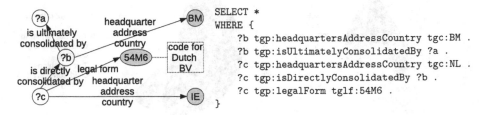

Fig. 8. Duck rabbit construction

Another analysis we conducted is related to addresses with high concentrations of companies using that address as a legal address. There are quite a few addresses which are used as legal addresses by thousands of companies. Examples for such addresses are shown in Fig. 9.

A particular observation of this analysis is that the two addresses most frequently used as legal addresses are in the state of Delaware, USA. We found that 36.7% of all US companies in our knowledge graph have their legal address in Delaware, whereas the state only accounts for 0.29% of the total US population. This phenomenon became known as the *Delaware Loophole* [32] and is a result of the Delaware tax legislation, which does not charge income tax on companies not *operating* in Delaware [4]. Consequently, only 15.3% of the companies having their legal address in Delaware also have their *headquarter* in that state.

3.3   Federated Querying

Although, as discussed above, federated queries for combining data from our knowledge graph with data from Wikidata are not very fast and scalable, they are still possible. One example is to use the area of cities – which is included in Wikidata but not in our KG – and compute the density of companies by headquarter and legal address in each city. The rationale is that cities exposing an overly large density are suspicious, similar to the analysis of addresses above.

Figure 10 depicts an example for a federated query using Wikidata. The inner query collects all cities with a minimum number of companies using that city in their address, the outer query retrieves the area for those cities from Wikidata to compute the density of companies in those cities. Table 4 shows the outcome

Table 3. Top 10 countries by companies per inhabitants (top) and per GDP (in Million USD, bottom). Germany and USA are listed for comparison.

Country	Population	Companies per capita
Liechtenstein	37,910	0.311
Cayman Islands	64,174	0.237
Luxembourg	607,728	0.063
Isle of Man	84,077	0.036
Bermuda	63,968	0.035
Monaco	38,682	0.017
Marshall Islands	58,413	0.017
Seychelles	96,762	0.011
Denmark	5,797,446	0.009
Saint Kitts and Nevis	52,441	0.008
Germany	82,927,922	0.002
USA	327,167,434	0.001

Country	GDP	Companies per 1M GDP
Marshall Islands	221,278	4.59
Cayman Islands	5,1413	2.96
Liechtenstein	6,214	1.90
Seychelles	1,590	0.66
Belize	1,871	0.61
Samoa	820	0.57
Saint Vincent and the Grenadines	811	0.54
Luxembourg	70,885	0.54
Saint Kitts and Nevis	1,011	0.46
Isle of Man	6,770	0.45
Germany	3,947,620	0.03
USA	20,544,343	0.01

of that query, showing the top 10 cities according to the density of headquarter and legal addresses registered. It can be observed that in both cases, Vaduz in Liechtenstein has the highest density of companies per square kilometer. For the density of legal addresses, Dover in Delaware shows up in the top list as another piece of evidence for the already mentioned *Delaware Loophole*.[12]

---

[12] The top 10 lists, however, have to be taken with a grain of salt. For a city to appear in the top 10 list, it requires that (a) we are able to link it to Wikidata using the approach sketched in Sect. 2, and (b) it has to have its area as a value in Wikidata. Therefore, those lists cannot be considered complete.

1209 Orange Street, Wilmington, Delaware, USA (14,551 companies)   251 Little Falls Drive, Wilmington, Delaware, USA (11,207 companies)

Eastwood House, Glebe Road, Chelmsford, CM1 1QW, Essex, United Kingdom (2,265 companies)   121 South Church Street, George Town, Cayman Islands (1,639 companies)

Fig. 9. Addresses with the highest frequency of being used as a legal address. Pictures from Google Street View.

4   Related Work

Parts of GLEIF, which we also used in this paper, have already been ported to an RDF representation and made available as a Linked Data endpoint [30]. However, the most important information for our use case – i.e., parent and child relations between companies – are not included in that representation.

Other approaches are restricted to single branches and/or countries, and thus would not allow for an analysis like the one conducted in this paper. An example for a branch specific solution is discussed in [9], where the authors build a populated ontology of bank holding companies and their ownership relations is introduced. The authors build an ontology and populated it from the Federal Reserve's public National Information Center (NIC) database[13]. Examples for country-specific solutions include a knowledge graph of Chinese companies [17], and a Linked Data endpoint of French business register data [6]. The euBusi-

---

[13] https://www.ffiec.gov/NPW.

```
SELECT ?c ?count ?a (?count/?a as ?density) WHERE {
  { SELECT COUNT(?x) AS ?count ?c WHERE {
  ?x tgp:headquartersAddressCityID ?c .
  }
  GROUP BY ?c
  HAVING(COUNT(?x)>1000)
  }
?c owl:sameAs ?wdc .
SERVICE <https://query.wikidata.org/bigdata/namespace/wdq/sparql> {
?wdc <http://www.wikidata.org/prop/direct/P2046> ?a}
} ORDER BY DESC(?density)
```

Fig. 10. Example for a federated query using Wikidata

nessGraph [28] project publishes data about businesses in the EU, but does not contain relationships between companies. Those datasets are often very detailed, but are of limited use for analyzing the taxation of multinational companies.

In addition to specific datasets, many cross-domain knowledge graphs also contain information about companies [26]. Hence, we also looked at such knowledge graphs as potential sources for the analysis at hand. However, since we need information not only for the main business entities, but also for smaller subsidiaries in order to identify tax compliance issues, we found that the information contained in those knowledge graphs is not sufficient for the task at hand. In Wikidata [31], DBpedia [15], and YAGO [18], the information about subsidiaries is at least one order of magnitude less frequent than in the graph discussed in this paper, as shown in Fig. 11: Especially longer chains of subsidiary relations, which are needed in our approach, are hardly contained in public cross-domain knowledge graphs.

In the tax accounting literature several scholars have already used data on multinational corporations to analyze the behavior of firms. It has been shown that some firms fail to publicly disclose subsidiaries that are located in tax havens [3]. In [1], the authors used very detailed data on the structure of multinational corporations to show that the introduction of public country by country reporting – the requirement to provide accounting information for each country a firm operates in to tax authorities – leads to a reduction in tax haven engagement.

A different strand of the tax literature has looked at networks of double tax treaties. Double tax treaties are in general bilateral agreements between countries that lower cross-border taxes in case of international transactions of multinational corporations. This literature on networks shows that some countries are strategically good choices for conduit entities to relocate profits and minimize cross-border taxation [12,25].

Table 4. Cities with largest densities of companies having registered their headquarter (upper half) and legal address (lower half)

City	Country	No. of companies	Area in sq. km	Density
Vaduz	Liechtenstein	9021	17.30	521.45
Puteaux	France	1334	3.19	418.18
Paris	France	16276	105.40	154.42
Geneva	Switzerland	2153	15.92	135.24
Brussels	Belgium	3756	33.00	113.82
Copenhagen	Denmark	7356	86.70	84.84
Barcelona	Spain	8305	101.30	81.98
Milan	Italy	12563	181.67	69.15
Zug	Switzerland	1027	21.61	47.52
Nicosia	Cyprus	2276	51.06	44.58
Vaduz	Liechtenstein	8460	17.30	489.02
Puteaux	France	1218	3.19	381.82
Dover	USA (Delaware)	11268	60.82	185.28
Paris	France	16253	105.40	154.20
Brussels	Belgium	4213	33.00	127.67
Geneva	Switzerland	1522	15.92	95.60
Copenhagen	Denmark	7432	86.70	85.72
Barcelona	Spain	8095	101.30	79.91
Milan	Italy	12599	181.67	69.35
Zug	Switzerland	1052	21.61	48.68

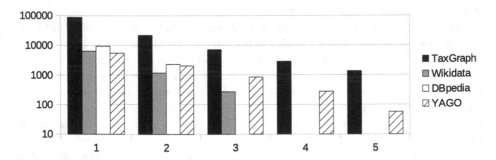

Fig. 11. Chains of subsidiaries by number of hops

5   Conclusion and Outlook

In this paper, we have introduced a knowledge graph for multinational companies and their interrelations. We have shown that the graph allows for finding companies using specific constructs, such as well-known aggressive tax planning strategies, as well as for identifying further anomalies.

Our current knowledge graph uses company data from GLEIF, which is openly available and encompasses about 1.5 M business entities. There are other (non-open) databases such as ORBIS [24], which contain even more than 40M business entities, but their licenses do not allow for making them available as a public knowledge graph. For the future, we envision the dual development of an open and a closed version of the graph, the latter based on larger, but non-public data. On the larger graph, we expect to find more evidence for known tax planning strategies and a larger number of interesting anomalies.

Another interesting source of information would be the mining of up to date information from news sites such as Reuters or Financial Times. This would allow feeding and updating the KG with recent information, and to directly rate events in the restructuring of multinational companies in the light of whether or not it is likely that those events happen for reasons of tax planning.

Apart from increasing the mere size of the graph, we also plan to include more diverse data in the graph. For example, adding branch information for companies would allow for more fine-grained analyses finding tax planning strategies, which are specific to particular branches. Further data about companies could include the size of companies (in terms of employees), or other quantitative revenue data mined from financial statements, and a detailed hierarchy of subsidiary relations describing the relations more closely (e.g., franchise, licensee, holding).

A particular challenge lies in the more detailed representation of taxation legislation. For the moment, we have only included average corporate tax rates as a first approximation, but having more fine grained representations in the knowledge graph would be a clear improvement. However, this requires some up-front design considerations, since the ontological representation of tax legislation is not straight forward.

References

1. De Simone, L., Olbert, M.: Real Effects of Private Country-by-Country Disclosure (ID 3398116) (2019). https://papers.ssrn.com/abstract=3398116
2. Devereux, M., Freeman, H.: A general neutral profits tax. Fiscal Stud. **12**(3), 1–15 (1991)
3. Dyreng, S., Hoopes, J.L., Langetieg, P., Wilde, J.H.: Strategic Subsidiary Disclosure (ID 3137138) (2018). https://doi.org/10.2139/ssrn.3137138, https://papers.ssrn.com/abstract=3137138
4. Dyreng, S.D., Lindsey, B.P., Thornock, J.R.: Exploring the role delaware plays as a domestic tax haven. J. Financ. Econ. **108**(3), 751–772 (2013)
5. Ehrlinger, L., Wöß, W.: Towards a definition of knowledge graphs. SEMANTiCS (Posters, Demos, SuCCESS) **48**, 1–4 (2016)
6. El Kader, S.A., et al.: Modeling and publishing French business register (sirene) data as linked data using the eubusiness graph ontology (2019)
7. Endres, D., Spengel, C.: International Company Taxation and Tax Planning. Kluwer Law International (2015)
8. Erling, O.: Virtuoso, a hybrid RDBMS/graph column store. IEEE Data Eng. Bull. **35**(1), 3–8 (2012)
9. Fan, L., Flood, M.D.: An ontology of ownership and control relations for bank holding companies. In: Proceedings of the Fourth International Workshop on Data Science for Macro-Modeling with Financial and Economic Datasets, pp. 1–6 (2018)

10. Fuest, C., Spengel, C., Finke, K., Heckemeyer, J.H., Nusser, H.: Profit shifting and "aggressive" tax planning by multinational firms: issues and options for reform. World Tax J. **5**(3), 307–324 (2013)
11. Heckemeyer, J.H., Spengel, C.: Maßnahmen gegen steuervermeidung: Steuerhinterziehung versus aggressive steuerplanung. Wirtschaftsdienst : Zeitschrift für Wirtschaftspolitik **93**(6), 363–366 (2013)
12. Hong, S.: Tax Treaties and Foreign Direct Investment: A Network Approach. Working Paper (2016)
13. Jacobs, O.H., Endres, D., Spengel, C., Oestreicher, A., Schumacher, A.: Internationale Unternehmensbesteuerung: deutsche Investitionen im Ausland; ausländische Investitionen im Inland. Beck-Online Bücher, C.H. Beck, 8, neu bearbeitete und erweiterte auflage edn. (2016). https://beck-online.beck.de/?vpath=bibdata/komm/JacobsHdbIntStR_8/cont/JacobsHdbIntStR.htm
14. Kleinbard, E.D.: Stateless income. Florida Tax Rev. **11**, 699–773 (2011)
15. Lehmann, J., et al.: Dbpedia-a large-scale, multilingual knowledge base extracted from wikipedia. Semant. Web **6**(2), 167–195 (2015)
16. Lüdicke, J.: "tax arbitrage" with hybrid entities: challenges and responses. Bull. Int. Taxation **68**(6), 309–317 (2014)
17. Ma, Y., Crook, P.A., Sarikaya, R., Fosler-Lussier, E.: Knowledge graph inference for spoken dialog systems. In: 2015 IEEE International Conference on Acoustics, Speech and Signal Processing (ICASSP), pp. 5346–5350. IEEE (2015)
18. Mahdisoltani, F., Biega, J., Suchanek, F.M.: Yago3: a knowledge base from multilingual wikipedias (2013)
19. OECD: Study into the Role of Tax Intermediaries. OECD Publishing, Paris (2008)
20. OECD: Model tax convention on income and on capital (2014)
21. OECD: Addressing the Tax Challenges of the Digital Economy, Action 1–2015 Final Report. OECD/G20 Base Erosion and Profit Shifting Project, OECD Publishing (2015). http://dx.doi.org/10.1787/9789264241046-en
22. OECD: Neutralising the Effects of Hybrid Mismatch Arrangements, Action 2–2015 Final Report. OECD Publishing (2015)
23. Paulheim, H.: Knowledge graph refinement: a survey of approaches and evaluation methods. Semant. Web **8**(3), 489–508 (2017)
24. Ribeiro, S.P., Menghinello, S., De Backer, K.: The OECD orbis database: Responding to the need for firm-level micro-data in the OECD. OECD Statistics Working Papers 2010(1), 1 (2010)
25. van't Riet, M., Lejour, A.: Profitable Detours: Network Analysis of Tax Treaty Shopping (2015)
26. Ringler, D., Paulheim, H.: One knowledge graph to rule them all? Analyzing the differences between DBpedia, YAGO, Wikidata & co. In: Kern-Isberner, G., Fürnkranz, J., Thimm, M. (eds.) KI 2017. LNCS (LNAI), vol. 10505, pp. 366–372. Springer, Cham (2017). https://doi.org/10.1007/978-3-319-67190-1_33
27. Ristoski, P., Paulheim, H.: Analyzing statistics with background knowledge from linked open data. In: SemStats@ ISWC (2013)
28. Roman, D., Soylu, A.: Enabling the European business knowledge graph for innovative data-driven products and services. ERCIM News **121**, 31–32 (2020)
29. Shakespeare, W.: Hamlet. Clarendon Press, Oxford (1912)
30. Trypuz, R., Kuzinski, D., Sopek, M.: General legal entity identifier ontology. In: JOWO@ FOIS (2016)
31. Vrandečić, D., Krötzsch, M.: Wikidata: a free collaborative knowledgebase. Commun. ACM **57**(10), 78–85 (2014)
32. Wayne, L.: How delaware thrives as a corporate tax haven. In: New York Times, vol. 30 (2012)

Turning Transport Data to Comply with EU Standards While Enabling a Multimodal Transport Knowledge Graph

Mario Scrocca[✉][iD], Marco Comerio[iD], Alessio Carenini[iD], and Irene Celino[iD]

Cefriel – Politecnico of Milano, Viale Sarca 226, 20126 Milan, Italy
{mario.scrocca,marco.comerio,alessio.carenini,irene.celino}@cefriel.com

Abstract. Complying with the EU Regulation on multimodal transportation services requires sharing data on the National Access Points in one of the standards (e.g., NeTEx and SIRI) indicated by the European Commission. These standards are complex and of limited practical adoption. This means that datasets are natively expressed in other formats and require a data translation process for full compliance.

This paper describes the solution to turn the authoritative data of three different transport stakeholders from Italy and Spain into a format compliant with EU standards by means of Semantic Web technologies. Our solution addresses the challenge and also contributes to build a multi-modal transport Knowledge Graph of interlinked and interoperable information that enables intelligent querying and exploration, as well as facilitates the design of added-value services.

Keywords: Transport data · Semantic data conversion · Multimodal transport knowledge graph · Transport EU regulation

1 Introduction

Semantic interoperability in the transportation sector is one of the European Commission challenges: establishing an interoperability framework enables European transport industry players to make their business applications 'interoperate' and provides the travelers with a new seamless travel experience, accessing a complete multi-modal travel offer which connects the first and last mile to long distance journeys exploiting different transport modes (bus, train, etc.).

With the ultimate goal of enabling the provision of multi-modal transportation services, the EU Regulation 2017/1926[1] is requiring transport service providers (i.e., transport authorities, operators and infrastructure managers) to give access to their data in specific data formats (i.e., NeTEx[2] and SIRI[3]) through the establishment of the so-called National Access Points (NAP).

---

[1] EU Reg. 2017/1926, cf. https://eur-lex.europa.eu/eli/reg_del/2017/1926/oj.
[2] NeTEx, cf. http://netex-cen.eu/.
[3] SIRI, cf. http://www.transmodel-cen.eu/standards/siri/.

© Springer Nature Switzerland AG 2020
J. Z. Pan et al. (Eds.): ISWC 2020, LNCS 12507, pp. 411–429, 2020.
https://doi.org/10.1007/978-3-030-62466-8_26

A survey conducted by the SNAP project[4] revealed that transport service providers have a poor knowledge of the EU Regulation 2017/1926 and its requirements and they do not use or even know the requested standards (NeTEx and SIRI). The transport stakeholders deem the conversion of their data to these standards as technically complex: they would need to dedicate a notable amount of resources to this effort, but often they lack such resources.

We designed and developed an innovative solution for data conversion, based on Semantic Web technologies, hiding the complexity of the conversion and enabling flexibility to address different scenarios and requirements. Our solution, described in this paper, has been validated within the SNAP project to enable the conversion of complex transportation data into EU-mandated standards. The proposed approach fits the needs of transport service providers rendering their legacy data interoperable and compliant with the regulation. Moreover, enabling a multi-modal transport knowledge graph, it also fosters data harmonization for the design and development of added-value travel services.

This paper is organized as follows: Sect. 2 presents our vision on how to address the transport data conversion using Semantic Web technologies; Sect. 3 describes our solution; Sect. 4 provides technical and business evaluations; Sect. 5 draws conclusions and ongoing works.

2   Challenges and Vision

Complying with the EU Regulation on multi-modal transportation services requires sharing data on the National Access Points in one of the standards indicated by the European Commission. This means that each affected organization – transport authority, transport operator and infrastructure manager – has to produce data in such formats. The goal of the European approach is clear and praiseworthy: limit the heterogeneity of data formats, require specific levels of quality and expressivity of data, and pave the way for interoperability.

As mentioned above, the mandated standards are complex and of limited practical adoption. NeTEx is a very broad and articulated XML schema and it was created as an interchange format rather than an operational format. Moreover, very few production system adopt it, so converting to NeTEx requires a deep understanding of its intricacies. NeTEx contains 372 XSD files with 2898 complex types, 7134 elements and 1377 attributes (without considering the national profiles that bring additional country-specific content).

This means that today existing datasets are natively expressed in other formats and require a data translation process for full compliance. Point-to-point conversion is of course possible, but it requires a complete knowledge of the target formats and the correspondences between the original schemata and the target standards. Our approach hides this complexity from the stakeholders, letting them keep using their current legacy systems.

---

[4] SNAP (Seamless exchange of multi-modal transport data for transition to National Access Points), cf. https://www.snap-project.eu.

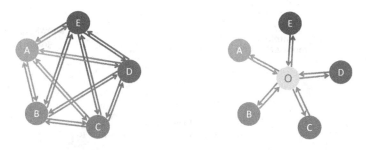

Fig. 1. Mappings required without and with a global conceptual model

The transportation domain is characterised by a proliferation of formats and standards which do not facilitate system interoperability. However, there are efforts that try to overcome the incompatibility challenge, such as Transmodel[5], which aims at representing a global conceptual model for this domain. Indeed, Transmodel is the rich and comprehensive reference schema (its data dictionary only contains 1068 concepts) on which other specific formats are based, such as NeTEx and SIRI themselves.

This is clearly an opportunity for Semantic Web technologies and for the adoption of an any-to-one centralized mapping approach [17], i.e., a global conceptual model (i.e. an ontological version of Transmodel) allowing for bilateral mappings between specific formats and the reference ontology [4].

This approach provides a twofold advantage. On the one hand, it reduces the number of mappings required for conversion and compliance, making the management of complexity manageable: if there are n different formats, the number of mappings is $2n$ instead of $2n(n-1)$ (cf. Fig. 1).

On the other hand, the two-step conversion process (i.e., lifting from the original format to the reference ontology and then lowering from the ontological version to the target format) has a positive "collateral effect": once lifted to the semantic conceptual level, data contributes to build a multi-modal transport knowledge graph of interlinked and interoperable information (cf. Fig. 2).

Taking the example of static transport information (e.g., timetables of public transportation), the target format required by the EU Regulation is the already mentioned NeTEx. Even when all data from different transport stakeholders are converted to NeTEx and shared on a National Access Point, still the challenges of data integration and reuse are not easily solved. However, the availability of a Semantic Web-enabled knowledge representation level, with data expressed with respect to a reference ontology, allows for seamless integration through automatic linking, for intelligent querying and exploration, and for the facilitation of added-value service design. Examples of such services are Intelligent Transport Systems – with new features for travelers such as planning of door-to-door journeys through the usage of multiple transportation modes – or Decision Support

---

[5] Transmodel, cf. http://www.transmodel-cen.eu/.

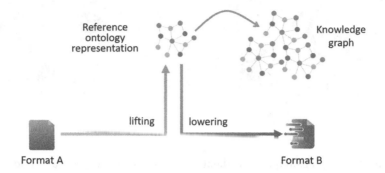

Fig. 2. Building a knowledge graph as a "collateral effect" of data conversion

Systems for government bodies e.g., for urban planners, providing statistics on mobility usage and supporting the analysis of territory coverage, to comply with the European Green Deal.

Moreover, this multi-modal transport knowledge graph can be implemented with an incremental approach, as suggested by [16], with the possibility to add additional data from multiple providers at different times, still highlighting the advantages and showcasing the opportunities of such a method.

In a nutshell, while addressing the challenge of compliance with the EU Regulation on transport information sharing, we advocate the adoption of a Semantic Web approach that not only solves the issue of data conversion, but also provides a strong and solid solution for data interoperability and for the creation of a new generation of Intelligent Transport Systems, which support the European Commission vision on seamless travel experience for travellers.

3   Our Solution

In this section, we show how we addressed the challenges and implemented the vision of Sect. 2. In particular, we designed a solution that implements the any-to-one centralized approach to semantic interoperability, with a global conceptual model to allow for bilateral mappings between specific formats and the reference ontology. After an overview on similar approaches and related efforts, in this section we explain the requirements, the architecture and the technical choices of Chimera, our open source framework to address the conversion challenge.

3.1   Related Works

Data transformations based on a common data model, such as the ones described in the previous section, require implementing two different processes. The former, which in the context of semantic technologies have been named "lifting", extracts knowledge from the data source(s), and represents it according to the common data model. The latter, which has been named as "lowering", accesses such information and represents it according to a different data model and serialisation format.

Different approaches have been proposed over the years to deal with data lifting to RDF. Most of them rely on code or format-specific transformation languages (such as XSLT applied to XML documents). Declarative languages have nonetheless been specified to deal with specific data sources, such as relational databases (R2RML [6]), XML documents (xSPARQL [2]) or tabular/CSV data (Tarql [5]). Other languages instead allow creating data transformations that extract information from multiple data sources, such as RML [9] (a generalisation of the R2RML language) and SPARQL-Generate [15]. All the cited approaches and languages support the representation of the source information according to a different structure, and the modification of values of specific attributes to adapt them to the desired data model.

Data lifting using semantic technologies has been also streamlined inside different semantic-based ETL ("Extract, Transform and Load") tools. Unified-Views [14] and LinkedPipes [13] have been implemented during the years, providing environments fully based on Semantic Web principles to feed and curate RDF knowledge bases. A different approach is used by Talend4SW[6], whose aim is to complement an already existing tool (Talend) with the components required to interact with RDF data. An ETL process is based not only on transformation steps. Other components, such as message filtering or routing, are usually involved. The main categorisation of the components and techniques that can be used in an integration process is the Enterprise Integration Patterns [12], which influenced heavily our work (cf. Sect. 3.3).

At the opposite side of the transformation process, lowering has received less attention from the academic community, and there is not a generic declarative language to lower RDF to any format. xSPARQL [2] provides a lowering solution specifically for XML, letting developers embed SPARQL inside XQuery. A downlift approach was proposed in [8], by querying lifting mappings to recreate a target CSV format. In our work, we adopt a declarative approach to implement lowering to XML, relying on SPARQL for querying data and on template engines to efficiently serialise data according to the target format.

Addressing both lifting and lowering is also possible by adopting an object-relational mapping approach (ORM), using and object-oriented representation of both the ontological and non-ontological resources through annotations and mashalling and unmashalling libraries. This approach is implemented in RDF-Beans[7] and Empire[8] and we also applied it in the past [3]. However, its main drawback is memory consumption, making it unsuitable for large data transformation; this is why we turned to the approach described in this paper.

The specific problem of producing datasets complying to the EU Regulation 2017/1926, converting GTFS datasets into NeTEx, is currently being addressed

---

[6] Talend4SW, cf. https://github.com/fbelleau/talend4sw.

[7] RDFBeans, cf. https://github.com/cyberborean/rdfbeans.

[8] Empire, cf. https://github.com/mhgrove/Empire.

by means of a non-semantic approach by Chouette [10]. Even if the implementation does not use Semantic Web languages, the conversion process is implemented in the same way. In Chouette, GTFS data is first loaded inside a relational database strongly based on the Neptune data model, and users can then export data from the database to NeTEx. Such database and related data model perform the same role than RDF and the Transmodel ontology in our approach. Chouette, as well as any direct translation solution, can address and solve the conversion problem; however, it does not build a knowledge graph nor it helps in enabling an easy integration or further exploitation of the converted data.

3.2   Chimera Requirements

The design of our software framework – that we named Chimera – aimed to address the data conversion challenge highlighted in Sect. 2 and, in the meantime, to offer a generic solution for data transformation adoptable in domains different than the transportation one. The requirements that guided our work can be summarized as follows: (i) the solution should address the conversion challenge following the any-to-one centralized mapping approach [17], by employing Semantic Web technologies; (ii) the conversion should support two different data transformation scenarios: batch conversion (i.e., conversion of an entire dataset, typical of static data) and message-to-message mediation (i.e., translation of message exchanges in a service-centric scenario, typical of dynamic data); (iii) the solution should minimise the effort required to the adopters to customize the data conversion process.

3.3   Chimera Architecture

The design of the Chimera architecture follows a modular approach to favour the conversion customization to different scenarios and formats, thus guaranteeing the solution flexibility. The decision of the modular approach is based on the assumption that a data conversion process could be broken down in a pipeline of smaller, composable and reusable elements.

We took inspiration from the already mentioned Enterprise Integration Patterns (EIP) approach, that offers best practices and patterns to break down a data processing flow into a set of building blocks. In this sense, Chimera is similar to the other data transformation solutions mentioned in Sect. 3.1. With respect to EIP terminology, a Chimera converter implements a *Message Translator* system, i.e., a pattern to translate one data format into another; in particular, a Chimera generic pipeline can be seen as a composition of specialized transformers called *Content Enricher*.

The generic pipeline for a data conversion process is illustrated in Fig. 3. The basic conversion includes the already mentioned lifting and lowering elements; however, to support a wider set of scenarios, the Chimera architecture includes additional building blocks.

Therefore, in the design of the Chimera architecture we identified the following typologies of building blocks:

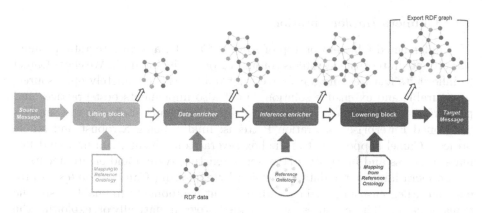

Fig. 3. A data conversion pipeline supported by Chimera.

- **Lifting block** (lifting procedure): this block takes structured data as input and, by means of mappings between the source data model and the reference ontology, transforms it into its ontological representation (an RDF graph);
- **Data enricher block** (merging/linking procedure): this (optional) block takes a set of additional RDF graphs and integrates them within the current graph; this block can be useful when additional data sources, already represented using the reference ontology, are available to enrich the current knowledge graph (i.e., information not present in input messages but needed for the conversion);
- **Inference enricher block** (inference procedure): this (optional) block takes the RDF graph and, based on a set of ontologies (containing axioms or rules), enables an inference procedure to enrich the graph with the inferred information;
- **Lowering block** (lowering procedure): this block queries the RDF graph and, by means of mappings between the reference ontology and the target data format, transforms the data into the desired output format.

In the conversion pipeline, the RDF graph is an intermediate product and it is not a necessary output of the data transformation, thus it could be discarded once the pipeline execution is completed. However, we believe that this "collateral effect" of building an RDF graph is one of the strengths of our approach, because it enables the incremental building of a knowledge graph. In our multi-modal transport scenario, this is of paramount importance, because it allows to build a knowledge graph covering different modes and therefore representing a precious source for multi-modal intelligent transport systems. The Chimera pipelines, therefore, can be configured to save the enriched RDF graph as an additional output of the transformation.

3.4   Chimera Implementation

We implemented Chimera on top of Apache Camel[9], a Java integration framework to integrate various systems consuming or producing data. We chose Camel to inherit its features and advantages: not only it is a completely open-source, configurable and extensible solution, but it also implements best practices and patterns to solve the most common integration problems, including the already-mentioned Enterprise Integration Patterns; finally, being a robust and stable project, Camel supports out-of-the-box several components, runtimes and formats to access and integrate a large set of existing system and environments.

Chimera implements data conversion by exploiting Camel's *Routes* to process data, i.e., custom pipelines with different components to enable a specific transformation. Those Routes can be defined programmatically or, exploiting the integration with Spring[10], they can be simply configured through an XML file. Thanks to the huge set of Camel's pre-defined components, Chimera can achieve both batch conversion and message-to-message mediation, and it can leverage multiple input and output channels: data can be acquired by polling directories or FTP servers, or by exposing REST services, Web APIs and SOAP services, by using different publish-subscribe technologies such as JMS or AMQP.

The Chimera framework implements the specific blocks illustrated in Sect. 3.3 with the help of Semantic Web technologies. On top of Chimera implementation, pipelines can be defined by simply providing a configuration file, without the need to add any code.

The basic idea of Chimera is to define a pipeline composed of different blocks and to "attach" an RDF graph to the specific "message" going through the Camel route. In this way, each block can process the incoming data by relying on a shared global RDF graph. All Chimera components use the RDF4J library[11] to process and handle RDF data and to operate on a Repository interface that can be in-memory, local or a proxy to a remote triplestore.

In the current release of Chimera, we provide the following blocks:

– **RMLProcessor**: a *lifting block* based on RML [9], exploiting our fork[12] of the rml-mapper library[13]; our modified version of the mapper extends RML to declare a set of InputStreams (i.e. a generic source stream rather than individual files[14]) as logical sources in the RML mapping file.

---

[9] Apache Camel, cf. https://camel.apache.org/.

[10] Spring, cf. https://spring.io/.

[11] RDF4J, cf. https://rdf4j.org/.

[12] Cf. https://github.com/cefriel/rmlmapper-cefriel.

[13] Cf. https://github.com/RMLio/rmlmapper-java.

[14] We borrowed the idea from the Carml implementation of RML, cf. https://github.com/carml/carml#input-stream-extension.

- **TemplateProcessor**: a *lowering block* based on Apache Velocity[15] to implement a template-based solution to query the RDF graph, process the result set and generate the output data in the desired format[16].
- **DataEnricherProcessor**: a *data enricher block* to add a set of RDF sources to the pipeline graph.
- **InferenceEnricherProcessor**: an *inference enricher block* loading a set of ontologies to enrich the pipeline graph with inferred information; it exploits the inferencing capabilities and configurations of the repository[17].
- A set of utility processors (**AttachGraph, DumpGraph**) to handle the initialization and export of the RDF graph.

Chimera is released as an open source software framework, with an Apache 2.0 license, and is available at https://github.com/cefriel/chimera. An example pipeline showing how to configure a semantic conversion pipeline using the described lifting, lowering and data enricher blocks is also available[18] and can be tested by following the instructions of the README file.

4   Real-World Evaluation of Our Solution

In this section, we describe the activities performed within the SNAP project to assess the proposed approach and the Chimera converter on concrete scenarios and real datasets. In Sect. 4.1, we describe the pilot scenarios, the involved stakeholders and the differences among the use cases. Then, we provide an evaluation of our proposed solution on four main dimensions:

- *Mappings*: in Sect. 4.2 we discuss the mappings' definition procedure, presenting a conceptualization of the different steps required to enable a conversion pipeline, and an assessment considering the pilots scenarios.
- *Flexibility*: in Sect. 4.3 we discuss the flexibility provided by Chimera in composing customized pipelines to address the specific requirements of the different pilots.
- *Performance*: in Sect. 4.4 we present the actual performance results and comment on conversion pipelines' execution using different input datasets.
- *Business Viability*: in Sect. 4.5 we present considerations emerged from pilots on the business viability of the proposed solution.

4.1   Evaluating the SNAP Solution

To evaluate our solution on both a technical and business side, in the context of the SNAP project we identified a set of stakeholders in the transport domain

---

[15] Cf. https://velocity.apache.org/.

[16] We implemented this approach both as a Chimera block and as a standalone tool available at https://github.com/cefriel/rdf-lowerer.

[17] Cf. for instance RDFS inference on in-memory/native RDF4J stores https://rdf4j.org/documentation/programming/repository/#rdf-schema-inferencing.

[18] Cf. https://github.com/cefriel/chimera/tree/master/chimera-example.

affected by the regulation and necessitating to convert their data. The identified pilots allowed us to test both functional and non-functional requirements of our conversion solution and let us generate valuable outputs for the involved stakeholders that not only obtained data compliant with the regulation, but they also received the RDF representation of their data, with respect to the common reference ontology based on Transmodel.

Three different pilots, covering the region of Madrid, the city of Milan, and the city of Conova, were executed. In each region, we engaged with the corresponding stakeholders to gather uses cases involving different actors identified by the EU regulation, i.e., transport operators, transport authorities and infrastructure managers.

In the region of Madrid, our *transport authority* pilot, we transformed into NeTEx the GTFS data sources of both the Consorcio Regional de Transportes de Madrid[19] (Light Railway, Metro Madrid, Regional Railway) and EMT Madrid[20] (the public bus company inside the city of Madrid). This pilot showcases the ability of our solution to deal with *large dataset conversion*.

In Milano, our *infrastructure manager* pilot, we transformed into NeTEx some data sources of SEA[21], the company managing both Linate and Malpensa airports. In particular, we focused on two different datasets: the airport facilities description (e.g., help desks, information points, ticket booths, lifts, stairs, entrances and exit locations, parking, gates) and the airport internal transport data (i.e., shuttle service between terminals). This pilot showcases the ability of our solution to build *custom conversion pipelines* that take multiple input sources into account and that involve *proprietary data formats*; indeed, we converted both datasets into an integrated NeTEx representation.

In Genova, our *transport operator* pilot, we involved AMT[22], the public transport service provider of the metropolitan area of Genoa. This pilot showcases the ability of our solution to take a generic GTFS-to-NeTEx conversion pipeline (similar to the one used for the Madrid pilot) and to customise it by adding an initial *data preparation* step to operate data cleaning and data enrichment of the original GTFS source provided by the transport operator.

4.2   Mappings' Definition

In all the pilots presented above, the expected outcome is data expressed in the NeTEx standard. Since NeTEx is a concrete XML serialization of Transmodel [1] and since Transmodel is a rich UML representation of the transportation domain, we took Transmodel as reference conceptualization. Therefore, together with the OEG group of UPM, we kick-started an effort to produce the ontological version of Transmodel; even if the ontology is far from being complete, the initial modules that allowed us to address our pilots' data conversion are already available[23].

---

[19] CRTM, cf. https://www.crtm.es/.
[20] EMT Madrid, cf. https://www.emtmadrid.es/.
[21] SEA Aeroporti, cf. http://www.seamilano.eu/.
[22] AMT Genova, cf. https://www.amt.genova.it/amt/.
[23] Cf. https://w3id.org/transmodel/.

At the time of writing, the following modules, encompassing 285 classes, 178 object properties and 66 data properties, are available: Organisations (information about the public transport organisations), Fares, Facilities, Journeys (trips and different types of journeys for passengers and vehicles), and Commons. It is important to note that the union of all these modules does not allow representing yet all the information that is expressed in the Transmodel specification, since the current focus was only to enable pilots' data conversion.

All data conversion pipelines share two steps: the *lifting* from the original data format to the Transmodel ontology and the *lowering* from the ontological version to NeTEx. The former requires the definition of lifting mappings in RML, the latter implies the creation of Apache Velocity templates to generate the final output.

To understand the complexity and the difficulties in defining the **lifting mappings**, we schematize the process in terms of its required steps, as follows:

1. *Assessing the input data model*, to identify the main concepts and relations in the input data.
2. *Aligning the identified concepts/relations with the respective ones in the reference ontology*, considering both the explicit information and the data that could be inferred. Some special cases should also be considered, in particular, cases where a one-to-one mapping cannot be identified.
3. *Extending the reference ontology*, in case some concepts or properties of the input data are not present in the reference ontology. In particular, this can be required if: (i) an information in the input format cannot be represented by the ontology but cannot be discarded, or (ii) the RDF representation of a given piece of information in the reference ontology requires the instantiation of intermediate entities and relations that cannot be materialized given a lack of information in the input data format.
4. *Coding the actual lifting mappings*, which implies creating the files that encode all the above, as well as specific configurations, such as custom functions applied in the process (e.g., to change the format for date and time). Considering the RML specification, a human readable text-based representation can be defined using YARRRML [11] then compiled to RML, and the Function Ontology can be used to declare functions in mappings [7] (FNO is very useful for conversion of specific formats, like date-time for example: if the source doesn't use xsd:dateTime, a custom conversion function is needed to generate the proper triples).

The definition of lifting mappings from GTFS to Transmodel (Madrid and Genova pilots) required the completion of all the above mentioned steps. The alignment was possible only through a deep study and understanding of the two data models and their complexity, since they are completely different, even if covering overlapping concepts. The need for ontology extension was very limited but the RML mapping coding required the definition of several custom functions to properly access and manipulate the input data.

In other words, this lifting activity was expensive, because it required a lot of manual work; however, the definition of mappings for a widespread format such as GTFS can be reused with a large set of potential customers, hence this activity cost can be recovered through economy of scale. The definition of lifting mappings for proprietary formats (Milano pilot) required a similar effort, which however cannot be leveraged with additional customers, hence it should be taken into account when valuing such custom data conversion activities.

Similarly to the lifting procedure, we schematize the steps of the **lowering mappings** definition as follows:

1. *Assessing the output data model,* to identify the main concepts and relations expected in the output format.
2. *Aligning with the reference ontology,* to identify how to query the knowledge graph expressed in the reference ontology to get the information required to produce the output format; the querying strategy should take into account both the explicit facts and the information that can be extracted with non trivial queries. Special cases should also be considered, in particular, cases requiring the definition of queries with complex patterns involving multiple entities.
3. *Extending the output format,* in case the knowledge graph contains relevant information that is not foreseen in the output format; of course, this is possible if those extensions are allowed by the output format, otherwise the additional knowledge should be discarded.
4. *Coding the actual lowering mappings,* which implies creating the actual templates that encode all the above to generate the output data. Considering Velocity templates, this coding phase includes: (i) identifying the output data structures to compose the skeleton for the Velocity template, (ii) defining the set of SPARQL queries to extract the relevant data from the knowledge graph, (iii) encoding the logics to iterate on the SPARQL results and properly populate the skeleton according to the output data format.

It is worth noting that the choice of Transmodel as reference conceptualization has a twofold advantage: on the one hand, as already said, Transmodel is a broad model that encompasses a rich spectrum of concepts and relations of the transport domain, and, on the other hand, it is the model on which the NeTEx XML serialization was defined, hence the correspondence between the ontology and the output format is quite large.

Therefore, the definition of lowering mappings between the Transmodel ontology and NeTEx was quite straightforward given the strict relation between the two models and the use of similar terminology. For this reason, the assessment of the output data model and the reference ontology alignment activities were simplified, and the output format extension activity was not necessary. Moreover, since all our pilots have NeTEx as output format, we reused the lowering mappings across the different scenarios.

4.3   Pipeline Composition

The three pilots illustrated in Sect. 4.1 have different requirements, hence they need different data conversion pipelines. In this section, we explain the three pipelines we built on top of Chimera, to show the flexibility of our approach. In general, setting up a pipeline is a matter of adapting and configuring Chimera blocks to address the specific scenario.

The basic pipeline, implemented for the *Madrid pilot*, executes the conversion from GTFS to NeTEx. This case demonstrates the main advantages of our solution in composing custom conversion pipelines. The first advantage is the possibility to create custom blocks in Chimera pipelines, besides the default ones. In this case, a custom *GTFS Preprocessing* block checks the file encoding (e.g., to handle UTF with BOM files) and generates input streams for the lifting procedure (e.g., creating different *InputStreams* from a single file to overcome a known limitation of the current RML specification, i.e. the impossibility to filter rows in CSV data sources). The second advantage is the possibility of including existing Camel blocks in the pipeline. In this case, the *ZipSplitter* block accesses the different files in the zipped GTFS feed, and other utility blocks deal with input/output management and routing. To design the final pipeline, the *lifting* and *lowering* blocks are configured with the specific mappings, and the additional blocks are integrated to define the intended flow: (i) an *AttachGraph* block to inizialize the connection with the remote RDF repository, (ii) a *ZipSplitter* block, (iii) the custom *GTFS Preprocessing* block, (vi) a *RMLProcessor* lifting block configured using RML mappings from GTFS to the Transmodel Ontology, (v) a *TemplateProcessor* lowering block configured using a Velocity template querying a knowledge graph described by the Transmodel Ontology and producing NeTEx output, and (vi) a *DumpGraph* block to serialize the content of the generated knowledge graph.

The pipeline implemented for SEA in the *Milano pilot* extends the Madrid pipeline showcasing how Chimera allows to easily integrate data in different formats. In this scenario, data on airports' facilities provided in a proprietary data format should be merged with GTFS data describing the shuttle service. To produce a unified pipeline for SEA, we configure Chimera to execute in parallel two lifting procedures: (i) a GTFS-to-Transmodel lifting portion, similar to the Madrid one, to obtain shuttle service data represented in the Transmodel ontology, and (ii) a custom SEA-to-Transmodel lifting portion, to obtain facility data represented in the Transmodel ontology. After the lifting procedures, a *DataEnricher* block is added to the pipeline to merge the RDF triples materialized in the two different lifting procedures. Finally, the Transmodel-to-NeTEx lowering templates are executed to map the shared graph to an integrated NeTEx representation.

The pipeline implemented for AMT in the *Genova pilot* extends the Madrid pipeline by integrating Chimera with functionalities of external systems. In this case, the conversion pipeline needed to interact with an external data preparation component. This external system, based on Fiware[24], enriches an initial GTFS feed with additional data sources and it can be configured to call a REST API whenever a new enriched feed is available. In the implemented pipeline, pre-existing Camel blocks are configured to accept POST calls containing an enriched

---

[24] Fiware, cf. https://www.fiware.org/.

GTFS feed. Each API call triggers the GTFS-to-NeTEx conversion pipeline, as in the other pipelines, and in the end also returns the results to the initial requester.

The pipelines of the three pilots demonstrate how the Chimera modular approach, inherited from Camel, allows to easily define customized conversion pipelines. The basic conversion pipeline can be configured with the default blocks for lifting and lowering. Complex scenarios can be addressed implementing additional blocks or employing already defined Camel blocks. Moreover, the same pipeline can be manipulated to fulfill different requirements with minimal mod ifications.

4.4 Performance Testing

In this section, we provide some statistics on the actual execution of the Chimera conversion pipelines presented in Sect. 4.3 and we comment on performance results and addressed issues.

We executed the conversion using Docker Containers on a machine running CentOS Linux 7, with Intel Xeon 8-core CPU and 64 GB Memory. Memory constraint is set to 16 GB using the Docker `--memory-limit` option on running containers, no limits are set on CPU usage. GraphDB 9.0.0 Free by Ontotext[25] was used as remote RDF repository by Chimera pipelines for storing the materialized knowledge graph and querying triples during the lowering phase.

In Table 1, we report the execution results. For each conversion case, we detail: the dataset size (for GTFS feeds, the total number of rows in all CSV files); the number of triples in the materialized knowledge graph; the lifting, lowering and total execution times; the NeTEx output dataset size.

The numbers show that NeTEx is much more verbose than GTFS and, since Transmodel has similar terminology, the knowledge graph and the output dataset are much bigger than the input data.

The conversion times showcase the ability of our solution to handle large knowledge graphs. Considering the size of CRTM, AMT and EMT datasets, we notice how conversion time grows almost linearly with respect to the input size. The biggest dataset conversion required one hour, which is reasonable for batch processing of static transport data that changes sporadically.

To make our solution efficient, we implemented in Chimera a set of optimizations as follows. The rml-mapper library, used by the Chimera lifting block, stores all triples generated in an in-memory object before serializing them at the end of the procedure. To reduce memory consumption, we implemented incremental upload of triples generated to a remote repository during the materialization. This modification reduced the memory consumption, but shifted the performance bottleneck to the triplestore. For this reason, we allowed to set in the pipeline the number of triples in each incremental *insert* query, we created a queue collecting pending queries and we spawn a thread pool of configurable size to consume the queue. Since the Free version of GraphDB allows for only two

---

[25] GraphDB, cf. http://graphdb.ontotext.com/documentation/9.0/free/.

Table 1. Conversion execution statistics; * indicates facilities in Malpensa and Linate airports, only a subset of this dataset is converted to NeTEx.

		CRTM	EMT	SEA	AMT
GTFS total no. rows		168,759	2,381,275	106	710,104
GTFS size	MB	12.46	100	0.005	45
Other data size	MB	-	-	1.62*	-
Lifting time	sec	121	1,697	6	546
Lowering time	sec	98	1,823	5	624
Conversion time	min:sec	3:39	58:40	00:11	19:30
No. triples		1,692,802	27,981,607	7,597	9,361,051
NeTEx size	MB	182.3	2,450	0.728	502

concurrent queries on a single core, the execution time is affected and directly influenced by the performances of the single core, slowing down the queue consumption.

We also experienced that a clear performance improvement in the RML materialization comes from a reduction of the number of *join* conditions in the RML mappings. In RML, triple generation is often achieved with joins between different data sources; this powerful RML construct, however, is computationally expensive, especially in case of large datasets and nested structures. In our tests, join conditions exponentially increase the required conversion time. Therefore, in our mappings we opted to limit the use of join conditions and instead use separate and simpler Triple Maps with the same IRI generation pattern, to achieve the same result without heavily affecting the conversion time.

Considering the lowering procedure, since a standard lowering approach has not emerged yet, we adopted a generic templating solution based on Apache Velocity, which is flexible to adapt to any target format. Any SPARQL SELECT query supported by the triplestore (that returns a table) can be included in the template; the processing of the SPARQL result table can be as complex as needed (using the Velocity Template Language). In the described scenario of Transmodel-to-NeTEx lowering, we implemented a set of generic templates, each with SPARQL queries with multiple *optional* patterns, to avoid assumptions on the available data. Of course, writing templates with focused queries (e.g. avoiding *optional* clauses) would improve the processing time, because the lowering performance is directly related to the query performance.

To further reduce the execution time, we generated supporting data structures to access data more efficiently, with queries avoiding nested loops. To reduce memory consumption, we removed white spaces and newlines in the Apache Velocity template, negatively impacting the output readability, which can however be improved at a later stage.

4.5   Business Viability

Finally, we add some considerations regarding the business evaluation of the presented work.

Initially, the goal of our Chimera framework was limited to provide a *data conversion solution*, to ensure the compliance with the mentioned EU Regulation. This objective was fully achieved, and all the involved stakeholders – representing transport authorities, transport operators and infrastructure managers – expressed their full satisfaction with the results obtained with our solution. We run a final meeting with each of them to illustrate the approach, the results, the pricing model and to ask for feedback. They all received their input data enriched, integrated and transformed into the required NeTEx format, with no effort on their side. Regarding the conversion effectiveness, they all agreed that we had successfully provided a solution to the problem of compliance to the European regulation and that they will adopt it, once the respective National Access Point will be setup, requiring them to supply their data in the NeTEx format.

With respect to the conversion issue, therefore, we showed the feasibility and viability of our solution. Furthermore, to preserve our business proposition, we keep the lifting and lowering mappings together with the pipeline configuration as our unique selling point, while we decided to release the Chimera framework as open source, as already mentioned, to facilitate its adoption and improvement by the larger community.

It is worth noting that developing a conversion solution on top of Chimera still requires all the technical skills to manually develop RML mappings (for the lifting step) and SPARQL queries to fill in the templates (for the lowering step), even when the full domain knowledge is available. This means that there is room for improvement in terms of supporting tooling by our community. At the moment, in business terms, this represents a competitive advantage for us and for Semantic Web-based solution providers.

While addressing the data conversion issue, however, we discovered that we can indeed offer a broader set of services to the target customers in the transport domain. As a result of our business modeling effort, we defined three offers:

- *Requirements analysis and compliance assessment*, to investigate the feasibility and complexity of converting the customer data in the desired format;
- *Conversion services*, including data preparation and enrichment, lifting to a reference ontology, data integration and lowering to the desired format;
- *Supporting services*, including long-term maintenance (e.g., repeated conversion, incremental addition of data sources), knowledge transfer and training.

In terms of pricing model, on the basis of the described experience, we now opt for different pricing depending on the *estimated difficulty of the conversion*, which we measure along three dimensions of the source data: (i) popularity of the data format (widespread or proprietary, because this impact the re-usability of the mappings), (ii) data volume, and (iii) input format complexity (again, because it impacts the mapping definition).

As a concluding remark, we highlight that the Madrid transport authority CRTM – which is already adopting Semantic Web solutions and managing RDF data – gave us a very positive feedback with respect to our solution's "collateral effect" of generating a nucleus of a *multimodal transport knowledge graph*: they perfectly understood and appreciated the enabled possibility to build additional added value services on top of the knowledge graph itself, well beyond the pure data conversion for EU Regulation compliance. Indeed, once National Access Points will be in place and transport stakeholder will provide their data, the actual exploitation of the multimodal transport knowledge graph will be enabled, e.g. by developing intelligent journey planning offering solutions based on the usage of different transport modes.

The availability of an integrated and harmonized knowledge graph can also pave the way for assessing the conversion completeness. However, in most of the pilot cases illustrated in this paper, the input data sources are in GTFS (a very simple and basic tabular data format), the target standard is NeTEx (a very complex and articulated XML format), and the Transmodel reference ontology is very close to NeTEx. Thus the coverage of the target standard (and the reference ontology) is always quite limited even if the entire information contained in the source data is completely used in the mapping. Therefore we leave this kind of analysis as future work to support and improve the mapping development (e.g. using shape validation to assess the quality/completeness of the mapping results with respect to the reference ontology).

5   Conclusions and Future Works

In this paper, we presented our solution to enable the conversion of transport data, into standards required by the European Commission, using Semantic Web technologies. To support the implementation, we designed the Chimera framework, providing a modular solution to build semantic conversion pipeline configuring a set of pre-defined blocks. The described solution has been employed within the SNAP project on concrete scenarios and real datasets involving transport stakeholders in Italy and Spain. The different pilots have been presented proposing an evaluation of the solution on different dimensions. The performed activities acknowledge the feasibility of the solution on a technological side, the desirability for stakeholders and the business viability of the approach.

We are also adopting Chimera in the ongoing SPRINT project[26] with different reference ontologies and source/target standards, to implement and evaluate conversion pipelines in both cases of batch transformation and message translation; we have already reached a clear scalability improvement with respect to our previous ORM-based solution [3] and we are proving the generalizability of our technological solution beyond the scenario offered in this paper.

As future works, we plan to: (i) explore additional tools to facilitate the mappings' definition (e.g. collaborative and visual tools), (ii) implement additional blocks for Chimera to offer more options in the pipeline definition (e.g.,

---

[26] Cf. http://sprint-transport.eu/.

approaches based on a virtualized graph to extract data from the original sources) and, (iii) perform an evaluation of the proposed solution considering dynamic data (e.g., DATEX II and SIRI formats) and requirements in real-time scenarios.

Acknowledgments. The presented research was partially supported by the SPRINT project (Grant Agreement 826172), co-funded by the European Commission under the Horizon 2020 Framework Programme and by the SNAP project (Activity Id 19281) co-funded by EIT Digital in the Digital Cities Action Line.

References

1. Arneodo, F.: Public transport network timetable exchange (NeTEx) - introduction. Technical report, CEN TC278 (2015). http://netex-cen.eu/wp-content/uploads/2015/12/01.NeTEx-Introduction-WhitePaper_1.03.pdf
2. Bischof, S., Decker, S., Krennwallner, T., Lopes, N., Polleres, A.: Mapping between RDF and XML with XSPARQL. J. Data Semant. **1**(3), 147–185 (2012)
3. Carenini, A., Dell'Arciprete, U., Gogos, S., Pourhashem Kallehbasti, M.M., Rossi, M.G., Santoro, R.: ST4RT - semantic transformations for rail transportation. Transp. Res. Arena TRA **2018**, 1–10 (2018). https://doi.org/10.5281/zenodo.1440984
4. Comerio, M., Carenini, A., Scrocca, M., Celino, I.: Turn transportation data into EU compliance through semantic web-based solutions. In: Proceedings of the 1st International Workshop on Semantics for Transport, Semantics (2019)
5. Cyganiak, R.: TARQL (SPARQL for tables): Turn CSV into RDF using SPARQL syntax. Technical report (2015). http://tarql.github.io
6. Das, S., Sundara, S., Cyganiak, R.: R2RML: RDB to RDF Mapping Language. W3C recommendation, W3C, September 2012. https://www.w3.org/TR/r2rml/
7. De Meester, B., Dimou, A., Verborgh, R., Mannens, E.: An ontology to semantically declare and describe functions. In: Sack, H., Rizzo, G., Steinmetz, N., Mladenić, D., Auer, S., Lange, C. (eds.) ESWC 2016. LNCS, vol. 9989, pp. 46–49. Springer, Cham (2016). https://doi.org/10.1007/978-3-319-47602-5_10
8. Debruyne, C., McGlinn, K., McNerney, L., O'Sullivan, D.: A lightweight approach to explore, enrich and use data with a geospatial dimension with semantic web technologies. In: Proceedings of the Fourth International ACM Workshop on Managing and Mining Enriched Geo-Spatial Data, pp. 1–6 (2017)
9. Dimou, A., Vander Sande, M., Colpaert, P., Verborgh, R., Mannens, E., Van de Walle, R.: RML: a generic language for integrated RDF mappings of heterogeneous data (2014)
10. Gendre, P., et al.: Chouette: an open source software for PT reference data exchange. In: ITS Europe. Technical Session (2011)
11. Heyvaert, P., De Meester, B., Dimou, A., Verborgh, R.: Declarative rules for linked data generation at your fingertips!. In: Gangemi, A., et al. (eds.) ESWC 2018. LNCS, vol. 11155, pp. 213–217. Springer, Cham (2018). https://doi.org/10.1007/978-3-319-98192-5_40
12. Hohpe, G., Woolf, B.: Enterprise Integration Patterns: Designing, Building, and Deploying Messaging Solutions. Addison-Wesley Professional (2004)

13. Klímek, J., Škoda, P., Nečaský, M.: LinkedPipes ETL: evolved linked data preparation. In: Sack, H., Rizzo, G., Steinmetz, N., Mladenić, D., Auer, S., Lange, C. (eds.) ESWC 2016. LNCS, vol. 9989, pp. 95–100. Springer, Cham (2016). https://doi.org/10.1007/978-3-319-47602-5_20

14. Knap, T., Kukhar, M., Macháč, B., Škoda, P., Tomeš, J., Vojt, J.: UnifiedViews: an ETL framework for sustainable RDF data processing. In: Presutti, V., Blomqvist, E., Troncy, R., Sack, H., Papadakis, I., Tordai, A. (eds.) ESWC 2014. LNCS, vol. 8798, pp. 379–383. Springer, Cham (2014). https://doi.org/10.1007/978-3-319-11955-7_52

15. Lefrançois, M., Zimmermann, A., Bakerally, N.: Flexible RDF generation from RDF and heterogeneous data sources with SPARQL-generate. In: Ciancarini, P., et al. (eds.) EKAW 2016. LNCS (LNAI), vol. 10180, pp. 131–135. Springer, Cham (2017). https://doi.org/10.1007/978-3-319-58694-6_16

16. Sequeda, J.F., Briggs, W.J., Miranker, D.P., Heideman, W.P.: A pay-as-you-go methodology to design and build enterprise knowledge graphs from relational databases. In: Ghidini, C., et al. (eds.) ISWC 2019. LNCS, vol. 11779, pp. 526–545. Springer, Cham (2019). https://doi.org/10.1007/978-3-030-30796-7_32

17. Vetere, G., Lenzerini, M.: Models for semantic interoperability in service-oriented architectures. IBM Syst. J. **44**(4), 887–903 (2005)

Enhancing Public Procurement in the European Union Through Constructing and Exploiting an Integrated Knowledge Graph

Ahmet Soylu[1]([envelope]), Oscar Corcho[2], Brian Elvesæter[1], Carlos Badenes-Olmedo[2], Francisco Yedro Martínez[2], Matej Kovacic[3], Matej Posinkovic[3], Ian Makgill[4], Chris Taggart[5], Elena Simperl[6], Till C. Lech[1], and Dumitru Roman[1]

[1] SINTEF AS, Oslo, Norway
{ahmet.soylu,brian.elvesaeter,till.lech,dumitru.roman}@sintef.no
[2] Universidad Politécnica de Madrid, Madrid, Spain
{ocorcho,cbadenes,fyedro}@fi.upm.es
[3] Jožef Stefan Institute, Ljubljana, Slovenia
{matej.kovacic,matej.posinkovic}@ijs.si
[4] OpenOpps Ltd., London, UK
ian@spendnetwork.com
[5] OpenCorporates Ltd., London, UK
chris.taggart@opencorporates.com
[6] King's College London, London, UK
elena.simperl@kcl.ac.uk

Abstract. Public procurement is a large market affecting almost every organisation and individual. Governments need to ensure efficiency, transparency, and accountability, while creating healthy, competitive, and vibrant economies. In this context, we built a platform, consisting of a set of modular APIs and ontologies to publish, curate, integrate, analyse, and visualise an EU-wide, cross-border, and cross-lingual procurement knowledge graph. We developed end-user tools on top of the knowledge graph for anomaly detection and cross-lingual document search. This paper describes our experiences and challenges faced in creating such a platform and knowledge graph and demonstrates the usefulness of Semantic Web technologies for enhancing public procurement.

Keywords: Public procurement · Knowledge graph · Linked data

1 Introduction

The market around public procurement is large enough so as to affect almost every single citizen and organisation across a variety of sectors. For this reason, public spending has always been a matter of interest at local, regional, and national levels. Primarily, governments need to be efficient in delivering services, ensure transparency, prevent fraud and corruption, and build healthy

© Springer Nature Switzerland AG 2020
J. Z. Pan et al. (Eds.): ISWC 2020, LNCS 12507, pp. 430–446, 2020.
https://doi.org/10.1007/978-3-030-62466-8_27

and sustainable economies [1,13]. In the European Union (EU), every year, over 250.000 public authorities spend around 2 trillion euros (about 14% of GDP) on the purchase of services, works, and supplies[1]; while the Organisation for Economic Co-operation and Development (OECD) estimates that more than 82% of fraud and corruption cases remain undetected across all OECD countries [19] costing as high as 990 billion euros a year in the EU [10]. Moreover, small and medium-sized enterprises (SMEs) are often locked out of markets due to the high cost of obtaining the required information, where larger companies can absorb the cost.

The availability of good quality, open, and integrated procurement data could alleviate the aforementioned challenges [11]. This includes government agencies assessing purchasing options, companies exploring new business contracts, and other parties (such as journalists, researchers, business associations, and individual citizens) looking for a better understanding of the intricacies of the public procurement landscape through decision-making and analytic tools. Projects such as the UK's GCloud (Government Cloud)[2] have already shown that small businesses can compete effectively with their larger counterparts, given the right environment. However, managing these competing priorities at a national level and coordinating them across different states and many disparate agencies is notoriously difficult. There are several directives put forward by the European Commission (e.g., Directive 2003/98/EC and Directive 2014/24/EU8) for improving public procurement practices. These led to the emergence of national public procurement portals living together with regional, local as well as EU-wide public portals [9]. Yet, there is a lack of common agreement across the EU on the data formats for exposing such data sources and on the data models for representing such data, leading to a highly heterogeneous technical landscape.

To this end, in order to deal with the technical heterogeneity and to connect disparate data sources currently created and maintained in silos, we built a platform, consisting of a set of modular REST APIs and ontologies, to publish, curate, integrate, analyse, and visualise an EU-wide, cross-border, and cross-lingual procurement knowledge graph [22,23] (i.e., KG, an interconnected semantic knowledge organisation structure [12,27]). The knowledge graph includes procurement and company data gathered from multiple disparate sources across the EU and integrated through a common ontology network using an extract, transform, load (ETL) approach [3]. We built and used a set of end-user tools and machine learning (ML) algorithms on top of the resulting knowledge graph, so as to find anomalies in data and enable searching across documents in different languages. This paper reports the challenges and experiences we went through, while creating such a platform and knowledge graph, and demonstrates the usefulness of the Semantic Web technologies for enhancing public procurement.

The rest of the paper is structured as follows. Section 2 presents the related work, while Sect. 3 describes the data sets underlying the KG. Section 4 explains the KG construction, while Sect. 5 presents the KG publication together with the overall architecture and API resources. Section 6 describes the use of the KG

---

[1] https://ec.europa.eu/growth/single-market/public-procurement_en.

[2] https://www.digitalmarketplace.service.gov.uk.

for anomaly detection and cross-lingual document search, while Sect. 7 presents the adoption and uptake. Finally, Sect. 8 concludes the paper.

2   Related Work

We focus on procurement data, related to tenders, awards, and contracts, and basic company data. We analyse relevant related works from the perspective of such types of data. Procurement and company data are fundamental to realising many key business scenarios and may be extended with additional data sources.

Public procurement notices play two important roles for the public procurement process: as a resource for improving competitive tendering, and as an instrument for transparency and accountability [15]. With the progress of eGovernment initiatives, the publication of information on contracting procedures is increasingly being done using electronic means. In return, a growing amount of open procurement data is being released leading to various standardisation initiatives like OpenPEPPOL[3], CENBII[4], TED eSenders[5], CODICE[6], and Open Contracting Data Standard (OCDS)[7]. Data formats and file templates were defined within these standards to structure the messages being exchanged by the various agents involved in the procurement process. These standards primarily focus on the type of information that is transmitted between the various organisations involved in the process, aiming to achieve certain interoperability in the structure and semantics of data. The structure of the information is commonly provided by the content of the documents that are exchanged. However, these initiatives still generate a lot of heterogeneity. In order to alleviate these problems, several ontologies including PPROC [16], LOTED2 [8], MOLDEAS [20], or PCO [17], as well as the upcoming eProcurement ontology[8] emerged, with different levels of detail and focus (e.g., legal and process-oriented). So far, however, none of them has reached a wide adoption mainly due to their limited practical value.

Corporate information, including basic company information, financial as well as contextual data, are highly relevant in the procurement context, not only for enabling many data value chains, but also for transparency and accountability. Recently, a number of initiatives have been established to harmonise and increase the interoperability of corporate and financial data. These include public initiatives such as the Global Legal Entity Identification System—GLEIS[9], Bloomberg's open FIGI system for securities[10], as well as long-established proprietary initiatives such as the Dun & Bradstreet DUNS number[11]. Other notable

---

[3] https://peppol.eu.
[4] http://cenbii.eu.
[5] https://simap.ted.europa.eu/web/simap/sending-electronic-notices.
[6] https://contrataciondelestado.es/wps/portal/codice.
[7] http://standard.open-contracting.org.
[8] https://joinup.ec.europa.eu/solution/eprocurement-ontology.
[9] https://www.gleif.org.
[10] https://www.omg.org/figi.
[11] http://www.dnb.com/duns-number.html.

initiatives include the European Business Register (EBR)[12], Business Register Exchange (BREX)[13], and the eXtensible Business Reporting Language (XBRL) format[14]. However, these are mostly fragmented across borders, limited in scope and size, and siloed within specific business communities. There are also a number of ontologies developed for capturing company and company-related data including the W3C Organisation ontology (ORG)[15], some e-Government Core Vocabularies[16], and the Financial Industry Business Ontology (FIBO) [4]. They have varying focuses, do not cover sufficiently the basic company information, or are too complex due to many ontological commitments [21].

There is so far no existing platform or KG (in whatever form) linking and provisioning cross-border and cross-language procurement and company data allowing advanced decision making, analytics, and visualisation.

3   Data Sets

The content of our KG is based on the procurement and company data that is provided by two main data providers extracting and aggregating data from multiple sources. The first one is OpenOpps[17], which is sourcing procurement data primarily from the Tenders Electronic Daily (TED)[18] data feed and from the procurement transparency initiatives of individual countries. TED is dedicated to European public procurement and publishes 520 thousand procurement notices a year. The second provider is OpenCorporates[19], which is collecting company data from national company registers and other regulatory sources. OpenOpps is the largest data source of European tenders and contracts, while OpenCorporates is the largest open database of companies in the world. Both OpenOpps and OpenCorporates gather relevant data using a range of tools, including processing API calls and Web scraping and data extraction.

Regarding procurement data, in the context of this work, OpenOpps provides gathered, extracted, pre-processed, and normalised data from hundreds of data sources completely openly through an API that can be used for research purposes. OpenOpps currently handles 685 data sources, with 569 of these being from Europe. This totals over 3 million documents dating back to 2010. All of the data for OpenOpps is gathered using a series of over 400 different scripts configured to collect data from each source. Each script is triggered daily and runs to gather all of the documents published in the last twenty-four hours. Each script is deployed on a monitored platform, giving the ability to check which scripts have failed, or which sources have published fewer than expected. Data is collected in

---

[12] http://www.ebr.org.
[13] https://brex.io.
[14] https://www.xbrl.org.
[15] https://www.w3.org/TR/vocab-org.
[16] https://joinup.ec.europa.eu/solution/e-government-core-vocabularies.
[17] https://openopps.com.
[18] https://ted.europa.eu.
[19] https://opencorporates.com.

the raw form and then mapped to the OCDS format after being cleansed. Where necessary, the data is processed, e.g., splitting single records into several fields, to comply with the data standard. Regarding company data, OpenCorporates provides more than 140 million company records from a large number of jurisdictions[20]. OpenCorporates also pre-processes and normalises data collected, maps collected data to its own data model, and makes data available through an API.

The data collected from OpenOpps and OpenCorporates is openly available under the Open Database License (ODbl)[21]. It is available on GitHub[22] in JSON format and is updated on a monthly basis. The data is also made available through Zenodo[23] with a digital object identifier (DOI) [26].

4   Knowledge Graph Construction

The KG construction process includes reconciling and linking the two aforementioned and originally disconnected data sets, and mapping and translating them into Linked Data with respect to an ontology network [24].

4.1   Ontology Network

We developed two ontologies, one for representing procurement data and one for company data, using common techniques recommended by well-established ontology development methods [6,18]. A bottom-up approach was used, including identifying the scope and user group of the ontology, requirements, and ontological and non-ontological resources. In general, we address suppliers, buyers, data journalists, data analysts, control authorities and regular citizens to explore and understand how public procurement decisions affect economic development, efficiencies, competitiveness, and supply chains. This includes providing better access to public tenders; spotting trends in spending and supplier management; identifying areas for cost cuts; and producing advanced analytics.

Regarding procurement data, we developed an ontology based on OCDS [25] – a relevant data model getting important traction worldwide, used for representing our underlying procurement data. The OCDS' data model is organised around the concept of a contracting process, which gathers all the relevant information associated with a single initiation process in a structured form. Phases of this process include mainly planning, tender, award, contract, and implementation. An OCDS document may be one of two kinds: a release or a record. A release is basically associated to an event in the lifetime of a contracting process and presents related information, while a record compiles all the known information about a contracting process. A contracting process may have many releases associated but only one record. We went through the reference specification of OCDS release and interpreted each of the sections and extensions

---

[20] https://opencorporates.com/registers.
[21] https://opendatacommons.org/licenses/odbl.
[22] https://github.com/TBFY/data-sources.
[23] https://zenodo.org.

(i.e., structured and unstructured). In total, there are currently 25 classes, 69 object properties, and 81 datatype properties created from the four main OCDS sections and 11 extensions. The core classes are `ContractingProcess`, `Plan`, `Tender`, `Award`, and `Contract`. A contracting process may have one planning and one tender stage. Each tender may have multiple awards issued, while there may be only one contract issued for each award. Other ontology classes include `Item`, `Lot`, `Bid`, `Organisation`, and `Transaction`. We reused terms from external vocabularies and ontologies where appropriate. These include Dublin Core[24], FOAF[25], Schema.org[26], SKOS[27], and the W3C Organisation ontology [28]. The OCDS ontology is available on GitHub in two versions[29]: one with the core OCDS terms and another with the extensions.

Regarding company data, one of the main resources used during the ontology development was data models provided by four company data providers: Open-Corporates, SpazioDati[30], Brønnøysund Register Centre[31], and Ontotext[32]. The data supplied by these data providers originally came from both official sources and unofficial sources. The need for harmonising and integrating data sets was a guiding factor for the ontology development process, since data sets have different sets of attributes and different representations with similar semantics. The resulting ontology, called euBussinessGraph ontology [21], is composed of 20 classes, 33 object properties, and 56 data properties allowing us to represent basic company-related data. The ontology covers registered organisations (i.e., companies that are registered as legal entities), identifier systems (i.e., a company can have several identifiers), officers (i.e., associated officers and their roles), and data sets (i.e., capturing information about data sets that are offered by company data providers). Registered organisations are the main entities for which information is captured in the ontology. The main classes include `RegisteredOrganisation`, `Identifier`, `IdentifierSystem`, `Person`, and `Dataset`. Three types of classifications are defined in the ontology for representing the company type, company status, and company activity. These are modelled as SKOS concept schemes. Some of the other external vocabularies and ontologies used are W3C Organisation ontology, W3C Registered Organisation Vocabulary (RegOrg)[33], SKOS, Schema.org, and Asset Description Metadata Schema (ADMS)[34]. The ontology, data sets and some examples are released as open source on GitHub[35].

---

[24] http://dublincore.org.
[25] http://xmlns.com/foaf/spec.
[26] https://schema.org.
[27] https://www.w3.org/2004/02/skos.
[28] https://www.w3.org/TR/vocab-org.
[29] https://github.com/TBFY/ocds-ontology/tree/master/model.
[30] http://spaziodati.eu.
[31] http://www.brreg.no.
[32] https://www.ontotext.com.
[33] https://www.w3.org/TR/vocab-regorg.
[34] https://www.w3.org/TR/vocab-adms.
[35] https://github.com/euBusinessGraph/eubg-data.

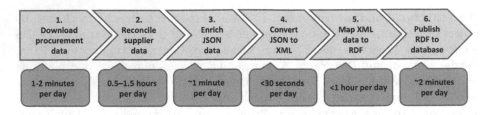

Fig. 1. The daily data ingestion process for the KG – on average 2500 OCDS releases are processed and 2400 suppliers (i.e., companies) are looked up per day.

4.2 Data Ingestion

The ingestion process extracts procurement and company data from the data providers, matches suppliers appearing in procurement data against company data (i.e., reconciliation), and translates the data sets into RDF using RML[36]. The daily process is composed of the following steps (see Fig. 1):

(1) **Download procurement data**: Downloads procurement data from the OpenOpps OCDS API[37] as JSON data files.
(2) **Reconcile suppliers**: Matches supplier records in awards using the Open-Corporates Reconciliation API[38]. The matching company data is downloaded using the OpenCorporates Company API[39] as JSON data files.
(3) **Enrich downloaded JSON data**: Enriches the JSON data files downloaded in steps 1 and 2, e.g., adding new properties to support the mapping to RDF (e.g., fixing missing identifiers).
(4) **Convert JSON to XML**: Converts the JSON data files from step 3 into corresponding XML data files. Due to limitations in JSONPath, i.e., lack of operations for accessing parent or sibling nodes from a given node, we prefer to use XPath as the query language in RML.
(5) **Map XML data to RDF**: Runs RML Mapper on the enriched XML data files from step 4 and produces N-Triples files.
(6) **Store and publish RDF**: Stores the RDF (N-Triples) files from step 5 to Apache Jena Fuseki and Apache Jena TBD.

We have been running the ingestion pipeline on a powerful server (see some performance metrics in Fig. 1), with the following hardware specifications: 2x Xeon Gold 6126 (12 Cores, 2.4 GHz, HT) CPU, 512 GB main memory, 1x NVIDIA Tesla K40c GPU, and 15 TB HDD RAID10 & 800 GB SSD storage. Python was used as the primary scripting language, RMLMapper was used as the mapping tool to generate RDF, and finally Apache Jena Fuseki & TDB was chosen as the SPARQL engine and triple store. The Python scripts operate on files (output and input) and services have been dockerised using Docker and made

---

[36] https://rml.io.
[37] https://openopps.com/api/tbfy/ocds.
[38] https://api.opencorporates.com/documentation/Open-Refine-Reconciliation-API.
[39] https://api.opencorporates.com/documentation/API-Reference.

available on Docker Hub[40] to ease deployment. All development work and results towards the creation of the knowledge graph are published and maintained as open source software on GitHub[41]. The data dumps of the KG, including more than 126M statements as of August 2020, are available on Zenodo [26].

5   Knowledge Graph Provisioning

We developed a platform and core API services for the KG ingestion and provisioning, using recent Linked Data and REST API design practices and principles.

5.1   Platform Architecture

Our platform follows state-of-the-art principles in software development, considering a low decoupling amongst all the software components.

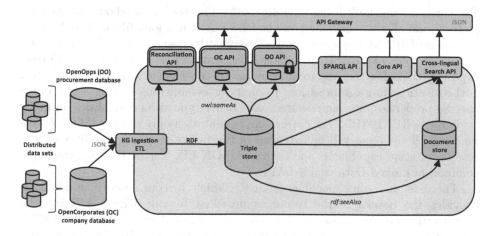

Fig. 2. The high-level architecture for KG ingestion and provisioning.

Figure 2 provides a high-level overview of the architecture. On the left-hand side, we include the ETL processes that are being used to incorporate the data sources into the KG. On the right-hand side we provide an overview of the main data storage mechanisms, including a triple store for the generated RDF-based data and a document store for the documents associated to public procurement (tender notices, award notices, etc.), whose URLs are accessible via specific properties of the KG (using `rdfs:seeAlso`). For those specific cases where a URI is also available in the original data sources (from OpenOpps and OpenCorporates), such URI is provided in the KG using a statement with `owl:sameAs`.

---

[40] https://hub.docker.com/r/tbfy/kg-ingestion-service.
[41] https://github.com/TBFY/knowledge-graph.

This would allow our data providers to provide additional information about tenders or companies with a different license or access rights (e.g., commercial use).

The KG is accessible via a core REST API. Our API catalogue is mostly focused on providing access mechanisms to those who want to make use of the knowledge graph, particularly software developers. Therefore, they are mostly focused on providing access to the KG through the HTTP GET verb and the API catalogue is organised around the main entities that are relevant for public procurement, as discussed in Sect. 4, such as contracting processes, awards, and contracts. Since the KG is stored as RDF in a triple store, there is also a SPARQL endpoint[42] for executing ad-hoc queries. Finally, there is a cross-lingual search API for searching across documents in various languages and an API Gateway providing a single-entry point to the APIs provided by the platform.

5.2   Core API

The core API was built using the R4R tool[43]. This tool is based on Velocity templates[44] and allows specifying how the REST API will look like and configure it by means of SPARQL queries, similarly to what has been proposed in other state of the art tools like BASIL (Building Apis SImpLy) [7] or GRLC [14]. Beyond exposing URIs for the resources available in the KG, it also allows including authentication and authorisation, pagination, establishing sorting criteria over specific properties, nesting resources, and other typical functionalities normally available in REST APIs. The current implementation only returns JSON objects for the API calls and will be extended in the future to provide additional content negotiation capabilities and formats (JSON-LD, Turtle, HTML), which are common in Linked Data enabled APIs.

There is an online documentation[45], which is continuously updated. It provides the details of the resources provided by our REST API in relation to the OCDS ontology. The core resources derived from the OCDS ontology are: (i) `ContractingProcess`, (ii) `Award`, (iii) `Contract`, (iv) `Tender`, and (v) `Organisation`. For all these resources, there is a possibility of paginating (e.g., `GET /award?size=5&offset=1`), sorting (e.g., `GET /contract?sort=-startDate`), and filtering (e.g., by the title of the award: `GET /award?status=active`).

6   Knowledge Graph in Use

We implemented a number of real-life use cases on the platform and KG: anomaly detection and cross-lingual document search.

---

[42] http://yasgui.tbfy.eu.
[43] https://github.com/TBFY/r4r.
[44] https://velocity.apache.org.
[45] https://github.com/TBFY/knowledge-graph-API/wiki.

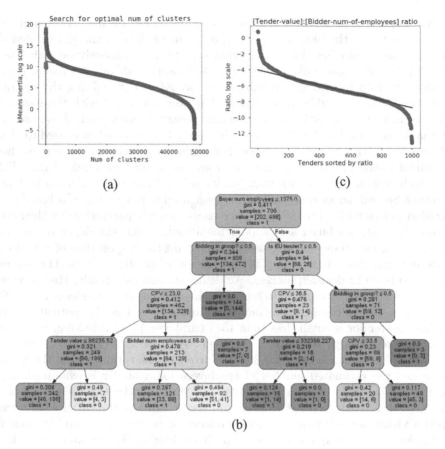

Fig. 3. (a) Anomaly detection in public procurement data with k-Means analysis. (b) The decision tree model for identifying successful tenders. (c) A graph showing interdependence between tender value and number of employees of bidder.

6.1   Anomaly Detection

Public procurement is particularly susceptible to corruption, which can impede economic development, create inefficiencies, and reduce competitiveness. At the same time, manually analysing a large volume of procurement cases for detecting possible frauds is not feasible. In this respect, using ML techniques for identifying patterns and anomalies, such as fraudulent behaviour or monopolies, in procurement processes and networks across data sets produced independently, is highly relevant [5]. For example, by building a network of entities (individuals, companies, governmental institutions, etc.) connected through public procurement events, one can discover exceptional cases as well as large and systematic patterns standing out from the norm, whether they represent examples of good public procurement practice or possible cases of corruption.

We applied several ML techniques, i.e., supervised, unsupervised, and statistical, on top of the Slovenian public procurement data in the KG to identify patterns and anomalies. First, clustering was used for anomaly detection (see Fig. 3 (a)), since one could quickly spot deviations with this approach. Every tender was transformed into a feature vector. After determining the optimal number of clusters, public procurement data were clustered with the K-means algorithm. Vectors deviating most from their centroids are identified and ordered by the deviation value (i.o., Cartesian distance). The approach was used to identify tenders with highest deviations. For example, among others, our method identified public procurement cases with an unusually high tender value. This approach gives a hint on data that "stick out" and are worth of more in-depth scrutiny. Second, supervised analysis implemented in our platform is based on a decision tree (see Fig. 3 (b)). We enabled users to select parameters by their own choice (for instance buyer size, bidder municipality, and the depth of decision tree model), and thus enabling users to compare the importance of subsets of various parameters contributing to the success of public tenders. The success definition is up to decision makers. We define success as a tender that received more than one bid. According to our preliminary analysis, a tender is successful – i.e. there will be competition (more than one bid) - if public institution who opened the tender is small (less than 1375 employees) and if bidding is done in group. Third, statistical approach was used to deal with various ratios between pre-selected parameters (see Fig. 3 (c)). Currently, the ratio between the tender value and the estimated number of employees for a bidder is examined. Bidders are then sorted by their ratio value and every bidder turned into a point: the x value is a consecutive number and the y value is the ration. We developed a visual presentation of interdependence of tender value and the number of employees. The graph shows deviating behaviour at the beginning as well as at the end of the list. On the upper left corner of the graph, we can see big companies with a high number of employees that won small tenders, and on the bottom right corner, there are companies with a small number of employees that won big tenders.

We implemented a system capable of processing tens of millions of records, based on the techniques mentioned and made it available online[46].

6.2   Cross-Lingual Document Search

Procurement processes are not only creating structured data, but also constantly creating additional documents (tender specifications, contract clauses, etc.). These are commonly published in the official language of the corresponding public administrations. Only some of these, for instance those published in TED, are multilingual, but the documents in the local language are typically longer and much more detailed than their translations into other languages. A civil servant working at a public administration on a contracting process may be interested in understanding how other public administrations in the same

---

[46] http://tbfy.ijs.si.

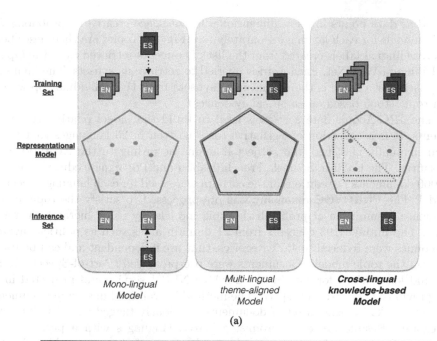

(a)

Topic3@EN	Topic3@ES	Topic26@FR	Topic10@PT
Communication Systems	Sistema de Comunicación	Systeme de Comunicacion	Communication Systems
radio	equipo	communications	rede
equipment	red	reseaux	comunicação
network	comunicación	electroniques	electrónico
communication	espectro	acces	acesso
regulatory	electromagnético	telecommunications	utilizador

(b)

Fig. 4. (a) Documents are represented in a unique space that relies on the latent layer of cross-lingual topics obtained by LDA and hash functions through hierarchies of synsets. (b) Theme-aligned topics described by top 5 words based on EUROVOC annotations.

country or in different countries (and with different languages) have worked on similar contexts. Examples may include finding organisations related to a particular procurement process, or search for tenders related to given procurement text.

We worked on an added-value service[47] in order to provide support to these types of users, with the possibility of finding documents that are similar to a given one independently of the language in which it is made available. We also generated a Jupyter notebook with some representative examples, so as to facilitate its use[48]. This service is based on the use of unsupervised probabilistic topic models, based on cross-lingual labels from sets of cognitive synonyms (synsets) to establish relations between language-specific topics [2]. Documents are repre-

---

[47] http://tbfy.librairy.linkeddata.es/search-api.
[48] http://bit.ly/tbfy-search-demo.

sented as data points in a low-dimensional latent space created by probabilistic topic models for each language separately (see Fig. 4). Topics are then described by cross-lingual labels created from the list of concepts retrieved from the Open Multilingual WordNet. Each word is queried to retrieve its synsets. The final set of synsets for a topic is the union of the synsets from the individual top-words of a topic (top5 based on empirical evidences).

The JRC-Acquis data set[49] was used to build the model relating the documents. It is a collection of legislative texts written in 23 languages that have been manually classified into subject domains according to the EUROVOC[50] thesaurus. The English, Spanish, French, Italian and Portuguese editions (about 20.000 documents per edition) of the corpora were used for each language-specific model. The EUROVOC taxonomy was pre-processed to satisfy the topic independence assumption of probabilistic topic models, by using hierarchical relations. The initial 7.193 concepts from 21 domain areas such as politics, law or economics were reduced to 452 categories, that are independent and can be used to train the topic models. Documents were pre-processed (Part-of-Speech filtering and lemmatized format) by the librAIry NLP[51] service and projected into the previously created topic space. The method is evaluated in several document retrieval tasks by using a set of documents previously tagged with EUROVOC categories. Results are quite promising across languages with a performance close to 0.8 in terms of accuracy, although a better performance is achieved with English texts.

7   Adoption and Uptake

We have used Semantic Web technologies to integrate disparate open data sources in a standardised way. They enabled us to ingest new data sources and integrate other relevant data sources (e.g., company and procurement data) without major restructuring efforts. Similar solutions could be provided using other technologies; however, without following the Linked Data and Semantic Web principles, they would rather remain ad-hoc and could not be easily scaled, given various independent data publishers and consumers.

The uptake of our platform and KG has been exemplified in four different cases so far. The core API and KG are used by the Spanish company OESIA[52] and by the city of Zaragoza, Spain. OESIA created a commercial tool for tender analysis, which is offered to SMEs. Zaragoza includes economic information in their transparency portal[53], including public procurement. Regarding advanced tools, the anomaly detection tool is used by the Ministry of Public Administration in Slovenia for detecting procurement anomalies, while the cross-lingual

---

[49] https://ec.europa.eu/jrc/en/language-technologies/jrc-acquis.
[50] http://eurovoc.europa.eu.
[51] http://librairy.linkeddata.es/nlp.
[52] https://grupooesia.com/en/.
[53] https://zaragoza.es/sede/servicio/transparencia.

similarity search is used by the Italian company CERVED[54] for finding tenders in other countries/languages and offering this as part of their services. The categories of users using the system include civil servants (i.e., Zaragoza and Slovenia), citizens (i.e., Zaragoza), and companies, especially SMEs (i.e., CERVED and OESIA). As of August 2020, over 3.700 queries have been submitted to the system APIs.

We plan to maintain the KG in the context of already funded innovation projects. Maintenance will include ingesting new data and operating the system. Agreements with data providers, i.e., OpenOpps and OpenCorporates, have been established to provide the KG with data on a continuous basis. Furthermore, the data and platform components are made available openly for the community to contribute (a catalogue is available[55]). We are also proposing our ontology network as the way to publish open data about procurement by governments. An example is the case of Zaragoza, which already adopted our ontology network.

8 Lessons Learned

There are plenty of lessons learned in the context of this work, which may be applicable to the construction of other KGs in similar or different domains. First, we provide a non-exhaustive list of major takeaways related to the whole process:

(i) The KG enabled easier and advanced analytics, which was otherwise not possible, by connecting companies (i.e., suppliers) appearing in the procurement data set to companies in company data set. However, getting and preprocessing the data (e.g., data curation) was a major and time-consuming task, requiring attention from national and EU-wide data providers.

(ii) The existing Semantic Web technologies and tools scaled well for ingesting and provisioning large amounts of data and RESTful approach was useful for bringing the Linked Data to non-Semantic Web application developers. However, more support is required such as visual editors for creating mapping definitions and specifying data transformations.

(iii) The process of building a high-quality KG that can be used extensively by users would be clearly improved if all data sources were providing their procurement data in a more structured manner. Data quality problems are still a relevant issue, as described in the followings, and reduce the result quality of ML processes such as anomaly detection and reconciliation.

(iv) There are still many documents associated to the procurement processes that are provided as PDFs (in some cases even scanned PDFs). Providing all documents in the form of raw texts as well would simplify the processing that needs to be done, and would allow applying more easily the techniques like the ones described for cross-lingual search.

(v) Data providers should also aim at publishing the information of all types of contracting processes that they are handling, independently of their size.

---

[54] https://company.cerved.com.
[55] https://tbfy.github.io/platform.

Currently, due to many types of regulations across countries, not all contracting processes (especially the smallest ones) are published.

We also faced a high number of data quality issues, even though there are mandates in place for buyers to provide correct data. This particularly applies to procurement data sources. These data quality issues could be classified as:

(i) *Missing data*: It is frequent that data is missing. Among others, the least frequently completed field in the tender and contracting data is the *value* field; it is usually completed in less than 10% of tender notices. One item of data that is particularly important to procurement transparency is the reference data required to link a contract award to a tender notice (very common in the TED data). We found that just 9% of award notices had provided a clear link between tenders and contracts. Subsequently, the majority of contract award notices had been orphaned and there was no link to the source tenders.

(ii) *Duplicate data*: Publishers frequently publish to multiple sources in order to meet the legal requirements of their host country and that of the European Union. This means that all over-threshold tenders are available at least twice. The task of managing duplicates is not always simple. It is common for different publishing platforms to have different data schemas and interoperability between schemas is not guaranteed.

(iii) *Poorly formed data*: Sources are frequently providing malformed data or data that cannot be reasonably parsed by code. The tender and contract value field can often include string values rather than numbers (same goes for the dates). Across the sources, approach to using character delimiters in value data is frequently heterogeneous, with different nationalities using different delimiters to separate numbers and to indicate decimals.

(iv) *Erroneous data*: Structured data such as numeric and date records are frequently a problem. Buyers often submit zero value entries in order to comply with the mandate and the lack of validation on date related data has allowed buyers to record inconsistent date data. There are some contracts where the date of publication exceeds the end date of the contract or the start date of the contract is greater than the end date of the contract.

(v) *Absent data fields*: In some cases, the sources lack core pieces of information, for instance, there is no value field in a number of European sources. A large number of sites also fail to publish the currency of their monetary values. In all cases, if a publisher sought to add the additional information, such as a different currency, there would be no capacity in the system to provide the information required in a structured form.

Most of these problems can be resolved through the use of standards and validation at the point of data entry. Requiring buyers to publish records to a standard would, in turn, require the platform providers to both mandate the field format and validate data entries. The usage of an ontology network for the development of the KG allowed us to inform public administrations willing to provide data on the minimum set of data items that are needed, and some of them are already adapting their information systems for this purpose [9].

Acknowledgements. The work reported in this paper is partly funded by EC H2020 TheyBuyForYou (780247) and euBusinessGraph (grant 732003) projects.

References

1. Alvarez-Rodríguez, J.M., et al.: New trends on e-Procurement applying semantic technologies. Comput. Ind. **65**(5), 797–799 (2014)
2. Badenes-Olmedo, C., et al.: Scalable cross-lingual similarity through language-specific concept hierarchies. In: Proceedings of K-CAP 2019, 147–153 (2019)
3. Bansal, S.K., Kagemann, S.: Integrating big data: a semantic extract-transform-load framework. Computer **48**(3), 42–50 (2015)
4. Bennett, M.: The financial industry business ontology: best practice for big data. J. Banking Regul. **14**(3), 255–268 (2013)
5. Chandola, V., et al.: Anomaly detection: a survey. ACM Comput. Surv. **41**(3) (2009)
6. Corcho, O., et al.: Ontological engineering: principles, methods, tools and languages. In: Calero, C., Ruiz, F., Piattini, M. (eds.) Ontologies for Software Engineering and Software Technology, pp. 1–48. Springer, Berlin (2006). https://doi.org/10.1007/3-540-34518-3_1
7. Daga, E., et al.: A BASILar approach for building web APIs on top of SPARQL endpoints. In: Proceedings of SALAD 2015. CEUR-WS.org (2015)
8. Distinto, I., et al.: LOTED2: an ontology of European public procurement notices. Semantic Web **7**(3), 267–293 (2016)
9. Espinoza-Arias, P., et al.: The zaragoza's knowledge graph: open data to harness the city knowledge. Information **11**(3) (2020)
10. The Cost of Non-Europe in the area of Organised Crime and Corruption. Technical report, European Parliment (2016). https://www.europarl.europa.eu/RegData/etudes/STUD/2016/579319/EPRS_STU%282016%29579319_EN.pdf
11. Futia, G., Melandri, A., Vetrò, A., Morando, F., De Martin, J.C.: Removing barriers to transparency: a case study on the use of semantic technologies to tackle procurement data inconsistency. In: Blomqvist, E., Maynard, D., Gangemi, A., Hoekstra, R., Hitzler, P., Hartig, O. (eds.) ESWC 2017. LNCS, vol. 10249, pp. 623–637. Springer, Cham (2017). https://doi.org/10.1007/978-3-319-58068-5_38
12. Giese, M., et al.: Optique: zooming in on big data. Computer **48**(3), 60–67 (2015)
13. Janssen, M., Konopnicki, D., Snowdon, J.L., Ojo, A.: Driving public sector innovation using big and open linked data (BOLD). Inf. Syst. Front. **19**(2), 189–195 (2017). https://doi.org/10.1007/s10796-017-9746-2
14. Meroño-Peñuela, A., Hoekstra, R.: grlc makes GitHub taste like linked data APIs. In: Sack, H., Rizzo, G., Steinmetz, N., Mladenić, D., Auer, S., Lange, C. (eds.) ESWC 2016. LNCS, vol. 9989, pp. 342–353. Springer, Cham (2016). https://doi.org/10.1007/978-3-319-47602-5_48
15. Miroslav, M., et al.: Semantic technologies on the mission: preventing corruption in public procurement. Comput. Industry **65**(5), 878–890 (2014)
16. Muñoz-Soro, J., et al.: PPROC, an ontology for transparency in public procurement. Semantic Web **7**(3), 295–309 (2016)
17. Necaský, M., et al.: Linked data support for filing public contracts. Comput. Industry **65**(5), 862–877 (2014)
18. Noy, N.F., McGuinness, D.L.: Ontology development 101: a guide to creating your first ontology. Technical report, Stanford Medical Informatics (2001)

19. OECD Principles for Integrity in Public Procurement. Technical report, OECD (2009). http://www.oecd.org/gov/ethics/48994520.pdf
20. Rodríguez, J.M.Á., et al.: Towards a pan-European E-procurement platform to aggregate, publish and search public procurement notices powered by linked open data: the moldeas approach. Int. J. Softw. Eng. Knowl. Eng. **22**(3), 365–384 (2012)
21. Roman, D., et al.: The euBusinessGraph ontology: a lightweight ontology for harmonizing basic company information. Semantic Web under Rev. (2020). http://www.semantic-web-journal.net/system/files/swj2421.pdf
22. Simperl, E., et al.: Towards a knowledge graph based platform for public procurement. In: Garoufallou, E., Sartori, F., Siatri, R., Zervas, M. (eds.) MTSR 2018. CCIS, vol. 846, pp. 317–323. Springer, Cham (2019). https://doi.org/10.1007/978-3-030-14401-2_29
23. Soylu, A., et al.: Towards integrating public procurement data into a semantic knowledge graph. In: Proceedings of EKAW 2018 Poster and Demonstrations. CEUR-WS.org (2018)
24. Soylu, A., et al.: An overview of the TBFY knowledge graph for public procurement. In: Proceedings of ISWC 2019 Satellite Tracks. CEUR-WS.org (2019)
25. Soylu, A., Elvesæter, B., Turk, P., Roman, D., Corcho, O., Simperl, E., Konstantinidis, G., Lech, T.C.: Towards an ontology for public procurement based on the open contracting data standard. In: Pappas, I.O., Mikalef, P., Dwivedi, Y.K., Jaccheri, L., Krogstie, J., Mäntymäki, M. (eds.) I3E 2019. LNCS, vol. 11701, pp. 230–237. Springer, Cham (2019). https://doi.org/10.1007/978-3-030-29374-1_19
26. TBFY: KG data dump (2020). https://doi.org/10.5281/zenodo.3712323
27. Yan, J., Wang, C., Cheng, W., Gao, M., Zhou, A.: A retrospective of knowledge graphs. Front. Comput. Sci. **12**(1), 55–74 (2018). https://doi.org/10.1007/s11704-016-5228-9

The OpenCitations Data Model

Marilena Daquino[1,2]([✉]) [iD], Silvio Peroni[1,2] [iD], David Shotton[2,3] [iD],
Giovanni Colavizza[4] [iD], Behnam Ghavimi[5] [iD], Anne Lauscher[6] [iD],
Philipp Mayr[5] [iD], Matteo Romanello[7] [iD], and Philipp Zumstein[8] [iD]

[1] Digital Humanities Advanced Research Centre (/DH.arc),
Department of Classical Philology and Italian Studies,
University of Bologna, Bologna, Italy
{marilena.daquino2,silvio.peroni}@unibo.it
[2] Research Centre for Open Scholarly Metadata,
Department of Classical Philology and Italian Studies,
University of Bologna, Bologna, Italy
[3] Oxford e-Research Centre, University of Oxford, Oxford, UK
david.shotton@opencitations.net
[4] Institute for Logic, Language and Computation (ILLC), University of Amsterdam,
Amsterdam, The Netherlands
g.colavizza@uva.nl
[5] Department of Knowledge Technologies for the Social Sciences,
GESIS - Leibniz-Institute for the Social Sciences, Mannheim, Germany
ghavimi.behnam@gmail.com, philipp.mayr@gesis.org
[6] Data and Web Science Group, University of Mannheim, Mannheim, Germany
anne@informatik.uni-mannheim.de
[7] École Polytechnique Fédérale de Lausanne, Lausanne, Switzerland
matteo.romanello@epfl.ch
[8] Mannheim University Library, University of Mannheim, Mannheim, Germany
philipp.zumstein@bib.uni-mannheim.de

Abstract. A variety of schemas and ontologies are currently used for
the machine-readable description of bibliographic entities and citations.
This diversity, and the reuse of the same ontology terms with different nuances, generates inconsistencies in data. Adoption of a single data
model would facilitate data integration tasks regardless of the data supplier or context application. In this paper we present the OpenCitations
Data Model (OCDM), a generic data model for describing bibliographic
entities and citations, developed using Semantic Web technologies. We
also evaluate the effective reusability of OCDM according to ontology
evaluation practices, mention existing users of OCDM, and discuss the
use and impact of OCDM in the wider open science community.

Authors' contributions specified according to the CrediT taxonomy. MD: Conceptualization, Data curation, Formal analysis, Investigation, Methodology, Software, Writing
– original draft, Writing – review & editing. She is responsible for section "Background"
and section "Analysis of OCDM reusability". SP and DS: Conceptualization, Investigation, Methodology, Software, Data curation, Supervision, Funding acquisition, Project
administration, Writing – original draft. GC, BG, AL, PM, MR and PZ: Investigation,
Resources, Validation, Writing – review & editing.

© Springer Nature Switzerland AG 2020
J. Z. Pan et al. (Eds.): ISWC 2020, LNCS 12507, pp. 447–463, 2020.
https://doi.org/10.1007/978-3-030-62466-8_28

Keywords: Open citations · Scholarly data · Data model

1 Introduction

In recent years, largely thanks to the Initiative for Open Citations (I4OC)[1], most major scholarly publishers have made their bibliographic reference data open, resulting, for example, in more than 700 million citations now being made openly available in the OpenCitations Index of Crossref open DOI-to-DOI citations (COCI) [17]. As a consequence, scholarly data providers and bibliometric analysis software have started to integrate open citation data into their services, thereby offering an alternative to the current reliance on proprietary citation indexes.

Open bibliographic and citation metadata are beneficial because they enable anyone to perform meta-research studies on the evolution of scholarly knowledge, and allows national and international research assessment exercises characterized by transparent and reproducible processes. Within this context, bibliographic citations are essential components of scholarly discourse, since they "remain the dominant measurable unit of credit in science" [12]. They carry evidence of scholarly networks and of the progress of theories and methods, and are fundamental aids in tenure evaluation and recommendation systems. To perform open bibliometric research and analysis, the publications upon which the work is based should be FAIR, namely Findable, Accessible, Interoperable, and Reusable [35]. Ideally, such data should be made available without any restrictions, licensed under a Creative Commons CC0 waiver[2], and the software for programmatically accessing and analysing them should be also released with open source licences.

However, data suppliers use a variety of licenses, technologies, and vocabularies for representing the same bibliographic information, or use ontology terms defined in the same ontologies with different nuances, thereby generating diversity in data representation. The adoption of a common, generic, open and documented data model that employs clearly defined ontological terms would ensure data consistency and facilitate integration tasks.

In this paper we present the OpenCitations Data Model (OCDM), a data model based on existing ontologies for describing information in the scholarly bibliographic domain with a particular focus on citations. OCDM has been developed by OpenCitations [29], an infrastructure organization for open scholarship dedicated to the publication of open bibliographic and citation data using Semantic Web technologies. Herein, we propose a holistic approach for evaluating the reusability of OCDM according to ontology evaluation methodologies, and we discuss its uptake, impact, and trustworthiness.

We compared OCDM to similar existing solutions and found that, to the best of our knowledge, OCDM (a) has the broadest vocabulary coverage, (b) is the best documented data model in this area, and (c) has already a significant uptake in the scholarly community. The main advantages of OCDM, in addition

---

[1] https://i4oc.org.
[2] https://creativecommons.org/publicdomain/zero/1.0/legalcode.

to the consistency of data description that it facilitates, are that it was designed from the outset to enable use by those who are not Semantic Web practitioners, as well as by those that are, that it is properly documented, and it is provided with accompanying software for managing the entire life-cycle of data created according to OCDM.

The paper is organized as follows. In Sect. 2 we clarify the scope and motivations for this work. In Sect. 3 we present the data model and its documentation, software and current early adopters. In Sect. 4 we present the criteria we have used to evaluate OCDM reusability and we present results, including figures about OCDM views, downloads and citations according to Figshare and Altmetrics, which are further discussed in Sect. 5.

2   Background

The OpenCitations Data Model (OCDM) [9] was initially developed in 2016 to describe the data in the OpenCitations Corpus (OCC). In recent years OpenCitations has developed other datasets while OCDM has been adopted by external projects, and OCDM has been expanded to accommodate these changes. We have recently further expanded the OpenCitations Data Model to accommodate the extended metadata requirements of the Open Biomedical Citations in Context Corpus project (CCC). This project has developed an exemplar Linked Open Dataset that includes detailed information on citations, in-text reference pointers such as "Berners-Lee et al. 2011", and identifiers of the citation contexts (e.g. sentences, paragraphs, sections) within which in-text reference pointers are located, to facilitate textual analysis of citation contexts. The citations are treated as first-class data entities [26], enriched with open bibliographic metadata released using a CC0 waiver that can be mined, stored and republished. This includes identifiers specifying the specific positions of the various in-text reference pointers within the text. However, the literal text of these contexts are not stored within the Open Biomedical Citations in Context Corpus, and regrettably in many cases the full text of the published entities cannot be mined from elsewhere in an open way, even for some (view only) Open Access articles, because of copyright, licensing and other Intellectual Property (IP) restrictions.

Table 1 shows the representational requirements (hereinafter, for the sake of simplicity, also called citation properties and numbered (P1–P8)) that we were interested in recording for each citation instantiated from within a single paper.

3   The OpenCitations Data Model

The OCDM permits one to record metadata about bibliographic references and their textual contexts, bibliographic entities (citing and cited publications) and the citations that link them, agents and their roles (e.g. author, editor), identifiers for the foregoing entities, provenance metadata and much more, as shown diagrammatically in Fig. 1. All terms described in the OCDM are brought together in the OpenCitations Ontology (OCO)[3]. OCO aggregates terms from

---

[3] https://w3id.org/oc/ontology.

Table 1. Representational requirements of the OpenCitations Data Model

ID	Description
P1	A classification of the type of citation (e.g. self-citation)
P2	The bibliographic metadata of the citing and cited bibliographic entities (e.g. type of published entity, identifiers, authors, contributors, publication date, publication venues, publication formats)
P3	The bibliographic reference, typically found within the reference list of the citing bibliographic entity, that references a cited bibliographic entity
P4	The separate identifiers of all the in-text reference pointers included in the text of the citing entity, that denote bibliographic references within the reference list
P5	The co-occurrence of in-text reference pointers within each in-text reference pointer lists (e.g. "[3, 5, 12]")
P6	The identifiers of structural elements (e.g. XPath of sentences, paragraphs, captions) that specify where, in the full text, an in-text reference pointer appears
P7	The function or purpose of the citation (e.g. to cite as background, extend, or agree with the cited entity) to which each in-text reference pointer relates
P8	Provenance information of the citation extraction process (e.g. responsible agents, data sources, extraction dates)

the SPAR (Semantic Publishing and Referencing) Ontologies [28] and other well-known ontologies, such as PROV-O [4] and Web Annotation Ontology [32].

Citations are instances of the class `cito:Citation` defined in CiTO, the Citation Typing Ontology[4]. Subclasses (not shown in Fig. 1), relevant for P1, include `cito:AuthorSelfCitation`, `cito:JournalSelfCitation`, `cito:FunderSelf-Citation`, `cito:AffiliationSelfCitation`, and `cito:AuthorNetworkSelf-Citation`. In addition, citations can be characterized with a purpose or function with respect to the related citation context, by means of the property `cito:hasCitationCharacterisation` and the use of one or more CiTO properties (e.g. `cito:usesMethodIn`) (P7).

Instances of the class `fabio:Expression`, defined in the FRBR-aligned Bibliographic Ontology (FaBiO)[5], can be linked to bibliographic metadata such as publication dates, authors, and venues. Instances of `fabio:Manifestation` aggregate information on specific editions and formats (P2).

Instances of `oa:Annotation`, defined in the Web Annotation Ontology (OA)[6], link instances of the class `cito:Citation` to instances of `biro:Bibliographic-Reference` (P3), defined in BiRO, the Bibliographic Reference Ontology[7], and individuals of `c4o:InTextReferencePointer` (P4), defined in C4O, the Citation

---

[4] http://purl.org/spar/cito.
[5] http://purl.org/spar/fabio.
[6] https://www.w3.org/ns/oa.
[7] http://purl.org/spar/biro.

Counting and Context Characterisation Ontology[8]. Lists of in-text reference pointers are represented by the class c4o:SingleLocationPointerList (P5).

Structural elements wherein in-text reference pointers appear are represented as individuals of deo:DiscourseElement, defined in DEO, the Discourse Element Ontology[9]. Elements are uniquely identified (P6) by means of instances of datacite:Identifier, defined in the DataCite Ontology[10].

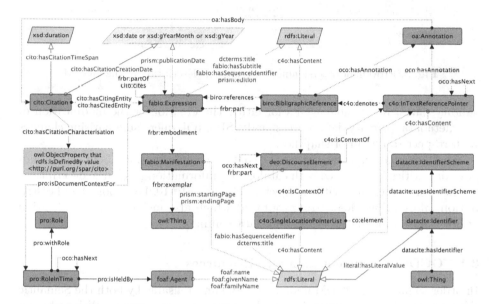

Fig. 1. Main classes and properties of the OpenCitations Ontology

Finally, as summarized in Fig. 2, OCDM provides guidance for describing the provenance and versioning of each entity under consideration, and also enables the specification of the main metadata related to the datasets containing such entities (P8). To this end, the OCDM reuses terms from PROV-O, the Provenance Ontology[11], VoID, the Vocabulary of Interlinked Datasets[12] [2], and DCAT, the Data Catalog Vocabulary[13] [24].

Each bibliographic entity described by the OCDM is annotated with one or more provenance snapshots (i.e. instances of prov:Entity, each snapshot intended as a specialisation of the bibliographic entity via prov:specializationOf) as defined in [30]. In particular, each snapshot records the set of statements having the bibliographic entity as its subject at a fixed point in time,

[8] http://purl.org/spar/c4o.
[9] http://purl.org/spar/deo.
[10] http://purl.org/spar/datacite.
[11] http://www.w3.org/ns/prov.
[12] http://rdfs.org/ns/void.
[13] http://www.w3.org/ns/dcat.

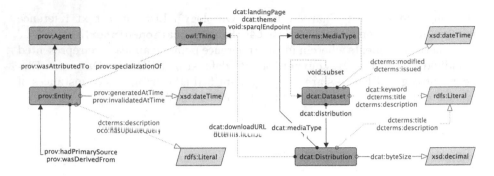

Fig. 2. Provenance, versioning, and dataset description in the OCDM

validity dates, responsible agents for either the creation or the modification of the metadata, primary data sources, and a SPARQL query summarising changes with respect to any prior snapshot.

Lastly, a dataset (`dcat:Dataset`) containing information about the bibliographic entities is described with cataloguing information (e.g. title, description, publication and change dates, subjects, webpage, SPARQL endpoint) and distribution information (`dcat:Distribution`) which also includes the specification of licenses, dumps, media types, and data volumes.

3.1   OCDM Documentation and Resources

In order to make the OCDM understandable and reusable by both the Semantic Web community and communities with no expertise in Semantic Web technologies, support material has been produced. All materials are available at http://opencitations.net/model and include the following resources (Fig. 3).

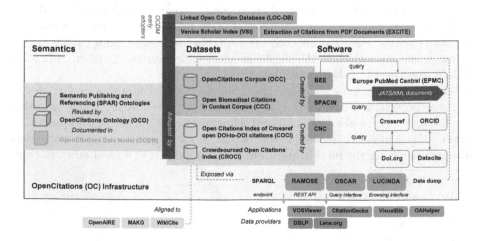

Fig. 3. Overview of OpenCitations ecosystem and acronyms used in this paper

Human-readable documentation. The OCDM documentation [9] provides (1) detailed definitions of terms characterising open citation data and open bibliographic metadata, (2) naming conventions and URI patterns, and (3) real-world examples. OCDM is supplemented by two additional specifications, i.e. the definition of the Open Citation Identifier (OCI) [26] and the definition of the In-Text Reference Pointer Identifier (InTRePID) [33].

OCDM-compliant data examples. All the data introduced in the OCDM documentation are expressed and provided in JSON-LD to make it easily understandable both to RDF experts and other Web users. In addition, CSV templates have been adopted so as to express and share parts of the OCDM – e.g. to store the citation data in COCI [17].

Ontology development documentation. The first version of the OCDM, released in 2016, addressed citation properties P1–P3 and P8, by directly reusing the SPAR Ontologies and other vocabularies [28]. Within the context of the CCC project described above, we used SAMOD [27], an agile data-driven methodology for ontology development, to extend OCO with terms relevant to P4–P7. Motivating scenarios, competency questions, and a glossary of terms of all the new entities included in the OCDM, are available for reproducibility purposes.

Open source software leveraging the data model. The source code of the knowledge extraction and data re-engineering pipeline for managing data according to OCDM is available at http://opencitations.net/tools. The pipeline includes software originally developed for creating the OpenCitations Corpus (BEE and SPACIN) and the OpenCitations Indexes (Create New Citations – CNC), and a user-friendly web application (BCite) [10] for creating OCDM-compliant RDF data from lists of bibliographic references. In addition, we have released tools to support the development of applications leveraging data organized according to OCDM: RAMOSE (to create RESTful APIs over SPARQL endpoints), OSCAR (to create user-friendly search interfaces for querying SPARQL endpoints [16]) and LUCINDA (a configurable browser for RDF data). Configuration files for setting up these tools are available in their GitHub repositories.

Licenses for reuse. OCDM (both the documentation and OCO) is released under a CC-BY license. Software solutions are released under the ISC license. The OCDM-compliant data served by OpenCitations are made open under CC0.

3.2 OCDM Early Adopters

To date, OCDM is central to the work of OpenCitations. The OpenCitations datasets modelled using OCDM include: the OpenCitations Corpus (OCC), including about 13 million citation links and the OpenCitations Indexes, which include more than 721 million citations. Forthcoming datasets, that will be released later in 2020, include OpenCitations Meta, which stores metadata of the citing and cited entities involved in the citations included in the Indexes, and the Open Biomedical Citations in Context Corpus (CCC), mainly derived from the Open Access corpus of biomedical articles provided by PubMed Central, that will include detailed information on in-text reference pointers denoting each reference in the reference list, and their textual contexts.

Moreover, OCDM has three external acknowledged early adopters. The Extraction of Citations from PDF Documents (EXCITE) project [20] is run by GESIS and the University of Koblenz. The aim of EXCITE is to extract and match citations from social science publications. To date, EXCITE has extracted around 1 million citations, has converted the data to RDF according to OCDM, and has then published it by ingestion into the OCC.

The Linked Open Citation Database (LOC-DB) [21] is a project which aims to demonstrate that it is possible for academic libraries to catalogue citation relations sustainably, accurately, and cooperatively. So far, the project has stored bibliographic and citation data for about 7000 published entities. LOC-DB has used a customisation of the OCDM as the data model for defining its data, and exports data in OCDM/JSON-LD so as to be ingested into the OCC.

The Venice Scholar Index (VSI)[14] is an instance of the Scholar Index, originated from the "Linked Books" project [8] founded by the Swiss National Science Foundation. The citation index includes about 4 million references to publications cited in the historiography of Venice. VSI exports data into RDF formats according to OCDM so as to be integrated into the OCC.

4   Analysis of OCDM Reusability

A holistic approach has been used to evaluate the OCO ontology and to infer properties relevant to OCDM. We adopted seminal definitions and classifications of ontology evaluation approaches [6, 14] and we selected the following dimensions and approaches that are representative with respect to OCDM reusability.

[E1] Lexical keyword similarity. This addresses the similarity of definitions (labels of terms) in OCO with respect to the real-world knowledge to be mapped. We adopted a data-driven evaluation [7] to map OCO definitions with terms included in a corpus of documents encoded in the Journal Article Tag Suite (JATS) XML schema[15]. JATS is used by Europe PubMed Central (EPMC)[16] to encode scholarly documents, that are in turn harvested by OpenCitations.

[E2] Vocabulary coverage. This addresses the coverage of concepts, instances, and facts of OCO with respect to the domain to be covered. [E2.1] We validated OCO coverage by comparing it with competing ontologies [25]. [E2.2] Secondly, we adopted an application-based approach [31] to address OCO coverage in four sources that leverage it: OpenCitations, EXCITE, LOC-DB, and ScholarIndex.

---

[14] https://scholarindex.eu/.
[15] https://jats.nlm.nih.gov/.
[16] https://europepmc.org/downloads/openaccess.

Also, we addressed aspects peculiar to OCDM reusability, namely:

[E3] Usability-profiling. This encompasses the communication context of OCDM, i.e. its pragmatics. We evaluated OCDM recognition level [13], i.e. the efficiency of access to OCDM ontologies, documentation, and software, by comparing it with competing ontologies [25].

Lastly, we addressed current uptake, potential impact, and trustworthiness of OCDM, including metrics about OCDM views, downloads and citations according to Figshare and Altmetrics.[17]

4.1 E1: Lexical Keyword Similarity

We created a randomized corpus of 2800 JATS documents taken from the Open Access Subset of biomedical literature hosted by Europe PubMed Central. We extracted the list of XML elements used in the documents within this corpus (117 elements), and we expanded element names with definitions scraped from the online XML schema guidelines (e.g. `<p>` became "Paragraph"). We manually pruned non-relevant elements such as MathML markup, text style elements (e.g. `<italic>`), redundant wrapping elements (`<keywordGroup>`) and elements that are out of scope (e.g. `<biography>`), resulting in a refined list of 45 terms.

Secondly, we extracted definitions from OCO (118). We manually pruned terms that were not relevant (e.g. annotation properties, provenance, and distribution related terms), terms that represent hierarchy, sequences, and linguistic aspects not available in XML (e.g. "partOf", "hasNext", "Sentence"), and terms dependent on post-processing activities (e.g. "self-citation", "hasCitationCharacterisation"), resulting in a refined list of 77 OCO definitions.

We then used Wordnet[18] to automatically expand both XML and ontology definitions with synonyms, and we matched synsets similarities. We used a symmetric similarity score to find best matches between the synsets. We considered two thresholds for the similarity match, 0.7 and 0.5, and we manually computed precision and recall. Table 2 shows the results.

Table 2. Lexical similarity between JATS/XML elements and OCO terms

Threshold	Matches	Precision	Recall
0.7	(25/45) 55.5%	(24/25) 96%	(24/45) 53.3%
0.5	(33/45) 73.3%	(31/33) 93.9%	(31/45) 68.8%

The coverage of JATS terms in OCO was 55.5% when the threshold was greater than 0.7, with high precision (96%) and average recall (53.3%). The coverage was 73.3% when the threshold was greater than 0.5, with still high

---

[17] Source code and results of this analysis are available at https://github.com/opencitations/metadata.

[18] https://wordnet.princeton.edu/.

precision (93.3%) and average recall (68.8%). False negative results included acronyms (e.g. "issn") that did not have a match in Wordnet, and terms of the taxonomy that were underrepresented in the corpus (e.g. "book"). Likewise, false positive results were due to acronyms used in XML definitions that were not correctly parsed (e.g. "URI for This Same Article Online" was incorrectly matched with "fabio:JournalArticle").

4.2   E2: Vocabulary Coverage

[E2.1] Vocabulary coverage in existing vocabularies. Since gold standard ontologies are not available, we referred to existing data models and relevant ontologies used by citation data providers. For the sake of completeness, we addressed both open and non-open citation data providers[19], and both graph data providers and others. We reviewed the vocabulary coverage with respect to P1–P8. We did not take into account discipline coverage or citation counting. The complete list of data models and references is available at https://github.com/opencitations/metadata. Table 3 summarizes the comparison of vocabularies coverage, an "x" indicating that the source had metadata of relevance to the citations properties P1–P8 (Table 1).

Table 3. Vocabulary coverage in existing vocabularies according to P1–P8

	P1	P2	P3	P4	P5	P6	P7	P8
Google Scholar		x						x
Scopus		x						x
Web of Science	x	x	x					x
CiteseerX	x	x	x			x		x
Dimensions		x	x					x
Crossref		x	x					x
EPMC		x	x					x
Datacite		x	x				x	x
DBLP		x						x
MAKG		x	x			x		
ORC		x						x
GORC		x	x	x	x	x		x
SciGraph		x						x
WikiCite		x						x
OpenCitations	x	x	x	x	x	x	x	x

Non-open citation data providers include Google Scholar, Scopus [1], Web of Science (WoS) [5], CiteSeerX [22] and Dimensions [19]. Their data models cover

---

[19] See the definition of "open" at https://opendefinition.org/licenses/.

a few aspects of bibliographic metadata (P2) and provenance data (P8). WoS, CiteSeerX, and Dimension also includes bibliographic references (P3). In addition, Wos and CiteSeerX also cover types of citations (P1), and only CiteSeerX includes citation context sentences (P6).

Open citation data providers include Crossref [18], Europe PubMed Central (EPMC), DataCite, DBLP, Microsoft Academic Knowledge Graph (MAKG) [11] (which is based on Microsoft Academic Graph [34] and which reuses the SPAR Ontologies and links to resources in Wikidata and OpenCitations), the Semantic Scholar Open Research Corpus (ORC) [3], the Semantic Scholar's Graph of References in Context (GORC) [23], Springer Nature's SciGraph [15] (which is based on Schema.org), WikiCite (which includes terms aligned to SPAR Ontologies and interlinks with the OpenCitations Corpus), and the OpenCitations datasets [29]. All data models cover P2, and all except MAKG also cover P8. Only OpenCitations covers P1. In addition, Crossref, Europe PMC, DataCite, MAKG, GORC, and OpenCitations cover P3. MAKG, GORC, and OpenCitations cover P6, while the latter two also includes in-text reference pointers (P4) and related lists (P5). DataCite and OpenCitations allow the tracking of citation functions (P7).

[E2.2] **Vocabulary coverage in early adopters.** We separately analysed the vocabulary coverage in acknowledged adopters of OCDM (Table 4).

Table 4. Vocabulary coverage in OCDM early adopters according to P1–P8

	P1	P2	P3	P4	P5	P6	P7	P8
EXCITE		x	x					x
LOC-DB		x	x					x
VSI		x	x	x		x		x

EXCITE data fully covers P2, P3 and P8. Its local data model also includes information about the data quality of extracted references, which is not currently mapped to OCDM. LOC-DB data fully covers P2, P3, and P8. The OCDM was extended in its local data model so as to cover information about its OCR activities performed on PDF scans. Venice Scholar Index (VSI) aligned data to OCDM terms so as to fully cover P2, P3, P4, P6, and P8. In order to cover peculiar needs of the project relevant to P2, the classes `fabio:Work` and `fabio:Expression` defined in the SPAR Ontologies (and reused in OCO) were specialized so as to include the following sub-classes: `fabio:ArchivalRecord`, `fabio:ArchivalRecordSet`, `fabio:ArchivalDocument`, and `fabio:ArchivalDocumentSet`[20].

---

[20] As documented at https://github.com/SPAROntologies/fabio/issues/1.

4.3 E3: Usability Profiling

We compared the documentation available for existing graph data providers, namely: MAKG, OC and GORC (Semantic Scholar), SciGraph, and WikiCite. We considered the same dimensions used to address OCDM documentation, namely: human-readable documentation, machine-readable data model and examples, ontology development documentation, open source software leveraging the model, and licenses for reuse (see Table 5).

Table 5. Usability of existing ontologies and data models

	HR docum.	MR data model	Ontology dev. docum.	Software	Licenses
MAKG	x				
ORC and GORC	x	x			
SciGraph	x	x			x
WikiCite	x			x	x
OCDM	x	x	x	x	x

The MAKG data model is graphically represented in [11]. Software for creating RDF data is available, but no machine-readable data model and examples are provided. Likewise, the development of the data model is not described. Moreover, according to Färber [11], the property `c4o:hasContext` is used to annotate instances of `cito:Citation`, rather than `c4o:InTextReferencePointer` as prescribed in C4O, preventing it from representing consistently P3, P4, and P7 in future works, and from merging third-party data with OpenCitations. Lastly, no license is specified for the data model.

The Semantic Scholar Open Research Corpus data model is described in [3]. A machine-readable example of the data model is presented in a dedicated web page[21]. No further documentation is available. Similarly, GORC is described in [23], where an example of JSON data is presented. Both datasets are released under OCD-BY (i.e. an open license), although programmatically accessing data through their APIs requires one to subscribe to a more restrictive and non-open license (comparable to CC-BY-NC-ND). No license associated with the data model is stated.

The Schema.org main classes reused in SciGraph are described in a dedicated web page[22]. While the ontology is reused as-is, the SciGraph data model[23] is released as a JSON-LD file and machine-readable examples are available under a CC-BY license. Development documentation of the data model is not available.

Sources addressing the Wikidata model used by WikiCite include templates[24] and examples[25]. However, no dedicated documentation nor a machine-readable

---

[21] http://s2-public-api-prod.us-west-2.elasticbeanstalk.com/corpus/.
[22] https://scigraph.springernature.com/explorer/datasets/ontology/.
[23] https://github.com/springernature/scigraph.
[24] https://www.wikidata.org/wiki/Template:Bibliographic_properties.
[25] https://www.wikidata.org/wiki/Wikidata:WikiProject_Source_MetaData.

version of the model having citations as a scope is separately available. Data, software, and the general data model are all released under the CC0 license.

Lastly, OCDM [9] is described in dedicated human-readable documentation, including machine-readable data model and examples, available under a CC-BY license. The ontology development documentation and the open source software leveraging the model are available on github (ISC licence). All materials are gathered in the official page of the OCDM data model[26].

4.4 OCDM Uptake, Potential Impact, and Trustworthiness

We can quantify current uptake of the OCDM documentation by using statistics provided by Figshare and Altmetrics, and the number of users' views of the model description page in the OpenCitations website. As of 18 August 2020, the Figshare document [9] has been viewed 10852 times, downloaded 1508 times, and cited 5 times. 100 tweets from 65 users include links to the document. The web page (http://opencitations.net/model) dedicated to the model has received 13,844 views from 8,202 unique users since 2018.

We can estimate the potential impact of OCDM by considering (a) different types of possible reuse of the model, (b) the number of current reusers of the data model, (c) projects and applications leveraging data created according to OCDM, and (d) the kind of users of data created according to OCDM.

In detail, OCDM can be reused 'as is', via alignment for interchange purposes, and as a JSON data model for non-Semantic Web users. Currently OCDM is used by OpenCitations for all its datasets, and by the three acknowledged early adopters, namely: EXCITE and LOC-DB, which reuse OCDM 'as is', and VSI, which aligned terms to OCDM. EXCITE data have been ingested in the OpenCitations Corpus, while LOC-DB and VSI data are going to be ingested soon. VOSViewer[27], CitationGecko[28], VisualBib[29], and OAHelper[30] are applications that leverage OpenCitations data conforming to OCDM retrieved via the OpenCitations REST APIs or directly through its SPARQL endpoints. Moreover, OpenAIRE[31], MAKG, and WikiCite align data to OpenCitations. Both DBLP and Lens.org[32] use citation data from OpenCitations to enrich their bibliographic metadata records.

Users of OpenCitations data include scholars in scientometrics, life sciences, biomedicine, the physical sciences, and the information technology domain. OpenCitations is currently expanding its coverage to include the social science and the arts and humanities disciplines. The main users of EXCITE data are researchers in the social sciences, while those of the data held by LOC-DB and the Venice Scholar Index include librarians and researchers in the humanities.

---

[26] http://opencitations.net/model.
[27] https://www.vosviewer.com/.
[28] https://citationgecko.com/.
[29] https://visualbib.uniud.it/en/project/.
[30] https://www.otzberg.net/oahelper/.
[31] https://www.openaire.eu/.
[32] https://lens.org.

Lastly, we address trustworthiness of OCDM. Long-term availability of ontologies is crucial for the development of the Semantic Web, and the trustworthiness of the ontology creators is important. OCDM, OCO, and the SPAR Ontologies are all maintained by OpenCitations, which has been recently selected by the Global Sustainability Coalition for Open Science Services (SCOSS)[33] as an open infrastructure deserving of crowdfunding support from the scholarly community, thereby helping to ensure its long-term sustainability.

Along with trustworthiness, another important factor is the general interest in the community towards research topics and outputs that can leverage OCDM So far, two OpenCitations projects dedicated to the enhancement of the OpenCitations Corpus and the creation of the Open Biomedical Citations in Context Corpus have been funded by the Alfred P. Sloan Foundation[34] and the Wellcome Trust respectively, as mentioned above in Section "Background". Moreover, the Internet Archive and Figshare have both offered to archive backup copies of the OpenCitations datasets without charge.

5   Discussion and Conclusions

First, we evaluated lexical similarity of OCO definitions over the knowledge included in data sources encoded in JATS/XML, a gold standard for academic publications [E1]. While the recall is only average, mainly due to mistakes in parsing of acronyms, for those terms that were correctly matched the lexical similarity precision is high, showing that OCO is appropriate for representing data sources organized according to the gold standard. One of the known limits of data-driven evaluation methodologies is that these do not address possible changes in the domain knowledge over time. To date, early adopters of OCO continuously contribute with new scenarios to be represented in the model, which is correspondingly expanded. As a result, OCO will remain a comprehensive reference point for future developments. Other statistical semantic approaches will be evaluated in the future.

Secondly, we evaluated OCO vocabulary coverage as compared with competing data models [E2.1] and in the context of early adopters [E2.2]. Only OCDM fully covers P1–P8. In particular, only one other provider covers P4 and P5 (identifiers for in-text references and groups of these), three providers cover property P6 (although they only store full-text sentences, and lack identifiers for in-text reference pointers), and only one other provider covers property P7 (citation function). Two graph-data providers reuse terms from SPAR Ontologies (either directly or by alignment) in different ways, generating heterogeneity in data.

Among early adopters, LOC-DB required extensions in order to represent special information related to the cataloguing of digital objects, and VSI required us to expand the FaBiO ontology to permit description of unpublished archival entities. While such changes can be deemed marginal, these are relevant hints for future developments in the humanities domain and will require further analysis.

---

[33] https://scoss.org/.
[34] See https://sloan.org/grant-detail/8017.

Nonetheless, the OCDM vocabulary coverage is satisfying and strengthens its reusability across domains and applications.

We showed how alternative citation data providers ensure access to their data models [E3]. Peer-reviewed articles are the main access point to descriptions of those data models, with additional information scattered across various web pages. While machine-readable data models and examples are mostly available, none of the other providers referenced detailed development documentation. Moreover, the licenses for reusing the data models are not always defined. In summary, OCDM appears to be the most documented and findable data model.

Again, no comparison was possible of the uptake of the alternative models in the community. We showed that OCDM has been relatively popular in community social networks, and that the documentation has been downloaded and read by many people. At the moment we cannot measure for what purpose the OCDM documentation has been reused, with the exception of the three early-adopter projects of which we are aware listed in this paper.

We have shown that OCDM is potentially of significant usefulness to several communities, and fosters reuse in combination with legacy technologies, and we have highlighted ongoing interest from several parties in the maintenance and ongoing development of OCDM in support of several projects.

In future works, we will (a) create SHeX shapes to facilitate reusers in mapping their data to OCDM, and (b) trace OCDM usage scenarios by asking users to fill in a form for statistical purposes.

Acknowledgements. This work was funded by the Wellcome Trust (Wellcome-214471_Z_18_Z). We thank Ludo Waltman (Centre for Science and Technology Studies - CWTS, Leiden University) and Vincent Larivière (École de bibliothéconomie et des sciences de l'information, l'Université de Montréal) for supervising aspects of this work, and Ivan Heibi (University of Bologna) for contributing with suggestions.

References

1. Aagaard, K., Kladakis, A., Nielsen, M.W.: Concentration or dispersal of research funding? Quant. Sci. Stud. **1**(1), 117–149 (2020)
2. Alexander, K., Cyganiak, R., Hausenblas, M., Zhao, J.: Describing linked datasets. In: Bizer, C., Heath, T., Berners-Lee, T., Idehen, K. (eds.) Proceedings of the WWW 2009 Workshop on Linked Data on the Web, LDOW 2009. CEUR Workshop Proceedings, Madrid, Spain, 20 April 2009, vol. 538. CEUR-WS.org (2009). http://ceur-ws.org/Vol-538/ldow2009_paper20.pdf
3. Ammar, W., et al.: Construction of the literature graph in semantic scholar. arXiv preprint arXiv:1805.02262 (2018)
4. Belhajjame, K., et al.: PROV-O: the PROV ontology. W3C Recommendation (2013)
5. Birkle, C., Pendlebury, D.A., Schnell, J., Adams, J.: Web of science as a data source for research on scientific and scholarly activity. Quant. Sci. Stud. **1**(1), 363–376 (2020)
6. Brank, J., Grobelnik, M., Mladenic, D.: A survey of ontology evaluation techniques. In: Proceedings of the Conference on Data Mining and Data Warehouses (SiKDD 2005), Ljubljana, Slovenia, pp. 166–170 (2005)

7. Brewster, C., Alani, H., Dasmahapatra, S., Wilks, Y.: Data driven ontology evaluation. In: 4th International Conference on Language Resources and Evaluation, LREC 2004 (2004)
8. Colavizza, G., Romanello, M.: Citation mining of humanities journals: the progress to date and the challenges ahead. J. Eur. Period. Stud. 4(1), 36–53 (2019)
9. Daquino, M., Peroni, S., Shotton, D.: The OpenCitations data model. Figshare (2018). https://doi.org/10.6084/m9.figshare.3443876.v7
10. Demidova, E., Zaveri, A., Simperl, E.: Creating open citation data with BCite. In: Emerging Topics in Semantic Technologies: ISWC 2018 Satellite Events, vol. 36, p. 83 (2018)
11. Färber, M.: The Microsoft academic knowledge graph: a linked data source with 8 billion triples of scholarly data. In: Ghidini, C., et al. (eds.) ISWC 2019. LNCS, vol. 11779, pp. 113–129. Springer, Cham (2019). https://doi.org/10.1007/978-3-030-30796-7_8
12. Fortunato, S., et al.: Science of science. Science 359(6379), eaao0185 (2018)
13. Gangemi, A., Catenacci, C., Ciaramita, M., Lehmann, J.: Modelling ontology evaluation and validation. In: Sure, Y., Domingue, J. (eds.) ESWC 2006. LNCS, vol. 4011, pp. 140–154. Springer, Heidelberg (2006). https://doi.org/10.1007/11762256_13
14. Gómez-Pérez, A.: Ontology evaluation. In: Staab, S., Studer, R. (eds.) Handbook on Ontologies. INFOSYS, pp. 251–273. International Handbooks on Information Systems, Springer (2004). https://doi.org/10.1007/978-3-540-24750-0_13
15. Hammond, T., Pasin, M., Theodoridis, E.: Data integration and disintegration: managing springer nature SciGraph with SHACL and OWL. In: International Semantic Web Conference (Posters, Demos & Industry Tracks) (2017)
16. Heibi, I., Peroni, S., Shotton, D.: OSCAR: a customisable tool for free-text search over SPARQL endpoints. In: González-Beltrán, A., Osborne, F., Peroni, S., Vahdati, S. (eds.) SAVE-SD 2017-2018. LNCS, vol. 10959, pp. 121–137. Springer, Cham (2018). https://doi.org/10.1007/978-3-030-01379-0_9
17. Heibi, I., Peroni, S., Shotton, D.: Software review: COCI, the OpenCitations Index of Crossref open DOI-to-DOI citations. Scientometrics 121(2), 1213–1228 (2019). https://doi.org/10.1007/s11192-019-03217-6
18. Hendricks, G., Tkaczyk, D., Lin, J., Feeney, P.: Crossref: the sustainable source of community-owned scholarly metadata. Quant. Sci. Stud. 1(1), 414–427 (2020)
19. Herzog, C., Hook, D., Konkiel, S.: Dimensions: bringing down barriers between scientometricians and data. Quant. Sci. Stud. 1(1), 387–395 (2020)
20. Hosseini, A., Ghavimi, B., Boukhers, Z., Mayr, P.: Excite-a toolchain to extract, match and publish open literature references. In: 2019 ACM/IEEE Joint Conference on Digital Libraries (JCDL), pp. 432–433. IEEE (2019)
21. Lauscher, A., et al.: Linked open citation database: enabling libraries to contribute to an open and interconnected citation graph. In: Proceedings of the 18th ACM/IEEE on Joint Conference on Digital Libraries, pp. 109–118 (2018)
22. Li, H., Councill, I., Lee, W.C., Giles, C.L.: CiteSeerx: an architecture and web service design for an academic document search engine. In: Proceedings of the 15th International Conference on World Wide Web, pp. 883–884 (2006)
23. Lo, K., Wang, L.L., Neumann, M., Kinney, R., Weld, D.S.: GORC: a large contextual citation graph of academic papers. arXiv preprint arXiv:1911.02782 (2019)
24. Maali, F., Erickson, J., Archer, P.: Data catalog vocabulary (DCAT). W3C recommendation (2014)
25. Maedche, A., Staab, S.: Comparing ontologies-similarity measures and a comparison study. AIFB (2001)

26. Peroni, S., Shotton, D.: Open citation identifier: definition. Figshare (2019). https://doi.org/10.6084/m9.figshare.7127816
27. Peroni, S.: A simplified agile methodology for ontology development. In: Dragoni, M., Poveda-Villalón, M., Jimenez-Ruiz, E. (eds.) OWLED/ORE -2016. LNCS, vol. 10161, pp. 55–69. Springer, Cham (2017). https://doi.org/10.1007/978-3-319-54627-8_5
28. Peroni, S., Shotton, D.: The SPAR ontologies. In: Vrandečić, D., et al. (eds.) ISWC 2018. LNCS, vol. 11137, pp. 119–136. Springer, Cham (2018). https://doi.org/10.1007/978-3-030-00668-6_8
29. Peroni, S., Shotton, D.: Opencitations, an infrastructure organization for open scholarship. Quant. Sci. Stud. **1**(1), 428–444 (2020)
30. Peroni, S., Shotton, D.M., Vitali, F.: A document-inspired way for tracking changes of RDF data **1799**, 26–33 (2016)
31. Porzel, R., Malaka, R.: A task-based approach for ontology evaluation. In: ECAI Workshop on Ontology Learning and Population, Valencia, Spain, pp. 1–6 (2004)
32. Sanderson, R., Ciccarese, P., Van de Sompel, H.: Designing the W3C open annotation data model. In: Proceedings of the 5th Annual ACM Web Science Conference, pp. 366–375 (2013)
33. Shotton, D., Peroni, S., Daquino, M.: In-text reference pointer identifier: definition. Figshare (2020). https://doi.org/10.6084/m9.figshare.11674032
34. Wang, K., Shen, Z., Huang, C., Wu, C.H., Dong, Y., Kanakia, A.: Microsoft academic graph: when experts are not enough. Quant. Sci. Stud. **1**(1), 396–413 (2020)
35. Wilkinson, M.D., et al.: The fair guiding principles for scientific data management and stewardship. Sci. Data **3**, 1–9 (2016)

Semantic Integration of Bosch Manufacturing Data Using Virtual Knowledge Graphs

Elem Güzel Kalaycı[1,3], Irlan Grangel González[2], Felix Lösch[2],
Guohui Xiao[1,4(✉)], Anees ul-Mehdi[2], Evgeny Kharlamov[5,6],
and Diego Calvanese[1,4,7]

[1] Free University of Bozen-Bolzano, 39100 Bolzano, Italy
{kalayci,xiao,calvanese}@inf.unibz.it
[2] Robert Bosch GmbH, Corporate Research, 70465 Stuttgart, Germany
{irlan.grangelgonzalez,felix.losch,anees.ul-mehdi}@de.bosch.com
[3] Virtual Vehicle Research GmbH, 8010 Graz, Austria
elem.guezelkalayci@v2c2.at
[4] Ontopic S.r.L., 39100 Bolzano, Italy
{guohui.xiao,diego.calvanese}@ontopic.biz
[5] Robert Bosch GmbH, Bosch Center for Artificial Intelligence,
71272 Renningen, Germany
evgeny.kharlamov@de.bosch.com
[6] University of Oslo, 0316 Blindern, Oslo, Norway
evgeny.kharlamov@ifi.uio.no
[7] Umeå University, 90187 Umeå, Sweden
diego.calvanese@umu.se

Abstract. Analyses of products during manufacturing are essential to guarantee their quality. In complex industrial settings, such analyses require to use data coming from many different and highly heterogeneous machines, and thus are affected by the data integration challenge. In this work, we show how this challenge can be addressed by relying on semantic data integration, following the Virtual Knowledge Graph approach. For this purpose, we propose the SIB Framework, in which we semantically integrate Bosch manufacturing data, and more specifically the data necessary for the analysis of the Surface Mounting Process (SMT) pipeline. In order to experiment with our framework, we have developed an ontology for SMT manufacturing data, and a set of mappings that connect the ontology to data coming from a Bosch plant. We have evaluated SIB using a catalog of product quality analysis tasks that we have encoded as SPARQL queries. The results we have obtained are promising, both with respect to expressivity (i.e., the ability to capture through queries relevant analysis tasks) and with respect to performance.

Electronic supplementary material The online version of this chapter (https://doi.org/10.1007/978-3-030-62466-8_29) contains supplementary material, which is available to authorized users.

© Springer Nature Switzerland AG 2020
J. Z. Pan et al. (Eds.): ISWC 2020, LNCS 12507, pp. 464–481, 2020.
https://doi.org/10.1007/978-3-030-62466-8_29

1   Introduction

The digitization trend in manufacturing industry, known as Industry 4.0, leads to a huge growth of volume and complexity of data generated by machines involved in manufacturing processes. These data become an asset of key relevance for enhancing the efficiency and efficacy of manufacturing. However, unlocking the potential of these data is a major challenge for many organizations. Indeed, often the data naturally reside in silos, which are not interconnected, but which contain semantically related data, possibly with redundant and inconsistent information. As a result, the effective use of data demands data integration, which includes cleaning, de-duplication, and semantic homogenization. As evaluated at Bosch, the integration effort needed for data integration is approximately 70–80%, in comparison to 20–30% required for data analysis [16], where the integration is mainly hampered by ad-hoc and manual approaches that are prone to data quality issues. In particular, this affects the reproducibility of analytical results as well as the consequent decision making [9,16,17].

A Bosch plant located in Salzgitter, Germany, that produces electronic control units, is not an exception in the digitization trend and the integration challenge [16]. Indeed, the product quality analysis that is performed at the plants requires integration of vast amounts of heterogeneous data. For instance, failure detection for Surface Mounting Process (SMT) fundamentally relies on the integration and analysis of data generated by the machines deployed in the different phases of the process. Such machines, e.g., for *placing electronic components* (SMD) and for *automated optical inspection* (AOI) of solder joints, usually come from different suppliers and they rely on distinct formats and schemata for managing the same data across the process. Hence, the raw, non-integrated data does not give a coherent view of the whole SMT process and hampers analysis of the manufactured products.

To address this problem, we propose to adopt an approach for semantic data integration and access based on virtual knowledge graphs (VKG)[1] [5,20,21], which we illustrate in Fig. 1. In such an approach, we use an ontology that exposes, in terms of a VKG, a conceptual view of the concepts and properties of relevance for the SMT manufacturing process. The log data of the process are extracted directly from JSON files generated by the different machines, and are loaded into a PostgreSQL database, without further conversion from JSON into the relational format. From there the data is connected to the ontology by making use of semantic mappings [1].

The strength of this approach comes on the one hand from the domain knowledge encoded in the ontology. This knowledge is used to enrich answers to user queries describing product analysis tasks. Another key advantage is the use of semantically rich VKG mappings. These bridge the impedance mismatch between the data layer and the ontology layer, by relying on the template-based mechanism of R2RML [4] to construct knowledge graph (KG) IRIs out of database values. Moreover, the mappings provide a solution to the integration

---

[1] Also known in the literature as Ontology-based Data Access/Integration (OBDA/I).

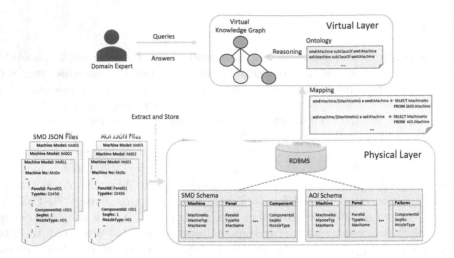

Fig. 1. Virtual Knowledge Graphs approach exemplified over Bosch SMT scenario.

challenge, since semantically homogeneous information coming from different log files, which use syntactically different representations, can be reconciled at the level of the KG. This makes it easy to query the overall data assets in an integrated way, by exploiting the semantics of the extracted information. For example, consider various types of machine failures, encoded in the different log files through different *magic numbers*, i.e., codes with a specific meaning that have to be known and interpreted by the Bosch engineers. Once the machine failures of the same type are reconciled in a single ontology class, Bosch engineers can effectively understand the meaning of failures, since the mappings guarantee that each failure data item (virtually) populates only the appropriate failure class. This is of crucial importance to allow engineers take the right decisions in the SMT process. In general, the approach enables to encode different product analysis tasks that are of importance in the Bosch SMT manufacturing process, by means of suitable SPARQL queries over the domain ontology. Notably, such queries make use of ontology terms to refer to the relevant information assets, and thus are very close to the natural language formulation of the analysis tasks, which in turn makes it easy for Bosch engineers to formulate them. We can then obtain the respective analysis data coming from the process logs, by simply executing such queries over the underlying database via a VKG engine.

The above solution has been implemented at Bosch in the VKG-based data integration framework called SIB (for Semantic Integration at Bosch) and deployed at Bosch. The purpose of this deployment has been to evaluate the feasibility of using semantic technologies and data integration based on them for supporting product quality analysis. More specifically, towards the development of SIB, we have overcome the technical challenges posed by semantic data integration, and we have provided the following contributions: *(i)* We developed the *SMT Ontology*, which is an OWL 2 QL ontology capturing the concepts and

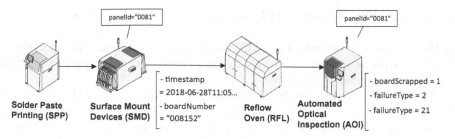

Fig. 2. Surface Mounting Process pipeline. The SMT process comprises four phases: SPP, SMD, RFL, and AOI. The machines, usually by different suppliers, rely on distinct formats and schemas for managing the same data across the process pipeline.

properties that are relevant for SMT manufacturing, together with important domain knowledge. This ontology is the basis of the VKG approach to integration for SMT manufacturing. *(ii)* We built a mapping layer that semantically connects the SMT ontology to a PostgreSQL database. Such a database in turn collects the relevant log data from JSON files produced by the machine components in the manufacturing pipeline. *(iii)* We encoded relevant product analysis tasks into a catalog of SPARQL queries formulated over the SMT Ontology. For processing such queries, taking into account both the SMT Ontology and mappings, we rely on the state-of-the-art VKG engine Ontop [2].

Moreover, we carried out an experimental evaluation of the SIB approach, with a two-fold aim. First, we wanted to assess its effectiveness in addressing significant product analysis tasks by relying on the answers returned by the SPARQL queries encoding such tasks. Second, we were interested in understanding the performance in query execution, and whether such performance is compatible with the requirements coming from product analysis. As a baseline for the evaluation, we have used SANSA-DL (where "DL" stands for Data-Lake), which is an alternative implementation of semantic integration based on SANSA [14], a distributed framework for scalable RDF data processing in turn based on Apache Spark. Similarly to Ontop, SANSA-DL supports SPARQL queries over an ontology connected via mappings to a data source. However, the data source consists of *parquet* files, loaded from the JSON data via a Scala script.

Our evaluation showed that, in this real-world use-case in industrial manufacturing, semantic data integration based on VKGs as realized in SIB is feasible, both from the point of view of semantics/expressiveness, and from the point of view of performance. Moreover, our results indicate that the VKG system Ontop adopted in SIB outperforms an alternative implementation based on Spark.

The rest of the paper is organized as follows. Section 2 describes the SMT manufacturing process and the challenges for integrating data generated by the different machines used for manufacturing. Section 3 describes the components of the VKG approach and its application to the SMT use case. In Sect. 4, we outline the evaluation results. In Sect. 5 we discuss the lessons learned, and we conclude the paper with an outlook and future work in Sect. 6.

2   Bosch Use Case: SMT Manufacturing Process

In this section we describe the SMT manufacturing process, the analytics it typically requires, and the challenges in implementing this analytics, which will be addressed further in the paper.

The SMT Process involves four main phases (cf. Fig. 2): *(1) Solder Paste Printing:* This phase consists of pasting solder paste on print circuit boards (PCBs), and is conducted by a Solder Paste Printing Machine; *(2) Surface Mounting:* This is the phase where electronic components are actually mounted on PCBs in a Surface Mounting Device (SMD); *(3) Heating:* To solder the mounted components properly, in this phase the boards are heated inside a Reflow Oven; and *(4) Automated Optical Inspection.* In the final phase, the boards are inspected by an AOI machine to find out whether during the previous phases any failure occurred, such as component misplacement or bad soldering. When the whole process is completed, the system generates log files, which contain two types of data: the *placement logs* by the SMD machine, include information on which component is mounted on which board; the *failure logs* generated by the AOI machine, comprise information where at a board and with what component a failure is encountered.

Product Analysis Tasks. Several product analysis tasks are carried out by Bosch engineers in order to be able to assess and evaluate the quality of the manufactured PCBs, and such tasks require in an essential way to access in an integrated way the data produced during the SMT process. A typical analytical request for information in this domain may look as follows: "For the *panels* processed in a given *time frame*, retrieve the number of *failures* associated with *scrapped boards* and grouped by *failure types*." To retrieve this information, data that reside in the SMD and AOI machines need to be semantically integrated. Referring again to Fig. 2, the SMD machine processes the panel, the board, and the timestamp information, i.e., the time at which a given panel was processed, while the AOI machine generates information to check whether a board was scrapped or not. The meaning of this information is encoded in numbers: a value of '1' when the board is scrapped and '0' when not. In addition, the AOI machine contains information about the different failure types, e.g., number 2 representing "false call" and 21 representing "misplaced component". The meaning of this information encoded in numbers is only available in internal documents and in the head of Bosch engineers. The panel provides the connection between the data generated by the machines. With this setup, the codes associated to various kinds of components need to be combined to create the ids for different types of objects, and this in turn complicates semantic interoperability between SMD and AOI data.

Technical Challenges. In this work, we focus on addressing the above challenges by means of semantic data integration. Specifically, we show how a VKG-based approach enables the semantic integration of the SMD and AOI data, and how the requests of domain experts can be answered on top of the semantically integrated data. Towards the objective of applying the VKG approach to the Bosch use case, we had to address the following technical challenges: **C1:** integration of syntactically heterogeneous machine log data involving semantic conflicts; **C2:** development of the SMT Domain Ontology capturing relevant domain knowledge; **C3:** transformation of product analysis questions into a catalog of SPARQL queries and evaluation of their effectiveness for product quality analysis; and **C4:** evaluation of the efficiency of the VKG approach in the execution of these queries, which are relevant to the Bosch use case. We discuss in this paper, how these challenges have been addressed and solved.

Related Work. The VKG-based integration approach has already been successfully applied in industrial settings [7]. These include, among many, the case of manufacturing [18], of assembling complex systems at Festo [6], of turbine diagnostics and other tasks at Siemens [8,10–12,17], and of exploration and analyses of geological data at Equinor [9,13]. In several of these, Ontop has been adopted as one of the key components. In this work we continue this line of applied research and bring semantic technologies into the Bosch corporate environment by adapting them to a concrete scenario.

3   Application of SIB at Bosch

In this section, we present the general VKG approach for semantically integrating data and its specific application at Bosch in SIB for the semantic integration of manufacturing data of the SMT process. SIB comprises the following components: *1)* data sources of the SMT manufacturing process; *2)* the SMT Ontology, i.e, a semantic model of the SMT domain; *3)* mappings between the data sources and the SMT Ontology; *4)* queries for expressing the requirements of the domain; and *5)* an implementation of the VKG approach using Ontop to semantically answer the queries while integrating the SMD and AOI data sources.

3.1   Overview of the SIB Framework

We gave a general overview of the Virtual Knowledge Graph (VKG) paradigm in Fig. 1 of Sect. 1. In this work we rely on the state-of-the-art VKG framework Ontop for developing SIB. Ontop computes answers end-user SPARQL queries by translating them into SQL queries, and delegating the execution of the translated SQL queries to the original data sources. We remark that with the VKG approach, there is no need to materialize into a KG all the facts entailed by the ontology. As seen in Fig. 3, the workflow of Ontop can be divided into an off-line and an online stage. As the first step at the *off-line stage*, Ontop loads the OWL 2 QL ontology and classifies it via the built-in reasoner, resulting in a directed acyclic graph stored in memory that represents the complete hierarchy of concepts and that of properties. In the second step, Ontop constructs a so-called *saturated mapping*, by compiling the concept and property hierarchies into the original VKG mapping. This aspect is important also in SIB, since the domain knowledge encoded in the ontology allows for simplifying the design of

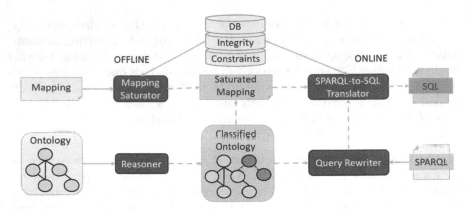

Fig. 3. The Ontop Workflow for query translation. The figure depicts the SPARQL-to-SQL query translation workflow of Ontop.

the mapping layer. During the offline stage, Ontop also optimizes the saturated mapping by applying structural and semantic query optimization [19].

During the *online stage*, Ontop takes a SPARQL query and translates it into SQL by using the saturated mapping. To do so, it applies a series of transformations that we briefly summarize here [2,22]: *(i)* it rewrites the SPARQL query w.r.t. the ontology; *(ii)* it translates the rewritten SPARQL query into an algebraic tree represented in an internal format; *(iii)* it unfolds the algebraic tree w.r.t. the saturated mapping, by replacing the triple patterns with their optimized SQL definitions; and *(iv)* it applies structural and semantic techniques to optimize the unfolded query. One of the key points in the last step is the elimination of self-joins, which negatively affect performance in a significant way. To perform this elimination, Ontop utilizes in an essential way the key constraints defined in the data sources. In those cases where it is not possible to define these key constraints explicitly in the data sources, or to expose them as metadata of the data sources so that Ontop can use them, Ontop allows one to define them implicitly, as part of the mapping specification. The data we have been working with in the Bosch use case was mostly log data and stored as separate tables containing often highly denormalized and redundant data. Consequently, there were a significant amount of constraints in the tables that are not declared as primary or foreign keys, which brought significant challenges to the performance of query answering. To address these issues, we had to declare these constraints manually, and supply them as separate inputs to Ontop.

3.2   Data Sources

Despite the fact that the SMT process comprises multiple data sources, we have focused in SIB only on the SMD and AOI log data generated by the respective machines, as these are the main data sources in the SMT process and are required to answer the queries we derived for product quality analysis. We extract the

Table 1. SMT Data. The main tables in the SMT schema are given on the left. The most frequently requested tables are given on the right, with a sample of data.

SMD Tables

smd_event

smd_location

smd_panel

smd_components

smd_panel

panelId	boardNo	machineName	processedTS	location
p01	b01	SMD Machine 1	24-04-2020	mes01

smd_components

panelId	boardNo	headId	nozzleId	turnNo	pickSeqNo	placeSeqNo
p01	b01	h01	n01	2	1	3

AOI Tables

aoi_event

aoi_location

aoi_panel

aoi_failures

aoi_failures

panelId	boardNo	refDesignator	windowNo	cPinNo	failureCode
p01	b01	rd01	w01	pn01	1

SMD and AOI log data from nested JSON files and without any pre-processing, we store them in a PostgreSQL database, which contains the relational tables shown in Table 1. The **event** suffixed tables contain information about mounting and inspection processes, while the **location** suffixed tables describe locations where processed panels are inspected. The **panel** suffixed tables contain information about processed and inspected panels and boards. The SMD data set tracks the information about the pick-and-place sequences, which is stored in the **smd_components** table. It contains the sequence numbers in which the component is picked and placed, and the turn number in which the pick-and-place is performed. Finally, the **aoi_failures** table encompasses information of the failures that are detected during the SMT process by visually inspecting the solder joints. It contains information about panels, boards, components, and pins associated with failures.

3.3 SMT Ontology

In the VKG approach, the domain of interest is described in a ontology in terms of *classes*, their *data properties* (i.e., attributes), and the *object properties* (i.e., relationships) in which they are involved. The ontology serves as an abstraction over the data sources and is used by the domain experts to formulate queries over the data. To apply the VKG approach, we developed the SMT Ontology (cf. Fig. 4), which is used as a domain model for semantic data integration and access in the domain, and is divided into three modules: *1) SMT Failure Ontology* (fsmt), modeling the SMT failures that can occur during the process; *2) SMT Product Ontology* (psmt), describing SMT products; and *3) SMT Machine Ontology* (msmt), modeling SMT machines. The SMT Ontology overall comprises 76 classes, 30 object properties, and 57 datatype properties.

The development of the SMT Ontology was the result of six workshops over a period of eight weeks, involving Bosch experts that include a line engineer, two line managers from the Bosch plant, an SMT process expert, an SMT data

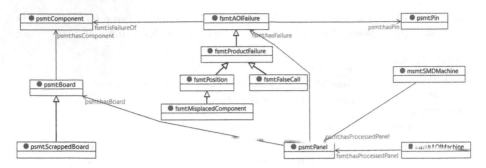

Fig. 4. The SMT Ontology. The SMT Ontology comprises the SMT Product, SMT Machine, and SMT Failure Ontologies. Above, we show a portion of the SMT Ontology.

manager, a project manager, two Big Data managers, and four semantic experts. The ontology creation process took several iterations that were required to understand the SMT process, the concepts that are relevant to it, and the relationships between these concepts. We combined a top-down approach in which we modeled classes and properties based on expert knowledge provided by the process and machine experts, with a bottom-up approach in which we looked at the JSON data together with the data engineers and experts to identify additional attributes of classes. We also had to study numerous Wiki pages with technical documentation created by the process experts of the Bosch plant.

Let us briefly discuss the case of AOI failure types, since these play a crucial role in the product quality assessment, and understanding their meaning is of crucial importance to take decisions in the SMT process. The failure types were partially described in textual form in the Wiki and partially they were only in the head of the process experts. Typically, the way in which failures are coded differs between different plants inside the same organization. To address this issue, we semantically codified the failures in the SMT Failure Ontology. For instance, the failure type *misplaced component*, meaning that a component was misplaced on top of the board, is represented in the data with the *magic number* 21. We created a class `fsmt:MisplacedComponent` to represent this type of failure. The class is a subclass of `fsmt:Position` as well as `fsmt:ProductFailure`, which semantically group all the failures that fall into these categories. Similarly, we created a class `psmt:ScrappedBoard`, for semantically representing all the boards that are scrapped, represented as subclass of `psmt:Board`.

3.4 Data-to-Ontology Mapping

A *mapping* is a set of assertions specifying how the classes and properties of the ontology are populated by the data from the sources. Each mapping assertion consists of an *identifier*, a *source part*, i.e., a SQL query over the data source schema, and a *target part*, i.e., an RDF triple template [3]. The standard language for representing a mapping is defined by the W3C R2RML specification [4].

The SMT VKG mapping contains 53 mapping assertions; they link the elements of the SMT data source to the basic ontological concepts and roles in the SMT Ontology. For the sake of readability and for lack of space, we present only a subset of the mappings that we developed. These mappings cover several important assertions relevant to the queries, and we formulate them using the Ontop mapping language [2], since we find it more compact and more end-user oriented than the R2RML mapping language.

The first mapping assertion we consider (with id `hasBoard`), instantiates the object property `psmt:hasBoard`, relating individuals of the classes `psmt:Panel` and `psmt:Board`, and also associates the individuals of `psmt:Panel` with their processing times.

```
mappingId  hasBoard
target     psmt:Panel/{panelId} psmt:hasBoard psmt:Board/{panelId}/{boardNo} ;
                               psmt:pTStamp {smdpTStamp}.
source     SELECT panelId, boardNo, smdpTStamp  FROM smd_panel
```

The next mapping assertion places some of the individuals of the `psmt:Board` class (namely those selected by the `WHERE` condition in the source query) also in the `psmt:ScrappedBoard` class.

```
mappingId  ScrappedBoard
target     psmt:Board/{panelId}/{boardNo} a psmt:ScrappedBoard .
source     SELECT panelId, boardNo FROM aoi_failures WHERE boardScrapped = 1;
```

As one can notice, this mapping assertion addresses the semantic interoperability problem, by carrying the information about scrapped boards that is encoded in numbers, from the data level to the conceptual level.

The following mapping assertion creates a bridge between SMD and AOI entities, by relating individuals of `fsmt:AOIFailures` with individuals of `psmt:Panel` via the `fsmt:hasFailure` object property.

```
mappingId  hasFailure
target     psmt:Panel/{panelId} fsmt:hasFailure
               fsmt:AOIFailure/{panelId}/{boardNo}/{refDesignator}/{windowNo}/{cPinNo}.
source     SELECT  panelId, boardNo, refDesignator, windowNo, cPinNo
           FROM aoi_failures
```

Finally, the following mapping assertions label the individuals of failures with a type, which may be either `"FalseCall"` or `"MisplacedComponent"`, depending on the numeric value of the `failureCode` attribute of the `aoi_failures` table, which identifies the failure types.

```
mappingId  FalseCall
target     fsmt:AOIFailure/{panelId}/{boardNo}/{refDesignator}/{windowNo}/{cPinNo}
               rdf:type fsmt:FalseCall ;
               fsmt:failureType "FalseCall" .
source     SELECT panelId, boardNo, refDesignator, windowNo, cPinNo
           FROM aoi_failures WHERE failureCode = 2

mappingId  MisplacedComponent
target     fsmt:AOIFailure/{panelId}/{boardNo}/{refDesignator}/{windowNo}/{cPinNo}
               rdf:type fsmt:MisplacedComponent ;
               fsmt:failureType "MisplacedComponent" .
source     SELECT panelId, boardNo, refDesignator, windowNo, cPinNo
           FROM aoi_failures WHERE failureCode = 21
```

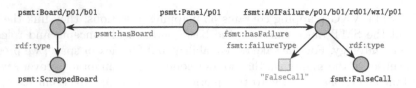

Fig. 5. A VKG. It is built from the mapping over the SMD and AOI data.

Based on the above mapping assertions defined over the SMD and AOI data, a VKG (cf. Fig. 5), is built. This graph contains instances of `psmt:Panel`, `psmt:Board`, and `fsmt:AOIFailure` as nodes, and also object properties `psmt:hasBoard` and `fsmt:hasFailure` as edges. Note that the special object property (i.e., edge) `rdf:type` is used to represent the membership of an object to a class.

4 Evaluation of SIB at Bosch

We have carried out an evaluation of SIB, both with respect to its effectiveness in supporting the formulation of typical product quality analysis tasks through SPARQL queries, and with respect to the efficiency with which such queries are executed by the underlying VKG system Ontop over distinct data sets with three different sizes.

4.1 Effectiveness of SIB

We start by describing our effectiveness evaluation. We measure the effectiveness be verifying whether typical product quality queries can be expressed over the ontologies that we developed. To this end, we developed a catalog of 13 queries that consolidate the requirements of Bosch experts. These queries were the result of a collaborative work and a careful selection during two visits to Bosch plants and meetings with Bosch line engineers and line managers. The queries offer a good balance among three dimensions: they are representative for product analyses, offer a good coverage of product analyses tasks, and they are complex enough to account for a reasonable number of domain terms. All these 13 queries were expressible over the ontologies of SIB.

Now we present three of these queries, formulating them both in natural language and in SPARQL. The queries are presented in increasing complexity, going from a simple query performing joins and applying `FILTER`s, to a more complex query involving a nested sub-query, up to a query involving complex aggregation. The complete query catalog is provided as supplemental material.

Query q2: "Return all panels that have been processed after a given panel P, and before `2018-06-28T11:05:42.000+02:00`."

This query fetches all the panels produced in a given time interval between the production time of panel *P* and the given time. Since timestamps recording processing time are associated to panels, we retrieve all panels processed after the given panel by comparing their timestamps. The query in SPARQL is formalized as follows:

```
SELECT DISTINCT ?pn2 ?ts2
WHERE {
    ?pn psmt:panelId ?panelId ;
        psmt:pTStamp ?ts .
    ?machine psmt:hasProcessedPanel ?pn2 .
    ?pn2 psmt:pTStamp ?ts2 .
    FILTER (?ts < ?ts2 && ?ts2 < '2018-06-28T11:05:42.000+02:00'^^xsd:dateTimeStamp)
    FILTER (?panelId = "0850799900252180622261041592") }
```

Query q3: *"Return all panels processed from a given time T up to the detection of a failure."*

This query is temporal in nature in the sense that we are interested in all panels that did not encounter failures in production *until* the first failure was encountered. The query can still be realized in SPARQL as shown below.

```
SELECT DISTINCT ?panel ?ts ?eventTime
WHERE {
    ?panel psmt:pTStamp ?ts . {
        SELECT ?eventTime
        WHERE {
            ?eventfailure fsmt:eTStamp ?eventTime .
            FILTER (?eventTime > '2018-06-01T00:06:00.000+02:00'^^xsd:dateTimeStamp)
        }
        ORDER BY (?eventTime) LIMIT 1
    }
    FILTER (?ts > '2018-06-01T00:06:00.000+02:00'^^xsd:dateTimeStamp && ?ts < ?eventTime) }
```

Query q9: *"For the panels processed in a given time frame, retrieve the number of failures associated with scrapped boards and grouped by failure types."*

This query is formalized in SPARQL using an aggregation operator, as shown below.

```
SELECT (COUNT(?failure) as ?f) ?type
WHERE {
    ?pn psmt:pTStamp ?ts ;
        psmt:hasBoard ?board ;
        fsmt:hasFailure ?failure .
    ?failure fsmt:failureType ?type ;
        rdf:type fsmt:AOIFailure .
    ?board rdf:type psmt:ScrappedBoard .
    FILTER (?ts > '2018-06-01T02:17:54+02:00'^^xsd:dateTimeStamp)
    FILTER (?ts < '2018-06-02T07:04:14+02:00'^^xsd:dateTimeStamp)
} GROUP BY ?type
```

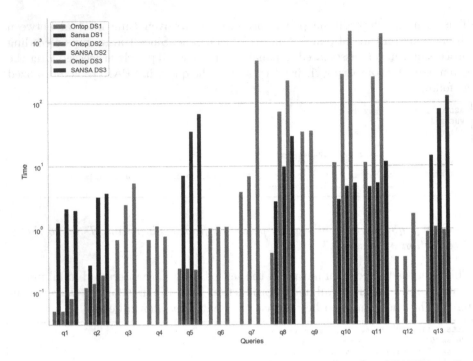

Fig. 6. Evaluation Results of executing the 13 queries using Ontop and SANSA on the three data sets used for the evaluation. The execution times are given in seconds. In general, Ontop outperforms SANSA. Ontop also supports more queries than SANSA.

4.2 Efficiency of SIB

In order to analyze the performance and scalability of the VKG approach for the SMT use case, we executed performance tests for the 13 queries in the query catalog. To evaluate the influence of the size of data on query performance, we executed all 13 queries on three SMD data sets of different sizes. These SMD data sets are combined with the AOI data set. Each line in the AOI data set describes when an error occurs. The three data sets in JSON, i.e., DS1, DS2, and DS3, have sizes of 3.15 GB, 31 GB, and 59 GB, respectively. Each line of the SMD data sets represents an event where a panel was produced. DS1 comprises 69995 lines, DS2 has 699962 lines, whereas DS3 has 1332161 lines. DS3 contains all the SMT data that was available for the experiment. The experiments were executed on a machine Intel(R) Xeon(R) Gold 6148 CPU @ 2.40GHz with 64 GB RAM and Red Hat Linux version 7 installed. As a database system we used PostgreSQL version 9.2.24. For the execution of the experiments, we used the Ontop v3.0 Plugin for Protégé, which provides an integrated framework for the definition of the ontology and the mappings to an underlying database, and gives also the possibility of executing SPARQL queries over such database.

We compared the results with SANSA [14], which is a distributed framework for scalable RDF data processing based on Apache Spark. SANSA allows for accessing both large RDF graphs, and heterogeneous large non-RDF data sources, thanks to its sub-component SANSA-DataLake (SANSA-DL) [15]. SANSA-DL can query out-of-the-box various types of sources ranging from plain files and file formats stored in Hadoop distributed file system, e.g., CSV and Parquet, as well as various other sources via a JDBC connector. We included SANSA because it is an open source solution, and also uses a virtualized approach.

Since neither Ontop nor SANSA natively support the JSON format, the three datasets used for the evaluation have been ingested into a database and into Parquet files, respectively. For Ontop, we created a Python script to read the JSON data and insert it into a PostgreSQL database. For SANSA, we have developed a Scala script to convert the data into Parquet files. Figure 6 reports on the query execution times in seconds for executing the queries in Ontop and SANSA with the different data sets. We have used gray bars to depict the execution times for Ontop, and black bars for those of SANSA.

The experimental results show that Ontop performs well in general and outperforms SANSA. For most of the queries (namely q1–q5, q6, q7, q12, and q13) the execution times scale sub-linearly, and most of them finish in less than one second even over the largest dataset DS3. For most of the complex queries, i.e., q5, q7, q9, q10, and q11, the execution lasts in the order of 10 s of seconds, which is considered a reasonable amount of time in the context of the SIB use case. We observe that SANSA does not support some queries, i.e., q3, q4, q6, q7, q9, and q12. This is because subqueries, queries containing variables at both object and subject position, i.e., joins on the object position, and GROUP BY clauses in SPARQL queries on variables appearing at subject position are currently not supported by SANSA. In two of the queries, i.e., q10 and q11 SANSA behaves better than Ontop. These queries comprise aggregation functions, i.e., *COUNT* in this case. The query processing in Ontop for aggregation functions require additional operations for query rewriting, while SANSA supports these functions as part of its SPARQL fragment. SANSA utilizes SPARK DataFrames to perform aggregation functions on top of the queried data. This seems to have an influence on the query performance as depicted in Fig. 6. To summarize our evaluation, Ontop shows quite fast query answering times for most of the queries. Even complex queries involving sub-queries could be answered at reasonable time for our use case. In contrast to Ontop, SANSA could not answer some of the queries. Based on that SANSA was not the first choice for our case as important queries for product analysis tasks could not be executed.

5   Lessons Learned

Involvement of Domain Experts. Involving the Bosch domain experts in the process of deploying the VKG approach to the SMT use case early on was important for the project. Although these experts had no knowledge of semantic technologies, they quickly understood that the SMT Ontology we developed together with them will support their product quality analysis tasks. Before we formally modelled the SMT domain as an Ontology, they started depicting a model of the process on a whiteboard.

Integrated View of Data provided by SMT Ontology. The SMT Ontology and VKG approach enabled complex product analysis tasks that require an integrated view of manufacturing data over multiple processes. Without this integrated view, it would have been very difficult for the Bosch domain experts to formulate their information needs as queries, as they were working on the level of incompatible raw data sources before.

Methodology and Tooling. The adoption of the VKG approach is a labor-intensive process. In particular, in the design phase of mappings three pieces of knowledge need to be combined: the domain knowledge of the SMT manufacturing process, the detailed database schemas, and how the VKG approach works. This makes it challenging to produce high-quality mappings. A proper methodology and tooling support will be important to improve the productivity. By applying the VKG approach we learned several guidelines that should be followed when designing the mappings, notably the following: *(i)* it is necessary to avoid joining multiple tables inside a mapping to reduce the complexity of query-rewriting; *(ii)* indexes in the database should be used for speeding up the processing of certain translated SQL queries; *(iii)* all necessary key constraints should ideally be defined and exported via JDBC, to enable the VKG system Ontop to perform extensive query optimization.

Handling Denormalized Data. The data we have been working with in this project is mostly log data. Each large log file was treated as a separate table. Such tables are often highly denormalized and contain a lot of redundancy. Consequently, there are a significant amount of constraints in the tables that are not declared as primary or foreign keys. These redundancies bring significant challenges to the performance of query answering. To address these issues, we have supplied the constraints as separate inputs to Ontop so as to avoid generating queries with redundancies. This optimization is critical and we have observed it turns many queries from timing out to almost finishing instantly. Currently, these constraints are declared manually, but in the future, we envision to automate this step as well.

Efficiency of VKG Approach. Bosch domain experts were skeptical in the beginning about the VKG approach, and they were confident that it could result in rather long query execution times for product quality analysis. After we presented our evaluation, they were quite enthusiastic about the results as their complex queries could be answered in a reasonable amount of time.

Impact of Our Approach to I4.0 at Bosch. The results of applying the VKG approach to the problem of integrating heterogeneous manufacturing data for the SMT manufacturing process are quite promising. We integrated the heterogeneous data sources to answer complex queries while at the same time achieving reasonable query execution times. We also showed that the SIB framework bridges the gap between the complex SMT domain and the underlying data sources, which is important for enabling the vision of I4.0. We see that the SIB framework can also be applied to other use cases at Bosch.

6   Conclusions and Outlook

Conclusions. We have presented the SIB Framework, in which we apply the VKG approach for the efficient integration of manufacturing data for product quality analysis tasks at Bosch. We introduced an ontology that models the domain knowledge and abstracts away the complexity of the source data for the *Surface Mounting Technology* Manufacturing Process. We mapped the data sources into ontological concepts and formulated queries that encode important product quality analysis tasks of domain experts at Bosch. We evaluated SIB over three different SMT data sets of different scales. We have presented the evaluation results and have shown that our framework retrieves the requested information mostly at reasonable time.

Outlook. This work is a first and important step towards adapting the VKG technology for integration of manufacturing data at Bosch, but there is a number of further steps to do. So far, we heavily involved domain experts from the Bosch Salzgitter plant in the development of the query catalog and ontologies and in the interpretation of results. This effort has shown that it is important to extend SIB with GUIs that will ease the interaction with domain experts, and will allow them to use SIB for various analytical tasks. This would enable us to conduct a user study that is needed for a more in depth evaluation of SIB. Moreover, this will permit us to scale the evaluation of the system from 13 queries to a much more substantial catalog. Then, it is important to show the benefits of SIB with analytical tasks beyond product analysis and evaluate the ability of the SMT ontology to accommodate a wider range of use cases. Another important next step is to extend SIB with a number of dashboards to facilitate aggregation and visualisation of query results. Moreover, an additional core step is popularization and validation of the SIB technology and its benefits with Bosch colleagues from various departments, where we already did the first step by presenting our results in an internal Bosch "Open Day" for internal projects. All of this should allow us to make SIB accessible and attractive to Bosch domain experts from factories and move us closer towards integrating SIB in the tool-set of Bosch analytical software. Finally, we plan to extend our solution and directly connect ontologies to the native JSON data, avoiding the intermediate step of materializing JSON into a relational DB. We also plan to compare the VKG approach to native triple stores.

Acknowledgements. This research has been partially supported by the Wallenberg AI, Autonomous Systems and Software Program (WASP) funded by the Knut and Alice Wallenberg Foundation, by the Italian Basic Research (PRIN) project HOPE, by the EU H2020 project INODE, grant agreement 863410, by the CHIST-ERA project PACMEL, by the Free Univ. of Bozen-Bolzano through the projects QUADRO, KGID, and GeoVKG, and by the project IDEE (FESR1133) through the European Regional Development Fund (ERDF) Investment for Growth and Jobs Programme 2014–2020.

References

1. Bienvenu, M., Rosati, R.: Query-based comparison of mappings in ontology-based data access. In: Proceedings of the 15th International Conference on Principles of Knowledge Representation and Reasoning (KR), pp. 197–206. AAAI Press (2016)
2. Calvanese, D., et al.: Ontop: answering SPARQL queries over relational databases. Semant. Web J. **8**(3), 471–487 (2017)
3. Cyganiak, R., Wood, D., Lanthaler, M.: RDF 1.1 concepts and abstract syntax. W3C Recommendation, World Wide Web Consortium, February 2014
4. Das, S., Sundara, S., Cyganiak, R.: R2RML: RDB to RDF mapping language. W3C Recommendation, World Wide Web Consortium, September 2012
5. De Giacomo, G., Lembo, D., Lenzerini, M., Poggi, A., Rosati, R.: Using ontologies for semantic data integration. In: Flesca, S., Greco, S., Masciari, E., Saccà, D. (eds.) A Comprehensive Guide Through the Italian Database Research Over the Last 25 Years. SBD, vol. 31, pp. 187–202. Springer, Cham (2018). https://doi.org/10.1007/978-3-319-61893-7_11
6. Elmer, S., et al.: Ontologies and reasoning to capture product complexity in automation industry. In: Proceedings of the ISWC 2017 P&D and Industry Tracks (2017)
7. Horrocks, I., Giese, M., Kharlamov, E., Waaler, A.: Using semantic technology to tame the data variety challenge. IEEE Internet Comput. **20**(6), 62–66 (2016)
8. Hubauer, T., Lamparter, S., Haase, P., Herzig, D.M.: Use cases of the industrial knowledge graph at siemens. In: Proceedings of the ISWC 2018 P&D/Industry/BlueSky Tracks (2018)
9. Kharlamov, E., et al.: Ontology based data access in Statoil. J. Web Semant. **44**, 3–36 (2017)
10. Kharlamov, E., et al.: An ontology-mediated analytics-aware approach to support monitoring and diagnostics of static and streaming data. J. Web Semant. **56**, 30–55 (2019)
11. Kharlamov, E., et al.: Semantic access to streaming and static data at Siemens. J. Web Semant. **44**, 54–74 (2017)
12. Kharlamov, E., Mehdi, G., Savkovic, O., Xiao, G., Kalayci, E.G., Roshchin, M.: Semantically-enhanced rule-based diagnostics for industrial Internet of Things: the SDRL language and case study for Siemens trains and turbines. J. Web Semant. **56**, 11–29 (2019)
13. Kharlamov, E., et al.: Finding data should be easier than finding oil. In: Proceedings of the IEEE International Conference on Big Data (Big Data), pp. 1747–1756. IEEE Computer Society (2018)
14. Lehmann, J., et al.: Distributed semantic analytics using the SANSA stack. In: d'Amato, C., et al. (eds.) ISWC 2017. LNCS, vol. 10588, pp. 147–155. Springer, Cham (2017). https://doi.org/10.1007/978-3-319-68204-4_15
15. Mami, M.N., Graux, D., Scerri, S., Jabeen, H., Auer, S., Lehmann, J.: Squerall: virtual ontology-based access to heterogeneous and large data sources. In: Ghidini, C., et al. (eds.) ISWC 2019. LNCS, vol. 11779, pp. 229–245. Springer, Cham (2019). https://doi.org/10.1007/978-3-030-30796-7_15
16. Mehdi, A., Kharlamov, E., Stepanova, D., Loesch, F., Gonzalez, I.G.: Towards semantic integration of Bosch manufacturing data. In: Proceedings of ISWC, pp. 303–304 (2019)
17. Mehdi, G., et al.: Semantic rule-based equipment diagnostics. In: d'Amato, C., et al. (eds.) ISWC 2017. LNCS, vol. 10588, pp. 314–333. Springer, Cham (2017). https://doi.org/10.1007/978-3-319-68204-4_29

18. Petersen, N., Halilaj, L., Grangel-González, I., Lohmann, S., Lange, C., Auer, S.: Realizing an RDF-based information model for a manufacturing company – a case study. In: d'Amato, C., et al. (eds.) ISWC 2017. LNCS, vol. 10588, pp. 350–366. Springer, Cham (2017). https://doi.org/10.1007/978-3-319-68204-4_31
19. Rodriguez-Muro, M., Rezk, M.: Efficient SPARQL-to-SQL with R2RML mappings. J. Web Semant. **33**, 141–169 (2015)
20. Xiao, G., et al.: Ontology-based data access: a survey. In: Proceedings of the 27th International. Joint Conference on Artificial Intelligence (IJCAI), pp. 5511–5519, IJCAI Org. (2018)
21. Xiao, G., Ding, L., Cogrel, B., Calvanese, D.: Virtual knowledge graphs: an overview of systems and use cases. Data Intell. **1**(3), 201–223 (2019)
22. Xiao, G., Kontchakov, R., Cogrel, B., Calvanese, D., Botoeva, E.: Efficient handling of SPARQL OPTIONAL for OBDA. In: Vrandečić, D., et al. (eds.) ISWC 2018. LNCS, vol. 11136, pp. 354–373. Springer, Cham (2018). https://doi.org/10.1007/978-3-030-00671-6_21

A Semantic Framework for Enabling Radio Spectrum Policy Management and Evaluation

Henrique Santos[1][(✉)], Alice Mulvehill[1,3], John S. Erickson[1,2],
Jamie P. McCusker[1], Minor Gordon[1], Owen Xie[1], Samuel Stouffer[1],
Gerard Capraro[4], Alex Pidwerbetsky[5], John Burgess[5],
Allan Berlinsky[5], Kurt Turck[6], Jonathan Ashdown[6],
and Deborah L. McGuinness[1]

[1] Tetherless World Constellation, Rensselaer Polytechnic Institute, Troy, NY, USA
{oliveh,mulvea,erickj4,mccusj2,gordom6,xieo,stoufs}@rpi.edu,
dlm@cs.rpi.edu
[2] The Rensselaer Institute for Data Exploration and Applications, Troy, NY, USA
[3] Memory Based Research LLC, Pittsburgh, PA, USA
[4] Capraro Technologies Inc., Utica, NY, USA
gcapraro@caprarotechnologies.com
[5] LGS Labs, CACI International Inc., Florham Park, NJ, USA
{a.pidwerbetsky,john.burgess,allan.berlinsky}@caci.com
[6] Air Force Research Laboratory, Rome, NY, USA
{kurt.turck,jonathan.ashdown}@us.af.mil

Abstract. Because radio spectrum is a finite resource, its usage and sharing is regulated by government agencies. These agencies define policies to manage spectrum allocation and assignment across multiple organizations, systems, and devices. With more portions of the radio spectrum being licensed for commercial use, the importance of providing an increased level of automation when evaluating such policies becomes crucial for the efficiency and efficacy of spectrum management. We introduce our Dynamic Spectrum Access Policy Framework for supporting the United States government's mission to enable both federal and non-federal entities to compatibly utilize available spectrum. The DSA Policy Framework acts as a machine-readable policy repository providing policy management features and spectrum access request evaluation. The framework utilizes a novel policy representation using OWL and PROV-O along with a domain-specific reasoning implementation that mixes GeoSPARQL, OWL reasoning, and knowledge graph traversal to evaluate incoming spectrum access requests and explain how applicable policies were used. The framework is currently being used to support live, over-the-air field exercises involving a diverse set of federal and commercial radios, as a component of a prototype spectrum management system.

Keywords: Dynamic spectrum access · Policies · Reasoning

Approved for public release (reference number: 88ABW-2020-1535).

© Springer Nature Switzerland AG 2020
J. Z. Pan et al. (Eds.): ISWC 2020, LNCS 12507, pp. 482–498, 2020.
https://doi.org/10.1007/978-3-030-62466-8_30

1   Introduction

Usable radio spectrum is becoming crowded[1] as an increasing number of services, both commercial and governmental, rely on wireless communications to operate. Techniques known as Dynamic Spectrum Access (DSA) [26] have been extensively researched as a way of promoting more efficient methods for sharing the radio spectrum among distinct organizations and their respective devices, while adhering to regulations.

In the United States, spectrum is managed by agencies that include the National Telecommunications and Information Administration[2] (NTIA) and the Federal Communications Commission[3] (FCC). The NTIA publishes revised versions of its Manual of Regulations and Procedures for Federal Radio Frequency Management[4] (commonly referred to as the NTIA Redbook) which is a compilation of regulatory policies that define the conditions that non-US, as well as federal and non-federal US, organizations, systems, and devices must satisfy in order to compatibly share radio spectrum while minimizing interference.

With the advent of 5G,[5] more parts of the radio spectrum are being licensed for commercial usage and, with the increased availability of cognitive radios [25] (devices that are able to automatically adjust their operating frequency), the importance of providing an increased level of automation when evaluating spectrum policies becomes crucial for the sustainability of spectrum management. This issue is currently under investigation by the National Spectrum Consortium[6] (NSC), a research and development organization that incubates new technologies to tackle challenges about radio spectrum management and utilization.

In this paper, we describe the Dynamic Spectrum Access Policy Framework (DSA Policy Framework). The DSA Policy Framework supports the management of machine-readable radio spectrum usage policies and provides a request evaluation interface that is able to reason about the policies and generate permit or deny results to spectrum access requests. This is accomplished via the utilization of a novel policy representation based on semantic web standards OWL and PROV-O that encodes its rules in an ontology. This ontology, combined with background knowledge originating from a number of relevant select sources, is stored in a Knowledge Graph that is used by a domain-specific reasoning implementation that mixes GeoSPARQL [22], OWL reasoning, and knowledge graph traversal to evaluate policies that are applicable to spectrum access requests. Effects (Permit/Deny/Permit with Obligations) are assigned and explanations are provided to justify why a particular request was permitted or denied access to the requested frequency or frequency range.

---

[1] http://bit.ly/FCC_AWS.
[2] http://ntia.doc.gov.
[3] http://fcc.gov.
[4] http://bit.ly/NTIA_Redbook.
[5] http://bit.ly/FCC_5G.
[6] http://nationalspectrumconsortium.org.

2   Sharing the Radio Spectrum

Radio spectrum policies specify how available spectrum should be used and shared. The applicability of existing policies must be checked for a variety of activities that use spectrum in many settings including various different types of requesters (e.g. systems and devices). For example, training exercises will request spectrum usage for a certain time frame and geographic region for potentially hundreds of radios with a wide variety of capabilities. During a training exercise, local policies are typically created to manage the spectrum that radios used in the exercise will require, and to minimize interference between federal and commercial (non-federal) radios that may be operating in the same area and within the same frequency range. Throughout this paper the following definitions are used:

– *High-level policies:* Policies as documented by authoritative agencies, including NTIA and FCC. These policies are not prone to change in the short term, although they may evolve when new versions of documents are released.
– *Local policies:* Specializations of high-level policies. These are created to locally manage the spectrum requests of various devices that want to operate in specific locations and/or for specific time periods.
– *Sub-policies:* A sub-section of a policy, sometimes referred to as "provisions" in NTIA Redbook policies.
– *Spectrum manager:* The role of a human who is responsible for managing policies. The spectrum manager verifies if existing policies are sufficient to support some activity and creates local policies to accommodate specific spectrum requirements.
– *Spectrum system:* This role represents some external system that is being used to generate spectrum requests on behalf of entities that require the use of a specific frequency or set of frequencies.

3   Dynamic Spectrum Access Policy Framework

The DSA Policy Framework supports the following objectives:

– Serve as a centralized machine-readable radio spectrum policy repository.
– Provide policy management features (including creation and customization) for a wide range of radio spectrum domain users.
– Use machine-readable policies as a basis for automatically evaluating radio spectrum access requests.

As shown on the top of Fig. 1, the DSA Policy Framework provides two major functions: *Policy Management* and *Request Evaluation. Policy Management* enables the spectrum manager to create local policies by referencing relevant higher level policies and adding customization. *Request Evaluation* utilizes all policies to automatically process spectrum requests. The results include references to any policy that was involved in the evaluation. As policies evolve, the underlying knowledge representation evolves, enabling the *request evaluation* module to use the most current policy information to reason and assign effects (permit, deny, permit with obligation) to spectrum requests. Spectrum managers can verify evaluation results in the presence of newly created policies through the *Request Builder* tool.

To support the DSA Policy Framework infrastructure we leveraged Whyis [20], a nanopublication-based knowledge graph publishing, management, and analysis framework. Whyis enables the creation of domain and data-driven knowledge graphs. The DSA Policy Framework takes advantage of the use of nanopublications [15] in Whyis, which allows it to incrementally evolve knowledge graphs while providing provenance-based justifications and publication credit for each piece of knowledge in the graph. This is particularly useful for the spectrum domain because policies do change and new policies can be created, potentially triggered by multiple sources. The DSA Policy Framework also makes use of the SETLr [19] Whyis agent, which enables the conversion of a number of the identified knowledge sources or derivatives to the Resource Description Framework (RDF), thereby bootstrapping the DSA Knowledge Graph.

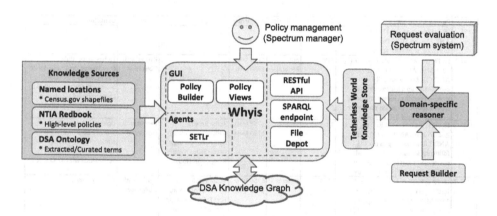

Fig. 1. DSA Policy Framework architecture

3.1 Knowledge Sources

In order to support the creation and interpretation of machine-readable radio spectrum policies, information from several sources was mined and incorporated into a Knowledge Graph. Spectrum policies and term definitions were obtained from the NTIA Redbook, an IEEE Standard Definitions and Concepts document [3], and from FCC 14–31 [4] (a FCC policy publication). Specific schemas associated with the spectrum domain were obtained from the Standard Spectrum Resource Format (SSRF) [5]. Finally, information about geographical locations was obtained from the Census.gov shapefiles.[7]

---

[7] http://bit.ly/Census_shapefiles.

During the *policy capture* process, Rensselaer Polytechnic Institute (RPI) collaborated with DSA domain experts from Capraro Technologies Inc. and LGS Labs of CACI International Inc. to select and analyze English text-based policies from the NTIA Redbook and from various FCC documents. In order to be used by the DSA Policy Framework, the English text was converted into a different representation, and many of the terms used in the English text were incorporated into a domain ontology. During this process, we observed that the text for many policies is logically equivalent to a conditional expression e.g., *IF* some device wants to use a frequency in a particular frequency range *AND* at a particular location, *THEN* it is either *PERMITTED* or *DENIED*.

More complex policies contain a set of conditional expressions, with each conditional expression focused on a particular request attribute, e.g., a request device type, frequency, frequency range. The spreadsheet displayed in Fig. 2 contains an example of a complex policy from the NTIA Redbook called US91. Due to space constraints, we have omitted some of the sub-policies for US91 and the columns that document policy metadata and provenance, including the original text, source document, URL, and page number.

RuleID	Parsed logical rule	Requester	Affiliation	Frequency	Location	Effect	Obligation
US91	IF RefFreq is ε (≥ 1755 MHz AND ≤ 1780 MHz) then the following provisions apply						
US91-1	IF TR is AWS AND RefFreq is ε (≥ 1755 MHz AND ≤ 1780 MHz) AND (TR has successfully coordinated on a nationwide basis prior to operation, unless otherwise specified by Commision rule, order , or notice) THEN TR is Primary	AWS	non Federal	1755 MHz - 1780 MHz		Permit with Obligation	TR has successfully coordinated on a nationwide basis prior to operation, unless otherwise specified by Commision rule, order, or notice and TR is Primary
US91-2	IF TR is Federal AND RefFreq is ε (≥ 1755 MHz AND ≤ 1780 MHz) AND TR is operating in one the following locations AND (Until reaccommodated in accordance with 47 CFR 301) THEN it is Primary	TR	Federal	1755 MHz - 1780 MHz	Yuma Proving Ground, Fort Irwin, Fort Polk, Fort Bragg, White Sands Missile Range, Fort Hood	Permit with Obligation	Operate on a co-equal, primary basis with AWS stationsTR can't transmit Until re-accommodated in accordance with 47 CFR 301
US91-3	IF TR is Federal AND JTRS AND RefFreq is ε (≥ 1755 MHz AND ≤ 1780 MHz) AND TR is operating in one the following locations THEN It is Primary	JTRS	Federal	1755 MHz - 1780 MHz	Yuma Proving Ground, Fort Irwin, Fort Polk, Fort Bragg, White Sands Missile Range, Fort Hood	Permit with Obligation	JTRS is Primary

Fig. 2. Spreadsheet excerpt showing the NTIA Redbook US91 policy capture

US91 regulates the usage of the 1755–1780 MHz frequency range and is an example of a spectrum range that must be efficiently shared by both federal and non-federal devices. Because spectrum usage can vary by device, affiliation and location, we decompose the policy into several sub-policies (US91-1, US91-2, and US91-3). A *parsed logical rule* is manually created for each policy by a domain expert. The elements of the logical rule are further expressed as attribute-value pairs, e.g., Requester = AWS. The attribute (column) names map into the following elements of the *policy logical expression* that is used by the framework:

- *Requester:* the device requesting access
- *Affiliation:* the affiliation of the requester (Federal, Non-Federal)
- *Frequency:* the frequency range or single frequency being requested
- *Location:* location(s) where the policy is applicable
- *Effect:* the effect a policy yields, if the rule is satisfied (Permit, Deny, Permit with Obligations)
- *Obligations:* the list of obligations the requester needs to comply with in order to be permitted

The framework utilizes several ontologies to support policy administration and spectrum request processing, including a domain ontology called the DSA Ontology. DSA ontology terms were collected during policy capture and/or derived from the NTIA Redbook, IEEE Standards, SSRF, or other domain source. All terms were curated by an ontology developer and linked to external ontologies including PROV-O and the Semanticscience Integrated Ontology (SIO) [7].

3.2   Representing Radio Spectrum Requests

Figure 3 shows the DSA request model and a sample request. The model is based on the World Wide Web Consortium's recommended standard for provenance on the web (PROV). The modeling of requests as activities and agents was influenced by the policy attributes described as columns in the policy capture spreadsheet (shown in Fig. 2) and extended to include the action associated with the requester in a spectrum request (currently we represent only the Transmission action). In the model, the requester (prov:Agent) is linked with an action prov:Activity using the prov:wasAssociatedWith predicate. The location attribute is represented as prov:Location and linked to the requester using the prov:atLocation predicate. The time attribute, which describes when the request action is to start and end, is represented using the literal data type xsd:dateTime and linked to the action using the prov:startedAtTime and prov:endedAtTime predicates.

For attributes not natively supported by PROV, we use SIO, which enables us to model objects and their attributes (roles, measurement values) with the use of the sio:Attribute class and sio:hasAttribute /sio:hasValue /sio: hasUnit predicates. In the DSA request model, the attributes Frequency, Frequency Range, and Affiliation are represented as sio:Attribute and linked to the requester using sio:hasAttribute. For attributes that can assume a value (currently only Frequencies), literal values (sio:hasValue) and units of measurement (sio:hasUnit) are used to express the quantification of an attribute as a specified unit from the Units of Measurement Ontology (UO) [13]. In Fig. 3, a sample radio spectrum request instance is shown, where the requester is a Generic Joint Tactical Radio System (JTRS) radio. The device is physically located at a location that is defined by the Well-Known Text (WKT) [2] string POINT(-114.23 33.20) and is requesting access to the frequency range 1755--1756.25 MHz using the FrequencyRange attribute, which is

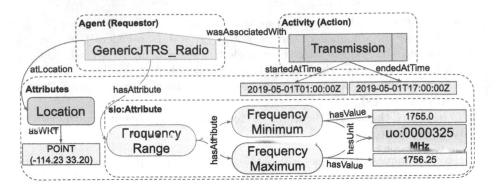

Fig. 3. The DSA request model and sample request

described by the composition of the `FrequencyMinimum` and `Frequency-Maximum` attributes, with their respective value and unit.

3.3 Representing Radio Spectrum Policies

The DSA policy model was created to enable (1) the unambiguous representation of rules as parsed during *policy capture*, (2) the reuse of an existing policy's rules for creating local policies, and (3) the implementation of request evaluation capabilities. The model is based on OWL, encoding policy rules as restrictions in OWL classes that represent policies. The OWL restrictions are constructed on the RDF properties of the DSA request model, presented in the previous subsection, while constraining their ranges to the expected literal value or class, as dictated by the policy rule being expressed. To demonstrate this, Listing 1.1 shows the OWL expression in Manchester syntax of an excerpt of the NTIA US91 policy that describes how a requester that is an example of a radio, categorized as JTRS, can use the spectrum regulated by US91 (1755–1780 MHz) in specific locations (US91-3 sub-policy in Fig. 2).

The model advocates for the creation of an OWL class for each rule that composes the complete policy rule expression. In the example, the OWL class `US91` has the restriction on the frequency range attribute, with the minimum value `1755` and maximum value `1780` (lines 4–11). This class is a subclass of `Transmission` (line 13), which is the action this policy regulates. In lines 15–20, a class `US91-3` restricts the requester to a `JointTacticalRadioSystem` and appends that rule to the rules of its super-class `US91`. The last class of this policy expression is `US91-3.1`, which encodes the location rule as a restriction on the `atLocation` property (lines 24–25), constraining it to a location class. When the composition of rules expressed by this last class is evaluated to be true, the policy should assign the permit effect. This is expressed by asserting `US91-3.1` as a subclass of `Permit` (line 27).

```
 1 Class: US91
 2   EquivalentTo:
 3     Transmission and
 4     (wasAssociatedWith some (hasAttribute some
 5      (FrequencyRange
 6        and (hasAttribute some
 7              (FrequencyMaximum and
 8                (hasValue some xsd:float[<= 1780.0f]))))
 9          and (hasAttribute some
10              (FrequencyMinimum and
11                (hasValue some xsd:float[>= 1755.0f]))))))
12   SubClassOf:
13     Transmission
14
15 Class: US91-3
16   EquivalentTo:
17     US91 and
18     (wasAssociatedWith some JointTacticalRadioSystem)
19   SubClassOf:
20     US91
21
22 Class: US91-3.1
23   EquivalentTo:
24     US91-3 and (wasAssociatedWith some
25                 (atLocation some US91-3.1_Location))
26   SubClassOf:
27     Permit, US91-3
```

Listing 1.1. OWL expression of part of the US91 policy in Manchester syntax

Many of the NTIA policies, including US91, contain location rules and, often, these rules are written in terms of location lists. To represent this, we defined OWL classes for the lists to be used in conjunction with the OWL expression of policies. Listing 1.2 shows the definition of the US91-3.1_Location class, which is used in the US91-3.1 restriction, as previously shown. We have used the GeoSPARQL predicate sfWithin in conjunction with OWL unions to express that the rule is satisfied if the location is specified in the list (lines 3–8). In this example, we leverage location information we imported from Census.gov shapes for US Federal locations.

```
 1 Class: US91-3.1_Location
 2   EquivalentTo:
 3     (sfWithin value White_Sands_Missile_Range) or
 4     (sfWithin value Ft_Irwin) or
 5     (sfWithin value Yuma_Proving_Ground) or
 6     (sfWithin value Ft_Polk) or
 7     (sfWithin value Ft_Bragg) or
 8     (sfWithin value Ft_Hood)
 9   SubClassOf:
10     Location
```

Listing 1.2. OWL expression of a location list in Manchester syntax

The creation of the OWL class hierarchy (and, therefore, the incremental addition of rules) maximizes the reuse of rules when spectrum managers create local policies. To illustrate this claim, Listing 1.3 shows a sample local policy which modifies the existing US91-3 sub-policy to Deny requests in a specific time window. Lines 4–5 contain the time restrictions on the PROV-O predicates. Local policies can have an explicit precedence level, as shown in line 7.

```
1 Class. US91 3 1-Local
2 EquivalentTo:
3   US91-3.1 and
4   endedAtTime some xsd:dateTime[>=2019-10-01T11:00:00Z] and
5   startedAtTime some xsd:dateTime[<=2019-10-01T17:00:00Z]
6 SubClassOf:
7   Deny, Priority_1, US91-3.1
```

Listing 1.3. OWL expression of a local policy in Manchester syntax

3.4 Policy Management

The DSA Policy Framework provides a web interface to allow spectrum managers to have a comprehensive understanding of the DSA Knowledge Graph, including policies, locations, and entities in the DSA Ontology. It leverages Whyis default "views" with some extensions for supporting a domain-specific display of pieces of the Knowledge Graph. The structure and content of the interface are driven by the DSA Knowledge Graph, which ensures that it displays relevant and contextualized information and features.

The *Policy Faceted Browser* that allows a user to quickly find policies based on the selection of attribute values. Spectrum managers can, for instance, find policies applicable to a list of select locations or policies applicable to a specific device, or even to a combination of both. This is accomplished by domain-specific SPARQL queries that retrieve and group attributes from the policy's OWL structure. The user can view details of a policy or reuse a policy's rules.

The *Policy Detail* view provides a display of policy metadata, including name, original text and identifier, and a human-readable version of the policy encoded rules. If the policy specifies locations, those locations will be displayed on a map.

The *Policy Builder* view can be used to build a policy from scratch or to create local policies by reusing existing policies' rules. The Policy Builder leverages the DSA Knowledge Graph to provide user support during policy creation. For instance, if a user wants to create a rule for an specific device, the Builder will display known devices as represented in the KG. More than that, the Builder "understands" the semantics of the rule which means that if the user wants to add a frequency range rule, for instance, the Builder will prompt the user to enter both lower and upper bound values. Users can also set policy effects, precedence, and obligations. In the back end, the policy is converted to the DSA policy model in OWL and stored as a new piece of knowledge in the DSA Knowledge Graph.

Knowledge curation is a time-consuming task and it might impact the pace at which updated knowledge becomes available for use when creating new policies.

To overcome this, the DSA Policy Framework supports policy additions that refer to terms not currently in the ontology, by allowing input of new terms along with the annotation that those terms need in order to be curated by the appropriate ontology owner. This allows a spectrum manager to input and test a new policy without having to wait for an ontology update first.

3.5 Request Evaluation: Domain-Specific Reasoning

To enable the evaluation of spectrum requests against policies, we have implemented a domain-specific reasoner that combines various Semantic Web computational approaches to assign `Permit`/`Deny` effects to requests, while fulfilling requirements, including geographical reasoning, policy precedence evaluation, and explanations for denied requests. The domain-specific reasoner follows a four-phase pipeline, with a set of requests as input and the assigned effect, a list of obligations, and a list of reasons for each request as output. The reasoner initiates by creating an in-memory RDF graph originated by the merge of the request RDF graph and the DSA Knowledge Graph.

Next, in the **geographical reasoning phase**, the reasoner elicits the geographical relationships among the requests' WKT locations, and the named locations present in the DSA KG, by using GeoSPARQL to infer triples like `:req_location geo:sfWithin :NAMED_LOCATION`. The inferred triples are then asserted back into the graph.

In the **OWL reasoning phase**, the reasoner makes use of the HermiT OWL reasoner [14] to perform Description Logic (DL) reasoning over the updated graph. The domain-specific implementation relies on the DL services of:

- *Classification* for computing all subclass relationships, allowing the inferred class hierarchy to be leveraged when querying policy Effect and Precedence.
- *Realization* of computing classes (policies) that individuals (requests) belong to. "Belonging" to a class means that an individual satisfies the constraints of that class; realization can be understood as determining *policy applicability.*
- *DL Query* for retrieving the individuals that were determined to belong to specific classes.

Using the list of applicable policies retrieved from HermiT, the domain-specific reasoner decides precedence, in the **precedence evaluation phase**, by a simple evaluation of which policy has the highest precedence level. Policies with no explicit precedence are assumed to have the lowest precedence. The highest precedence policy effect is then assigned to the request. The reasoner follows these heuristics to explain the assignment of a `Deny` effect to a request, in the **explanation phase**:

- Requests can be denied by a specific applicable policy with a `Deny` effect. In this case, the "reason" for the denial is the identity of this policy and the specification of what attributes of the request fulfill the rules contained within. The reasoner retrieves the policy's rules and presents them as reasons for denial.

– Requests can be denied due to a lack of a policy with a `Permit` effect (i.e. there is no applicable policy with either a `Deny` or `Permit` effect). In these cases the reasoner finds the rules that the request did not satisfy in order to be assigned a `Permit`. The reasoner calculates the paths in the OWL class hierarchy from those policies that the request was reasoned to belong to, to the policies that would result in a `Permit`. These paths contain classes that the request was determined not to belong to. Unfulfilled rules for each policy found to be in the path are retrieved and presented as reasons for the `Deny`.

As example, the request in Fig. 3 would ultimately be determined to belong to the `US91`, `US91-3`, and `US91-3.1` classes displayed in Listing 1.1 and, therefore, assigned the `Permit` effect. Nevertheless, when the local policy shown in Listing 1.3 exists in the graph, the request is then reasoned to belong to the `US91-3.1-Local` class as well. Since this local policy contains a time constraint rule with a higher precedence, the reasoner assigns the effect to be a `Deny` and returns the reason, "the request is in a prohibited time window."

Conversely, if the request is modified to a different location outside the locations expressed in Listing 1.2, it would be reasoned to belong only to `US91` and `US91-3` classes, based on the frequency range and requester attributes of the request. These applicable classes don't express an *explicit* `Permit` or `Deny` effect, so the reasoner must default to `Deny`. The reasoner then calculates the path to the class `US91-3.1` that would `Permit` it *if* the conditions had been met, and retrieves the rules of each class in the path (in this case, only the rules of the `US91-3.1` class) and returns the reason, "the request is not in a permitted location."

4   Evaluation of the Framework

In order to evaluate the DSA Policy Framework's semantic representation of the domain, we identified several fundamental spectrum policy constructs. They are displayed in bold in the first column of Table 1. For each of them, we worked with domain experts to identify the elements that were required in order to effectively represent policies and support request evaluation. The table contains columns for Policy Representation (high-level and local policies) and Request Evaluation. "Yes" in the columns indicates that the policy construct is either *Relevant* or it has been fully addressed and *Implemented*. "Partial" indicates that the current implementation meets a simplified version of the requirement, while "uc" means that the construct element is currently under consideration.

The basic structure of a policy varies among source documents. The NTIA Redbook uses provisions to distinguish policy behavior with regards to attributes (specific device or location, for instance). Provisions are described as sub-policies in the policy capture spreadsheet and in the Framework, and they are leveraged during request evaluation. Policies can be specified in the *Parser logical rule* format, which is enabled in the Policy Builder. The framework does not currently represent the individual elements of an obligation. Instead, it is represented as text or as a canned identifier to support request evaluation.

Table 1. DSA Policy Framework policy semantics coverage

Domain policy construct	Policy representation		Request evaluation	
	Relevant	Implemented	Relevant	Implemented
Provision	yes	yes	yes	yes
Parsed logical rule	yes	yes	no	no
Obligation	yes	partial	yes	partial
Requesters				
Device	yes	yes	yes	yes
Organization	yes	yes	yes	yes
Dependency	yes	uc	uc	uc
Licensee	yes	partial	yes	partial
Affiliations				
Federal/Non-Federal	yes	partial	yes	partial
Named requester	yes	yes	yes	yes
Frequencies				
Frequency range	yes	yes	yes	yes
Single frequency	yes	yes	yes	yes
Named bands	yes	yes	uc	uc
Units	yes	yes	yes	yes
Time				
Timezones	yes	yes	yes	yes
Policy validity	yes	yes	yes	yes
Locations				
Named locations	yes	yes	yes	yes
Relative locations	yes	uc	yes	uc
Polygons/Circles	yes	yes	yes	yes
Geographical rules				
Specific location	yes	yes	yes	yes
List of locations	yes	yes	yes	yes
Precedence				
Levels	yes	yes	yes	yes
Explanations				
Policy triggered	yes	yes	yes	yes
Rules not satisfied	yes	yes	yes	yes

The text in the source policies specifies regulations for a variety of requester types. While all requester types are relevant for policy representation, requests only express the device. However, the DSA KG does represent several types including Organization and Licensee, so if a request was received for one of these types, theoretically, it could be evaluated. If a policy specifies a dependency between requesters, this dependency is only currently treated as text. Policies can also regulate the spectrum usage by affiliation (Federal systems are permitted to use some frequency, for instance). The Framework allows affiliation rules to be specified in policies, however, affiliation reasoning is currently limited by the expressiveness of affiliations in the DSA Ontology, where only requesters with an explicit affiliation (named requester) are effectively reasoned.

Frequency rules are specified in terms of a range or single frequency in different units (MHz, GHz). Sometimes, a range is described as a band, which is a named frequency range. Requests do not express named bands. The framework supports all of these constructs, using the `Frequency` and `FrequencyRange` attributes, and the `sio:hasUnit` property. The time attribute exists in local policies only, and they specify the validity of a policy. Time and timezones are supported using `xsd:dateTime` literals and PROV-O time predicates.

Most policies refer to locations by names or by coordinates (points, polygons, and circles), but sometimes a location is expressed in relation to another location. The framework uses Census.gov shapes to refine named locations and WKT literals to represent polygons and circles. Relative locations are still under development. Geographical rules are defined in terms of the requester being in a location or in a list of locations. The framework implements all of these constructs using `geo:sfWithin` predicate and OWL unions.

The framework implements all of the identified precedence needs. It can define and evaluate precedence levels. For explaining evaluation results, the framework implements all current explanation requirements. It outputs which policy was triggered by a request and presents reasons based on the presented heuristics. Finally, the constructs identified as "uc" are areas for future work.

5 System Adoption and Deployment

The DSA Policy Framework is being used in simulated scenarios, where it supports the research & development of other components of a dynamic spectrum management system. It currently contains approximately 165 high-level policies from the NTIA Redbook (including their sub-policies). The DSA Ontology contains 695 classes and is constantly evolving to address new domain constructs and support more precise request evaluation.

The framework is transitioning to support live, over-the-air field exercises involving a diverse set of federal and commercial radios. During these exercises, the Framework supports (1) the creation, deletion, and revision of local policies, (2) the real-time processing of numerous spectrum requests, and (3) the generation of explanations that describe how the spectrum requests were processed. The public released assets developed during the course of the project can be accessed at https://github.com/tetherless-world/dsa-open/.

6 Related Work

Kirrane [17] offers a comprehensive survey of access control models, well-known policy languages, proposed frameworks that utilize ontologies and/or rules to express policies, and a categorization of policy languages and frameworks against access control requirements. XACML 3.0, the eXtensible Access Control Markup Language [1], is a well-known policy language and *de facto* standard for representing attribute-based access control (ABAC) [16] policies and requests. Importantly, XACML provides a reference architecture for centralizing access control and a process model for evaluating requests against existing policies that inform the design of access control systems across domains and technologies.

Thi [24] proposes an OWL-based extension to XACML to support a generalized context-aware role-based access control (RBAC) model providing spatio-temporal restrictions and conforming with the NIST RBAC standard [9]. Their work augments the XACML architecture with new functions and data types.

Muppavarapu [21] identifies the limitations of identity-based access control schemes used in the Open Grid Services Architecture (OGSA) and proposes the use of OWL to represent the ontology of an organization's resources and users. They further propose the use of semantics in conjunction with the XACML standard for better interoperability and reduced administration overhead.

Our approach combines OWL, PROV-O, and the HermiT OWL reasoner with an ontology, represented as a knowledge graph, to support the representation of policies governing access to available spectrum. Relevant related research is described in Dundua [8], where previous work is described that uses OWL for modeling and analyzing access control policies, especially ABAC, and considers how the ABAC model can be integrated into ontology languages. In addition, Sharma [23] describes how OWL can be used to formally define and process security policies that can be captured using ABAC models. This work demonstrates how models, domains, data and security policies can be expressed in OWL and how a reasoner can be used to decide what is permitted.

Kolovski [18] maps the web service policy language, WS-Policy [6], to the description logic fragment species of OWL and demonstrates how standard OWL reasoners can check policy conformity and perform policy analysis tasks.

Garcìa [11,12] and Finin [10] offer important contributions on how end-to-end usage rights and access control systems may be implemented in OWL and RDF. Garcìa proposes a "Copyright Ontology" based on OWL and RDF for expressing rights, representations that can be associated with media fragments in a web-scale "rights value change." Finin describes two ways to support standard RBAC models in OWL and discusses how their OWL implementations can be extended to model attribute-based RBAC or, more generally, ABAC.

Our policy representation approach builds on previous work by matching the cross-domain policy expression semantics of XACML. It extends these semantics with the capacity to express rich spatio-temporal restrictions, enabling the implementation of a wide variety of attribute-based policies across domains. It leverages background knowledge from domain-specific knowledge graphs that are structured with a domain-derived ontology, enabling the inference of policy applicability based on attributes and constraints. Our approach uniquely conceptualizes policy requests as PROV activities and request evaluations as realizations. Finally, our approach provides a novel reasoner-based explanation in request evaluation results, enabling domain policy developers to understand the precise reasons for policy decisions.

7 Conclusion

We described a policy Framework that leverages a machine-readable, radio spectrum policy representation to support policy management and enable a domain-specific reasoner to efficiently process spectrum requests. The DSA policy model uses OWL restrictions on PROV-O properties to represent policy rules in a hierarchical approach that maximizes the reuse of rules when local policies are created, therefore facilitating the creation of local policies. The hierarchical nature of this policy representation also supports the explanation of evaluation results, by traversing the graph to find rules that were not satisfied. Because it leverages a domain Knowledge Graph, built from multiple knowledge sources, the domain-specific reasoner allows rich semantics which otherwise would be difficult to achieve with approaches that rely on a "flat" representation of attributes.

During the course of the project, we encountered OWL reasoning performance issues when multiple requests are simultaneously received. To decrease reasoning time, we partitioned the request graph into smaller graphs and evaluated each in parallel, using multiple processor cores. This allowed us to match the required response time of under 10 s.

Future work includes additional support for enforcement of the DSA policy model. Although the DSA policy model's OWL hierarchy maximizes the reuse of rules, there is currently no enforcement. Overlapping rules can be created, which can lead to multiple policies with the same precedence level being triggered during request evaluation. The DSA Ontology is constantly changing as additional policies are added to the framework with terms that have yet to be defined. Existing policies can be affected by changes in the DSA Ontology as their rules reference entities in it and they may need to be reviewed with regards to the updated representation of the domain. We plan to provide spectrum managers a way of tracking policies that are subject to review due to ontology changes.

Acknowledgement of Support and Disclaimer. This work is funded in support of National Spectrum Consortium (NSC) project number NSC-17-7030. Any opinions, findings and conclusions or recommendations expressed in this material are those the authors and do not necessarily reflect the views of AFRL.

References

1. eXtensible Access Control Markup Language (XACML) Version 3.0. http://docs.oasis-open.org/xacml/3.0/xacml-3.0-core-spec-os-en.html
2. ISO/IEC 13249–3:2016 Information technology – Database languages - SQL multimedia and application packages - Part 3: Spatial

3. IEEE Standard Definitions and Concepts for Dynamic Spectrum Access: Terminology Relating to Emerging Wireless Networks, System Functionality, and Spectrum Management. IEEE Std 1900.1-2008, pp. 1–62 (2008)

4. Federal Communications Commission, FCC 14–31. Technical report (2014)

5. Standard Spectrum Resource Format (SSRF), Data Exchange Standard, Version 3.1.0 (MC4EB Pub 8). Technical report (2014)

6. Curbera, F., Hallam-Baker, P., Hondo, V.M., Nadalin, A., Nagaratnam, N., Sharp, C.: Web services policy framework (ws-policy) (2006). https://www.w3.org/Submission/WS-Policy/

7. Dumontier, M., et al.: The semanticscience integrated ontology (SIO) for biomedical research and knowledge discovery. J. Biomed. Semant. **5**, 14 (2014)

8. Dundua, B., Rukhaia, M.: Towards integrating attribute-based access control into ontologies. In: 2019 IEEE 2nd Ukraine Conference on Electrical and Computer Engineering (UKRCON), pp. 1052–1056 (2019)

9. Ferraiolo, D.F., Kuhn, D.R.: Role-based access controls (2009)

10. Finin, T., et al.: ROWLBAC: representing role based access control in OWL. In: Proceedings of the 13th ACM symposium on Access Control Models and Technologies, SACMAT 2008, pp. 73–82. Association for Computing Machinery, Estes Park (2008)

11. García, R., Castellà, D., Gil, R.: Semantic copyright management of media fragments. In: DATA 2013 - Proceedings of the 2nd International Conference on Data Technologies and Applications, pp. 230–237 (2013)

12. García, R., Gil, R., Delgado, J.: A web ontologies framework for digital rights management. Artif. Intell. Law **15**(2), 137–154 (2007)

13. Gkoutos, G.V., Schofield, P.N., Hoehndorf, R.: The Units Ontology: a tool for integrating units of measurement in science. Database **2012** (2012)

14. Glimm, B., Horrocks, I., Motik, B., Stoilos, G., Wang, Z.: HermiT: an OWL 2 reasoner. J. Autom. Reasoning **53**(3), 245–269 (2014). https://doi.org/10.1007/s10817-014-9305-1

15. Groth, P., Gibson, A., Velterop, J.: The anatomy of a nanopublication. Inf. Serv. Use **30**(1–2), 51–56 (2010). publisher: IOS Press

16. Hu, V.C., Kuhn, D.R., Ferraiolo, D.F., Voas, J.: Attribute-based access control. Computer **48**(2), 85–88 (2015)

17. Kirrane, S., Mileo, A., Decker, S.: Access control and the resource description framework: a survey. Semant. Web **8**(2), 311–352 (2017)

18. Kolovski, V., Parsia, B., Katz, Y., Hendler, J.: Representing web service policies in OWL-DL. In: Gil, Y., Motta, E., Benjamins, V.R., Musen, M.A. (eds.) ISWC 2005. LNCS, vol. 3729, pp. 461–475. Springer, Heidelberg (2005). https://doi.org/10.1007/11574620_34

19. McCusker, J.P., Chastain, K., Rashid, S., Norris, S., McGuinness, D.L.: SETLr: the semantic extract, transform, and load-r. Technical report, e26476v1, PeerJ Inc. (2018)

20. McCusker, J.P., Rashid, S.M., Agu, N., Bennett, K.P., McGuinness, D.L.: The Whyis knowledge graph framework in action. In: International Semantic Web Conference (P&D/Industry/BlueSky) (2018)

21. Muppavarapu, V., Chung, S.M.: Semantic-based access control for grid data resources in open grid services architecture - data access and integration (OGSA-DAI). In: 2008 20th IEEE International Conference on Tools with Artificial Intelligence, vol. 2, pp. 315–322 (2008). iSSN: 2375–0197

22. Perry, M., Herring, J.: OGC GeoSPARQL-a geographic query language for RDF data. OGC Implement. Stand. **40** (2012)

23. Sharma, N.K., Joshi, A.: Representing attribute based access control policies in OWL. In: 2016 IEEE Tenth International Conference on Semantic Computing (ICSC), pp. 333–336 (2016)
24. Tran Thi, Q.N., Dang, T.K.: X-STROWL: a generalized extension of XACML for context-aware spatio-temporal RBAC model with OWL. In: Seventh International Conference on Digital Information Management (ICDIM 2012), pp. 253–258 (2012)
25. Zhang, Y.: Dynamic spectrum access in cognitive radio wireless networks. In: 2008 IEEE International Conference on Communications, pp. 4927–4932 (2008). iSSN: 1938–1883
26. Zhao, Q., Sadler, B.M.: A survey of dynamic spectrum access. IEEE Signal Process. Mag. **24**(3), 79–89 (2007)

Leveraging Linguistic Linked Data
for Cross-Lingual Model Transfer
in the Pharmaceutical Domain

Jorge Gracia[1(✉)], Christian Fäth[2], Matthias Hartung[3], Max Ionov[2],
Julia Bosque-Gil[1], Susana Veríssimo[3], Christian Chiarcos[2],
and Matthias Orlikowski[3]

[1] Aragon Institute of Engineering Research, University of Zaragoza, Zaragoza, Spain
{jogracia,jbosque}@unizar.es
[2] Goethe University Frankfurt, Frankfurt, Germany
faeth@em.uni-frankfurt.de, ionov@cs.uni-frankfurt.de,
chiarcos@informatik.uni-frankfurt.de
[3] Semalytix GmbH, Bielefeld, Germany
{hartung,susana.verissimo,matthias.orlikowski}@semalytix.com

Abstract. We describe the use of linguistic linked data to support a cross-lingual transfer framework for sentiment analysis in the pharmaceutical domain. The proposed system dynamically gathers translations from the Linked Open Data (LOD) cloud, particularly from Apertium RDF, in order to project a deep learning-based sentiment classifier from one language to another, thus enabling scalability and avoiding the need of model re-training when transferred across languages. We describe the whole pipeline traversed by the multilingual data, from their conversion into RDF based on a new dynamic and flexible transformation framework, through their linking and publication as linked data, and finally their exploitation in the particular use case. Based on experiments on projecting a sentiment classifier from English to Spanish, we demonstrate how linked data techniques are able to enhance the multilingual capabilities of a deep learning-based approach in a dynamic and scalable way, in a real application scenario from the pharmaceutical domain.

Keywords: Apertium RDF · Cross-lingual model transfer · Fintan

1 Introduction

One of the biggest challenges faced by international companies in Europe and worldwide is that markets are spread across countries and languages. Thus, their ability to adapt to new markets is of vital importance. To that end, language technologies (LT) and linked data (LD) have been recognised as core technologies to reduce language barriers between different national markets [16].

A major challenge faced by suppliers of LT services and products in global markets arises from the complexity of business use cases, technical components

© Springer Nature Switzerland AG 2020
J. Z. Pan et al. (Eds.): ISWC 2020, LNCS 12507, pp. 499–514, 2020.
https://doi.org/10.1007/978-3-030-62466-8_31

needed to address them, and input data that comes from multiple languages. Approaching this challenge by attempting to build dedicated Natural Language Processing (NLP) stacks for each new language from scratch is not scalable, due to generally high on-boarding costs for initial model development and refinement.

As an alternative, *cross-lingual model transfer* methods are based on the idea that NLP models readily existing for a source language can be transferred to a new target language of interest without language-specific supervision in terms of manually created training data being required in this target language [17]. As a primary source of cross-lingual information, many transfer approaches rely on bilingual lexical resources in order to bridge the language gap.

A growing number of lexical resources is made publicly available as part of the Linguistic Linked Open Data (LLOD) cloud[1] [4]. In this paper, we demonstrate the strong potential of LLOD resources to be used as catalysers of cross-lingual transfer of NLP models in deep learning frameworks. This is illustrated by way of the multilingual lexical resource Apertium RDF v2.0 that has been created and published as LLOD in order to meet the requirements of an LT-based real-world evidence platform for the pharmaceutical industry in a software-as-a-service setting. The specific use case aims at rapidly and cost-effectively increasing the multilingual capabilities of the NLP components underlying the platform, which we demonstrate here for the case of a pharma-specific sentiment detection model that is transferred from English to Spanish. In order to allow for a flexible and automated way of transforming the Apertium data into the LLOD formats, we rely on Fintan [8], a newly developed RDF transformation platform.

The remainder of this paper is organised as follows. In Sect. 2 we describe the background and technological context of this research. In Sect. 3 we give an overview of the overall architecture. Then, Sect. 4 describes the transformation and linking steps carried out to convert the Apertium original data into RDF. In Sect. 5 the role of the Apertium RDF data to improve bilingual sentiment embeddings is explained, and Sect. 6 reports on some experimental validations. Finally, Sect. 7 contains conclusions and future work.

2   Background and Related Work

In this section we describe some core technologies needed to better understand our approach, namely the Ontolex-lemon model and the Apertium initiative. We also report on the recent advancements in cross-lingual transfer learning.

2.1   OntoLex-Lemon

In the context of LLOD, OntoLex-lemon[2] is the primary community standard for representing lexical data in RDF [13]. This was originally developed with the aim to provide a rich linguistic grounding for ontologies, meaning that the

---

[1] http://linguistic-lod.org/llod-cloud.

[2] https://www.w3.org/2016/05/ontolex/.

natural language expressions used in labels, definitions or comments of ontology elements are equipped with an extensive linguistic description.

The main class for linguistic description in OntoLex is `LexicalEntry`, which corresponds to a word, a multi-word expression, or an affix. Lexical entries have different lexical forms (through the `Form` class) with their corresponding written and/or phonetic representations. The connection of a lexical entry to an ontological entity is marked mainly by the `denotes` property or is mediated by the `LexicalSense` or the `LexicalConcept` classes.[3]

Other modules extend the core module such as the variation and translation (*vartrans*) module, which introduces the representation of translations as a subtype of `SenseRelation`, i.e., a relation established between lexical senses.[4]

2.2 Apertium

Apertium[5] is a free/open-source machine translation platform [6] that mostly relies on the use of symbolic methods and currently includes over 50 language pairs.[6] It provides NLP components for many languages, as well as transfer rules and bilingual dictionaries for their respective translation.

A subset of the family of bilingual dictionaries developed in Apertium was converted to the LMF [7] ISO standard as part of the METANET4U Project.[7] From that subset of Apertium dictionaries, only the entries in Apertium which were annotated as nouns, proper nouns, verbs, adjectives and adverbs were considered (from a long list of heterogeneous parts of speech present across datasets). This LMF subset constituted the basis for the first RDF representation of the Apertium dictionaries [10], which was released as LLOD[8] (we will refer to it as `Apertium RDF v1.0` in the rest of this paper). Such an RDF version of the Apertium dictionary data was based on the *lemon* model, the predecessor of Ontolex-lemon, and its translation module [9].

Given that Apertium RDF v1.0 only covered the language pairs for which an LMF version was available, and that the initiative to convert Apertium dictionaries into LMF was not continued, we decided to expand Apertium RDF by accessing the Apertium source data directly and converting them into the more recent OntoLex version of the *lemon* model.

2.3 Cross-Lingual Transfer Learning

Cross-lingual induction of resources for multilingual text analytics, instead of creating them from scratch, has attracted much attention in the NLP literature over the last decades, dating back at least to [19]. These early works are

---

[3] See https://www.w3.org/2016/05/ontolex/#core for a diagram and complete description of the OntoLex-lemon core module.

[4] See the whole diagram of the vartrans module at https://www.w3.org/2016/05/ontolex/#variation-translation-vartrans.

[5] https://www.apertium.org/.

[6] http://wiki.apertium.org/wiki/Main_Page.

[7] http://www.meta-net.eu/projects/METANET4U/.

[8] http://linguistic.linkeddata.es/resource/id/apertium.

comparatively resource-intensive themselves, as they assume the availability of parallel or aligned corpora, which is a requirement that is hard to meet for many language pairs, and even more so in technical domains.

In more recent work, these requirements are substantially alleviated by representation learning approaches capitalizing on bilingual word embeddings which can be induced from parallel and non-parallel corpora (cf. [17] for an overview). Due to their generality, bilingual embedding approaches are sufficiently versatile in order to be applied to a variety of cross-lingual text classification problems [15]. Cross-lingual sentiment analysis, as a special case, is investigated in multiple studies from a representation learning perspective [1,20,21].

UBiSE [5] presents a projection approach based on bilingual sentiment-specific word embeddings without any cross-lingual supervision, thus reducing resource requirements to a minimum: Only relying on a labeled sentiment corpus in the source language, as well as monolingual embeddings for both languages, their method outperforms Bilingual Sentiment Embeddings (BLSE) [1] on online customer reviews. In light of our results presented in this paper, it remains to be evaluated as to whether UBiSE can be scaled to technical domains as well.

Our assumption is that the use of the LD version of the Apertium data (and in general of any linguistic data) for cross-lingual model transfer has a number of advantages: It does not rely on proprietary formats and APIs but on standard representation mechanisms and access means (RDF, ontologies, SPARQL, etc.), which also makes linkage and combination with other LD resources easier. Further, the continuous enrichment and growth of the LLOD cloud (e.g., more translations among new language pairs are available) can lead to the improvement of the NLP stack exploiting them with little or no extra effort.

3   Overall Architecture

In this section we describe the whole pipeline that the multilingual data traverse: the Apertium source data is taken as input and converted into RDF based on the OntoLex-lemon model. Then, it is linked to the LexInfo[9] catalogue of linguistic categories and published as LD. In the next step, the RDF data is consumed by a sentiment analysis component in a user application, to support cross-lingual model transfer. Such a pipeline is illustrated in Fig. 1. Two main components implement such a pipeline, namely Fintan and Pharos®:

Fintan, the Flexible, Integrated Transformation and Annotation engineering platform [8] has been developed in the context of the Prêt-à-LLOD project[10] and allows for creating complex transformation pipelines between widely used formats for representing linguistic resources. Fintan allows existing RDF converters to be integrated with stream-based graph processing steps which modify the resulting data to comply with standard data models such as OntoLex-lemon

---

[9] https://www.lexinfo.net/ontology/2.0/lexinfo.
[10] https://www.pret-a-llod.eu/.

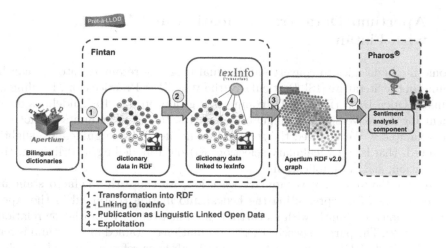

Fig. 1. Apertium RDF v2.0 pipeline: from source data to exploitation

and interlink it with existing LD resources. Fintan thus poses an ideal framework for mapping the Apertium XML data as RDF.

Pharos® is marketed by Semalytix[11] as a Pharma Analytics Platform that provides actionable real-world evidence (RWE)[12] for customers from the global pharmaceutical industry. Since its inception in 2019, the platform has been adopted in more than 10 projects by pharma companies from three countries. RWE extraction requires to analyse large volumes of heterogeneous content, including subjective assessments of patients and medical experts, which is typically available as unstructured text in multiple languages. Underlying Pharos, there is a complex stack of NLP components and modules, comprising entity and concept recognition, relation extraction, sentiment analysis, among others. In this paper, we focus on cross-lingual transfer of an RWE-tailored sentiment model from English to Spanish using an LLOD-based transfer learning framework.

Our final goal is running the whole Apertium pipeline in a fully automated way, therefore periodically gathering updates in Apertium, running the transformation and linking scripts through Fintan, and serving the produced LD to Pharos®, in an automated manner. Manual intervention is only necessary if adjustments to the data model or the mapping of annotation schemes are required (see Sects 4.1 and 4.2). Then, the whole pipeline can be run automatically since such a modelling and mapping design is common for all the Apertium data and dictionaries.

---

[11] https://www.semalytix.com.

[12] RWE is evidence for the effectiveness and safety of a drug product, gathered outside of the controlled settings of clinical trials, in order to demonstrate added value of a drug in terms of improvements in quality of life in specific patient populations.

4   Apertium Data Transformation and Linking with Fintan

Some methodologies to convert multilingual language resources into LD can be found in the literature [18]. Particularly, the W3C Best Practices for Multilingual Linked Open Data (BPMLOD) community group proposed a guidelines document for the conversion of bilingual dictionaries, taking Apertium as a motivating example.[10] We followed the steps recommended in such guidelines, slightly adapted, that is: (i) vocabulary selection, (ii) data modelling, (iii) linking, (iv) generation, and (v) publication.

As for the first step, vocabulary selection, we chose the de-facto standard Ontolex-lemon for representing the lexical information contained in the Apertium dictionaries, jointly with its *vartrans* module to specify translation relations (see Sect. 2). The part of speech (POS) information contained in Apertium is represented, in its RDF counterpart, by using LexInfo as reference model, which is a registry of linguistic categories widespread in the linked data community [3]. In the rest of this section we review the remaining steps for the RDF conversion.

4.1   Data Modelling

Following the Apertium RDF v1.0 approach, three files are generated for each language pair in a source Apertium dictionary: one for each dictionary (source and target lexicons), and the third one for the translation relations between the corresponding lexical senses. Figure 2 shows the RDF representation of the translation relation between the senses of the entries *safety* in English and *seguridad* in Spanish based on the *vartrans* module of OntoLex-lemon.

To represent the POS tags of Apertium as RDF, a URI in the Apertium namespace is associated to each tag, using the string value of every tag as its local name, e.g. `apertium:n` for the tag n (noun), and we assign it as a morphosyntactic property to the lexical entry: `:safety-n-en lexinfo:morphosyntactic-Property apertium:n`.

4.2   Mapping to LexInfo

As a part of the conversion explained in the previous section, a list of approx. 700 category abbreviations used for morphosyntactic description across the datasets in Apertium source files were extracted and gathered under the same namespace (e.g. `apertium:def` for `definite`, `apertium:dat` for `dative case`). However, the tags in Apertium to indicate POS and other morphosyntactic properties are not normalised and sometimes are not very informative. In order to allow for the integration of the Apertium dictionaries among themselves and with external resources, a normalisation process is necessary. To that end we have mapped the Apertium POS tags to LexInfo, resulting in a homogeneous tagging across all the Apertium dataset family and facilitating its querying and reuse.

---

[13] https://www.w3.org/2015/09/bpmlod-reports/bilingual-dictionaries/.

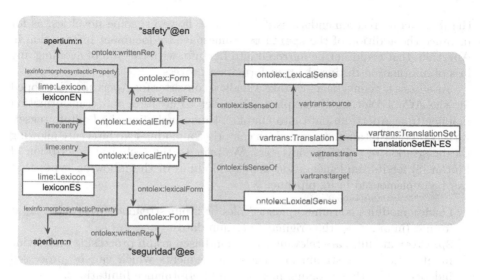

Fig. 2. Modelling example of the translation between "safety" @en and "seguridad" @es, prior to the linking to LexInfo.

Table 1. Apertium-LexInfo mapping examples, for *adjective* and *determiner*.

Apertium tag	Lexinfo property	Lexinfo tag
apertium:adj	lexinfo:partOfSpeech	lexinfo:adjective
apertium:A	lexinfo:partOfSpeech	lexinfo:adjective
apertium:det	lexinfo:partOfSpeech	lexinfo:determiner

We use `lexinfo:morphosyntacticProperty` to account for the Apertium POS individuals initially. The mapping between Apertium POS and LexInfo is defined as a CSV file, which provides `predicate - object` pairs for each of those Apertium tags acting as objects (e.g. `lexinfo:partOfSpeech`, `apertium:vblex`, `lexinfo:verb`). In total, the initial number of Apertium categories identified as POS was 104, which were mapped into 28 different LexInfo categories. Table 1 shows three mapping examples. The initial mapping between the POS Apertium tags and LexInfo was performed manually by the authors and made available online for the review and validation by the larger community of linguists and lexicographers.[14]

4.3   RDF Generation

Prior to generate the RDF data, a URI naming strategy has to be defined. To that end, we follow the same approach as in Apertium v1.0, which follows

---

[14] The mapping is available as CSV and TSV in GitHub and open to comments and modification by the community. See https://github.com/sid-unizar/apertium-lexinfo-mapping.

the ISA Action recommendations.[15] There are, however, some novelties, as for instance the addition of the `apertium`: namespace to document information in Apertium that could not be mirrored into LexInfo, with the aim of avoiding any loss of information during the transformation process.

As a basis for conversion, we take a shallow converter for Apertium, developed for the ACoLi Dictionary Graph [2]. In order to create a full transformation pipeline from Apertium data into OntoLex-lemon including the LexInfo tagset, we rely on the Fintan platform, which comprises a modular architecture allowing the integration of existing converters. We refer to the technical description of Fintan [8] for its implementation details. Modules of the following types that can be implemented in its pipeline:

- `Loader` modules consume uploaded files or input streams of a specific input format (in our case, the original Apertium data).
- `Splitter` modules are relevant for stream-based graph processing and divide input data into a stream of RDF data segments which can be processed independently, thus avoiding memory and performance limitations.
- `Update` modules consume a stream of RDF data segments and use multi-threading to process multiple segments in parallel.
  The transformation steps are rendered as SPARQL updates which are sequentially executed and optionally iterated to allow recursive operations.
- `Writer` modules export graph data into RDF serializations or other formats. Fintan currently supports the native export of TSV data, however, a custom `Writer` module can be integrated in the same way as a `Loader`.

All modules can be built into complex pipelines using a graphical workflow manager, or be directly called using a command-line interface. Figure 3 shows the Apertium pipeline within the Fintan workflow manager. The Apertium transformation pipeline in Fintan consists of the following steps:

1. The current Apertium repositories are checked out,
2. Morphological properties are extracted from all the source files,
3. With XSLT, each dictionary is transformed from XML to OntoLex-lemon,
4. Using the LexInfo mapping table, a dynamically built SPARQL update script inserts LexInfo morphological categories into the RDF, removing raw Apertium ones where possible,
5. The output is the Apertium RDF data in turtle and TSV formats, for the NLP application to choose the most suitable format for consuming the data.

Given the iterative nature of the conversion, Fintan is a suitable choice for making the workflow more reproducible, user-friendly and less resource-intensive, since some of the Apertium dictionaries are quite large and applying the update to the whole dataset can pose a bottleneck.

---

[15] http://ec.europa.eu/isa/actions/01-trusted-information-exchange/1-1action_en. htm.

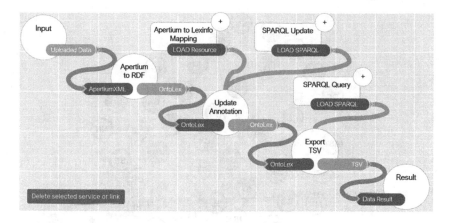

Fig. 3. Conversion pipeline for Apertium in the Fintan Workflow Manager

4.4 Publication

A result of the previously described pipeline, a graph of dictionary linked data, interconnected at the level of lexical entries and linked to an external resource such as LexInfo, has been created. It allows for a seamless exploration of the Apertium data, moving them beyond its original data silos (bilingual dictionaries in XML) and application domain (Machine Translation) to enable other usages like the one illustrated in this paper (sentiment analysis in a multilingual setting). Figure 4 illustrates the new Apertium RDF graph, covering 44 languages and 53 translation sets among them (compare with the 16 languages of the previous Apertium RDF version). It contains a total of 1,535,853 translations among different lexical entries and 1,838,295 links to LexInfo.

A preliminary version of the Apertium RDF v2.0 dictionaries, provided under GPL license (like the original data), is available via https://github.com/acoli-repo/acoli-dicts. The release contains the build scripts, such that the data can be locally re-built if new Apertium dictionaries are being published or existing dictionaries are being updated. The build scripts provide an implicit versioning via the time-stamp provided with every RDF dump they create.[16]

5 RDF Exploitation

In this section, we demonstrate how the RDF workflow presented above can be exploited in a real-world industry use case. We address the problem of transferring a domain-specific model for sentiment prediction that exists for

---

[16] Access to a testing SPARQL endpoint, as well as a number of example queries to the Apertium RDF v2.0 dataset, can be found at 10.6084/m9.figshare.12355358. A stable version of Apertium RDF v2.0 will be uploaded to http://linguistic.linkeddata.es/ apertium/ and hosted by Universidad Politécnica de Madrid (UPM) as part of the Prêt-à-LLOD project and documented through https://lod-cloud.net/.

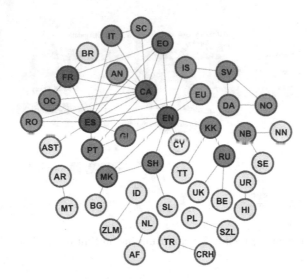

Fig. 4. New Apertium RDF graph. The nodes represent monolingual lexicons and edges the translation sets among them. Darker nodes correspond to more interconnected ones.

pharmaceutical text in a source language (here: English) to a target language (here: Spanish) for which no labeled training data is available. Our approach capitalizes on a deep learning transfer framework based on BLSE (*Bilingual Sentiment Embeddings*) [1].

In comparison to other approaches, BLSE is relatively parsimonious in terms of language data and resources required, as training signals for the learning algorithm need to be provided only in terms of ground-truth sentiment labels in the source language and a bilingual lexicon which contains translation pairs of words in both languages. In our use case scenario, we can assume that ground-truth labels are available in terms of manual annotations, whereas the selection of the most appropriate bilingual lexicon(s) is subject to empirical evaluation.

In the following, we describe the BLSE architecture, the lexical resources that are acquired using the RDF workflow presented above, and methods to combine such resources in order to increase their domain- and task specificity.

5.1 BLSE Architecture

As can be seen from the high-level architecture displayed in Fig. 5, BLSE requires (i) monolingual word embeddings in both the source and target language, (ii) ground-truth annotations in the source language, and (iii) a bilingual dictionary that maps words from the source language to their translations in the target language. These resources provide the foundation for learning mappings M and M' into a bilingual task-specific embedding space. The learning procedure is

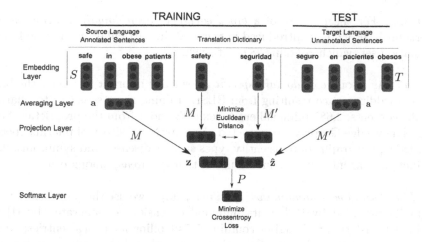

Fig. 5. Overview of BLSE architecture (slightly adapted from [1])

guided by a composite loss function based on the cross-entropy between sentiment predictions and ground truth labels in the source language and the spatial proximity of source/target pairs from the bilingual dictionary in the bilingual embedding space. The latter part enables the model to tailor target-language embeddings such that they can be used as input to a softmax classification layer that returns target-language predictions without any direct supervision in this language being available (see [1] for more details of BLSE).

5.2 Lexical Resources Used in BLSE

Monolingual Word Embeddings used in this study are selected along the two axes of *language* and *domain*: For English, we use *google*[17] open-domain and *PMC*[18] biomedical embeddings. For Spanish, we use *sg_300_es*[19] as open-domain embeddings, and *scielo_wiki*[20] as domain-specific representations. All embeddings were pre-trained on the respective corpus using word2vec [14].

Bilingual Dictionaries. In order to inform the cross-lingual projection in BLSE, we apply three different lexicons that provide Spanish translations for English lexical entries. These lexicons were selected according to the criteria of *domain-* and *task specificity*.

---

[17] Trained on news text, available from https://drive.google.com/open?id=1GpyF2h 0j8K5TKT7y7Aj0OyPgpFc8pMNS.

[18] Trained on the PubMed Central corpus, available from http://bio.nlplab.org.

[19] Trained on Wikipedia text, available from https://drive.google.com/open?id=1Gpy F2h0j8K5TKT7y7Aj0OyPgpFc8pMNS.

[20] Trained on the concatenation of the Scielo corpus and a medical subset of Wikipedia text, available from https://zenodo.org/record/2542722#.XeUOo5NKjUK.

Apertium. For the purpose of a *broad-coverage, open-domain* lexicon, we use Apertium RDF v2.0, as introduced in Sect. 2.2. In particular, we use the EN-ES translation set, which contains 28,611 translations.

Pharma. As a source of *domain-specific* knowledge in order to render the bilingual embedding space resulting from BLSE training more sensitive to pharma-specific contents, 2,687 bilingual entity lexicalizations from the proprietary Semalytix Knowledge Graph were extracted. As a large repository of pharma-specific knowledge, the graph contains entity types such as diseases and symptoms, drug products and agents, drug manufacturers, therapy areas, among others.

BingLiu. As an *open-domain, task-specific* resource, we use the sentiment lexicon originally provided by [12] in its bilingual extension as generated by [1] via machine translation.[21] BingLiu contains 5,749 bilingual lexical entries; we do not make use of the polarity information that is provided alongside each entry.

Before being used in BLSE, each resource undergoes a procedure of (i) *deduplication* (removing duplicate entries), (ii) *disambiguation* (in case of translation ambiguities, selecting the translation candidate that occurs most frequently in the target-language corpus) and (iii) *filtering* (removing all entries with translations that do not occur in the target-language corpus). This results in 5,084 processed entries for Apertium, 277 for Pharma, and 1,362 for BingLiu.

Lexicon Extension Procedures. In order to exploit complementarities in the lexical content of the previously described lexical resources,[22] we generate three extensions as summarized in Table 2. For each extension, the individual source lexicons are composed successively in the given order by either adding novel entries or overwriting existing ones (in case of conflicting translations in the source lexicons). After composition, each extension undergoes the same post-processing procedure described above.

Table 2. Overview of extensions generated by composing individual source lexicons, with numbers of original and processed entries (i.e., translation pairs) per extension

	Source lexicons	#Entries (original)	#Entries processed
Domain extension	Apertium + Pharma	31,192	5,307
Task extension	Apertium + BingLiu	34,254	5,799
Full extension	Apertium + Pharma + BingLiu	36,941	6,018

---

[21] Available from https://github.com/jbarnesspain/blse/tree/master/lexicons/bingliu.

[22] The overlap between these resources amounts to 647 processed entries between Apertium and BingLiu, but only 54 between Apertium and Pharma, and only 12 between Pharma and BingLiu.

6 Experiment: Impact of Lexical Resources on Cross-Lingual Transfer of Sentiment Detection Models

In this section, we report on an experimental evaluation of the different configurations of lexical resources as regards their performance in the cross-lingual sentiment projection task.

6.1 Corpus

The corpus used in our experiments consists of a non-parallel sample of comparable English and Spanish transcripts of summaries of conversations between pharma representatives and medical experts. In these conversations, the medical experts are asked to state their opinions and assessments about particular aspects of medical treatments (e.g., safety and effectiveness of a drug, among others). The following examples denote positive and negative assessments of safety and effectiveness, respectively:

(1)   DRUG can be safely used in elderly patients with renal failure. – SAFETY; positive

(2)   No effect on glycaemic control when using DRUG as add-on. – EFFECTIVENESS; negative

A collection of 21,400 English summaries is manually annotated with binary sentiment labels at the document level (11,069 positives vs. 10,331 negatives) and subsequently used for training the cross-lingual transfer model in a cross-validation setting. A set of 1,001 Spanish summaries is annotated likewise (559 positives, 442 negatives) in order to provide a test set in the target language which is used for evaluation purposes only.

6.2 Results

Table 3. Accuracy scores in the target language obtained from BLSE when different lexicons are used in isolation (upper part) or in combination (lower part).

	Monolingual embeddings	Target language accuracy
Apertium	google; scielo_wiki	**0.768**
Pharma	google; sg_es_300	0.434
BingLiu	PMC; sg_300_es	0.711
Apertium & Pharma	google; scielo_wiki	0.763
Apertium & BingLiu	google; scielo_wiki	**0.773**
Apertium & Pharma & Bing Liu	google; scielo_wiki	0.767

Performance of Individual Lexicons. The upper part of Table 3 shows the results of cross-lingual projection using BLSE when each of the lexicons introduced in Sect. 5.2 above is injected into the BLSE framework as the only source of bilingual information. We clearly observe that the best accuracy[23] in the target language is due to Apertium (Acc = 0.768). The considerable margin over Pharma and BingLiu confirms the status of Apertium as a linguistically rich, general-purpose source of bilingual lexical knowledge. Even though the underlying data set is highly pharma-specific, the relative individual performance of Pharma and BingLiu suggests that task-specific sentiment information benefits cross-lingual projection approaches more than technical domain knowledge.

With respect to the monolingual word embeddings involved, a clear pattern of complementarity can be observed: Apertium benefits most[24] from domain-specific embeddings in the target language, whereas the domain-specific Pharma lexicon is best complemented by open-domain embeddings in both the source and the target language. For BingLiu, using sentiment-specific knowledge from the lexicon and domain knowledge from the (source) embeddings works best.

Performance of Extended Lexicons. The impact of lexicon extensions as generated through the procedure described in Sect. 5.2 can be seen from the lower part of Table 3. In comparison to using Apertium as the only source of bilingual information, we find that lexicon composition in individual configurations can be effective in generating richer bilingual lexical representations that result in more accurate cross-lingual projection of sentiment classifiers. Apparently, this is due to a certain degree of complementarity among the original lexicons, given that extending the general-purpose lexicon Apertium by task-specific knowledge from BingLiu yields the best performance overall (Acc = 0.773).

6.3 Discussion

The present case study clearly demonstrates the value of Apertium as an example of a bilingual LLOD resource for cross-lingual transfer of NLP models in practical application scenarios. In our experiments on sentiment projection from English to Spanish in the pharmaceutical domain, we found Apertium to excel both in terms of its individual linguistic richness (being the most informative source of bilingual lexical information among the different resources compared) and resource interoperability (facilitating additional performance gains when being combined with complementary task-specific resources).

The good results are not inherent to the LD nature of the data, but illustrate how high quality resources can be gathered from the LLOD cloud and dynamically combined with other data sources and plugged into NLP pipelines.

---

[23] Accuracy is defined as the proportion of correct labels in all labels predicted by the model on the test set.

[24] For Apertium, Pharma, and Bing Liu, Table 3 displays only the best-performing configurations of monolingual embeddings.

The induced sentiment model meets an excellent trade-off on the cost-effectiveness spectrum: Without any supervision being required in the target language, its predictive performance is (i) reasonably close to the one of a supervised source language classifier,[25] and (ii) largely superior to a sequential machine translation pipeline, as reported in our previous work [11].

7 Conclusions and Future Work

This paper outlines the strong potential of LLOD resources to be used as catalyzers of cross-lingual transfer of NLP models in deep learning frameworks. We were able to demonstrate this with the Apertium RDF data processing pipeline for a real-world industry use case involving cross-lingual transfer of a sentiment model in the pharmaceutical domain.

In our experiments, we observed a beneficial effect of composing different lexical resources in order to achieve optimal transfer performance. This underlines the great potential of LLOD-based pipelines for setting up flexible and fully automated transfer workflows which could exploit regular updates of the underlying lexical resources in a dynamic manner.

Despite these encouraging findings, we believe that our approach has not exhausted its full potential and that some challenges remain. For instance, the extension to other language pairs will need additional validation. Further, we plan to extend the current workflow into a fully automated pipeline in order to (i) exploit bilingual lexical information in a fully dynamic manner, thus benefiting from regular updates and extensions of the Apertium data, and (ii) rapidly scale the transfer approach to numerous other languages already available as LLOD.

Acknowledgements. This work was funded by the Prêt-à-LLOD project within the European Union's Horizon 2020 research and innovation programme under grant agreement no. 825182. This work is also based upon work from COST Action CA18209 – NexusLinguarum "European network for Web-centred linguistic data science", supported by COST (European Cooperation in Science and Technology). It has been also partially supported by the Spanish projects TIN2016-78011-C4-3-R (AEI/FEDER, UE) and DGA/FEDER 2014–2020.

References

1. Barnes, J., Klinger, R., Schulte im Walde, S.: Bilingual sentiment embeddings: joint projection of sentiment across languages. In: Proceedings of ACL (2018)
2. Chiarcos, C., Fäth, C., Ionov, M.: The ACoLi dictionary graph. In: Proceedings of LREC, pp. 3281–3290. ELRA, Marseille (2020)
3. Cimiano, P., Buitelaar, P., McCrae, J., Sintek, M.: LexInfo: a declarative model for the lexicon-ontology interface. J. Web Semant. 9(1), 29–51 (2011)

---

[25] Despite not being exactly comparable due to non-parallel evaluation data, the classifiers resulting from the Task Extension setting differ by only 4.3 points in source vs. target language accuracy (0.816 vs. 0.773, respectively).

4. Cimiano, P., Chiarcos, C., McCrae, J.P., Gracia, J.: Linguistic Linked Data: Representation Generation and Applications. Springer International Publishing, Switzerland (2020)
5. Feng, Y., Wan, X.: Learning bilingual sentiment-specific word embeddings without cross-lingual supervision. In: Proceedings of NAACL:HLT (2019)
6. Forcada, M.L., et al.: Apertium: a free/open-source platform for rule-based machine translation. Mach. Transl. **25**(2), 127–144 (2011)
7. Francopoulo, G., et al.: Lexical Markup Framework (LMF) for NLP multilingual resources. In: Proceedings of the Workshop on Multilingual Language Resources and Interoperability, pp. 1–8, Sydney (2006)
8. Fäth, C., Chiarcos, C., Ebbrecht, B., Ionov, M.: Fintan - flexible, integrated transformation and annotation engineering. In: Proceedings of LREC (2020)
9. Gracia, J., Montiel-Ponsoda, E., Vila-Suero, D., Aguado-de Cea, G.: Enabling language resources to expose translations as linked data on the web. In: Proceedings of LREC, pp. 409–413 (2014)
10. Gracia, J., Villegas, M., Gómez-Pérez, A., Bel, N.: The apertium bilingual dictionaries on the web of data. Semant.Web **9**(2), 231–240 (2018)
11. Hartung, M., Orlikowski, M., Veríssimo, S.: Evaluating the impact of bilingual lexical resources on cross-lingual sentiment projection in the pharmaceutical domain. https://doi.org/10.5281/zenodo.3707940 (2020)
12. Hu, M., Liu, B.: Mining and summarizing customer reviews. In: Proceedings of KDD, pp. 168–177 (2004)
13. McCrae, J.P., Bosque-Gil, J., Gracia, J., Buitelaar, P., Cimiano, P.: The OntoLex-Lemon Model: development and applications. In: Proceedings of eLex 2017 Electronic lexicography in the 21st century, pp. 587–597 (2017)
14. Mikolov, T., Chen, K., Corrado, G., Dean, J.: Efficient estimation of word representations in vector space, January 2013
15. Mogadala, A., Rettinger, A.: Bilingual word embeddings from parallel and non-parallel corpora for cross-language text classification. In: Proceedings of NAACL:HLT (2016)
16. SRIA-Editorial-Team: Strategic Research and Innovation Agenda for the Multilingual Digital Single Market. Technical report, Cracking the Language Barrier initiative (2016)
17. Søgaard, A., Vulic, I., Ruder, S., Faruqui, M.: Cross-lingual word embeddings. Morgan Claypool (2019)
18. Vila-Suero, D., Gómez-Pérez, A., Montiel-Ponsoda, E., Gracia, J., Aguado-de-Cea, G.: Publishing linked data on the web: the multilingual dimension. In: Buitelaar, P., Cimiano, P. (eds.) Towards the Multilingual Semantic Web, pp. 101–117. Springer, Heidelberg (2014). https://doi.org/10.1007/978-3-662-43585-4_7
19. Yarowsky, D., Ngai, G., Wicentowski, R.: Inducing multilingual text analysis tools via robust projection across aligned corpora. In: Proceedings of HLT (2001)
20. Zennaki, O., Semmar, N., Besacier, L.: Inducing multilingual text analysis tools using bidirectional recurrent neural networks. In: Proceedings of COLING (2016)
21. Zhou, X., Wan, X., Xiao, J.: Cross-lingual sentiment classification with bilingual document representation learning. In: Proceedings of ACL (2016)

Reasoning Engine for Support Maintenance

Rana Farah[1]([⊠]), Simon Hallé[1], Jiye Li[1], Freddy Lécué[1,3]([⊠]), Baptiste Abeloos[1],
Dominique Perron[1], Juliette Mattioli[2], Pierre-Luc Gregoire[1], Sebastien Laroche[1],
Michel Mercier[1], and Paul Cocaud[1]

[1] Thales Digital Solution, Québec, QC G1P4P5, Canada
rana.farah@polymtl.ca, {simon.halle,jiye.li,baptiste.abeloos,
dominique.perron,pierre-luc.gregoire,sebastien.laroche,
michel.mercier,paul.cocaud}@ca.thalesgroup.com,
freddy.lecue@inria.fr
[2] Thales, 92098 Paris La défense, France
juliette.mattioli@thalesgroup.com
[3] Inria Sophia Antipolis, 06902 Valbonne, France

Abstract. This paper presents a reasoning system deployed for supporting the maintenance of IT devices in use by a leading broadcasting and cable television company in North America. We describe a reasoning engine pipeline relying on semantic data representation and some machine learning approaches such as clustering. The engine derives problems on a telecommunication network from a textual description and uses structured historical data of problems, error codes and proposed solutions to prescribe potential solutions. The engine is capable of proposing solutions to unseen problems by using analogical reasoning on structured representations. When a problem happens on the network or more precisely on one of the devices, these devices generate error codes. We addressed two scenarios; (i) we assumed that the list of error codes that we captured is complete, (ii) we assumed, more realistically, that this list is incomplete. In the first case, we suggested solutions for seen and new problems and reported results on real data. In the second case, we proposed a method to infer the complete list of errors, tested that method on synthetic data and showed results with high accuracy. Although both scenarios are in-use, the first scenario is more usual than the second one, but both need to be considered.

Keywords: Knowledge graphs · Graph embedding · Reasoning engine

1 Introduction and Literature Review

Facilities-based providers are looking into tools that preempt problems on their service platform, network and their customers' equipment. To support their customers, they employ a huge amount of resources being the Network Operation Centers (NOCs), support personnel, maintenance technicians, software, equipment, and fleets of vehicles. These providers are starting to look into preempting any potential failures if possible. Or, in the worst case, try to solve the problem from the first attempt, as soon as possible, with

© Springer Nature Switzerland AG 2020
J. Z. Pan et al. (Eds.): ISWC 2020, LNCS 12507, pp. 515–530, 2020.
https://doi.org/10.1007/978-3-030-62466-8_32

the least human interaction and, especially, client impact. Thus, call influx and duration in support centers can be reduced, money saved and client satisfaction is enhanced [1].

When it comes to maintenance solutions, we are interested in two kinds predictive-maintenance and prescriptive-maintenance. The difference is that predictive-maintenance is pre-occupied with predicting when faults may occur while prescriptive-maintenance builds on that and proposes solutions and actions that should be taken to correct the faults or even prevent them.

First, we looked at predictive maintenance works [2]. The techniques used covered knowledge based approaches [3–7], traditional machine learning approaches [8–10] and deep learning approaches [11–14].

While predictive-maintenance techniques are widely documented in the literature, prescriptive-maintenance is still a very new discipline and very poorly documented. The methods used include probabilistic models [15], rule based [16] and machine learning models [17, 18]. The works mentioned in this review are representative and not comprehensive.

Even though the techniques employed varied, the entirety of these works rely only on sensor data in the form of digital measurement such as images, vibration signals, temperature measurements among others. However, in some cases as in the NOCs operations, data can be presented in the form of natural language. Indeed, even though part of that data is collected from sensors and machines on the network, a large part of that data is generated from the conversation between the clients and the support agent and recorded in natural language.

The work that we will describe in this paper is a knowledge based approach and more precisely an ontology based approach. It differs from other ontology based approaches [3, 19, 20] in that it uses semantics not only to capture the relations between the entities in the domain context knowledge but also to capture the information and establish relations from natural language data. It also differs in that it can use historical fault description to proposing solutions for new unseen fault description.

The work that we present in this paper describes the system we implemented toward a prescriptive-maintenance operation at Thales. The paper is organized as follows; First we describe the use case we worked on, the challenges that this use case presented and the objectives we worked toward achieving. Second, we describe the data used and the semantic representation that we designed to integrate this data into our solution. Then we describe the methods that we used to leverage the semantic representation of the data to create new useful data to prescribe solutions to old and new maintenance problems. Last, we conclude with our take-always from implementing such a solution in a NOC environment.

2   Use Case and Challenges

2.1   Thales Reflex Platform for Support Maintenance

The Thales Reflex Platform is ensuring data management services to telecommunication providers. Among this data, the platform aims at managing error codes logs from the customer devices: mainly internet, phones and TV services. As part of the services provided to clients, the platform exposed a new capability to better exploit historical and real-time data of problems and solutions, to allow technicians in NOCs to rapidly, and even before the fact, provide solutions to problems that are not in the customer support textbook but may have been successful in resolving problems by some in the past or recently (from historical data) [1]. As a first target, the Thales Reflex Platform's aim is to reduce the average call duration from an average of 7 min to 5 min and save 10% of operating cost that amounts to a 56 million dollars savings. This operation touches on tens of million eventual operator customers and about 27.5 million customers. The Reflex Platform main operation consists of collecting multimodal data from the different devices on the network, organizing this data in a semantic framework, using machine learning and a reasoning model to identify faults on this network and prescribing actions either to the devices themselves or to the customers to correct the fault. The platform also reports these faults and the prescribed solutions in the NOC (see Fig. 1.)

Fig. 1. A schema showing the existing and targeted functionalities on the Thales Reflex Platform. Image adapted from [1] for illustration purposes.

2.2 Use Case

As a use case, we will present the work done to extend the Reflex platform. In this use case we exploited two types of data (textual and code) to put together a pipeline that prescribes resolutions to problems that happen on a telecom network. The problems can happen at the customer end or on the network itself. In a normal operation case, the customer (a subscriber to the service or another technician) calls the NOC to report a problem. This prompts the issue of a ticket. The ticket is a record of the problem details and the suggested solutions. The agent at the NOC, transcribes the information in a natural language format and suggests actions to the customer (ex: reboot the device) or takes actions to resolve the problem. More than one action can be needed to resolve the problem. These are also recorded in a natural language format in the ticket record. To drive this process efficiently, and reduce the call duration and the time it takes to propose the right solution, we looked at ways to augment the process with semantic representation of tickets and problems as well as machine learning to expose prescriptive actions.

2.3 Challenges

When looking at the available data and analyzing the requirement of the problem we realized that we are facing several challenges. First, the search space is extremely wide. The number of problems that can be faced on the platform is not trivial and the syntax to express these problems is also largely varied. The data is noisy. The problem description is seldom accurate and the syntax used to express these problems could also be vague. In addition, the solution proposed and documented by the support agent is not guaranteed to be the "right" solution. There is no indication if the problem was definitely solved during the interaction. We also realized that there is no standard taxonomy or ontology for this particular type of data. The latter is crucial to structure the information, the tickets, challenges and potential solutions which could be re-usable among similar semantic cases.

2.4 Objectives and Solution Motivation

Our ultimate objective is to build a prescriptive-maintenance system which will rely on semantic representation of information, and then exploit structure representation through a hybrid combination of machine learning and reasoning paradigms. The system will prescribe potential solutions when presented with certain symptoms or similarities with other cases. In particular, the engine should prescribe fixes when presented with the description of a problem on a digital telecommunication platform and other relevant information.

The solution is not diagnostic. We do not particularly aim at understanding the problem in itself or validate it when it is presented. We are more interested in proposing solutions when certain symptoms are presented such as a problem description (which could be accurate or not). The reasoning engine is examining historical relational data (the information collected in the available ticket reports as described in Sect. 3) and uses analogical reasoning (on top of semantically augmented data) to derive conclusions. The engine is not limited to proposing solutions only for problems that are described in historical data or some given guidelines. It is also required to proposing solutions for unseen problems.

Given that the data is multimodal, the solution encodes this data in a unified format that captures its semantics and the relationships between the entities of the context domain.

The number of faults that occur on networks covered by the Reflex Platform is extremely large. On a ticket, each of these faults is non-uniformly expressed given that they are recorded in natural language. This creates a large domain space and, in practice, the data available to represent this space will be relatively sparse. For this reason, the solution that we adopted is not enormously data hungry.

The reasoning engine pipeline that we implemented is shown in Fig. 2.

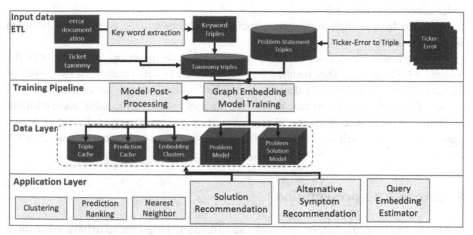

Fig. 2. The proposed reasoning engine pipeline. The blue blocks refer to data instances or repositories and the grey blocks are processes. The upper most layer describes the data preprocessing and structuring. The following layer describes the data parsing into vectors in the embedding space that encode semantic relational information. The third layer describes the different cache used to allow the pipeline to be used in real time. The most bottom layer describes the solution recommendation part of the reasoning engine process. (Color figure online)

3 Data

A three months period of ticket and error codes data was collected for model training purposes. This reaches a total of 3.6 million ticket records, and 61 million error code records, which amount to approximately 30 GB. We report the experiments on the ticket/error code data set.

The ticket dataset contains individual entries summarizing details of a problem reported to a support agent. The tickets, in this dataset, have been logged by support agents.

The error code dataset contains codes collected in the network. These error codes are generated at the device level. A problem can generate a collection of error codes, and also error codes that are not unique to one problem. The ideal solution would be to guess the combination list of error codes generated by a certain problem. However, the association between such a list and a problem is not straight forward. Also, the problem reports are not punctual. A problem is usually detected much later than its occurrence. It is the same for the error codes generated. Also devices frequently generate "benign" errors codes. Given that there is no indication of when the problem(s) happened or a way to distinguish "benign" from "problematic" codes, the association between the problem and its combination of generated error codes is far from straightforward. For the purposes of this project, we associated a problem to the list of error codes that occurred in the interval of the 7 days of which the reporting time is in the middle. This assumption is adopted by the technicians themselves. Of course, this highly inaccurate assumption creates a highly noisy association between the reported problems and the list of error codes associated with each one.

We also have a dictionary that associates each error code to potential problem descriptions.

We constructed a dataset of 600 k ticket-error code pairs by sampling over the original 61 M records. This resulted in 35 k reported tickets relating to two types of services ("hsd service" and "vidoo service"). We limited our analysis to these services because they had enough coverage in the ticket dataset and are of the most interest for our predictive-maintenance platform. These ticket-error code pairs covered a set of device manufacturers and each ticket (problem) report is associated on average with 20 error codes.

4 Semantic Representation

Due to the specialized nature of our problem and dataset, it was crucial to have a dedicated ontology (see Fig. 3) to populate a knowledge graph of tickets, related by potential symptoms and solutions (see Fig. 4).

4.1 Domain Ontology

The domain ontology consisted mainly of ten major high-level classes:

Ticket: represents a certain ticket and each instance expressed using a unique ID (ex: DI0618002227)

Date/Time: indicating the date time that a ticket was reported.

Account: represents the customer account. Each instance is identified by a unique ID.

Error: represents the error code generated by the hardware and potentially assigned to a problem. (ex: b20f8ee03140218f)

Problem: describes the problem reported by a customer (ex: video quality problem)

Manufacturer: represents the manufacturer of a certain device potentially affected by a reported problem (ex: dwalin).

Device: represents a device that is potentially affected by a reported problem. (ex: 1:px022mine)

Resolution: represents the solution proposed by the agent. This is however not a guarantee appropriate solution. (ex: power cycled device)

ProblemDescription: This is the description provided by the ticket (ex: unable to play VOD assets)

Network: The network affected or that the device is branched to. (ex: eriador-minhriath)

The seven classes are related by the following relations *fromAccount, causedBy, causeType, affectsDevice, manufacturer, Network*. Using these relationships we formed 820 k triples after filtering the duplicate errors codes. Figure 3 shows a simplified version of the ontology graph.

When looking at the *Problem*, *Resolution* and *Error* classes, each had an enormous extended taxonomy (the tree of subclasses was very shallow) that was not exploitable in this case. Keeping up with this degree of detail was not advantageous. For this reason, we decided to create a certain hierarchy and group these subclasses into new mega-subclasses according to their semantic similarities. These mega-subclasses were related to their respective classes via the relationship (Problem, Resolution and Error) *isA* thus forming 20 k triples in total.

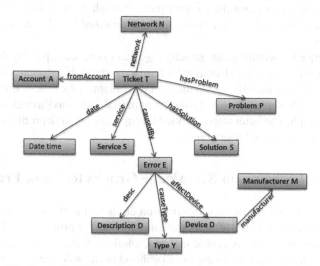

Fig. 3. A simplified version of the ontology (encoded in RDF).

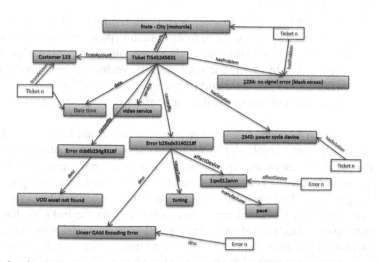

Fig. 4. A local snapshot of a neighborhood in the training data graph. Some values are fictitious for client privacy consideration.

4.2 Knowledge Graph

Once all data is structured following ontology in Fig. 3, the result is a knowledge graph that contains the entire relational information in the dataset. Figure 4 shows a snapshot of the knowledge graph. As our task is to offer solutions to, not only seen requests, but also new or similar ones using the knowledge that we collected and structured, it is important to structure the domain information in a format that could be linked with external vocabularies such as DBPedia [21]. This is crucial to link entities and potentially identify commonalities among them through external links. This ensures to obtain a domain knowledge graph which is contextualized but also interpretable across other domains.

By deriving a knowledge graph structure we can 1) encode explicitly domain knowledge through relations, 2) add external knowledge through external vocabularies, 3) design the problem of identifying solutions of a problem as the task of link prediction in a knowledge graph, 4) identify exact solution or partial solutions through semantic relatedness in the graph. The latter step is tackled through missing link prediction, combining machine learning on knowledge graphs.

5 Machine Learning on Knowledge Graphs for Link Prediction

To be able to exploit the relational information contained in the knowledge graph, and then derive potential solutions of new problems, we map the problem as a link prediction task in a knowledge graph. A problem P and solutions S are classes in the graph, and weighted links Rs between them are the likelihood of the solutions in S to solve P. We addressed this problem by exploiting machine learning on knowledge graphs through graph node embedding [22, 23].

The goal of a graph node embedding is to parse the nodes of a graph in a low-dimensional vector space such that the optimized vector space encodes the structure or relations of the original graph. In this space, also, the relationship, or the edges between the nodes are represented by geometric relationships. Vectors in the embedding space (to which we will simply refer to as embeddings in the rest of the paper) do not only encode information about the entity itself but also about its position and relationships in the graph.

We evaluated four embedding algorithms on the same dataset; TransE [24], DistMult [25], and HolE [26] and ComplEx [27]. We based our judgment on quantitative assessment of the result of each of the algorithms. Results of this particular analysis are not reported in the paper due to page limits[1]. ComplEx was retained for converging to better results during training. These results are consistent with reported results in the literature [23, 27]. The ComplEx algorithm represents each node and the links Rs that connect them with complex-valued embeddings. The algorithm then calculates the score for each triple (*head, tail, relation*) as the real part of the trilinear product of the corresponding embeddings. Overall, the algorithm optimizes the values of the embeddings so that the

---

[1] The details and results of the embedding algorithms comparison are at http://www-sop.inria.fr/members/Freddy.Lecue/thales/iswc-2020-in-use-PrescriptiveMaintenance-extra-results.pdf.

nodes that are semantically related are attributed a high score while the nodes that are not related are attributed low scores.

We used the implementation of ComplEx as is in the platform Ampligraph [28].

6   Solution Reconstruction for Support Maintenance

In this context, given a problem, predicting its solution would have been to add a corresponding node in the graph and predicting a link to a "best" solution node in the graph. However, we did not adopt this method for two reasons. 1) Generating embedding space for new nodes in a graph is not a straightforward task. Adding a new node to the graph will affect its neighborhood in the least and will relatively affect the embedding values of all the nodes of the graph. Repeating the complete space embedding calculation process for each new query is not a practical solution. It is computationally exhaustive and it will not suit real time applications. 2) We wanted to emulate the reasoning process of the support technician. We wanted to look at the error code space and attempt to find patterns or deviations from those patterns that can be leveraged not only to propose a one shot solution but to be able to create a reasoning pipeline that is able to propose a solution and alternative solutions that are not similar but complimentary. In case the first solution failed, we wanted to propose other options that are independent and not similar and that can solve that same problem. That is why the knowledge graph structure is crucial in our settings. We suspected that the best scored solutions would be similar and propose similar approaches.

6.1   Vector Approximation in the Embedding Space

We evaluate the embedding vector of a new problem as an approximation of semantically close problem embeddings. To this end we calculate the embedding of a new problem node as the mean embedding of problem nodes having similar linked error codes. The rationale is that similar problems produce similar error codes. A problem generated on a given network, related to given devices by given manufacturers, manifesting in a given manner and so on, also manifest a similar set of error codes every time. Following this logic, problems with a similar list of error codes have similar attributes and, as a result, similar embeddings.

```
errors_PN = the list of errors associated with problemN
for error_n in errors_PN
    problems_en = list of problems associated with error_n
  for problem_y in problems_en
        errors_y= the list of errors associated with problem_y
        score_y = |errors_PN ∩ errors_y|/|errors_y|
embedding_ProblemN = average of the five problem embeddings with highest
scores
```

Fig. 5. A pseudo code that describes the procedure to approximate new problem embeddings.

Also, in the context of this project a new problem, *problemN* (represented by an embedding, *embedding_N*), does not create new attributes (i.e., network, device, customer, error codes, etc.). A new problem has a new ID and a new combination of individual attributes. Those individual attributes are already represented by nodes in the graph and have their already computed embeddings. Then, for each of the error codes in the error code list attributed to *problemN*, we calculate the problems, in the graph, that are linked to that error code. We score the problems by the number of error codes that they share with *problemN* and average the embeddings of the best five scored problems to approximate the *embedding_N*. This procedure is described in the pseudo code of Fig. 5.

6.2 Solution Recommendation

Our aim is to emulate the process of problem troubleshooting that an agent has during an interaction with a customer facing a problem. We proposed a process formed of a set of condition-based instructions that aims at maximizing the confidence in each consequent proposed solution. Our stepwise approach (see Fig. 6) is as follows:

Step1. Verify if the exact problem exists in our dataset. We do so by calculating the Euclidean distance between *embedding_N* and the other problem embeddings. If the best distance is lower than a threshold, **propose the solution associated with the problem that corresponds to that distance.**

Step 2. Use K-means [29, 30] and pre-cluster all the problem embeddings in the dataset into K clusters using the Euclidean Distance.

Step 3. Calculate the *Nearest Neighbors (NN)* problem embeddings of the *problemN* embedding. Then, group these embeddings according to the clusters calculated in step 1 (see Fig. 5). Our justification for clustering the problems is that we considered the shared solutions of problems that are similar to *problemN* to constitute the recommended list of solutions to *problemN*. Also we considered multiple problems instead of just one (the nearest one) to minimize the impact of accumulated error codes throughout the complete pipeline.

When computing the neighbors, we considered three metrics to calculate the distance between the embeddings: the Euclidean distance, the cosine similarity distance and the hyperbolic distance. The Euclidean distance and the cosine distance returned almost similar results. However, we adopted the hyperbolic distance because it better emphasized the relevant differences between the problems (ex: one additional error code between two problems with the same list or error codes).

Step 4. Use ComplEx to perform link prediction between the *NN* problem nodes and solution nodes, and recommend solutions for each of the *NN* problems (see Fig. 6). Each of the solutions is predicted with a confidence score.

Step 4.1. If all the *NN* problems agree on one solution, **then propose this solution.**
Step 4.2. If all the *NN* problems fall under one cluster (cluster$_x$), then we consider that the proposed solution is the synthesis of the total of the solutions proposed by this cluster (*solution_nnList*). The solution embedding (embedding$_S$) to *problemN* is calculated as such

$$embedding_S = embedding_N - weightedAvg(cluster_x) + avg(solution\_nnList) \quad (1)$$

The average is weighted by the confidence associated with each predicted solution by ComplEx. **Propose the solution with the closest embedding to** *embedding_solution*.

Step 4.3. If the conditions in Step 4.1 and Step 4.2 are not met, our confidence in the one possible solution is decreased and we aim at proposing a list of two possible solutions as follows

Step 4.3.1. Calculate the average embedding of each group of problem embeddings calculated in Step 2

Step 4.3.2. Maximize the following distance for each combination possible

$$Distance_{S_M^{GA}, S_N^{GB}} = argmax_{S_M^{GA}, S_N^{GB}}\left(|GA - GB| + \left|S_M^{GA} - S_N^{GB}\right|\right) \quad (2)$$

where GA and GB are the two average embeddings associated with each of the two groups, respectively. Also, S_M^{GA} is one of the solutions associated to one of the problems in one of the problem groups GA and S_N^{GB} is one of the solutions associated with one of the problems in another of the problem groups GB. Where $S_M^{GA} - S_N^{GB}$ is the distance between S_M^{GA} and S_N^{GB}.

Propose the List of Two Solutions S_M^{GA} And S_N^{GB}
In choosing such a combination, we aim to ensure that the solutions we are proposing are alternative solutions and not similar solutions.

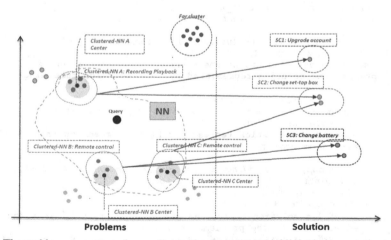

Fig. 6. The problems are grouped according to the clusters that highlight their semantic similarity. Also, each group has its mean embedding calculated. We used ComplEx to estimate the solutions (right plane) associated to each *Nearest Neighbors(NN) problems* (left plane)

Experiment and Results. We tested the proposed solution recommendation algorithm on a dataset that included problem samples spanning 76 solution classes. The dataset also consisted of 5,000 problems and was unbalanced with respect to the solution classes (realistic case). We considered a hit when the recommender returned the real solution in a list of 5. It is important to keep in mind that when we query the ComplEx algorithm to establish a link (make an association), the algorithm calculates the probability of a link between the given entity and others in the graph. For these results we considered the 5 most probable links to solution nodes. The results are summarized in Table 1.

Lessons Learnt. The results achieved in this work reach the level of readiness for integration in the platform for deployment and in-use in the context of the telecommunication scenario. The limitation factors of the approach are: (1) the high uncertainty in the relations established in the training dataset, in particular the association between the list of

```
Let embeddingN be the embedding that represents problemN in the embed-
ding space
Let embeddingList be the list of embeddings in the embedding space that
belong to the problems on the database
Embeddingy ∈ embeddingList
Distancemin = argminy({Euclidean_Distance(embeddingy, embeddingN})

Let embeddingmin = the embedding in embeddingList that is associated with
distancemin
Let solutionmin = the solution represented by embeddingmin
if distancemin < thresholdn
  propose the solution represented by solutionmin
else
    clusterlist = k_means(embeddingList)
    nn_neighbourList = nearest_neighbours(embeddingN, embeddingList)

    for neighbournn in solution_nnList
    solutionnn = ComplEx(neighbournn)

Let solution_nnList = {solutionnn| solutionnn is unique}
Let clusterx ∈ clusterList
if solution_nnList contains one element solution embedding
  propose the solution represented by that solution embedding

elif solution_nnList ⊂ clusterx
    embeddings = (embeddingN - weighted_avg(clusterx)) +
    avg(solution_nnList)
    propose the solution represented by the embedding that is clos-
    est to embeddings
else
    Let groupg = {embeddingx ∈ clusterx and neighbournn |clusterx
    clusterList, neighbournn ∈ solution_nnList }
  Let G = {avg(groupg)}
  Let GA ∈ G and GB ∈ G
    Let SGA_M ∈ solution_nnList and groupA ∈ groupg | SGA_M is associat-
    ed with one of the problems in groupA
    Let SGB_N ∈ solution_nnList and groupB ∈ groupg | SGB_N is associat-
    ed with one of the problems in groupB
Distance_SGA_M,SGB_N=argmaxGA,GB,M,N({|GA - GB| + |SGA_M - SGB_N|})

propose the two solutions that corresponds to SGA_M and SGB_N
```

Fig. 7. Pseudocode describing the solution recommendation process

Table 1. The solution recommendation results.

Accuracy	Precision	Recall	F1 score
0.58	0.69	0.58	0.60

error codes and the problem and the relation between the problem and the solution, (2) initial inaccuracy in problems reporting.

6.3 Alternative Solution Recommendation

Usually the problem description in the ticket is not accurate, or the list of associated error codes is missing a key error code that should be associated with the given problem. This results in a solution recommendation with low confidence that manifests as low scores in the solution proposal in 3.1, 3.2 or 3.3.4. Normally, the agent has to ask a long series of questions to resolve the problem. The situation usually annoys the customer and costs a lot of money in time spent over the phone or necessitating the dispatching of a technician on location. We address this issue and theorize that similar problems are represented by similar embeddings and also have similar error code lists.

Our method consists of proposing a list of P (in this case 5) solutions as follows:

Step 1. Among the nearest neighbors of *problemN*, consider the one with the highest score associated to its solution (the correlation between the solution and the problem is high)

Step 2. Look at the error codes associated with the neighbor and not associated with *problemN*. We assume that these are the error codes that were missing from the original list of error codes associated with *problemN*.

Step 3. Add the solutions associated with these error codes to the list of alternative recommendation solutions.

Step 4. Relax the threshold on the nearest neighbors calculation (widen the neighborhood) and reiterate steps 1, 2 and 3 until the number P of items in the list is met.

Experiment and Results. To test the alternative solutions recommendation procedure we had to build a special dataset for evaluation purpose. We needed problems that are replicas of existing problems (having the same problem description and having the same list of error codes associated with them) except for one error code missing from the list associated with one problem and existing in the other. These cases were hard to find in our dataset so we built such problems by randomly selecting 15 thousand problems and cloning them and their associated list of error codes. For each of the clones, we removed one error code from the list. We made sure that no replicas of the clones exist in the dataset or if a clone exists the solutions associated with it are different from the solutions associated with the original problem (parent problem of the clone). We considered a hit when the right solution was among the list of 5. See the results in Table 2.

Table 2. The results for the alternative solution recommendation method

Accuracy	Precision	Recall	F1 score
0.82	0.84	0.82	0.83

These results are very encouraging. If a given query (*problemN*) has a low solution prediction score, the alternative solution recommendation algorithm is of interest. Considering other related non-occurring errors codes would help the technician finding more related problems and further investigating more plausible solutions.

Lessons Learnt. When looking at these results we can conclude that the algorithm achieved high accuracy in encoding the semantic relation between a given list of error codes and its corresponding problem. These results were even robust with respect to incomplete or noisy (missing or extra codes) lists.

6.4 Automated Chatbot Support Agent

A chatbot was developed for a customer support agent to exploit the platform and quickly understand/differentiate the caller's problem and recommend solutions. The chatbot uses the customer's problem description to initiate the interaction. Through an iterative process of question and answers, the chatbot refines its understanding of the current problem and begins suggesting solutions when it reaches the appropriate level of certainty in the problem and applicable solution. A video[2] is available to demonstrate the functionalities of the chatbot.

7   Conclusion

We presented work we deployed to reduce phone call durations in NOC and to increase customer satisfaction. We implemented and deployed a reasoning engine pipeline that, based on structured historical data, was able to establish connections between new unseen problems and historical documented problems and propose a list of potential solutions. All information has been encoded in a knowledge graph for encoding its semantics, and then we expose the problem as a link prediction problem in a knowledge graph. Even though the data captured uncertainty the method reached level ready for deployment in our platform and testing with a client for support maintenance. We identified several exploration venues to improve further the results. First, we need to address the issue of identifying the list of malign error codes associated with a certain problem. We are aware that this list will never be deterministic or accurate otherwise the problem of finding the solution will become straight forward. However efforts should be made to

---

[2] http://www-sop.inria.fr/members/Freddy.Lecue/thales/iswc-2020-in-use-PrescriptiveMaintenance.mp4.

lower the uncertainty in selecting the error codes. Second, we are also aware that the quality of the embedding calculated for new entities is not sufficient. Efforts should be made to construct new knowledge graph embeddings for new entities without having to execute the complete embedding generation algorithm. This is still an emerging research question and we plan to address it in our future work.

Finally, the work described in this paper consists of a very valuable feature of the Reflex platform. It also gives the platform an edge in the domain of prescriptive-maintenance and customer support.

Acknowledgements. The authors would like to thank Dr. Roger Brooks for his support along the duration of the project.

References

1. Watson, J., Brooks, R., Colby, A., Kumar, P., Malhotra, A., Jain, M.: Predicting service impairments from set-top box errors in near real-time and what to do about It. In: the 2018 Fall Technical Forum (2018)
2. Ran, Y., Zhou, X., Lin, P., Wen, Y., Deng, R.: A survey of predictive maintenance: systems, purposes and approaches. arXiv:1912.07383 (2019)
3. Konys, A.: An ontology-based knowledge modelling for a sustainability assessment domain. Sustainability **10**(2), 300 (2018)
4. Peng, Y., Dong, M., Zuo, M.J.: Current status of machine prognostics in condition-based maintenance: a review. Int. J. Adv. Manuf. Technol. **50**(1), 297–313 (2010)
5. Zhang, Z., Si, X., Hu, C., Lei, Y.: Degradation data analysis and remaining useful life estimation: a review on Wiener-process-based methods. Eur. J. Oper. Res. **271**(3), 775–796 (2018)
6. Olde Keizer, M.C.A., Flapper, S.D.P., Teunter, R.H.: Condition-based maintenance policies for systems with multiple dependent components: a review. Eur. J. Oper. Res. **261**(2), 405–420 (2017)
7. Hong, H.P., Zhou, W., Zhang, S., Ye, W.: Optimal condition-based maintenance decisions for systems with dependent stochastic degradation of components. Reliab. Eng. Syst. Saf. **121**, 276–288 (2014)
8. Rao, B.K.N., Pai, P.S., Nagabhushana, T.N.: Failure diagnosis and prognosis of rolling - element bearings using artificial neural networks: a critical overview. J. Phys: Conf. Ser. **364**, 012023 (2012)
9. Susto, G.A., Schirru, A., Pampuri, S., McLoone, S., Beghi, A.: Machine learning for predictive maintenance: a multiple classifier approach. IEEE Trans. Ind. Inform. **11**(3), 812–820 (2015)
10. Chen, X., Wang, P., Hao, Y., Zhao, M.: Evidential KNN-based condition monitoring and early warning method with applications in power plant. Neurocomputing **315**, 18–32 (2018)
11. Guo, L., Lei, Y., Li, N., Yan, T., Li, N.: Machinery health indicator construction based on convolutional neural networks considering trend burr. Neurocomputing **292**, 142–150 (2018)
12. Zhu, J., Chen, N., Peng, W.: Estimation of bearing remaining useful life based on multiscale convolutional neural network. IEEE Trans. Ind. Electron. **66**(4), 3208–3216 (2019)
13. Wu, Y., Yuan, M., Dong, S., Lin, L., Liu, Y.: Remaining useful life estimation of engineered systems using vanilla LSTM neural networks. Neurocomputing **275**, 167–179 (2018)
14. Deutsch, J., He, D.: Using deep learning-based approach to predict remaining useful life of rotating components. IEEE Trans. Syst. Man Cybern.: Syst. **48**(1), 11–20 (2018)

15. Ansari, F., Glawar, R., Sihn, W.: Prescriptive maintenance of CPPS by integrating multimodal data with dynamic bayesian networks. Machine Learning for Cyber Physical Systems. TA, vol. 11, pp. 1–8. Springer, Heidelberg (2020). https://doi.org/10.1007/978-3-662-59084-3_1
16. Matyas, K., Nemeth, T., Kovacs, K., Glawar, R.: A procedural approach for realizing prescriptive maintenance planning in manufacturing industries. CIRP Ann. – Manuf. Technol. **66**, 461–464 (2017)
17. Goyal, A., et al.: Asset health management using predictive and prescriptive analytics for the electric power grid. IBM J. Res. Devel. **60**(1), 4:1–4:14 (2016)
18. Ansari, F., Glawar, R., Nemeh, T.: PriMa: a prescriptive maintenance model for cyber-physical production systems. Int. J. Comput. Integr. Manuf. **32**(4:5), 482–503 (2019)
19. Schmidt, B., Wang, L., Galar, D.: Semantic framework for predictive maintenance in a cloud environment. In: Proceedings of the 10th CIRP, vol. 62, pp. 583–588 (2017)
20. Medina-Oliva, G., Voisin, A., Monnin, M., Leger, J.-B.: Predictive diagnosis based on a fleet-wide ontology approach. Knowl.-Based Syst. **68**, 40–57 (2014)
21. Auer, S., Bizer, C., Kobilarov, G., Lehmann, J., Cyganiak, R., Ives, Z.: DBpedia: a nucleus for a web of open data. In: Aberer, K., et al. (eds.) ASWC/ISWC -2007. LNCS, vol. 4825, pp. 722–735. Springer, Heidelberg (2007). https://doi.org/10.1007/978-3-540-76298-0_52
22. Zhang, S., Tay, Y., Yao, L., Liu, Q.: Quaternion knowledge graph embeddings. In: NIPS, pp. 2735–2745 (2019)
23. Wang, Q., Mao, Z., Wang, B., Guo, L.: Knowledge graph embedding: a survey of approaches and applications. IEEE Trans. Knowl. Data Eng. **29**(12), 2724–2743 (2017)
24. Bordes, A., Usunier, N., Garcia-Duran, A., Weston, J., Yakhnenko, O.: Translating embeddings for modeling multi-relational data. In: NIPS, pp. 2787–2795 (2013)
25. Yang, B., Yih, W., He, X., Gao, J., Deng, L.: Embedding entities and relations for learning and inference in knowledge bases. In: 3rd International Conference on Learning Representations, (ICLR 2015) San Diego (2015)
26. Nickel, M., Rosasco, L., Poggio, T.: Holographic embeddings of knowledge graphs. In: Proceedings of the Thirtieth AAAI Conference on Artificial Intelligence, pp. 1955–1961, Phoenix, Arizona (2016)
27. Trouillon, T., Welbl, J., Riedel, S., Gaussier, É., Bouchard, G.: Complex embeddings for simple link prediction. In: Proceedings of the 33rd International Conference on International Conference on Machine Learning, vol. 48, pp. 2071–2080. New York (2016)
28. Costabello, L., Pai, S., Van, C. L., McGrath, R., McCarthy, N.: AmpliGraph: a Library for Representation Learning on Knowledge Graphs, Zenodo (2019)
29. Forgy, E.W.: Cluster analysis of multivariate data: efficiency versus interpretability of classifications. Biometrics **21**(3), 768–769 (1965)
30. Lloyd, S.: Least squares quantization in PCM. IEEE Trans. Inf. Theory **28**(2), 129–137 (1982)

Ontology-Enhanced Machine Learning: A Bosch Use Case of Welding Quality Monitoring

Yulia Svetashova[1,2]([✉]), Baifan Zhou[1,2]([✉]), Tim Pychynski[1], Stefan Schmidt[1], York Sure-Vetter[2], Ralf Mikut[2], and Evgeny Kharlamov[3,4]

[1] Bosch Corporate Research, Renningen, Germany
svetashova@gmail.com,
{Baifan.Zhou,Tim.Pychynski,Stefan.Schmid5}@de.bosch.com
[2] Karlsruhe Institute of Technology, Karlsruhe, Germany
{York.Sure-Vetter,Ralf.Mikut}@kit.edu
[3] Bosch Center for Artificial Intelligence, Renningen, Germany
Evgeny.Kharlamov@de.bosch.com
[4] University of Oslo, Oslo, Norway
Evgeny.Kharlamov@ifi.uio.no

Abstract. In the automotive industry, welding is a critical process of automated manufacturing and its quality monitoring is important. IoT technologies behind automated factories enable adoption of Machine Learning (ML) approaches for quality monitoring. Development of such ML models requires collaborative work of experts from different areas, including data scientists, engineers, process experts, and managers. The asymmetry of their backgrounds, the high variety and diversity of data relevant for quality monitoring pose significant challenges for ML modeling. In this work, we address these challenges by empowering ML-based quality monitoring methods with semantic technologies. We propose a system, called SemML, for ontology-enhanced ML pipeline development. It has several novel components and relies on ontologies and ontology templates for task negotiation and for data and ML feature annotation. We evaluated SemML on the Bosch use-case of electric resistance welding with very promising results.

1 Introduction

Industry 4.0 [16] and technologies of the Internet of Things (IoT) [13] behind it lead to unprecedented growth of data generated during manufacturing processes [3,35]. Indeed, modern manufacturing machines and production lines are equipped with sensors that constantly collect and send data and with control units that monitor and process these data, coordinate machines and manufacturing environment and send messages, notifications, requests. Availability of

Y. Svetashova and B. Zhou—Contributed equally to this work as first authors.

© Springer Nature Switzerland AG 2020
J. Z. Pan et al. (Eds.): ISWC 2020, LNCS 12507, pp. 531–550, 2020.
https://doi.org/10.1007/978-3-030-62466-8_33

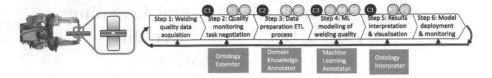

Fig. 1. A machine for automated welding (left) and an ML workflow enhanced with our semantic modules for welding quality monitoring (right). C stands for *challenges*, and R for *requirements*, see Sect. 1. ETL stands for *Extract, Transform, Load*. (Color figure online)

these voluminous data has led to a large growth of interest in data analysis for a wide range of industrial applications [25,26,41], especially the use of Machine Learning (ML) approaches for *monitoring* manufacturing processes, machines, and products by predicting machines' down-times or the quality of manufactured products [37].

Consider an example of *welding quality monitoring* at Bosch, where welding is performed with machines as shown in Fig. 1 (left) to connect pieces of metal together by pressing them and passing high current electricity through them [6]. For the purpose of developing ML approaches for welding quality monitoring, Bosch adopts the workflow slightly adjusted from [7,27] as schematically depicted in Fig. 1 (right). The workflow is iterative and includes data collection (Step 1), task negotiation, to define feasible and economic tasks (Step 2), data integration, to integrate data from different conditions and factories (Step 3), ML model development (Step 4), result interpretation and model selection (Step 5), and finally, model deployment in production (Step 6).

Development of such ML approaches is a complex and costly process where the following three challenges are of high importance for Bosch since they consume more than 80% of the overall time of development. The first challenge (C1) is *communication*: Steps 2 and 5 of welding quality monitoring require collaborative work of experts from different areas, including data scientists, engineers, process experts, and managers that have asymmetric backgrounds, which makes communication time consuming and error-prone. The second challenge (C2) is *data integration*: Step 3 requires to integrate data from dozens of sources with highly manual modification. The third challenge (C3) is *generalisability* of ML quality models: each ML model developed in Step 4 is typically tailored to a specific dataset and one welding process. Thus, reuse of this ML model for other data or processes requires a significant effort, while the reuse is highly desired, considering Bosch's wide spectrum of processes, equipment, and locations. In Fig. 1 we annotated Steps 2–5 with the challenges as C1–C3.

In this work we address these three challenges by enhancing machine learning development for quality monitoring with semantic technologies that have recently gained a considerable attention in industry for a wide range of applications and automation tasks such as modelling of industrial assets [18] and industrial analytical tasks [21], integration [11,19,20] and querying [32] of production data, and for process monitoring [29] and equipment diagnostics [22].

In particular, we developed a system, called SemML, that extends the conventional ML workflow with four semantic components that are depicted with grey boxes in Fig. 1. These components rely on ontologies, ontology templates, and reasoning. In particular, SemML exploits upper-level and concrete domain ontologies and the ML-ontology that captures machine learning tasks. The four semantic components of SemML are:

- **Ontology extender** that allows domain experts to describe domains in terms of an upper-level ontology by filling in templates. Data scientists then also use templates to annotate domain terms with quality related information. Then, they use the ontologies they jointly developed as a "lingua franca" for task negotiation.
- **Domain knowledge annotator** that enables data integration by annotating, mapping raw data to the terms in domain ontologies with ontology-to-data mappings.
- **Machine learning annotator** that uses automated reasoning to infer ML-relevant information from ontology-to-data mappings and creates the mappings between ML ontologies and data for each raw data source.
- **Ontology interpreter** that facilitates uniform and explainable inspection of ML models and raw data.

Ontology extender and interpreter help us to address the communication challenge, domain knowledge annotator addresses the data integration challenge, and ML annotator addresses the generalisability challenge.

We evaluated SemML with a group of domain users. In particular, we conducted two experiments with data scientists, measurement experts, and domain experts from two welding processes: resistance spot welding (RSW) and hot-staking (HS). To this end, we developed a set of templates, domain ontologies, and welding quality monitoring tasks. The users were first asked to create their domain ontologies using the ontology extender, and then map the variable names in raw data to the datatype properties of their created ontologies. After each task, they answered questionnaires to provide information on subjective satisfaction. The time and accuracy of these tasks and the scores of the questionnaires were recorded, analysed, and evaluated with promising results.

In Sect. 2 we introduce the Bosch use case of electric resistance welding. Section 3 describes the architecture and functions of SemML. Section 4 reports our user study.

2   Use Case: Quality Monitoring in Electric Resistance Welding

We now discuss the Bosch welding quality monitoring use case, the corresponding ML workflow for predictive quality monitoring, and then enumerate challenges that we address with semantic technologies.

Bosch Welding Quality Monitoring. Bosch is one of the global manufacturing leaders in the automotive industry. Welding is heavily used in industry for numerous applications including car production. Indeed, a typical car body can contain up to 6000 welding spots [38] where pieces of metal are connected. Bosch welding solutions include welding equipment (Fig. 1 for RSW) software, service, development support, etc. These solutions are used in Bosch plants and many customers worldwide, e.g. Daimler, BMW, Volkswagen, Audi, Ford. Enabled by the abundant data and computing resources behind the IoT technologies, Bosch is developing ML methods as depicted in Fig. 1 to predict the welding quality of next spots, before the actual welding happens. This allows to take necessary measures beforehand, like automatic adjustment of welding parameters, to improve the expected welding quality and avoid potential quality failure.

In the example illustrated in Fig. 2, the developed ML approaches should be able to predict the quality of the 6th welding spot, based on data of previous welded spots, including sensor measurements, welding configurations, past spot quality, etc. This requires acquisition of welding data from welding processes and measurements, as in *Step 1* of Fig. 1. The process, data and possible interesting tasks need to be explained to

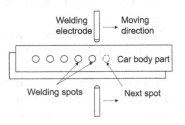

Fig. 2. Resistance spot welding

data scientists. The latter have to understand the process, the data, translate the task description from engineering languages into machine learning languages, evaluate the feasibility and cost for solving these questions. Some preliminary data analysis and visualisation are done in this step, known as Exploratory Data Analysis. Managers also have to participate in negotiating prioritising activities and goals from a view of strategic interest and make decisions of defining the task. The task negotiation is highly iterative, comprising *Step 2* of Fig. 1. This step is very time consuming and error-prone.

Then, in *Step 3* of Fig. 1 the Bosch RSW production data collected from various monitoring software in at least 4 locations and 3 original equipment manufacturers are integrated. These data may have different names for the same variables, or have some variables missing in one source but present in another, or measured with different sampling rate, etc. Besides the production data, Bosch has data collected from laboratory and simulation for process development. Extra sensors are installed in the laboratory, and the simulation data are generated with different mechanisms. Integration of all these data requires collaborative work of data scientists, data managers, process experts, and measurement experts. Thus, *Step 3* is essential but laborious and time-consuming.

After the data are integrated, the ML modelling in *Step 4* of Fig. 1 starts. It includes feature engineering which is time-consuming [1] since different datasets/domains may have different features and require different feature engineering strategies. Finally, after the heavy work of ML modelling in *Step 5* of Fig. 1, the data scientists present and visualise ML results and models, and together with other stakeholders discuss and interpret them. Managers then have to choose the best model to be deployed in *Step 6* of Fig. 1.

Use Case Requirements. Summing up, the time and effort required for ML development is heavily affected by *(C1)* the necessity of multiple iterations of communication by different stakeholders, *(C2)* the complexity of the data integration process, and *(C3)* generalisability of the developed ML models to similar processes and datasets. In order to enhance the ML workflow adopted by Bosch in a way that it addresses C1–C3 we derived the following five system requirements:

- *R1: Uniform communication model for various stakeholders*: The system should rely on a common vocabulary, with unambiguously defined relations between the terms. This vocabulary should be machine-readable and minimally controversial.
- *R2: Uniform data format and ML vocabulary*: the results of the ETL process are the input to ML modelling. Thus, the system should offer a uniform format for the data storage and a uniform naming of variables.
- *R3: Mechanism for generalising ML models*: the system should offer a mechanism for machine learning methods developed on one dataset to be reused or generalised to other datasets and manufacturing processes.
- *R4: Data enrichment mechanism*: the system should enable the enrichment of data with some task-specific information so that the integrated data can be linked to the generalisable machine learning approaches.
- *R5: Flexibility, extensibility, maintainability*: the system and its functionalities should enable accommodation of new data sources and ML tasks.

Note that the requirements R1 and R5 address the challenge C1, then R2 and R5 address C2, and R3–R5 address C3; we depict it in Fig. 1 with yellow circles.

3 SemML: Ontology-Enhanced Machine Learning Development

In this section, we present our SemML system that has a modular and multilayered architecture and illustrated in Fig. 3. In order to simplify for the reader the understanding of how SemML works, we overlay the architecture with the workflow from Fig. 1 where the steps are indicated with blue arrows. SemML has three layers: *Industry Applications Layer* where the welding monitoring, diagnostics, and analyses happen, *System Layer* that contains machine learning modules enhanced with our semantic modules, where orange circles indicate the requirements from Sect. 2 that are addressed by the corresponding modules, and *Data and Knowledge Layer* that contains ontologies, ontology templates, data, ML models and other relevant artifacts. The *Semantic Artifacts* of the latter layer serve as a bridge between the data sources and the modules in the system layer. We now discuss the layers and their interactions.

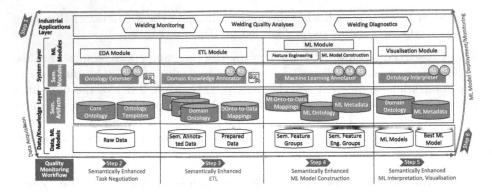

Fig. 3. An architectural overview of our semantically enhanced ML solution SemML for welding quality monitoring, where we overlay the welding quality monitoring workflow of Fig. 1 and the use-case requirements. EDA: explorotary data analysis, Sem.: semantics, Eng.: engineered. (Color figure online)

3.1 SemML Workflow

The arrows in Fig. 3 indicate the data flow from the raw sources generated by the welding monitoring and diagnostics applications through the machine learning modules back to the top layer where the developed quality models are deployed and monitored. We now walk the reader through these workflow steps.

Semantically Enhanced Task Negotiation. Once the raw data is acquired, data scientists and process experts align their backgrounds and specify the task of quality analysis. To this end, we enriched the traditional ML module for *Exploratory Data Analysis* (EDA) with the semantic *Ontology Extender*. Its graphical user interface allows experts to describe their domain in terms of an upper-level ontology, *Core Ontology* that we developed, by filling in *Ontology Templates* that we also developed. The users thus create domain ontologies that reflect the specificity of the raw data and a manufacturing process. Templates are also used by data scientists to annotate domain terms with quality-related information. Thus, the ontology extender, as well as the core ontology and templates, addresses the R1 and R5 requirements: the core ontology serves as a common communication model and the templates make the system flexible and extensible to new data sources.

Semantically Enhanced ETL. Our *Domain Knowledge Annotator* enables data integration via the mapping of the raw data to the terms in the domain ontologies. For mappings, we introduce a compact graphical user interface with browsing functionalities which is linked to the ontology extender. In case when a required term is missing, the user can switch to the ontology extender, and the newly introduced term immediately becomes available for use. The resulting domain knowledge mapping (*DOnto-to-Data Mapping*, where DOnto stands for Domain Ontology) is used by the *Extract-Transform-Load* (ETL) module to prepare the data for machine learning. Thus, the domain knowledge annotator

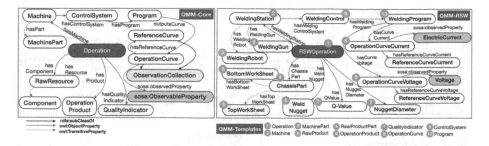

Fig. 4. QMM-Core, QMM-RSW and Templates where prefixes such as `qmm-core` are omitted.

addresses the R2 and R5 requirements: it allows to represent data in a uniform agreed format.

Semantically Enhanced ML Model Construction. The data prepared after the semantically enhanced ETL go through the *ML-Annotator* module. It relies on an ontology reasoner and the ML ontology to infer machine learning-relevant information from the DOnto-to-Data mappings. It creates the *MLOnto-to-Data Mapping* (where MLOnto stands for ML Ontology) for each raw data source. The resulting two kinds of mappings store different relationships. Indeed, consider for example a sensor measurement feature named as "CurrentAmp" that contains a series of observations of electric current values with time stamps. This feature will be mapped to the domain term "operationCurveCurrentValue" with a DOnto-to-Data mapping and to the ML term "TimeSeries" with an MLOnto-to-Data mapping. The latter indicates that this column will be treated as time-series (a special feature group) in machine learning. MLOnto-to-Data mappings enable the uniform handling of the prepared data by ML algorithms in the *Feature Engineering* module. This module performs various transformations of data categorised as feature groups and can also add new *Engineered Groups* of features. After feature engineering, several machine learning models are constructed in the *ML Model Construction* module. Information about the used feature engineering algorithms and engineered features are stored in the data layer as the *ML-Metadata*, that is, an application ontology which facilitates visualisation of the machine learning modelling. Our ML-Annotator addresses the requirements R3–R5.

Semantically Enhanced ML Interpretation and Visualisation. In order to conduct ML interpretation, data scientists discuss the ML models with other stake-holders through the *Visualisation Module*. Our *Ontology Interpreter* module facilitates a uniform and explainable inspection of ML models and raw data using ontologies, and thus, addresses the requirements R1 and R5. After the inspection, a selected ML model, and insights provided by ML analysis are deployed in the industrial applications layer.

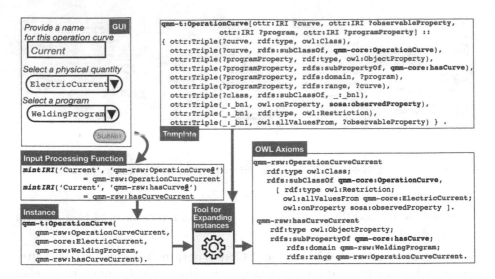

Fig. 5. Template, template instance and serialised OWL axioms.

3.2 Semantic Artifacts of SemML for Automated Welding Monitoring

We now give more details on the semantic artifacts of SemML that we developed for automated welding monitoring. In particular, we will discuss QMM-Core, the upper-level ontology for Quality Monitoring in Manufacturing, the library of templates, and show how they were used to construct one of the domain ontologies, QMM-RSW, for the manufacturing process of resistance spot welding. We then describe QMM-ML, the ontology for machine learning that powers the machine learning components of the system, and show how the automated reasoning and generalisability are enabled for different domains.

QMM-Core **ontology** is an OWL 2 ontology and can be expressed in the Description Logics $\mathcal{S}(\mathcal{D})$. With its 1170 axioms, which define 95 classes, 70 object properties and 122 datatype properties, it models the processes of discrete manufacturing with an emphasis on quality analysis. The left part of Fig. 4 displays the main classes and relations between them. This ontology has been developed through a series of workshops, taking inputs from various Bosch experts of engineering and machine learning. It can serve as one solution to reflect the consensus terminology for a common base of discussion. The ontology takes an operation-centred perspective: this orientation naturally follows from the analytical task of quality prediction described in Sect. 2. In particular, a qmm-core:Operation is performed by a qmm-core:Machine on a qmm-core:RawProduct. It results in a qmm-core:OperationProduct. Sensor observations are stored as qmm-core:OperationCurves and represent series of observation results with their corresponding timestamps. This class is our lightweight adaptation of ssn-ext:ObservationCollection from the proposed extensions to the Semantic Sensor Network Ontology [4]. We thus align

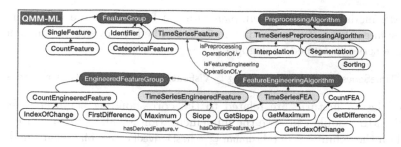

Fig. 6. A fragment of the QMM-ML ontology.

QMM-Core with the established way to model and query sensor observations – the SOSA/SSN ontology [10].

QMM-T Template Library. Our templates can be seen as parametrised ontologies and they rely on the Reasonable Ontology Templates (OTTR) framework [31]. By providing values (arguments) for each parameter and a user can create an instance of a template, which is then serialised as OWL axioms. Templates guarantee uniformity of the updates and the consistency of the updated ontology, as well as the relative simplicity of the ontology extension process. For our use cases, we created a template library QMM-T that relies on the classes and relationships of QMM-Core and has 30 templates. We also implemented a GUI that exposes the QMM-T to end-users. In Fig. 5 we exemplify our template instantiation process with one template qmm-t:OperationCurve.

QMM-RSW Resistance Spot Welding Ontology. By applying our templates, Bosch domain specialists created the QMM-RSW – ontology for the resistance spot welding process. It features 1542 axioms, which define 84 classes, 123 object properties and 246 datatype properties and can be expressed using $\mathcal{SH}(\mathcal{D})$ Description Logics. QMM-RSW and templates are partially shown in the right part of Fig. 4.

QMM-ML Ontology is partially depicted in Fig. 6. It has classes to categorise features as qmm-ml:FeatureGroups: time series, categorical features, identifiers, etc. It contains 62 classes, 4 object properties, 2 datatype properties as well as 210 axioms and 122 annotation assertions; it can be expressed using $\mathcal{ALH}(\mathcal{D})$ Description Logics. When QMM-ML is mapped to data, then MLOnto-to-Data mappings store qmm-ml:FeatureGroups for all columns in the prepared data. These mappings can be created automatically using reasoning and then modified by users. The ML module of SemML, in turn, has generic operations and algorithms with the behaviour specified on the level of qmm-ml:FeatureGroups of QMM-ML. Then, the ML Module retrieves the pre-processing and feature engineering algorithms for each group of features. To this end, it relies on the corresponding class definitions in the QMM-ML. For instance, the pre-processing algorithm for time series is defined as follows:

TimeSeriesPreprocessingAlgorithm $\sqsubseteq \forall$isPreprocessingAlgorithmOf.TimeSeries.

The ML module contains the implementation for each of the algorithm's sub-classes: Interpolation, Segmentation and Sorting. The feature engineering algorithm for time series is defined analogously:

TimeSeriesFeatureEngineeringAlg ⊑ ∀isFeatureEngineeringAlgOf.TimeSeries.

Its subclasses, in turn, are related to the corresponding engineered features, and the ML module will have the implementation for all of them. For example, based on the definition: GetMaximum ⊑ ∀ hasDerivedFeature.Maximum, the Feature Engineering module of SemML will apply the implemented GetMaximum algorithm to all time series features and generate new features with the token "Maximum" in their name for all of them.

In ML terms, the way how our semantically enhanced ML module works is: $h : \mathcal{X} \xrightarrow{M} \{\{FG_1\} \cdots \{FG_N\}\} \xrightarrow{\text{QMM-ML}} \{\{FEG_1\} \cdots \{FEG_K\}\} \rightarrow \hat{\mathcal{QI}}$, where h is a hypothesis that maps raw input features $\mathcal{X}$ into an estimation $\hat{\mathcal{QI}}$ of a welding quality indicator $\mathcal{QI}$. This mapping has two intermediate steps: (1) using MLOnto-to-Data Mapping M it fetches a set of standardised Feature Groups FGs and (2) using QMM-ML it turns them into a set of Feature Engineered Groups FEGs. This makes the developed ML approaches easily extendable to similar tasks and datasets. Moreover, this enables non-ML-experts to better understand the ML approaches, and even to modify the ML approaches with minimal training of ML expertise. Note that the classical ML module starts with $\mathcal{X}$ and may develop different ad hoc feature processing strategies for different tasks and data sources to estimate $\hat{\mathcal{QI}}$, or schematically: $h : \mathcal{X} \rightarrow \hat{\mathcal{QI}}$.

3.3   Related Work

Survey [30] extensively covers the usage of semantic technologies in data mining and knowledge discovery, and in particular in the facilitation of machine learning workflows. Still, to the best of our knowledge, existent approaches and system solutions, including the recent developments of digital twins for manufacturing [14], only partially meet our requirements R1–R5. Thus we had to develop our own ontologies and templates as well as ontology-based, highly customised and configurable solution, integrated into the workflow to support quality analysis in manufacturing. The users of our system are the different experts responsible for the task of developing machine learning methods. Indeed, none of the ontologies for manufacturing (e.g., [2,8,23,24,33,34]) fully serve as the communication model for our use cases and sufficiently cover our domains. The mapping-based data integration solutions like Ontop [17] are not particularly targeted towards our aim of minimising the involvement of ontologists into the model maintenance processes. Moreover, the role of mappings in our context is not limited to the transformation of data into the RDF format. Firstly, we integrate data sources for machine learning, secondly and in line with these tools, we transform some parts of it to RDF to explore the data. In the metadata management solutions for data lakes like Constance [9] and GEMMS [28], the metadata descriptions are used to integrate the raw sources. As the mapping-based data integration solutions, these systems lack the extensibility aspect. We found that the existing tools for ontology extension, e.g. template-driven systems (Webulous [15], TermGenie [5], Ontorat [36]) required considerable adjustments (including but not limited to the development of the new graphical user interface) and could not be easily integrated with the machine learning workflow and our infrastructure. Thus, we developed our own ontology extension tooling.

4   User Study

Our user study evaluates how well SemML addresses the challenges C1 on communication by evaluating the Ontology Extender and C2 on data integration by evaluating Domain Knowledge Annotator. Evaluation for C3 is our future work. To this end, we organised a workshop with three parts. First, we organised a thirty-minutes crash course to explain the ontology QMM-Core and templates. Then, we conducted two experiments: Experiment 1 on *Ontology Extension*, where the users were asked to describe their domains in terms of QMM-Core by filling in the proposed templates, and Experiment 2 on *Data Mapping*, where the users were asked to map the variables in the raw data sources to the datatype properties in the ontologies they created. Note that our experiments do not aim at comprehensive coverage of the welding domains and data sources relevant for welding quality: in our evaluation tasks we tried to balance the coverage and the time required to accomplish them.

4.1   Design of Experiments

We give further details on experiments and participants.

Users. Two target user groups (with the roles of domain experts and data scientists) participated in the experiments with two welding processes: resistance spot welding (RSW) and hot-staking (HS). The users could choose to participate only in Experiment 1 or in both. Some of them took part in the experiments with more than one domain or role. This is the case, e.g., for users who are domain experts both for RSW and HS, and some users who are domain experts but are learning data analysis or vice versa. In total, from 14 participants 25 result instances were collected in Experiment 1, and 19 instances in Experiment 2. Before the experiments, the participants rated their domain expertise (E1), experience with semantic technologies (E2) and experience with data mapping tools (E3) on a Likert scale (1: Beginner, 2: Developing, 3: Competent, 4: Advanced, 5: Expert).

Experiment 1: Ontology Extension. The users were asked to use Ontology Extender to create their ontologies. As illustrated in Fig. 7.1.1, for each term highlighted with the blue background in the short descriptions for the welding processes on the left side, the users selected a template on the right side, and then made choices to link the created class to its dependencies (drop-down list in Fig. 7.1.2). The resulting ontology terms (classes and properties) were then visualised (Fig. 7.1.3). Note that domain experts and data scientists did their

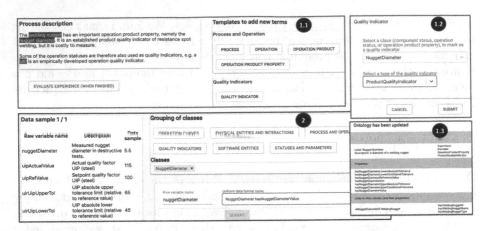

Fig. 7. Graphical user interfaces for (1.1–1.3) ontology extension and (2) Data mapping. (Color figure online)

tasks sequentially: the former created an ontology, and then the latter inspected their ontologies and extended them with quality indicators.

Experiment 2: Data Mapping. As illustrated in Fig. 7.2, the users were asked to use Domain Knowledge Annotator to map data. For each term in the column of raw variable names on the left side, they clicked the group of classes from the right top panel, selected a class, and then chose the datatype properties where the class is a domain from a drop-down list (in the right bottom panel).

4.2   Evaluation Metrics

According to ISO 9241-11 system usability has 3 dimensions: *effectiveness*, *efficiency*, and *satisfaction* [12]. We rely on them and their correlations with user expertise.

Effectiveness shows to which extent the intended goal is achieved [12]. We use *correctness*, the percentage of successfully completed tasks, as the metric for it. We are fully aware that there is no absolute correctness for these tasks because the domains or data can be understood in different ways. This issue is however not critical in our experiments since we carefully designed the tasks so that the answers are minimally controversial across the experts. In Experiment 1, the correctness is defined as the percentage of correctly chosen templates for a given term (Template Correctness, TC), the percentage of correct choices linking the dependencies between classes (Choice Correctness, ChC), and the percentage of fully correctly created classes, for which the correct template is chosen and all dependencies are correctly specified (Final Correctness, FC). In Experiment 2, the correctness is defined as the percentage of correctly chosen classes (Class Correctness, ClC) and the percentage of correctly mapped datatype properties for each item of raw variable names (Item Correctness, IC).

Table 1. Satisfaction metrics: Questionnaires and aggregated quality dimensions.

Experiment 1: Ontology Extension		Experiment 2: Data Mapping		Dimension
Domain Experts	Data Scientists	Domain Experts	Data Scientists	
Q1: I felt very confident using the system				D1: User
Q2: I found the system unnecessarily complex				Friendliness
Q3: I needed to learn many things before I became productive with this system				D2: Self-
Q4: I needed support of a technical person to be able to use this system				Explainability
Q5: I thought there was too much inconsistency in this system				D3: Consistency
Q6: I think the system covers most of my fundamental requirements for the *				D4:
* description of the process	* understanding of the process	* description of the data	* understanding of the data	Complete-ness
Q7: I think the system/resulting model allows me to unambiguously *				D5: Descriptive Power
* describe the process	* understand the process	* describe the data	* understand the data	
Q8: I think the relevant aspects of process quality are well presented in the *		I think the mapping system saves me effort of *		
* system	* system and the resulting model	* describing the data	* understanding and completing the data mapping	
Q9: I think the resulting ontology/mapping would be very easy to understand for *				D6: Communication Easiness
* data scientists	* domain experts	* data scientists	* domain experts	
Q10: I think the resulting ontology/mapping provides a good common base for discussion with *				
* data scientists	* domain experts	* data scientists	* domain experts	

Efficiency corresponds to "the resources (such as time or effort) needed by users to achieve their goals" [12]. We use *time spent on tasks* as the metric of efficiency.

Satisfaction was evaluated with the questionnaires after each experiment on 6 dimensions (see Table 1): [D1] *User Friendliness*: the system is easy to use; [D2] *Self-Explainability*: the system does not require extra knowledge or support; [D3] *Consistency*: the system is consistent in format, workflow, wording, etc.; [D4] *Completeness*: the system covers the domain/data to describe/understand; [D5] *Descriptive Power*: the system allows to describe the domain/data effectively, clearly; [D6] *Communication Easiness*: the system eases the communication between experts. The questions are presented in Table 1 where the first five are inspired by System Usability Scale (SUS). Questions 1–5 (and the corresponding Dimensions 1–3) target the usability of the graphical user interface. They are identical for all roles and experiments. Questions 6–10 (Dimensions 4–6) address more specific issues and differ slightly with respect to the role or the experiment. The users were asked to give scores ranging from 1 to 5 with a Likert scale (1: Strongly disagree, 2: Disagree, 3: Neither agree or disagree, 4: Agree, 5: Strongly agree). Note that Questions 2–5 are negatively formulated. Their scores are reversed in the later analysis to make the representation of the results more intuitive and consistent. E.g. if a user scores Q2 with 1, which means the user strongly disagrees that the system is complex, the corresponding score is reversed to 5, indicating the system is not complex.

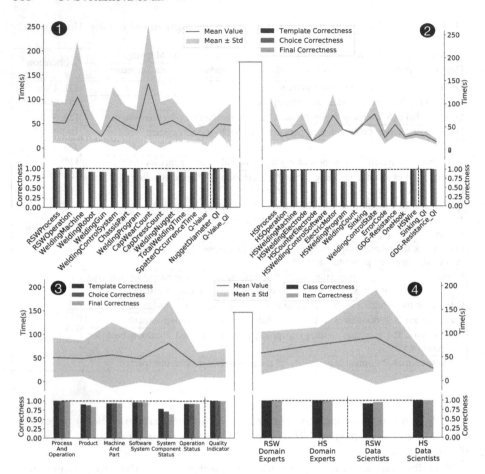

Fig. 8. 1 and 2: User performance of time and correctness for RSW in 1 and for HS in 2, aggregated on template groups for both in 3. Time and correctness for data mapping in 4.

4.3   Evaluation Results and Discussion

The results of the Effectiveness and Efficiency metrics are summarised in Fig. 8, which shows the user performance on Ontology Extender (Fig. 8.1–8.3) and Domain Knowledge Annotator (Fig. 8.4).

Results for Experiment 1: Ontology Extension. Domain experts in RSW created 14 terms and those in HS created 15 terms. Data scientists created 2 terms for both processes. On average, the users needed about 50s to create a new term. Note that the description of one term adds from 4 to 25 classes and properties to the ontology (see the process exemplified in Fig. 5 of Sect. 3.2).

Some users needed extra time for the terms WeldingMachine, CapWearCount and CapDressCount (with high standard deviation shown in the figures). The potential reasons are that the users needed to understand the complex structure of

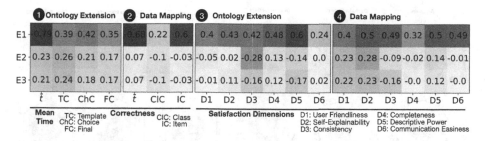

Fig. 9. Heatmaps of correlation coefficients between the usability metrics and self-accessed expertise. E1: Domain expertise, E2: Experience with semantic technologies, E3: Experience with data mapping tools.

machine and its multiple parts; for the latter two terms (both are created by the SystemComponentStatus template in Fig. 8.3) the users specified two dependencies, which is one more than the normal case of one choice. Another reason could be that the users moved to a new template group, which increased the cognitive complexity of the task and thus, the time spent on the task. In line with this tendency, we observe a gradual decline in time for subsequent terms created with the same or similar templates. For example, WeldingRobot and WeldingGun are machine parts directly following the WeldingMachine, and CapDressCount directly follows CapWearCount. This strongly supports the learnability of the system: having experience with a template increases efficiency and effectiveness.

The average correctness for applying a template is 93%, for making choices of the dependencies is 92%, and for both (final correctness) is 90%. The terms, e.g. CapWearCount, that required more time to create often have a relatively low correctness ratio. The high average correctness strongly demonstrates the usability and the error prevention potential of the system.

One of the goals of Ontology Extender was to serve as the communication platform between domain experts and data scientists. In our experimental setup, the data scientists were supposed to (1) inspect the domain ontologies created by the domain experts and (2) add the terms relevant for quality analysis. In particular, they had to add or find a term, and characterise it as a quality indicator. We separate these two parts by the vertical dashed lines in Fig. 8.1–8.3. All data scientists achieved 100% correctness with an average time of 39 s.

We now analyse the correlations between the self-reported expertise of our users and their performance. Figure 9.1 shows a strong negative correlation between domain expertise (E1) and time t and relatively strong positive correlation between E1 and the three types of correctness (template, choice and final). Not surprisingly, the users with higher domain expertise provided more correct modeling solutions and were faster than the beginners. The figure also suggests insignificant correlation between the performance of users and their experience in semantic technologies and mapping tools. This is encouraging since it suggests that the usage of our system requires no or little prior training in these disciplines and activities.

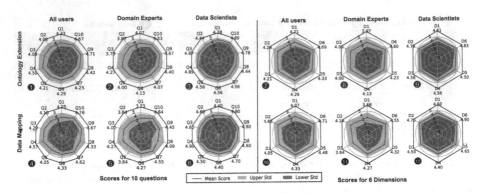

Fig. 10. Radar charts of questionnaires scores on 10 questions (1–6) and aggregated to 6 dimensions (7–13) defined in Table 1. Std is standard deviation. The blue lines indicate the mean scores, the light blue shadow the mean + std, and the dark blue shadow the mean - std. (Color figure online)

Results for Experiment 2: Data Mapping. The majority of users correctly mapped column names in the suggested files to the newly introduced terms, achieving 100% correctness (Fig. 8.4). The average time they spent for each term is about 50s. The correlations in Fig. 9.2 support the idea that domain expertise will ease the work and the other two parameters, including the self-accessed experience with mapping tools, have almost no effect. We interpret it as the evidence that the system is able to serve as a solution for both tasks – data modeling and data mapping – and does not require any prior experience with similar technologies. We complement our analysis with the results of the satisfaction questionnaires in the following section.

Satisfaction. We report the satisfaction results in the radar charts in Fig. 10 separately for data scientists and domain experts, and aggregated for all users. The charts in Fig. 10.1 and 10.4 represent the average scores for both user groups. These scores are higher than 4, which indicates a general good impression of users. The mean scores on Questions 4, 9 and 10 are very high (>4.5): The users evaluate the tool as easy to use without support of a technical person, and they think their working results will be easy to understand for other experts. This supports our vision that an ontology can serve as a good communication base.

The comparison of the scores on questions by the domain experts and data scientists (Fig. 10.2 vs. 10.3, 10.5 vs. 10.6), reveals that the data scientists evaluate the system with higher scores in average, and smaller standard deviation. This indicates that the data scientists have better and more uniform opinions on the system, while users taking the role of domain experts have more diverse opinions. One reason for that could be that the tasks for data scientists were only related to quality indicators and thus more clearly defined, while the tasks for domain experts who needed to describe complicated processes were more demanding.

In Fig. 10.7–10.12 the scores on questions are aggregated to six dimensions (see the meanings of dimensions in Table 1). This aggregation makes it easier to draw conclusions. Firstly, all six dimensions have average scores over 4, which means the users are satisfied with the system in general. The scores for D6 (communication easiness) and D5 (descriptive power) are the highest, indicating the users appreciate the ease of communication. Dimensions more related to the system usability (D1–D3) have scores around four, which means there is improvement space for the user interface.

Correlations in Fig. 9.3 and 9.4 reveal similar results as for the performance analysis: domain expertise correlates with high satisfaction scores, while the other two areas of expertise have little effect, which supports that the tool requires little prior training.

5 Conclusion, Lessons Learned, And Outlook

Conclusion. In this work we presented a Bosch use case of automated welding, challenges with ML-based welding quality monitoring, and requirements for a system to address them. To address the challenges we proposed to enhance the welding quality monitoring ML development with four semantic components: Ontology Extender, Domain Knowledge Annotator, Machine Learning Annotator, and Ontology Interpreter. We implemented the enhancement as the SemML system. We then evaluated SemML by focusing on the first two semantic modules. To this end, we conducted a user study with 14 Bosch experts. The evaluations show promising results: SemML can indeed help in addressing the challenges of communication and data integration.

Lessons Learned. First, an ontology as a formal language can be very effective to provide a lingua franca for communication between experts with different knowledge backgrounds. The process of developing the Core model was onerous, time-consuming and cognitively demanding at the initial phase. After that, we had a basis of core model and a set of templates. It revealed the development process became much easier because the developed ontologies facilitated communication. Second, the technology of templates enables non-ontologists to describe their domains and data in a machine-readable and unambiguous way. In contrast to the simple, tabular interfaces which exist for the template-based ontology construction, we needed to address the new requirements for our use case. In the use case, the users need to generalise a series of classes, while the later generated classes have dependencies on the older ones. This has two requirements: (1) the newly generated classes need to be accessible to later generated classes; (2) sequences of templates need to be applied in a particular order because the later classes presuppose the existence of their depending classes. Furthermore, the users need assistance like drop-down lists and visualisation of changes. Third, the users that are unfamiliar with the domains will need more time for some tasks and yield lower correctness. This indicates that we need to split the iterative process of task negotiation to smaller units so that different experts can digest each other's information more smoothly.

Outlook. We have an on-going study [39] that informally evaluates SemML's third semantic module and provides a user-interface [40] and we plan to further extend it to a larger scale user evaluation. Then, SemML is currently deployed in a Bosch evaluation environment, and we plan to push it into the production. This in particular requires to show the benefits of SemML with more Bosch users and in other use cases. This also requires to further improve the usability of SemML with more advanced services such as access control as well as with various ontology visualisation modules.

References

1. Bengio, Y., Courville, A., Vincent, P.: Representation learning: a review and new perspectives. IEEE Trans. Pattern Anal. Mach. Intell. **35**, 1798–1828 (2013)
2. Borgo, S., Leitão, P.: The role of foundational ontologies in manufacturing domain applications. In: Meersman, R., Tari, Z. (eds.) OTM 2004. LNCS, vol. 3290, pp. 670–688. Springer, Heidelberg (2004). https://doi.org/10.1007/978-3-540-30468-5_43
3. Chand, S., Davis, J.: What is smart manufacturing. Time Mag. Wrapper **7**, 28–33 (2010)
4. Cox, S.: Extensions to the semantic sensor network ontology. W3C Working Draft (2018)
5. Dietze, H., et al.: TermGenie-a web-application for pattern-based ontology class generation. J. Biomed. Semant. **5** (2014). https://doi.org/10.1186/2041-1480-5-48
6. DIN EN 14610: Welding and allied processes - definition of metal welding processes. German Institute for Standardisation (2005)
7. Fayyad, U., Piatetsky-Shapiro, G., Smyth, P.: From data mining to knowledge discovery in databases. AI Mag. **17**(3), 37 (1996)
8. Fiorentini, X., et al.: An ontology for assembly representation. Technical report. NIST (2007)
9. Hai, R., Geisler, S., Quix, C.: Constance: an intelligent data lake system. In: SIGMOID 2016 (2016)
10. Haller, A., et al.: The SOSA/SSN ontology: a joint WEC and OGC standard specifying the semantics of sensors observations actuation and sampling. In: Semantic Web (2018)
11. Horrocks, I., Giese, M., Kharlamov, E., Waaler, A.: Using semantic technology to tame the data variety challenge. IEEE Internet Comput. **20**(6), 62–66 (2016)
12. ISO: 9241–11.3. Part II: guidance on specifying and measuring usability. ISO 9241 ergonomic requirements for office work with visual display terminals (VDTs) (1993)
13. ITU: Recommendation ITU - T Y.2060: Overview of the Internet of Things. Technical report. International Telecommunication Union (2012)
14. Jaensch, F., Csiszar, A., Scheifele, C., Verl, A.: Digital twins of manufacturing systems as a base for machine learning. In: 2018 25th International Conference on Mechatronics and Machine Vision in Practice (M2VIP), pp. 1–6. IEEE (2018)
15. Jupp, S., Burdett, T., Welter, D., Sarntivijai, S., Parkinson, H., Malone, J.: Webulous and the Webulous Google Add-On-a web service and application for ontology building from templates. J. Biomed. Semant. **7**, 1–8 (2016)
16. Kagermann, H.: Change through digitization—value creation in the age of industry 4.0. In: Albach, H., Meffert, H., Pinkwart, A., Reichwald, R. (eds.) Management of Permanent Change, pp. 23–45. Springer, Wiesbaden (2015). https://doi.org/10.1007/978-3-658-05014-6_2

17. Kalaycı, E.G., González, I.G., Lösch, F., Xiao, G.: Semantic integration of Bosch manufacturing data using virtual knowledge graphs. In: Pan, J.Z., et al. (eds.) ISWC 2020. LNCS, vol. 12507, pp. 464–481. Springer, Cham (2020) (2020)

18. Kharlamov, E., et al.: Capturing industrial information models with ontologies and constraints. In: Groth, P., et al. (eds.) ISWC 2016. LNCS, vol. 9982, pp. 325–343. Springer, Cham (2016). https://doi.org/10.1007/978-3-319-46547-0_30

19. Kharlamov, E., et al.: Ontology based data access in Statoil. J. Web Semant. **44**, 3–36 (2017)

20. Kharlamov, E., et al.: Semantic access to streaming and static data at Siemens. J. Web Semant. **44**, 54–74 (2017)

21. Kharlamov, E., et al.: An ontology-mediated analytics-aware approach to support monitoring and diagnostics of static and streaming data. J. Web Semant. **56**, 30–55 (2019)

22. Kharlamov, E., Mehdi, G., Savković, O., Xiao, G., Kalaycı, E.G., Roshchin, M.: Semantically-enhanced rule-based diagnostics for industrial Internet of Things: the SDRL language and case study for Siemens trains and turbines. J. Web Semant. **56**, 11–29 (2019)

23. Krima, S., Barbau, R., Fiorentini, X., Sudarsan, R., Sriram, R.D.: OntoSTEP: OWL-DL ontology for STEP. Technical report. NIST (2009)

24. Lemaignan, S., Siadat, A., Dantan, J.Y., Semenenko, A.: MASON: a proposal for an ontology of manufacturing domain. In: IEEE DIS (2006)

25. Mikhaylov, D., Zhou, B., Kiedrowski, T., Mikut, R., Lasagni, A.F.: High accuracy beam splitting using SLM combined with ML algorithms. Opt. Lasers Eng. **121**, 227–235 (2019)

26. Mikhaylov, D., Zhou, B., Kiedrowski, T., Mikut, R., Lasagni, A.F.: Machine learning aided phase retrieval algorithm for beam splitting with an LCoS-SLM. In: Laser Resonators, Microresonators, and Beam Control XXI, vol. 10904, p. 109041M (2019)

27. Mikut, R., Reischl, M., Burmeister, O., Loose, T.: Data mining in medical time series. Biomed. Tech. **51**, 288–293 (2006)

28. Quix, C., Hai, R., Vatov, I.: GEMMS: a generic and extensible metadata management system for data lakes. In: CAiSE Forum (2016)

29. Ringsquandl, M., et al.: Event-enhanced learning for KG completion. In: Gangemi, A., et al. (eds.) ESWC 2018. LNCS, vol. 10843, pp. 541–559. Springer, Cham (2018). https://doi.org/10.1007/978-3-319-93417-4_35

30. Ristoski, P., Paulheim, H.: Semantic web in data mining and knowledge discovery: a comprehensive survey. J. Web Semant. **36**, 1–22 (2016)

31. Skjæveland, M.G., Lupp, D.P., Karlsen, L.H., Forssell, H.: Practical ontology pattern instantiation, discovery, and maintenance with reasonable ontology templates. In: Vrandečić, D., et al. (eds.) ISWC 2018. LNCS, vol. 11136, pp. 477–494. Springer, Cham (2018). https://doi.org/10.1007/978-3-030-00671-6_28

32. Soylu, A., et al.: OptiqueVQS: a visual query system over ontologies for industry. Semant. Web **9**(5), 627–660 (2018)

33. Usman, Z., Young, R.I.M., Chungoora, N., Palmer, C., Case, K., Harding, J.: A manufacturing core concepts ontology for product lifecycle interoperability. In: van Sinderen, M., Johnson, P. (eds.) IWEI 2011. LNBIP, vol. 76, pp. 5–18. Springer, Heidelberg (2011). https://doi.org/10.1007/978-3-642-19680-5_3

34. Šormaz, D., Sarkar, A.: SIMPM - upper-level ontology for manufacturing process plan network generation. Robot. Comput. Integr. Manuf. **55**, 183–198 (2019)

35. Wuest, T., Weimer, D., Irgens, C., Thoben, K.D.: Machine learning in manufacturing: advantages, challenges, and applications. Prod. Manuf. Res. **4**, 23–45 (2016)

36. Xiang, Z., Zheng, J., Lin, Y., He, Y.: Ontorat: automatic generation of new ontology terms, annotations, and axioms based on ontology design patterns. J. Biomed. Semant. **6** (2015). https://doi.org/10.1186/2041-1480-6-4

37. Zhao, R., Yan, R., Chen, Z., Mao, K., Wang, P., Gao, R.X.: DL and its applications to machine health monitoring. MS&SP **115**, 213–237 (2019)

38. Zhou, B., Pychynski, T., Reischl, M., Mikut, R.: Comparison of machine learning approaches for time-series-based quality monitoring of resistance spot welding (RSW). Arch. Data Sci. Ser. A **5**(1), 13 (2018). (Online first)

39. Zhou, B., Svetashova, Y., Byeon, S., Pychynski, T., Mikut, R., Kharlamov, E.: Predicting quality of automated welding with machine learning and semantics: a Bosch case study. In: CIKM (2020)

40. Zhou, B., Svetashova, Y., Pychynski, T., Kharlamov, E.: SemFE: facilitating ML pipeline development with semantics. In: CIKM (2020)

41. Zhou, B., Chioua, M., Bauer, M., Schlake, J.C., Thornhill, N.F.: Improving root cause analysis by detecting and removing transient changes in oscillatory time series with application to a 1, 3-butadiene process. Ind. Eng. Chem. Res. **58**, 11234–11250 (2019)

Revisiting Ontologies of Units of Measure for Harmonising Quantity Values – A Use Case

Francisco Martín-Recuerda[1(✉)], Dirk Walther[1], Siegfried Eisinger[1],
Graham Moore[2], Petter Andersen[3], Per-Olav Opdahl[3], and Lillian Hella[3]

[1] DNV GL, Oslo, Norway
{francisco.martin-recuerda,dirk.walther,siegfried.eisinger}@dnvgl.com
[2] Sesam.io, Oslo, Norway
graham.moore@sesam.io
[3] ix3 – an Aker Solutions Company, Fornebu, Norway
{petter.andersen,per-olav.opdahl,lillian.hella}@akersolutions.com

Abstract. Processing quantity values in industry applications is often arduous and costly due to different systems of units and countless naming and formatting conventions. As semantic technology promises to alleviate the situation, we consider using existing ontologies of units of measure to improve data processing capabilities of a cloud-based semantic platform for operating digital twins of real industry assets. We analyse two well-known ontologies: OM, the Ontology of units of Measure, and QUDT, the Quantities, Units, Dimensions and Types ontology. These ontologies are excellent resources, but do not meet all our requirements. We discuss suitable modelling choices for representing quantities, dimensions and units of measure and we outline the process we followed to adapt relevant definitions from OM and QUDT into a new ontology of units of measure that better meets our needs. Compared with the alternative of manually creating the ontology from scratch, the development and maintenance costs and duration were reduced significantly. We believe this approach will achieve similar benefits in other ontology engineering efforts.

Keywords: Digital twin · Ontology engineering · Units of measure

1 Introduction

Aker Solutions,[1] DNV GL[2] and Sesam[3] are three companies based in Norway that are collaborating on the development of services and IT infrastructure for the design, implementation, operation and quality assurance of digital twins

---

[1] https://www.akersolutions.com/.
[2] https://www.dnvgl.com/.
[3] https://sesam.io/.

This work has been supported by Aker Solutions, DNV GL, Sesam and SIRIUS labs. A special thanks to Kaia Means (DNV GL) for reviewing this document.

© Springer Nature Switzerland AG 2020
J. Z. Pan et al. (Eds.): ISWC 2020, LNCS 12507, pp. 551–567, 2020.
https://doi.org/10.1007/978-3-030-62466-8_34

(DTs).[4] To this end, Aker Solutions has founded a dedicated software company, ix3,[5] responsible for the implementation, management and commercialisation of the *ix3's digital twin platform Integral* (*Integral platform* for short), a cloud-based digital infrastructure for operating DTs. This platform facilitates the deployment, integration and orchestration of multiple types of digital services including data analytics and (multi-physics and process) simulation, which are essential when operating DTs. In particular, the Integral platform collects, harmonises, contextualises, analyses, visualises and distributes large amounts of heterogeneous data. This includes not only operational data from sensors and control systems but also engineering and enterprise data from production systems. In this paper, we define *data harmonisation* as the process of providing a common representation of quantities and units (including the use of unique identifiers), and *data contextualisation* as the process of enriching data with additional information that facilitates its understanding and processing.

Collecting, harmonising, contextualising and distributing data are tasks delegated to *Sesam Datahub*, enriched with an ontology library, developed by ix3 in cooperation with DNV GL, the *Information Model* (IM). The IM includes more than 100 OWL 2 (Direct Semantics) ontologies organised in a strict dependency hierarchy using the OWL 2 import mechanism. On top of the hierarchy are domain-independent ontologies, such as the ISO 15926-14 ontology,[6] SKOS[7] and PAV.[8] The next level of ontologies describes generic concepts in the engineering domain and mappings between system codes from Aker Solutions and from its customers. The lower level of ontologies represents oil and gas assets and related technical and commercial documentation. The IM provides Sesam with a controlled language with unique identifiers (IRIs), definitions for each term and mappings that explicitly state how terms defined by external sources are related to terms defined by the IM. Sesam uses this information to contextualise and harmonise the data propagated through *pipes*, which are internal processes defined using a declarative language and supported by specific built-in connectors and microservices for consuming, transforming and exporting data from/to external data-sources.

Data consumed and produced by DTs are often in the form of quantity values, which are products of numbers and units. Processing quantity values is still arduous and costly, mainly due to different systems of units and countless naming and formatting conventions. For instance, the following data items represent the same quantity value: '392.75 FAHR', '392,75 F', '39,275,231e−5 oF', '200,42 C', '200.416.667E−6 DegC', '473,57 K' and '4.7357e2 KLV'. A unit of measure, such as Fahrenheit, has several different abbreviations (i.e. FAHR, F and oF), punctuation may vary, or other units may be used, such as Celsius (i.e. C and DegC) or Kelvin (i.e. K and KLV). To support Sesam when harmonising these examples of quantity values, the IM provides a preferred label and a unique

---

[4] https://en.wikipedia.org/wiki/Digital_twin.
[5] https://www.ix3.com/.
[6] https://www.iso.org/standard/75949.html.
[7] https://www.w3.org/TR/skos-reference/.
[8] https://github.com/pav-ontology/pav/.

identifier for units of measure such as Fahrenheit, Celsius and Kelvin. The IM also includes the information to convert quantity values into different units of measure (i.e. from Fahrenheit to Kelvin). In addition, the IM maintains mappings between its terms and external terms using SKOS semantic properties (i.e. the terms FAHR, F and oF represents the unit Fahrenheit in the IM). Additional context of a quantity value, such as a related quantity kind (i.e. thermodynamic temperature), can be also obtained from the IM.

Main Results and Organisation. In this paper, we describe how we extended the IM to better support Sesam when harmonising and contextualising quantity values in the Integral platform for digital twins. To minimise development costs, we tried to reuse existing ontologies of units of measure. In Sect. 2, we provide a summary of the analysis of two well-known such ontologies, the Ontology of units of Measure (OM)[9] and the Quantities, Units, Dimensions and Types ontology (QUDT).[10] Although both ontologies are excellent resources, we concluded that they must be revised to meet all of our requirements. Therefore, we built a new ontology of units of measure based on QUDT and OM that is better adapted to our needs. In Sect. 3, we discuss suitable modelling choices and ontology design patterns [2] that we considered for representing quantities, dimensions and units. We revisited the implementation and build processes of the IM to accommodate the new ontology of units of measure (cf. Sect. 4). In particular, we tried to reduce manual effort, facilitate collaborative development and ensure uniform modelling by implementing and applying ontology templates [6] and specific SPARQL queries. We conclude the paper in Sect. 5.

Notation and Terminology. In this paper, we use notation and terminology related to units of measure, ISO standards, W3C recommendations and ontologies. Terms such as dimensional analysis, units of measure, quantities (or quantity kinds), dimensions and systems of units are well introduced by Wikipedia[11] and the standard ISO/IEC 80000,[12] which describes the International System of Quantities (ISQ)[13] and also includes all units defined by the International System of Units (SI).[14] The various W3C recommendations for OWL 2, including syntaxes and semantics are accessible from the W3C website.[15] This is relevant when we refer to IRIs, named and anonymous individuals, classes, object and data properties, class expressions (or restrictions), assertions and punning. SKOS is a W3C recommendation that includes an OWL ontology. The notions of SKOS concepts, SKOS concept schemes and SKOS semantic and mapping properties are relevant for this paper. In addition to the ontologies OM, QUDT, SKOS, ISO 15926-14 and PAV we also refer to the ontologies DTYPE v1.2,[16]

---

[9] https://github.com/HajoRijgersberg/OM.
[10] https://github.com/qudt/qudt-public-repo/releases.
[11] https://en.wikipedia.org/wiki/Dimensional_analysis.
[12] https://www.iso.org/committee/46202/x/catalogue/.
[13] https://en.wikipedia.org/wiki/International_System_of_Quantities.
[14] https://en.wikipedia.org/wiki/International_System_of_Units.
[15] https://www.w3.org/TR/owl2-overview/.
[16] http://www.linkedmodel.org/doc/2015/SCHEMA_dtype-v1.2.

and VAEM v2.0.[17] When terms defined by these ontologies are explicitly mentioned, we use the following prefixes: om, qudt, skos, dtype, iso, pav, dtype and vaem, respectively. QUDT includes additional prefixes when referring to quantity kinds, dimensions and units such as quantitykind, qkdv and unit. Due to space limitations, we do not include the namespaces that correspond to these prefixes.

2   Assessing Ontologies of Units of Measure

In this section, we discuss the suitability of the ontologies OM and QUDT, based on a selection of the requirements that we considered in our use case, and we refrain from reexamining existing work on assessing ontologies of units of measure, such as Keil et al. [3]. OM and QUDT are outstanding candidates among the ontologies of units of measure that we considered. Both ontologies are prominent and explicitly referenced, for instance by Wikidata[18] and W3C SOSA/SSN.[19] In addition, these ontologies fulfil many of the requirements discussed in this section.

The selection of requirements includes a mix of functional and non-functional requirements that are also relevant for other use cases aiming to contextualise and harmonise quantity values using ontologies of units of measure. In fact, we noticed that many of the requirements of the Integral platform are similar to other projects where we applied ontologies to improve interoperability of applications. The key words *MUST* and *SHOULD* are to be interpreted as described in RFC 2119.[20] Table 1 summarises the assessment of the requirements considered. We use the following codes: +, - and *, which mean the requirement is satisfied, not satisfied, and nearly satisfied, respectively.

Table 1. Requirements assessment for the ontologies OM and QUDT

Requirement	Key words	OM	QUDT
R01	Public licence	+	+
R02	Active maintenance	+	+
R03	Coverage	+	+
R04	OWL 2 Direct Semantics	+	-
R05	(Web) Protégé & OWLAPI	+	-
R06	HermiT reasoner	-	-
R07	Ontology design patterns	*	*
R08	Dimensional analysis & SPARQL	-	+
R09	Compatibility with ISO 15926-14	+	*
R10	Compatibility with SKOS	-	-
R11	Modularity	-	+

---

[17] http://www.linkedmodel.org/doc/2015/SCHEMA_vaem-v2.0.
[18] https://www.wikidata.org/wiki/Wikidata:Main_Page.
[19] https://www.w3.org/TR/vocab-ssn/.
[20] https://tools.ietf.org/html/rfc2119.

It is obvious that QUDT and OM meet the requirements **R01**, **R02** and **R03**. Both ontologies are distributed under public licence CC BY 4.0[21], new releases of these ontologies are made available every few months, and they define a broad range of units, quantities and dimensions. For the rest of the requirements, we provide a more detailed discussion in the following paragraphs.

R04: *A candidate ontology MUST be specified using OWL 2 language under direct, model-theoretic semantics.* This requirement is a consequence of implementing the Information Model in OWL 2 and applying OWL 2 (DL) reasoners such as HermiT[22] to verify the consistency of the ontologies and to compute their class and property hierarchies (which would be impractical when using OWL 2 RDF-based Semantics[23]) during the build process (cf. Sect. 4). The requirement is fulfilled by OM but not by QUDT. We identified several issues in the QUDT ontologies, as well as imported ontologies such as DTYPE v1.2. These issues are mostly related to the fact that DTYPE and QUDT follow OWL 2 RDF-based Semantics instead of OWL 2 Direct Semantics. For instance, under OWL 2 Direct Semantics, it is not possible to implement cardinality restrictions on transitive properties.[24] The property qudt:isScalingOf is defined as being transitive and it is used in the definition of the class qudt:ScaledUnit together with a cardinality restriction. Additional typing restrictions defined under OWL 2 Direct Semantics[25] are not followed by QUDT and DTYPE. For instance, a property cannot be an object and a data property in the same set of axioms. We observed that DTYPE and QUDT ontologies define several properties of type rdf:Property which is a superclass of owl:ObjectProperty and owl:DataProperty. In addition to the issues regarding semantics, we also detected several syntactic errors. For instance, the definition of the class qudt:PhysicalConstant includes a cardinality restriction over the annotation property qudt:latexDefinition, which is not valid OWL 2 syntax.

R05: *A candidate ontology MUST be compliant with "de facto" reference tools for OWL 2 such as (Web-)Protégé and OWLAPI.* The requirement is needed to support the team responsible of developing and maintaining the Information Model. They use (Web-)Protégé[26] to manually inspect ontologies and the OWLAPI[27] to support some of the services responsible for building and validating new releases of the Information Model (cf. Sect. 4). The requirement is fulfilled by OM but not by QUDT. In the case of QUDT, it is not a surprise given the problems we reported with requirement R04.

R06: *A candidate ontology MUST ensure acceptable reasoning performance using the state-of-the-art reasoner HermiT.* The requirement is relevant for releasing, (manually) inspecting and applying our ontologies (cf. requirement

---

[21] https://creativecommons.org/licences/by/4.0/legalcode.

[22] http://www.hermit-reasoner.com/.

[23] https://www.w3.org/TR/owl2-primer/#OWL_2_DL_and_OWL_2_Full.

[24] https://www.w3.org/TR/owl-ref/#TransitiveProperty-def.

[25] https://www.w3.org/TR/owl2-syntax/#Typing_Constraints_of_OWL_2_DL.

[26] https://protege.stanford.edu/.

[27] https://github.com/owlcs/owlapi.

R05 and Sect. 4). Acceptable reasoning performance should be achieved on a commodity PC. Reasoning tasks are not limited to consistency checking and classification, they also include materialisation of inferred class and property assertions. Since QUDT does not fulfil requirements R04 and R05, it does not meet this requirement. Neither does OM. Classifying OM takes more than 25 min for HermiT (v1.4.3.456) running on a Dell Latitude laptop with i5-6300U CPU (2.50 Ghz), 16 GB RAM and Windows 10 (64 bit). Materialisation of class and property assertions is even more time consuming. After more than five hours, HermiT could not complete these reasoning tasks. The problem is related to the definition of classes for units. Instead of explicitly asserting units (represented by named individuals) into classes of units, this is done by defining complex class restrictions involving nominals and disjunctions. In particular, we noticed that it is very difficult to reason with subclasses of compound and prefixed units due to the definition of specific equivalent-classes restrictions. After simplifying the definitions of the unit classes in OM, Hermit completed classification and materialisation of class and property assertions in less than one minute.

R07: *A candidate ontology MUST provide well-defined ontology design patterns for quantity values, quantity kinds and units.* Well-defined ontology design patterns [2] are instrumental when implementing and maintaining large industrial ontologies such as the Information Model. Definition of classes for quantity values, quantity kinds, units and dimensions should include class restrictions that clearly state the expected class and property assertions for individuals of these classes. OM and QUDT nearly achieve this requirement. In the case of OM, it includes well-defined ontology design patterns for quantity values, quantity kinds and dimension vectors. However, we missed the expected class restrictions in the classes om:Measure, om:Quantity, and om:Dimensions. Class restrictions can be found in the definition of the subclasses of om:Quantity and om:Dimensions. The subclasses of om:Quantity also provide the necessary class restrictions for individuals of the class om:Measure. The definition of the class om:Unit does not include any class restrictions either. This is due to the different nature of the subclasses of units defined by the designers of OM.

QUDT includes well-defined ontology design patterns for quantity values, units, quantity kinds and dimensions. These design patterns are clearly specified by the respective classes qudt:QuantityValue, qudt:Unit as well as the classes qudt:QuantityKind and qudt:QuantityKindDimensionVector. QUDT also includes the class qudt:Quantity that represents a relation between a phenomenon (object or event), a quantity kind and a collection of quantity values of the same kind (similar to individuals of type om:Quantity). The specification is not clear about which property relates a phenomenon with a quantity, but we suspect it is qudt:hasQuantity.

R08: *A candidate ontology MUST be optimised for dimensional analysis using SPARQL.* The requirement is essential for practical applications based on semantic technologies. This includes conversion of units, verification of dimensional homogeneity and discovery of base quantities (and units) from derived quantities (and units). QUDT fulfils this requirement, OM does not (not entirely). OM and QUDT provide an excellent support for the verification of dimensional homogeneity and the identification of base quantities (and units) from derived quantities (and units). This is possible thanks to the notion of *dimension vector*, representing a derived dimension, defined by ISQ as the product of powers of the seven base dimensions. Both ontologies define each dimension vector as a named individual asserted to the exponents of each base dimension. OM and QUDT define seven data properties, each related to one base dimension. By adding the exponents of each base dimension, it is possible to verify if two derived quantity kinds or two complex expressions involving powers of products of derived quantity kinds are commensurable. This can be done with a relatively simple ASK SPARQL query that will return true if the base dimensions of two expressions are the same.

Unit conversion using SPARQL is relatively simple in QUDT but not in OM. Computing the conversion of two commensurable units defined by QUDT is done by using the values of two data properties: qudt:conversionMultiplier and qudt:conversionOffset. The values of these properties determine how the magnitude of a quantity value can be converted to a base (or reference derived) unit. For instance, the conversion multiplier and offset of the unit kilowatt hour (unit:KiloW-HR) are $3.6e^6$ and 0.0. These values are defined with respect to the reference derived unit Joule (unit:J), which is of the same kind, and has as conversion multiplier 1.0 and offset 0.0. Therefore, converting a magnitude defined in kilowatt hour into joules is as simple as multiplying by $3.6e^6$.

The case for unit conversion using SPARQL on OM is not as simple, for two main reasons. The first is that the conversion factor (or multiplier) and offset are not available for all derived units. Consequently, to obtain the necessary conversion factors and offsets for a given unit, we must traverse the RDF graph until we find a unit that has one or two of these properties. The second reason is that OM defines several types of units with different properties. This means that we must add several optional graph patterns to deal with all these cases in a SPARQL query. This can be easily verified by trying to convert again a magnitude defined in kilowatt hour (om:kilowattHour) into joules (om:joule). The unit om:kilowattHour does not include any conversion factor in its definition. Therefore, we need to find the information in the definition of its constituent units starting from om:kilowatt and om:hour.

R09: *A candidate ontology MUST be compatible with ISO 15926-14 upper ontology.* The requirement is the result of using ISO 15926-14 as our reference upper ontology, which is influenced by BFO. The ontologies OM and QUDT are both compatible with ISO 15926-14. The ontology design patterns defined by OM for representing quantity values, quantity kinds and units fit well to ISO 15926-14. Dimensions are not explicitly covered by the ISO 15926-14 ontology, but they can be represented as individuals of a subclass of iso:InformationObject. QUDT does not meet the requirement as well as OM, because it uses a different ontology design pattern for representing quantity kinds. Contrary to ISO 15926-14 and OM, QUDT models quantity kinds only as individuals and not as classes. These individuals do not represent a relation between a particular phenomenon (such as an object or event) and a collection of quantity values, as in OM and ISO 15926-14.

R10: *A candidate ontology SHOULD enable mapping definitions with related terms defined by external data sources using SKOS.* We often use SKOS to define mappings between locally defined terms and related terms defined by external sources. Neither OM, nor QUDT provide appropriate support for SKOS. However, it is possible to adapt these ontologies to fulfil the requirement. In fact, QUDT uses SKOS properties to define semantic relations between QUDT terms and to specify mappings with terms defined by Wikipedia and DBpedia. However, we observed that these properties are not always consistently applied and, in general, the use of SKOS must be revised to make QUDT fully compliant with the W3C recommendation. For instance, the quantity kind quantitykind:LuminousEnergy is related to the quantity kind quantitykind:RadiantEnergy, using the SKOS mapping property skos:closeMatch. None of these terms are defined as SKOS concepts and are not grouped in specific SKOS concept schemes, which is expected when using SKOS properties.

R11: *A candidate ontology SHOULD be developed as a coherent collection of modules.* The requirement is related to well-known best practices when developing large industrial ontologies such as the Information Model. QUDT meets this requirement, but OM does not. The latter is released as a single ontology, whereas QUDT is released as a collection of 52 ontologies. We appreciate the strict separation between the QUDT upper ontology, SCHEMA_QUDT-v2.1.ttl, and the remaining files. Units, quantity kinds, dimensions and physical constants are also defined in separate ontologies.

The result of our analysis of OM and QUDT shows that both ontologies are, in principle, suitable for the harmonisation and contextualisation of quantity values and the implementation of services for dimensional analysis. However, we also concluded that neither of these ontologies fulfil all of our requirements. This appears to be a common situation that private and public organisations may frequently face when attempting to incorporate external resources into their ontology libraries. This usually ends up in the decision to build a new ontology from scratch, where certain term definitions are manually copied from existent ontologies, edited and finally inserted into the new ontology. As reported by Skjæveland et al. [6], the manual creation and maintenance of ontologies is expensive and error-prone. The problem is exacerbated by the fact that the ontologies QUDT and OM are updated often. To reduce time and development costs of creating and maintaining a new ontology of units of measure based on QUDT and OM that better suits our needs, and at the same time, while preserving interoperability, we must consider carefully the design, implementation and build phases. In the design phase, we must identify, which modelling choices and ontology design patterns better suit our needs. This is the aim of the next section.

3   Modelling Choices and Design Patterns

Ontology design patterns (ODPs) [2] define relevant and best-practice conceptual building blocks for implementing high-quality ontologies. ODPs ensure consistent and uniform modelling of similar terms and facilitate understanding of large ontologies, which is instrumental for reducing maintenance costs. As reported during the assessment of requirement R07, QUDT and OM provide well-defined modelling patterns (with few exceptions) for each of their fundamental concepts (i.e. quantity values, dimensions, quantity kinds and units of measure). Therefore, QUDT and OM modelling patterns represent a good starting point when defining the modelling patterns for the new ontology of units of measure. The assessment of requirement R08 indicates that the modelling pattern for representing dimensions (very similar in both ontologies) facilitates the verification of dimensional homogeneity using SPARQL. The same is not true when computing unit conversion using SPARQL, where the modelling pattern of QUDT for representing units of measure has a clear advantage. In addition, the definition of classes of units in OM, particularly subclasses of prefixed and compound units must be updated to avoid inefficient reasoning (cf. assessment of requirement R06).

Before discussing specific modelling patterns for representing quantity values (Sect. 3.2), dimensions (Sect. 3.3), quantity kinds (Sect. 3.4) and units of measure (Sect. 3.5), we will briefly introduce some general design decisions regarding modular structure and the use of upper ontologies (Sect. 3.1).

3.1   General Considerations

The Information Model follows a strict dependency hierarchy, where the ontologies ISO 15926-14 and SKOS are at the top. It is important to take this into consideration when defining the modular structure of the new ontology of units of measure and how it is integrated with the upper and domain independent ontologies. These decisions will contribute to a better fulfilment of the requirements R09, R10 and R11 discussed in the previous section.

Modular Structure. As with QUDT, we stored the definitions of quantity kinds, (types of) quantity values, dimensions and units of measure in different ontologies (with different namespaces). Moreover, we placed common term definitions (of classes and properties) in a dedicated *core* ontology, that is imported by other ontologies. These ontologies were released as an ontology library (UoM library, for short) organised following a strict dependency hierarchy, where ontologies can only refer to terms defined in ontologies located in upper levels. Respect a strict dependency hierarchy in the UoM library may be difficult because a term in one ontology, such as a specific unit, may refer to terms in different ontologies, such as a particular quantity kind or dimension. To avoid cross references between ontologies at the same level, we separate term declarations from term definitions and we store them in different files. Files containing the term declarations import the core ontology and they are, in turn, imported by the ontologies that include the definitions of these terms. In the remaining document, we refer to defined terms using the following prefixes: uomcore (common terms), uomqd (types of quantity values), uomqk (quantity kinds), uomqkdv (dimensions) and uomunit (units).

ISO 15926-14 and SKOS Ontologies. It is important that the ontologies of the UoM library are well aligned with the ISO 15926-14 and SKOS ontologies. The alignment with the ISO 15926-14 ontology is relatively easy to achieve, because QUDT and OM are already quite compatible. For instance, the ISO 15926-14 classes iso:PhysicalQuantity, iso:QuantityDatum and iso:UnitOfMeasure correspond well with the classes in QUDT and OM that define quantity kinds, (types of) quantity values (or measurements) and units of measure. The case of quantity kinds is somewhat more challenging as the design pattern adopted for the ISO 15926-14 ontology is similar to its counterpart in OM but not to the one in QUDT. The class for defining dimension vectors (defined similarly by QUDT and OM) is not supported by the ISO 15926-14 ontology, but it can be specified as a subclass of iso:InformationObject.

Just below the ISO 15926-14 ontology, the Information Model includes the ontology SKOS as part of the ontology (import) hierarchy. SKOS has been adopted by Aker Solutions, ix3, Sesam and DNV GL to facilitate interoperability with external resources. To define SKOS mappings between terms in the ontology library (local terms) and terms used by external resources (external terms), we apply SKOS semantic properties such as skos:Broader. This requires that all mapped terms must be defined as SKOS concepts and associated to one or more SKOS concept schemes, which implies that these terms are being declared as OWL individuals. When they are not, we apply *punning* to produce the necessary OWL individuals.

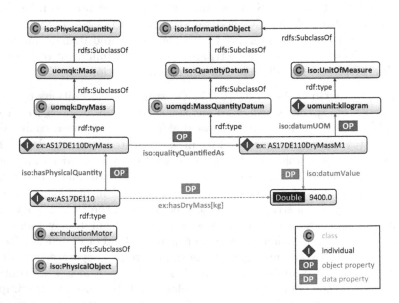

Fig. 1. Modelling example

The modelling example in Fig. 1 describes a simplified scenario illustrating some of the modelling choices we considered when building a new ontology of units of measure. The individual ex:AS17DE110 represents an induction motor that has the physical quantity of type uomqk:DryMass associated with it (the prefix ex: refers to a namespace for examples). As an example of the modelling pattern for representing quantity kinds as classes (cf. Sect. 3.4), the relation is defined using the ISO 15926-14 object property iso:hasPhysicalQuantity and the individual ex:AS17DE110DryMass. A quantity value representing a measurement of the dry mass of the induction motor is represented by the individual ex:AS17DE110DryMassM1. Notice that we used named individuals instead of anonymous individuals for physical qualities and values to make the example easier to understand. We will discuss this example in the following subsections.

3.2   Quantity Values

A quantity value is represented as a relation that contains at least one numerical value (or magnitude) and one unit of measure. This is illustrated by the individual ex:AS17DE110DryMassM1 in Fig. 1, where the data property iso:datumValue states the numerical value and the object property iso:datumUOM indicates the unit of measure. In addition, we extended this relation to include a timestamp that indicates when the quantity value was generated and the uncertainty of the quantity value commonly represented by the (relatively) standard uncertainty (both properties are defined in the core ontology and are not included in Fig. 1 for the sake of simplicity). Therefore, a suitable approach for representing n-ary relations in OWL 2 was required. Among the different available modelling approaches, we followed Noy et al. [4] and defined quantity values as individuals. Notice that QUDT and OM represent quantity values using the same modelling approach.

There are two issues to note. First, the question of whether individuals representing quantity values should be defined as named or anonymous individuals. This is not clearly stated in ISO 15926-14, OM and QUDT, and the decision may depend on the application. In the case of the Integral platform, named individuals make sense, because the application Sesam is responsible for detecting changes in the external data sources and updating the data accordingly. Therefore, the numerical value does not determine the identity of a quantity value. However, tracking quantity values is not a simple task, and it requires substantial contextual information around each quantity value. This can be costly in terms of storage and computing resources. The second issue is determining the suitability of this modelling approach, i.e. modelling quantity values as individuals, when there are many values to be represented or when values stem from time series data. In the former case, there are simpler modelling approaches, where quantity values are represented using data property assertions (cf. Fig. 1, ex:hasDryMas[kg]) as done similarly in the SOSA/SSN ontology using the data property sosa:hasSimpleResult.[28]. We also considered the latter case, time series data, where different modelling patterns can be applied, but this is not discussed here.

------

[28] https://www.w3.org/TR/vocab-ssn/#SOSAhasSimpleResult.

3.3   Dimension Vectors

A dimension vector, representing the product of powers of the seven base dimensions, is defined using an individual and seven data property assertions, where each data property corresponds to one base dimension. This is similar to the way dimension vectors are represented in OM and QUDT. This representation facilitates the verification of dimensional homogeneity using SPARQL queries, as discussed earlier (cf. requirement R08).

We decided to adapt the naming conventions specified by QUDT for labelling dimension vectors. For instance, the acceleration dimension, defined by the expression LT^{-2}, is represented by the label 'N0I0L1J0M0H0T-2', where the numbers correspond to the powers of each base dimension. For the data properties stating the power of each base dimension, we followed the naming convention adopted by OM. All individuals representing dimension vectors are asserted to the class uomcore:Dimension which is defined as a subclass of iso:InformationObject. For reasons of simplicity, we do not include dimensions in the modelling example depicted in Fig. 1.

3.4   Quantity Kinds

The question of how to model quantity kinds has created a significant controversy between the authors of this work. This is also a source of disagreement for the designers of OM and QUDT, as they followed different design patterns for representing quantity kinds. We considered the following design patterns: *quantity kinds as individuals*, *quantity kinds as classes* and *quantity kinds as properties*. For a more in-depth discussion about these three design patterns, we recommend interested readers the overview given by Rijgersberg et al. [5].

Quantity Kinds as Individuals. This is the preferable modelling choice in QUDT. In our ontology, we also defined quantity kinds as individuals, as in QUDT, but these individuals are defined as SKOS concepts and related to specific SKOS concept schemes. The purpose of this approach is to map our quantity kinds to similar terms defined by external data sources using SKOS semantic properties (cf. requirement R10). This is also expected by ontologies following ISO 15926-14 (cf. requirement R09).

All quantity kinds in QUDT (e.g. quantitykind:DRY-MASS) are defined using named individuals and asserted to the class qudt:QuantityKind. To indicate that a quantity kind, such as quantitykind:DRY-MASS, is more general (or more specific) than other quantity kinds, such as quantitykind:Mass, QUDT uses the properties qudt:generalization and skos:broader (or qudt:specialization and skos:narrower). In our example, the individual ex:AS17DE110DryMass would be asserted to the class qudt:Quantity and it would be related to the quantity kind quantitykind:DRY-MASS using the object property qudt:hasQuantityKind.

Quantity Kinds as Classes. This is the preferable modelling choice in OM and also implemented in our ontology of units of measure. This is also the recommended design pattern for representing quantities in ISO 15926-14 and BFO upper ontologies (cf. requirement R09). Both upper ontologies define qualities as classes representing *dependent continuants* [1], meaning that the existence of a quality or a physical quantity depends on the existence of its bearer. For instance, in the modelling example depicted in Fig. 1, the induction motor represented by the individual ex:AS17DE110 has the physical quantity dry mass and this is stated by a property assertion with the individual ex:AS17DE110DryMass.

As a result of this modelling approach, we created a hierarchy of classes representing quantity kinds, where iso:PhysicalQuantity is the top class. Subclass relations for quantity kinds are determined by the notion of dimensional homogeneity. For instance, the class uomqk:DryMass is defined as a subclass of uomqk:Mass. This is because both have the same dimension, represented by the individual (not included in Fig. 1), and uomqk:DryMass is more specific than uomqk:Mass.

Quantity Kinds as Properties. This design pattern was supported in earlier releases of OM [5] and is the preferred representation in the Sesam datahub. As in the case of quantity kinds represented as individuals, this design pattern does not follow ISO 15926-14 (or BFO) but it does not prevent integration with this upper ontology (cf. requirement R09). As with quantity kinds implemented as classes, we can create a hierarchy of properties representing quantity kinds, where iso:hasPhysicalQuantity is the top property. Subproperty relations for quantity kinds are also determined by the notion of dimensional homogeneity. For instance, the object property uomqk:hasDryMass is defined as a subproperty of uomqk:hasMass, as both have the same dimension. A clear advantage of this pattern is that many users are more comfortable working with properties than classes, where properties define meaningful relations between objects (or events) and quantity values. For instance, an object representing a particular induction motor is related to a quantity value representing a measurement of its dry mass using the property uomqk:hasDryMass.

3.5   Units of Measure

As discussed earlier, unit conversion using SPARQL is easier in QUDT than in OM (cf. requirement R08). Therefore, we decided to define units of measure using the same conversion multipliers and offsets. We explicitly indicated which unit is related to the conversion values in the definition of each unit, using the object property uomcore:hasReferenceUnit. QUDT is also used as the main reference for the identifiers of the units even though OM defines more intuitive identifiers, especially for simple units. This is because the identifiers defined in QUDT are more compact, particularly for compound units.

The designers of QUDT made a reasonable attempt at defining a generic class of units of measure, qudt:Unit. We adapted this by improving or removing some class restrictions. This includes, for instance, deleting unneeded references to latex or MathML expressions, simplifying the definition of units. In addition, we included all the subclasses of qudt:Unit defined by QUDT and we extended this class hierarchy with the OM class om:CompoundUnit and its direct subclasses.

The definitions have been simplified to improve reasoning performance, as discussed (cf. requirement R06).

To facilitate the verification of dimensional homogeneity (i.e. if two units are commensurable) using SPARQL (cf. requirement R08), we adopted the approach implemented by OM that relates each unit to a suitable dimension vector. In addition, we related each unit of measure to quantity kinds of the same dimensions and to relevant base quantities and units (when defining derived units only).

4   Implementation and Build Process

After analysing and selecting suitable modelling choices for the design of the UoM library, we focus on how to optimise its implementation by reducing manual efforts, facilitating collaborative development and ensuring uniform modelling. One of the key ingredients to achieving this is the use of ontology templates for representing ontology design patterns in the form of parameterised ontologies [6]. In this section, we discuss how we produced a tailor-made UoM library for the Integral platform, based on QUDT and OM ontologies, where it is instrumental to fulfil requirements R04, R05 and R06. Figure 2 outlines the main phases of the process, including *Preprocessing*, *Creation* and *Build & Deployment*. The input and output (results) of each phase are represented in the upper part and the tools used to produce the output in the lower part.

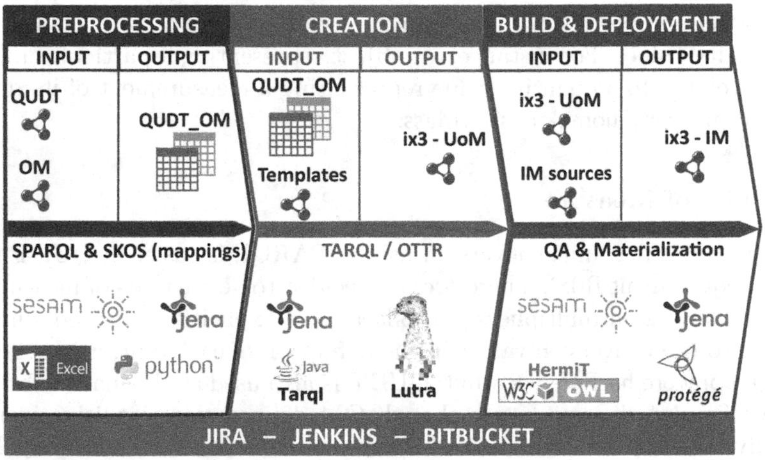

Fig. 2. Implementation and build process

Pre-processing. The assessment of the requirement R07 helped us to identify the relevant design patterns in QUDT and OM. These patterns provide a useful guideline when defining the graph patterns of the SELECT SPARQL queries we executed to extract information from QUDT and OM. The new design patterns, defined for the UoM Library, also helped us to establish the output of these queries.

The results of the execution of these SELECT SPARQL queries were stored in several Comma-separated values (CSV) files (one for each design pattern and represented in Fig. 2 by the tables *QUDT_OM*). These files were processed further to make them suitable for populating the UoM ontology library using ontology templates. For instance, each term stored in the CSV files received an identifier, an IRI. Then new columns were added to enrich the information and to state, for instance, that each relevant term is defined as an OWL named individual and as a SKOS concept. This process was done using common tools such as Microsoft Excel, Python and Jena.

The main challenge was, and still is, to efficiently align and combine information from the ontologies OM and QUDT. To do this, we created a specific CSV file where similar terms in OM and QUDT had been identified and mapped using SKOS semantic properties. We applied algorithms (in Python) for detecting similar strings in the labels of each term. The resulting mapping list was manually refined, which is an expensive process. We expect that new versions of Sesam data hub can soon take care of this process and significantly reduce manual work.

Creation (from Tarql to OTTR). After we collected, refined and stored QUDT and OM data in CSV files, we generated the various ontologies of the UoM library by instantiating ontology templates. Currently, we are using Tarql,[29] to define templates (as SPARQL queries) and instantiate these templates (by running these queries). We adopted Tarql due to its simplicity, but discovered several limitations when building a library or reusable templates. Therefore, we turned our attention to Reasonable Ontology Templates (OTTR) [6], a language for representing ontology templates. In combination with the tool Lutra,[30] it is possible to instantiate OTTR templates from tabular data and produce RDF graphs and OWL ontologies. In the future, it may be possible to define OTTR templates for querying data from QUDT and OM ontologies and avoid the use of CSV files. We defined several OTTR templates to represent the design patterns associated to quantity kinds, units, dimension vectors and quantity values. We are currently testing these OTTR templates with data from QUDT and OM, and the preliminary results look promising. We hope we can shortly introduce OTTR templates when building the Information Model at ix3.

Build and Deployment. ix3 has designed and implemented an infrastructure and a build process to deliver and deploy new releases of the Information Model that includes the UoM library. Changes in the source files of the Information Model are requested and discussed in Jira,[31] a tool for issue tracking and project

---

[29] https://tarql.github.io/.

[30] https://gitlab.com/ottr/lutra/lutra.

[31] https://www.atlassian.com/software/jira.

management. Jira is integrated with Jenkins[32] and Bitbucket,[33]. Jenkins takes care of the distribution and deployment of new releases of the Information Model, and Bitbucket hosts the different files needed to build the Information Model and facilitating the collaborative edition of these files.

During the build process of a new release of the Information Model, several quality assurance and refining tasks are executed. This includes, for instance, the validation of the syntax of the different RDF files produced using the tool Jena. Consistency of the OWL ontologies delivered in the release is tested using the HermiT reasoner, and the classification of the ontology (class and property hierarchies) is computed and materialised (cf. requirements R04, R05 and R06). Materialisation of subclass and subproperty relations becomes valuable when executing SPARQL in the triple stores where the Information Model resides. In addition, class and property assertions are inferred and materialised to speed up SPARQL queries. Manual inspection using (Web) Protégé is also done as part of the quality assurance process (cf. requirement R05).

5   Conclusion

In this paper, we discuss how we extended the Information Model, the ontology library supporting the ix3 Integral platform for digital twins, to better support the contextualisation and harmonisation of quantity values and units of measures. To minimise development costs, we tried to reuse the ontologies QUDT and OM. Although they are excellent resources, they did not meet all our requirements. Therefore, we decided to create a new ontology based on QUDT and OM. We started by analysing suitable ontology design patterns and we implemented the selected modelling patterns using ontology templates. By applying these templates and specific SPARQL queries as part of the implementation and build processes at ix3, we reduced manual effort and we ensured uniform modelling of the new ontology of units of measure. The latest release of this ontology is currently integrated and deployed as part of the Integral platform. Despite the resulting ontology not being released under a public licence, we can share our experience gained with this work. This is to help practitioners who aim to incorporate and adapt external resources to their ontology libraries, and who seek to benefit from using ontologies for contextualisation and harmonisation of quantities and units of measure.

References

1. Arp, R., Smith, B., Spear, A.D.: Building Ontologies with Basic Formal Ontology. MIT Press, Cambridge (2015)
2. Gangemi, A., Presutti, V.: Ontology design patterns. In: Staab, S., Studer, R. (eds.) Handbook on Ontologies. IHIS, pp. 221–243. Springer, Heidelberg (2009). https://doi.org/10.1007/978-3-540-92673-3_10

---

[32] https://www.jenkins.io/.
[33] https://bitbucket.org/.

3. Keil, J.M., Schindler, S.: Comparison and evaluation of ontologies for units of measurement. Semant. Web J. **10**(1), 33–51 (2019)
4. Noy, N., Rector, A., Hayes, P., Welty, C.: Defining N-ary relations on the semantic web (W3C working note). Technical report, World Wide Web Consortium (W3C) (2006). https://www.w3.org/TR/swbp-n-aryRelations/
5. Rijgersberg, H., van Assem, M., Top, J.: Ontology of units of measure and related concepts. Semant. Web J. **4**(1), 3–13 (2013)
6. Skjæveland, M.G., Lupp, D.P., Karlsen, L.H., Forssell, H.: Practical ontology pattern instantiation, discovery, and maintenance with reasonable ontology templates. In: Vrandečić, D., et al. (eds.) ISWC 2018. LNCS, vol. 11136, pp. 477–494. Springer, Cham (2018). https://doi.org/10.1007/978-3-030-00671-6_28

NEO: A Tool for Taxonomy Enrichment with New Emerging Occupations

Anna Giabelli[1,3] (ID), Lorenzo Malandri[1,3(✉)] (ID), Fabio Mercorio[1,3(✉)] (ID),
Mario Mezzanzanica[1,3] (ID), and Andrea Seveso[2,3] (ID)

[1] Department of Statistics and Quantitative Methods, University of Milano Bicocca,
Milan, Italy
[2] Department of Informatics, Systems and Communication,
University of Milano-Bicocca, Milan, Italy
[3] CRISP Research Centre, University of Milano-Bicocca, Milan, Italy
{anna.giabelli,lorenzo.malandri,fabio.mercorio,
mario.mezzanzanica,andrea.seveso}@unimib.it

Abstract. Taxonomies provide a structured representation of semantic relations between lexical terms, acting as the backbone of many applications. This is the case of the online labour market, as the growing use of Online Job Vacancies (OJVs) enables the understanding of how the demand for new professions and skills changes in near-real-time. Therefore, OJVs represent a rich source of information to reshape and keep labour market taxonomies updated to fit the market expectations better. However, manually updating taxonomies is time-consuming and error-prone. This inspired NEO, a Web-based tool for automatically enriching the standard occupation and skill taxonomy (ESCO) with new occupation terms extracted from OJVs. NEO - which can be applied to any domain - is framed within the research activity of an EU grant collecting and classifying OJVs over all 27+1 EU Countries.

As a contribution, NEO (i) proposes a metric that allows one to measure the pairwise semantic similarity between words in a taxonomy; (ii) suggests new emerging occupations from OJVs along with the most similar concept within the taxonomy, by employing word-embedding algorithms; (iii) proposes GASC measures (Generality, Adequacy, Specificity, Comparability) to estimate the adherence of the new occupations to the most suited taxonomic concept, enabling the user to approve the suggestion and to inspect the skill-gap. Our experiments on 2M+ real OJVs collected in the UK in 2018, sustained by a user-study, confirm the usefulness of NEO for supporting the taxonomy enrichment task with emerging jobs. A demo of a deployed instance of NEO is also provided.

The research leading to these results is partially supported within the EU Project AO/DSL/VKVET-GRUSSO/Real–time LMI 2/009/16 granted by the EU Cedefop Agency, in which some authors are involved as PI and researchers. *All authors equally contributed to this work.*

© Springer Nature Switzerland AG 2020
J. Z. Pan et al. (Eds.): ISWC 2020, LNCS 12507, pp. 568–584, 2020.
https://doi.org/10.1007/978-3-030-62466-8_35

1   Introduction and Motivation

Over the past several years, the growth of web services has been making available a massive amount of structured and semi-structured data in different domains. An example is the web labour market, with a huge number of Online Job Vacancies (OJVs)[1] available through web portals and online applications. The problem of processing and extracting insights from OJVs is gaining researchers' interest in the recent years, as it allows modelling and understanding complex labour market phenomena (see, e.g. [8,9,11,16,20,33,34]). At the same time, the ability to extract valuable knowledge from these resources strongly depends on the existence of an *up-to-date* target taxonomy. Those resources are essential for machine understanding and many tasks in natural language processing. In the European labour market, the key resource is ESCO.[2] Organisations and governments are making a great effort in keeping ESCO up-to-date with the labour market through expert knowledge. This challenging task needs an automated, scalable method capable of enriching ESCO with new terms from the job market.

Unlike the automated construction of new taxonomies from scratch, which is a well-established research area [31], the augmentation of existing hierarchies is gaining in importance, given its relevance in many practical scenarios (see, e.g. [30]). Human languages are evolving, and online contents are constantly growing. As a consequence, people often need to enrich existing taxonomies with new concepts and items, without repeating their whole construction process every time. To date, the most adopted approach to enrich or extend standard *de-jure* taxonomies lean on expert panels and surveys, that identify and validate which term has to be added to a taxonomy. The approach totally relies on human knowledge, with no support from the AI-side, and this makes the process costly, time-consuming, and unable to consider the peculiarities of country-specific labour markets. To extract semantic information from the OJVs, we resort to *distributional semantics*, a branch of linguistics based on the hypothesis that words occurring in similar context tend to have similar meaning [17]. Words are represented by semantic vectors, which are usually derived from a large corpus using co-occurrence statistics or neural network training, and their use improves learning algorithms in many NLP tasks. Semantic word vectors have empirically shown to preserve linguistic regularities [22], demonstrating their ability to enrich existing knowledge structures as well [12,25].

Contribution. In this paper we design and develop NEO, a novel system for enriching the ESCO taxonomy with mentions as they appear in the real labour market demand. A *novel occupation*, indeed, is a term that deserves to be represented within the taxonomy, that might represent either an emerging job (e.g.,

---

[1] An Online Job Vacancy (OJV, *aka,* job offers, job ads) is a document containing a *title* - that shortly summarises the job position - and a *full description*, usually used to advertise the skills a candidate should hold.

[2] European Commission: ESCO: European skills, competences, qualifications and occupations, available at https://ec.europa.eu/esco/portal/browse(2019).

SCRUM master) or a new alternative label characterising an existing job (e.g., *Android developer*). The novelty of NEO goes toward three directions:

- Define a domain-independent metric, i.e., the *Hierarchical Semantic Related-ness* (HSR) to measure the pairwise semantic similarity between words in a taxonomy;
- Synthesise and evaluate, with the supervision of the HSR, vector-space models for encoding the lexicon of both the taxonomy and the OJVs, in order to extract from the latter potential new emerging occupations, and define a set of measures, namely GASC (Generality, Adequacy, Specificity, Comparability) to estimate their suitability as entities of different taxonomic concepts;
- Provide to final users a Web-based tool to vote suggested mentions, supporting the experts in the taxonomy extension activity, and explaining the rationale behind each suggested new occupation through a skill-gap analysis.

The project - in which NEO is framed within - aims at realising a European system to collect and classify OJVs for the whole EU, including 27+1 EU country members and all the 32 languages of the Union [10] through machine learning. The use of classified OJVs and skills, in turn, enables a number of third-party research studies to understand complex labour market phenomena. Just to give a few examples, in [11] authors used OJVs to estimate the impact of AI in job automation and to measure the impact of digital/soft skills within occupations, validating theoretical results from [15]; in reaction to the COVID-19 emergency, the EU Cedefop Agency has been using those OJVs to build the *Cov19R* index for identifying workers with a higher risk of COVID exposure, who need greater social distancing, affecting their current and future job performance capacity.[3]

Though the ESCO taxonomy is a standard, it encodes neither the peculiarities nor the characteristics of countries' labour markets, that vary in terms of skills requested, the lexicon used for advertising similar jobs, and country-related economic conjunctures. This, as a consequence, sharply limits the usability of ESCO as a system for understanding and representing different labour markets.

The Relevance of NEO for the Labour Market. The following example should help in clarifying the matter. Figure 1 shows the structure of the ESCO occupation pillar: it is built on top of the ISCO taxonomy down to the fourth digit (i.e., 436 occupation codes). Then, it continues with 2,942 ESCO occupation concepts and up to 30,072 alternative labels as well. *How to maintain the taxonomy up-to-date to adhere to labour market lexicon? How to enrich the taxonomy with those mentions representing novel occupations? How to estimate similarities between occupations?* Those are just few questions with which economists and policymakers have to deal. The inspiring idea of NEO is to build a Web-based tool for supporting labour market specialists in enriching the taxonomy to better fit the labour market demand of new occupations.

The remainder of the paper is organised as follows. In Sect. 2 we survey related works and formalise the taxonomy enrichment problem and solution. In

---

[3] https://tinyurl.com/covid-r.

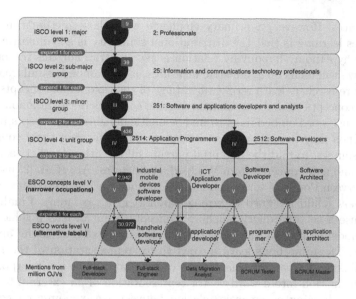

Fig. 1. *Motivating example.* Representation of the ESCO taxonomy, with new mentions representing novel jobs not yet included in ESCO as they emerge from the UK Web Labour Market Demand (2M+ vacancies processed in 2018).

Sect. 3 we describe NEO in three conceptual steps: (i) learn word embeddings while preserving taxonomic relations, (ii) suggest new entities for ESCO and (iii) evaluate the fitting of the new entities through GASC measures and by expert evaluation. Those steps are implemented in a real case scenario presented in Sect. 4. Finally, Sect. 5 concludes the paper and draws future work; a demo of NEO is also provided.

2 Related Work and Preliminaries

In the past literature, despite the automatic construction of taxonomies from scratch has received a considerable attention [31], the same cannot be said for augmentation of existing hierarchies. Most of the work in the area of automated taxonomy enrichment relies heavily on or domain specific knowledge [5,14] or lexical structures specific to an existing resource, like the WordNet synset [18,26,29] or Wikipedia categories [28]. In recent years few scholars tried to overcome those limitations developing methodologies for the automated enrichment of generic taxonomies. Wang et al. [32] use a hierarchical Dirichlet model to complete a hierarchy with missing categories. Then they classify elements of a corpus with the supervision of the complete taxonomy. For our purposes, this work has two shortcomings: the authors i) modify the structure of the taxonomy, while we want to preserve the ESCO framework and ii) do not update the hierarchical categories with new entities, which is the main goal of our tool. Other scholars [3,13] exploit semantic patterns between hypernyms and hyponyms in word vector spaces. However, a primary challenge with those methods in semantic hierarchies learning is that the distributional similarity is a symmetric measure, while the hypernomy-hyponymy relation is asymmetric. For this reason, in this research we will focus on symmetric measures, like taxonomic similarity and *rca*.

Other researchers learn term embeddings of the taxonomic concepts and connect new concepts to the most similar existing concepts in the taxonomy. Vedula et al. [30] use word embeddings to find semantically similar concepts in the taxonomy. Then they use semantic and graph features, some of them coming from external sources, to find the potential parent-child relationship between existing concepts and new concepts from Wikipedia categories. Aly et al. [2] use the similarity between Poincaré term embeddings to find *orphans* (disconnected nodes) and *outliers* (child assigned to wrong parents) in a taxonomy. Finally, in [27] authors use a set of <query, anchor> concepts from an existing hierarchy to train a model to predict the parent-child relationship between the anchor and the query. Those methods learn a word vector representation of the taxonomy, without linking it to an external corpus of web data, while we incorporate taxonomic information into a word vector representation of an external text corpus. This allows drawing a semantic relation between a taxonomic concept and a mention.

2.1   Setting the Stage

In this section we introduce a formal definition of taxonomy and we formulate the problem of taxonomy enrichment, relying on the formalisation proposed by [19].

Definition 1 (Taxonomy).   *A taxonomy T is a 4-tuple $T = (C, W, \mathcal{H}^c, \mathcal{F})$. C is a set of concepts $c \in C$ (aka, nodes) and W is a set of words (or entities) belonging to the domain of interest; each word $w \in W$ can be assigned to none, one or multiple concepts $c \in C$. $\mathcal{H}^c$ is a directed taxonomic binary relation existing between concepts, that is $\mathcal{H}^c \subseteq \{(c_i, c_j) | (c_i, c_j) \in C^2) \wedge i \neq j\}$. $\mathcal{H}^c(c_1, c_2)$ means that c_1 is a sub-concept of c_2. The relation $\mathcal{H}^c(c_1, c_2)$ is also known as $IS - A$ relation (i.e., c_1 $IS - A$ sub-concept of c_2). $\mathcal{F}$ is a directed binary relation mapping words into concepts, i.e. $\mathcal{F} \subseteq \{(c, w) | c \in C \wedge w \in W\}$. $\mathcal{F}(c, w)$ means that the word w is an entity of the concept c. Notice T might be represented as a DAG.*

Given an existing taxonomy T, the goal of NEO is to expand T with new *mentions* (entities) coming from a text corpus. Each mention is assigned to one or multiple candidate destination nodes of T, along with a score value and a set of measures. More formally we have the following.

Definition 2 (Taxonomy Enrichment Problem (TEP)).   *Let T be a taxonomy as in Definition 1, and let $\mathcal{D}$ be a corpus; a Taxonomy Enrichment Problem (TEP) is a 3-tuple $(\mathcal{M}, \mathcal{H}^m, \mathcal{S})$, where:*

- $\mathcal{M}$ is a set of mentions extracted from $\mathcal{D}$, i.e., $m \in \mathcal{M}$;
- $\mathcal{S} : \mathcal{W} \times \mathcal{M} \rightarrow [0, 1]$ is a scoring function that estimates the relevance of m with respect to an existing word w. Ideally, the scoring function might consider the frequency of m, as well as the similarity between m and w according to $\mathcal{D}$.
- $\mathcal{H}^m \subseteq \{(c, m) | c \in \mathcal{C} \wedge m \in \mathcal{M}\}$ is a taxonomic relation (edge) existing between a $<concept, mention>$ pair. Intuitively, $\mathcal{H}^m$ models the pertinence of m to be an entity of the existing concept c;

A solution to TEP computed over $\mathcal{D}$ is a 7-tuple $\mathcal{T}^{\mathcal{D}} = (\mathcal{C}, \mathcal{W}, \mathcal{H}^c, \mathcal{F}, \mathcal{M}, \mathcal{H}^m, \mathcal{S})$.

3   How Does NEO Work?

In this section we describe our overall approach to enriching ESCO with new emerging occupations, that is mainly composed by three steps: i) *learn word embeddings* ii) *suggest new entities* iii) *vote and enrich* as shown in Fig. 2.

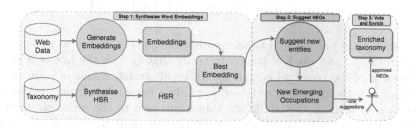

Fig. 2. A representation of the NEO workflow highlighting the main modules.

3.1   Step 1: Synthesise Word Embeddings

The first step requires to learn a vector representation of the words in the corpus to preserve the semantic relationships expressed by the taxonomy itself. To do that, we rely on three distinct sub-tasks, that are the following.

Step 1.1: Generation of embeddings. NEO employs and evaluates three of the most important methods to induce word embeddings from large text corpora. One, GloVe [23], is based on co-occurrence matrix factorisation while the other two, Word2Vec [21], and FastText [6], on neural networks training. Notice that FastText considers sub-word information, and this allows one to share information between words to represent rare words, typos and words with the same root.

Step 1.2: Hierarchical Semantic Relatedness (HSR). Semantic relatedness is a well-known measure for the intrinsic evaluation of the quality of word embeddings, developed in [4]. It evaluates the correlation between a measure of similarity between two terms used as the gold standard and the cosine similarity between their corresponding word vectors. A common way to build the gold standard is by human evaluation [4]. In many cases, however, this task is difficult, time-consuming, and it might be inaccurate.

Below we introduce a measure of relatedness in a semantic hierarchy, based on the concept of information content. Intuitively, the lower the rank of a concept c which contains two entities, the higher the information content the two entities share. Building on [24], we can supplement the taxonomy with a probability measure $p : C \to [0, 1]$ such that for every concept $c \in C$, $p(c)$ is the probability of encountering an instance of the concept c. From this definition, p (i) is monotonic and (ii) decreases with the rank of the taxonomy, that is if c_1 is a sub-concept of c_2, then $p(c_1) \leq p(c_2)$. According to information theory, the self-information of a concept $c \in C$ can be approximated by its negative log-likelihood defined as:

$$\mathcal{I}(c) = -\log p(c) \tag{1}$$

We can define the relatedness between concepts of the semantic hierarchy as:

$$sim(c_1, c_2) = \max_{c \in Z(c_1, c_2)} \mathcal{I}(c) = \mathcal{I}(\ell_{c_1, c_2}) \tag{2}$$

where $Z(c_1, c_2)$ is the set of concepts having both c_1 and c_2 as sub-concepts. Given (i), (ii) and Eq. 1, it is easy to verify that ℓ_{c_1, c_2} is the Lowest Common Ancestor of the concepts c_1 and c_2. To estimate the values of p, in [24] the author uses the frequency of the concepts in a large text corpus. Anyway, our purpose is to infer the similarity values intrinsic to the semantic hierarchy. Since we want to extend a semantic hierarchy built by human experts, we adopt those values as a proxy of human judgements. As a consequence, we use the frequencies of the concepts in the taxonomy to compute the values of p.

$$\hat{p}(c) = \frac{N_c}{N} \tag{3}$$

where N is the cardinality, i.e. the number of entities (words), of the taxonomy, and N_c is the sum of the cardinality of the concept c with the cardinality of all its sub-concepts. Note that $\hat{p}(c)$ is monotonic and increases with granularity, thus respects our definition of p. Now, given two words $w_1, w_2 \in W$, we define $Z(w_1)$ and $Z(w_2)$ as the sets of concepts containing w_1 and w_2 respectively, i.e. the *senses* of w_1 and w_2. Therefore, given a pair of words w_1, w_2, there are $N_{w_1} \times N_{w_2}$ possible combinations of their word senses, where N_{w_1} and N_{w_2} are the cardinality of $Z(w_1)$ and $Z(w_2)$ respectively. We refer to $\mathcal{L}$ as the set of all the Lowest Common Ancestor ℓ_{c_1, c_2} for all the combinations of $c_1 \in Z(w_1), c_2 \in Z(w_2)$.

Hence, the hierarchical semantic relatedness between the words w_1 and w_2 is:

$$HSR(w_1, w_2) = \sum_{\ell \in \mathcal{L}} \frac{N_{<w_1, w_2> \in \ell}}{N_{w_1} \times N_{w_2}} \times \mathcal{I}(\ell) \tag{4}$$

where $N_{<w_1,w_2>\in\ell}$ is the number of pairs of senses of word w_1 and w_2 which have ℓ as lowest common ancestor.

Step 1.3. Word Embedding Selection. Finally, the performance of each word vector model generated in Step 1.1 is assessed by the Spearman Correlation of the HSR between all the pairs of words in the taxonomy with the cosine similarity between their vectors in the model space. The Spearman Correlation coefficient can be interpreted as a measure of fidelity of the vector model to the taxonomy.

3.2   Step 2: Suggest New Entities

Step 2 is aimed at extracting new occupation terms from the corpus of OJVs, and at suggesting the most suitable concepts under which they could be positioned in $\mathcal{T}$. To do this, NEO works in two distinct steps shown in PseudoCode 1: first, it extracts a set of mentions $\mathcal{M}$ from the corpus $\mathcal{D}$ of OJVs; then, it proposes a set of measures, namely GASC (Generality, Adequacy, Specificity, Comparability) to estimate the suitability of a mention $m \in \mathcal{M}$ as an entity of the concepts in $\mathcal{C}$.

PseudoCode 1 NEO

Require: $\mathcal{T}(\mathcal{C},\mathcal{W},\mathcal{H}^c,\mathcal{F})$ as in Def. 1; $\mathcal{D}$ dataset;
1: $\mathcal{E} \leftarrow$ best_embedding$(\mathcal{D},\mathcal{T})$ ❭ *Step1*
2: $\mathcal{M} \leftarrow \varnothing$ //init the set of mentions
3: **for all** $w \in \mathcal{W}$ **do**
4:     $\mathcal{M} \leftarrow \mathcal{M} \cup most\_similar(\overrightarrow{\mathcal{E}[w]},\mathcal{S})$ //ordered according to $\mathcal{S}$ of Eq.5
5: **for all** $m \in \mathcal{M}$ **do**   *Step2*
6:     $\mathbf{G}_m \leftarrow$ compute Eq.6
7:     **for all** $c \in \mathcal{C}$ **do**
8:        $\mathbf{S}_{m,c}, \mathbf{A}_{m,c}, \mathbf{C}_{m,c} \leftarrow$ compute Eq.7, 8, 9
9:        $\mathcal{H}^m \leftarrow$ user_eval$(m, c, \mathbf{G}_m, \mathbf{A}_{m,c}, \mathbf{S}_{m,c}, \mathbf{C}_{m,c})$ ❭ *Step3*
10: **return** $(\mathcal{M}, \mathcal{H}^m)$

Step 2.1: Extract new mentions from the corpus. First we select a starting word w_0 from the taxonomy. Then we consider the top-5 mentions in $\mathcal{D}$ with associated the highest *score* value $\mathcal{S}$ with w_0, where the score $\mathcal{S}(m,w)$ of the mention m w.r.t. the generic word w is defined as:

$$\mathcal{S}(m,w) = \alpha \cdot cos\_sim(m,w) + (1 - \alpha) \cdot freq(m) \tag{5}$$

where $cos\_sim(m,w)$ is the cosine similarity between the mention m and the word w in the word embedding model $\mathcal{E}$ selected in Sect. 3.1, while $freq(m)$ is the frequency of the mention m in the corpus. We concentrate on the most important terms, (i) computing the score value only for the top-k most similar mentions, (ii) filtering out the words which are rarely used in the OJVs. To do this, we compute the cumulative frequency of $freq(m)$ and we keep only the mentions determining the 95% of the cumulative.[4]

Step 2.2: Suggest the best entry concepts for the new mention.
Once $\mathcal{M}$ is synthesised, the most suitable concepts are identified on the basis of four measures, namely GASC (_Generality_, _Adequacy_ _Specificity_, and _Comparability_), that estimate the fitness of a concept c for a given mention m.

Generality and Specificity. The _Generality_ (**G**) of a mention m measures to which extent the mention's embedding is similar to the embeddings of all the words in the taxonomy $\mathcal{T}$ as a whole, in spite of the concept. Conversely, the _Specificity_ (**S**) between the mention m and the concept c measures to which extent the mention's embedding is similar to the embeddings that represent the words associated to concept c in $\mathcal{E}$. They are defined as follows.

$$(6)\ \mathbf{G}_m = \frac{1}{|\mathcal{W}|} \sum_{w \in \mathcal{W}} \mathcal{S}(m,w) \qquad\qquad \mathbf{S}_{m,c} = \frac{1}{|\mathcal{F}(c,w)|} \sum_{w \in \mathcal{F}(c,w)} \mathcal{S}(m,w)\ (7)$$

Adequacy. The _Adequacy_ (**A**) between m and c estimates the fitting of the new mention m, extracted from the corpus, to the ESCO concept c, on the basis of the vector representation of m and the words $w \in \mathcal{F}(c,w)$, i.e. their use in the OJVs corpus. **A** is computed as:

$$\mathbf{A}_{m,c} = \frac{e^{\mathbf{S}_{m,c}} - e^{\mathbf{G}_m}}{e-1} \in [-1,1] \tag{8}$$

On one side, the _Adequacy_ of a mention m to the concept c is directly proportional to the similarity with the other words $w \in \mathcal{F}(c,w)$ (i.e., the _Specificity_ to the concept c). On the other side, the _Adequacy_ is also inversely proportional to the similarity of m with all the words $w \in W$ (i.e., its _Generality_). The _Adequacy_ is defined to hold the following:

$$\mathbf{A}_{m,c} \begin{cases} \overset{\geq}{<} 0 & \text{if } \mathbf{S}_{m,c} \overset{\geq}{<} \mathbf{G}_m \\ > \mathbf{A}_{m_2,c_2} & \text{if } \mathbf{S}_{m,c} - \mathbf{G}_m = \mathbf{S}_{m_2,c_2} - \mathbf{G}_{m_2} \wedge \mathbf{S}_{m,c} > \mathbf{S}_{m_2,c_2} \end{cases}$$

The first property guarantees zero to act as a threshold value, that is, a negative value of **A** indicates that the mention is related more to the taxonomy, rather than that specific concept c. Conversely, a positive **A** indicates the mention m might be a sub-concept of c. The second property guarantees that given two pairs of concepts and mentions - e.g. (m, c) and (m_2, c_2) - if the difference between their _Specificity_ and _Generality_ values is the same, then the pair having the higher _Specificity_ will also have a higher value of _Adequacy_, still allowing NEO to distinguish between the two.

Comparability. To better investigate the comparability of the new mention m with the existing ESCO concepts, we consider their required skills. The skills

---

[4] k is set to $1,000$ whilst α is set to 0.85 to weight the frequency less than the similarity.

are identified in the context of [10] in the OJVs' descriptions, and classified using the ESCO skills/competencies pillar. Let us consider a set K_c of skills associated to the occupations belonging to the concept c in the OJVs, and a set K_m of skills associated to the mention $m \in \mathcal{M}$ in the OJVs. Given the set $K_U = K_c \cup K_m$ of the L skills associated with at least one out of m and c, we define two L-dimensional vectors $\mathbf{t_c} = (t_{c1}, \ldots, t_{cL})$ and $\mathbf{t_m} = (t_{m1}, \ldots, t_{mL})$ where the generic elements t_{cl} and t_{ml} represent the *revealed comparative advantage* (*rca*) [5] of skill k_l for c and m respectively. If $k_l \notin K_c, t_{cl} = 0$, and similarly if $k_l \notin K_m, t_{ml} = 0$. Given these vectors $\mathbf{t_c}$ and $\mathbf{t_m}$, the *Comparability* (**C**) between the concept c and the mention m is defined as:

$$\mathbf{C}_{m,c} = \frac{\sum_{l=1}^{L} \min(t_{ml}, t_{cl})}{\sum_{l=1}^{L} \max(t_{ml}, t_{cl})} \tag{9}$$

The *Comparability* represents a method to assess the similarity between an ESCO occupation and a potentially new one not on the basis of their vector representation, but on the basis of their characteristics in the domain of interest.

3.3   Step 3: Vote and Enrich

Finally, we engage labour market experts to validate the outcome of Sect. 3.1 and Sect. 3.2, which are fully automated. The user evaluation is composed of two questions. We ask to evaluate *Q1*) if the mentions extracted from the corpus in Step 2.1 are valid emerging occupations and *Q2*) to which extend the concepts suggested as entry for a new mention are appropriate for it, basing on the name of the mention and the concepts and their skill-gap. We recall that a *novel occupation* is a term that deserves to be represented within the taxonomy, as it might represent either an emerging job or a new alternative label characterising an existing job. For *Q1* the user is asked to give a yes/no answer, while *Q2* is evaluated using a 1–6 Likert scale (from 1: *Completely disagree*, to 6: *Completely agree*). The user feedback is used as a judgement of response quality, meaning that a high evaluation of the best proposed suggestion implies a high quality of suggestion. In the study, we select 12 of the most representative ESCO ICT occupations, i.e. taxonomic entities. For each of them NEO, according to Step 2.1, suggests 5 new mentions, for a total of 60, and the expert evaluates whether they can be considered terms representing emerging occupations or not (*Q1*). Then, for each one of the 60 suggestions, NEO proposes three candidate concepts where to place the new mention. The first is the concept of starting word, and the other two are those with the highest *Adequacy*, as computed in Step 2.2, among the remaining. The experts evaluate the appropriateness of the proposed mentions for those three concepts (*Q2*).

---

[5] The *rca* $\in [0, +\infty]$ was introduced in 2018 in [1] to assess the relevance of skills in the US taxonomy O*Net. We adapted the *rca* to work on ESCO as well.

4   Experimental Results

Experimental Settings. The corpus contains 2,119,025 OJVs published in the United Kingdom during the year 2018, referring to the ESCO ICT positions reported in Table 1, and classified as we specified in [7,9]. OJV's titles were preprocessed applying the following pipeline: (1) tokenisation, (2) lower case reduction, (3) punctuation and stopwords removal (4) n-grams computation.

We deployed NEO over the UK dataset following the workflow of Sect. 3.1.

Table 1. OJVs collected from UK in 2018. Only Information and Communication Technology (ICT) occupation codes are shown.

ISCO code	Occupation description	OJVs number
1330	ICT service managers	176,863
2511	Systems analysts	402,701
2512	Software developers	740,112
2513	Web and multimedia developers	225,784
2514	Applications programmers	30,383
2519	Software and applications developers and analysts	44,339
2521	Database designers and administrators	42,305
2522	Systems administrators	45,542
2523	Computer network professionals	15,411
2529	Database and network professionals	110,210
3511	ICT operations technicians	44,585
3512	ICT user support technicians	168,705
3513	Computer network and systems technicians	55,222
3514	Web technicians	5,708
3521	Broadcasting and audiovisual technicians	11,121

4.1   Step 1: Synthesise Word Embeddings

We trained space vector models using various architectures: *Word2Vec*, *GloVe* and *FastText*, generating 260 models. Hyperparameter selection for each architecture was performed with a grid search over the following parameter sets:

- *Word2Vec* (80 models): Algorithm $\in$ {SG, CBOW} $\times$ HS $\in$ {0, 1} $\times$ embedding size $\in$ {5, 20, 50, 100, 300} $\times$ number of epochs $\in$ {10, 25, 100, 200};
- *GloVe* (20 models): embedding size $\in$ {5, 20, 50, 100, 300} $\times$ number of epochs $\in$ {10, 25, 100, 200};
- *FastText* (160 models): Algorithm $\in$ {SG, CBOW} $\times$ embedding size $\in$ {5, 20, 50, 100, 300} $\times$ number of epochs $\in$ {10, 25, 100, 200} $\times$ learning rate $\in$ {0.01, 0.05, 0.1, 0.2}

Average training times (with std) in seconds were 890±882 for *Word2Vec*, 55±74 for *GloVe* and 246±333 for *fastText*, running on an Intel i-7 CPU equipped with 32GB RAM. An intrinsic evaluation - as detailed in Step 1.3 - has been performed to select the embedding that better preserves taxonomic relations, by computing the Spearman correlation of the cosine similarity between each couple of occupations and their corresponding HSR. The model with highest correlation, with $\rho = 0.29$ and $p\_value \simeq 0$, has the following parameters: architecture $= fastText$, algorithm $=$ CBOW, size $= 300$, epochs $= 100$, learning rate $= 0.1$. Figure 5 provides a scatter plot produced over the best embedding model - as emerges from Table 1 - generated by means of UMAP. Each icon is assigned to one ISCO level 4 group, as in Fig. 1. The ESCO concepts and words belonging to each group are showed, distinguishing between narrower occupations (shallow shape) and alternative labels (filled shape). Focusing on Fig. 5 one might observe that though a *data engineer* and a *data scientist* were designed to be sub-concepts in ESCO, as they belong both to the ▼*2511: System Analyst* ISCO group, their meaning is quite different in the real-labour market, as any computer scientist knows. The former indeed shares much more with ■ *2521: Database designers and administrators* rather than its theoretical group. Conversely, in many practical cases, the taxonomy perfectly adheres to the real labour market demand for occupations. This is the case of ◆ *3521: Broadcasting and audio-visual technicians*, that composes a very tight cluster in the map, showing a perfect match between de-facto and de-jure labour market occupations. This also applies to ∗ *3513: Computer network and systems technicians*, even though in a lesser extent.[6]

4.2   Step 2: Suggest New Emerging Occupations

As a first step, the user selects the starting word w_0 among the occupations already in ESCO (*data analyst* in the example in Fig. 3). Then, NEO prompts the 5 mentions with associated the highest *score* with w_0 (Fig. 3a). The user can therefore select a mention m (*business system analyst*) to evaluate to which extent the mention fits as an entity of the starting word's ESCO concept c_j and as an entity of other two ESCO concepts $c_l, c_k \in C \setminus c_j$, that are those with associated the highest value of *Adequacy* with the mention m (*ict business analyst* and *ict system analyst* in Fig. 3). For each one of these three pairs mention m and ESCO concept, NEO provides the GASC measures (see Fig. 3b). For each pair NEO provides comparison of the *rca* of skills for both the mention and the concept (Fig. 3b). These skills, together with the GASC measures, support the domain expert in evaluating if the suggested entry is appropriate or not as an entity of a concept, as thoroughly explained in Sect. 4.3.

---

[6] Both best/worst embeddings are available at https://tinyurl.com/worst-neo and https://tinyurl.com/best-neo respectively.

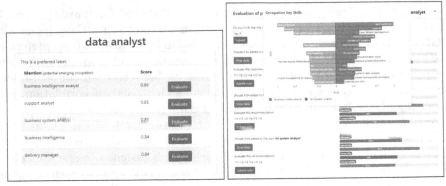

(a) New mentions from OJVs starting from the word *data analyst*.

(b) Vote the best class for the new mention selected (*business system analyst*).

Fig. 3. NEO suggests new mentions from the OJV corpus.

4.3   Step 3: Vote and Enrich with User Evaluation

In order to evaluate the effectiveness of NEO we recruited 10 among ML engineers and labour market experts involved in the development of the ML-based system within [10], but not in this research activity. We asked to ten experts to evaluate the system as detailed in Sect. 3.3.

Q1: **Does NEO suggest valid new emerging occupations?** In *Q1* we ask to the voters whether a suggested mention can be considered an occupation or not. Out of 60 proposed mentions, 11 are repeated starting from different words. For the remaining 49 unique mentions, 6 of them were evaluated to not be proper occupations, according to the majority of the votes. This means that 88% of the occupations were successfully evaluated to be new occupations. Though 6 out of 49 mentions did not pass the test, they are strongly related to the starting concept, referring to skills requested by those job profiles.[7] Figure 4e shows the new occupations found by NEO and the median of Likert scores of experts along with the ESCO concept suggested by NEO and approved by experts.

Q2: **To which extent the new mentions fit the suggested taxonomic concepts?** To assess the significance of our GASC measures, we use two well-known hypothesis tests, the Spearman's ρ and the Kendall's τ coefficient, that proved to be effective in the labour market domain (see, e.g. [33]). The correlation values are shown in Table 2 while the distribution of the GASC measures grouped according to values of the Likert scale is shown in Fig. 4(a–d). The association between the Likert values and the corresponding *Adequacy, Specificity*, and *Comparability* is positive, and hypothesis tests indicate that it is statistically significant. The strongest correlation is between Likert values

---

[7] The mentions evaluated not to be proper occupations are: data analytics, business intelligence, penetration testing, operation, data management, drupal.

and *Comparability*, indicating this is the measure on which the experts relied more. Conversely, the association between the Likert values and the *Generality* isn't statistically significant, coherently with the nature of *Generality* that does not aim to rank concepts with respect to a mention.

In summary, our results - sustained by an expert user study - show that NEO is able (i) to accurately identify novel occupations and (ii) to put them in the right place within the taxonomy. This, in turn, makes NEO a tool for supporting the process of identification of new emerging occupations enriching the taxonomy accordingly, taking into account the real labour market demand.

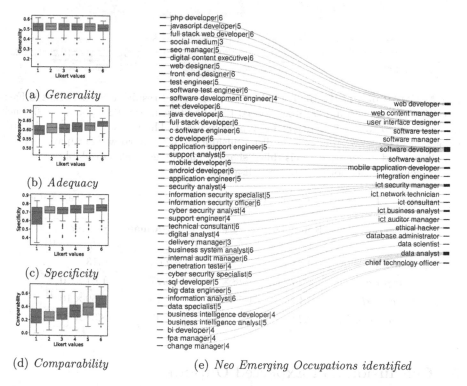

(a) *Generality*

(b) *Adequacy*

(c) *Specificity*

(d) *Comparability*

(e) *Neo Emerging Occupations identified*

Fig. 4. *(left-side)* Box-plots representing the distribution of *Generality*, *Adequacy*, *Specificity* and *Comparability* grouped for each value of the Likert scale. *(right-side)* Alluvial diagram showing the mentions recognised as New Emerging Occupations with the median of Likert values (i.e., neo|score) and the corresponding ESCO concept suggested by NEO and validated by experts.

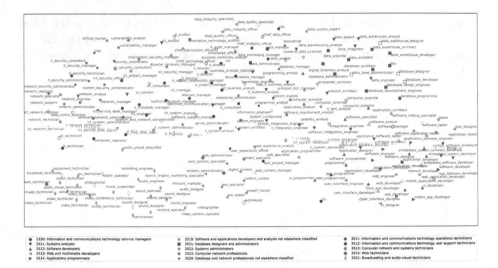

Fig. 5. UMAP plot of the **best** word-embedding model resulting Step 1, that is Fast-Text, CBOW algorithm, Learning rate $= 0.1$, embedding size $= 100$, epochs $= 100$. Each icon is assigned to one ISCO level 4 group, as in Fig. 1. The ESCO concepts and words belonging to each group are shown distinguishing between narrower occupations (shallow shape) and alternative labels (filled shape). The image is also available at https://tinyurl.com/best-neo for a better visualisation.

Table 2. The results of correlation analysis between GASC and Likert values.

Measure	Kendall's τ	p-value $(H_0 : \tau = 0)$	Spearman's ρ	p-value $(H_0 : \rho = 0)$
Generality	-0.03	0.14	-0.04	0.13
Adequacy	0.20	2.48×10^{-21}	0.27	1.61×10^{-21}
Specificity	0.14	1.59×10^{-11}	0.19	2.59×10^{-11}
Comparability	0.34	2.21×10^{-60}	0.45	8.33×10^{-62}

5   Conclusion and Expected Outlook

In this paper, we proposed NEO, a tool framed within the research activities of an ongoing EU grant in the field of Labour Market Intelligence. NEO has been deployed on a set of 2M+ real OJVs collected from UK in 2018 within the project. NEO synthesised and evaluated more than 240 vector space models, identifying 49 novel occupations, 43 of which were validated as novel occupations by a panel of 10 experts involved in the validation of the system. Two statistical hypothesis tests confirmed the correlation between the proposed GASC metrics of NEO and the user judgements, and this makes the system able to accurately identify novel occupations and to suggest an IS-A relation within the taxonomy.

We are working to scale NEO over multiple country-datasets and occupations, as well as to apply the approach proposed by NEO to other domains.

DEMO. A demo video is provided at https://tinyurl.com/neo-iswc-demo.

References

1. Alabdulkareem, A., Frank, M.R., Sun, L., AlShebli, B., Hidalgo, C., Rahwan, I.: Unpacking the polarization of workplace skills. Sci. Adv. **4**(7), eaao6030 (2018)
2. Aly, R., Acharya, S., Ossa, A., Köhn, A., Biemann, C., Panchenko, A.: Every child should have parents: a taxonomy refinement algorithm based on hyperbolic term embeddings. arXiv preprint arXiv:1906.02002 (2019)
3. Anh, T.L., Tay, Y., Hui, S.C., Ng, S.K.: Learning term embeddings for taxonomic relation identification using dynamic weighting neural network. In: EMNLP (2016)
4. Baroni, M., Dinu, G., Kruszewski, G.: Don't count, predict! A systematic comparison of context-counting vs. context-predicting semantic vectors. In: ACL (2014)
5. Bentivogli, L., Bocco, A., Pianta, E.: ArchiWordNet: integrating wordnet with domain-specific knowledge. In: International Global Wordnet Conference (2004)
6. Bojanowski, P., Grave, E., Joulin, A., Mikolov, T.: Enriching word vectors with subword information. TACL **5**, 135–146 (2017)
7. Boselli, R., et al.: WoLMIS: a labor market intelligence system for classifying web job vacancies. J. Intell. Inf. Syst. **51**(3), 477–502 (2018). https://doi.org/10.1007/s10844-017-0488-x
8. Boselli, R., Cesarini, M., Mercorio, F., Mezzanzanica, M.: Classifying online job advertisements through machine learning. Future Gener. Comput. Syst. **86**, 319–328 (2018)
9. Boselli, R., Cesarini, M., Mercorio, F., Mezzanzanica, M.: Using machine learning for labour market intelligence. In: Altun, Y., et al. (eds.) ECML PKDD 2017. LNCS (LNAI), vol. 10536, pp. 330–342. Springer, Cham (2017). https://doi.org/10.1007/978-3-319-71273-4_27
10. CEDEFOP: Real-time labour market information on skill requirements: setting up the EU system for online vacancy analysis (2016). https://goo.gl/5FZS3E
11. Colombo, E., Mercorio, F., Mezzanzanica, M.: AI meets labor market: exploring the link between automation and skills. Inf. Econ. Policy **47**, 27–37 (2019)
12. Efthymiou, V., Hassanzadeh, O., Rodriguez-Muro, M., Christophides, V.: Matching web tables with knowledge base entities: from entity lookups to entity embeddings. In: d'Amato, C., et al. (eds.) ISWC 2017. LNCS, vol. 10587, pp. 260–277. Springer, Cham (2017). https://doi.org/10.1007/978-3-319-68288-4_16
13. Espinosa-Anke, L., Camacho-Collados, J., Delli Bovi, C., Saggion, H.: Supervised distributional hypernym discovery via domain adaptation. In: EMNLP (2016)
14. Fellbaum, C., Hahn, U., Smith, B.: Towards new information resources for public health–from WordNet to MedicalWordNet. J. Biomed. Inform. **39**(3), 321–332 (2006)
15. Frey, C.B., Osborne, M.A.: The future of employment: how susceptible are jobs to computerisation? Technol. Forecast. Soc. Change **114**, 254–280 (2017)
16. Giabelli, A., Malandri, L., Mercorio, F., Mezzanzanica, M.: GraphLMI: a data driven system for exploring labor market information through graph databases. Multimed. Tools Appl. (2020). https://doi.org/10.1007/s11042-020-09115-x. ISSN 1573-7721

17. Harris, Z.S.: Distributional structure. Word **10**(2–3), 146–162 (1954)
18. Jurgens, D., Pilehvar, M.T.: Reserating the awesometastic: an automatic extension of the WordNet taxonomy for novel terms. In: ACL, pp. 1459–1465 (2015)
19. Maedche, A., Staab, S.: Ontology learning for the semantic web. IEEE Intell. Syst. **16**(2), 72–79 (2001)
20. Mezzanzanica, M., Boselli, R., Cesarini, M., Mercorio, F.: A model-based approach for developing data cleansing solutions. J. Data Inf. Qual. (JDIQ) **5**(4), 1–28 (2015)
21. Mikolov, T., Chen, K., Corrado, G., Dean, J.: Efficient estimation of word representations in vector space. arXiv preprint arXiv:1301.3781 (2013)
22. Mikolov, T., Sutskever, I., Chen, K., Corrado, G.S., Dean, J.: Distributed representations of words and phrases and their compositionality. In: NIPS (2013)
23. Pennington, J., Socher, R., Manning, C.: GloVe: global vectors for word representation. In: EMNLP, pp. 1532–1543 (2014)
24. Resnik, P.: Semantic similarity in a taxonomy: an information-based measure and its application to problems of ambiguity in natural language. JAIR **11**, 95–130 (1999)
25. Ristoski, P., Paulheim, H.: RDF2Vec: RDF graph embeddings for data mining. In: Groth, P., et al. (eds.) ISWC 2016. LNCS, vol. 9981, pp. 498–514. Springer, Cham (2016). https://doi.org/10.1007/978-3-319-46523-4_30
26. Schlichtkrull, M., Alonso, H.M.: MSejrKu at SemEval-2016 Task 14: taxonomy enrichment by evidence ranking. In: SemEval, pp. 1337–1341 (2016)
27. Shen, J., Shen, Z., Xiong, C., Wang, C., Wang, K., Han, J.: TaxoExpan: self-supervised taxonomy expansion with position-enhanced graph neural network. In: WWW, pp. 486–497 (2020)
28. Sumida, A., Torisawa, K.: Hacking Wikipedia for hyponymy relation acquisition. In: IJCNLP (2008)
29. Toral, A., Monachini, M.: Named entity wordnet. In: LREC (2008)
30. Vedula, N., Nicholson, P.K., Ajwani, D., Dutta, S., Sala, A., Parthasarathy, S.: Enriching taxonomies with functional domain knowledge. In: SIGIR (2018)
31. Wang, C., He, X., Zhou, A.: A short survey on taxonomy learning from text corpora: issues, resources and recent advances. In: EMLP, pp. 1190–1203 (2017)
32. Wang, J., Kang, C., Chang, Y., Han, J.: A hierarchical dirichlet model for taxonomy expansion for search engines. In: WWW, pp. 961–970 (2014)
33. Xu, T., Zhu, H., Zhu, C., Li, P., Xiong, H.: Measuring the popularity of job skills in recruitment market: a multi-criteria approach. In: AAAI (2018)
34. Zhang, D., et al.: Job2Vec: job title benchmarking with collective multi-view representation learning. In: CIKM, pp. 2763–2771 (2019)

Domain-Specific Customization of Schema.org Based on SHACL

Umutcan Şimşek(✉) ⑩, Kevin Angele⑩, Elias Kärle⑩, Oleksandra Panasiuk⑩,
and Dieter Fensel

University of Innsbruck, Innsbruck, Austria
{umutcan.simsek,kevin.angele,elias.kaerle,
oleksandra.panasiuk,dieter.fensel}@sti2.at

Abstract. Schema.org is a widely adopted vocabulary for semantic annotation of web resources. However, its generic nature makes it complicated for publishers to pick the right types and properties for a specific domain. In this paper, we propose an approach, a domain specification process that generates domain-specific patterns by applying operators implemented in SHACL syntax to the schema.org vocabulary. These patterns can support annotation generation and verification processes for specific domains. We provide tooling for the generation of such patterns and evaluate the usability of both domain-specific patterns and the tools with use cases in the tourism domain.

Keywords: SHACL · Schema.org · Semantic annotation · Domain-specific patterns

1 Introduction

schema.org [7] is currently the de facto standard for annotating resources on the web. The vocabulary is maintained by the schema.org initiative, and it contains 821 types and 1328 properties[1]. The schema.org vocabulary has a highly generic nature. On the one hand, the vocabulary covers many domains (e.g., events, media, accommodation) superficially, on the other hand, it does not cover individual domains in detail.

The data model of schema.org vocabulary is quite flexible in terms of type hierarchy and inheritance of properties by specialized types (e.g., a Waterfall can have a telephone number). Moreover, there are multiple ways to represent the same information (e.g., the address of a Place can be represented with three different properties in two different ways). This flexibility and heterogeneity come with two side effects for data publishers: (a) The generic nature of the vocabulary can make the creation of correct, complete and concise annotations quite challenging, especially within a specific domain. The publishers may not know which types and properties to select for their domain or may use different

---

[1] See https://schema.org/docs/schemas.html Last accessed: 16.04.2020.

© Springer Nature Switzerland AG 2020
J. Z. Pan et al. (Eds.): ISWC 2020, LNCS 12507, pp. 585–600, 2020.
https://doi.org/10.1007/978-3-030-62466-8_36

properties to represent the same information, and (b) the schema.org vocabulary may not contain specific types and properties for their domain. These two issues may have harming implications on the overall quality of the annotations and consequently the applications built on top of them. The completeness would be harmed by missing important domain-relevant information. Similarly, the conciseness of the annotations would be harmed since publishers could be more prone to use different properties to represent same information on an annotation (see coverage and succinctness quality dimensions [9, Sect. 7.2 and 7.4]).

One way to guide data publishers to create semantically annotated data and content for a domain is to provide domain-specific patterns of the schema.org vocabulary. They would serve as an agreement between publishers and consumers to ease the annotation generation and verification processes. In this paper, we present an approach for generating such patterns. Figure 1 depicts the domain specification process. We apply domain specification operators implemented with Shapes Constraint Language (SHACL) [10] syntax to the schema.org vocabulary to have *an extended subset of schema.org* for a domain[2]. The main reason for utilizing (a subset of) SHACL is to benefit from its widespread adoption and tool support. It is a W3C recommendation that would have a substantial impact on the uptake of our approach.

The remainder of the paper is structured as follows: In Sect. 2, we give a brief introduction to schema.org and our usage of SHACL. Section 3 describes the domain specification process and gives a running example. Section 4 briefly introduces the tool support for domain specifications and their usage in annotation generation. We evaluate the usability and benefit of domain-specific patterns and its tool support in Sect. 5 with use cases in tourism domain. Section 6 gives an overview of the related work and Sect. 7 concludes the paper with final remarks and future directions.

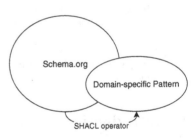

Fig. 1. The domain specification process

2   Preliminaries

In this section, we give an introduction to the data model of schema.org as well as SHACL, the syntax of which we use to define the domain specification operators.

---

[2] There is an automatically generated version of entire schema.org in SHACL maintained by TopQuadrant http://datashapes.org/schema.

2.1   Schema.org

The data model of schema.org is quite simple and "very generic and derived from RDF Schema."[3]. It does not have a formal semantics[4]. Nevertheless, it provides informal definitions and guidelines. Similar to an RDFS vocabulary, the schema.org data model organizes types in a multiple inheritance hierarchy. There are two disjoint hierarchies, namely the item types that are more specific than s:Thing and data types that are more specific than s:DataType[5]. The vocabulary contains properties that have one or more types in their domains and one or more types in their range definitions. A significant deviation from RDFS comes with the properties, mainly how their domains and ranges are defined. schema.org defines the domains and ranges with s:domainIncludes and s:rangeIncludes properties, respectively. The semantics of these properties are not formally defined. However, the documentation indicates that domain and range definitions are disjunctive, which is not the case for domain and range definitions with RDFS.

The data model of schema.org allows global ranges, meaning each range is valid for each domain of a property. However, in many domain-specific cases, local ranges could be instrumental (cf. [12]). For instance, a SportsEvent is in the domain of the location property. The property has a global range of Place. In a domain-specific scenario, it is not hard to imagine that a sports event takes place in a sports activity location. Therefore, defining the range of the location property on SportsEvent type as SportsActivityLocation may be a better modeling practice. With the domain specification approach, we allow the definition of schema.org types with local ranges on their properties.[6]

When creating annotations on the web, the range of a property may need to be altered with the conjunction of multiple types. A prominent example for this is the so-called multi-typed entity (MTE) practice for the annotation of hotel rooms[7]. The s:HotelRoom type contains properties for describing beds and amenity features. However, to describe a daily room price, the hotel room must also be defined as an instance of s:Product, which allows the usage of schema:offers property. The MTE practice is an excellent example of why customization of schema.org for specific domains is needed. For data publishers, it may be complicated to find out which types should be used for an MTE. The domain-specific patterns created by domain experts can enforce the conjunction of multiple types as the range of a property to guide data publishers.

---

[3] https://schema.org/docs/datamodel.html - accessed on 16.04.2020.

[4] Patel-Schneider gives an analysis of the vocabulary and a possible formal semantics in [12].

[5] s is a prefix for the http://schema.org/ namespace.

[6] This is also why we ignore the property hierarchy of schema.org since this is already implemented via local properties.

[7] See MTE documentation: https://tinyurl.com/s2l3btw.

2.2   Our Take on SHACL

SHACL is a W3C recommendation for defining constraints over RDF graphs. For the domain specification process proposed in this paper, we use a subset of SHACL-CORE[8] elements. We adopt three main types of SHACL elements, namely shapes, target selectors, and constraint components [10]:

- *Class based target selector (sh:targetClass)*, to specify the type on which the pattern is based
- *Node Shape with a target selector*, to specify the domain-specific pattern
- Shape-based Constraint Components to define local properties (sh:property) and range restrictions (sh:node)
- *Value Type Constraint Components (sh:class and sh:datatype)*, to define ranges on local properties
- *Logical Constraint Components (sh:or)* to define disjunctive ranges
- and various other constraint components to define constraints on local properties beyond range restrictions (see Sect. 3 for a complete list).

Despite adopting most of the SHACL-CORE components, the domain specification operators have stricter syntax rules. For instance, SHACL allows the usage of many constraint components on both node and property shapes, but we allow all constraint components only on property shapes. Moreover, we allow target definitions only on the node shapes that are not a value of sh:node property, in other words, only on the node shape that specifies the pattern. The property shapes are only allowed as a Shape based-constraint and not as standalone shapes[9]. In the next section, we explain how each SHACL element is used in the domain specification process.

3   Domain Specification

A domain expert creates a domain-specific pattern from schema.org and its extensions through the following actions:

1. Removing types and properties from schema.org that are not relevant for a specific domain.
2. Defining local properties and their ranges on the remaining types
3. Optionally, further restricting the types in the ranges of defined local properties
4. Optionally, extending the ranges with new types from schema.org and its extensions
5. Optionally, defining additional constraints on the local properties.

---

[8] *sh* prefix is used for the SHACL-Core namespace.
[9] An abstract syntax for our domain specification operators based on SHACL can be found online: https://tinyurl.com/qmkb3ln.

We explain the process mentioned above more concretely with a running example. A domain expert creates a domain specification operator that generates a domain-specific pattern for the accommodation domain. The process starts by eliminating all irrelevant types from schema.org. In this example, this action leaves us with the s:LodgingBusiness, s:Text, s:DateTime, schema:Place and s:PostalAddress and properties s:name, s:checkInTime, s:checkOutTime, s:containsPlace and s:location. Listing 1 shows the domain specification operator in the SHACL syntax. This operator defines five local properties and ranges on s:LodgingBusiness type[10]. The first two steps eliminate more than 100 properties that are allowed on s:LodgingBusiness by schema.org.

Additionally, some of the types in the property ranges are eliminated. A domain expert may want to describe the location of a lodging business with its address. The property has a disjunction of four types in its range, including the s:Place, s:PostalAddress, s:VirtualLocation and s:Text. The s:VirtualLocation type may not be desired for describing the location of an accommodation provider. s:Text is not expressive enough to make a granular description of an address. Both s:Place and s:PostalAddress types can be used to describe a postal address, but s:Place requires more properties than s:PostalAddress. Therefore a domain expert may eliminate the types other than s:PostalAddress from the range of s:location property.

A domain expert may choose to restrict further the types in the ranges of the local properties. Listing 2 shows such a restriction on two properties. The s:PostalAddress type in the range of the s:location property defined on s:LodgingBusiness is restricted further to a type that allows only s:addressLocality and s:addressCountry properties. Similarly, the range of s:containsPlace is restricted to a type that is a conjunction of s:HotelRoom (subtype of s:Place) and s:Product. The property s:containsPlace is defined by schema.org vocabulary very generically, to define a place that contains other places. In a specific domain like accommodation, domain experts may want to describe only hotel rooms and their offers. Therefore, they restrict the s:Place type in the range to s:HotelRoom and to allow the definition of offers, they create a conjunction with the type s:Product (see also the explanation of MTEs in Sect. 2.1).

A domain-specific pattern is an extended subset of schema.org; however, so far, we only defined a subset of the vocabulary. The schema.org vocabulary can be extended externally[11]. The extensions of schema.org are built following the same data model as schema.org and assumed to be hosted externally. The domain specification process can use external extensions of schema.org for:

- using a type from an extension to specify the pattern
- defining a property from an extension as a local property on a type
- adding types from an extension to the ranges of local properties

---

[10] Datatypes like s:Text and s:DateTime are mapped to xsd:string and xsd:datetime respectively for compatibility to SHACL syntax.

[11] See External Extensions: https://schema.org/docs/schemas.html.

```
@prefix sh: <http://www.w3.org/ns/shacl#>.
@prefix schema: <http://schema.org/>.
@prefix smtfy: <https://semantify.it/ds/> .
@prefix xsd: <http://www.w3.org/2001/XMLSchema#> .

smtfy:149vQ318v a sh:NodeShape;
    sh:targetClass schema:LodgingBusiness;
    schema:name "LodgingBusiness";
    sh:property [
        sh:path schema:name;
        sh:datatype xsd:string;
    ];
    sh:property [
        sh:path schema:checkInTime;
        sh:datatype xsd:datetime;
    ];
    sh:property [
        sh:path schema:checkOutTime;
        sh:datatype xsd:datetime;
    ];
    sh:property [
        sh:path schema:containsPlace;
        sh:class schema:Place;
    ];
    sh:property [
        sh:path schema:location;
        sh:class schema:PostalAddress;
    ].
```

Listing 1: A domain specification operator for defining a domain-specific pattern
in the accommodation domain

In Listing 3, we define a new local property n:totalNumberOfBeds[12] from an
external schema.org extension. Similarly, the range of an existing property can
be extended with a type from an extension. Here we add the n:Sauna type to
the range of the s:containsPlace property.

A domain-specific pattern can apply constraints on properties beyond the
type restrictions in their range definitions. A domain specification operator supports the definition of following types of constraint (based on the naming convention of SHACL constraint components):

- Cardinality constraints to enforce arbitrary cardinalities on property values
  (e.g., via sh:minCount)
- Value-range constraints to restrict the ranges of numerical property values
  and time values (e.g., via sh:minInclusive)
- String-based constraints to enforce length ranges, patterns or language tags
  on string literal values (e.g., via sh:pattern)
- Property pair constraints to enforce constraints involving the values of two
  properties (e.g., via sh:lessThan)
- Enumeration constraints to restrict the range of a property to a specific set
  of values (via sh:in parameters)

---

[12] n prefix is used for the namespace of a schema.org extension.

```
...
    sh:property [
        sh:path schema:containsPlace;
        sh:class schema:HotelRoom;
        sh:class schema:Product;
    ];
    sh:property [
        sh:path schema:location;
        sh:class schema:PostalAddress;
        sh:node [
            a sh:NodeShape;
            sh:property [
                sh:path schema:addressCountry;
                sh:datatype xsd:string;
            ];
            sh:property [
                sh:path schema:addressLocality;
                sh:datatype xsd:string;
            ];
        ]
    ].
```

Listing 2: The domain specification operator further restricts the local ranges

Listing 4 shows the addition of a property pair constraint that ensures that check-in time is always earlier than the check-out time. Additionally, it adds a minimum cardinality constraint to the s:checkinTime property to indicate that the property is required.

In this section, we described the domain specification process that applies a domain specification operator to schema.org vocabulary in order to create extended subsets of the vocabulary. As shown in the running example, the first two steps are enough to have the simplest form of a domain specification operator since it already creates a subset of schema.org. A full domain specification operator with further restrictions on ranges (e.g. further restrictions on the range of s:containsPlace and more cardinality constraints) can be found online[13].

The patterns can be used as machine-readable guidelines for annotation creation, but also as a specification to verify existing annotations (see also Sect. 4). In order to focus on the in-use aspects, we have left out the formal definitions in this section. A detailed abstract syntax for domain specification operators and formal semantics of the verification process can be found online[14].

4   Tools

We developed a set of tools to help domain experts with the creation of domain-specific patterns and to help data publishers with the usage of the patterns for creating annotations with schema.org. The tools are integrated with seman-

---

[13] https://semantify.it/ds/l49vQ318v.
[14] https://tinyurl.com/qmkb3ln.

```
...
    sh:property [
        sh:path n:totalNumberOfBeds;
        sh:datatype xsd:integer;
    ];
    sh:property [
        sh:path schema:containsPlace;
        sh:or ([sh:class schema:HotelRoom;
            sh:class schema:Product;] [sh:class n:Sauna])
    ];
...
```

Listing 3: A property from an external extension is defined on s:LodgingBusiness. The range of s:containsPlace is extended with a type from an extension

```
...
    sh:property [
        sh:path schema:checkInTime;
        sh:datatype xsd:datetime;
        sh:minCount 1;
        sh:lessThan schema:checkOutTime;
    ];
...
```

Listing 4: The domain specification operator extended with property pair and cardinality constraints

tify.it[15], a platform for creation, publication, and evaluation of semantic annotations. In this section, we will briefly introduce the domain specification tools, namely the domain specification editor and visualizer, the annotation verifier, and the annotation editors that utilize the domain specifications.

4.1   Domain Specification Editor and Visualizer

The domain specification editor provides a user interface for generating domain-specific patterns[16]. Figure 2 shows a part of the interface. A user first fills some metadata about the domain-specific pattern and selects a target type. Alternatively, an existing pattern can be used as a template. At this stage, extensions of schema.org can also be included in the editor via the advanced settings. After the target type is selected, the local properties and their ranges must be selected recursively. In Fig. 2, the user can click the blue pencil icon to restrict the range of the location property based on the PostalAddress type. Further constraints for a property can be specified by clicking the advanced settings button next to

---

[15] https://semantify.it.

[16] https://semantify.it/domainspecifications/create.

that property[17]. After a domain-specific pattern is created, it can be visualized in multiple forms, such as tabular, tree, and graph representation[18,19].

Using a subset of SHACL-CORE components helps us to keep the tool simple with a rather straightforward workflow. Unlike more generic shape editors (e.g. [4]), the domain specification editor supports a linear workflow. For the case of semantic annotations, this is more intuitive since the annotations are typically single sourced graphs with a main type[20].

Fig. 2. A screenshot of the domain specification editor

4.2   Annotation Verifier

The semantify.it platform hosts a verifier implementation that extracts schema.org annotations from a web page and verifies them against domain-specific patterns[21,22]. An annotation is first syntactically verified against the schema.org vocabulary, then verified against the selected domain-specific pattern. A report is then produced for each annotation, describing the errors and their sources[23].

---

[17] Not all advanced constraints (e.g. property-pair consraints for dates) are supported by the editor at the submission of this paper.

[18] We use a customized version of VOWL library http://vowl.visualdataweb.org/webvowl.html.

[19] An example can be found in https://semantify.it/domainspecifications/public/l49vQ318v.

[20] See the formal definition: https://tinyurl.com/qmkb3ln.

[21] https://semantify.it/validator/.

[22] https://semantify.it/domainSpecifications Currently, there are 400 domain-specific patterns created by semantify.it users.

[23] Comparable to a SHACL validator.

4.3 Annotation Editors

The domain-specific patterns are used to guide the data publishers. We automatically generate editors to create annotations based on domain-specific patterns. An example of such generated editor is our Wordpress extension "Instant Annotation"[24]. The plugin uses a predefined set of domain-specific patterns to create minimal editors to annotate Wordpress content. These domain-specific patterns typically target Search Engine Optimization.

5   Use Cases and Evaluation

In this section, we first describe different use cases where the domain-specific patterns are being applied. The use cases are currently dominantly from the tourism domain. However, our approach can be applied to any domain since the syntax of domain specification operators (i.e., SHACL) and the approach itself are domain-independent[25]. We also provide a preliminary evaluation of the domain specification editor's usability and the benefit of domain-specific patterns.

5.1   Web Content Annotations in Tourism

Domain-specific patterns have been used in the tourism domain by Destination Management Organizations (DMOs). The work in [1] describes a use case regarding the generation of schema.org annotated tourism-related data such as events, accommodation, and infrastructure from raw data sources. The patterns have been used to guide the mapping process of the metadata of the IT solution provider of DMO Mayrhofen[26],[27]. The domain-specific patterns are also used for manual generation of annotated data, through an editor that dynamically builds forms for annotation creation based on domain-specific patterns. The generated annotations have been used to annotate web pages for Search Engine Optimization and as a data source for a chatbot. The annotations created with the help of the domain-specific patterns are evaluated qualitatively by observing search engine results.

5.2   Schema-Tourism and DACH-KG Working Groups

The schema-tourism working group was founded as a place for the experts in the tourism domain and researchers to commonly work on a) a unified way to

---

[24] https://wordpress.org/plugins/instant-annotation/.

[25] See BioSchemas [6] for a potential use case in another domain.

[26] https://mayrhofen.at.

[27] DMOs are organizations that promote touristic services in a region. Similar annotation projects have been conducted with other DMOs such as DMO Fügen (https://best-of-zillertal.at), Seefeld (https://seefeld.com), Wilder Kaiser (https://wilderkaiser.info), and promotion agencies like Tirol Werbung (https://tirol.at).

use schema.org in the tourism domain and b) to identify the shortcomings of schema.org and extend the vocabulary when needed. The findings of the working group are published online as domain-specific patterns[28]. There are currently 81 patterns created by the working group. The documentation follows the schema.org documentation style and provides a list of all available domain-specific patterns, including a short description. By clicking on the title of a pattern, its description page is shown. This page lists all mandatory and optional properties with their respective ranges. If a type in the range is restricted in the pattern, then a link leads to its description. Otherwise, the link leads to schema.org's description of the type. Alongside the human-readable representation, the domain specification operator in SHACL syntax can also be obtained for each pattern.

The DACH-KG working group was founded with the primary goal of building a touristic knowledge graph for the region of Austria, Germany, South Tyrol, and Switzerland. Several stakeholders in the consortium[29], including but not limited to the German National Tourist Board[30], Austria Tourism Agency[31] and IDM South Tyrol[32] are currently working on a unified schema to represent their data in the knowledge graph. Hence the current focus of this working group lies on identifying the mappings from heterogeneous data sources to schema.org and defining best practices for further use of schema.org. The domain-specific patterns help the DACH-KG to formalize their findings and disseminate their best practice patterns to data and IT solution providers.

As a starting point, DACH-KG working group analyzes the existing patterns developed by the Schema-Tourism Working Group and develops additional patterns or suggests modifications. This process also includes the development of a schema.org extension, in order to increase the domain-specific coverage of schema.org for tourism. During this process, DACH-KG consortium uses schema.org extensions from different IT solution providers in tourism sector as an input. An example pattern developed by the working group can be found online[33].

The domain-specific patterns are already being adopted by the individual participants of the DACH-KG group in their internal semantic annotation processes. For example, Thüringen Tourism is using semantify.it, particularly the domain specification editor to develop their own patterns[34]. Additionally, the German Center For Tourism announced an open call for a feasibility study for

---

[28] https://ds.sti2.org.

[29] Full list can be found at https://www.tourismuszukunft.de/2018/11/dach-kg-auf-dem-weg-zum-touristischen-knowledge-graph/.

[30] https://germany.travel.

[31] https://austriatourism.com.

[32] https://www.idm-suedtirol.com/.

[33] https://ds.sti2.org/TQyCYm-r5.

[34] see slide 5 at https://thueringen.tourismusnetzwerk.info/download/pdf-Veranstaltungen/ThueCAT.pdf.

the upcoming German Tourism Knowledge Graph and required applicants to follow the domain-specific patterns on https://ds.sti2.org[35].

5.3 Preliminary Evaluation

We conducted two user studies in order to evaluate the usability of the domain specification editor and the benefit of domain-specific patterns for creating semantic annotations with schema.org.

Domain Specification Editor Usability Study. The usability of the domain specification editor has been evaluated with a System Usability Scale (SUS) [3] survey. The survey contains ten questions with 5-point Likert Scale answers. We added the eleventh question, "Overall, I would rate the user-friendliness of this editor as:" with options such as "Worst Imaginable, Awful, Poor, Good, Excellent and Best Imaginable", in order to validate the quantitative SUS score with the qualitative perception of usability. This question has been adapted from the study in [2].

The survey has been conducted anonymously among domain experts and researchers from the DACH-KG working group, who are specialized in tourism (8 participants) and senior bachelor students of Economy, Health, and Sports Tourism (WSGT) program (29 participants). The members of each target group have a basic understanding of schema.org, domain-specific patterns and their purpose.

Table 1 shows the results of the study. The leftmost column represents different target groups. The main difference is that the DACH-KG members are more experienced in the tourism sector than the WSGT students. Judging from the mean ($\bar{x} = 55.27$) and median ($\tilde{x} = 55$) SUS values in comparison with the mean values of the SUS score for each qualitative perception class, the editor has overall "Good" ($\bar{x} = 49.86$) usability. The high difference between the median score of two groups may also indicate that experienced tourism experts appreciate the domain specification editor more than the tourism students[36].

The meaning of SUS scores has also been investigated in [2] with different adjective scales. The adjective scale between Worst Imaginable and Best Imaginable is mapped to minimum mean SUS scores with a study made over 1000 SUS surveys. This adjective scale in this survey also has a central value "OK", but it is observed that the intended meaning of the word OK (whether it is acceptable) is not clear; therefore this option is left out in our study. The minimum mean score for OK in [2] is 50.9, and Good is 71.4; therefore in the presence of the central value, overall usability can be seen as "OK"[37]. The more experienced DACH-KG users' qualitative judgment is more consistent with their SUS scores, while for inexperienced users, participants were tending towards more "Good" (58.6%) than "Poor" (17.2%), even though they have given a low SUS score.

---

[35] See page 4 at https://tinyurl.com/vlnxu76.

[36] $p = 0.01$. The difference in the median of SUS scores of two groups is significant at $p < 0.05$ based on Mann-Whitney U Test.

[37] Marginally acceptable in acceptability scale. See [2].

Table 1. SUS survey results

	$\bar{x}$	σ	$\tilde{x}$	Awful		Poor		Good		Excellent	
				$\bar{x}$	%	$\bar{x}$	%	$\bar{x}$	%	$\bar{x}$	%
DACH-KG	75	25	82.5	–	0	27.5	12.5	77.5	12.5	82.5	75
WSGT-students	49.82	16.5	50	20	3.4	36.5	17.2	48.23	58.6	70.41	20.7
ALL	55.27	21.24	55	20	2.7	35	16.2	49.86	48.6	76.45	32.4

Domain-Specific Pattern Usage Survey. This survey aims to see how the domain-specific patterns can help the data publishers and IT solution providers with creating semantic annotations with schema.org. We gave 14 computer science master students and software developers who had at least some experience in creating annotations with schema.org and the task of creating annotations with the help of domain-specific patterns hosted on https://ds.sti2.org.

They were first asked to create annotations without any domain-specific patterns; then, they created one with the help of a pattern. After they created their annotations, we have asked two main questions: (a) "Did you find the domain-specific patterns understandable?" (b) "Did domain-specific patterns help you with creating annotations?"

Only 21.4% of the participants found the domain-specific patterns difficult to understand, while 78.6% found it either easy or very easy. As for the second question, we provided answer options like "It made it more complicated", "It did not make any difference" or "It helped by saving time, reducing the complexity or enabling the usage of types and properties that do not exist in schema.org vocabulary"[38]. All participants reported that the domain-specific patterns helped them in some way. More than 50% of the participants reported that the patterns helped them to save time and reduce the complexity of schema.org. About 50% stated that they also helped them to discover and use new types and properties.

6   Related Work

Although there are not many tools and approaches specifically targeting customization of schema.org for semantic annotation, an informal definition of domain specification is made in [14]. This approach only takes a subset of schema.org by removing types and properties. We define the domain specification process formally and make a clear conceptual distinction between the process of domain specification and resulting domain-specific patterns. Additionally, we do not only restrict but also allow extension of the schema.org vocabulary.

A plethora of approaches for verifying RDF data has been proposed. RDFUnit [11] is a framework that mimics the unit tests in software engineering for RDF data by checking an RDF graph against test cases defined with SHACL

---

[38] Multiple answer selection was possible.

and SPARQL. Integrity Constraints in OWL [15] describes an integrity constraint semantics for OWL restrictions to enable verification. The constraints are then checked with SPARQL. Recent efforts on standardization of RDF verification has lead to approaches that revolve around the notion of *shapes*. A shape typically targets a set of nodes in an RDF graph and applies certain constraints on them. The shape specifications have already found various usage scenarios in the literature: They can be used to assess whether an RDF graph is valid with regards to a specification (o.g , set of shapes) or as an explicit knowledge to determine the concept of completeness for a knowledge graph, typically to be used for symbolic learning[39]. For specifying shapes, two languages, namely the W3C recommendation SHACL [10] and a W3C community group effort Shape Expressions (ShEx) [13] have emerged. SHACL facilitates the creation of data shapes for validating RDF data. Following the specification, SHACL shapes are informally converted into SPARQL queries. Similarly, the ShEx provides mechanisms to create shapes and has an abstract syntax and formal semantics.

Domain-specific patterns can also be seen from the perspective of Content Ontology Design Patterns (CPs). The CPs are used to solve recurrent content modelling problems [8]. A high-level CP can be taken as a reference to create a pattern with our approach for a specific domain. The CPs typically target OWL ontologies while domain-specific patterns are fit to schema.org characteristics (see Sect. 2.1). Therefore, they can be seen as a Content ODP language for semantic annotations on the web, given that the schema.org is the de facto vocabulary for this task.

Perhaps the most relevant work to the domain-specific patterns is BioSchemas[40] [6], which provides extensions to schema.org for the life sciences domain together with simple constraints for their usage such as cardinality. To the best of our knowledge, they currently provide only human-readable specifications publicly. We see BioSchemas as a good potential use case for the domain specification approach and will investigate the possible cooperation in the future work.

The domain specification approach proposed in this paper aims to bring a compact, formal solution for customizing a very large and heterogeneous vocabulary, the de facto content annotation standard schema.org, for specific domains. We utilize SHACL to benefit from its widespread adoption, but the portability of this approach to another language like ShEx is rather straightforward. Even though our approach is related to RDF verification, the main focus here is not to verify a graph but rather to create machine-readable patterns for content and data annotations. The domain-specific patterns can have further interesting usage scenarios. For example, they can be used as a template to determine which subgraph should be returned in case of node lookups, as they would represent the relevant data for a domain (cf. [9, Sect. 9.2.2]).

---

[39] We refer the reader to the survey by Hogan et al. [9].
[40] https://bioschemas.org.

7   Conclusion and Future Work

Schema.org is the de facto industrial standard for annotating content and data. Due to its design, it is not very straightforward for data publishers to pick the right types and properties for specific domains and tasks. In order to overcome this challenge, machine-understandable domain-specific patterns can guide the knowledge generation process.

In this paper, we proposed an approach, the domain specification process for generating domain-specific patterns. The domain specification process applies an operator implemented in SHACL to schema.org vocabulary, in order to tackle the issues that come with its generic nature and make it more suitable for specific domains. We presented our approach by defining the domain specification process and with a running example of domain specification operators based on SHACL. We demonstrated the utility of our approach via various use cases. The common benefit of machine-understandable domain-specific patterns observed in all use cases is to have a rather standardized way of encoding the domain expert's knowledge for publishing annotations. The patterns can support the generation of user interfaces and mappings for annotation creation, as well as verification of annotations against a specification. Thanks to the adoption of (de-facto) standardized semantic technologies, the patterns are reusable and inter-operable, which is very important, especially for providing common schemas for a domain such as tourism, where many different data providers and application developers are involved. This brings a considerable advantage over non-semantic approaches such as ad-hoc excel sheets and vendors-specific non-reusable solutions. Given the concrete scenarios and the results of the preliminary[41] user studies, the domain-specific patterns and their tools address an important challenge standing in front of the publication of semantic data by bridging the gap between data publishers and domain experts, as well as the application developers.

For the future work, we will work on expanding the application of domain-specific pattern patterns on verification of instances in knowledge graphs [5]. We will work on the formalization of different types of relationships (e.g., subsumption) between domain-specific patterns. Additionally, we will address a second dimension alongside domain, namely the task dimension with the patterns of schema.org. We will use such patterns to make restrictions and extensions of schema.org to make it suitable for certain tasks such as Web API annotation. Another interesting direction to go would be extracting domain-specific patterns from Knowledge Graphs by using the graph summarization techniques (see [9, Sect. 3.1.3] for details).

References

1. Akbar, Z., Kärle, E., Panasiuk, O., Şimşek, U., Toma, I., Fensel, D.: Complete semantics to empower touristic service providers. In: Panetto, H., et al. (eds.) OTM 2017. LNCS, vol. 10574, pp. 353–370. Springer, Cham (2017). https://doi.org/10.1007/978-3-319-69459-7_24

---

[41] Larger studies will be conducted in the future work.

2. Bangor, A., Kortum, P., Miller, J.: Determining what individual SUS scores mean: adding an adjective rating scale. J. Usability Stud. **4**(3), 114–123 (2009)

3. Brooke, J.: SUS-a quick and dirty usability scale. Usability Eval. Ind. **189**(194), 4–7 (1996)

4. De Meester, B., Heyvaert, P., Dimou, A., Verborgh, R.: Towards a uniform user interface for editing data shapes. In: Ivanova, V., Lambrix, P., Lohmann, S., Pesquita, C. (eds.) Proceedings of the 4th International Workshop on Visualization and Interaction for Ontologies and Linked Data. CEUR Workshop Proceedings, vol. 2187, October 2018. http://ceur-ws.org/Vol-2187/paper2.pdf

5. Fensel, D., et al.: Knowledge Graphs - Methodology, Tools and Selected Use Cases. Springer, Cham (2020). https://doi.org/10.1007/978-3-030-37439-6

6. Gray, A.J., Goble, C.A., Jimenez, R., et al.: Bioschemas: from potato salad to protein annotation. In: International Semantic Web Conference (Posters, Demos & Industry Tracks), Vienna, Austria (2017)

7. Guha, R.V., Brickley, D., Macbeth, S.: Schema.org: evolution of structured data on the web. Commun. ACM **59**(2), 44–51 (2016). http://dblp.uni-trier.de/db/jour nals/cacm/cacm59.html#GuhaBM16

8. Hammar, K.: Content Ontology Design Patterns: Qualities, Methods, and Tools, vol. 1879. Linköping University Electronic Press, Linköping (2017)

9. Hogan, A., et al.: Knowledge graphs, March 2020. http://arxiv.org/abs/2003.02320

10. Knublauch, H., Kontokostas, D. (eds.): Shapes Constraint Language (SHACL). World Wide Web Consortium, July 2017. https://www.w3.org/TR/shacl/

11. Kontokostas, D., et al.: Test-driven evaluation of linked data quality. In: WWW 2014 - Proceedings of the 23rd International Conference on World Wide Web, pp. 747–757 (2014)

12. Patel-Schneider, P.F.: Analyzing schema.org. In: Mika, P., et al. (eds.) ISWC 2014. LNCS, vol. 8796, pp. 261–276. Springer, Cham (2014). https://doi.org/10.1007/ 978-3-319-11964-9_17

13. Prud'hommeaux, E., Labra Gayo, J.E., Solbrig, H.: Shape expressions: an RDF validation and transformation language. In: Proceedings of the 10th International Conference on Semantic Systems, pp. 32–40. ACM (2014)

14. Şimşek, U., Kärle, E., Holzknecht, O., Fensel, D.: Domain specific semantic validation of schema.org annotations. In: Petrenko, A.K., Voronkov, A. (eds.) PSI 2017. LNCS, vol. 10742, pp. 417–429. Springer, Cham (2018). https://doi.org/10.1007/ 978-3-319-74313-4_31

15. Tao, J., Sirin, E., Bao, J., McGuinness, D.L.: Integrity constraints in OWL. In: Fox, M., Poole, D. (eds.) Proceedings of the Twenty-Fourth AAAI Conference on Artificial Intelligence, AAAI 2010, Atlanta, Georgia, USA, 11–15 July 2010. AAAI Press (2010)

Understanding Data Centers from Logs: Leveraging External Knowledge for Distant Supervision

Chad DeLuca, Anna Lisa Gentile$^{(\boxtimes)}$, Petar Ristoski, and Steve Welch

IBM Research Almaden, San Jose, CA, USA
{delucac,welchs}@us.ibm.com,
{annalisa.gentile,petar.ristoski}@ibm.com

Abstract. Data centers are a crucial component of modern IT ecosystems. Their size and complexity present challenges in terms of maintaining and understanding knowledge about them. In this work we propose a novel methodology to create a semantic representation of a data center, leveraging graph-based data, external semantic knowledge, as well as continuous input and refinement captured with a human-in-the-loop interaction. Additionally, we specifically demonstrate the advantage of leveraging external knowledge to bootstrap the process. The main motivation behind the work is to support the task of migrating data centers, logically and/or physically, where the subject matter expert needs to identify the function of each node - a server, a virtual machine, a printer, etc - in the data center, which is not necessarily directly available in the data and to be able to plan a safe switch-off and relocation of a cluster of nodes. We test our method against two real-world datasets and show that we are able to correctly identify the function of each node in a data center with high performance.

1 Introduction

Understanding the functions implemented within a data center (which software processes are running, and where) is often an extremely challenging problem, especially in situations where good documentation practices are not in place, and machine re-configurations, software updates, changing software installation, failures, malevolent external attacks, etc. make the ecosystem difficult to understand. When data center migration is offered as a third party service, it is important to enable the practitioners to quickly and precisely characterize the nature, role, and connections - which are often not explicitly declared - of the multitude of nodes in the data center to migrate, in order to offer a reliable service.

Migrating data centers, either physically relocating machines or logically moving applications to the cloud, is a time and resource intensive task. Preparing a migration plan, especially in the absence of well documented information about the inner workings of the datacenter, involves intensive data analysis. Often the practitioners have to rely heavily on logs and network activities of each node in

© Springer Nature Switzerland AG 2020
J. Z. Pan et al. (Eds.): ISWC 2020, LNCS 12507, pp. 601–616, 2020.
https://doi.org/10.1007/978-3-030-62466-8_37

the data center to understand its cartography. Discovering and understanding connections and dependencies can be very laborious and missing any component of a dependency can result in unplanned outages during, or after, a migration. Traditional data analysis tools offer little support during the planning phase, which typically requires a significant amount of labor.

In this work, we propose a data exploration solution that allows the subject matter expert (SME) to interactively augment the collected data with structured knowledge and semantic information which is not initially present in the data. We combine traditional Information Extraction techniques together with human-in-the-loop learning to construct a semantic representation of the functions provided by the data center.

The main contribution of this work is a novel technique to create a semantic representation of a data center. Knowledge extraction is performed with a human-in-the-loop model: we first collect available knowledge about software processes from the Linked Open Data (LOD) cloud and use it in a distant supervision fashion to generate initial tags for each node in the data center; an SME validates (accept/reject/correct) the proposed tags; the validated tags are used to train several learning models and label all the processes from each node in the data center. The SME validates new annotations and the process can be repeated until the coverage is considered satisfying.

The nature of this problem is unique in the sense that, while for many processes in the data center we have useful textual information (long process name strings captured from logs), for many others we only have information about ports, connections, etc, without textual logs. Using all available textual content from those processes with logs, we generate initial tags using knowledge from the LOD. The main advantage of our solution is that we effectively exploit external knowledge to create tags and bootstrap the annotation process, training the model using both the textual information as well as the graph structure. We then apply and refine the models on the entire dataset representing the data center, including nodes where no textual content is available. Finally, all the enriched data is rendered with graph visualization tools. The SMEs have access to combined information about the nodes' logs, together with iteratively added knowledge, which creates an intelligible cartography of the data center.

In the following, we give an overview of related work (Sect. 2), formally define the problem (Sect. 3), introduce our approach (Sect. 4), followed by an evaluation (Sect. 5). We draw conclusions and discuss potential future work in Sect. 7.

2   State of the Art

The literature about data center analysis is quite broad, in the following we will highlight the differences of our proposed work in terms of (i) methods, (ii) scope - the ultimate task they are tackling - and (iii) utilized features.

In terms of methods, the common ground for knowledge discovery from log-like data is the usage of rules - in the form of regular expressions, filters, etc. [16] which can in turn rely on different features - e.g. tokenization, dictionaries or timestamp filters. While we previously investigated the use of regular expressions for this task [1], in this work we successfully explore the use of external knowledge resources and match them against the data with vector space based methods, followed by iteratively assigning validated annotations with a human-in-the-loop model.

In terms of scope, there are different classes of work in this field: some of the efforts have the goal of identifying *user sessions*, e.g., in Web based search [14] or within e-commerce frameworks [2]; some focus on identifying *anomalies and attacks* [13] while others, more similar in scope to this work, attempt to generate a *semantic view of the data* [5,9,17]. The Winnower system [13] is an example of a tool for monitoring large distributed systems, which parses audit records (log files) into provenance graphs to produce a behavioral model for many nodes of the distributed system at once. The main difference with our work is that while Winnower builds signatures of "normal behavior" for nodes - to be able to identify an anomaly - our goal is to semantically characterize the function of each node in the data center, rather than produce a binary classification of normal vs anomalous behavior. The majority of works that focus on generating a semantic view of data centers are substantially ontology-based solutions [5,9], where the logs are aligned to a specific available knowledge representation of the data center. While in this work we also create a semantic representation of the data, we propose a knowledge selection and refinement approach: we use pre-existing target knowledge but we let the final semantic representation emerge from the human interaction. The SME can add, discard, and modify semantic tags to characterize the logs, and we collect and organize them in a growing consolidated domain knowledge set. The work by Mavlyutov et al. [17] is the most similar to ours, in the sense that they analyze logs without using any pre-existing target knowledge, rather letting the semantic representation emerge from the data. They propose Dependency-Driven Analytics (DDA), which infers a compact dependency graph from the logs, and constitutes a high level view of the data that facilitates the exploration. The main difference is that the purpose of DDA is to offer compact access to the log files for tracking provenance information and allowing for different levels of inspection, e.g. to retrieve all the logs related to certain particular job failures or the debug recurring jobs etc. Differently from DDA, we generate high level, semantic views of the data center and the implemented functions, rather than semantically indexing the logs files. The logs are only exploited to generate semantic tags for the nodes in the data center, with the aid of a human-in-the-loop paradigm.

Last, but not least, logs are not the only source of information available in data centers. A plethora of sources of data are available, including network data, configuration management, databases, data from monitoring devices and appliances, etc., all of which can be leveraged to generate a model of the data center [4,7,11]. While we use all this data, this is not the focus of this paper as the major novelty of our work is a methodology to leverage unstructured data (the logs) and make sense of them using external Knowledge Resources and active interaction with the subject matter experts.

Finally, it is worth noting that there exist works in the literature that focus on understanding data centers in terms of work load, memory utilization, optimization of batch services etc. [12,21,23]. These works are also encouraged by benchmarking initiatives, such as the Alibaba Cluster Trace Program[1] that distributes data about some portion of their real production data center data. The purpose of these types of work is entirely different from this work, as we do not look at performance optimization but we aim to produce a detailed "semantic inventory" of a data center, that depicts all the applications running on each node, which we make consumable in a visual interface via (i) graph exploration, (ii) summary bar that offers contextual and global information about the data center and (iii) the knowledge cards that describe each node. The ultimate purpose is that facilitating data center migration plans. Moreover, we tackle the problem of understanding a data center, even in the absence of structured and detailed data, relying on logs collected from each node. It should also be specified that large, homogeneous data centers (i.e., Google data centers) fall outside the scope of this work. These types of data centers tend to treat each node as an anonymous worker unit with management policies that are similar to cluster or hypervisor management systems.

3   Problem Statement

A data center refers to any large, dedicated cluster of computers that is owned and operated by a single organization [3]. We formally define it as follows:

Definition 1. *A data center D is a tuple $<H, P, C>$ where H is a set of hosts, i.e. physical or virtual machines, $P = p_1 \ldots p_n$ is a list of processes running on each host $h \in H$, and C is a list of directed links between the processes P. Each host h has a set of features $s \in S$. Each s is a datatype property of the host, such as the operating system, the hardware platform, available network interfaces, etc. Each process p has a set of features $d \in D$, where each d can be either a datatype property of the process, such as port number, IP address, consumer/service status etc. or a relational property of the process, such as parent/child/sibling processes within the host. C contains the links between all processes in P, which express a directional relation among them: each process is either a consumer or a provider for another processes.*

Each data center D can be transformed into a knowledge graph $G = (V, E)$.

Definition 2. *A knowledge graph is a labeled, directed graph $G = (V, E)$ where V is a set of vertices, and E is a set of directed edges, where each vertex $v \in V$ is identified by a unique identifier, and each edge $e \in E$ is labeled with a label from a finite set of link edges.*

To transform a given data center D into a knowledge graph G, we convert each process p into a graph vertex v_p. Then, we generate a list of labels for each

---

[1] https://github.com/alibaba/clusterdata.

feature of the existing processes. For each feature of each process we generate a triple in the form $<v_p\ e_n\ e_p>$, where e_n is the feature predicate, e_p is the value of the feature for the given process v_p. Then we transform the set of links C between all the processes to triples in the form $<v_p\ c_n\ v_{pi}>$, where each type of link is represented with a set of triples, i.e., one triple for each feature of the link $<v_c\ e_n\ e_c>$.

Given a data center $D = <H, P, C>$, our objective is to use the information provided for each process, and the links to other processes, to assign a set of labels $L_p = l_1, l_2, ..., l_n$ to each process p and - by inheritance - to each host h. In the literature, this task is also known as knowledge graph type prediction.

4   Understanding Data Centers

In the following, we describe the end-to-end methodology to create a semantic representation of a data center starting from scratch, i.e. from collecting raw data from the individual nodes of the data center to creating a representation which can be fully visualized and explored as a graph.

The workflow of the system consists of four subsequent steps. The *Data Collector* process (Sect. 4.1) collects log data from a portion of nodes in the data center. The *Knowledge Matcher* component (Sect. 4.2) obtains relevant knowledge from the LOD cloud and produces candidate labels for each node in the data center, in a distant supervision fashion. The candidate labels are validated by the SME (Sect. 4.3) before being used to train several classification models (Sect. 4.4). All augmented data is made available to the users via a visual tool, the Data Center Explorer (covered in Sect. 6) which allows the SMEs to efficiently visualize data center structure and perform complex queries.

4.1   Collecting the Data

Data is collected by running a script on each machine, virtual or physical, that is deemed important. This means deploying the script to at least 10%–20% of the nodes in a data center. Broader deployments are preferable, but not required, as they reduce the human effort required in the final stages of a migration to ensure success. A typical configuration is to run the script for 2 weeks, taking a data snapshot every 15 min. Upon the first snapshot, the script records attributes considered "static", including operating system, hardware, and network interface information. Every snapshot includes a list of all active connections at that point in time, along with every running process. An active connection includes a process (either the consumer or provider of a service), a direction indicator, a target port, and the IP address of the other side of the connection. If the collection script is running on the node at the other side of this connection, we'll be able to combine both sets of collected data to construct a full, end-to-end definition of a particular dependency. In cases where the other side of the dependency is not running the collection script, we'll only have a partial understanding of the dependency, given the fact that we cannot know the exact process communicating on the other node.

When running processes are captured, we record the process log string (which we refer to as *process name*) along with its Process ID (PID) and its Parent Process ID (PPID). These elements allow us to construct a process tree as it existed when that snapshot was taken. We can combine the process tree from each snapshot into a meta-tree such that, at any point-in-time, the process tree represents a subset of the process meta-tree with all relationships intact. The relationship between processes can be helpful in process identification, for example when a particular communicating process is unknown, but its parent is known. In such cases, cascading attributes from a parent process to its children may be beneficial. All collected data is transformed into a knowledge graph representation, following Definition 2 (Sect. 3).

4.2   Bootstrapping the Labelling Process: The Knowledge Matcher

The goal of this step is twofold: to identify an initial set of labels, i.e. the inventory of possible software that can run on a node, and to create an initial set of annotated processes, i.e. assign some of these labels to a portion of the processes in the data center, that is going to be used as initial training set. We tackle the task as a distant supervision problem [18]: we leverage pre-existing, structured knowledge to identify potential labels and annotate the data center processes. The idea is to use external knowledge to construct dictionaries for target concepts and use them to label the target data. The approach is not bound to any specific ontology or resource: the only assumption is to have a dictionary that contains instances of the target concept. The dictionary does not have to be exhaustive either: as we will show in Sect. 4.3, we allow the SME to expand the knowledge with any missing concept.

For this work, we construct a dictionary for the concept *Software* following the same approach as [10]. Given a SPARQL endpoint,[2], we query the exposed Linked Data to identify the relevant classes. We manually select the most appropriate classes and properties that describe the concept of interest and craft queries to obtain instances of those.

We build a vector space representation of this constructed knowledge, where each software item is represented as a tf-idf vector and the vocabulary of features is built using the software names, labels and, optionally, any of the properties which is of text type (e.g., one can collect features such as the operating system, the type of software, etc.). Similarly, we build a vector space representation of all the processes $p \in P$, and the vocabulary of features is built using all the text in the collected logs strings. We discard any annotation with similarity below a certain threshold[3] and for each $p \in P$ we select the top similar instance from the dictionary. We then use standard cosine similarity to assign one label from the software dictionary to each process $p \in P$. The intuition for using the

---

[2] For example the DBpoedia endpoint http://dbpedia.org/sparql.
[3] For this work the similarity threshold has been set to 0.6.

vector space model and cosine similarity is that some of the words in a log string running a certain software, will be similar to some of the information that we collected from the knowledge base - for example the path were the software is installed or the name of executable files or some of the parameters can contain the name of the software itself. The annotations produced in this step are not expected and not meant to be exhaustive, and our only purpose is to create a reliable initial training set. For this reason we want the annotations to be as accurate as possible so that they can be used to train machine learning models (Sect. 4.4).

4.3   Human-in-the-Loop: Validation and Knowledge Expansion

We start by ranking all retained annotations by their similarity score. During the validation process, if the SME does not accept any of the candidate annotations, they can browse the reference dictionary and manually select the appropriate one - if available. We also give the SME the possibility to manually add an entry to the dictionary, if they deem it missing, or delete any dictionary entry that they consider spurious, redundant, or incorrect. All the validated annotations are immediately added as tags for the processes and can be used as exploring dimensions to query and visualize the data (the visual tool for the graph exploration is discussed in Sect. 6).

4.4   Training the Models

The goal of this step is to leverage the validated tags to train a classification model in order to extend the tags to the whole data center. To this end, we experiment with several different neural network models, spanning from simple architectures that only exploit string information about the text in the logs to more complex graph embeddings that capture all the relations among the nodes in the data center.

For String-based approaches we use multi-label Logistic Regression (LR), multi-label Support Vector Machines (SVM) and Convolutional Neural Network (CNN) text classifiers, trained to classify log entries. The architecture of the CNN is inspired by Kim et al. [15], which has shown high performance in many NLP tasks.

In many cases of analyzing data center logs, a significant portion of the processes do not include a name or string description, i.e., either only the process ID is available or the string description is encoded, which cannot be used to infer the label of the process. To be able to correctly label such processes we use multiple graph embedding approaches. Graph embedding approaches transform each node in the graph to a low dimensional feature vector. The feature vector embeds the graph characteristics of the node and it can be used to predict the node's label, i.e., similar nodes in the graph should have the same label. To build graph embedding vectors on a data center D, we transform it to a knowledge graph $G = (V, E)$. In this work we consider 5 state-of-the-art graph embedding approaches: RDF2Vec [20], TransE [6], DistMult [24], ComplEx [22], HolE [19].

608   C. DeLuca et al.

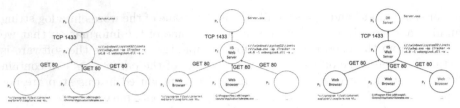

n) *Input graph of communicating*   b) *Inferred process labels after ap-*   c) *Inferred process labels after ap-*
processes.   *plying the string-based model.*   *plying the graph-based model.*

Fig. 1. Example of a sub-graph of communicating processes being labeled using the string-based model then the graph-based model.

The output of each graph embedding approach, which is a numerical vector for each node, is used to build a classification model for node label prediction, i.e., multi-label Logistic Regression, multi-label Support Vector Machines, and CNN network.

The string-based model and the graph-based model are complementary to each other. The string-based model is used to label processes when the process name is available and the model can predict the process label with confidence higher than a specified threshold. In all the other cases we apply the graph-based model, which is able to set a label based on the structure of the graph. Figure 1 shows a simplified example of the labeling steps when using both models. The input is a sub-graph of 5 processes $p1$ to $p5$ (Fig. 1a). The process $p3$ doesn't have a name, while process $p5$ has a very generic name "Server.exe". As such, after applying the string-based model, the system can label the nodes $p1$, $p2$ and $p4$ with high confidence, as the process names are rather informative, e.g., from the processes' names we can easily infer that both $p1$ and $p2$ are web browsers, "Google Chrome" and "Internet Explorer", while $p4$ is "Internet Information Services", which is a "Web Server" (Fig. 1b). However, the string-based model cannot predict the label for process $p3$ as there is no process name provided, and for process $p5$ can only make a prediction with a very low confidence, as the name of the process is very general. In such cases, the graph-based model can help identify the missing labels. As the graph-based model is able to capture the graph structure and the relations between the processes, the model can predict the labels with high confidence when only a couple of labels are missing. In this case, the embedding vector for process $p3$ is going to be quite similar to the one from $p1$ and $p2$ as they all communicate with $p4$ with the same type of relation, which means that the model can label this process as "Web Browser" with high confidence. Regarding process $p5$, we expect that the model should be able to learn that in many other cases a process labeled as "Web Server" communicates with a process "Database Server" over a TCP link (Fig. 1c).

5   Experiments

5.1   Description of the Data

We ran experiments against data collected from 2 different data centers: a large US financial institution (*US-Financial*) and a large Canadian healthcare organization (*Canadian-Healthcare*), both representing industries with the most extreme demands on data center operations, in terms of security, resiliency, and availability. Data is collected following the methodology described in Sect. 5.1 for both data centers. The *US-Financial* dataset contains data about 2,420 discovered hosts and additional 24,127 inferred hosts (5,993 of which are external to the institution's data center). The *Canadian-Healthcare* dataset contains data about 2,139 discovered hosts, as well as 44,169 inferred hosts (526 of which external). While the data collected for *discovered hosts* is richer and contains the strings of all running processes, the information about *inferred hosts* only concerns network connections, ports, etc. For this reason, when it comes to process identification, we focus solely on discovered hosts, since these are the only hosts where we can see the names of running processes and their position within a host's process tree, nonetheless the information about *inferred hosts* is leveraged in the learning steps, especially in the graph embeddings approach (Sect. 4.4). In fact, the graph based methods largely benefit of all the information collected about links and communications among hosts. Specifically, from the 2,420 discovered hosts in the *US-Financial* dataset, we collected 1,375,006 running processes. Of these running processes, 186,861 processes are actually communicating. The *Canadian-Healthcare* dataset derived from 2,139 discovered hosts contains data on 339,555 running processes, 98,490 of which are communicating on the network. Each unique dependency forms a link between two hosts.[4] A unique dependency consists of five elements: Source Host, Source Process, Target Host, Target Process, Target Port. While both hosts and the target port are required to form the link, it is perfectly acceptable for one or both processes to be null. Given this definition of a dependency, the *US-Financial* dataset contains 2,484,347 unique dependencies and the *Canadian-Healthcare* dataset contains 22,016,130 unique dependencies, or links, between nodes.

5.2   Human-in-the-Loop Distant Supervision

As target knowledge for the tags, we use a dictionary of Software constructed from DBpedia, which initially contains 27,482 entries. Each entry in the dictionary has a unique identifier, a name and, when available, a software type. We run the distant supervision module on the 1,375,006 running processes. For each process we select the highest scoring tag, with a similarity score of at least 0.6.[5] These gave us 639,828 tagged processes, annotated with 632 different tags, which are passed to the SME for validation. At the end of this step, we have a

---

[4] The words host and node are used interchangeably.
[5] The threshold was selected based on empirical observation.

gold standard of 442,883 annotated processes, meaning that more than 69% of the annotations produced with distant supervision where considered correct by the SME. After this validation, 122 unique tags remained in the pool of useful ones. Some of these tags classify processes that are crucial to be identified when planning a migration, such as "Internet Information Services", "Microsoft SQL Server", and "NetBackup". During the validation process, the SME pointed out that some of the proposed tags identified software tools that they wouldn't have thought about looking for, if not otherwise prompted by the system. We provided the ability to remove entries based on softwareType, and our SME decided to remove entries of type "Fighting game", "Shooter game", "Sports video game", and others. One example of a manually added entry in this experiment was Internet Explorer, which was not initially included in the dictionary as the DBpedia type for http://dbpedia.org/page/Internet_Explorer is Television Show (http://dbpedia.org/ontology/TelevisionShow).

5.3   Training the Models

String-Based Approach. We developed two baseline string-based machine learning approaches, i.e., multi-label Logistic Regression (LR) and multi-label Support Vector Machines (SVM). We use standard bag-of-word with tf-idf weights to represent each log entry. We use the scikit-learn multi-label implementation with standard parameters.[6] Furthermore, we use a multi-label Convolutional Neural Network (CNN) text classifier and we train it to classify log entries. We selected the following parameters for the CNN model: an input embedding layer, 4 convolutional layers followed by max-pooling layers, a fully connected sigmoid layer, rectified linear units, filter windows of 2, 3, 4, 5 with 100 feature maps each, dropout rate of 0.2 and mini-batch size of 50. For the embedding layer, we use word2vec embeddings trained on 1.2 million log entries, with size 300 using the skip-gram approach. We train 10 epochs with early stopping. Using a sigmoid activation function in the last dense layer, the neural network models the probability of each class as a Bernoulli distribution, where the probability of each class is independent from the other class probabilities. This way we can automatically assign multiple labels to each log entry.

Graph Embedding Approaches. We experiment with 5 different graph embeddings approaches, i.e. RDF2Vec, TransE, DistMult, ComplEx and HolE. For the RDF2Vec approach, we use the implementation provided in the original paper.[7] For the rest of the approaches, we are using the implementation provided by AmpliGraph [8].[8] For the RDF2vec approach, for each entity in the DCE graph we generate 500 random walks, each of depth 8. We use the generated sequences to build a Skip-Gram model with the following parameters: window size = 5; number of iterations = 10; negative sampling for optimization;

---

[6] https://scikit-learn.org/.

[7] http://data.dws.informatik.uni-mannheim.de/rdf2vec/code/.

[8] https://github.com/Accenture/AmpliGraph.

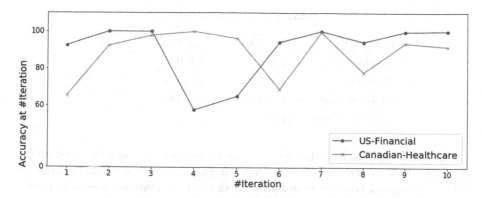

Fig. 2. Accuracy of the string-based CNN per iteration.

negative samples = 25; vector size = 300. For TransE, DistMult, ComplEx and HolE, we learn embeddings with 300 dimensions and maximum of 50 epochs.

The learned embeddings from the 5 approaches are then used to train a machine learning model for type prediction. The output of the embeddings is used to train a multi-label Logistic Regression (LR), multi-label Support Vector Machines (SVM) and a 1-dimensional CNN network. The architecture of the CNN is as follows: two one-dimensional convolutional layers followed by one max-pooling layer and a fully connected sigmoid layer. Each convolutional layer has 100 filters with size 10, using ReLU activation function.

5.4 Human-in-the-Loop Gold Standard Generation

We adapted a human-in-the-loop approach to generate a gold standard in order to evaluate our approaches on the whole data center. We used the dataset from the distant supervision module to bootstrap the labeling process. The labeling process runs in iterations, where in each iteration we (i) train a CNN model using the available labeled data, (ii) apply the model on the unlabeled data, (iii) present the candidates to the SME with confidence above 0.8,[9] and (iv) add the labeled data by the SME in the training set. For the *US-Financial* dataset, we run the process in 10 iterations. In each iteration we present the SME with the top 100, 000 candidates sorted by the model's confidence and calculate the accuracy at each iteration. Note that this step of the pipeline is the equivalent of generating a gold standard dataset, hence the accuracy measure: at each iteration we calculate the percentage of the 100, 000 candidates automatically labeled by the model that are deemed correct by the SME. Figure 2 plots the classification accuracy for both datasets: at each iteration the y-axis shows the percentage of processes correctly labeled by the model and manually validated by the SME. We perform the experiment first on *US-Financial* and then on *Canadian-Healthcare*.

---

[9] The threshold was selected based on empirical evaluation.

Table 1. Datasets statistics

Dataset	*US-Financial*	*Canadian-Healthcare*
#nodes	2,420	2,139
#processes	1,375,006	339,555
#labeled processes	1,324,759	264,496
#unique labels	153	64

For *US-Financial*, iteration number one represents the model trained on distant supervision data and all subsequent iterations use the feedback from the SME as additional training material. We can notice a relative decrease in performance in iterations 4 and 5. At iteration 4, the SME started validating processes where the classification confidence of the model was lower and identified that the knowledge base did not contain any appropriate tags for them. The SME added 31 additional tags to the knowledge base in iterations 4 and 5, and we can then notice an improvement of performance on the next iterations. At the end of the annotation process, the complete dataset consists of 1,324,759 processes, including those labeled in the bootstrapping step. The SME was unable to validate only 50,247, which is less than 4% of the data, as a result of insufficient information about the processes. Finally, the total number of used tags to label all the processes is 153.

For *Canadian-Healthcare*, the bootstrapping phase is done using the model from *US-Financial*, and then proceed updating the model with the newly labeled (and validated) instances at each subsequent iteration. In the plot (Fig. 2), iteration one of *Canadian-Healthcare* represents results obtained with the previously trained model (on the *US-Financial* dataset): the accuracy is rather low at this first iteration as the overlap with the previous dataset is not very high, but the performance improves with the subsequent interactions with the human. At the end of the annotation process, the complete dataset consists of 264,496 processes. The SME was unable to validate 75,059 as a result of insufficient information about the processes. Tthe total number of used tags to label all the processes is 64. The statistics for both datasets are given in Table 1.

5.5   Results

We evaluate the approaches using the standard measures of precision, recall and F-score. Our datasets are imbalanced: while some of the tags are only used a handful of times, others classify many processes. We therefore weight all the evaluation metrics by the class size. The results are calculated using stratified 20/80 split validation, i.e., we use stratified random 20% of the data to train the model and we test the model on the remaining 80%. We have to note that the results significantly improve when using 10-fold cross validation, however that is not a realistic representation of the problem we are trying to solve, i.e. in our application we are trying to use as little labeled data as possible to automatically label the reset of the dataset. The graph embedding approaches are built on the

Table 2. Results on the *US-Financial* dataset.

Method	Precision	Recall	F-Score
TF-IDF - LR	89.46	73.2	75.85
TF-IDF - SVM	90.11	72.64	76.86
word2vec - CNN	**91.51**	**92.43**	**89.58**
RDF2Vec - LR	47.31	24.58	33.04
RDF2Vec - SVM	43.01	34.95	26.86
RDF2Vec - CNN	72.36	59.17	62.43
TransE - LR	46.98	37.57	28.95
TransE - SVM	41.35	22.86	29.29
TransE - CNN	54.65	32.40	40.64
DistMult - LR	74.97	71.51	71.73
DistMult - SVM	79.39	79.07	79.08
DistMult - CNN	82.44	81.78	82.29
ComplEx - LR	75.86	72.11	72.44
ComplEx - SVM	79.59	79.16	79.19
ComplEx - CNN	84.19	84.97	83.57
HolE - LR	76.51	74.97	75.18
HolE - SVM	78.62	81.36	80.24
HolE - CNN	89.28	88.21	87.93

Table 3. Results on the *Healthcare* dataset.

Method	Precision	Recall	F-Score
TF-IDF LR	94.12	79.06	85.53
TF-IDF SVM	95.30	91.99	92.98
word2vec CNN	**97.38**	**98.17**	**97.63**
RDF2Vec - LR	45.67	21.93	29.51
RDF2Vec - SVM	41.29	32.14	36.04
RDF2Vec - CNN	69.12	57.36	62.57
TransE - LR	12.78	13.01	12.89
TransE - SVM	29.80	17.16	17.02
TransE - CNN	34.48	14.56	4.90
DistMult - LR	43.84	39.59	30.96
DistMult - SVM	41.33	34.92	36.97
DistMult - CNN	47.65	32.49	38.58
ComplEx - LR	75.45	71.48	73.33
ComplEx - SVM	76.35	73.51	74.83
ComplEx - CNN	87.95	79.48	80.17
HolE - LR	76.69	73.31	74.43
HolE - SVM	76.63	72.39	73.39
HolE - CNN	89.46	74.20	81.11

complete graph, using all available information, without the tags of the processes that are part of the test set in the current fold.

Table 2 and Table 3 show a summary of the results for both datasets, and compares figures for each model - Logistic Regression (LR), Support Vector Machine (SVM) or Convolutional Neural Network (CNN) - when using either string-based features (tf-idf/word2vec) or graph embeddings features (RDF2Vec/TransE/ DistMult/ComplEx/HolE).

Among the string-based approaches, the CNN with word2vec embedding is the best performer on both datasets. For the graph based approaches, the CNN using HolE embedding outperforms the rest. While DistMult and ComplEx perform comparably to HolE, RDF2vec and TransE show worse performance. On the *Canadian-Healthcare* the TransE model cannot learn a good representation of the graph, which leads to rather bad results. When comparing string-based methods against graph-based approaches, it is important to notice that, despite the fact that word2vec-CNN outperforms HolE-CNN, all the graph-based approaches completely ignore the *process name*, i.e. the text content, and solely use the graph structure to generate the graph embeddings. The result is significant in this particular domain, because much of the data comes without any text content. This means that, while we can successfully use string-based approaches for a portion of the data, we can supplement with graph-based approaches for datasets lacking string descriptions. Finally, one of the major advantages of our

Fig. 3. Data Center Explorer - example view of a portion of the datacenter graph.

methodology was creating the pool of tags and the initial training data in a distant supervision fashion, using LOD as external knowledge. The validation of this initial data took less than a handful of hours for the SME, but enabled a granular understanding of the data center.

6   System Deployment and User Engagements

The machine learning pipeline described in this work is currently deployed on a machine with 100 cores and 1 TB of RAM. The models are continuously updated in the background, and the new results are used to produce data center graphs for any new engagement. To preserve data privacy, for each client engagement we build a separate set of models, trained only on the client's data, without transfer learning from other engagements. Each produced data center graph is used to feed the User Interface used by our SMEs, called *Data Center Explorer* (DCE). The DCE visual tool builds a color-coded representation of all the nodes, as well as their incoming and outgoing connections. We then overlay the produced semantic representation: each node is characterized by all its entities of interest, i.e. the processes running on the machine and their associated tags. An example view produced by DCE is shown in Fig. 3.

We create a knowledge card for each node that summarizes all the information about the node itself, including all the semantic tags associated with all processes running on the node. Moreover, all the knowledge collected in the enriched data center graph can be used to query the data center via GraphQL[10] queries, as well as using the tags as active facets.

DCE is currently used internally by one of the IBM business units working on datacenter migrations engagements with different clients. The team involved in our pilot consists of about 15 practitioners that have the mission of designing and implementing datacenter migration tasks. Since its deployment on our internal cloud, the DCE tool has been used by the team to process 8 large data

---

[10] https://graphql.org/.

center migrations (over 15000 hosts). The team was previously only relying on manually populated spreadsheets and client interviews, but since the deployment of our pipeline they use our solution to bootstrap any new datacenter migration engagement. From the informal feedback that we gathered from the team, they identified 3 striking benefits of the solution. First, information provided via client interviews is often refuted by DCE reports, with DCE being far more reliable. Second, enormous amounts of time can be saved or redirected to other tasks, as the majority of upfront labor is relegated as unnecessary by DCE's semantic bootstrapping process. Finally, the immediate cost savings provided by DCE (reduced manual labor and fewer migration failures) tends to be between 1 and 2 orders of magnitude higher than the cost to develop, maintain, and train DCE. The major lesson learned for us was the importance of being able to quickly add initial semantic annotations to the nodes of the datacenter, by transparently mining labels from Linked Data, but giving the SMEs full control on which label to add and use in the graph. The combination of automatic semantic bootstrapping with human-in-the-loop achieves the right level of representation to improve the quality and effectiveness of the datacenter migration task.

7   Conclusions and Future Work

Understanding the structure of a data center is crucial for many tasks, including its maintenance, monitoring, migration etc. In this work, we presented a distant supervision approach that bootstraps the annotation of logs using knowledge from the Linked Open Data Cloud. We then train neural models and refine them with a human-in-the-loop methodology. We performed quantitative experiments with two real-world dataset and showed that the approach can classify nodes with high performance.

While in this work we focus on the task of type prediction, in the future we will extend the pipeline to also perform link prediction, i.e. identifying missing links between nodes in the data center, and identifying the type of the relation. Furthermore, using graph embeddings to represent the whole data center and having the whole data into one single feature space opens the possibility for many useful applications, e.g., clustering nodes by type, identifying dependency clusters (clusters of nodes that must be migrated together), ranking nodes and clusters based on importance, and finally building ad-hoc machine learning models for custom data analytic.

References

1. Alba, A., et al.: Task oriented data exploration with human-in-the-loop. A data center migration use case. In: Companion Proceedings of the 2019 World Wide Web Conference, WWW 2019, pp. 610–613. ACM, New York (2019)
2. Awad, M., Menasc, D.A.: Automatic workload characterization using system log analysis. In: Computer Measurement Group Conference (2015)
3. Benson, T., Akella, A., Maltz, D.A.: Network traffic characteristics of data centers in the wild. In: 10th ACM SIGCOMM (2010)

4. Benzadri, Z., Belala, F., Bouanaka, C.: Towards a formal model for cloud computing. In: Lomuscio, A.R., Nepal, S., Patrizi, F., Benatallah, B., Brandić, I. (eds.) ICSOC 2013. LNCS, vol. 8377, pp. 381–393. Springer, Cham (2014). https://doi.org/10.1007/978-3-319-06859-6_34

5. Bernstein, D., Clara, S., Court, N., Bernstein, D.: Using Semantic Web Ontology for Intercloud Directories and Exchanges (2010)

6. Bordes, A., Usunier, N., Garcia-Duran, A., Weston, J., Yakhnenko, O.: Translating embeddings for modeling multi-relational data. In: Advances in Neural Information Processing Systems, pp. 2787–2795 (2013)

7. Bourdeau, R.H., Cheng, B.H.C.: A formal semantics for object model diagrams. IEEE Trans. Softw. Eng. 21(10), 799–821 (1995)

8. Costabello, L., Pai, S., Van, C.L., McGrath, R., McCarthy, N.: AmpliGraph: a Library for Representation Learning on Knowledge Graphs (2019)

9. Deng, Y., Sarkar, R., Ramasamy, H., Hosn, R., Mahindru, R.: An Ontology-Based Framework for Model-Driven Analysis of Situations in Data Centers (2013)

10. Gentile, A.L., Zhang, Z., Augenstein, I., Ciravegna, F.: Unsupervised wrapper induction using linked data. In: Proceedings of the Seventh International Conference on Knowledge Capture, pp. 41–48 (2013)

11. Grandison, T., Maximilien, E.M., Thorpe, S., Alba, A.: Towards a formal definition of a computing cloud. In: Services. IEEE (2010)

12. Guo, J.: Who limits the resource efficiency of my datacenter: an analysis of Alibaba datacenter traces. In: IWQoS 2019 (2019)

13. Hassan, W.U., Aguse, L., Aguse, N., Bates, A., Moyer, T.: Towards scalable cluster auditing through grammatical inference over provenance graphs. In: Network and Distributed Systems Security Symposium (2018)

14. Jiang, Y., Li, Y., Yang, C., Armstrong, E.M., Huang, T., Moroni, D.: Reconstructing sessions from data discovery and access logs to build a semantic knowledge base for improving data discovery. ISPRS 5, 54 (2016)

15. Kim, Y.: Convolutional neural networks for sentence classification. In: EMNLP 2014, pp. 1746–1751. ACL, October 2014

16. Lemoudden, M., El Ouahidi, B.: Managing cloud-generated logs using big data technologies. In: WINCOM (2015)

17. Mavlyutov, R., Curino, C., Asipov, B., Cudre-mauroux, P.: Dependency-driven analytics: a compass for uncharted data oceans. In: 8th Biennial Conference on Innovative Data Systems Research (CIDR 2017) (2017)

18. Mintz, M., Bills, S., Snow, R., Jurafsky, D.: Distant supervision for relation extraction without labeled data. In: ACL (2009)

19. Nickel, M., Rosasco, L., Poggio, T.: Holographic embeddings of knowledge graphs. In: Thirtieth AAAI Conference on Artificial Intelligence (2016)

20. Ristoski, P., Rosati, J., Di Noia, T., De Leone, R., Paulheim, H.: RDF2Vec: RDF graph embeddings and their applications. Semant. Web 10, 1–32 (2018)

21. Shan, Y., Huang, Y., Chen, Y., Zhang, Y.: LegoOS: a disseminated, distributed {OS} for hardware resource disaggregation. In: 13th Symposium on Operating Systems Design and Implementation, pp. 69–87 (2018)

22. Trouillon, T., Welbl, J., Riedel, S., Gaussier, É., Bouchard, G.: Complex embeddings for simple link prediction. In: ICML (2016)

23. Wu, H., et al.: Aladdin: optimized maximum flow management for shared production clusters. In: 2019 IEEE PDPS, pp. 696–707. IEEE (2019)

24. Yang, B., Yih, W., He, X., Gao, J., Deng, L.: Embedding entities and relations for learning and inference in knowledge bases. arXiv preprint arXiv:1412.6575 (2014)

Assisting the RDF Annotation
of a Digital Humanities Corpus
Using Case-Based Reasoning

Nicolas Lasolle[1,2](✉) [iD], Olivier Bruneau[1], Jean Lieber[2], Emmanuel Nauer[2],
and Siyana Pavlova[2]

[1] Université de Lorraine, CNRS, Université de Strasbourg, AHP-PReST,
54000 Nancy, France
{nicolas.lasolle,olivier.bruneau}@univ-lorraine.fr
[2] Université de Lorraine, CNRS, Inria, LORIA, 54000 Nancy, France
{nicolas.lasolle,jean.lieber,emmanuel.nauer,siyana.pavlova}@loria.fr

Abstract. The Henri Poincaré correspondence is a corpus composed of
around 2100 letters which is a rich source of information for historians of
science. Semantic Web technologies provide a way to structure and publish data related to this kind of corpus. However, Semantic Web data editing is a process which often requires human intervention and may seem
tedious for the user. This article introduces RDFWebEditor4Humanities,
an editor which aims at facilitating annotation of documents. This tool
uses case-based reasoning (CBR) to provide suggestions for the user which
are related to the current document annotation process. These suggestions are found and ranked by considering the annotation context related
to the resource currently being edited and by looking for similar resources
already annotated in the database. Several methods and combinations
of methods are presented here, as well as the evaluation associated with
each of them.

Keywords: Semantic Web · Content annotation · Case-based
reasoning · RDF(S) · SPARQL query transformation · Digital
humanities · History of science · Scientific correspondence

1 Introduction

Born in Nancy, France, in 1854, Jules Henri Poincaré is considered one of the
major scientists of his time. Until his death in 1912, he relentlessly contributed to
the scientific and social progress. Most known for his contribution in mathematics
(automorphic forms, topology) and physics (3-Body problem resolution), he also
played a significant role in the development of philosophy. His book *La Science et
l'Hypothèse* [15] had a major international impact for the philosophy of science.

His correspondence is a corpus composed of around 2100 letters which
includes sent and received letters. It gathers scientific, administrative and private correspondence which are of interest for historians of science. The *Archives*

© Springer Nature Switzerland AG 2020
J. Z. Pan et al. (Eds.): ISWC 2020, LNCS 12507, pp. 617–633, 2020.
https://doi.org/10.1007/978-3-030-62466-8_38

Henri-Poincaré is a research laboratory located in Nancy, in which one of the important works is the edition of this correspondence. In addition to the physical publication, a keen interest is devoted to the online publishing of this correspondence. On the Henri Poincaré website[1] are available the letters associated with a set of metadata. Different search engines may be used by historians or, more globally, by those who show an interest in the history of science. This has been achieved by the use of Semantic Web technologies: RDF model to represent data, RDFS language to represent ontology knowledge and SPARQL language to query data. During the human annotation process, several difficulties have emerged. Indeed, the annotation is a tedious process which requires the constant attention of the user to avoid different kinds of mistakes. The *duplication mistake* is encountered when the user inserts data that is already in the database. The *ambiguity mistake* happens when the user does not have enough information to distinguish items. It occurs when the same description or label is used for different items or when an item identifier does not give explicit information about its type and content. For instance, if a search is made based on the string "Henri Poincaré", different types of resources may be returned. Indeed, the most plausible expected answer should refer to the famous scientist, but this term also refers to different institutes, schools and, since 1997, a mathematical physics prize exists named in his memory. The *typing mistake* is encountered when the user wants to write an existing word to refer to a specific resource but inadvertently mistypes it. If not noticed, this error can lead to the creation of a new resource in the database instead of referring to an existing resource. In addition to these possible mistakes, the cognitive load effect associated with the use of an annotation system should not be neglected. Depending on the volume of the corpus to annotate, this process could be a long-term project. Keeping the users motivated when performing the associated tasks is a real issue.

This article intends to present an efficient tool currently in use for content annotation. A suggestion system is proposed to the user to assist her/him during the annotation process. Four versions of the system have been designed: the *basic editor*, the *deductive editor*, the *case-based editor* and the last version, the *combination editor* which is a combination of the methods used in the two previous versions of the system. The last two versions use case-based reasoning (CBR) to find resources presenting similarities with the one currently being edited and thus take advantage of the already indexed content. The following hypotheses are made and are evaluated in the evaluation section:

Hypothesis 1 the use of CBR improves the suggestion list provided to the user.
Hypothesis 2 the combination of the use of CBR and RDFS entailment improves the suggestion list with respect to the use of CBR alone and of RDFS entailment alone.

Section 2 shortly introduces Semantic Web notions that are considered in this article, and presents a brief reminder of CBR. Section 3 explains the current

---

[1] http://henripoincare.fr.

infrastructure related to the Henri Poincaré correspondence edition and summarizes the previous work about the annotation tool. Section 4 focuses on how the use of CBR improves the annotation process and addresses the issues mentioned above. Section 5 describes the evaluation. Section 6 details some of the choices that have been made through the development of this editor and situates this work by comparing it with related works. Section 7 concludes and points out future works.

2 Preliminaries on Semantic Web Technologies and CBR

This section introduces the CBR methodology and terminology. A brief reminder of the Semantic Web notions and technologies that are RDF, RDFS and SPARQL is provided afterwards.

2.1 CBR: Terminology and Notation

Case-based reasoning (CBR [17]) aims at solving problems with the help of a *case base* CB, i.e., a finite set of cases, where a case represents a problem-solving episode. A case is often an ordered pair (x, y) where x is a problem and y is a solution of x. A case (x^s, y^s) from the case base is called a *source case* and x^s is a *source problem*. The input of a CBR system is a problem called the *target problem* and is denoted by x^{tgt}.

The 4 Rs model decomposes the CBR process in four steps [1]. (1) A $(x^s, y^s) \in$ CB judged similar to x^{tgt} is selected (*retrieve step*). (2) This retrieved case (x^s, y^s) is used for the purpose of solving x^{tgt} (*reuse step*): the proposed solution y^{tgt} is either y^s (reused as such) or modified to take into account the mismatch between x^s and x^{tgt}. (3) The pair (x^{tgt}, y^{tgt}) is then tested to see whether y^{tgt} correctly solves x^{tgt} and, if not, y^{tgt} is repaired for this purpose (*revise step*); this step is often made by a human. (4) Finally, the newly formed case (x^{tgt}, y^{tgt}) is added to CB if this storage is judged appropriate (*retain step*). In some applications, the retrieve step returns several source cases that the reuse step combines.

Consider an example related to a case-based system in the cooking domain [10]. This system lets users formulate queries to retrieve cooking recipes. It provides adapted recipes when the execution of a query does not give any result. For instance, a user may want to find a recipe for an *apple pie with chocolate*. There is no such recipe in the case base but there exists a recipe of a *chocolate pear pie* which is the best match to the initial query (*retrieve step*). The *reuse step* may consist in adapting the recipe by replacing pears with apples, with an adjustment of the ingredient quantities This adapted recipe is proposed to a user who may give a feedback useful to improve this recipe. (*revise step*). Finally, the newly formed case is added to the case base (*retain step*).

2.2 RDF(S): a Semantic Web Technology

The term RDF(S) refers to the combined use of RDF, RDF Schema and SPARQL technologies.

RDF. Resource Description Framework [12] (RDF) provides a way for representing data, by using a metadata model based on labeled directed graphs. Information is represented using statements, called triples, of the form ⟨*subject predicate object*⟩. The subject s is a resource, the predicate p is a property, and the object o is either a resource or a literal value.

RDFS. RDF Schema [5] (RDFS) adds a logic upon the RDF model. A new set of specific properties are introduced: `rdfs:subclassof` (resp `rdfs:subpropertyof`) allows to create a hierarchy between classes (resp. properties). `rdfs:domain` (resp. `rdfs:range`) applies for a property and adds a constraint about the type of the resource which is in *subject* (resp. *object*) position for a triple. In the remainder of the article, short names will be used for these properties: a for `rdf:type`, domain for `rdfs:domain` and range for `rdfs:range`. The inference relation ⊢ is based on a set of inference rules [5].

SPARQL. SPARQL is the language recommended by the World Wide Web Consortium (W3C) to query RDF data [16]. Consider the following informal query:

$$\mathcal{Q} = \left| \begin{array}{l} \text{``Give me the letters sent by Henri Poincaré to} \\ \text{mathematicians between 1885 and 1890''} \end{array} \right.$$

This query can be represented using SPARQL:

```
        SELECT ?l
        WHERE {?l a letter .
               ?l sentBy henriPoincaré .
Q =            ?l sentTo ?person .
               ?person a Mathematician .
               ?l sentInYear ?y .
               FILTER(?y >= 1885 AND ?y <= 1890)}
```

For the sake of simplicity, in the remainder of the article, queries are presented in an informal way though all of them correspond to actual SPARQL queries.

3   The Henri Poincaré Correspondence: Edition and Online Publishing

3.1   Context and Works

Through the Henri Poincaré website, anyone can access the letters of the Henri Poincaré correspondence. About 60% of the letters are associated with a plain text transcription (in *XML* or LaTeX), a critical apparatus and a set of metadata. Metadata can either refer to the letter as a physical object (writing date, place of expedition, sender, recipient, etc.) or to the content of the letter (scientific topics discussed, people quoted, etc.). The content management system Omeka S [4] has been used to create this website and to publish data related to this correspondence. This platform enables the web-publishing and sharing of cultural collections from institutions (museums, archives, etc.). It operates with a MySQL database. A search engine using the Solr tool, which allows plain text search, has been implemented to retrieve transcribed letters. Although appropriate to some situations, this search engine suffers from a lack of expressiveness. In practice, historians often need to express more complex and structured queries. As an example, one can be interested in finding the letters sent by Henri Poincaré to members of his family in which he mentioned his classmates at the time he was a student of the "École polytechnique". To address this issue, an RDFS base has been initialized and is daily updated by translating the Omeka S content to Turtle files.[2] This database is structured using the *Archives Henri-Poincaré Ontology* which, in particular, describes resources and relations in the context of scientific correspondences. It gathers classes and properties related to persons, institutions, places, documents (e.g. books, articles, letters, etc.). This ontology is aligned with the use of several standard vocabularies (dcterms, bibo, rel, etc.). Three SPARQL querying modes have been created and are available on the website [7]. The *classical mode* requires to directly write SPARQL queries. The *form-based mode* proposes a set of input fields to assist the user in the generation of the query. The *graphical mode* presents a graph-based interface which lets the user manipulate nodes and edges to formulate the query.

As described in the introduction of this article, the annotation process is a tedious work which justifies the need of a dedicated editor to assist the user. This system should enable an efficient interactive update of an RDFS base, by visualizing the already edited statements and providing suggestions appropriate to the current annotation context.

3.2   Proposal of an Annotation Tool

A suggestion system has been developed to assist the user during the annotation process. This system has been implemented in Java. It comes with a set of parameters in order to connect to any given RDF base. It can be a file system base or a base reachable through a SPARQL endpoint. Different engines can be used to dialog with the RDF base (e.g. Jena [13], Corese [9]).

Associated to this system, a web user interface has been developed to use and compare the different versions of the system. This interface proposes an autocomplete mechanism that uses the suggestion system for providing values. The interface is common to all versions of the system. The tool enables the visualization and update of RDF databases. Three fields are available to set the values of subject, predicate and object. The use of prefixes has been implemented to improve the readability of the tool. The list of existing namespaces and associated prefixes is accessible through the "Prefixe" tab. When editing a triple, the editor displays the associated *context*. This corresponds to the set of already edited triples related to the current subject resource. For example, if the current subject is letter11, the editor displays the results of the execution of

---

[2] Turtle is a RDF serialization which is easily readable [8].

the query SELECT ?p ?o WHERE {letter11 ?p ?o}. This context is refreshed each time the value of the subject field is updated. When a new triple is created and inserted into the database, it is added to the current context. An excerpt of this interface is presented in Fig. 1. The full interface associated with several use cases is the subject of a presentation video accessible online.[3] The first two versions of the suggestion engine are presented here.

The *basic editor* assists the user by proposing an autocomplete mechanism in which the suggestions are ranked using the alphabetical order. The proposed suggestions do not depend neither on the current annotation problem nor on the available data and knowledge.

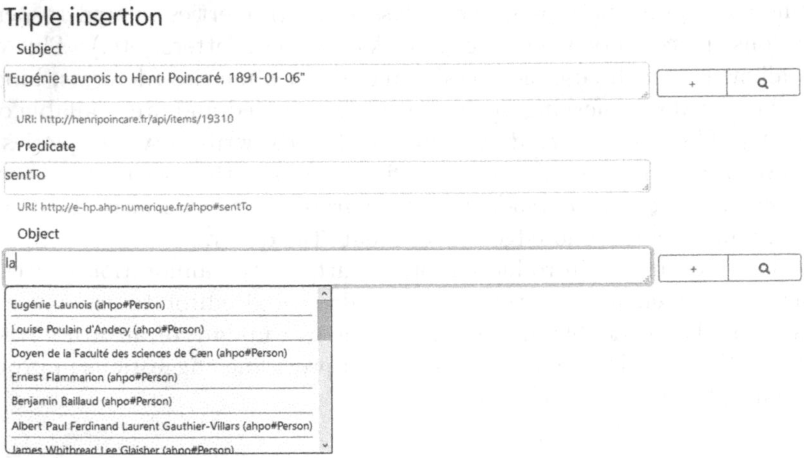

Triple insertion

Subject

"Eugénie Launois to Henri Poincaré, 1891-01-06"   +   Q

URI: http://henripoincare.fr/api/items/19310

Predicate

sentTo

URI: http://e-hp.ahp-numerique.fr/ahpo#sentTo

Object

la   +   Q

Eugénie Launois (ahpo#Person)
Louise Poulain d'Andecy (ahpo#Person)
Doyen de la Faculté des sciences de Caen (ahpo#Person)
Ernest Flammarion (ahpo#Person)
Benjamin Baillaud (ahpo#Person)
Albert Paul Ferdinand Laurent Gauthier-Villars (ahpo#Person)
James Whitbread Lee Glaisher (ahpo#Person)

Fig. 1. An excerpt of the RDFWebEditor4Humanities interface.

The *deductive editor* benefits from the use of RDFS knowledge for ranking the suggestions provided to the user. The notion of *annotation question* is introduced: this corresponds to a triple for which 1, 2 or the 3 fields are unknown, and for which a field is currently being edited. This field is represented by using a frame around an existential variable (i.e. $\boxed{?p}$, $\boxed{?o}$). For instance, $\langle s \ \boxed{?p} \ ?o \rangle$ corresponds to an annotation question type for which the subject is known, the predicate is currently being edited and the object is unknown. There exist twelve different annotation question types. For each of them, the knowledge about the **domain** and **range** can be used to rank the potential values for the targeted field.

Consider the annotation question $\langle \text{letter11} \ \text{sentBy} \ \boxed{?o} \rangle$ which is of the type $\langle s \ p \ \boxed{?o} \rangle$. The objective here is to provide appropriate suggestions by listing and ranking the potential values for the object field. For this version of the

---

[3] https://videos.ahp-numerique.fr/videos/watch/0d544e5b-b4be-423e-9497-216f29ab44f3.

editor, the top suggestions are resources of the classes **Person** and **Institution**. In the ontology, **Person** and **Institution** are **range** of the property **sentBy**. This knowledge is used to favor the instances of these classes. However, the resources that do not explicitly belong to these classes are still suggested because RDFS works with "open world assumption".[4]

There exist different rules which can be used to retrieve a list of potential values and whose applications depend on the type of the annotation question. For instance, the rule used to answer the annotation question presented in the example below is called **rangePred** and may apply for the annotation questions of the type $\langle s \ p \ \boxed{?o} \rangle$ and $\langle ?s \ p \ \boxed{?o} \rangle$. To answer an annotation question, a count is computed for each potential value ?v based on the number of rule applications which retrieved this value. The final list of suggestions is ranked according to this count (in decreasing order). For the potential values with the same count, the alphabetical order is used. 6 different rules have been defined. **domainPred** uses the knowledge about the domain of a given property p and may apply for the annotation questions of the type $\langle \boxed{?s} \ p \ o \rangle$ and $\langle \boxed{?s} \ p \ ?o \rangle$. The application of **predProperty** increases the count of each candidate solution which is defined as an **rdf:Property** for the annotation questions in which the target is the value in predicate position. **subjectInDomain** uses the value s defined in subject position: if s is an instance of a class D, each candidate value having D as domain will have its count incremented. It may apply for the annotation questions of the type $\langle s \ \boxed{?p} \ ?o \rangle$ and $\langle s \ \boxed{?p} \ o \rangle$. In a symmetrical way, **objectInRange** uses the range of the value o and may apply for the annotation questions of the type $\langle ?s \ \boxed{?p} \ o \rangle$ and $\langle s \ \boxed{?p} \ o \rangle$. For the annotation questions of the type $\langle \boxed{?s} \ ?p \ o \rangle$, each candidate value which is in the domain of a property whose range contains o will have its count incremented. This rule is called **subjImRel** and may apply in a symmetrical way for the annotation questions of the type $\langle s \ ?p \ \boxed{?o} \rangle$ by using the value s.

4   A Case-Based Editor

4.1   Case-Based Content Annotation

The use of RDFS deduction (as described in Sect. 3.2) brings a first improvement to the suggestion system by using the knowledge about the **domain** and **range** of the properties defined in the base. However, in some situations, this is not enough to propose the most appropriate resources for the current editing question. As an example, consider a triple currently being edited for which the subject is an instance of **Letter** (**letter2100**), the predicate is **sentTo** and for which suggestions for the object field are expected. As the class **Person** is **range** of

---

[4] If a fact is not asserted, it does not imply that this fact is false. In this situation, there may exist a resource r that is intended to represent a person (resp. institution) but is such that the triple $\langle r \ \texttt{a} \ \texttt{Person} \rangle$ (resp. $\langle r \ \texttt{a} \ \texttt{Institution} \rangle$) cannot be entailed by the *current* RDFS base. Therefore, r can be suggested as well, though further in the suggestion list.

the property `sentTo`, the system will favor the instances of this class in the suggestion list. But the problem is that there are many instances of this class in the base,[5] and there is no guarantee that the appropriate value will be among the first suggestions in the list. Indeed, for the values with the same count, the alphabetical order is used.

An alternative way to obtain a relevant ranking of the suggestion list would be to follow the CBR methodology: in the current situation, pieces of information from similar situations can be reused. In this framework, an annotation problem x^{tgt} is composed of an editing question and a context. For editing questions of the type $\langle s \; p \; \boxed{?o} \rangle$, it is defined as follows:

$$x^{tgt} = \begin{array}{|l|} \hline \text{question: } \langle subj^{tgt} \; pred^{tgt} \; \boxed{?o} \rangle \\ \text{context: the set of triples related to } subj^{tgt} \\ \hline \end{array}$$

For the running example, this gives:

$$x^{tgt} = \begin{array}{|ll|} \hline \text{question: } & \langle \texttt{letter2100 sentTo} \; \boxed{?o} \rangle \\ \text{context: } & \langle \texttt{letter2100 sentBy henriPoincaré} \rangle \\ & \langle \texttt{letter2100 hasTopic écolePolytechnique} \rangle \\ & \langle \texttt{letter2100 quotes paulAppell} \rangle \\ \hline \end{array}$$

The case base is the RDF database $\mathcal{D}_{HP}$. A source case is given by a triple $\langle subj^s \; pred^s \; obj^s \rangle$ of $\mathcal{D}_{HP}$, considered among all the triples of $\mathcal{D}_{HP}$, and which, in relation to x^{tgt}, can be decomposed into a problem x^s and a solution y^s:

$$x^s = \begin{array}{|l|} \hline \text{question: } \langle subj^s \; pred^s \; \boxed{?o} \rangle \\ \text{context: the database } \mathcal{D}_{HP} \\ \hline \end{array}$$

and the solution $y^s = obj^s$. For the purpose of this example, consider an excerpt $\mathcal{D}_{ex}$ of the Henri Poincaré correspondence database $\mathcal{D}_{HP}$ composed of the letters related to the following instances of the class `Person`: göstaMittagLeffler, alineBoutroux, eugénieLaunois, felixKlein and henriPoincaré. How should the list composed of these 5 resources be ordered? To propose a solution to this problem, the method consists in retrieving the cases which correspond the best to the current annotation problem. At each source case x^s is associated a value y^s which is used as a candidate solution to x^{tgt}. A count is computed for ranking the candidate solutions. This corresponds to the number of letters having this value associated with the property `sentTo`. An initial SPARQL query Q based on x^{tgt} is defined to compute this count. For the running example, it gives:

$$Q = \left| \begin{array}{l} \text{``Give me, for each potential value } \texttt{?o}\text{, the number of letters} \\ \text{having this value associated with the } \texttt{sentTo} \text{ property} \\ \text{where letters have } \texttt{écolePolytechnique} \text{ as topic,} \\ \text{quote } \texttt{paulAppell} \text{ and have been sent by } \texttt{henriPoincaré''} \end{array} \right.$$

---

[5] At the time of writing this article, there are around 1800 persons defined within the database.

The execution of Q on $\mathcal{D}_{ex}$ returns an empty set of results. Indeed, it is uncommon to find two different letters having exactly the same context. So, the question is to find a method to retrieve the most similar source cases. This issue can be addressed by using SPARQL query transformations. An engine has already been designed to manage transformation rules and has proven useful in different contexts including the search in the Henri Poincaré correspondence corpus and the case-based cooking system Taaable [6]. Rules are configured by the user and can be general or context-dependent. To each rule is associated a cost, corresponding to a query transformation cost. Two rules are considered in the running example:

- r_{exchange}: exchanges the sender and recipient of the letter (cost of 2);
- $r_{\text{genObjInst}}$: generalizes a class instance in object position (cost of 3).

A search tree can be explored starting from the initial query Q by applying one or several successive transformation rules. A maximum cost is defined to limit the depth of the search tree exploration. For this application, this maximum cost is set to 10.

At depth 1, the application of the rule r_{exchange} on Q generates the query Q_1 with a cost of 2 (the modified part of the query is underlined):

$$Q_1 = \left| \begin{array}{l} \text{``Give me, for each potential value ?o, the number of letters} \\ \text{having this value associated with the \textbf{sentBy} property} \\ \text{where letters have écolePolytechnique as topic,} \\ \text{quote \textbf{paulAppell} and have been \underline{received} by \textbf{henriPoincaré}''} \end{array} \right.$$

The result of the execution of Q_1 on $\mathcal{D}_{ex}$ is: [{eugénieLaunois : 2}, {alineBoutroux : 1}]. Three applications of the rule $r_{\text{genObjInst}}$ exist at depth 1, each of them for a cost of 3. The first one applies for the people quoted, by replacing **paulAppell** by any instance of the class **Mathematician** (because Paul Appell belongs to that class), the second one applies for the sender of the letter, and the last one applies for **écolePolytechnique** by replacing the value by any instance of the class **Topic**. The generated queries are Q_2, Q_3 and Q_4:

$$Q_2 = \left| \begin{array}{l} \text{``Give me, for each potential value ?o, the number of letters} \\ \text{having this value associated with the \textbf{sentTo} property} \\ \text{where letters have écolePolytechnique as topic,} \\ \text{quote \underline{a Mathematician} and have been sent by \textbf{henriPoincaré}''} \end{array} \right.$$

$$Q_3 = \left| \begin{array}{l} \text{Give me, for each potential value ?o, the number of letters} \\ \text{having this value associated with the \textbf{sentTo} property} \\ \text{where letters have écolePolytechnique as topic,} \\ \text{quote \textbf{paulAppell} and have been sent by a \underline{Mathematician}} \end{array} \right.$$

$$Q_4 = \left| \begin{array}{l} \text{Give me, for each potential value ?o, the number of letters} \\ \text{having this value associated with the \textbf{sentTo} property} \\ \text{where letters have a \underline{defined topic},} \\ \text{quote \textbf{paulAppell} and have been sent by \textbf{henriPoincaré}} \end{array} \right.$$

The execution of Q_2 on $\mathcal{D}_{ex}$ gives: [{eugénieLaunois : 138}, {alineBoutroux : 4}]. The executions of Q_3 and Q_4 give no result. As the maximum cost has been set to 10, it is possible to continue the tree exploration on the different branches to find new possible suggestions. At depth 2, the application of $r_{\texttt{genObjInst}}$ on Q_2 (applied for the topic) generates the query:

$$Q_{21} = \left| \begin{array}{l} \text{Give me, for each potential value ?o, the number of letters} \\ \text{having this value associated with the } \texttt{sentTo} \text{ property} \\ \text{where letters have a } \underline{\text{defined topic,}} \\ \text{quote a } \texttt{Mathematician} \text{ and have been sent by } \texttt{henriPoincaré} \end{array} \right.$$

The execution of Q_{21} on $\mathcal{D}_{ex}$ gives: [{eugénieLaunois : 280}, {göstaMittagLeffler : 74}, {alineBoutroux : 17}]. At depth 3, the application of $r_{\texttt{genObjInst}}$ on Q_{21} (applied for the sender of the letter) generates the query:

$$Q_{211} = \left| \begin{array}{l} \text{Give me, for each potential value ?o, the number of letters} \\ \text{having this value associated with the } \texttt{sentTo} \text{ property} \\ \text{where letters have a defined topic,} \\ \text{quote a } \texttt{Mathematician} \text{ and have been sent by a } \underline{\texttt{Mathematician}} \end{array} \right.$$

The execution of Q_{211} on $\mathcal{D}_{ex}$ gives: [{eugénieLaunois : 305}, {henriPoincaré : 219}, {göstaMittagLeffler : 141}, {felixKlein : 25}, {alineBoutroux : 21}]. The other possible rule applications (considering maximum cost) generate queries already generated by other combinations or which give the same resources but with a greater cost.

The final list of suggestions is ranked by ordering the resources based on the required minimal transformation cost. For resources with the same minimal cost, the count related to the execution of the query associated to this cost is used (in a decreasing order). For the running example, this gives, for the first 5 suggestions from number 1 to number 5: eugénieLaunois, alineBoutroux, göstaMittagLeffler, henriPoincaré[6] and felixKlein. The remainder of the suggestions is composed of all the resources of the database ranked using the alphabetical order.

This approach constitutes the *retrieve* step of the CBR model. The *reuse* step is a *reuse as such* approach: there is no modification of the proposed resources. After this, the user chooses the appropriate resource, which could be considered as a *revise* step. Then the edited triple is inserted into the database (*retain* step).

4.2 Combining RDFS Deduction with CBR

The last version of the editor combines the use of RDFS deduction with CBR. It takes advantage of both the knowledge about the resources similar to the

---

[6] This suggestion could be removed if the system knows that the recipient of a letter cannot be its sender. This is considered again in the future work part of this article conclusion.

one being edited and the **domain** and **range** of the properties used during the editing. The resources found using CBR are on top of the suggestion list. For the other resources, a count is computed for ranking the potential values as presented in Sect. 3.2. Consider the example presented above, in which the editing question was ⟨letter2100 sentTo ⌷?o⌷⟩ and suggestions for the object field were expected. Using the CBR version of the system, the first five suggestions are resources which seem to be pertinent considering the current editing context and by looking for the similar objects in the database. But for the remainder of the suggestions, only the alphabetical order is used for ranking. This can be addressed by using the **range** of the property **sentTo** (as explained in Sect. 3.2) for ranking the second part of the suggestions list (from the 6th value). As **Person** is **range** of the property **sentBy**, all the instances of this class would be higher in the suggestions list than instances of other classes.

5   Evaluation

The goal of the evaluation is to compare the efficiency of the different versions of the system for concrete annotation situations. The first evaluation is human-based through a user who will test and compare the four versions. A second evaluation is managed through a dedicated program and will provide objective measures. Both evaluations focus on a subset of 7 properties among the most frequently used when editing letters: **sentBy** defines the sender; **sentTo** defines the recipient; **hasTopic** gives one of the topics; **archivedAt** specifies the place of archive; **hasReply** gives a letter responding to the current letter; **replies** gives a letter to which the current letter responds; **citeName** refers to a person mentioned in the letter transcription.

5.1   Human Evaluation

This evaluation involves a single user who is one of the people in charge of the editing of the Henri Poincaré correspondence corpus. He was using Omeka S before moving on to the system presented in this article. He had no previous experience with this tool at the time he carried out the evaluation. The test set is composed of 10 letters which have been randomly chosen from a set of 30 unpublished letters from the Henri Poincaré correspondence corpus. This set constitutes a real annotation case with respect to the already edited letters in the corpus database. The new items from the evaluation corpus have been edited using Omeka S before the start of the evaluation, so as to ensure that no version of the system would suffer from being the first one to be evaluated. For each version (presented in a random order), the user edits (i.e. create the triples) the same 10 letters using the interface provided with the tool. Before switching to the next version, the RDF database is reset to correspond to the initial state.

   After having edited the complete set of letters with one version, the user is invited to complete a survey and to provide feedback about this version. This survey insists on the appreciation of the autocomplete mechanism efficiency (but

Table 1. The average score (on a 1 to 7 scale) associated with the suggestions provided by the four versions of the system.

	Basic editor	Deductive editor	Cased-based editor	Combination editor
Average score	3.4	5.7	5.3	7

experience feedback about the user interface is also expected). For each property, the user is invited to attribute a score by using a *Likert scale* [2], from 1 (not at all relevant) to 7 (very relevant) to characterize the relevance of the suggestions provided for annotation questions linked to that property.

What emerges of this evaluation is that the combination editor has been perceived as the most efficient, and this for all the properties of the evaluation. The average scores of all property evaluations are given on Table 1. The basic editor is the version that obtained the lowest score. It has been perceived as "not assisting the annotation", but still not causing any problems to the user. The deductive and case-based editors got high average scores. However, in situations in which the retrieval of source cases leads to an empty set of cases, the CBR engine only uses the alphabetical order for ranking the list of suggestions, and may provide irrelevant resources. This caused frustration for the user and explains why the average score of the CBR engine is lower than the deductive engine. Combining the two engines is a good method to counter these situations.

Furthermore, the interface associated with the tool helped avoiding the mistakes described in the introduction: it prevents the insertion of triples which already exist in the database (*duplication mistake*), the type of the selected resource is always visible (*ambiguity mistake*), and the use of labels simplify the management of resources for the user (*typing mistake*). The user felt in control when he was performing actions to alter the database. At the end of the evaluation, the user has proceed with a few more tests to compare Omeka S with the last version of the editor. He has estimated that the time required for the annotation of a letter using the combination editor was about half the time he needed with Omeka S.

5.2   Automatic Evaluation

The aim of the automatic evaluation is to compare the performances of the different versions of the tool by computing measures. The chosen measures are related to the rank of the expected value $\texttt{rank}(aq)$ where aq is the current annotation question. $\texttt{rank}(aq) = 1$ means that the associated value is the first in the suggestion list. In others words, the lower the rank is, the better the version of the system is.

The RDF graph of the Henri Poincaré correspondence $\mathcal{G}_{\text{HP}}$ has been used as a test set. This graph is formed by the union of the database $\mathcal{D}_{\text{HP}}$ and the ontology $\mathcal{O}_{\text{HP}}$: $\mathcal{G}_{\text{HP}} = \mathcal{D}_{\text{HP}} \cup \mathcal{O}_{\text{HP}}$. At the time of writing this article, the RDF database $\mathcal{D}_{\text{HP}}$ is composed of around 220 000 triples. The database and ontology triples are stored in Turtle files. For this evaluation, the application of the inference

Table 2. Rank measures for the suggestions provided by the four versions of the system.

	Basic editor	Deductive editor	Case-based editor	Combination editor
$rank \leq 15$	11.3%	22.11%	49.0%	49.5%
$rank \leq 10$	7.1%	21.15%	43.2%	43.2%
$rank \leq 5$	2.7%	19.23%	33.6%	34.7%

rules mentioned in Sect. 2 has been considered. Different classes of items exist in the database (e.g. `Letter`, `Person`, `Article`, etc.) but, for this evaluation, the focus is put on the editing of letters. A set of 100 letters is randomly extracted from the existing set of annotated letters. For each letter of this set, the related triples part of the context are used to simulate annotation questions for which the answer is already known. For each triple, the order of editing of the 3 fields (subject, predicate and object) is considered in a random order so as to include various annotation question types in the evaluation. For each annotation question aq, the four suggestion engines are called to provide an ordered suggestion list. The rank of the expected value $rank(aq)$ in the list is saved for each version and is added to the related multi-set $Ranks(system)$. At the end of the evaluation, measures related to the elements of $Ranks(system)$ are computed. These measures correspond to the percentage of annotation questions for which the expected value was given among the n first propositions $(rank \leq n)$.

The results of this evaluation are given in Table 2 for each version of the system. Different values of n have been chosen (5, 10 and 15) but the evolution of the efficiency of the versions is the same in all situations. This shows that the combination editor provides the best results for the different annotation questions related to this evaluation because it suggests the appropriate value more often. It is thus more likely to assist the user during the annotation process.

Although not used as a measure during the evaluation, the computing time has been considered. It corresponds to the time needed to provide the suggestion list for an annotation question. Indeed, the reaction time to requests should be considered in a human interaction system especially since this system is using an autocomplete mechanism for which a user expects no latency. The computation time is more important when using the combination editor but this stays low enough not to impact the user (around 1 s for a standard laptop).

6   Discussion and Related Work

The method presented in Sect. 4 is inspired from the UTILIS system [11]. This system introduces the idea of looking for resources similar to the one being edited to suggest values that might be appropriate to the current annotation problem. But the form of query relaxation proposed is different as it mainly uses generalization rules. The engine presented in this article allows the users to define their own rules to correspond to a specific database. Combined with the use of RDFS knowledge, it allows to propose suggestions appropriate to various annotation question types.

One of the most frequently used tools for editing Semantic Web data is Protégé [14]. When editing an instance of a specific class, Protégé uses the domain and range of the properties to make suggestions for predicate and object values. But these suggestions do not have profit from the already edited triples. Other approaches exist to assist the editing of RDF databases, several of them being based on natural language processing. The GINO editing tool [3] proposes the use of a guided and controlled natural language which lets the user specify sentences corresponding to statements. The main idea is that the principles of Semantic Web are sometimes not easily apprehended by non specialists, and thus should be encapsulated within a more user-friendly system. The syntax of this language is close to English syntax (e.g. "There is a mount named Everest", "The height of mount Everest is 29 029 feet", etc.). A suggestion mechanism proposes classes, instances and properties to complete the current annotation. These suggestions are ranked alphabetically and are consistent with the defined ontology. The main challenge in this system is the interpretation of the user request to build triples from sentences.

More generally, the tool that is presented in this article could be categorized as a recommender system. These systems intend to assist the user by presenting information likely to interest her/him. Different recommender systems, such as the one presented here, use case-based recommendation [18]. A great variety of methods exists, and the tool presented in this article could benefit from several of them. As an example, the involvement of the user in the suggestion proposal mechanism is considered. The explainability of the tool could be reinforced as it may be important to understand why some resources are more favored than others. This system could also benefit from the use of a preference-based feedback system which could improve the results of the tool in several situations, make the user feel included and thus reinforce the positive view about the tool. On the other hand, the query transformation mechanism which has been used in this application framework could be reused in other recommender systems.

7   Conclusion

The use of Semantic Web technologies has proven useful for the corpus of the correspondence of Henri Poincaré. A manual annotation process has been chosen to edit data related to items of this corpus (e.g. *letters*, *persons*, *places*, etc.) This process has been identified tedious for users in charge of the editing. To deal with this issue, a tool providing a suggestion system has been proposed. It intends to be a general tool for the editing of Semantic Web data. It uses some inferences with RDFS entailment combined with a CBR methodology. Different versions of the system have been implemented. The first version ranks the potential values by using the alphabetical order. The second version takes advantage of the knowledge about the domain and range for the properties of the base. The third version uses CBR to exploit the knowledge about similar edited resources. The last version is a combination of the two latter versions. Two different evaluations have been conducted. The human evaluation allowed to compare the different versions of the system between them and with the current existing annotation system (Omeka S). The automatic evaluation brought metrics by comparing, for a selected set of annotation questions, the suggestions of the different versions of the system. The relevance of the suggestions and the computing time have been taken into account. As explained in Sect. 5, while the metrics computed by the automatic evaluation show that the use of CBR alone brought better results than the use of RDFS deduction alone, the case-based version is sometimes insufficient and can provide irrelevant resources in some situations. The hypothesis 1 is validated by the automatic evaluation but not by the human evaluation. For both evaluations, the results show that the last version of the system combining the use of RDFS deduction with CBR is the most efficient and thus validates the hypothesis 2 stated in the introduction. However, in some situations, this system tends to show some limitations. For instance, consider the annotation question presented in Sect. 4. Both third and fourth versions of the system proposes `henriPoincaré` as a plausible answer although he is already defined as the sender of the current edited letter. A way to deal with this issue would be the use of some domain knowledge used as integrity constraints. For this example, the piece of knowledge "A resource cannot be at the same time the recipient and the sender of letter" could be used to prevent the suggestion of `henriPoincaré` as the recipient of a letter he sent. Another point observed during both human and automatic evaluations is that the order of editing of the different properties affects greatly the efficiency of the suggestion engine. Indeed, some properties values give more information about the resource than others, and thus having these values filled first should improve the ranking of the suggestions. Main challenge is to find the best order of editing for the properties of the base. This constitutes a future work. Another future work is related to the use of a more expressive logic than RDFS such as OWL-DL. A logic containing a form of negation would enable to remove some values from the list of potential values. However, such an extension could affect the computation time and its implementation should be investigated.

Although the RDFWebEditor4Humanities tool is now used for the editing of RDF data, Omeka S still provides some useful functionalities. It forms a stable environment for both editing and publishing of the items related to that corpus. Two solutions are considered: the first one consists in integrating some functionalities of Omeka S in the new annotation tool; the second one considers the creation of a new Omeka S module which would call the suggestion system to assist the user during the annotation process.

Acknowledgments. This work was supported partly by the french PIA project "Lorraine Université d'Excellence", reference ANR-15-IDEX-04-LUE. It was also supported by the CPER LCHN (Contrat de Plan État-Région Lorrain "Langues, Connaissances et Humanités Numériques") that financed engineer Ismaël Bada who participated to this project. We greatly thank Mickaël Smodis who is a final user of the tool and who participated to the human evaluation and Laurent Rollet who provided us with unpublished letters of the Henri Poincaré correspondence.

References

1. Aamodt, A., Plaza, E.: Case-based reasoning: foundational issues, methodological variations, and system approaches. AI Commun. **7**(1), 39–59 (1994)
2. Allen, I.E., Seaman, O.A.: Likert scales and data analyses. Qual. Prog. **40**(7), 64–65 (2007)
3. Bernstein, A., Kaufmann, E.: GINO – a guided input natural language ontology editor. In: Cruz, I., et al. (eds.) ISWC 2006. LNCS, vol. 4273, pp. 144–157. Springer, Heidelberg (2006). https://doi.org/10.1007/11926078_11
4. Boulaire, C., Carabelli, R.: Du digital naive au bricoleur numérique: les images et le logiciel Omeka. In: Cavalié, É., Clavert, F., Legendre, O., Martin, D. (eds.) Expérimenter les humanités numériques. Des outils individuels aux projets collectifs. Les Presses de l'Université de Montréal, Montréal, Québec, pp. 81–103 (2017)
5. Brickley, D., Guha, R.V.: RDF Schema 1.1, W3C recommendation (2014). https://www.w3.org/TR/rdf-schema/. Accessed Aug 2020
6. Bruneau, O., Gaillard, E., Lasolle, N., Lieber, J., Nauer, E., Reynaud, J.: A SPARQL query transformation rule language—application to retrieval and adaptation in case-based reasoning. In: Aha, D.W., Lieber, J. (eds.) ICCBR 2017. LNCS (LNAI), vol. 10339, pp. 76–91. Springer, Cham (2017). https://doi.org/10.1007/978-3-319-61030-6_6
7. Bruneau, O., Lasolle, N., Lieber, J., Nauer, E., Pavlova, S., Rollet, L.: Applying and developing semantic web technologies for exploiting a corpus in history of science: the case study of the Henri Poincaré correspondence. Semant. Web. IOS Press Journal (2020)
8. Carothers, G., Prud'hommeaux, E.: RDF 1.1 Turtle (2014). http://www.w3.org/TR/2014/REC-turtle-20140225/. Accessed Aug 2020
9. Corby, O., Dieng-Kuntz, R., Faron Zucker, C.: Querying the semantic web with corese search engine. In: López de Mántaras, R., Saitta, L. (eds.) European Conference on Artificial Intelligence, Valence, Spain, pp. 705–709, August 2004
10. Cordier, A., et al.: Taaable: a case-based system for personalized cooking. In: Montani, S., Jain, L.C. (eds.) Successful Case-Based Reasoning Applications-2. SCI, vol. 494, pp. 121–162. Springer, Cham (2014). https://doi.org/10.1007/978-3-642-38736-4_7
11. Hermann, A., Ferré, S., Ducassé, M.: An interactive guidance process supporting consistent updates of RDFS graphs. In: ten Teije, A., et al. (eds.) EKAW 2012. LNCS (LNAI), vol. 7603, pp. 185–199. Springer, Heidelberg (2012). https://doi.org/10.1007/978-3-642-33876-2_18
12. Manola, F., Miller, E., McBride, B., et al.: RDF Primer (2004). https://www.w3.org/TR/rdf-primer. Accessed Aug 2020
13. McBride, B.: Jena: a Semantic Web toolkit. IEEE Internet Comput. **6**(6), 55–59 (2002)
14. Noy, N.F., Sintek, M., Decker, S., Crubézy, M., Fergerson, R.W., Musen, M.A.: Creating semantic web contents with Protégé-2000. IEEE Intell. Syst. **16**(2), 60–71 (2001)
15. Poincaré, H.: La Science et l'Hypothèse. Flammarion, Paris (1902)
16. Prud'hommeaux, E.: SPARQL Query Language for RDF, W3C Recommendation (2008). http://www.w3.org/TR/rdf-sparql-query/. Accessed Aug 2020

17. Riesbeck, C.K., Schank, R.C.: Inside Case-Based Reasoning. Lawrence Erlbaum Associates Inc., Hillsdale (1989)
18. Smyth, B.: Case-based recommendation. In: Brusilovsky, P., Kobsa, A., Nejdl, W. (eds.) The Adaptive Web. LNCS, vol. 4321, pp. 342–376. Springer, Heidelberg (2007). https://doi.org/10.1007/978-3-540-72079-9_11

A First Step Towards a Streaming Linked Data Life-Cycle

Riccardo Tommasini[2]([⊠])(ID), Mohamed Ragab[2](ID), Alessandro Falcetta[1],
Emanuele Della Valle[1](ID), and Sherif Sakr[2](ID)

[1] DEIB, Politecnico di Milano, Milan, Italy
{alessandro.falcetta,emanuele.valle}@polimi.it
[2] DataSystem Group, University of Tartu, Tartu, Estonia
riccardo.tommasini@polimi.it, {mohamed.ragab,sherif.sakr}@ut.ee

Abstract. Alongside with the ongoing initiative of FAIR data management, the problem of handling *Streaming Linked Data* (SLD) is relevant as never before. The Web is changing to tame Data Velocity and fulfill the needs of a new generation of Web applications. New protocols (e.g. Web-Sockets and Server-Sent Events) emerge to grant continuous and reactive data access. Under the Stream Reasoning initiative, the Semantic Web community has been actively working on query languages, engines, and vocabularies to address the scientific and technical challenges of taming Data *Velocity* without neglecting Data *Variety*. Nevertheless, a set of guidelines that showcase how to *reuse* existing resources to produce and consume streams on the Web is still missing. In this paper, we walk through the life-cycle of streaming linked data. We discuss the challenges of applying FAIR principles when publishing data streams. Moreover, we contextualise the usage of prominent Semantic Web resources, i.e., (i) `TripleWave`, R2RML/RML, `VoCaLS`, `RSP-QL`. We apply the guidelines to three representative examples of *real-world* Web streams: `DBpedia` Live changes, `Wikimedia` EventStreams, and the Global Database of Events, Language and Tone (`GDELT`). Last but not least, we open-sourced our code at https://w3id.org/webstreams.

Keywords: RDF Streams · Streaming linked data · RDF Stream Processing · Stream reasoning

1 Introduction

Alongside with the ongoing initiative of FAIR data management [30], the Semantic Web community, under the Stream Reasoning umbrella [12,19], started investigating the problem of Streaming Linked Data (SLD) management.

The Web is changing to fulfill the needs of a new generation of Web applications that demands real-time data access [9]. On the one hand, new protocols like `HTTP Long Polling`[1] and `WebSockets`(see footnote 1), and `Server-Sent`

---

[1] https://www.pubnub.com/learn/glossary/what-is-http-streaming/.

© Springer Nature Switzerland AG 2020
J. Z. Pan et al. (Eds.): ISWC 2020, LNCS 12507, pp. 634–650, 2020.
https://doi.org/10.1007/978-3-030-62466-8_39

Events[2] are becoming popular. On the other hand, the Stream Reasoning initiative is actively working to extend the Semantic Web Stack for RDF Stream Processing (RSP)[3].

RSP aims at taming Data Velocity, i.e., the challenge of processing data as soon as they are produced and before they are no longer valuable, without neglecting Data Variety [10]. RSP extends the Semantic Web stack with a data model (i.e., RDF Stream), continuous extensions of SPARQL (e.g., RSP-QL [11]), and a number working prototypes that enable *on-the-fly* analysis of vast and heterogeneous data streams coming from complex domains [5,24]. The literature also includes middleware systems like LSM and Ztreamy [3,24], resources for publishing like TripleWave [20], and vocabularies to describe data streams [28].

Nevertheless, the quest for streams on the Web is still not FAIR. The Streaming Linked Data life-cycle is still unclear since a set of guidelines that explain how to produce and consume Streaming Linked Data is still missing. In this paper, we fill this gap presenting a Streaming Linked Data life-cycle. Additionally, we bootstrap a catalog of linked streams publishing three Real-World Web Streams [20]: DBPedia Live[4] that shares the RDF changes on DBPedia, EventStreams[5] that streams the changes across all the WikiMedia projects, and the Global Database of Events, Languages, and Tone (GDELT)[6]. We choose these streams to illustrate a comprehensive set of challenges that need to be tackled when publishing streams on the Web. In summary, the main contributions of this paper are:

1. We discuss the challenges to publish FAIR streaming data on the Web.
2. We present a publication lifecycle that is suitable for streaming linked data.
3. We include a set of representative examples for producing and consuming Streaming Linked Data that *reuse* existing resources (i.e., RML, TripleWave, VoCaLS), and RSP-QL.

This study is relevant for the Semantic Web community, in particular the RSP community, given the growing interest in Streaming Linked Data [23]. Moreover, the selected streams are relevant for those whose interests includes Knowledge Graph, news analysis, and Fact-Checking.

The remainder of the paper is organized as follows: Sect. 2 presents the Streaming Linked Data life-cycle, it discusses FAIR principles for streaming data management, and positions related resources. Sect. 3 presents the three real-world Web Streams, i.e., DBPedia Live, Wikimedia EventStreams, and GDELT. Sect. 4 applies the life-cycle to the selected streams. It highlights the assumptions, the design choices, and the open problems. Moreover, it shows examples of RDF Stream Processing using RSP-QL. Finally, Sect. 5 draws to conclusions, and presents future works.

---

[2] https://developer.mozilla.org/en-US/docs/Web/API/EventSource.
[3] https://www.w3.org/community/rsp/.
[4] https://wiki.dbpedia.org/online-access/DBpediaLive.
[5] https://stream.wikimedia.org.
[6] https://gdeltproject.org.

2 Streaming Linked Data Life-Cycle

In this section, we present the publication life-cycle for Streaming Linked Data. A premise to data publication is the identification of relevant sources. Figure 1 shows the three situations a practitioner might find when they publish Streaming Linked Data, i.e., our ultimate goal, which is identified by the *lower-right* quadrant.

The other quadrants present possible starting points, i.e., (upper-left) Web Data published in batches; (upper-right) Linked Data published in batches; and (lower-left) Web Data published as streams. Before introducing the life-cycle, we present a requirement analysis. In particular, we start from the FAIR principles [30] for data management, which aim at publishing data in a format that is Findable, Accessible, Interoperable, and Reusable, and we discuss their applicability to streaming data.

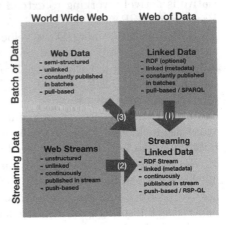

Fig. 1. Publication starting points.

2.1 FAIR Principles for Streaming Data

Findable. (F1) data should be assigned unique and persistent identifiers, e.g., DOI or URIs. (F2) data should be assigned metadata that includes descriptive information, data quality, and context. (F3) metadata should explicitly name the persistent identifier since they often come in a separate file. (F4) Identifiers and metadata should be indexed or searchable.

Requirements related to findability immediately apply to Streaming Data. It is not hard to imagine a Web Stream as a resource that is uniquely identified, thoroughly described with rich metadata, indexed, searched.

For effective machine-readable stream representation, Tommasini et al. proposed The Vocabulary for Cataloging and Linking Streams (VoCaLS) [28]. VoCaLS builds on DCAT[7] and is organized into three modules: Core, Service Description, and Provenance. In particular, *VoCaLS Core* follows the F-principles, i.e., a (i) *vocals:Stream* represents a Web stream, while a (ii) *vocals:StreamDescriptor* is an HTTP-accessible document that contains the stream's metadata.

Accessible. (A1) Data and metadata should be accessible via (a) free and (b) open-sourced, and (c) standard communication protocols, e.g., HTTP or FTP. Nonetheless, authorization and authentication are possible. (A2) metadata should be accessible even when data is no longer available.

---

[7] https://www.w3.org/TR/vocab-dcat/.

Fig. 2. Streaming linked data publication lifecycle.

Streaming data introduces a paradigm-shift for what concerns data access. Data are no longer put at rest and analysed. Instead, data are processed as soon as they arrive and before they are no longer useful. This change reached the Web architecture, which includes new protocols like *Server-Sent Events* for reactive processing and WebSocket for continuous consumption. These standardized solutions full-fill velocity-related requirements [26] as well as A1 and A2. Nevertheless, the Web is still founded on HTTP. Thus, Tommasini et al. proposed to decouple the stream description from the endpoint that enables data distribution. This approach ensures scalability and interoperability because a stream might have multiple endpoints. VoCaLS, in its current version, allows defining a *vocals:StreamEndpoint* that points to actual data sources.

Interoperable. (I1) Data and metadata must be written using formal languages and shared vocabularies that are accessible to a broad audience. (I2) Such vocabularies should also fulfill FAIR principles. (I3) Data and metadata should use qualified references to other (meta-)data.

In the literature, there is a general agreement on *RDF Stream* as the datamodel of choice for making streaming data interoperable (cf Definition 1).

Definition 1. *An RDF Stream is a data stream where the data items o are RDF graph, i.e., a set of RDF triples of the form (subject, predicate, object) [11].*

Prominent vocabularies to use for the stream content include but are not limited to FrAPPE [4], SAO [14], SSN [7] ,and SIOC [22]. Recently, Schema.org includes the classes relevant for streams representation, i.e., DataFeed[8] and DataFeedItem[9], but their adoption has not been estimated yet.

Reusable. (R1) Data should adopt an explicit license for access and usage. (R2) Data provenance should be documented and accessible. (R3) Data and metadata should comply with community standards.

Requirements related to reusability apply straightforwardly to streaming data. In particular, *VoCaLS* provenance modules allow modeling streaming data generation and transformation (R2). Additionally, requirements and commandments recognized by the stream reasoning community are available (R3).

2.2 Five-Steps into Streaming Linked Data

In this section, we present our publication life-cycle for Streaming Linked Data. Figure 2 summarises our proposal that takes into account FAIR principles, W3C

---

[8] https://schema.org/DataFeed.

[9] https://schema.org/DataFeedItem.

best practices[10], and the seminal of Hyland et al. [17], Hausenblas et al. [8] Villazon-Terrazas et al. [29]. Figure 2 shows the following steps: (0) Name (1) Model, (2) Describe, (3) Convert, (4) Publish, and (5) Processing. We elaborate on steps (0)–(4) in the following, while we postpone step (5), which is usually a corollary to publication, to Sect. 4.2.

Step 0 Name Things with (HTTP) URIs. The goal of Step 0 is to design (HTTP) URIs that identify the relevant resources [17]. To this extent, W3C presents a set of best practices for publishing[11], i.e., do not bind URIs to any implementation, keep them stable, and opaque[12]. Moreover, Linked Data Principles prescribe to use (HTTP) URIs to identify resources.

In the RSP community, The general trend is decoupling the access to the stream resource, which is available via HTTP, and the stream content, which is not a resource and can use an arbitrary Web protocol (e.g. WebSocket) [6,28]. Moreover, Sequeda et al. [25] proposed a URI-based mechanism to identify and access stream data items that take into account temporal and spatial aspects.

Step 1 Model the Streams. The goals of Step 1 are: (i) understanding and capturing the domain knowledge for enriching the raw [17] as well as (ii) identifying relevant resources.

In practice, this process requires collecting and reviewing applications, documentation, and data samples Then, with the help of knowledge engineers, domain knowledge, is captured into ontological models using a knowledge representation languages like RDFS or OWL2. The reuse of existing authoritative vocabularies is of paramount importance for FAIR data management.

During Step 1, the stakeholders ask questions that are translated into information needs [17]. For data streams, some information needs imply continuous semantics and their answer change over time, e.g., *What is the stream rate? What are the most relevant resources in the stream in the last 5 min?* Moreover, standard information needs are meaningful, e.g., *What is the stream main topic? For what statistics are useful? What problem you aim to solve?* The identified information needs should be formulated in natural language, and then *post-hoc* formalized using an appropriate formal language.

Step 2 Describe the Stream. Step 2 aims at providing representations for the streams [17] that can be consumed by both human and software agents.

To this extent, VoCaLS is the vocabulary of choice [28]. It allows writing machine-readable stream descriptions to share via HTTP. Moreover, it enables continuous data access via specified stream endpoints with streaming protocols.

In practice, describing a stream requires identifying relevant metadata such as publishers, the domain of provenance, documentation, and related resources. If the streaming data re-uses ontologies or schemas, they should be linked alongside the license for accessing the streaming data. Additionally, time-varying statics such as the stream rate and the number of active consumers should be available.

---

[10] https://www.w3.org/TR/ld-bp/.

[11] https://www.w3.org/TR/cooluris/#cooluris.

[12] I.e., do not let the user understand the underlying infrastructure.

Step 3 Convert Data to RDF Streams. The goal of Step 3 is fostering interoperability by enabling RDF (Stream) provisioning. Indeed, data are frequently shared using document-based format (e.g., JSON, XML), APIs, or using (semi)-structured data formats (e.g., TSV, CSV).

In practice, the conversion mechanism occurs automatically using approaches based on mapping languages that decouple conversion and modeling [17]. For instance, R2RML[13], and its extension RML [13], allow converting almost any non-RDF data source. Nevertheless, existing mapping engines were successfully used only for *batch* data conversion and, thus, they must be adapted or extended to work with Streaming Linked Data. The conversion mechanism involving data stream, due to their infinite nature, calls for *continuous* semantics, i.e., producing an infinite output from an infinite input [12]. Although this research problem is still open, a suitable solution is converting one stream-element a time [20].

Step 4 Serve Streams on the Web. The goal of this step is to provide the data to the audience of interest. Static Datasets – opportunely described with contextual vocabularies [1] – are either added to the Linked Open Data (LOD) Cloud, shared using REST APIs, or exposed via SPARQL endpoints. Critical aspects of Step 6 are licensing, audit, and access control [8,17].

Currently, one cannot add Streaming Linked Data as they are to the LOD Cloud [20,28], nor one can query them via SPARQL endpoints [12]. However, *Triple-Wave* [20] is a resource for publishing Streaming Linked Data that exposes an

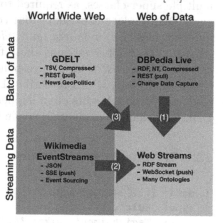

Fig. 3. Stream examples overview.

S-Graph via HTTP, i.e., a "static" named graph called that acts as stream descriptor. The S-Graph can be added to the LOD Cloud or indexed by search engine, making the stream findable and accessible via available endpoints.

3 Real-World Web Streams

In this section, we present three representative examples of real-world Web Streams published as Streaming Linked Data, i.e., DBPedia Live, Wikimedia EventStream, and the Global Dataset of Event, Language and Tone (GDELT). Our goal is to provide guidelines that (i) motivate the samples of RDF streams we selected to bootstrap our catalog, and (ii) facilitate future extensions of the catalog. Therefore, we selected three examples that cover all the publication scenarios. Figure 3 positions each of them w.r.t. the scenarios from Fig. 1.

---

[13] https://www.w3.org/2001/sw/rdb2rdf/r2rml/.

3.1   DBPedia Live

DBPedia is a community project to extract structured data from Wikipedia in the form of an Open Knowledge Graph. DBPedia is served as linked data using Entity-Pages, REST APIs, or SPARQL endpoints. The latest version of DBPedia counts more than 470 millions of RDF triples[14].

DBPedia Live (DBL) is a changelog stream continuously published to keep DBPedia replicas in-synch. A Synchronization tool designed to consume the batches and update the local DBPedia copy is available [21] DBL uses a pull-based mechanism for data provisioning. DBL shares RDF data using DBPpedia ontology (DBO), which is a cross-domain ontology, manually created from the most used Wikipedia info-boxes. The latest version of DBO covers 685 classes and 2795 different properties. It is a direct-acyclic graph, i.e., classes may have multiple super-classes, as required to map it to schema.org.

A DBL update consists of four compressed N-Triples (NT) files. Two main different files, i.e., *removed*, and *added*, determine the insertion and deletion stream. Two further streams share files for clean updates that are optional to execute: *reinserted*, which corresponds to unchanged triples that can be reinserted, and *clear*, which prescribe the delete queries that clear all triples for a resource. Although edits often happen in bulk, this information is not present in DBL, i.e., changes come with no timestamps.

3.2   Wikimedia EventStreams

```
1  meta: [...], timestamp: 1554284688, id: 937929642, bot: false,
2  type: 'edit', title: 'Q31218558',
3  user: 'Tagishsimon', wiki: 'wikidatawiki'
4  comment:   '...', parsedcomment: [...],
5  minor: false, namespace: 0, patrolled: true,
6  length: { new: 5530, old: 5445 },
7  revision: { new: 901332361, old: 756468340 },
8  server_name: 'www.wikidata.org',
9  server_script_path: '/w', server_url: 'https://www.wikidata.org',
```

Listing 1.1. Wikimedia EventStream recentchanges example data.

Wikimedia EventStream (WES) is a web service created at Wikimedia Foundation, i.e., an American non-profit organization that hosts, among the other, open-knowledge projects like Wikipedia. In particular, Wikimedia invests financial and technical resources for the maintenance of projects that foster free and open knowledge. WES was originally used for internal data analysis and was open-sourced in 2018. It exposes streams of structured data using SSE.

WES data is gathered from the internal *Kafka* cluster. it includes logs and change-data captures. In practice, WES refers to eight distinct *JSON* streams: (i) recentchanges (ii) revision-create (iii) page-create (iv) page-property-change

---

[14] http://dbpedia-live.openlinksw.com/live/.

(v) page-links-change (vi) page-move (vii) page-delete (viii) page-undelete. In this section, we focus only on the *recentchanges* stream, which is the one with the most complex content.

Listing 1.1 shows an example of a *rechentchange* data item. In WES's recent changes, the event title links to the Wikidata entity entity, e.g., in Listing 1.1 title points https://www.wikidata.org/entity/Q31218558.ttl. Each item is timestamped (line 1) and typed (line 2), helping us to introduce the concept of event [18]. Four kinds of events are possible: *"edit"* (cf. Listing 1.1) for existing page modification; *"new"*, for new page creation, *"log"* for log action, *"external"* for external changes, and *"categorize"* for category membership change.

3.3   Global Database of Events, Language and Tone (GDELT)

Table 1. Example of GDELT event data.

GLOBALEVENTID	...	Actor1Code	...	EventCode	...	AvgTone	...	DATEADDED
35209457		GOV		020		-3.4188		20190401203000
835209458		LEG		120		1.39860		20190401203000
835209459		USA		040		1.39860		20190401203000

The GDELT is the largest open-access *spatio-temporal* archive for human society. Its Global Knowledge Graph spans more than 215 years and connects people, organizations, locations worldwide. GDELT captures themes, images, and emotions into a single holistic global network. GDELT data come from a multitude of news sources using Natural Language Processing techniques.

GDELT consists of three different streams, i.e., Events, Mentions[15], and Global Knowledge Graphs (GKG)[16], delivered as compressed TSV every 15 min: The *Event stream* shares Geo-Political events appearing in the news. Each event refers to two Actors, e.g., nations or public figures, participating in each event and the action they perform, e.g., a *diplomatic visit*. The *Mention stream* records every mention of a particular event in the news over time, along with the article's timestamp. The mention stream is linked to the Event stream by the Global Event ID field. The *GKG stream* connects people, organizations, locations, themes, news sources, and events across the planet into a massive network. GKG stream is rich and multidimensional. Fields like the GCAM, Themes, or Persons include more than one entry, using a CSV or similar formats.

GDELT streams use the dyadic format for Conflict and Mediation Event Observations (CAMEO) [15], and they contain Global Content Analysis Measures (GCAM). Table 1 shows few relevant fields for the Event stream, i.e., GLOBALEVENTID and DATEADDED that identify the event uniquely and over time; Actors and Event code that link the event to CAMEO, and as an example of GCAM dimension, AvgTone.

---

[15] Link to GDELT-Event_Codebook-V2.0.pdf.
[16] Link to GDELT-Global_Knowledge_Graph_Codebook-V2.1.pdf.

4 Putting the Life-Cycle into Practice

In this section, we walk through the life-cycle steps for DBL, WES, and GDELT.

4.1 Publication

The Streaming Linked Data life-cycle starts with data publication, i.e., Steps (0)-(4) described in Sect. 2.

Step 0 Name. Regarding *Step 0*, we opted for re-naming the published linked streams. Our base-uri is http://linkeddata.stream. Moreover, we adopt the DBPedia best practice for URI design, i.e.,

- /resource/stream indicates Web stream URI and its RDF representation
- /page/stream provides an HTML representation of the stream resource;
- /ontology/onto provides the PURL to vocabulary of the stream resource.

Step 1 Model the Streams. Regarding *Step 1*, we opted for using OWL 2 and RDFS as knowledge representation languages.

DBL provisions data in RDF make use of DBPpedia ontology (DBO), therefore we did not need to take care of any further modelling effort.

For *WES recentchange*, we collected the information about the streams schemas into an OWL 2 ontology(see footnote 19). WES are designed around the notion of event. Thus, we ported related classes from contextual vocabularies like the Event Ontology[17]. Regarding the *recentchanges* stream, we emphasizes the modeling of the events types in our ontology, i.e., "edit", "new", "log", "categorize", or "external". Similarly, we take into account what could be represented as external resources like Wikidata. Since data items are timestamped individually, and we used this timestamp to name the graph containing all the event data.

For GDELT, we collected all the information regarding the streams schemas into an OWL 2 Ontology that involves both CAMEO[18] and GCAM[19]. The CAMEO ontology is a coding scheme designed for the study of third-party mediation in international disputes. It contains a hierarchical coding scheme for dealing with sub-state actors, event types, and an extensive taxonomy for religious groups and ethnic groups. For CAMEO, we focus on event types and actors, creating a comprehensive hierarchy for the stream. GCAM is a pipeline of 18 content analysis tools. Each news article monitored by GDELT goes through GCAM pipeline that captures over 2230 dimensions, reporting density, and value scores for each. Using GCAM, you can assess the density of "*Anxiety*" speech via Linguistic Inquiry and Word Count (LIWC), or "*Smugness*" via WordNet Affect. The GCAM Master Codebook lists of all of the dimensions available[20]. We converted the Codebook in RDF, and we refined it manually to link it to DBPedia and WordNET.

---

[17] http://motools.sourceforge.net/event/event.122.html.
[18] http://linkeddata.stream/ontologies/cameo.owl.
[19] http://linkeddata.stream/ontologies/gcam.owl.
[20] http://data.gdeltproject.org/documentation/GCAM-MASTER-CODEBOOK. TXT.

Step 2 Describe the Stream. For *Step 2*, we opted for using VoCaLS and DCAT as vocabularies for writing stream descriptions. Relevant metadata about the streams include license, data formats, documentations, and human-readable descriptions. We collected all this information for each of the aforementioned Web streams into a *vocals:StreamDescriptor*.

```
:dbl a vocals:StreamDescriptor ;
 dcat:title "DPedia Live" ; dcat:publisher <http://www.dbpedia.org> ;
 rdfs:seeAlso <https://wiki.dbpedia.org/services-resources/ontology>
 dcat:license <https://creativecommons.org/licenses/by-nc/4.0/> ;
 dcat:publisher <http://www.linkeddata.stream/about> ;
 dcat:dataset :dblstream .
:dblstream a vocals:RDFStream .
 vocals:hasEndpoint :dblendpoint, :dblendpointold .
```

Listing 1.2. A DBL Stream VoCaLS description. Prefixes omitted.

Listing 1.2 presents the *vocals:StreamDescriptor* that contains necessary information about the DBL[21], which contains basic information about the publisher and the license (line 6) as well as links to other relevant datasets (lines 5-6).

```
wes:recentchange a vocals:StreamDescriptor ;
 dcat:dataset :recentchange ;
 dcat:title "Wikimedia Recentchanges Event Stream"^^xsd:string ;
 dcat:publisher <http://www.linkeddata.stream/about> ;
 dcat:license <https://foundation.wikimedia.org/wiki/Terms_of_Use/en> .
:recentchange a vocals:Stream ;
    vocals:hasEndpoint :wesendpoint, :wesendpointold .
```

Listing 1.3. Wikimedia EventStream recentchanges sGraph. Prefixes Omitted.

Listing 1.3 shows the *vocals:StreamDescriptor* for the *recentchanges* stream[22], which includes the Wikimedia terms of use as license, and the OWL 2 ontology we designed to describe WES' domain.

```
:events a vocals:StreamDescriptor ;
 dcat:title "GDELT Event Stream"^^xsd:string ;
 dcat:publisher <http://www.linkeddata.stream/about> ;
 dcat:description "GDELT Events Stream"^^xsd:string ;
 dcat:license <https://creativecommons.org/licenses/by-nc/4.0/> .
 dcat:dataset :eventstream .
:eventstream a vocals:Stream ;
 vocals:windowType vocals:logicalTumbling ;
 vocals:windowSize "PT15M"^^xsd:duration ;
 vocals:hasEndpoint :eventEndpoint , :oldEndpoint .
```

Listing 1.4. A GDELT Stream description using VoCaLS. Prefixes omitted.

---

[21] http://linkeddata.stream/resource/dbl.
[22] http://linkeddata.stream/resource/recentchanges.

Listing 1.4 shows a VoCaLS description for the GDELT Event Stream. GDELT does not use a license format, thus we include a license that is compliant with the terms of use. We also linked ontologies and mappings using `rdfs:seeAlso`. Since the stream is published regularly as 15 min batch, we include metadata about the rate, i.e., `vocals:windowType` and `vocals:windowSize` at lines 6, 7.

Step 3 Convert Data to RDF Streams. For *Step 3*, we faced difference cases, RDF triples, JSON, and TSV. Thus, we discuss the best choice for each scenarios to the extent of providing useful guidelines.

DBPedia Live already provides RDF Data. Thus, we do not have to apply any conversion mechanism. Nevertheless, DBL updates are shared as compressed NT triple files, which allow discussing the alternative RDF Streams serializations, i.e., triple-based (TB) and graph-based (GB). Although the two serializations have been proven to be semantically equivalent, some trade-offs are worth unveil. Table 2 lists the most important differences: (i) Triple-based RDF Stream serialization will require to extend the RDF format to enable Event-Time processing. Indeed, no additional information like the timestamp can be carried when streaming a single RDF triple. On the other hand, graphs named after their time annotation. For instance, using the Time-Ontology, can be used as a form of punctuation. (ii) While TP is optimized for triple-pattern evaluations [23], GB simplifies BGPs'. (iii) Finally, GB simplifies time-based provenance tracking as graphs are annotated with timestamps. On the other hand, TP simplifies the window maintenance as new items arrive as triple to add/remove already.

On the other hand, in WES and GDELT stream, the elements provisioned rich document formats, i.e., JSON and TSV. Therefore, we must set up a conversion pipeline to foster interoperability (I).

Table 2. Triple-based vs Graph-based RDF Stream Serialization

Triple	Graph
Custom RDF format for Event Time	Use named graph for punctuation
Optimised for triple patterns evaluation	Optimised for (named) basic graph patterns
Simplifies window maintenance	Naturally supports provenance

```
1   "@context": {
2    "@vocab":"http://linkedata.stream/ontology/wes/",
3    "xsd": "http://www.w3.org/2001/XMLSchema#",
4    "rdfs": "http://www.w3.org/2000/01/rdf-schema#",
5    "timestamp": {  "@type": "xsd:dateTime" },
6    "title": {"@id":"rdfs:about"},
7    "type": "@type",    "id": {"@id":"@id", "@type":"Event"}}
```

Listing 1.5. JSON-LD Context for Event in Listing 1.1

In WES's recent changes, the event title links to the Wikidata entity, e.g., in Listing 1.1 title points https://www.wikidata.org/entity/Q31218558.ttl. The first approach for *recentchange* conversion will require de-reference this information at conversion time and stream out the stream content with contextual domain data. However, such an approach is discouraged as we have the size of the referred entity might be inappropriate for stream provisioning (E.g., the entity above contains 6166 triples). Alternatively, being WES's data JSON, it is sufficient to attach a valid JSON-LD context to each document cf Listing 1.5.

In GDELT, data arrive in TSV format. Thus, we cannot count on the *JSON-JSON-LD* compatibility, we must set up a mapping-based conversion. We opt for the RML mapping language [13] and for a modified version of `CARML` mapping engine that handles the annotation process incrementally to minimize the translation latency.

```
<GEM> a rr:TriplesMap ; rml:logicalSource <source> ;
  rr:subjectMap [
    rr:template "http://linkedata.stream/resource/gdelt/{GLOBALEVENTID
      }";
    rr:class gdelt:Event;
    rr:graphMap
    [ rr:template"http://linkedata.stream/resource/time/{DATEADDED}"
      ]];
  rr:predicateObjectMap [ rr:predicate cameo:type;
    rr:objectMap
  [rr:template "http://linkedata.stream/ontologies/cameo/{EventCode}";
    ]];
  rr:predicateObjectMap [
    rr:predicate cameo:actor;
    rr:objectMap [ rr:parentTriplesMap <Atr1TM> ]]; [...] .
```

Listing 1.6. RML Mapping for GDELT Event Stream Conversion (Subset).

Listing 1.6 shows a portion of the GDELT event RML mapping. We created a named graph after the event timestamp using `rr:graphMap` at line 5. We used `rr:class` to assign the `sem:Event` type to the data at line 4. Moreover, we also assign the CAMEO type using `cameo:type` at line 7. To model the actors and their hierarchy of types, we include a separate triple-map at line 12.

Step 4 Serve Streams on the Web. Regarding *Step 4*, we enable continuous data access to the content of the streams using an customized version of `TripleWave` that connects to the streams of choice and provision the RDF Streams via WebSocket. Nevertheless, to make this access sustainable over time, we require the interested user to run, locally to their machine, a docker image that executes the conversion pipeline accessing data at the source.

```
:dblendpoint a vocals:StreamEndpoint ;
    dcat:format frmt:JSON-LD ;
    dcat:accessURL "ws://localhost:8081/dbl" .
:dblendpointold  a vocals:StreamEndpoint ;
    dcat:format frmt:NT ;
    dcat:accessURL "http://downloads.dbpedia.org/live/changesets/
        lastPublishedFile.txt".
```

Listing 1.7. DBL Stream Endpoints, Prefixes omitted.

To enable data access, we added *vocals:StreamEndpoint* to each stream description that links to the exposed WebSocket. Moreover, for provenance reasons we included a *vocals:StreamEndpoint* that points to the original data source. Listing 1.7, Listing 1.8, and Listing 1.9 show such endpoints for DBL, WES recentchange, and GDELT Events, respectively.

```
:wesendpoint a vocals:StreamEndpoint ;
    dcat:format frmt:JSON-LD ;
    dcat:accessURL "ws://localhost:8081/wes/recentchanges" .
:wesendpointold a vocals:StreamEndpoint ;
    dcat:format frmt:JSON ;
    dcat:accessURL "https://stream.wikimedia.org/v2/stream/recentchange".
```

Listing 1.8. WES recentchange Stream Endpoints. Prefixes omitted.

```
:eventEndpoint . a vocals:StreamEndpoint ;
    dcat:format frmt:JSON-LD;
    dcat:accessURL "ws://localhost:8080/gdelt/events" .
:oldEndpoint . a vocals:StreamEndpoint ;
    dcat:format frmt:TSV;
    dcat:accessURL "http://data.gdeltproject.org/gdeltv2/lastupdate.txt"
        .
```

Listing 1.9. GDELT Event Stream Endpoints. Prefixes omitted.

4.2 Processing

This last step, which is a corollary to the publication cycle, aims at making sense of the published streams using standard languages and protocols.

Although many of SPARQL dialects for continuous query exist [6,23,27], RSP-QL is the reference model of choice for RDF Stream Processing (RSP). Moreover, it is also query language that includes all the existing SPARQL extensions [11] RSP-QL allows defining Stream-to-Stream transformations using a simple query model that is backward-compatible with SPARQL 1.1. and includes the operators' families: *Stream-to-Relation* (S2R) operators, which bridge the world of unbounded streams with the world of finite stream portions. Listing 1.10 shows in line 4 a typical operator of this kind, i.e., a Time Window. *Relation-to-Relation* (R2R) operators, which can be executed over the finite outputs of the S2R. In the RSP context, R2R coincides with SPARQL algebraic operators. *Relation-to-Stream* (R2S), which returns to infinite data streams. Listing 1.10 shows in line 2 the RStream, which is an example of R2S that append a timestamp to the output of the R2R operator.

```
PREFIX dbl <http://linkeddata.stream/resource/dbl/>
SELECT (COUNT (?entity) AS ?count) ?entity
FROM NAMED WINDOW <wa> ON dbl:added [RANGE PT1M STEP PT10S]
FROM NAMED WINDOW <wr> ON dbl:removed [RANGE PT1M STEP PT10S]
WHERE { WINDOW ?w { ?entity ?p ?o . } }
GROUP BY ?entity ORDER BY DESC(?count)
```

Listing 1.10. RSP-QL Query counting the most edited entities in a minute.

DBL was successfully used to analyze DBPedia evolution. In particular, DBL was used to satisfy information needs like *How many entities are updated in last 5 min?*. DBPedia live statistics is a daily updated Web page that shows analyses like top-k entity changes. We decided to compute the aforementioned statistics using RSP-QL. Listings 1.10 shows how the query that counts the top-20 most edited entities in the last minute looks like.

A recent challenge triggered a real-time data analysis of WES data, most of which focus on the data visualization aspect[23]. The projects empower simple statistical analyses such as comparison of *bot vs. human* editor, *minor vs. major* changes detection or categorizing the type of events. Listing 1.11 shows an example of RSP-QL query calculating the stream rate every minute.

```
PREFIX wes : <http://linkeddata.stream/resource/wes>
SELECT (COUNT{*}/60) ?ratesec
FROM NAMED WINDOW <w> ON wes:recentchange [RANGE PT60S STEP PT60S]
WHERE { WINDOW <w> { ?s a wes:Event } }
```

Listing 1.11. WES Recentchange rate

Several studies have been running using GDELT. In particular, data visualization techniques that take into account the spatio-temporal metadata of GDELT extracted events. GDELT offers different APIs to run analysis and a `Google Big Query`[24] access to the database. GDELT exposes examples of pre-configures analyses via the analysis service. For instance, the Event `TimeMapper` visualizes events matching a given search over time.

---

[23] Wikimedia EventStream Terms Of Service.
[24] https://cloud.google.com/bigquery/.

```
PREFIX gdelt : <http://linkeddata.stream/resource/gdelt>
SELECT (COUNT(?newsa) AS ?tot)
FROM NAMED WINDOW <e>  ON gdelt:events    [RANGE 30MPT STEP 15MPT]
FROM NAMED WINDOW <m>  ON gdelt:mentions  [RANGE 30MPT STEP 15MPT]
FROM NAMED WINDOW <g>  ON gdelt:gkg       [RANGE 30MPT STEP 15MPT]
WHERE {
  WINDOW <e> { ?event :quadClass cameo:4 ; :actionGeo_cc "IZ". }
  WINDOW <m> { ?event :mentions ?newsa. }
  WINDOW <g> { ?newsa :theme gcam:kill ; :location geo:iraq. } ]
GROUP BY ?newsa
```

Listing 1.12. RSP-QL Crossing GDELT Streams. Perfixes omitted.

5   Discussion and Conclusion

In this paper, we walked through the life-cycle of Streaming Linked data discussing how FAIR principles for data management applies and showcasing three examples of the Web Streams, i.e., DBPedia Live Stream, Wikimedia EventStream, and GDELT, that we work as open-source at https://w3id.org/webstreams.

While Naming, Describing, and Serving can count to sufficient examples that ease the user experience, *Modelling* and *Data Conversion* still demand a huge amount of manual work. For the former most of the complexity lies in the data domain understanding. Making sense of domain data like GDELT requires very specific knowledge. Moreover, knowledge representation still hides some challenges when it concern streaming data. For latter, best practices are emerging from the community around R2RML/RML.

We consider an interesting research direction investigating how knowledge engineering changes when data analysis is bound to specific temporal constraints. To the best of our knowledge, current works focus on temporal data modeling for historical data management [16]. Moreover, we consider the following exciting future works: (i) extending and maintaining the catalog of streams, the released ontologies, and used resources; (ii) extending the Streaming Linked Data processing step of the life-cycle with dedicated guidelines, i.e., identifying canonical problems for streaming linked data processing. (iii) the introduction of performance assessment and benchmarking as an explicit part of the life-cycle, taking into account the ongoing work of the RSP community and LDBC [2].

Acknowledgments. Dr. Tommasini acknowledges support from the European Social Fund via IT Academy program.

References

1. Alexander, K., Cyganiak, R., Hausenblas, M., Zhao, J.: Describing linked datasets. In: Proceedings of the WWW2009 Workshop on Linked Data on the Web, LDOW 2009, Madrid, Spain, 20 April 2009 (2009)

2. Angles, R., et al.: The LDBC social network benchmark. CoRR abs/2001.02299 (2020)
3. Arias-Fisteus, J., García, N.F., Fernández, L.S., Fuentes-Lorenzo, D.: Ztreamy: a middleware for publishing semantic streams on the web. J. Web Semant. **25**, (2014)
4. Balduini, M., Della Valle, E.: FraPPE: a vocabulary to represent heterogeneous spatio-temporal data to support visual analytics. In: Arenas, M., et al. (eds.) ISWC 2015. LNCS, vol. 9367, pp. 321–328. Springer, Cham (2015). https://doi.org/10.1007/978-3-319-25010-6_21
5. Barbieri, D.F., Braga, D., Ceri, S., Della Valle, E., Grossniklaus, M.: Querying RDF streams with C-SPARQL. SIGMOD Rec. **39**(1), 20–26 (2010)
6. Barbieri, D.F., Della Valle, E.: A proposal for publishing data streams as linked data - a position paper. In: Proceedings of the WWW 2010 Workshop on Linked Data on the Web, LDOW 2010, Raleigh, USA, 27 April 2010 (2010)
7. Compton, M., et al.: The SSN ontology of the W3C semantic sensor network incubator group. J. Web Sem. **17**, 25–32 (2012)
8. Consortium, W.W.W., et al.: Best practices for publishing linked data (2014)
9. Della Valle, E., Balduini, M.: Listening to and visualising the pulse of our cities using social media and call data records. In: Abramowicz, W. (ed.) BIS 2015. LNBIP, vol. 228, pp. 3–14. Springer, Cham (2015). https://doi.org/10.1007/978-3-319-26762-3_1
10. Della Valle, E., Dell'Aglio, D., Margara, A.: Taming velocity and variety simultaneously in big data with stream reasoning: tutorial. In: DEBS (2016)
11. Dell'Aglio, D., Della Valle, E., Calbimonte, J., Corcho, Ó.: RSP-QL semantics: a unifying query model to explain heterogeneity of RDF stream processing systems. Int. J. Seman. Web Inf. Syst. **10**(4), 17–44 (2014)
12. Dell'Aglio, D., Della Valle, E., van Harmelen, F., Bernstein, A.: Stream reasoning: a survey and outlook. Data Sci. **1**(1–2), 59–83 (2017)
13. Dimou, A., et al.: Mapping hierarchical sources into RDF using the RML mapping language. In: 2014 IEEE International Conference on Semantic Computing, Newport Beach, CA, USA, 16–18 June 2014, pp. 151–158 (2014)
14. Gao, F., Ali, M.I., Mileo, A.: Semantic discovery and integration of urban data streams. In: Proceedings of the Fifth Workshop on Semantics for Smarter Cities a Workshop at the 13th International Semantic Web Conference (ISWC 2014), Riva del Garda, Italy, 19 October 2014, pp. 15–30 (2014)
15. Gerner, D.J., Schrodt, P.A., Yilmaz, O., Abu-Jabr, R.: Conflict and mediation event observations (cameo): a new event data framework for the analysis of foreign policy interactions. International Studies Association, New Orleans (2002)
16. Gottschalk, S., Demidova, E.: Eventkg - the hub of event knowledge on the web - and biographical timeline generation. Semantic Web **10**(6), 1039–1070 (2019)
17. Hyland, B., Wood, D.: The joy of data-a cookbook for publishing linked government data on the web. In: Wood, D. (ed.) Linking Government Data, pp. 3–26. Springer, Heidelberg (2011). https://doi.org/10.1007/978-1-4614-1767-5_1
18. Luckham, D.: The power of events: an introduction to complex event processing in distributed enterprise systems. In: Bassiliades, N., Governatori, G., Paschke, A. (eds.) RuleML 2008. LNCS, vol. 5321, pp. 3–3. Springer, Heidelberg (2008). https://doi.org/10.1007/978-3-540-88808-6_2
19. Margara, A., Urbani, J., van Harmelen, F., Bal, H.E.: Streaming the web: reasoning over dynamic data. J. Web Sem. **25**, 24–44 (2014)
20. Mauri, A., et al.: TripleWave: spreading RDF streams on the Web. In: Groth, P., et al. (eds.) ISWC 2016. LNCS, vol. 9982, pp. 140–149. Springer, Cham (2016). https://doi.org/10.1007/978-3-319-46547-0_15

21. Morsey, M., Lehmann, J., Auer, S., Stadler, C., Hellmann, S.: DBpedia and the live extraction of structured data from wikipedia. Program **46**(2), 157–181 (2012)
22. Passant, A., Bojārs, U., Breslin, J.G., Decker, S.: The SIOC Project: semantically-interlinked online communities, from humans to machines. In: Padget, J., et al. (eds.) COIN -2009. LNCS (LNAI), vol. 6069, pp. 179–194. Springer, Heidelberg (2010). https://doi.org/10.1007/978-3-642-14962-7_12
23. Phuoc, D.L., Dao-Tran, M., Tuán, A.L., Duc, M.N., Hauswirth, M.: RDF stream processing with CQELS framework for real-time analysis. In: Proceedings of the 9th ACM International Conference on Distributed Event-Based Systems, DEBS 2015, Oslo, Norway, 29 June-3 July 2015, pp. 285–292 (2015)
24. Phuoc, D.L., Nguyen-Mau, H.Q., Parreira, J.X., Hauswirth, M.: A middleware framework for scalable management of linked streams. J. Web Semant. **16**, 42–51 (2012)
25. Sequeda, J.F., Corcho, Ó.: Linked stream data: a position paper. In: Proceedings of the 2nd International Workshop on Semantic Sensor Networks (SSN09), Collocated with the 8th International Semantic Web Conference, Washington DC, USA 2009
26. Stonebraker, M., Çetintemel, U., Zdonik, S.B.: The 8 requirements of real-time stream processing. SIGMOD Rec. **34**(4), 42–47 (2005)
27. Tommasini, R., Della Valle, E.: Yasper 1.0: towards an RSP-QL engine. In: Proceedings of the ISWC 2017 Posters & Demonstrations and Industry Tracks co-located with 16th International Semantic Web Conference (ISWC) (2017)
28. Tommasini, R., et al.: VoCaLS: vocabulary and catalog of linked streams. In: Vrandečić, D., et al. (eds.) ISWC 2018. LNCS, vol. 11137, pp. 256–272. Springer, Cham (2018). https://doi.org/10.1007/978-3-030-00668-6_16
29. Villazón-Terrazas, B., Vilches-Blázquez, L.M., Corcho, O., Gómez-Pérez, A.: Methodological guidelines for publishing government linked data. In: Wood, D. (ed.) Linking Government Data, pp. 27–49. Springer, New York (2011). https://doi.org/10.1007/978-1-4614-1767-5_2
30. Wilkinson, et al.: The fair guiding principles for scientific data management and stewardship. Sci. Data **3**(1), 160018 (2016). https://doi.org/10.1038/sdata.2016.18

AWARE: A Situational Awareness Framework for Facilitating Adaptive Behavior of Autonomous Vehicles in Manufacturing

Boulos El Asmar[1]([✉])(iD), Syrine Chelly[1,2](iD), Nour Azzi[1,4], Lynn Nassif[1,4], Jana El Asmar[1,4], and Michael Färber[3](iD)

[1] BMW Group, Munich, Germany
boulos.el-asmar@bmw.de
[2] Technical University of Munich (TUM), Munich, Germany
syrine.chelly@tum.de
[3] Karlsruhe Institute of Technology (KIT), Karlsruhe, Germany
michael.faerber@kit.edu
[4] Faculty of Engineering (ESIB), Saint Joseph University of Beirut, Beirut, Lebanon
{nour.azzi,lynn.nassif,jana.asmar}@net.usj.edu.lb

Abstract. In this paper, we introduce AWARE, a knowledge-enabled framework for robots' situational awareness. It is designed to support autonomous logistics vehicles operating in automobile manufacturing plants. AWARE comprises an ontology grounding robots' observations, a knowledge reasoner, and a set of behavioral rules: The AWARE ontology models data streams of proprioceptive and exteroceptive sensors into high-level semantic representations. The knowledge reasoner infers adequate policy by reasoning over a sliding window of observations, presumably depicting the robot's perceptions and actual state of knowledge. The behavioral rules, in analogy to road traffic rules and common sense, regulate the operation of autonomous robots in a manufacturing environment despite their obvious peculiarity. Our rules are the first ones facilitating the orderly and timely flow of vehicles. We show the applicability of AWARE in an industrial set up. Overall, we posit that situational awareness is a fundamental element towards functional autonomy and argue that it can provide a reliable basis for organizing and controlling robots in a smart factory in the near future.

Keywords: Knowledge graphs · Autonomous vehicles · Semantics-based smart factory · Internet of Things

1 Introduction

A plethora of autonomous and automated guided vehicles[1] are increasingly engaging in logistics operations. With the diversity of autonomous robots such

---

[1] We use vehicle and robot interchangeably throughout the course of this paper.

This work was supported by BMW Group.

© Springer Nature Switzerland AG 2020
J. Z. Pan et al. (Eds.): ISWC 2020, LNCS 12507, pp. 651–666, 2020.
https://doi.org/10.1007/978-3-030-62466-8_40

as transport robots, autonomous forklifts and autonomous tugger trains, substantial challenges emerge to optimize operations within a smart factory [1]. The German Association of the Automotive Industry published the communication interface VDA5050[2], between automated guided vehicles (AGVs) and a central master controller within the automotive manufacturing plants. VDA5050 enables the implementation of parallel and complementary operations of AGVs through a central master controller. Further, the American National Standards Institute/Industrial Truck Safety Development Foundation released the safety standard ANSI/ITSDF B56.5-2019[3] for driverless, automatically guided industrial vehicles. However, less work has been done towards governing interactions with other agents encountered on the shop floor that are not monitored by the same master controller, such as manned industrial vehicles and autonomous vehicles. In an analogous domain, in road autonomous driving, vehicles' interactions are typically governed by established priors like traffic rules and common sense. However, to the best of our knowledge, such priors over operational conduct of industrial vehicles, besides the safety-related priors [2,3], have not yet been considered in research and standards efforts. Examples of operational priors include right of way or overtaking behavior on divided aisles. Autonomous robots in logistics are currently able to perform complex tasks such as transportation, goods pick up and goods drop off. In a case study on logistics vehicles in an automobile manufacturing plant, we deduce that autonomous vehicles, whereas operating safely without being controlled directly by humans, still lack behavioral grounding. We observe operational impediments caused by: (1) low agility of autonomous vehicles compared to manned vehicles caused by rigorous safety regulations implemented on the autonomous vehicle, (2) autonomous vehicles possibly getting into bottlenecks in various situations such as in intersections, or narrow aisles.

Current autonomous robots are equipped with various sensors like depth cameras, LiDAR, indoor localization tags and ultrasonic sensors. Despite the development of artificial intelligence approaches depicting the streams of data published from the sensors, such as object detection [4] and 3D pose estimation [5], we posit that such representation is not solely sufficient to ensure timely and orderly operation, and must be complemented with situational awareness. Vehicles cannot understand and reason over their environment without a high-level semantic representation of the data. In [6], we introduced the AWARE ontology eliciting the knowledge of the moment as perceived by the autonomous robot, including its telemetry, priors on its environment, its sensed surrounding, and the rules governing the relations between the perceived assets. AWARE was developed in Web Ontology Language (OWL)[4] [7], which is particularly advantageous when reasoning and handling data from heterogeneous data streams. In

---

[2] VDA5050 – Schnittstelle zur Kommunikation zwischen Fahrerlosen Transportfahrzeugen (FTF) und einer Leitsteuerung, https://www.vda.de/en.

[3] ANSI/ITSDF B56.5-2019 Safety standard for driverless, automatic guided industrial vehicles and automated functions of manned industrial vehicles.

[4] https://www.w3.org/OWL/.

this paper, we introduce the Aware framework to close the gap between the perceptions of the robot and further knowledge processing. Aware represents processed data streams as what we refer to as observations using timestamp-based temporal RDF representation [8]. The Aware decision module uses a set of rules to reason over observations and priors in order to adapt the behavior of the robot. The proposed approach can be easily extended to deal with further situations requiring robots' awareness by adding more rules and adapting the ontology accordingly to cover the application domain. Overall, our main contributions of this paper are:

1. We introduce Aware, a knowledge-enabled framework for robots' reasoning that is, for the first time, specifically designed for enhancing situational awareness of autonomous robots operating in a manufacturing plant. Aware includes an ontology, a set of rules, and a reasoner.
2. We publish the first set of rules governing autonomous logistics vehicles in a manufacturing plant, resolving traffic bottlenecks and facilitating orderly and timely operations within a smart factory.
3. We show the applicability of the Aware ontology to ground robots' proprioceptive and exteroceptive perceptions, as well as priors on the environment, for the purpose of situational awareness.

The remainder of this paper is organized as follows: In Sect. 2, we review work related to the manufacturing domain's ontologies, knowledge processing frameworks, and situational awareness applications. Further, we review impediments we identified by observing operational autonomous robots deployed in a manufacturing environment. Next, in Sect. 3, we introduce the Aware framework. In Sect. 4, we describe how we evaluated the framework and its comprised ontology. We describe the lessons learned from developing an industrial application based on Semantic Web technologies in Sect. 5. We summarize the paper in Sect. 6.

2   Related Work and Priors

In this section, we provide background information about the two areas whose intersection this work resides in: ontologies supporting robotics applications, and knowledge processing frameworks. Then we present work related to situational awareness of autonomous robots and provide insights on impediments observed on the shop floor that can be resolved through situational awareness.

Ontologies. Ontologies have been applied in robotics applications to describe the semantic knowledge of robots. Low-level data streams from sensors are transformed into high-level semantic representations following ontology grounding. Most previous research related to robotics cognition, such as [9–13], adopted knowledge models focused on task planning. In [13], the knowledge schema represents robots' actions and perceptions but does not address knowledge of intrinsic states. Further, [9–11,13] lack the use of common terminologies provided by IEEE 1872 [14], W3C[5] or OGC[6] standards. Moreover, the operational environ-

---

[5] https://www.w3.org/.

[6] https://www.opengeospatial.org/.

ment represented in these knowledge models is not relevant to manufacturing plants. In [6], we introduced the AWARE ontology to represent the prevailing state of knowledge of the autonomous robot operating within an automobile manufacturing plant.

Knowledge Processing Frameworks. Ontology-based approaches for robots' autonomy are thoroughly discussed in a recently published review [15]. OMRKF [9] was designed to enable service robots in household environments. The framework enhances the robot's navigation and task planning capabilities. KnowRob [16] also focuses on household environments. Knowledge in KnowRob is organized in an action-centric way to support reasoning about action and task planning. OUR-K [10] is oriented towards robot intelligence for service robot use cases. It builds up rich knowledge for the robot to allow the completion of tasks even if the information at its disposal is incomplete. Open-Robots (ORO) [11] focuses on the implementation of a knowledge representation and reasoning for autonomous robots deployed in complex environments where they need human-machine interaction capabilities. Perception and Manipulation Knowledge (PMK) [12] is an ontological-based reasoning framework to enhance a robot's task- and motion-planning capabilities in the manipulation domain.

Situational Awareness. Situational awareness, as defined in the Oxford dictionary, is knowing that something exists and is important. Endsley [17] defined the scientific term "Situational Awareness" (SA) as the perception of relevant elements in the environment, the comprehension of their significance, and the projection of their future status. Thus, in this context, achieving situational awareness in robotics goes beyond ensuring basic functionalities such as navigation or task planning to decide on the most favorable course of action. Awareness has been essential in a wide range of domains such as urban autonomous driving [18–21] where SA helps to understand the interactions between perceived entities and empowers decision making in traffic situations. In air traffic control [22,23], SA helps ensuring efficiency and safety during take-off and landing by assessing locations of the aircrafts and projecting their future locations. In [24], SA increases cell phone profitableness by improving its functionalities. According to Endsley [17], good SA still does not ensure good performance, however, good performance becomes more likely with SA.

Impediment Situations in Manufacturing Environments. The complexity of manufacturing environments where autonomous robots are deployed is mainly due to their heterogeneity: they comprise both static and moving objects, humans as well as machines, autonomous and non-autonomous vehicles. The complexity is increased by the difficulty of establishing a traffic rule book that regulates the traffic within the manufacturing plant. In a case study conducted on autonomous transport robots deployed in an automobile manufacturing plant, we observe impediments in multiple situations such as at intersections. Although the robots are equipped with the required hardware and software to operate safely without human supervision, their behavior still demonstrates a lack of smoothness due to their shortness in cognition abilities. In Fig. 1, we illustrate

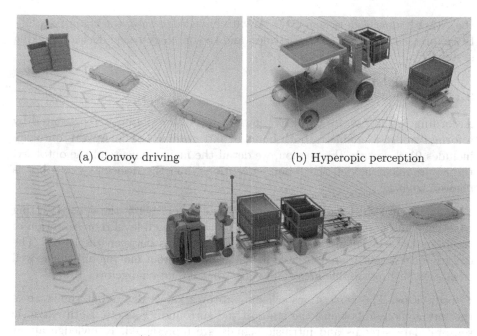

(a) Convoy driving   (b) Hyperopic perception

(c) Oncoming traffic with long obstacle

Fig. 1. Pictures of impediments encountered in a case study on autonomous robots deployed in an automobile manufacturing plant.

some operational drawbacks encountered on the shop floor case study. These drawbacks are illustrated to motivate the necessity of situational awareness for a functional autonomy: In a convoy driving situation, as shown in Fig. 1a, a desired behavior for the rear robot is to mimic the behavioral pattern of the front robot, without attempting to overtake, since the latter has an anticipated insight and thus a more reliable judgment. In Fig. 1b, the field of view of the autonomous robot is deficient because of proximity. The autonomous robot is required to perceive the loaded forklift while approaching and to reason and deduce potential collision of loads in order to increase separation distance. In Fig. 1c, overtaking an obstacle on a two-ways aisle might lead to a bottleneck with oncoming traffic. A more suited behavior is to avoid overtaking long obstacles.

3   AWARE: Situational Awareness Framework

AWARE is the first situational awareness framework specifically conceived for the purpose of augmenting autonomous robots in automobile manufacturing plants with awareness capabilities. The framework can be easily extended to support other applications of autonomous vehicles. In the following, we present the AWARE knowledge schema, the AWARE knowledge base, the cognitive abilities expressed through behavioral rules, and the ontology-based decision-making system.

3.1 Knowledge Schema

According to the AWARE ontology[7] [6], low-level information is represented by a format understandable to both humans and machines. The AWARE ontology includes 91 classes. In this section, we detail the main elements of the ontology: the environment model, the robot perceptions, and the decisions the robot is allowed to make.

Environment Model. The environment model represents the spatial setting where the autonomous robot operates. It includes a high-level representation of the manufacturing plant, its assets, and the relations between the assets. The environment assets are represented in classes for different moving objects as well as classes for topographic areas and zones that do not change over time.

Perceptions. The perceptions schema captures knowledge about the state of the autonomous robot and the state of the surrounding assets. Both extrinsic sensors' data streams and intrinsic signals are represented. Knowledge about the surrounding assets is modeled using the class *Observation*. An observation instance is used to link all elements of a perception: (1) the observed element, (2) its observed property, (3) the sensor that made the observation, (4) the procedure or algorithm used to extract the property, and (5) the timestamp of the observation. The classes *TransitwayObstacle* and *ObjectOfFocus* are used to identify objects that are of particular relevance in the robot's field of focus. It is out of the scope of this paper to detail the AWARE data acquisition module and the related fields of focus. The class *TransitwayObstacle* is used to describe objects treated as obstacles to be avoided by the robot. *ObjectOfFocus* indicates objects perceived by the camera within the robot's field of focus, that may represent an obstacle in the future.

3.2 Knowledge Base

Apart from the ontology, we create a knowledge base (KB) containing instances of the concepts in our ontology. Our knowledge base contains both time-invariant instances and instances that do vary over time. As time-invariant instances we have instances of the class *Decision*, such as *pause*, *adjustSafetyRange*, and instances of the class *Procedure*, such as object detection models like *YOLO* [25] or *DetectNet* [4]. Furthermore, instances of the class *OperationalArea* and of the class *ConstraintZone* are time-invariant. *OperationalArea* represents parts of the plant with a particular functionality such as aisles or drop-off areas, while

---

[7] https://w3id.org/AWARE/ontology.

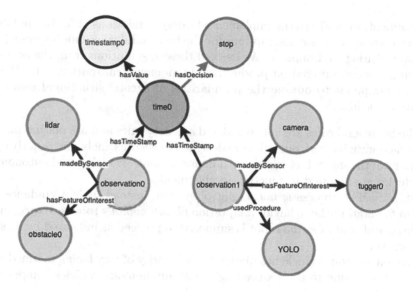

Fig. 2. Observations modeled in a knowledge base.

the class *ConstraintZone* refers to delimited surfaces in the plant where specific behavioral regulations apply such as zones with limited speed or limited capacity zones. Time-variant instances are characterized by a timestamp. They represent processed data from different intrinsic and extrinsic sensor streams. Data extracted from the autonomous robot's internal state and its surrounding environment is inserted into the KB as instances of the class *Observation*. An example of two instances from class *Observation* is shown in Fig. 2. One observation is concerned with one feature of interest only. Multiple observations can be characterized by the same timestamp. If multiple features of interest appear simultaneously, such as multiple detections within a single frame, an observation is created for each independently.

3.3 Basic Assumptions

AWARE reasons over behavioral rules to enhance the performance of autonomous transport robots. To the best of our knowledge, unlike in road traffic, the functioning of autonomous vehicles in closed environments is not standardized by an established code of conduct. No published standard regulates traffic within a manufacturing plant, such as intersections right of way or rules on when an autonomous vehicle is supposed to yield way. Nevertheless, autonomous robots are able to safely perform complex tasks without human intervention. The observed right of way allocation is often following a first-come first-served basis, or an it-fits-I-pass policy. We set forth the need to introduce behavioral rules to govern the behavior of autonomous vehicles deployed in a production environment, similarly to the implemented safety regulations [2, 3]. In the following, we present our basic assumptions for the behavioral rules guiding vehicles in a manufacturing environment. We derived these assumptions from the observed impediments encountered on productive autonomous transport robots. We list these assumptions to outline the peculiarities of agents' situational awareness and its modeling.

1. The behavioral rules are not considered as safety rules and are not intended to replace such; instead, our rules are designed to ensure timely and orderly operations of the smart factory, where humans, manned vehicles, and autonomous vehicles are required to function in alignment.
2. Situational awareness is not a control system; instead, it is a guidance system facilitating the behavior adaptation of *autonomous* robots. Hence, in the absence of guidance, the robot is supposed to proceed as indicated by its state machine.
3. The autonomous vehicle has always lower priority of way facing manned vehicles. This is due to the reduced agility of autonomous vehicles compared to manned vehicles.
4. Autonomous vehicles interact with each other following right of way rules similar to road traffic rules. That requires the ability for autonomous vehicles to recognize other autonomous vehicles and differentiate them from manned vehicles.
5. All autonomous vehicles deployed in the same operations environment are expected to follow the same traffic rules.
6. All autonomous vehicles deployed in the same operations environment are expected to have the same priors on the environment. Priors examples are intersections, driveway side, main and secondary aisles.
7. Autonomous vehicles cannot communicate between each others. To the best of our knowledge, no standard has been published to enforce lateral communication between autonomous vehicles. Thus, we do not enforce that constraint to solve eventual traffic congestion. AWARE identifies unsolvable congestions and notifies the cloud master controller. We predicate the need for such standardized vehicle-to-vehicle communication to guarantee a complete autonomy.

3.4   AWARE Architecture

The decision making system is built on top of the knowledge base schema enriched by a set of behavioral rules. Reasoning over the statements in the knowledge base, the robot adapts its behavior and avoids different bottleneck situations by applying the inferred decisions such as 'pause' or 'increaseSafetyRange'. An architecture diagram of the framework is shown in Fig. 3. In the following, we outline the components of the architecture in more detail.

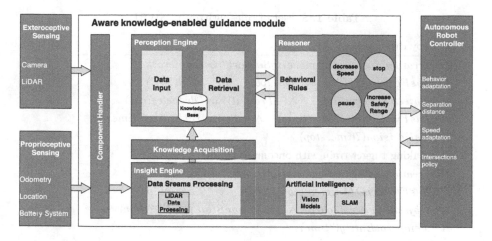

Fig. 3. An overview of the AWARE architecture.

Component Handler. The component handler is the data extraction module, adapting frame rate of data streams, and ensuring alignment of data and timestamps.

Insight Engine. The insights engine is the central data processing component through artificial intelligence, and diverse and redundant data analysis processes. Thus, real-world knowledge extracted by the Components Handler is structured according to the ontology. For example, images captured by the camera are fed to a trained neural network for object detection.

Knowledge Acquisition. The knowledge acquisition layer applies masks on the processed data to narrow down the insights to the area of focus. The area of focus varies with every sensor: for camera input for example, we filter out detected objects following a trapezium of interest as in [26].

Perception Engine. The perception engine handles data input and data retrieval into and from the knowledge base. This module manages a time window of observations in memory.

Reasoner. On knowledge insertion, a rule written in Prolog [27] automatically checks all the defined behavioral rules to trigger the ones that match the current instantiated state. Depending on the observations in the time window, the inferred guidance is published to the control system to adapt the ego's behavior according to the perceived environment.

Table 1. Subset of rules written in SWRL

Convoy driving
$Observation(?obs) \land madeBySensor(?obs, camera)$
$\land hasFeatureOfInterest(?obs, ?obj) \land STR(?obj)$
$\land ObjectOfFocus(?obj) \land TransitWayObstacle(?obj)$
$\land hasTimeStamp(?obs, ?time) \land TemporalEntity(?time)$
$\rightarrow hasDecision(?time, stop)$

Overtaking tugger train with oncoming traffic
$Observation(?obs) \land madeBySensor(?obs, camera)$
$\land hasFeatureOfInterest(?obs, ?obj) \land Tugger(?obj)$
$\land ObjectOfFocus(?obj) \land hasTimeStamp(?obs, ?time)$
$\land TemporalEntity(?time)$
$\rightarrow hasDecision(?time, stop)$

Hyperopic perception
$isLoaded(ego, True) \land Observation(?obs)$
$\land madeBySensor(?obs, camera) \land hasFeatureOfInterest(?obs, ?obj)$
$\land Forklift(?obj) \land ObjectOfFocus(?obj)$
$\land hasTimeStamp(?obs, ?time) \land TemporalEntity(?time)$
$\rightarrow hasDecision(?time, increaseSafetyRange)$

3.5 AWARE Implementation

We developed the ontology with the latest version of the Web Ontology Language OWL2 using Protégé, a free open-source ontology editor developed by Stanford[8]. Despite their simplicity, SWRL rules have the disadvantage of being computationally expensive when reasoning over a large number of rules [20]. Hence, we implemented ontology and rules in SWI-Prolog[9] [28], which is a computational effective logic programming language. We store both the ontology and the rules using Prolog [27]. We load the knowledge schema, represented in RDF triples in .owl format, into the konwledge base (KB) of Prolog. The reasoner inspects the data available in the KB and checks it over the rules in order to infer the best course of action following the prevailing situation. We continuously update the KB to keep a 1-minute-duration of knowledge history. We store observations and inferences over a window of time in order to ensure a smooth decision making and filter out erroneous decisions generated by noisy data. The rules, defined in Prolog language, are stored in .pl format. In Table 1, for the sake of expressiveness, we show the rules in SWRL corresponding to the impediments illustrated in Fig. 1.

---

[8] https://protege.stanford.edu/.
[9] https://www.swi-prolog.org/.

4   Evaluation

Evaluating a knowledge processing framework for situational awareness and behavior adaptation of autonomous vehicles is not a trivial task. The reason behind such challenge is that no evaluation methods or benchmarks are established so far as it is the case in other areas of artificial intelligence research. To assess our framework, we perform a quantitative evaluation by measuring its scalability and responsiveness, and we perform a qualitative evaluation by testing our system on various scenarios.

Quantitative Evaluation. We evaluate the framework's performance by measuring the scalability and the responsiveness. Specifically, we evaluate scalability by computing the time consumed to store a set of RDF triples resulting from observations. We wrote a test script to generate up to 45,000 mock observations from different data streams in a loop, which resulted in 497,104 RDF triples in total. We measure the responsiveness by evaluating the inference time of 200 randomly-generated rules over the RDF triples. All the measurements have been taken on an Intel(R) Core(TM) i7-8565U with a speed 1.80 GHz and 16.0 GB of RAM.

We report the results of our evaluation in Fig. 4. The time consumed scales linearly with the number of generated observations to reach 2.45 s for 45,000 observations. The response time remains almost constant. The reduced response time is due to the optimization of the rules' structure that accelerates the querying process. These results demonstrate that a reasonable amount of knowledge, as expected in the Aware observations time window, can be stored and processed efficiently by our framework. The scalability of the system is currently limited to what a single machine can handle. Overall, the framework's capability for responsiveness appears to be sufficient for modeling situational awareness.

Qualitative Evaluation. To evaluate Aware qualitatively, competency situations were implemented in a simulation environment using the Unity[10] game engine. We collected competency situations by analyzing the behavior of autonomous transport robots deployed in automobile manufacturing plants. We documented the behavior of the deployed robots via onsite observations and expert feedback in three production manufacturing plants in Germany. The observed fleet of deployed autonomous transport robots comprises 100 robots operating during two 8-hour-shifts per day. The study to collect the competency situations was conducted over 10 months.

In Table 2, we list the situations encountered by the autonomous robot that we refer to as *Ego* vehicle, and the corresponding guidance output. For example, on intersections, referred to as *crossingArea*, a desired behavior of autonomous robots is to yield way to manned vehicles. In such case Aware would return a *stop* guidance. Our qualitative evaluation was conducted in an iterative manner

---

[10] https://unity.com/.

Table 2. List of competency situations and expected output of AWARE

Situation	Aware guidance
Ego vehicle is located in a *CrossingArea*	*decreaseSpeed*
Ego vehicle is on *MainAisle* in a *CrossingArea* and detects a manned vehicle in field of focus	*stop*
Ego vehicle is on *MainAisle* in a *CrossingArea* and is driving *straight*	-
Ego vehicle is on *MainAisle* in a *CrossingArea* and is turning *right*	-
Ego vehicle is on *MainAisle* in a *CrossingArea* and is turning *left*. Ego vehicle detects an autonomous vehicle on the opposite *lane* in field of focus	*pause*
Ego vehicle is in a *CrossingArea* with *Decision* of previous timestamp is *pause* and the detected autonomous vehicle on the opposite *lane* in field of focus is stationary	-
Ego vehicle is on *MainAisle* in a *CrossingArea* and is turning *left* Ego vehicle does not encounter an autonomous vehicle on the opposite *lane* in field of focus	-
Ego vehicle is on *SecondaryAisle* in a *CrossingArea* and detects a manned vehicle in field of focus	*stop*
Ego vehicle is on *SecondaryAisle* in a *CrossingArea* while *MainAisle* is not clear	*stop*
Ego vehicle is on *SecondaryAisle* in a *CrossingArea* and is driving *straight* while *MainAisle* is clear	-
Ego vehicle is on *SecondaryAisle* in a *CrossingArea* and is turning *left*. Ego vehicle detects an autonomous vehicle in field of focus	*stop*
Ego vehicle is on *SecondaryAisle* in a *CrossingArea* and is turning *left*	-
Ego vehicle is on *SecondaryAisle* in a *CrossingArea* and is turning *right* while left *Aisle* is clear	-
Ego vehicle is in a *TwoWayAisle* and detects a *Tugger* as *ObjectOfFocus*	*stop*
Ego vehicle is in a *CrossingArea* and detects a *TransitwayObstacle*	*stop*
Ego vehicle detected in the last timestamp a manned vehicle as a *TransitwayObstacle* and enabled obstacle avoidance. Ego vehicle detects another *Entity* as a *ObjectOfFocus* in the field of focus	*stop*
Ego vehicle detects an autonomous vehicle as a *TransitwayObstacle* and as a *ObjectOfFocus*	*stop*
Ego vehicle detects a *Forklift* as a *ObjectOfFocus*	*increaseSafetyRange*

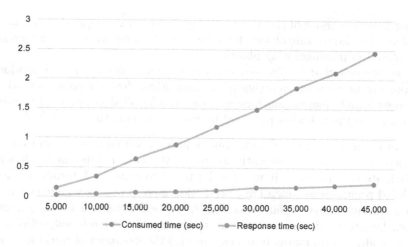

Fig. 4. Scalability and responsiveness per number of observations

during the ontology and framework development process in order to identify missing terms in the ontology and to ensure that AWARE satisfies all competency situations in the end. Evaluating the framework also next to the framework development significantly helped the AWARE ontology and framework to become mature.

5   Lessons Learned

The adoption of semantic technologies in industrial robotics applications is still facing the challenge of bridging the gap between robotics and semantics disciplines. We observed that, heretofore, the impact made by semantic technologies in robotics is limited in industry. Established productive robotics solutions, including route planning, task planning, and manipulation problems, use traditional optimization approaches. Through the work presented in this paper, we pave the way for a productive application of semantic technologies to enhance operations of autonomous robots. For example, the ability to dynamically adapt behavior of the robot has always been a requested feature reported by onsite robots fleet operators to avoid bottlenecks in ways seeming trivial to the human operators. Human operators have priors from road traffic rules, and expect robots to operate similarly. Also, drivers of manned vehicles on the shop floor request that autonomous robots avoid overtaking them. Such behavioral adaptation requires understanding and reasoning capabilities, besides knowledge acquisition and storage.

Knowledge acquisition is challenging for modalities like images where low-level pixels data need to be interpreted into world concepts: to recognise encountered agents through computer vision, a labeled dataset of all possible assets on the shop floor is required, similarly to existing benchmarks for roads autonomous driving [29]. As a result of this work, an object detection images dataset was collected and labeled to train object detection models to recognize and detect assets encountered in manufacturing plants.

It is planned that a pilot AWARE robot fleet is deployed in a productive environment of car manufacturing in autumn 2020. Hence, a policy for bringing awareness to autonomous machine was clearly identified as crucial. Overall, we have observed the following practical findings from our study:

1. Creating an ontology is doable, but requires good communication and best practices. Besides a systematic approach to avoid redundant work and to eliminate design errors, it was crucial to us to consider Internet of Things-related peculiarities which have been addressed in ontology engineering only to a limited degree so far. Specifically, we paid attention to (a) *perception* (i.e., how to establish a connection to the world), (b) *intersubjectivity* (i.e., how to align world representations), and (c) the *dynamics* of world knowledge (i.e., how to model events). For more information about these aspects, we can refer to [30].
2. Reasoning based on a rule-engine and an ontology has been applied in various scenarios. In the light of having a well-functioning and scalable Internet of Things scenario, using RDF triples and Prolog turned out to be a valid choice.
3. Rules need to be created by domain experts in order to cover all situations sufficiently. Also, time- and location-related constraints need to be taken into account. For instance, similar to varying traffic rules from country to country, robots operating in one environment (e.g., plant A) might need to cope with different observations and rules in another environment (e.g., plant B).
4. Deploying the framework in production also requires robust knowledge acquisition components adapted to the robots' sensors. In the case of diverse AI solutions, labeled datasets are needed (e.g., for object detection). This aspect should not be underestimated.

6   Conclusion and Prospects

In this paper, we introduced AWARE, a situational awareness framework adapted to the perception of autonomous robots operating in automobile production intralogistics. AWARE is the first knowledge-enabled framework designed to advance robot cognition within manufacturing environments. AWARE incorporates an ontology, a knowledge reasoner, and behavioral rules. The presented knowledge schema integrates proprioceptive and exteroceptive observations. Thus, AWARE models the intrinsic and extrinsic perceptions, framing low-level multi-dimensional data streams into high-level semantic representations. Furthermore, the knowledge reasoner provides guidance to the robot state machine based on the set of rules governing the interaction of autonomous robots with other agents operating in the same closed manufacturing environment. We predicate the lack of standards towards managing traffic within a smart factory, since only safety-related priors have been considered in research and standardization efforts so far.

Our future work orientations are two-fold: first, we will develop late fusion components to enable sensor fusion at the knowledge representation level. Therefore, for example, single objects detected by LiDAR as *TransitWayObstacle* and by camera as *ObjectOfFocus* will have their respective *Observation* entries linked to the same feature of interest. Secondly, we will focus on modeling projected future observations based on the observations recorded in a given time window. Ultimately, such projections will enable autonomous robots to distinguish between approaching and receding vehicles.

References

1. Wang, S., Wan, J., Li, D., Zhang, C.: Implementing smart factory of industrie 4.0: an outlook. Int. J. Distrib. Sensor Netw. **12**(1), 3159805:1–3159805:10 (2016)
2. DIN, E.: 1525, Safety of Industrial trucks-Driverless Trucks and their Systems, vol. 12. Beuth-Verlag, Berlin (1998)
3. ANSI/ITSDF: B56.5: 2019, Safety Standard for Driverless, Automatic Guided Industrial Vehicles and Automated Functions of Manned Industrial Vehicles EFFECTIVE. Accessed 08 Dec 2020
4. Tao, A., Barker, J., Sarathy, S.: DetectNet: deep neural network for object detection in DIGITS. Parallel Forall **4**, 1–8 (2016). https://devblogs.nvidia.com/parallelforall/detectnet-deep-neural-network-object-detection-digits/
5. Sundermeyer, M., Marton, Z.-C., Durner, M., Brucker, M., Triebel, R.: Implicit 3D orientation learning for 6D object detection from RGB images. In: Ferrari, V., Hebert, M., Sminchisescu, C., Weiss, Y. (eds.) ECCV 2018. LNCS, vol. 11210, pp. 712–729. Springer, Cham (2018). https://doi.org/10.1007/978-3-030-01231-1_43
6. El Asmar, B., Chelly, S., Färber, M.: AWARE: An Ontology for Situational Awareness of Autonomous Vehicles in Manufacturing. (2020) https://people.aifb.kit.edu/mfa/AWARE/AWARE-Ontology.pdf
7. Welty, C., McGuinness, D.L., Smith, M.K.: OWL Web Ontology Language Guide. W3C recommendation, W3C, February 2004. http://www.w3.org/TR/2004/REC-owl-guide-20040210
8. Gutierrez, C., Hurtado, C.A., Vaisman, A.: Introducing time into RDF. IEEE Trans. Knowl. Data Eng. **19**(2), 207–218 (2006)
9. Suh, I.H., Lim, G.H., Hwang, W., Suh, H., Choi, J.H., Park, Y.T.: Ontology-based multi-layered robot knowledge framework (OMRKF) for robot intelligence. In: 2007 IEEE/RSJ International Conference on Intelligent Robots and Systems, pp. 429–436. IEEE (2007)
10. Lim, G.H., Suh, I.H., Suh, H.: Ontology-based unified robot knowledge for service robots in indoor environments. IEEE Trans. Syst. Man Cybern.-Part A: Syst. Hum. **41**(3), 492–509 (2010)
11. Lemaignan, S., Ros, R., Mösenlechner, L., Alami, R., Beetz, M.: ORO, a knowledge management platform for cognitive architectures in robotics. In: 2010 IEEE/RSJ International Conference on Intelligent Robots and Systems, pp. 3548–3553. IEEE (2010)
12. Diab, M., Akbari, A., Ud Din, M., Rosell, J.: PMK-A knowledge processing framework for autonomous robotics perception and manipulation. Sensors **19**(5), 1166 (2019)

13. Beetz, M., Beßler, D., Haidu, A., Pomarlan, M., Bozcuoğlu, A.K., Bartels, G.: Know Rob 2.0 - a 2nd generation knowledge processing framework for cognition-enabled robotic agents. In: 2018 IEEE International Conference on Robotics and Automation (ICRA), pp. 512–519. IEEE (2018)

14. IEEE Robotics and Automation Society: IEEE standard ontologies for robotics and automation. IEEE Stan. **1872–2015**, 1–60 (2015)

15. Olivares-Alarcos, A., et al.: A review and comparison of ontology-based approaches to robot autonomy, Knowl. Eng. Rev. **34**, e29 (2019)

16. Tenorth, M., Beetz, M.: KNOWROB–knowledge processing for autonomous personal robots. In: 2009 IEEE/RSJ International Conference on Intelligent Robots and Systems, pp. 4261–4266. IEEE (2009)

17. Endsley, M.R.: Toward a theory of situation awareness in dynamic systems. Hum. Factors **37**(1), 32–64 (1995)

18. Wardziński, A.: The role of situation awareness in assuring safety of autonomous vehicles. In: Górski, J. (ed.) SAFECOMP 2006. LNCS, vol. 4166, pp. 205–218. Springer, Heidelberg (2006). https://doi.org/10.1007/11875567_16

19. Buechel, M., Hinz, G., Ruehl, F., Schroth, H., Gyoeri, C., Knoll, A.: Ontology-based traffic scene modeling, traffic regulations dependent situational awareness and decision-making for automated vehicles. In: 2017 IEEE Intelligent Vehicles Symposium (IV), pp. 1471–1476. IEEE (2017)

20. Armand, A., Filliat, D., Ibañez-Guzman, J.: Ontology-based context awareness for driving assistance systems. In: 2014 IEEE intelligent vehicles symposium proceedings, pp. 227–233. IEEE (2014)

21. McAree, O., Aitken, J.M., Veres, S.M.: Towards artificial situation awareness by autonomous vehicles. IFAC-PapersOnLine **50**(1), 7038–7043 (2017)

22. Endsley, M.R., Smolensky, M.W.: Situation Awareness in Air Traffic Control: The Picture. Academic Press, Cambridge (1998)

23. Hagemann, T., Weber, P.: Situational awareness in air traffic management (ATM). Hum. Factors Aerosp. Saf. **3**, 237–243 (2003)

24. Siewiorek, D.P., et al.: SenSay: a context-aware mobile phone. In: 7th International Symposium on Wearable Computers (ISWC 2003), pp. 248–249 (2003)

25. Redmon, J., Divvala, S., Girshick, R., Farhadi, A.: You only look once: unified, real-time object detection. In: Proceedings of the IEEE conference on computer vision and pattern recognition, pp. 779–788 (2016)

26. Tapu, R., Mocanu, B., Zaharia, T.: DEEP-SEE: joint object detection, tracking and recognition with application to visually impaired navigational assistance. Sensors **17**(11), 2473 (2017)

27. Bratko, I.: Prolog Programming for Artificial Intelligence. Pearson Education, London (2001)

28. Wielemaker, J., Schrijvers, T., Triska, M., Lager, T.: SWI-Prolog. Theor. Pract. Logic Program. **12**(1–2), 67–96 (2012)

29. Geiger, A., Lenz, P., Stiller, C., Urtasun, R.: Vision meets robotics: the KITTI dataset. The Int. J. Rob. Res. **32**(11), 1231–1237 (2013)

30. Färber, M., Svetashova, Y., Harth, A.: Theories of meaning for the Internet of Things. In: Bechberger, L., Kühnberger, K.U., Liu, M. (eds.) Concepts in Action - Representation, Learning and Application. Springer (2020)

Google Dataset Search by the Numbers

Omar Benjelloun[iD], Shiyu Chen[iD], and Natasha Noy[(✉)][iD]

Google Research, New York, USA
benjello@google.com, shiyuc@google.com, noy@google.com

Abstract. Scientists, governments, and companies increasingly publish datasets on the Web. Google's Dataset Search extracts dataset metadata—expressed using schema.org and similar vocabularies—from Web pages in order to make datasets discoverable. Since we started the work on Dataset Search in 2016, the number of datasets described in schema.org has grown from 500K to almost 30M. Thus, this corpus has become a valuable snapshot of data on the Web. To the best of our knowledge, this corpus is the largest and most diverse of its kind. We analyze this corpus and discuss where the datasets originate from, what topics they cover, which form they take, and what people searching for datasets are interested in. Based on this analysis, we identify gaps and possible future work to help make data more discoverable.

1 Dataset Search as a Snapshot of Datasets on the Web

We live in a data-driven world. Scientists, governments, journalists, commercial companies, and many others publish millions of datasets online. There are thousands of Web sites that publish datasets—some publish a handful, some publish hundreds of thousands [3]. Google's Dataset Search[1] is a search engine for datasets on the Web [16]. It relies on schema.org and similar open standards to extract the semantics of dataset metadata and to make it searchable.

Arguably, the mere existence of Dataset Search and its reliance on semantic markup provided a strong incentive for dataset providers to add such markup to their Web pages. Indeed, we have seen an explosive growth of dataset metadata on the Web since we started the work on Dataset Search. In the Fall of 2016, there were about 500K Web pages that included schema.org/Dataset markup, with half of them coming from data.gov, the US Open Government portal [10]. Today, there are tens of millions of such pages, from thousands of sites.

A recent comprehensive survey highlights a variety of approaches to help users find datasets [3]: these approaches range from searching within a collection of tables with different schemas [14], to finding data in repositories, such as Figshare, Zenodo, or DataDryad, to using metadata search engines, like Dataset Search. There are a number of well respected directories of dataset publishers (e.g., DataCite [19], re3data [12], Scientific Data in Nature [15]), but they inevitably miss new datasets or repositories [4]. To the best of our knowledge,

---

[1] http://datasetsearch.research.google.com.

© Springer Nature Switzerland AG 2020
J. Z. Pan et al. (Eds.): ISWC 2020, LNCS 12507, pp. 667–682, 2020.
https://doi.org/10.1007/978-3-030-62466-8_41

Dataset Search is the only collection of dataset metadata that includes all semantically annotated datasets on the Web.

We chose to rely primarily on schema.org for describing dataset metadata because both search engines and open-source tools have used it successfully to build an open ecosystem for various types of content [8]. In recent years, the scientific community has also embraced it for publishing data, by creating mappings from other metadata standards to schema.org. For example, Sansone and colleagues define a mapping from the DATS standard in the biomedical community to schema.org [20]. Wang and colleagues use schema.org to describe research-graph data, comprised of researchers, datasets and scholarly articles [24]. Efforts such as bioschemas.org [6] extend schema.org to include domain-specific terminology and relationships.

In this paper, we analyze the Dataset Search corpus of metadata. As of March 2020, the corpus contained 28 million datasets from more than 3,700 sites. While limited to the dataset metadata that is available in schema.org or DCAT, this corpus contains a sizable snapshot of the datasets on the Web. And because many researchers and scientists rely on search engines to find datasets [7], learning from this corpus can inform both the work to improve search engines for datasets and, more important, highlight the gaps in representation and coverage for the community at large. Specifically, in this paper, we make the following contributions:

- We present methods for analyzing an organically created corpus of metadata for 28 million datasets on the Web (Sect. 2).
- We identify a set of research questions that such a corpus can help analyze and present results of the analysis of the corpus (Sect. 3).
- We discuss lessons learned from the corpus analysis (Sect. 4).

2   Data Collection Methods

In this section, we describe the methods that we used to collect the metadata and to prepare it for the analysis in Sect. 3. In the remainder of this paper, we abbreviate the `schema.org` namespace as `so#` and the `DCAT` namespace as `dct#`.

2.1   From schema.org and DCAT on the Web to the Corpus

We described the details of the Dataset Search architecture elsewhere [16]. In brief, Dataset Search relies on the Google Web crawl to find pages that contain dataset metadata and to extract the corresponding triples. A post-processing of the Web crawl data parses RDFa, Microdata, and JSON-LD into a common graph data model, broadly equivalent to W3C's RDF triples [18]. We keep `so#Dataset`, `dct#Dataset`, and all the related entities and their properties.

We enhance, normalize, and augment this corpus in a variety of ways in order to provide users with a meaningful search experience. In this section, we focus only on those processing steps that are relevant to the subsequent data analysis. The processing happens at multiple levels of granularity: At the corpus level, we ensure that datasets are unique and attempt to remove non-datasets (i.e., pages that include dataset markup, but do not describe datasets). At the dataset level, we augment the metadata with inferred properties. Finally, at the property level, we clean up and normalize values.

The **corpus-level analysis** starts by removing duplicates within each site [16]. We found that many dataset repositories add markup both to the dataset landing pages and to the pages that list search results within that repository. We keep only the former in the corpus through simple heuristics: When the same dataset (according to values of key properties) appears on multiple pages, we keep the page that contains only one dataset. We also remove dataset metadata that does not have values for basic properties such as title and description [5].

At the **dataset level**, we process the values for properties such as title and description as well as the terms on the Web page itself in order to identify the main topics covered by the dataset. We use the topics from `re3data.org` [12] and a similar set of topics from the Google Knowledge Graph [17] as our vocabulary.

In addition, our **page-level** analysis collects information from the Web page that the dataset originated from, such as the domain of the page and its language.

For **individual properties**, we normalize, clean, and reconcile values for:

- **Data downloads and formats:** We identify the patterns that data providers use to represent download information and normalize them to a single representation [16]. Providers may specify file formats through the `so#fileFormat` or `so#encodingType` properties. When both of these properties are missing, we extract a file extension from the data-download URL.
- **DOIs and compact identifiers:** Persistent citable identifiers, such as Digital Object Identifiers (DOIs) and Compact Identifiers [26], may appear in several properties, such as `so#identifier`, `so#url`, or even `so#sameAs`, and `so#alternateName`. We use regular expressions to find patterns that correspond to these identifiers, and look for known prefixes from `identifiers.org` in all of these properties.
- **Level of access:** Two properties define the level of access of a dataset: `so#isAccessibleForFree` is a boolean value that indicates whether or not the dataset requires a payment. `so#license` links to a license or specifies one inline. We normalize the license information into known classes of licenses, such as Creative Commons and open government licenses. Any license that allows redistribution essentially makes the dataset available for free. We count datasets with these licenses as well as datasets with `so#isAccessibleForFree` set to true as the datasets that are "open."
- **Providers:** There is some ambiguity in schema.org on how to specify the the source of a dataset. We use the `so#publisher` and `so#provider` properties to identify the organization that provided the dataset. As with other properties, the value may be a string or an `Organization` object. Wherever possible, we reconcile the organization to the corresponding entity in the Google Knowledge Graph.
- **Updated date:** There are several date properties associated with a dataset: `so#dateCreated`, `so#datePublished`, `so#dateModified` (and similar

properties in DCAT). There is little consistency in how dataset publishers distinguish between them. However, the most recent value across these dates is usually a reliable approximation on when a dataset was last updated. We use several parsers to understand dates expressed in common formats and to normalize them. If there is no date in the metadata, we use the date when the Web page itself was last updated as a proxy.

Finally, to analyze the usage of datasets in the Dataset Search application, we look at logs for two weeks in May 2020. We extract the identifiers of the datasets that appeared in search results, and join them with their metadata to analyze search behavior in aggregate.

All the data and analyses in this paper are based on a snapshot of the Dataset Search corpus from March 26, 2020. We also compare the status of the corpus with a version from ten months prior, in June 2019.

2.2 Limitations of the Analysis

While we believe that our corpus is a reasonably representative snapshot of the datasets published on the Web, we recognize that it has limitations. Indeed, we have no way of measuring how well the corpus covers all the datasets available on the Web.

First, the corpus contains only the datasets that have semantic descriptions of their metadata in schema.org or DCAT. If a dataset page does not have metadata in a machine-readable format and in a vocabulary that we recognize, it will not be in our corpus (and will not be discoverable in Dataset Search).

Second, if a dataset page is not accessible to the Google crawler or is not being crawled for some reason (e.g., because of `robots.txt` restrictions), it will not be in our corpus. When the crawler processes the page, it often needs to execute JavaScript to get the metadata. If a page is slow to render, we may not obtain dataset metadata from it.

Third, our methods for inferring new values, such as dataset topics, may be imprecise, and we have not formally evaluated their accuracy yet.

Fourth, in our analysis, we "trust" what the metadata says. For instance, if a dataset's metadata says that the dataset is accessible for free, we count it among the open datasets. In some cases, the reality may be different when users try to download the dataset.

Finally, a significant amount of pages on the Web are designed for Search Engine Optimization or are simply spam. A page may have `so#Dataset` on it but not actually describe any dataset metadata. While we do our best to weed out such pages, our techniques are not perfect, and we cannot be certain that all the datasets in the corpus that we describe are indeed datasets.

3   Results and Observations

We start with the results of a corpus-level analysis (Sect. 3.1), then look at specific metadata properties (Sect. 3.2) and finally present our observations on the usage of datasets in Dataset Search (Sect. 3.3).

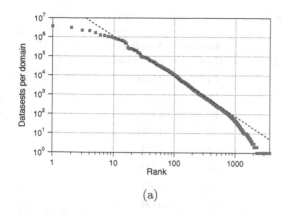

Domain	Datasets
ceicdata.com	3.7M
data.gov	3.1M
hikersbay.com	2.3M
tradingeconomics.com	2.2M
knoema.com	1.7M
figshare.com	1.3M
stlouisfed.org	1.2M
datacite.org	1.1M
thermofisher.com	1.0M
statista.com	0.9M

(a)   (b)

Fig. 1. Number of datasets per domain: (a) A log-log plot of domains and their relative sizes in terms of the number of datasets. The dotted line shows a fit to power-law for most of the range with the coefficient of 2.08 ± 0.01, which is very close to quadratic fit. (b) Domains with the largest number of datasets. These ten domains are responsible for 65% of the datasets in the corpus.

3.1 Corpus-Level Analysis

Looking at the corpus as a whole, as well as characteristics of the Web pages that we extracted metadata from, enables us to answer the following questions.[2]

Datasets and domains: How many datasets does each domain have? What is the distribution of the number of datasets by domain? What are the domains contributing the largest number of datasets? What are the most common top-level domains with datasets, and what fraction of datasets do they contribute? We know that many of the datasets come from open-government initiatives across the world. But just how many?

Dynamics of the corpus: How has the corpus of dataset metadata grown over time? Pages with datasets inevitably go offline, get moved, change their URLs. Can we trace this churn in our corpus? How prevalent is it? These numbers give a sense of how metadata on the Web is changing.

Metadata on metadata: Which fraction of datasets use schema.org vs DCAT? Which metadata fields are frequently populated, and which ones rarely have any values? These numbers give us probably the most actionable items in terms of improving metadata quality.

Datasets and Domains. The snapshot that we analyze in the rest of this section, taken on March 26, 2020, contains 28M datasets from 3,700 domains. The number of datasets per domain mostly follows a power law distribution, as the logarithmic scale plot in Fig. 1a shows: A small number of domains publish millions or hundreds of thousands of datasets, while the long tail of domains

---

[2] Here and elsewhere, "domain" refers to "internet domain."

Table 1. Number of datasets for the top twenty top-level Internet domains.

Top-level domain	Number of datasets	Top-level domain	Number of datasets
.com	14,956K	.ru	243K
.org	4,696K	.co	218K
.gov	3.386K	.nl	181K
.at	010K	.au	160K
.net	760K	.pl	152K
.es	524K	.uk	144K
.de	366K	.ca	139K
.edu	293K	.io	79K
.fr	281K	.world	56K
.eu	263K	.info	54K

hosts just a handful of datasets. The two domains with the largest number of datasets (`ceicdata.com` and `data.gov`) have more than 3 million datasets each. The ten largest domains (Fig. 1b) account for 65% of all datasets.

While "typical" Web pages about datasets correspond to a single dataset, some pages may have multiple datasets on them. For instance, a page may describe a large dataset and break down its components as multiple datasets; or a page may be dynamically generated in response to a search in a dataset repository. In our corpus, we found that over 90% of datasets come from pages that contain exactly one dataset. Still, more than 1.6M datasets come from pages with more than ten datasets.

Table 1 shows the distribution of datasets by top-level internet domains. The vast majority of the datasets come from .com domains, but both .org and government domains are well represented. For the country-specific domains, Austria, Spain, Germany, and France are at the top of the list. If we combine all government domains across the world (.gov, .gouv.*, .gv.*, .gov.*, .gob.*, etc.), we find 3.7M datasets on these government domains.

To get a more complete picture of the international coverage of datasets, Table 2 breaks them down by language, as specified by or extracted from the Web pages that contain them. More than 18M datasets, or 64% are in English, followed by datasets in Chinese and Spanish, both of which are growing faster than datasets in English. Note that these numbers do not capture the nuance of specific schema.org property values using multiple languages.

Dynamics of the Corpus. The next question that we study is the change in the corpus over time. Figure 2a shows the growth in the number of datasets since the beta launch of Dataset Search, in September 2018. The number of datasets has grown steadily, from about 6M then to 28M in March, 2020. We have reported earlier [16] that, day-to-day, about 3% of the datasets are deleted

Table 2. Number of datasets by language and the % change between June 2019 and March 2020.

Language	Number of datasets	% increase
English	18,650K	67%
Chinese	1,851K	82%
Spanish	1,485K	70%
German	743K	74%
French	492K	76%
Arabic	435K	75%
Japanese	404K	72%
Russian	354K	65%
Portuguese	304K	69%
Hindi	288K	70%

from our index while 7–10% new datasets are added. Enterprise data repositories have a similarly large level of churn [9]. Figure 2b shows the results of comparing the URLs between snapshots from June 2019 and March 2020, when the corpus almost doubled in size: Only 8.8M of the 14M URLs in the June 2019 corpus, or 63%, are still there in March 2020. The other 5.4M are no longer at the same location—or may no longer be in the corpus at all. This dynamic indicates a very high level of churn.

Figure 3 captures the updates to individual datasets. 14M datasets, or 50%, have a value for at least one of the date properties in metadata (Sect. 2). For an additional 10.7M datasets, we were able to determine the date when the Web page was modified. Out of these 24.7M datasets with a known date of last update, 21M datasets, or 85%, have this date within the last 5 years (Fig. 3). The short-term distribution (Fig. 3a) shows that more datasets were last updated within the last month than in any other month in the past year. Looking at the long-term distribution (Fig. 3b), 49% of datasets were last updated within the past year.

Metadata on Metadata. While schema.org is our primary semantic vocabulary for dataset description, we also understand and map basic DCAT properties. However, we found that fewer than 1% of datasets use the DCAT vocabulary.

Table 3 shows which fraction of datasets have values for specific properties. We require datasets to have a title and description [5]; hence, their 100% coverage in our corpus. Because we normalize and reconcile values from different properties, the properties in Table 3 do not always directly correspond to schema.org or DCAT predicates. For instance, we combine so#publisher and so#creator into "provider" because we observed that data owners do not really distinguish between the different semantics.

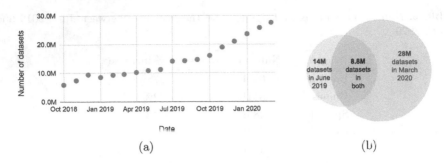

(a)   (b)

Fig. 2. Corpus dynamics: (a) Growth of datasets since the beta launch in September 2018. The number of datasets grew from 6M to 28M (b) Changed URLs in the corpus between June 2019 and March 2020: 8.8M URLs of the 14M in June 2019 remained in the corpus. The rest have either disappeared or changed.

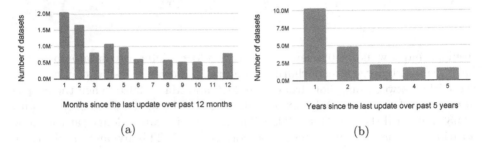

(a)   (b)

Fig. 3. Distribution of the date when a dataset was last updated: (a) at monthly granularity over the past year; (b) at yearly granularity over the last five years. Note that we have this information only for 85% of datasets.

3.2 Inside the Metadata

We focus on the properties that we found most informative in understanding the corpus—or that we got asked about most often (e.g., what are the formats for the available datasets). We do not try to provide an exhaustive description of the ranges of values for every property.

Topics. Figure 4 shows the distribution of the topics that datasets are associated with. We generate the topics automatically, based on dataset titles and descriptions as well as the text on the page where the dataset came from. The two largest topics are geosciences and social sciences—the areas that we focused on specifically before the beta launch in September 2018.

Data Downloads. Most users who search for datasets ultimately want the data itself and not just its metadata. Datasets can specify means to download their data via the so#distribution property.

Table 3. Percentage of datasets with specific properties. Column 2 lists the source predicates for each property. Properties not listed in the table have values in fewer than 1% of the datasets.

Property	Source predicates	Percentage
Description	so#description, purl#description	100.00%
Title	so#name, purl#title	100.00%
Provider	so#publisher, so#provider, purl#publisher	84.59%
Keywords	so#keywords, dct#keyword, purl#keyword	80.08%
URL	so#url, dct#accessurl, dct#landigpage	68.30%
Temporal coverage	so#temporalCoverage, so#temporal, purl#temporal	45.41%
Data download	so#distribution, dct#distribution	44.34%
Spatial coverage	so#spatialCoverage, so#spatial, purl#spatial	38.69%
Date modified	so#dateModified, purl#modified	37.46%
License	so#license and so#license on so#distribution	34.80%
Date published	so#datePublished, purl#published	30.83%
Catalog	so#includedInCatalog	29.74%
Variable	so#variableMeasured, dct#theme	20.90%
Authors	so#author, so#creator	14.12%
Same_as	so#sameAs, rdf#same_as	12.72%
Date created	so#dateCreated	9.62%
Alternate name	so#alternateName, rdf-schema#label	3.40%
Is accessible for free	so#isAccessibleForFree	3.04%

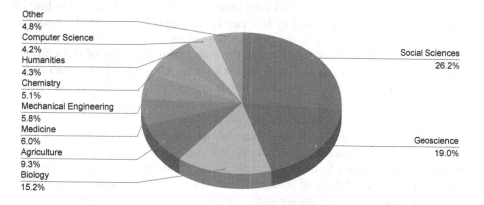

Fig. 4. Distribution of datasets by broad coverage topic, inferred from dataset metadata and the Web page itself.

Availability: Only 44% of datasets specify a data download link in their metadata (Table 3). Looking at the origin of datasets with downloads, we found that 85% of them are provided by just 10 domains.

Table 4. Number of datasets by content type. The counts are based on `so#fileFormat` or `so#encodingType` properties and file extensions. Note that some datasets have multiple distribution formats. Therefore, the total number of entries here is larger than the 12M datasets with data downloads.

Category	Number of datasets	% of total	Sample formats
Tables	7,822K	37%	CSV, XLS
Structured	6,312K	30%	JSON, XML, OWL, RDF
Documents	2,277K	11%	PDF, DOC, HTML
Images	1,027K	5%	JPEG, PNG, TIFF
Archives	659K	3%	ZIP, TAR, RAR
Text	623K	3%	TXT, ASCII
Geospatial	376K	2%	SHP, GEOJSON, KML
Computational biology	110K	<1%	SBML, BIOPAX2, SBGN
Audio	27K	<1%	WAV, MP3, OGG
Video	9K	<1%	AVI, MPG
Presentations	7K	<1%	PPTX
Medical imaging	4K	<1%	NII, DCM
Other categories	2,245K	11%	

Data Types and Formats: Zooming in on the subset of datasets that specify a data download, what are the broad categories of content and their relative prevalence? To answer this question, we first extract the file format of data downloads, and then bucket them into categories. For bucketing, we defined a high-level classification inspired by Elsevier DataSearch,[3] and created a mapping from the file formats found in the data to the target categories (Table 4).

Tables in CSV or XLS formats are the most common type of data (37%), followed by structured formats such as JSON and XML (30%) and documents in PDF or DOC format (11%). The latter category is problematic for many applications, as it is not machine readable. Audio, video, and medical imaging formats all constitute less than 1% of the datasets.

A subset of datasets that is of interest to the Semantic Web community is the datasets that contain graph data. We can approximate this number by summing over common formats: owl, rdf, xml+rdf, sparql, and so on. Together, they represent only 0.54% of datasets with downloads. Semantic web data is largely under-represented among datasets that use Semantic Web methods to describe metadata.

Making Data Citable. As we hope that datasets themselves become first-class citizens of the scientific discourse, we must develop mechanisms to reference and cite them. Scientists commonly use digital object identifiers (DOIs) and compact

---

[3] http://datasearch.elsevier.com/.

Table 5. Datasets with DOIs and compact identifiers

(a) Top ten providers of datasets with DOIs.

Domain	Datasets with DOIs
figshare.com	1,300,745
datacite.org	1,070,066
narcis.nl	118,210
openaire.eu	109,149
datadiscoverystudio.org	72,063
osti.gov	62,923
zenodo.org	49,622
researchgate.net	41,494
da-ra.de	39,318

(b) Providers with more than 100 datasets with compact identifiers.

Domain	Datasets with compact identifierss
neurovault.org	73,869
alliancegenome.org	29,204
datacite.org	14,982
openaire.eu	4,262
scicrunch.org	1,522
mcw.edu	517
duke.edu	306

identifiers (provided by services such as identifiers.org) for this purpose. We extract DOIs and compact identifiers from the dataset URLs or the values of the so#url property, as well as from so#sameAs and so#identifier properties. About 11% of the datasets in the corpus (or ~3M) have DOIs; about 2.3M of those come from two sites, datacite.org and figshare.com (Table 5a). Only a tiny fraction, 0.45% of the datasets, have compact identifiers (Table 5b).

Data Providers. While internet domains provide the conduit that brings datasets to users, the semantic provenance of datasets is more accurately captured by the notion of "provider." Domains and providers often align, but they also may differ when a provider hosts their datasets on a platform different from their own Web site. About 84% of all datasets specify a provider and there are about 100k distinct data providers in the corpus. The top 3 providers are CEICdata.com, Knoema, and the U.S. Geological Survey. The top 20 providers account for 78% of the total datasets. Most of them are the hosts of the top domains in Figure 1b. However, 87% of the providers are "small" providers, who publish fewer than 10 datasets each.

How Open is the Data. Finally, we analyze the licenses and availability of datasets. Dataset Search does not require the data to be open; only the metadata must be accessible to the crawler. Publishers specify access requirements for a dataset via the so#license property, the so#isAccessibleForFree boolean property, or both. About a third of the datasets (34%) specify license information and 3% of the datasets have the value for so#isAccessibleForFree (Table 3). Of the datasets that specify a license, we were able to recognize a known license in 72% of the cases. Those licenses include Open Government licenses for the US and Canada, Creative Commons licenses, and several Public Domain licenses. We found that for 89.5% of these datasets either the so#isAccessibleForFree bit is set to true or their license is a license that allows redistribution, or both.

In other words, almost 90% of these datasets are available for free. And of these open datasets, 5.6M, or 91%, allow commercial reuse.

3.3   What Do Users Search for

Finally, we look at what Dataset Search users search for. Overall, 2.1M unique datasets 2.6K domains appeared in the top 100 search results over 14 days in May 2020. Figure 5 shows the topics for these datasets. Note that the distribution of topics in Fig. 5 is different from the one for the corpus as a whole (Fig. 4), with geosciences, for instance, taking up a much smaller fraction; conversely, biology and medicine take a larger fraction relative to their share of the corpus. We are writing this paper during the Coronavirus pandemic; this timing likely explains the increased share of the biology and medicine datasets.

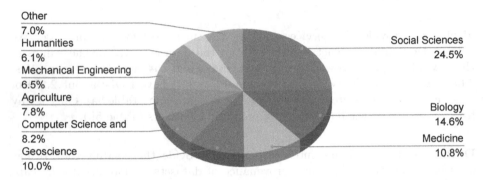

Fig. 5. Topic distribution of datasets that appeared in search results over 14 days in May 2020.

4   Discussion

We start our discussion by highlighting the results we found surprising or counter-intuitive. We then focus on the results that point to future work around improving the quality of metadata in general and provenance information in particular.

4.1   What Surprised Us in the Data

We do not attempt to discuss every table and graph in Sect. 3. Rather, we focus on the results that require some explanation or discussion.

Licenses and Access: Only 34% of the datasets provide any licensing information, through the so#license property for dataset or distribution. When no license is specified, the user technically must assume that they cannot reuse the data. Thus, adding licensing information, and, ideally, adding as open a license as possible, will greatly improve the reusability of the data [2]. At the same time, we were encouraged to see that in the vast majority of datasets that specify a license were available for free, allowed redistribution under certain conditions, and almost always allowed reuse for both commercial and non-commercial purposes. While 34% is still relatively low, this number is significantly higher than the 9% of linked-data datasets with licenses that Schmachtenberg and colleagues found in their 2014 survey [21].

Availability of Data Downloads: We found that only 44% of datasets specify a data download link in their metadata. This number is surprisingly low, because datasets are merely containers for data. A possible explanation is that Webmasters (or dataset-hosting platforms) fear that exposing the data-download link through Schema.org metadata may lead search engines or other applications to give their users direct access to downloading the data, thus "stealing" traffic from their Web site. Another concern may be that data needs the proper context to be used appropriately (e.g., methodology, footnotes, license information), and providers feel that only their Web pages can give the complete picture. In Dataset Search, we made the decision not to show download links as part of dataset metadata so that users can get the full context from the publisher's Web site before downloading the data.

Quality and Completeness of the Metadata: As Table 3 shows, the majority of metadata properties have under 50% coverage. While properties such as spatial or temporal coverage apply only to specific domains, others such as authors, variables being measured, updated date, licensing and download information are general, and need to become much more prevalent for metadata to be truly descriptive. Meusel and colleagues analyzed schema.org adoption across the Web in 2015 [13] and found that data providers both adopt and drive changes in the schema.

Effect of Outreach: We discussed elsewhere [16] that reaching out to specific communities and finding influencers there was key to bootstrapping the corpus in the first place. At the time, we focused on geosciences and social sciences. Since then, we have allowed the corpus to grow organically. We were surprised to see that, even after the corpus has grown manifold, the communities that we reached out to early on are still dominating the corpus: 45.2% of the datasets are from these two disciplines. Of course, this dominance may be due to other factors, such as differences in culture across communities. For instance, geosciences have been particularly successful in making their data FAIR [22].

Persistent Identifiers and URLs: Many scientific disciplines have come to a consensus (or have been compelled by funding agencies and academic publishers) that it is important to publish data and to cite it [25]. There are, for example, peer-reviewed journals dedicated to publishing valuable datasets, such as Nature Scientific Data [15], and efforts such as DataCite [19], that provide digital object identifiers (DOIs) for datasets and both encourage and enable scientists to publish their datasets. For datasets to become first-class citizens in scientific discourse, they must be citable. Unfortunately, Fig. 2b shows that

URLs of datasets are not persistent: 37% of URLs that had dataset metadata in June 2019 either do not have the metadata or are no longer accessible in March 2020. This high level of churn argues strongly for the use of persistent identifiers for datasets, such as DOIs and compact identifiers. This practice is now widespread for publications, and we argue that it should become just as widespread for datasets.

(Not) Eating our Own Dogfood. Fewer than 1% of datasets in our corpus are in linked-data formats. The Dataset Search approach relies on semantic-web technologies such as DCAT and schema.org. At the same time, the Semantic Web community is either not producing enough data, not sharing it, or not adding semantic metadata to it. From profiling efforts (e.g., [1,13]), it seems that the problem is the latter: there is plenty of shared data that researchers produce, but the final step of describing it appears to be less common.

4.2   Future Work

Improving Metadata Quality. Throughout this analysis, we found many places where metadata was missing, formatted wrongly, not normalized, and so on. We discuss a few possible approaches to improve the quality of metadata:

Automated techniques: We continue to develop better techniques to automatically clean, normalize, and reconcile dataset metadata. These techniques are hard to do at scale given the heterogeneity of the data, and they will never be perfect. The benefits of these techniques are often application-specific, and there are no easy mechanisms to share them back with the community.

Feedback to publishers: We can let the owners of datasets know that their dataset metadata can be improved. Google tools such as Search Console and the Structured Data Testing tool already highlight some of these issues. One could also consider developing interactive tools to create and validate dataset metadata, which could be integrated into popular dataset management CMSs or hosting platforms.

Crowdsourcing: Why not let users of datasets fix their metadata or point to possible improvements? One option would be to provide that functionality in the Dataset Search tool, which would then funnel suggestions to the publishers. Vrandečić [23] proposed using WikiData to crowdsource the definition of metadata for popular datasets that the community cares about.

Improving Provenance Information. Many datasets are duplicates of other datasets, or derived from other datasets. Knowing these relationships is important not just for Dataset Search, but to any user of datasets who cares about data provenance. These relationships are also critical to giving dataset publishers credit when their data is reused: derivative datasets are akin to citations of papers. While schema.org provides properties to describe these relationships, namely so#sameAs and so#isBasedOn, the usage of these properties is low. How can we improve the coverage of this lineage information?

Automated detection: We already detect duplicate datasets and cluster them [16]. Identifying that a dataset is derived from other datasets is a much more difficult problem, but it may be feasible in restricted cases (e.g., specific forms of data, such as tables or images, and limited transformation operations).

License requirements: Most data licenses require citation when re-using a dataset, however there is no obligation to make that citation machine readable. There would be huge value in requiring the usage of schema.org properties when citing datasets.

Community incentives: Data provenance can help complete the picture of the usefulness and impact of datasets, together with paper citations, application usage, etc. How can we incentivize broad adoption of data provenance in the scientific and open data communities [11]?

5   Conclusions

In this paper, we analyzed the corpus of dataset metadata used in Google's Dataset Search, a search engine over datasets on the Web. While it has limitations (Sect. 2.2), it is a large snapshot of datasets on the Web in a variety of disciplines. Our analysis shows that datasets on the Web are very diverse, with no one discipline truly dominating; there are datasets with semantic markup in Web sites from any country and in any language. We have observed an explosive growth over the last three years.

Yet, metadata still leaves a lot to be desired if data is truly to become a first-class citizen in scientific discourse: We need tools to ensure that the metadata is more complete and mechanisms to encourage the use of licensing information for data and persistent identifiers. And the Semantic Web community needs to eat its own dogfood by adding semantic metadata to its datasets.

References

1. Ben Ellefi, M., et al.: RDF dataset profiling–a survey of features, methods, vocabularies and applications. Semant. Web **9**(5), 677–705 (2018)
2. Carbon, S., Champieux, R., McMurry, J.A., Winfree, L., Wyatt, L.R., Haendel, M.A.: An analysis and metric of reusable data licensing practices for biomedical resources. PLOS ONE **14**(3) (2019). https://doi.org/10.1371/journal.pone.0213090
3. Chapman, A., et al.: Dataset search: a survey. VLDB J. **29**(1), 251–272 (2019). https://doi.org/10.1007/s00778-019-00564-x
4. Fenner, M., Crosas, M., Grethe, J., et al.: A data citation roadmap for scholarly data repositories. bioRxiv (2017). https://doi.org/10.1101/097196
5. Datasets: Search for developers. https://developers.google.com/search/docs/data-types/dataset
6. Gray, A.J., Goble, C.A., Jimenez, R.: Bioschemas: from potato salad to protein annotation. In: International Semantic Web Conference (Posters, Demos & Industry Tracks) (2017)

7. Gregory, K., Groth, P., Scharnhorst, A., Wyatt, S.: Lost or found? Discovering data needed for research. Harvard Data Sci. Rev. (2020). https://doi.org/10.1162/99608f92.e38165eb

8. Guha, R.V., Brickley, D., Macbeth, S.: Schema.org: evolution of structured data on the web. Commun. ACM **59**(2), 44–51 (2016)

9. Halevy, A., et al.: Goods: organizing Google's datasets. In: ACM SIGMOD (2016)

10. Hendler, J., Holm, J., Musialek, C., Thomas, G.: US government linked open data: Semantic data.gov. IEEE Intell. Syst. **27**(3), 25–31 (2012). https://doi.org/10.1109/MIS.2012.27

11. Herschel, M., Diestelkämper, R., Lahmar, H.B.: A survey on provenance: what for? what form? what from? VLDB J. **26**(6), 881–906 (2017)

12. Kindling, M., et al.: The landscape of research data repositories in 2015: a re3data analysis. D-Lib Mag. **23**(3/4) (2017). https://doi.org/10.1045/march2017-kindling

13. Meusel, R., Bizer, C., Paulheim, H.: A web-scale study of the adoption and evolution of the schema.org vocabulary over time. In: International Conference on Web Intelligence, Mining and Semantics. ACM, New York (2015). https://doi.org/10.1145/2797115.2797124

14. Nargesian, F., Zhu, E., Pu, K.Q., Miller, R.J.: Table union search on open data. VLDB J. **11**(7) (2018). https://doi.org/10.14778/3192965.3192973

15. Nature scientific data (2018). https://www.nature.com/sdata

16. Noy, N., Burgess, M., Brickley, D.: Google dataset search: building a search engine for datasets in an open web ecosystem. In: The Web Conference, pp. 1365–1375. ACM (2019). https://doi.org/10.1145/3308558.3313685

17. Noy, N., Gao, Y., Jain, A., Narayanan, A., Patterson, A., Taylor, J.: Industry-scale knowledge graphs: lessons and challenges. Commun. ACM **62**(8), 36–43 (2019). https://doi.org/10.1145/3331166

18. RDF 1.1 Concepts and Abstract Syntax. https://www.w3.org/TR/rdf11-concepts/

19. Rueda, L., Fenner, M., Cruse, P.: Datacite: lessons learned on persistent identifiers for research data. IJDC **11**(2), 39–47 (2016). https://doi.org/10.2218/ijdc.v11i2.421

20. Sansone, S.A., et al.: DATS, the data tag suite to enable discoverability of datasets. Sci. Data **4**, 170059 (2017)

21. Schmachtenberg, M., Bizer, C., Paulheim, H.: Adoption of the linked data best practices in different topical domains. In: Mika, P., et al. (eds.) ISWC 2014. LNCS, vol. 8796, pp. 245–260. Springer, Cham (2014). https://doi.org/10.1007/978-3-319-11964-9_16

22. Stall, S., et al.: Make scientific data FAIR (2019)

23. Vrandečić, D.: Describing datasets in Wikidata. In: Advanced Knowledge Technologies for Science in a FAIR World, IEEE eScience Conference (2019)

24. Wang, J., Aryani, A., Wyborn, L., Evans, B.: Providing research graph data in JSON-LD Using Schema.org. In: 26th International Conference on World Wide Web Companion, pp. 1213–1218 (2017). https://doi.org/10.1145/3041021.3053052

25. Wilkinson, M.D., Dumontier, M., Aalbersberg, I.J., et al.: The FAIR guiding principles for scientific data management and stewardship. Sci. Data **3**, 1–9 (2016)

26. Wimalaratne, S.M., Juty, N., Kunze, J., Janée, G., et al.: Uniform resolution of compact identifiers for biomedical data. Sci. Data **5**, 180029 (2018)

Dynamic Faceted Search for Technical Support Exploiting Induced Knowledge

Nandana Mihindukulasooriya$^{(\boxtimes)}$, Ruchi Mahindru,
Md Faisal Mahbub Chowdhury, Yu Deng, Nicolas Rodolfo Fauceglia,
Gaetano Rossiello, Sarthak Dash, Alfio Gliozzo, and Shu Tao

IBM Research, T.J. Watson Research Center, Yorktown Heights, NY, USA
{nandana.m,nicolas.fauceglia,Gaetano.Rossiello}@ibm.com
{rmahindr,mchowdh,dengy,sdash,gliozzo,shutao}@us.ibm.com

Abstract. IT support is a vital and integral part of technology adoption. Conventionally, IT support service providers heavily rely on human effort and expertise to respond to user queries. Given the cost-benefit and 24×7 availability for answering user questions, Virtual Assistants (VA) are highly applicable in the technical support domain. In this paper, we describe a novel methodology for building interactive virtual assistants for IT support using Dynamic Faceted Search (DFS). Given a question, dynamic facets are generated automatically, enabling the user to refine and narrow down their intent. To do so we leverage knowledge automatically induced from textual content and existing Semantic Web resources such as Wikidata. Such knowledge is then used to dynamically generate facets interactively based on the user's responses as shown in the demo video (https://ibm.box.com/v/iswc2020-dfs). The experiments on two real-world datasets in the IT support domain show the effectiveness of DFS in refining the user's queries and efficiently identifying possible solutions to their technical problems.

Keywords: Faceted search · Knowledge induction · IT support · Virtual assistant

1 Introduction

IT support is committed to help customers identify solutions for issues occurring in hardware and software products, and suggest respective solutions. In this domain, a large amount of support documentation is available, typically consisting of user manuals and troubleshooting (*how-to* or *what-is*) documents. Such documentation contains solutions to common problems that the customers may face with the products.

Information retrieval methods, *e.g.* keyword-based search, could be applied in seeking the relevant document which provides the solution to user's query. However, this strategy is not adequate in domains like IT support, where the

---

N. Mihindukulasooriya and R. Mahindru—Equal contributions.

© Springer Nature Switzerland AG 2020
J. Z. Pan et al. (Eds.): ISWC 2020, LNCS 12507, pp. 683–699, 2020.
https://doi.org/10.1007/978-3-030-62466-8_42

complexity of problems often leads to several challenges. First, user's problems are often too complex to be expressed with a single query, as the user may be observing several different issues at the same time, e.g. `fan making noise` and `battery not charging` can occur simultaneously. In some cases, problems observed could be related but in other cases they may be completely unrelated. Second, user may craft the question based on their observations, while the content in technical documentation is formally written by Subject Matter Experts (SMEs). Therefore, leaving a tremendous gap in wording and rhetorical structure used for asking a question versus content writing. Third, a user may not know a priori the context (relevant elements) that should be provided for the system to efficiently find the target result.

Faceted Search (FS) [15] is a prevalent technique for interactive information retrieval, e.g. in e-commerce. It involves augmenting a document retrieval system with facets to narrow down search results[1]. Users looking for IT support often find it challenging to formulate complete query for search. FS can **guide** users to refine initial queries and **navigate** to the target result. Traditional facet generation approaches present several drawbacks. Documents must be tagged with an existing taxonomy, adding overhead in content curation and management. Moreover, static facets are not based on the matching documents or queries. DFS overcomes such limitations [7].

In attempting to solve the limitations discussed above, we propose VA for IT Support based on DFS. Using DFS, our system allows refinement of the initial user query, in an interactive way, by presenting to the user multiple choices (for augmenting and refining the query) in the form of facets. Use of VAs to help answering customer issues is on a rise [6], as it is becoming a necessity for service providers to provide 24 × 7 IT support at a lower cost.

Typically, building a VA for a specific domain requires a considerable effort as it often relies on curated knowledge consisting of hand-crafted databases, problem determination flows and predefined question/answer pairs developed by SMEs. Such effort could typically range from anywhere between a few weeks to months which hinders automatic and rapid initialization of VAs to new domains.

Additionally, our experience shows that typically support issues cover 10–20% common questions and, remaining are long-tail questions addressed using documents. VA uses dialog technology for the common questions, redirecting the long tail to regular search on technical documentation. In this work, we address the long tail questions using Dynamic Faceted Search. DFS provides an interactive search experience to guide the user to form a more complete query.

In the absence of any kind of annotated training data (which is common for many IT service providers), exploiting the textual content of technical documentation in an unsupervised (or minimally supervised) manner is the most viable choice to avoid the laborious effort from the SMEs and delay to market.

The main features of our proposed DFS based VA system are as follows:

- The facets are dynamically generated from the textual content (in IT support documents) during runtime. In other words, they are not static, i.e.,

---

[1] A VA capable of fully automatic dialog generation is out of scope for DFS in the context of this work.

pre-tagged for every document. Instead, they are dynamically generated based on the user's query and facets selected so far.

– Our system generates two kinds of facets, namely `flat facets` and `typed facets` (more details in Sect. 3.3). The system can be configured to show the kind of facets a user prefers.

– Unlike conventional faceted search systems (often used for e-commerce), our facets are not specific to particular items (e.g. "price" or "brand" products) or documents (e.g. "topics" or "categories" of a document). Such facets are not sufficient for all domains. Due to this flexibility, even though we are presenting our VA system here for the IT support use case, it can be easily adapted to other domains that share similar or subset of characteristics as IT Support domain.

In addition to the above, this paper has two additional contributions. Firstly, we propose an automatic evaluation setting to simulate human users. Secondly, we empirically show that our FS based approach improves the results of document retrieval of a popular IR based approach.

In the next section, we define the use case relevant in scope for information retrieval over textual documents. An overview of our system and the architectural design are described in Sect. 3. In Sect. 4, we present our experimental protocol with the discussion relative to the results. Section 5 presents an overview regarding the progress made by the community in conversational search. Finally, Sect. 6 concludes the paper along with directions in our research agenda.

2 Use Case

IT Support is contracted to deliver several business key performance indicators (KPIs), e.g. first-time-fix (FTF). FTF is a measure indicating, whether the provided solution resolved the user's issue in the first time, they contacted the support or used the troubleshooting tools (e.g. search, virtual assistant etc.) directly to resolve their problem. For any customer side support whether from technical domain or not, customer satisfaction is typically one of the most important business success metrics. Therefore, it is critical to increase FTF improve customer satisfaction. In fact, it was found that the average first response time for support tickets is $5.32\,h^2$.

In the IT domain, user questions typically fall in three major categories: (1) `troubleshooting`, e.g. *"no power"*, *"jextract utility OutOfMemory error"* or *"key is stuck"*; (2) `how-to`, e.g *"password reset"*, *"replace disk drive"*, *"secure DataPower MQ"*; (3) `what-is`, e.g. *"what version of java is supported in DataPower Appliance 6.0+"*, *"what feature does my Android have"*.

End users' problems, expressed with queries, are primarily biased by the terminology that they are familiar with. While the technical documents are written by experts and the technical description of the problem may be quite

---

[2] https://www.jitbit.com/news/255-lessons-learned-from-analyzing-7-million-customer-support-tickets/.

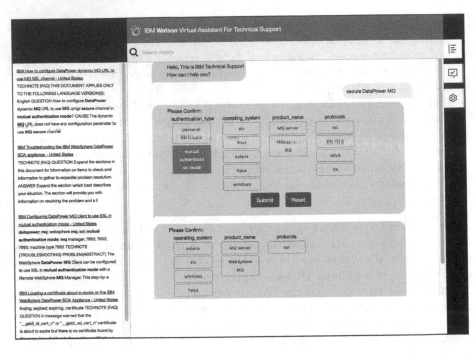

Fig. 1. Virtual assistant interface

remote from how the user may describe the issue. This disconnect in user's query and content along with complexity of technical problems often leads to several challenges, as discussed below.

- The user may express multiple problems in a single query. For example, *keyboard not typing* and *battery keeps discharging*. Here, there are two independent problems described. In another example, *ac adaptor faulty* and *battery is not charging*. In this case, *battery not charging* could be a side effect of *ac adaptor faulty*, hence there's a cause and effect relation.
- User may not know a-priori the context (relevant elements) that should be provided to efficiently find the target result. For instance, a user may experience that there is *"no audio"* output. There can be many different documents (*e.g.*, driver-related issues, connectivity issues, hardware issues) pertaining to a user query such as *"no audio"*. Such a query is under-expressed leaving a large room for the system to determine the precise answer document.
- User may craft the question based on their observations, while the content in technical documentation is formally written by Subject Matter Experts (SMEs). For example, user may express their query as *storage failure* while the document may express and describe the problem as *Hard Disk Drive Failure, Solid State Drive Failure*. Both *Hard Disk Drive* and *Solid State Drive* are storage related but neither document may directly use the word storage in them, while there could be many other unrelated documents using

the term *storage failure* in them but not necessarily related to resolving user's problem. Therefore, leaving a tremendous gap in wording used for asking a question versus content writing.

To address the challenges described above, our goal is to build a VA that given a user query, determines whether the user query is under-specified, contains multiple problems based on the rhetorical structure, or it is missing context.

In this paper, our focus is to tackle under-specified queries. Given the user query, VA fetches the relevant documents and presents them to the user, along with the facets. Thus interactively nudging the user to further refine their query, so that the augmented query can be used to retrieve more relevant documents (see Fig. 1). One can argue that this may also be accomplished to a certain extent by a non-interactive search system. Based on our observation from the real-world application, interactive VA helps reduce the cognitive burden of the user in coming up with a more complete query. And thus, guide them towards the target answer by presenting the contextually relevant facets. VA for technical support may cover other use cases[3], which are more close to dialog style (i.e. conversational) interaction, e.g. chatbots rather than search-oriented (long tail). In this paper, our focus is primarily on leveraging textual documents to 1) perform content driven and interactive user query augmentation, and 2) improve the ranking of the results retrieved.

3   End-to-End System Overview

We present an end-to-end system composed of two phases: 1) *Virtual Assistant Initialization (VAI)*, an offline process, which consists of two main steps – unsupervised knowledge induction and curation; and 2) *Virtual Assistant Runtime (VAR)*, also consists of two main steps, an online VA application for user interaction, integrated with the DFS component (see Fig. 2). In each phase, there are dedicated personas interacting with the system in the VAI and VAR phases.

3.1   Personas

The two personas that are involved in the end-to-end use case (described earlier) are – the solution designer (SD), e.g. an SME, who is specialized in the problem troubleshooting and the required domain content, is interested in creating a VA and leads the VAI, and the user of the VA is involved in the VAR.

Solution Designer. The SD uploads relevant documents (which would be used later for resolving users' questions) in the system. The system automatically induces knowledge (e.g. domain specific terms, taxonomy, etc) from the documents. The SD can fine-tune the induced taxonomy using the smart spreadsheet

---

[3] For example, interactive problem diagnosis containing test and action steps; and process automation to invoke enterprise endpoints. for common questions.

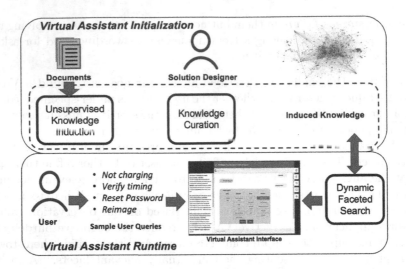

Fig. 2. End-to-end system overview

editor available in the system (see Sect. 4 in [10]). Lastly, the VA uses this curated knowledge together with other artifacts to generate facets dynamically to help the user refine their query.

Virtual Assistant User. The user persona uses the VA to find a solution to a problem that they are facing. In the IT domain, it could be either a support agent responsible for handling customer service requests, or it could be an end user directly interacting with the VA on the vendor's web-site.

Users of popular search engines, such as Google, generally have the tendency to formulate fragment queries [14] which are often under-specified. Therefore, by presenting dynamically generated facets relevant to the user's query, the VA can interactively solicit additional information from the user. Such information augments the query and enables the VA to retrieve updated results along with a new set of ranked facets. This interaction is repeated until the user is satisfied with the results presented and no further refinement is needed.

The VA interface drives the conversation and handles the user queries. Given a user query, it performs the tasks as shown in the *Facet Generation* component in Fig. 3. The refined facets and re-ranked search results are presented to the user in the VA after every interaction.

3.2   Virtual Assistant Initialization

This phase includes two main components as illustrated in Fig. 3:

- A component for automatic knowledge induction.
- A component for curating automatically induced knowledge where the induced knowledge is presented to the SD in a `smart spreadsheet` [10], so that they *quickly* remove spurious results to enhance it.

Unsupervised Knowledge Induction. The knowledge induction (KI) service trains term embedding models, extracts domain specific terms and induces taxonomies (types and their instances) from input documents. One of the embedding models trained is a type model (see Sect. 3.2 in [10]). The embedding models are used by all the components in Fig. 3. The state-of-the-art automatic taxonomy induction component uses an ensemble of a pattern-based approach, symbolic knowledge-based approach, and a neural network-based approach [8,10].

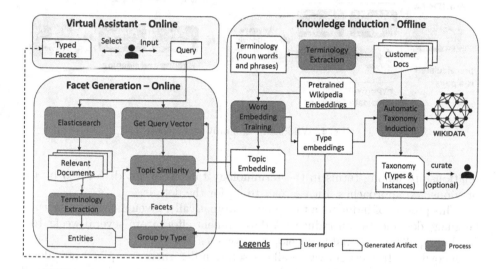

Fig. 3. Architectural overview

In the pattern-based approach, twenty-four *lexico-syntactic* patterns (e.g., "NP_y is a NP_x"), aka Hearst-like patterns [11] are used for *is-a* pair extraction. Then, several post-processing steps are carried out to clean the extracted *is-a* pairs by removing cycles and proper nouns as types [10] and to expand them using super and nested terms [5].

Figure 4 illustrates the usage of symbolic knowledge from Wikidata in knowledge induction. Wikidata is used to create a dictionary of pre-known *is-a* using *instance of* (P31) and *subclass of* (P279) relations. For each corpus, terminology is extracted as shown in Fig. 3 and linked to the entities in Wikidata. This process extracts an *intermediate domain taxonomy* from the dictionary. Finally, to mitigate any entity linking error propagation, we perform a cleaning step using cosine similarity (a) each *type* and *instance* and (b) between all *instances* of a given type. This produces a domain taxonomy for the given corpus.

In the neural network-based approach, we use STRICT PARTIAL ORDER NETWORKS (SPON) [8], a neural network architecture comprising of *non-negative* activations and *residual* connections designed to enforce strict partial order as a soft constraint. The union of the *is-a* pairs extracted from previous two approaches are used as the input for the neural model. SPON finds the *is-a*

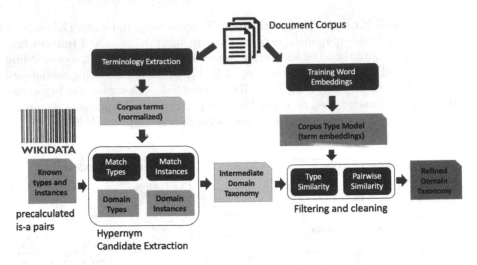

Fig. 4. Wikidata taxonomy generation

relationships among terms in the terminology that were not discovered by the previous two approaches. Please refer to [8] for the details.

This process of inducing a taxonomy automatically, given a corpus of natural language documents can reduce VA development effort from several weeks to a few hours, allowing scalable development of such VAs.

Knowledge Induction service allows solution designers to upload new documents and rerun the knowledge induction process to support evolving corpora. It will update the knowledge artefacts such as the terminology, embedding models, and taxonomies. Once updated, these new artefacts will be used seamlessly by `Dynamic Facet Generation` so that the knowledge from new documents will be used in DFS.

Knowledge Curation. As mentioned earlier, the SD, if needed, can curate the induced domain specific terms and taxonomy, to further refine them in a human-in-the-loop manner; based on the real-world use cases (with corpora up 800 K documents), this process can be generally performed in a few hours for a reasonably large corpora.

This curation process generally starts with presenting the SD with a list of types from the generated taxonomy. For making the inspection of SD efficient, the list can be sorted by the frequency they appear in the corpus or the number of the instances a type has. Once the relevant facets are selected, SD can export them to a `smart spreadsheet` (Fig. 5). In this spreadsheet, SD can fine-tune the instances in the types shown or add brand new types. Furthermore, SD can expand the instances using existing ones as seeds and finding the nearest neighbours of those in the type embedding model or by performing Wikidata lookups.

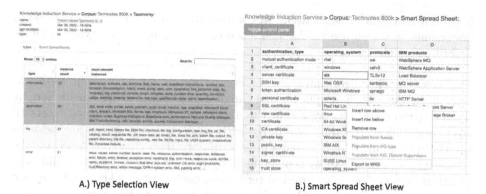

A.) Type Selection View   B.) Smart Spread Sheet View

Fig. 5. Smart spreadsheet for knowledge curation

3.3 Virtual Assistant Runtime

This phase includes two main components as illustrated in Fig. 2 –

- A component with a chat interface (see Fig. 1). It receives user questions and shows the most relevant documents containing the answer.
- A component for facet generation and search (see Fig. 3). It dynamically generates facets based on information retrieval result for the query and induced knowledge.

Virtual Assistant Interface. First, the user uses the VA interface to submit the initial query. Next, the system submits the query to the dynamic facet generation component. As described in Sect. 3.3, these facets are generated from the terminology present in the corresponding search results of the initial query. We would like to emphasize that such content driven guidance is critical, as the user may not know off hand the terminology relevant to retrieve the target answer. The user selects one or more relevant facets to refine the query, hence helping the user find the most relevant document faster (Fig. 1).

For example, suppose the user has a problem with "*database password running on Microsoft SQL Server*". But they may under specify a query such as "*password issue*", which is typical for the support agents. In this case, the VA starts by returning a set of initial results based on the query, as shown on the left hand side of the interface. In addition, the VA nudges the user to refine their query, if needed. The user is able to further clarify the query as their problem is related to a *database error* running on *SQL Server*. Finally, the user finds the relevant document per the refinements.

Dynamic Facet Generation. We automatically generate facets from the document corpus and show the relevant facets dynamically per the semantic matching of the user query and search results (Fig. 3). As a first step, the user query is submitted to the ElasticSearch[4] IR component, which returns a ranked list of documents as search results. As mentioned earlier, there are two types of facets described as follows.

Flat facets – This algorithm starts by extracting all domain specific terms, $\mathcal{E}$, present in the search results for the input query, which then are filtered by removing the ones – (i) present in the query, (ii) previously selected facets (if any), and (iii) whose topic similarity with respect to the query are below a threshold (default, 0.5). The remaining terms are then clustered based on the similarity score predicted by type embedding model. This is done to avoid near identical terms. The highest ranked term from each cluster is kept and they are ranked according to their topic similarity with respect to the query. The top n remaining terms are then returned as facets.

Typed facets – In this case, the system ranks each term in the induced taxonomy by topic similarity with respect to the query, looks for their corresponding types and collects weights for each type following a k-nn (k-nearest neighbours) based approach. This allows the system to select a certain number of types and rank them. Following that, the system looks for other terms (in the documents in search results for the query) that are similar to the terms in the type set selected in previous step. This is done by selecting the top n other terms if their type similarity is above a certain threshold. Finally, the top k types containing maximum top x facets per type are returned as typed facets.

4   Experiments

We evaluate our VA system through a quantitative and qualitative analysis. Quantitative evaluation aims to measure the impact of the proposed dynamic facet generation in retrieving the target (i.e. most relevant) document after the facet selection(s). We perform qualitative evaluation to investigate the quality and usefulness of those facets from the user experience perspective. A manual evaluation with SMEs is not viable for large amounts of data and for repeating experiments with different settings. Therefore, we have devised an automated evaluation for quantitative analysis using two datasets for experiments. Each dataset consists of document corpus and question answer pairs. The question answer pairs are actual questions posted by real-world users in forums and answer is link to the document in the corpus.

TechQA Dataset. The first dataset is the TechQA dataset [4] which contains real-world user questions posted on IBM DeveloperWorks[5] forums in the domain of technical customer support. This dataset is created by extracting forum questions with their accepted answers where a link to an *IBM Technote*, a technical

---

[4] https://elastic.co/enterprise-search.
[5] https://www.ibm.com/developerworks/community/forums/.

document, appears in the text. We use both the 160 question answer pairs in the development set and the full set of 610 question answer pairs for our benchmarks. The TechQA dataset also contains a corpus of 801,998 publicly available IBM Technotes documents.

Private Dataset. The second dataset is a private dataset, which contains a corpus of 4,000 technical troubleshooting documents from a Personal Computer domain. Additionally, it has 50 question answer pairs from forum corresponding to the same domain and product. This dataset comes from a real technical support customer engagement that we have used for internal evaluation purposes and due to confidentiality, the dataset cannot be shared publicly.

For each dataset, we use the documents to induce their own knowledge artifacts as described in Sect. 3.2. In our experiments, we only use the title of the forum post as the query to simulate under-specified user queries. Then we simulate the refinement of the query using the facets generated by the system in an interactive manner. The criteria used in the evaluation is how efficiently (with minimum number of clicks) a user can find the answer with the help of facets generated by our VA.

4.1   Quantitative Evaluation

In order to evaluate the impact of the facets, we compare our VA system against the vanilla ElasticSearch (i.e. baseline), which is used by our system as the IR component. We evaluate if the results are improved after the queries are augmented with the generated dynamic facets. We use four evaluation metrics: *Mean Reciprocal Rank (MRR), Hits@1, Hits@5* and *Hits@10*. For *Hits@K* metrics, we share the ratio of the number of queries where the expected document is ranked within top-K results compared to the total number of queries.

User facet selection is simulated using an algorithm that we call **Oracle**, discussed below.

1. First, the n-grams are extracted from the user query.
2. Second, the facet generation component is used to obtain the search results and generate facets with scores (either typed facets or flat facets).
3. Third, the Oracle first selects the top-K[6] facets (regardless the facets are typed or flat) based on the scores associated with relevance of each facet with the query. Here, by selecting top-K, we assume that in a live setting, a human user might not read more than the top-K facets.
4. Next, out of the top-K facets, the Oracle selects the best facet which retrieves the target document at the highest possible rank. Here, we make a simplified assumption that in a live setting a human user (*e.g.* SME) will be always able to identify the best facet (with respect to her query) among the top-K facets presented.

---

[6] K = 5 in our experiments.

Table 1 shows the performance comparison for the TechQA development dataset of 160 question answer pairs and full dataset of 610 question answer pairs. As shown in the table, both DFS approaches, i.e. search enhanced by flat facets and typed facets, have improved the baseline results for both cases by refining user queries. Interestingly, *DFS using typed facets* performs better than *DFS using flat facets*. We observe similar performance improvements in the private dataset as shown in Table 2. This underpins the value of our unsupervised induced taxonomy for organizing dynamically generated facets.

In case of VA use case, our target is to present the best possible document result in the top 5 (*Hits@5*) to improve user experience with the minimum number of facet clicks. Based on our interaction with the real-world users, we used 3 as the maximum number of rounds for facet selection (i.e. up to 3 user clicks to find the answer document). With the typed facets, we observe that the *Hits@1* and *Hits@5* are improved by up to 9% (as shown in Table 1). Additionally, as observed, flat facets follow the similar trend in improvement, where *Hits@1* and *Hits@5* are improved by 6% (as shown in Table 1). In the second data, similar hyper parameters were used and confirms the same observation in result improvements, as shown in Table 2).

4.2   Qualitative Evaluation

For the qualitative evaluation, we selected a sample set of random queries from the 160 queries. We manually inspected the facets using a *human-in-the-loop* approach. A facet is considered useful, if it (i) contains terminology (that is contextually related but not already mentioned in the user's actual query) from the fully specified query, and (ii) appears in the target answer document.

Table 3 shows a few fully specified queries, actual (under-specified) queries entered by the users in the VA, and the corresponding dynamically generated flat facets.

For example, in the first row of Table 3, the ideal query is *"How to configure SSL using mutual authentication in DataPower MQ client?"*, while the user may enter *"DataPower MQ client security"*. This under-specified actual query leaves room for fetching many generic documents containing "security" and "DataPower MQ".

Table 1. MRR and Hits@K results for the development set of 160 questions and the full set of 610 questions in the TechQA dataset using flat and typed facets.

Metric	Dev 160 questions			Full 610 questions		
	Baseline (ES)	DFS (Flat)	DFS (Typed)	Baseline (ES)	DFS (Flat)	DFS (Typed)
MRR	0.47	0.53	**0.55**	0.47	0.53	**0.54**
Hits@1	0.38	0.44	**0.47**	0.39	0.46	**0.47**
Hits@5	0.58	**0.64**	**0.64**	0.55	**0.62**	**0.62**
Hits@10	0.64	0.68	**0.69**	0.60	0.65	**0.66**

Table 2. MRR and Hits@K results for the questions in the private dataset using flat and typed facets.

Metric	Baseline (ES)	DFS (Flat)	DFS (Typed)
MRR	0.29	0.34	**0.39**
Hits@1	0.24	0.28	**0.34**
Hits@5	0.32	0.38	**0.42**
Hits@10	0.34	0.42	**0.46**

Table 3. Sample queries with the dynamically generated flat facets from the first dataset.

Fully specified query	Actual user query	Ranked (by similarity score) dynamically generated (flat) facets
How to configure SSL using mutual authentication in DataPower MQ client?	DataPower MQ client security	**Personal certificate, mutual authentication mode, aix, windows, linux, hpux**, MQ Server, Websphere MQ, **SSL, SSLV3**, ftp server, application server
NMAagent installation failure	NMA failure	**TPC Data Agent, CANDLEHOME directory, configuration failure, GSKit files**, configuration phase, installed agents, **failed probe**, Agent Builder agents, **N4 agent, failed install**
Mismatched MQ jars in my application server	MQ jar mismatch	**Java EE server**, Java Compute Nodes, jar package, Restart SBI, **JVM classpath**, com.ibm.mq.allclient.jar, client jar file, MQ WMQ, MQ client jars, **Java EE application server**, dhbcore.jar
How can I export a private key from DataPower Gateway Appliance?	Export key datapower	**OpenSSH format, Personal Certificates section**, key database password, **PFX file, certificate object**, key ring file, stash password, unencrypted file, **pkcs12 certificates, tklmKeyImport command**
How to refresh a DataPower WebService Proxy when WSDL got changed?	Refresh webservice proxy	Proxy recorder, **WSDL description, proxy gateway**, DNS Static Host, **Multi - Protocol Gateway**, Service Gateway, web service gateway, **service endpoint, MPGW service**

The 3rd column shows the top n dynamically generated facets, *personal certificate, mutual authentication mode, aix, windows, linux, hpux*, MQ Server, Websphere MQ, **SSL, SSLV3**, ftp server, application server. Facets in **BOLD** were considered useful, as they augmented the actual user query.

We performed similar qualitative evaluation on the private dataset as shown in Table 4. It appears from these random queries that on average 60% facets generated by the system are useful, based on the aforementioned criteria.

4.3 SME User Evaluation

In order to further validate the usefulness of the generated facets with real users. We conducted formal evaluation with 2 SMEs to capture their feedback on the relevance of the flat facets in context of the query. We provided them 50 actual user queries (collected from the production usage logs for the private dataset) and their respective top 20 flat facets. The SMEs' task was to evaluate and provide feedback on the facet relevance in context of the query.

SMEs decided whether the facets are useful based on the query and context that they have based on their skill and expertise. Among the annotations of the two SMEs, there was a Cohen's kappa coefficient of 0.49 inter-annotator agreement, which is a moderate agreement[7]. This is understandable because feedback of the users is subjective and it depends on their skills and experience.

Table 5 shows the average percentage of facets that the SME users found relevant for the given query is about 50% (out of top 20 facets per query). It is important to note that the standard deviation is high, which is indicative of large variance in the quality of the facets for a given query. This has opened additional avenues of investigation to further analyze, predict and understand the cause the under performing queries.

Table 4. Sample queries with the dynamically generated flat facets from the second dataset.

Fully specified query	Actual user query	Ranked (by similarity score) dynamically generated (flat) facets
machine is not powering but there is beep	no power	start up, **HDD**, fan, **AC power**, card, touchpad, **CHKDSK**, beeps, **optical drive, post**
audio is not coming from the speaker my machine	no audio	power, **headphone, music, HDMI, audio output, Bluetooth Headset, dock**, projector, **failed probe**, sound
Help increase the battery life, it keeps on running out fast	increase battery life	power button, **charger, performance, battery pack,** usage, time, **Battery saver**, level, batteries
Card not detected on new system using a specific adaptor	card not detected	**memory**, carrier, power, **adapter, HDD, SIM card**, case, **video card, SD Card, SIM**
Dock may fail to connect to a network when USB booting or PXE booting	dock failure	**USB-C, dock driver, docking station, USB**, basic USB, VGA, **Ultra Dock, 3.0 Dock, 3 Dock**

---

[7] 0.41– 0.60 is considered as moderate [12].

5 Related Work

Given that the goal of our system is to improve document retrieval for user's query, our system recommends facets (some of which are named entities) merely to refine user's problem. With this goal, we briefly highlight research that are closely relevant to our work.

The idea of using the conversational paradigm for information seeking has been originally proposed in [3]. Recently, Radlinski *et al.* [13] propose a theoretical framework for conversational search by suggesting that the interaction based on information retrieval interfaces makes the user experience more natural and convenient. The use of knowledge bases, such as DBpedia [2], has been proven to be effective in measuring the semantic coherence during the dialog sessions [18]. Moreover, structured data models as background knowledge allow the systems to handle complex user queries [16], i.e. retrieving answers which require multi-hop steps. However, the formulation of complex queries requires a considerable cognitive effort for the user. Also, there are certain answers that may not be found with a single query on a given data model. Conversational browsing [17] is proven to be a valuable interaction model in order to satisfy complex users' information needs by iteratively refining the initial query with relevant new concepts. This consideration motivated the idea behind our VA.

Yang *et al.* [19] proposed a learning framework with Deep Matching Networks for response ranking leveraging external knowledge. Knowledge is incorporated using Pseudo-relevance feedback (PRF) (i.e., using top N documents from the initial retrieval of an external collection as feedback to improve the query) and using QA correspondence knowledge distillation. In contrast, in our approach we leverage the knowledge automatically induced from a domain corpus to provide the user dynamically generate facets to refine under-specified queries iteratively.

Table 5. Average and Standard Deviation of flat facet SME Evaluation.

User	Average of relevant facets	Standard Deviation of Relevant Facets
User1	10.90	5.22
User2	10.85	4.84

Aliannejadi et al. [1] describe a conversational search system, which selects clarifying questions for a given query from a large pool of questions. The system uses the BERT [9] model to learn both query and question representations, based on which, it matches questions to a query and conversation context. Different from our approach, it requires a set of manually curated candidate questions.

6 Conclusions and Future Work

In this paper, we proposed a novel and general approach for initializing VAs by leveraging knowledge automatically induced from technical documentations with the aim to provide a better user experience than a conventional keyword-based retrieval paradigm. Our current system is driven by a dynamic faceted search algorithm. One of the building blocks of our proposed approach is an unsupervised automatic taxonomy induction method.

The evaluation on two real-world datasets in the IT domain has shown the effectiveness of our approach in better refining the users' information need about solving their technical issues. Though it is out of scope for this paper, we have tried out the proposed system in other domains such as medical, oil and gas, legal and the early feedback from the users is positive. The proposed system does not have any domain-specific methods. Thus, it can be easily applied to any domain.

Based on the real-world user evaluation, we learnt that certain user queries do not perform well for facet generation. Though it may be challenging, it will be an interesting exercise to predict the queries that will tend to perform badly.

One of the challenges we identified in the real-world SME evaluation is that it is hard to evaluate the usefulness of the facets automatically as it is highly dependent on the users skill set, domain knowledge and experience. In the SME evaluation, some users annotated some facets as relevant because they had experience of the exact problem in the question whereas as some others did not identify the connection between the problem and the facet.

In our vision, this approach is an early yet fundamental step toward an unsupervised conversational VA for technical support: a system that is able to be easily adapt to new domains, and which does not require any training data.

As future work, we plan to evolve our system by augmenting and strengthening the generation and selection of facets with automatically generated natural languages questions guided by a knowledge graph with a wider range of relation types. Furthermore, at the moment, we are only using taxonomic relations in text for generation and organisation of facets. We plan to induce knowledge graphs with all other relations in the text using techniques such as OpenIE and use them for facet generation.

References

1. Aliannejadi, M., Zamani, H., Crestani, F., Croft, W.B.: Asking clarifying questions in open-domain information-seeking conversations. In: Proceedings of the 42nd International ACM SIGIR Conference on Research and Development in Information Retrieval SIGIR 2019, pp. 475–484 (2019)
2. Auer, S., Bizer, C., Kobilarov, G., Lehmann, J., Cyganiak, R., Ives, Z.: DBpedia: a nucleus for a web of open data. In: Aberer, K., et al. (eds.) ASWC/ISWC -2007. LNCS, vol. 4825, pp. 722–735. Springer, Heidelberg (2007). https://doi.org/10.1007/978-3-540-76298-0_52
3. Belkin, N.J.: Anomalous states of knowledge as a basis for information retrieval (1980)
4. Castelli, V., et al.: The TechQA Dataset. ArXiv abs/1911.02984 (2019)
5. Chowdhury, M.F.M., Farrell, R.: An efficient approach for super and nested term indexing and retrieval. CoRR abs/1905.09761 (2019). http://arxiv.org/abs/1905.09761

6. Cui, L., Huang, S., Wei, F., Tan, C., Duan, C., Zhou, M.: Superagent: a customer service chatbot for e-commerce websites. In: ACL : System Demonstrations (2017)
7. Dash, D., Rao, J., Megiddo, N., Ailamaki, A., Lohman, G.: Dynamic faceted search for discovery-driven analysis. In: Proceedings of the 17th ACM Conference on Information and Knowledge Management (2008)
8. Dash, S., Chowdhury, M.F.M., Gliozzo, A., Mihindukulasooriya, N., Fauceglia, N.R.: Hypernym detection using strict partial order networks. In: AAAI (2020)
9. Devlin, J., Chang, M.W., Lee, K., Toutanova, K.: BERT: pre-training of deep bidirectional transformers for language understanding. In: NAACL 2019 (2019)
10. Fauceglia, N.R., Gliozzo, A., Dash, S., Chowdhury, M.F.M., Mihindukulasooriya, N.: Automatic taxonomy induction and expansion. In: EMNLP : System Demonstrations (2019)
11. Hearst, M.A.: Automatic acquisition of hyponyms from large text corpora. In: COLING (1992)
12. McHugh, M.L.: Interrater reliability: the kappa statistic. Biochemia Med. Biochemia Med. **22**(3), 276–282 (2012)
13. Radlinski, F., Craswell, N.: A theoretical framework for conversational search. In: CHIIR 2017, pp. 117–126. ACM (2017)
14. Safran, N.: Psychology of the searcher, April 2015. https://www.immagic.com/eLibrary/ARCHIVES/GENERAL/BNILE_US/B150428S.pdf
15. Tunkelang, D.: Faceted search. Synth. Lect. Inf. Concepts Retrieval Serv. **1**(1), 1–80 (2009)
16. Usbeck, R., Gusmita, R.H., Ngomo, A.N., Saleem, M.: 9th challenge on question answering over linked data. In: ISWC 2018. CEUR, vol. 2241, pp. 58–64 (2018)
17. Vakulenko, S.: Knowledge-based conversational search. CoRR abs/1912.06859 (2019). http://arxiv.org/abs/1912.06859
18. Vakulenko, S., de Rijke, M., Cochez, M., Savenkov, V., Polleres, A.: Measuring semantic coherence of a conversation. In: Vrandečić, D., et al. (eds.) ISWC 2018. LNCS, vol. 11136, pp. 634–651. Springer, Cham (2018). https://doi.org/10.1007/978-3-030-00671-6_37
19. Yang, L., et al.: Response ranking with deep matching networks and external knowledge in information-seeking conversation systems. In: SIGIR 2018, pp. 245–254 (2018)

Linking Ontological Classes
and Archaeological Forms

Vincenzo Lombardo$^{(\boxtimes)}$ ⓘ, Rossana Damiano ⓘ, Tugce Karatas ⓘ,
and Claudio Mattutino ⓘ

Università di Torino, Torino, Italy
{vincenzo.lombardo,rossana.damiano,
tugce.karatas,claudio.mattutino}@unito.it

Abstract. Archaeological studies are a trans-disciplinary endeavor, where a number of different scientists collaborate to get a reasonable account of material artefacts, through the various phases of recovery, analysis, and, recently, also exhibition. A large amount of digital data support the whole process, and there is a high value of keeping the coherence of the information and knowledge contributed by each discipline. The paper introduces a modular computational ontology, which is in use in a comprehensive archaeological project, Beyond Archaeology. The ontology provides the information structure to all the phases of the project, from the excavation phase, to the archaeometric analyses, up to the design and the implementation of the exhibition. The computational ontology is compliant with CIDOC-CRM reference model and introduces a number of novel properties and classes to link the description of the archaeological world with the forms traditionally used by the archaeologists to record the excavation and data about findings on the field and in the lab. The forms are implemented through a CMS structured site, for the creation of a data base, that is also filled with multimedia items that are to be employed in interpretation and exhibition, respectively.

Keywords: Archaeology · CRMarchaeo model · CMS

1 Introduction

Archaeological projects are more and more digital, in many accounts, as it happens in many areas of cultural heritage: data collection, curation, and visualization (see, e.g. [6,11], among others), analysis (e.g., GIS [2]), exhibition (starting from the virtual archeological reconstructions of the 1990s [1,9] and addressing general public outreach and participation [10]).

The scientific community of the archaeologists has been realizing the importance of the digital data curation, alongside with physical artefacts. Projects such as the Digital Archaeological Record[1], the catalogue section of the Central Institute of Cataloguing and Documentation of the Italian Ministry of Cultural

---

[1] http://www.tdar.org/.

© Springer Nature Switzerland AG 2020
J. Z. Pan et al. (Eds.): ISWC 2020, LNCS 12507, pp. 700–715, 2020.
https://doi.org/10.1007/978-3-030-62466-8_43

Heritage[2], and the Archaeology Data Service[3] are archiving and making available a number of archeological data for quantitative testing and processing. These data can be reused by people other than the original creators, in ways that they had not even envisioned before (see, e.g., [12]).

There also are projects that have been carried out with the goal of maintaining the data as long as possible. The Çatalhöyük Database and the Çatalhöyük Image Collection Database[4] make available the documentation of the Çatalhöyük excavation site. These are custom platforms, indeed searchable data management systems, updated during every excavation season, which have been made available through the Çatalhöyük Living Archive[5], an experimental web application that provides access to the data from two decades of excavation and analysis at a Neolithic settlement in Turkey. They also provide an API to query the database.

However, although languages and tools seem to be available and effective, there exist, in general, many limits concerning sharing and standardization of data [3]. A very recent survey made within the AriadnePlus project[6] reports that researchers are not very aware of the issues of data sharing and Linked Data. They also find useful to raise the competence in the alignment of terminologies through the usage of international thesauri (e.g., Getty AAT[7]) and to promote the usage of domain computational ontologies (e.g., CIDOC-CRM collaboration family[8]).

In this scenario, the Semantic Web approach has been invoked to support the sharing of data, particularly in the trans-disciplinary endeavours [5], as in the case of archaeology. Though we agree that a thorough development of the need for data sharing goes with the growth of awareness that is achieved through pervasive data modeling, training, and knowledge (see AriadnePlus report above), we believe that a boost in this direction can come up by the successful implementation of truly trans-disciplinary projects, where research questions arise through the collaboration and peer-to-peer cross-fertilization of several disciplines [8]. Archaeology is an ideal testbed, especially in its multiple relations with archaeometry and laboratory science, philosophy and social sciences, activities on the field (including the negotiation with contractors and the public) and in the university rooms, where "ordering and reconstructing the past" co-exist with "articulating activist political positions in the present" [13].

This paper describes a Semantic Web approach to the conduction of an ongoing EU project named Beyond Archaeology (BeArchaeo)[9], which consists in an

---

[2] http://www.iccd.beniculturali.it.

[3] http://archaeologydataservice.ac.uk/.

[4] http://www.catalhoyuk.com/research/database (last visited on 15 May 2020).

[5] http://catalhoyuk.stanford.edu.

[6] D2.1 Initial Report on Community Needs https://ariadne-infrastructure.eu/wp-content/uploads/2019/11/ARIADNEplus_D2.1_Initial-Report-on-Community-Needs-1.pdf, dated 31 October 2019.

[7] https://www.getty.edu/research/tools/vocabularies/aat/.

[8] http://www.cidoc-crm.org/collaborations.

[9] https://www.bearchaeo.com/ (last visited on 15 May 2020).

archaeological excavation, the consequent archaeometric analyses of the site and the excavated materials, the interpretation of the findings, and the dissemination of the results through physical and virtual exhibitions. The whole project depends on the creation, maintenance, and employment of digital data documentation, that ambitiously supports all the project phases, from the excavation to the exhibitions. The effort aims to overcome some of the limits that have been raised for IT applications in archaeology, which, on the one hand have been appointed to bring, notwithstanding a number of criticisms, some data-driven theory-neutrality to archaeological investigations (together with data recording and visualization), while, on the other, have been appraised as "unrealized 'great expectation"' [7]. In this paper, we introduce the core of BeArchaeo ontology that encodes the conceptual model of the data base. The ontology is publicly available at /purl.org/beArchaeo. In many cases, the ontology classes and properties specialize the entities of the well-known CRMarchaeo model[10] and represent a concrete realization of the application of the ontology from the initial phases of a project. In particular, the ontology captures the entities that are necessary to encode the knowledge that supports the archaeologists in filling the forms that document the excavation and the interpretation phases in an archaeological project. With respect to CRMarchaeo, we have addressed the descriptive issues that are recorded in the documentation rather than the processes that cause the existence of some encountered object.

In the next section, we introduce the BeArchaeo project. Then, we illustrate how we encoded the knowledge about the forms and how it is related to the knowledge of the archaeological entities and processes. Finally, we describe how we have implemented the forms in a CMS structure to allow the filling operation in the field. Some comment on the lesson learnt and conclusions end the paper.

2   The Bearchaeo Project

Project Beyond Archaeology (BeArchaeo) is a RISE European project that consists in the archaeological excavation, archaeometric analyses, interpretation of the findings, and eventually dissemination of the results about the Tobiotsuka Kofun (Soja city in Okayama Prefecture), together with other Kofun burial mounds and the related archaeological material in ancient Kibi and Izumo areas (present Okayama and Shimane Prefectures), in Japan[11]. Archaeologists and archaeometrists (e.g., chemists, physicists, ...) from both Europe and Japan work on a major period (the Kofun period) of the Japan history with a truly transdisciplinary vision of archaeology combined with archaeometry; the project activities and outcomes are accessible to the general public through engaging media communication along the project development and two final exhibitions in Italy and Japan, to be held at the end of the project.

---

[10] http://www.cidoc-crm.org/crmarchaeo/, (last visited on 15 May 2020).

[11] BeArchaeo website https://www.bearchaeo.com/ (last visited on 15 May 2020) and RISE programme https://ec.europa.eu/research/mariecurieactions/news/research-and-innovation-staff-exchange-rise-bridging-ri-sectors-europe-and-worldwide_en.

A preliminary achievement of this research has been the design and implementation of a semantic database for the encoding and storing of the digital data concerning the documentation of the archaeological excavation and the account of the metadata for the several disciplines[12]. We have developed a domain ontology centred around the major classes that appear in the archaeological projects, according to the major forms that are in use, namely the forms for the stratigraphic units and the archaeological findings, respectively. We have analyzed the major documentation sources about the forms in use by the European and the Japanese archaeological teams and we have encoded the related knowledge into the ontology, while keeping the alignment with the CRMarchaeo model. The documentation sources are mostly published by the national organizations of cultural heritage (see, e.g., the documentation records of the Italian Central Institute for the Catalogue and the Documentation[13]). This documentation is rarely related to some shared knowledge source, although in some cases there has been some post-alignment of relevant data bases. For example, the NIOBE database (concerning the Colosseum, the Roman age National Museum, and the Rome Archaeological Area) has been recently mapped onto CRMarchaeo[14] in the context of the ArcheoSITAR project[15].

As far as we know, BeArchaeo is the first archaeological project that assumes a Semantic Web approach from the start. In fact, the multi-disciplinary, multi-cultural, and multi-lingual characters of Be-Archaeo raise a high demand of interoperability of knowledge and data. The alignment with CIDOC-CRM is pursued at the disciplinary level, by aligning the archaeologic and the archaeometric descriptions through the CRMarchaeo and CRMsci models, wherever possible. These issues are particularly relevant in the mapping of the forms to be filled by the researchers onto the ontology classes and properties; so, we designed both a practical workflow and the form interfaces for collecting the data as the excavation goes on, to be continued in the analysis labs, and eventually with the design of the exhibition.

In the next section, we describe the modeling of the BeArchaeo ontology, by highlighting both the encoding of the forms and the alignment with CIDOC-CRM and then we see how it is interfaced on the documentation website.

3   The Modeling of the BeArchaeo Ontology

There have been two major guiding principles in the development of the BeArchaeo ontology. The first is that it should capture concepts and properties in archaeology and archaeometry and how these are connected. The long term aim is a truly trans-disciplinary approach between the archaeologic and

---

[12] https://bearchaeo.unito.it/omeka-s (last visited on 15 May 2020).

[13] http://www.catalogo.beniculturali.it/sigecSSUFE/, in Italian (last visited on 15 May 2020).

[14] http://www.archeositarproject.it/wp-content/uploads/2017/06/Mapping-NIOBE-towards-CIDOC-CRM-final_12.10.2016.pdf, (last visited on 15 May 2020).

[15] http://www.archeositarproject.it, (last visited on 15 May 2020).

the archaeometric disciplines mediated by the formal ontology. There has been an improvement from the 1990s, when the natural sciences were to deliver data to be interpreted by the archaeological theories, to nowadays, when the data emerging from scientific analysis are viewed as more objective and a stable foundation for interdisciplinary analysis [13]. The second is that the ontology must align with the forms currently used by the archaeologists. In particular, the ontology implementation must provide a detailed account of the archaeological knowledge that can lead to the publishing of the record forms that are typically filled by the archaeologists on the field and in the lab, and are the object of a continuous review and interpretation. This, in turn, requires that the interface should recall the traditional forms filled by the archaeologists, in order to match their standard working practices and consequently achieve their full collaboration in the filling of the database. These forms, though sharing a number of features (see below), are usually provided by the individual national institutions, and can be more or less rich. Also, in this project, these forms are, for the first time, augmented with fields that encode the archaeometric analyses, so to achieve a transparent management of the interpretations.

Finally, as an add-on requirement, the ontology must capture features that can support the work of the designers and professionals that will work on the main exhibition, which will be both physical and virtual, at the end of the project. This requirement has been currently limited to the storage of the media (3D models and images) that are associated with the items, together with the algorithms and the procedures used to achieve them.

The modeling has worked in parallel between the encoding already realized by CRMarchaeo and the forms provided by the Italian Ministry of Culture and translated into English for one excavation mission in Pompei. In the following, we introduce the two knowledge sources, the RDF-formalized CRMarchaeo and the forms of the Italian Ministry of Culture, respectively, and then we address the modeling of the BeArchaeo ontology.

3.1   CRMarchaeo

CRMarchaeo, an extension of CIDOC–CRM, was developed to support the activities concerning any archaeological project (actually, mainly the excavation process). Also, it has been growing from standards and models already in use by national and international cultural heritage institutions as well as from the metadata contained in the archaeological documentation. The Conceptual Reference Model is a formal ontology for the integration and interchange of cultural heritage information, which displays an heterogeneous nature. It provides the semantic definitions that underlie the construction of coherent resources, with a super-institutional perspective, to enable semantic interoperability.

The CIDOC-CRM family of models (Fig. 1, top, right) extends the general documentation model through specialised thematic models for the needs of projects and organisations. In particular CRMdig is a model for provenance metadata, CRMgeo is a model spatio-temporal entities. As super-models of CRMarchaeo, and of particular interest for the BeArchaeo project, are

- CRMinf, a formal ontology about argumentation and inference making, provides a formal description of the semantic relationships between premises, reasoning activities, and conclusions. For example, class *CRMinf/I2 Belief* encodes the fact that some associated proposition set is held to have a particular belief value about some subject on behalf of some actor (e.g., "Italian team believes that Archaeological finding AF29 is of 6th Century AD").
- CRMsci, a formal ontology about scientific observation, measurements and processed data in descriptive and empirical sciences, provides a formal description of the causal relationships in scientific investigations. For example, class *CRMsci/S3 Measurement by Sampling* encodes activities of taking a sample and measuring or analyzing it, in which the sample is typically not identified and preserved (e.g., "S3 Metabarcoding of microbial taxonomic diversity in sample SU202A has observed presence of Rhodosporidiobolus.").

CRMarchaeo, as well as BeArchaeo, take inspiration from Harris' model [4], which takes into account the stratified arrangement of an archaeological excavation. The excavation model includes the description of the dichotomy between the (natural or human) phenomena that produced the stratification (centred around the class *CRMarchaeo/A1 Excavation Process Unit*) and the units that are the outcome of the generation/modification process (centred around the class *CRMarchaeo/A8 Stratigraphic Unit*). Stratigraphic units contain some remains, classified as physical objects (centred around the class *CIDOC CRM/E18 Physical Object* of the core ontology). Stratifications and their contents are analyzed and interpreted to determine the relative chronological order of the strata, together with the classification and functionality of the objects therein, till the high-level reconstruction of the beliefs and behaviors of some group of people in the past in that place.

Figure 1 illustrates the major relationships between BeArchaeo ontology and CRMarchaeo, as well as the reference to the two thesauri (BeArchaeo Archaeological Finding Thesaurus – AFT, for a taxonomy of Japanese history materials, built within the project, and Getty Art and Architecture Thesaurus – AAT). While the class *CRMarchaeo/A8 Stratigraphic Unit* has been imported as it is, the class *bearchaeo/ArchaeologicalFinding* specializes the generic *CIDOC CRM/E18 Physical Object*, in order to connect it to the corresponding catalogue record (see below). We also see that the stratigraphic relation, existing between stratigraphic units, is specialized into several subproperties, as reported by the catalogue record forms below, namely the following spatial relations:

- isEqualTo, for two stratigraphic units that are claimed to be the same;
- isBoundTo, for a stratigraphic unit that is a limit for another one;
- Abuts/isAbuttedTo, for a stratigraphic unit that edges another one;
- Cuts/isCutBy, for a stratigraphic unit that introduces a discontinuity into another one;

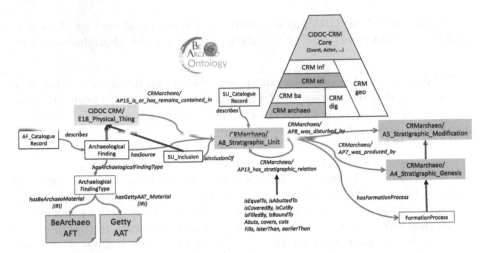

Fig. 1. Major relationships between BeArchaeo and CIDOC-CRM. Colors are employed to distinguish provenances. (Color figure online)

- Covers/isCoveredBy, for a stratigraphic unit that covers (stands over) another one;
- Fills/isFilledBy, for a stratigraphic unit that has filled a cut (see above);

Also, there are two temporal relations, laterThan and earlierThan, resulting from the interpretation of the stratigraphy.

3.2   The Archaeological Forms

When on the field as well as when in the lab, archaeologists fill forms that are prepared by the national authorities to track all the entities that have been recognized and to update the interpretations of the findings. So, it is important, for the practical application of the encoded ontological knowledge, that knowledge and forms are connected. The solution devised in BeArchaeo has been to also encode the form fields as properties of the ontology, developed as modules that included the archaeologic knowledge, the archaeometric knowledge, and the catalogue record knowledge (see below). The forms we have encoded, with some adjustments after long and productive discussions with the archaeological team of the project, have been the ones distributed by the Italian Ministry of Culture, and in particular, the forms of the Stratigraphic Unit and the Archaeological Finding.

In Fig. 2 we see an excerpt of the Stratigraphic Unit form, in an English translation (for the sake of understanding). The upper left part is the registry part of the Stratigraphic Unit, reporting identifiers and locations; then going down, after the informal definition and position, we find, among others,

- the distinguishing criteria (three-valued multiple choice), which were employed by the archaeologist to identify such a stratum in the soil;

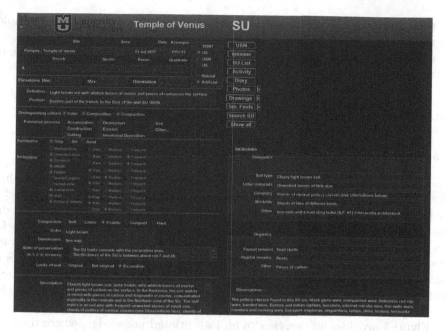

Fig. 2. The form provided by the Italian Ministry of Culture, in the English translation provided by an excavation mission carried out by the universities of Missouri and Mount Alison in Pompei (Courtesy of Ivan Varriale).

- the formation process (connected to the genesis or modification of the unit), which can be valued with a number of common processes, together with the possibility of free insertion ("other", in the form);
- the type of soil matrix, which can also be a combination of values;
- the inclusions, i.e. the generic types of physical objects found in the unit (a list is provided, together with the freedom of some further insertion), with their (three-valued) frequency of occurrences;
- the (five-valued) compaction attribute;
- the color of the unit.

The form for the Archaeological Finding record that we have taken into account is an extract from the very articulate documentation reported in the web platform of the Italian Ministry of Culture, named SigecWeb[16]. The reduction was due to the fact that the original format is designed to track the existence of the finding in passing through various institutions during its entire life cycle (restoration centres, museums, churches, ...), which is possibly very long and departs from the goals of BeArchaeo. The major fields of the record concern:

- the source, i.e. the stratigraphic unit that contains it (actually a subproperty of CRMarchaeo property *AP15 is or contains remains of*);

---

[16] http://www.sigecweb.beniculturali.it.

- the type of the finding (in terms of materials and functions), which we implemented as a reference to the two thesauri mentioned above;
- an indication of the guessed (or confirmed) chronology, together with a motivation for it.

Now we see how both the CRMarchaeo model and the archaeological excavation forms have contributed to the BeArchaeo ontology.

3.3 The BeArchaeo Ontology

The BeArchaeo ontology is geared to the description of the objects rather than the forming processes, and merges the general classes and properties provided by CRMarchaeo with the fields of the archaeological catalogue records. We did not employ specific design patterns because we were not aware of patterns for connecting knowledge and forms, and because our solution was straightforward (see below). The result is an application ontology that connects three types of knowledge: the archaeologic knowledge (lower left part of Fig. 3), the archaeometric knowledge (lower right part of Fig. 3), and the catalogue record knowledge (upper part of Fig. 3).

Figure 3 provides an overview of an instantiated knowledge, where the rectangles in grey or black are the individuals, and the white rectangles are the classes; object properties are depicted as blue lines, while datatype properties are depicted as green lines; the three elements in Courier font, highlighted in yellow, are the strings that are actually written in the final form interface. Going left to right: the stratigraphic unit "SU 202" (content of the title field of the catalogue record for this unit) is the source of the archaeological finding "AF 59" (content of the title field of the catalogue record for this finding); the type of the finding is "Sue (ceramics style)", as selected from the Getty AAT thesaurus and "sekki", as selected from the BeArchaeo thesaurus; the finding body[17] has undergone some chemical test, which has produced a composition descriptor (the individual is actually a table reporting the presence of substances), which in turn is input to a Data Evaluation process (related to CRMsci ontology), which assigns some dimension, namely an attribute for the body composition ("Calcareous").

In the Figs. 4 and 5 there are the major relations of the stratigraphic unit class and the archaeological finding class, respectively. Going clockwise, a stratigraphic unit has inclusions (i.e., entities that are contained in stratum), which are of some type, that can be generic or specific, and has a frequency of occurrence in the unit, qualitatively valued as rare, medium, or frequent. Inclusions have types that are taken from partially overlapping vocabularies, based on the practical experience of the archaeologists (these may change and should be aligned with the types included in the thesauri for the archaeological findings). Some informal properties, noted as free text, are the state of preservation of the unit

---

[17] Usually, for chemical tests, an archaeological finding is considered as composed a body, a coating, and an embellishment.

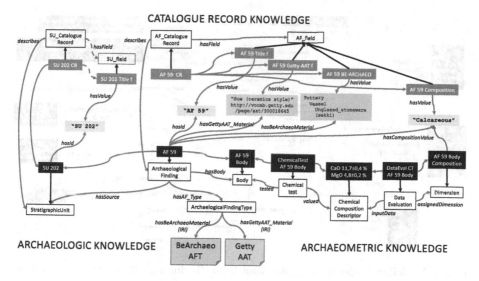

Fig. 3. Modeling of the archaeological finding, exemplifying archaeologic and archaeometric knowledge, respectively, and the corresponding fields in the archaeologic finding record.

and the measurements taken during the excavation, with a particular concern for Elevation. The distinguishing criterion determines how this unit has been identified: the terms that concern this attribute are three (Color, Composition and Compaction) and there are other three properties that possibly specify the actual values for such attributes (namely 6-valued soil/matrix term for composition, 5-valued term for compaction, and a free string for color). Color, in the relationship with archaeometrists (specifically, the soil scientists) has been augmented with the encoding provided by the well-known Munsell color system, in use in pedological studies[18]. Finally, the formation process concerns a specialization of the processes that are responsible for the creation and modification of the unit, with a frequent term vocabulary, which can be further augmented with free text insertion. The properties in the center of the figure specialize the stratigraphic relations, in spatial and temporal terms (see above).

An archaeological finding (Fig. 5) can be part of another archaeological finding (frequent is the case of fragments to be composed afterwards) and is sourced by some stratigraphic unit as well as museum collection or other places. This variety of sources concerns the goals of the BeArchaeo project, because of the employment of the ontology into the design of the final exhibition. The archaeological finding has a type, referring to terms in the widely acknowledged Getty

---

[18] Munsell color system is based on the three-dimensional model, where each color is defined by a triple of hue (the color of the color), value (how light or dark is the color), and chroma (or saturation/brilliance of the color), set up as a numerical scale with visually uniform steps https://munsell.com/about-munsell-color/how-color-notation-works/.

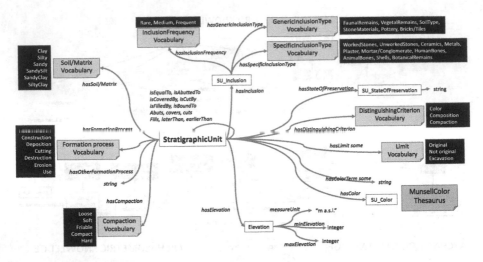

Fig. 4. Modeling of the stratigraphic knowledge (including references to thesauri and vocabularies (with list of terms)).

AAT thesaurus and the BeArchaeo AF thesaurus, the latter encoding knowledge from an authoritative Japanese reference [14]. Finally, an archaeological finding is marked with its chronology, currently limited to a free text insertion, together with its motivation, but with the idea of linking to a time ontology in the LOD panorama.

From a technical point of view, the model has been described as a number of subontologies that address different sections of the forms. In particular, there are five subontologies for the stratigraphic unit catalogue record: "registry" (which contains identifiers and formal issues concerning the location, the trench, the section, ...), "description" (concerning inclusions and soil attributes), "stratigraphy" (concerning the relations with other stratigraphic units), "dating" (where dating elements and chronology are represented), "sampling" (concerning some data about the excavation process). Then, there is one ontology for the archaeological finding record. And, finally, there is the encoding of the archaeological knowledge. The several subontologies for the sections are connected to the main ontology for the record through the properties *hasField* and *hasSection*, while the ontologies for the records are connected to the archaeological knowledge are connected through the property *arco/describes*, as introduced by project ArCo[19] for the relationship between an entity that describes another entity in the field of cultural heritage. The ontology is expressed in OWL/RDF formats and published at the permanent address /purl.org/beArchaeo[20].

---

[19] http://wit.istc.cnr.it/arco/.
[20] File "BeArchaeo_merge_all.owl" merges all the other sub-ontologies.

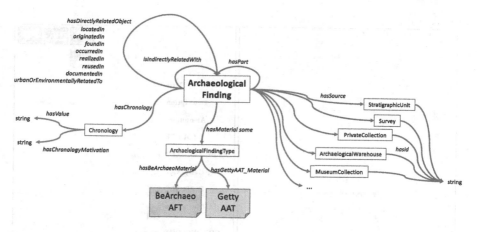

Fig. 5. Modeling of the archaeological finding. Exemplifying archaeologic and archaeometric knowledge, respectively, and the corresponding fields in the archaeologic finding record.

4 Ontology BeArchaeo in Use: CMS Approach to Form Filling and Lesson Learnt

In this section, we describe how the ontology has been employed for the excavation campaign carried out by the BeArchaeo team in August 2019 and reported on the project website[21]. In order to make the knowledge available to the archaeologists on the field, we built a website, based on an installation of a Content Management System (CMS), for achieving an immediate deployment of the forms. The CMS Omeka-S[22] is particularly suited for the import of semantic properties defined in a RDF file, the definition of customized vocabularies, and the construction of templates for the instantiation of filling forms. Also, from the inserted items, one can easily build a website for immediate check of the data base content, sharing of the data, and the development of specific functions, based on the native API.

We exploited the possibility of the fast prototyping of a user interface for the back-end of the system, accessible by the people on the field, and a quick front-end, where supervisors and stakeholders could explore the development of the archive and the related findings. Also, we have started uploading a number of rich media materials (currently photos and 3D models acquired from photogrammetry and scanning), that are being used for interpretation as well as will be the base for the final exhibition. Figure 6 reports two images, from the back end and the front end, respectively, of the production website[23]. The archaeologists have used the back end on the field, introducing data through tablets (stratigraphic units and archaeological findings), and afterwards in the labs through laptop

---

[21] https://www.bearchaeo.com.
[22] https://omeka.org/s/ (last visited on 15 May 2020).
[23] https://bearchaeo.unito.it/omeka-s.

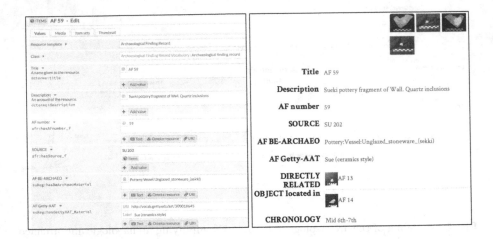

Fig. 6. Screenshot from the BeArchaeo resources website, concerning the Archaeological finding no. 59, with the related fields and media. On the left, the back end; on the right, the front end. Elements in red are links to other elements of the documentation (e.g., Stratigraphic Unit 202) or to some external knowledge source (e.g., Getty AAT thesaurus).

and desktop computers (archaeological findings). They found the tool useful, especially for the digital support (usually, archaeological teams without an IT support notate item data on paper and then transfer data on Excel files); the organization provided by the CMS to split into Authors, Reviewers, and Editors, the roles of the archaeologists with respect to the platform has eased the check and revision of the inserted data.

Differences in interpretations have been annotated through the possibility offered by Omeka-S to insert more than one value for a property; this also happens with functional properties, and should be solved through the reasoning operated by the semantic server. Also, fields with free text have allowed to include motivations for the choices made and thus enable the communication within the team. Indeed, this feature of a centralized database has been particularly appreciated by the team, and editors have made a specific pressure on the rest of the team for an extensive use the platform. The database-based approach has been useful to start the collaboration with the archaeometric team in the interpretation process. Indeed, some archaeometric analyses, such as the geophysical ones and the biological ones, have started before the archaeological excavation process. However, the encoding of the archaeometric knowledge, which requires a mapping with the CRMsci model, is under development.

The splitting of the five sections for the stratigraphic units, as structured in the original forms, has been seen as a complication of the filling work: so, the archaeological team have required for a unique form to be navigated through scrolling for the access to a large number of fields. This will require some programming effort to build an external website (with a specific connection with the database) and will be released in the next version of the platform. Also, the interface to the thesauri (both Getty AAT and BeArchaeo AFT) requires some modification. It is currently implemented as a keyword search/completion, in one case (Getty AAT), and a drop-down menu, in the other (BeArchaeo AFT). In particular, the terms are compiled from a number of features, namely Material, Object function, Sub-material, Object generic class, subclass, shape, subshape. For example, the following path can be built for a specific finding:

```
-- Metal
---- Objects Facet
------ Bronze (d ki)
-------- Weapon (buki)
---------- Sword (katana)
------------ Long sword (tachi)
-------------- Sword with a bulbous pommel (kabutsuchi no tachi)
```

The drop-down menu navigates the hierarchical classes by proposing allowed combinations of feature values; also, sometimes not all the features are overtly expressed. Archaeologists have found some difficulties in the navigation and prefer an interface arrangement that allows for the direct setting of the specific feature values. In this case, the system should then propose some terms from the thesaurus that are consistent with the setting provided. Again, a programmatic solution to be devised in the future.

Finally, two notes concerning the context of the BeArchaeo intercontinental project, because of the different excavation techniques that pertain the two traditional schools of archaeology and the linguistic issues for the interfaces. In the first case, we have that similar terms, such as trenches, sections, and rooms of the excavation call to slightly different definitions according to the two traditions; so, the interface must accomodate both methods and interpretations. At the current stage of development, the two teams are still looking for a common arrangement and the ontological vocabulary will be updated accordingly. In the second case, the interface in English was a limitation for the Japanese archaeologists in the form filling process, because the data insertion process could not be done in the native language (also because of the different interpretations above). On a development site[24], we are experiencing a number of innovations in preparation of the second excavation campaign (originally scheduled for August 2020, but now, due to the pandemic, postponed to 2021). In particular, we are addressing the encoding of the forms into Japanese: there are some Japanese resource templates for the Archaeological Finding and Stratigraphic Unit records, respectively, as well as a front end website in Japanese[25].

5   Conclusions

The paper has described an ontological approach to the encoding of the archaeological knowledge, in its relation with the archaeometric knowledge and the

---

[24] https://bearchaeo.di.unito.it/omeka-s.

[25] https://bearchaeo.di.unito.it/omeka-s/s/jtoppage/page/welcome.

practical forms for filling the archaeological/archaeometric information on the field and in the lab. In particular, we have presented how we have encoded an ontology of archaeological knowledge that is compliant to CRMarchaeo ontology, which is in use in a EU project concerning an excavation process in Japan. Also, we have seen how the ontology is the base for a CMS-based web platform for supporting the archaeologist's work in recording the excavation and interpretation activities. The encoding of the archaeological knowledge in an ontology that is compliant with CRMarchaeo, and so to CIDOC-CRM, and the implementation of a CMS-based solution for a concrete project can have a deep impact on fostering projects that adhere to the Semantic Web paradigm and address data sharing effectively. The publication of the ontology and the availability of a widespread CMS can be easily replicated in further projects. The ontology and the derived database will also be employed in the definition of the contents of the exhibition that will present the outcomes of the BeArchaeo project to large audiences in collaboration with museum institutions in Europe and Japan.

The BeArchaeo ontology has also been extended with the investigation processes and the related outcomes that concern the archaeometric part. However, we are working on how to build the interface forms to have them operational on the field and mostly in the labs afterwords. Also, we are going to connect the ontology with other resource for the cataloguing of cultural heritage assets (e.g. the Knowledge Graph for the Italian cultural heritage ArCo. We also want to improve the ontology interoperability, by replacing a number of customized vocabularies with domain ontologies (e.g., for chronology and formation processes).

6   Author Statement and Acknowledgements

Vincenzo Lombardo carried out the design and implementation of the ontology and wrote the core sections of this paper. Rossana Damiano worked on the system design and implementation of the web platform. Tugce Karatas worked on the project digital curation strategy and the usage of the system interface. Claudio Mattutino worked on the platform implementation and maintenance.

The Be-Archaeo project is funded by the European Union's Horizon 2020 research and innovation programme under the Marie Skłodowska-Curie, Grant Agreement No 823826. We thank all the team members for the common discussions during the illustration of the ontology and the database schema. In particular, we thank Ivan Varriale for sharing his insights on the archaeological knowledge, and Daniele Petrella (with the collaboration of Naoko Matsumoto and Akira Seike) for having enlisted the BeArchaeoAF thesaurus.

We are particularly grateful to the reviewers (especially the Reviewer 1) who made a detailed analysis of the work and greatly contributed to the readability of the paper. We also thank Carmine Montefusco and Angelo Saccà for the UniTo hosting service of BeArchaeo database.

References

1. Barcelo, J., Forte, M., Sanders, D.: Virtual Reality in Archaeology. ArcheoPress, Oxford (2000)
2. Conolly, J., Lake, M.: Geographical Information Systems in Archaeology. Cambridge University Press, Cambridge (2006)
3. Costopoulos, A.: Digital archeology is here (and has been for a while). Front. Digit. Humanit. 3 (2016). https://doi.org/10.3389/fdigh.2016.00004
4. Harris, E.: Principles of Archaeological Stratigraphy. Academic Press, London (1989)
5. Lampe, K.H., Riede, K., Doerr, M.: Research between natural and cultural history information: benefits and IT-requirements for transdisciplinarity. ACM J. Comput. Cult. Heritage 1(1) (2008)
6. Lercari, N., Shiferaw, E., Forte, M., Kopper, R.: Immersive visualization and curation of archaeological heritage data: Çatalhöyük and the Dig@IT app. J. Archaeol. Method Theory 25(2), 368–392 (2017). https://doi.org/10.1007/s10816-017-9340-4
7. Niccolucci, F., Hermon, S., Doerr, M.: The formal logical foundations of archaeological ontologies. In: Barcelo, J., Bogdanovic, I. (eds.) Mathematics and Archaeology, pp. 86–99. CRC Press, Boca Raton (2015)
8. Nicolescu, B.: Methodology of transdisciplinarity - levels of reality, logic of the included middle and complexity. Transdisciplinary Eng. Sci. 1(1), 19–38 (2010)
9. Reilly, P.: Towards a virtual archaeology. In: Lockyear, K., Rahtz, S. (eds.) Computer Applications in Archaeology, pp. 133–139. Oxford: BAR 565 (1990)
10. Richardson, L.: A Digital Public Archaeology? Papers from the Institute of Archaeology. UCL, London (2013)
11. Roosevelt, C.H., Cobb, P., Moss, E., Olson, B.R., Ünlüsoy, S.: Excavation is digitization: advances in archaeological practice. J. Field Archaeol. 40, 325–46 (2015). https://doi.org/10.1179/2042458215Y.0000000004
12. Silva, F., Linden, M.V.: Amplitude of travelling front as inferred from 14c predicts levels of genetic admixture among European early farmers. Sci. Rep. 7 (2017). https://doi.org/10.1038/s41598-017-12318-2
13. Stutz, L.N.: A future for archaeology: in defense of an intellectually engaged, collaborative and confident archaeology. Norw. Archaeol. Rev. 51(1–2), 48–56 (2018). https://doi.org/10.1080/00293652.2018.1544168
14. Tadanao, Y.: Dictionary of Japanese Archaeological terms. Tokyo Bijutsu Publishing, Tokyo (2001)

Correction to: KGTK: A Toolkit for Large Knowledge Graph Manipulation and Analysis

Filip Ilievski(iD), Daniel Garijo(iD), Hans Chalupsky(iD),
Naren Teja Divvala, Yixiang Yao(iD), Craig Rogers(iD),
Rongpeng Li(iD), Jun Liu, Amandeep Singh(iD), Daniel Schwabe(iD),
and Pedro Szekely

Correction to:
Chapter "KGTK: A Toolkit for Large Knowledge Graph
Manipulation and Analysis" in: J. Z. Pan et al. (Eds.):
The Semantic Web – ISWC 2020, LNCS 12507,
https://doi.org/10.1007/978-3-030-62466-8_18

In the originally published version of chapter 18 the name of Rongpeng Li was misspelled. This has been corrected.

The updated version of this chapter can be found at
https://doi.org/10.1007/978-3-030-62466-8_18

© Springer Nature Switzerland AG 2021
J. Z. Pan et al. (Eds.): ISWC 2020, LNCS 12507, p. C1, 2021.
https://doi.org/10.1007/978-3-030-62466-8_44

Correction to: Transparent Integration and Sharing of Life Cycle Sustainability Data with Provenance

Emil Riis Hansen, Matteo Lissandrini, Agneta Ghose,
Søren Løkke, Christian Thomsen, and Katja Hose

Correction to:
Chapter "Transparent Integration and Sharing
of Life Cycle Sustainability Data with Provenance"
in: J. Z. Pan et al. (Eds.): *The Semantic Web – ISWC 2020*,
LNCS 12507, https://doi.org/10.1007/978-3-030-62466-8_24

Chapter "Transparent Integration and Sharing of Life Cycle Sustainability Data with Provenance" was previously published non-open access. It has now been changed to open access under a CC BY 4.0 license and the copyright holder updated to 'The Author(s)'. The book has also been updated with this change.

The updated version of this chapter can be found at
https://doi.org/10.1007/978-3-030-62466-8_24

© The Author(s) 2022
J. Z. Pan et al. (Eds.): ISWC 2020, LNCS 12507, p. C2, 2022.
https://doi.org/10.1007/978-3-030-62466-8_45

Author Index

Printed in the United States
by Baker & Taylor Publisher Services

Printed in the United States
by Baker & Taylor Publisher Services